D0154421

Power System Analysis

McGraw-Hill Series in Electrical and Computer Engineering

SENIOR CONSULTING EDITOR
Stephen W. Director, University of Michigan, Ann Arbor

Circuits and Systems
Communications and Signal Processing
Computer Engineering
Control Theory
Electromagnetics
Electronics and VLSI Circuits
Introductory
Power
Radar and Antennas

Previous Consulting Editors
Ronald N. Bracewell, Colin Cherry, James F. Gibbons, Willis W. Harmon, Hubert Heffner, Edward W. Herold, John G. Linvill, Simon Ramo, Ronald A. Rohrer, Anthony E. Siegman, Charles Susskind, Frederick E. Terman, John G. Truxal, Ernst Weber, and John R. Whinnery

Power

SENIOR CONSULTING EDITOR
Stephen W. Director, University of Michigan, Ann Arbor

Chapman: *Electric Machinery Fundamentals*
Elgerd: *Electric Energy Systems Theory*
Fitzgerald, Kingsley, and Umans: *Electric Machinery*
Gonen: *Electric Power Distribution System Engineering*
Grainger and Stevenson: *Power System Analysis*
Krause and Wasynczuk: *Electromechanical Motion Devices*
Nasar: *Electric Machines and Power Systems: Volume I, Electric Machines*
Stevenson: *Elements of Power System Analysis*

Also Available from McGraw-Hill

Schaum's Outline Series in Electronics & Electrical Engineering

Most Outlines include basic theory, definitions, and hundreds of example problems solved in step-by-step detail, and supplementary problems with answers.

Related titles on the current list include:

Analog & Digital Communications
Basic Circuit Analysis
Basic Electrical Engineering
Basic Electricity
Basic Mathematics for Electricity & Electronics
Digital Principles
Electric Circuits
Electric Machines & Electromechanics
Electric Power Systems
Electromagnetics
Electronic Communication
Electronic Devices & Circuits
Feedback & Control Systems
Introduction to Digital Systems
Microprocessor Fundamentals
Signals & Systems

Schaum's Electronic Tutors

A Schaum's Outline plus the power of Mathcad® software. Use your computer to learn the theory and solve problems—every number, formula, and graph can be changed and calculated on screen.

Related titles on the current list include:

Electric Circuits
Feedback & Control Systems
Electromagnetics
College Physics

Available at most college bookstores, or for a complete list of titles and prices, write to:

The McGraw-Hill Companies
Schaum's
11 West 19th Street
New York, New York 10011-4285
(212-337-4097)

Power System Analysis

Hadi Saadat

Milwaukee School of Engineering

Boston Burr Ridge, IL Dubuque, IA Madison, WI New York San Francisco St. Louis
Bangkok Bogotá Caracas Lisbon London Madrid
Mexico City Milan New Delhi Seoul Singapore Sydney Taipei Toronto

WCB/McGraw-Hill
*A Division of The **McGraw-Hill** Companies*

Power System Analysis

This book is printed on acid-free paper.

1 2 3 4 5 6 7 8 9 0 DOC DOC 9 4 3 2 1 0 9

P/N 012235-0
Set ISBN 0-07-561634-3

Publisher: Kevin Kane
Executive editor: Betsy Jones
Sponsoring editor: Lynn Cox
Marketing manager: John Wannemacher
Project manager: Eve Strock
Production supervisor: Rich DeVitto
Compositor: York Graphic Services, Inc.
Printer: R. R. Donnelley & Sons, Inc.

Library of Congress Cataloging-in-Publication Data

Saadat, Hadi.
Power system analysis / Hadi Saadat
p. cm.
Includes bibliographical references and index.
ISBN 0-07-012235-0
1. Electric power systems. 2. System analysis. I. Title.
TK1011.S23 1999
621.31--dc21

http://www.mhhe.com

CONTENTS

PREFACE

This book is intended for upper-division electrical engineering students studying power system analysis and design or as a reference for practicing engineers. As a reference, the book is written with self-study in mind. The text has grown out of many years of teaching the subject material to students in electrical engineering at various universities, including Michigan Technological University and Milwaukee School of Engineering.

Prerequisites for students using this text are physics and mathematics through differential equations and a circuit course. A background in electric machines is desirable, but not essential. Other required background materials, including *MATLAB* and an introduction to control systems, are provided in the appendixes.

In recent years, the analysis and design of power systems have been affected dramatically by the widespread use of personal computers. Personal computers have become so powerful and advanced that they can be used easily to perform steady-state and transient analysis of large interconnected power systems. Modern personal computers' ability to provide information, ask questions, and react to responses have enabled engineering educators to integrate computers into the curriculum. One of the difficulties of teaching power system analysis courses is not having a real system with which to experiment in the laboratory. Therefore, this book is written to supplement the teaching of power system analysis with a computer-simulated system. I developed many programs for power system analysis, giving students a valuable tool that allows them to spend more time on analysis and design of practical systems and less on programming, thereby enhancing the learning process. The book also provides a basis for further exploration of more advanced topics in power system analysis.

MATLAB is a matrix-based software package, which makes it ideal for power system analysis. *MATLAB*, with its extensive numerical resources, can be used to obtain numerical solutions that involve various types of vector-matrix operations.

In addition, *SIMULINK* provides a highly interactive environment for simulation of both linear and nonlinear dynamic systems. Both programs are integrated into discussions and problems. I developed a power system toolbox containing a set of M-files to help in typical power system analysis. In fact, all the examples and figures in this book have been generated by *MATLAB* functions and the use of this toolbox. The power system toolbox allows the student to analyze and design power systems without having to do detailed programming. Some of the programs, such as power flow, optimization, short-circuit, and stability analysis, were originally developed for a mainframe computer when I worked for power system consulting firms many years ago. These programs have been refined and modularized for interactive use with *MATLAB* for many problems related to the operation and analysis of power systems. These software modules are versatile, allowing some of the typical problems to be solved by several methods, thus enabling students to investigate alternative solution techniques. Furthermore, the software modules are structured in such a way that the user may mix them for other power system analyses.

This book has more than 140 illustrative examples that use *MATLAB* to assist in the analysis of power systems. Each example illustrates a specific concept and usually contains a script of the *MATLAB* commands used for the model creation and computation. Some examples are quite elaborate, in order to bring the practical world closer. The *MATLAB* M-files on the accompanying diskette can be copied to the user's computer and used to solve all the examples. The scripts can also be utilized with modifications as the foundation for solving the end-of-chapter problems.

The book is organized into 12 chapters and 3 appendixes. Each chapter begins with a introduction describing the topics students will encounter. **Chapter 1** is a brief overview of the development of power systems and a description of the major components in the power system. Included is a discussion of generating stations and transmission and subtransmission networks that convey the energy from the primary source to the load areas. **Chapter 2** reviews power concepts and three-phase systems. Typical students already will have studied much of this material. However, this specialized topic of networks may not be included in circuit theory courses, and the review here will reinforce these concepts. Before going into system analysis, we have to model all components of electrical power systems. **Chapter 3** addresses the steady-state presentation and modeling of synchronous machines and transformers. Also, the per unit system is presented, followed by the one-line diagram representation of the network.

Chapter 4 discusses the parameters of a multicircuit transmission line. These parameters are computed for the balanced system on a per phase basis. **Chapter 5** thoroughly covers transmission line modeling and the performance and compensation of the transmission lines. This chapter provides the concepts and tools necessary for the preliminary transmission line design. **Chapter 6** presents a comprehensive coverage of the power flow solution of an interconnected power system during normal operation. First, the commonly used iterative techniques for the solution of nonlinear algebraic equation are discussed. Then several approaches to the solution of power flow are described. These techniques are applied to the solution of practical systems using the developed software modules.

Chapter 7 covers some essential classical optimization of continuous functions and their application to optimal dispatch of generation. The programs developed here are designed to work in synergy with the power flow programs. **Chapter 8** deals with synchronous machine transient analysis. The voltage equations of the synchronous machine are first developed. These nonlinear equations are transformed into linear differential equations using Park's transformation. Analytical solution of the transformed equations can be obtained by the Laplace transform technique. However, *MATLAB* is used with ease to simulate the nonlinear differential equations of the synchronous machine directly in time-domain in matrix form for all modes of operation. Thus students can observe the dynamic response of the synchronous machine during short circuits and appreciate the significance and consequence of the change of machine parameters. The ultimate objective of this chapter is to develop simple network models of the synchronous generator for power system fault analysis and transient stability studies.

Chapter 9 covers balanced fault analysis. The bus impedance matrix by the *building algorithms* is formulated and employed for the systematic computation of bus voltages and line currents during faults. **Chapter 10** discusses methods of symmetrical components that resolve the problem of an unbalanced circuit into a solution of a number of balanced circuits. Included are graphical displays of the symmetrical components transformation and some applications. The method is applied to the unbalanced fault, which once again allows the treatment of the problem on simple per phase basis. Algorithms have been developed to simulate different types of unbalanced faults. The software modules developed for unbalanced faults include single line-to-ground fault, line-to-line fault, and double line-to-ground fault.

Chapter 11 covers power system stability problems. First, the dynamic behavior of a one-machine system due to a small disturbance is investigated, and the analytical solution of this linearized model is obtained. *MATLAB* and *SIMULINK* are used conveniently to simulate the system, and the model is extended to multimachine systems. Next, the transient stability using equal area criteria is discussed, and the result is represented graphically, providing physical insight into the dynamic behavior of the machine. An introduction to nonlinear differential equations and their numerical solutions is given. *MATLAB* is used to obtain the numerical solution of the swing equation of a one-machine system. Simulation is also obtained using the *SIMULINK* toolbox. A program compatible with the power flow programs is developed for the transient stability analysis of the multimachine systems.

Chapter 12 is concerned with power system control and develops some of the control schemes required to operate the power system in the steady state. Simple models of the essential components used in control systems are presented. The automatic voltage regulator (AVR) and the load frequency control (LFC) are discussed. The automatic generation control (AGC) in single-area and multiarea systems, including tie-line power control, are analyzed. For each case, the responses to the real power demand are obtained. The generator responses with the AVR and various compensators, such as rate feedback and *Proportional Integral Derivative* (PID) controllers, are obtained. Both AGC and AVR systems are illustrated by several examples, and the responses are obtained using *MATLAB*. These analyses are supplemented by constructing the *SIMULINK* block diagram, which provides a highly interactive environment for simulation. Some basic materials of modern control theory are discussed, including the pole-placement state feedback design and the optimal controller designs using the linear quadratic regulator based on the *Riccati* equation. These modern techniques are then applied for simulation of the LFC systems.

Appendix A is a self-study *MATLAB* and *SIMULINK* tutorial focused on power and control systems and coordinated with the text. **Appendix B** includes a brief introduction to the fundamentals of control systems and is suitable for students without a background in control systems. **Appendix C** lists all functions, script files, and chapter examples. Answers to problems are given at the end of the book. The instructor's manual for this text contains the worked-out solutions for all of the book's problem.

The material in the text is designed to be fully covered in a two-semester undergraduate course sequence. The organization is flexible, allowing instructors to select the material that best suits the requirements of a one-quarter or a one-semester course. In a one-semester course, the first six chapters, which form the basis for power system analysis, should be covered. The material in Chapter 2 contains power concepts and three-phase systems, which are usually covered in circuit courses. This chapter can be excluded if the students are well prepared, or it can be used for review. Also, for students with electrical machinery background, Chapter 3 might be omitted. After the above coverage, additional material from the remaining chapters may then be appropriate, depending on the syllabus requirements and the individual preferences. One choice is to cover Chapter 7 (optimal dispatch of generation); another choice is Chapter 9 (balanced fault). The generator reactances required in Chapter 9 may be covered briefly from Section 8.7 without covering Chapter 8 in its entirety.

After reading the book, students should have a good perspective of power system analysis and an active knowledge of various numerical techniques that can be applied to the solution of large interconnected power systems. Students should find *MATLAB* helpful in learning the material in the text, particularly in solving the problems at the end of each chapter.

I would like to express my appreciation and thanks to the following reviewers for their many helpful comments and suggestions: Professor Max D. Anderson, University of Missouri-Rolla; Professor Miroslav Begovic, Georgia Institute of Technology; Professor Karen L. Butler, Texas A&M University; Professor Kevin A. Clements, Worcester Polytechnic Institute; Professor Mariesa L. Crow, University of Missouri-Rolla; Professor Malik Elbuluk, University of Akron; Professor A. A. El-Keib, University of Alabama; Professor F. P. Emad, University of Maryland; Professor L. L. Grigsby, Auburn University; Professor Kwang Y. Lee, Pennsylvania State University; Professor M. A. Pai, University of Illinois-Urbana; Professor E. K. Stanek, University of Missouri-Rolla.

My sincere thanks to Lynn Kallas, who proofread the early version of the manuscript. Special thanks goes to the staff of McGraw-Hill: Lynn Cox, the editor, for her constant encouragement, Nina Kreiden, editorial coordinator, for her support, and Eve Strock, senior project manager, for her attention to detail during all phases of editing and production.

I wish to express my thanks to the Electrical Engineering and Computer Science Department of Milwaukee School of Engineering and to Professor Ray Palmer, chairman of the department, for giving me the opportunity to prepare this material.

Last, but not least, I thank my wife Jila and my children, Dana, Fred, and Cameron, who were a constant and active source of support throughout the endeavor.

CHAPTER 1

THE POWER SYSTEM: AN OVERVIEW

1.1 INTRODUCTION

Electric energy is the most popular form of energy, because it can be transported easily at high efficiency and reasonable cost.

The first electric network in the United States was established in 1882 at the Pearl Street Station in New York City by Thomas Edison. The station supplied dc power for lighting the lower Manhattan area. The power was generated by dc generators and distributed by underground cables. In the same year the first water-wheel driven generator was installed in Appleton, Wisconsin. Within a few years many companies were established producing energy for lighting – all operated under Edison's patents. Because of the excessive power loss, RI^2 at low voltage, Edison's companies could deliver energy only a short distance from their stations.

With the invention of the transformer (William Stanley, 1885) to raise the level of ac voltage for transmission and distribution and the invention of the induction motor (Nikola Tesla, 1888) to replace the dc motors, the advantages of the ac system became apparent, and made the ac system prevalent. Another advantage of the ac system is that due to lack of commutators in the ac generators, more power can be produced conveniently at higher voltages.

The first single-phase ac system in the United States was at Oregon City where power was generated by two 300 hp waterwheel turbines and transmitted at 4 kV to Portland. Southern California Edison Company installed the first three-phase system at 2.3 kV in 1893. Many electric companies were developed throughout the country. In the beginning, individual companies were operating at different frequencies anywhere from 25 Hz to 133 Hz. But, as the need for interconnection and parallel operation became evident, a standard frequency of 60 Hz was adopted throughout the U.S. and Canada. Most European countries selected the 50-Hz system. Transmission voltages have since risen steadily, and the extra high voltage (EHV) in commercial use is 765 kV, first put into operation in the United States in 1969.

For transmitting power over very long distances it may be more economical to convert the EHV ac to EHV dc, transmit the power over two lines, and invert it back to ac at the other end. Studies show that it is advantageous to consider dc lines when the transmission distance is 500 km or more. DC lines have no reactance and are capable of transferring more power for the same conductor size than ac lines. DC transmission is especially advantageous when two remotely located large systems are to be connected. The dc transmission tie line acts as an asynchronous link between the two rigid systems eliminating the instability problem inherent in the ac links. The main disadvantage of the dc link is the production of harmonics which requires filtering, and a large amount of reactive power compensation required at both ends of the line. The first $\pm$400-kV dc line in the United States was the Pacific Intertie, 850 miles long between Oregon and California built in 1970.

The entire continental United States is interconnected in an overall network called the *power grid*. A small part of the network is federally and municipally owned, but the bulk is privately owned. The system is divided into several geographical regions called *power pools*. In an interconnected system, fewer generators are required as a reserve for peak load and spinning reserve. Also, interconnection makes the energy generation and transmission more economical and reliable, since power can readily be transferred from one area to others. At times, it may be cheaper for a company to buy bulk power from neighboring utilities than to produce it in one of its older plants.

1.2 ELECTRIC INDUSTRY STRUCTURE

The bulk generation of electricity in the United States is produced by integrated investor-owned utilities (IOU). A small portion of power generation is federally owned, such as the Tennessee Valley Authority and Bonneville Power Administration. Two separate levels of regulation currently regulate the United States electric system. One is the Federal Energy Regulatory Commission (FERC), which regulates the price of wholesale electricity, service terms, and conditions. The other is the Securities and Exchange Commission (SEC), which regulates the business structure of electric utilities.

The transmission system of electric utilities in the Unites States and Canada is interconnected into a large power grid known as the North American Power Systems Interconnection. The power grid is divided into several pools. The pools consist of several neighboring utilities which operate jointly to schedule generation in a cost-effective manner. A privately regulated organization called the North American Electric Reliability Council (NERC) is responsible for maintaining system standards and reliability. NERC works cooperatively with every provider and distributor of power to ensure reliability. NERC coordinates its efforts with FERC as well as other organizations such as the Edison Electric Institute (EEI). NERC currently has four distinct electrically separated areas. These areas are the Electric Reliability Council of Texas (ERCOT); the Western States Coordination Council (WSCC); the Eastern Interconnect, which includes all the states and provinces of Canada east of the Rocky Mountains (excluding Texas), and Hydro-Quebec, which has dc interconnects with the northeast. These electrically separate areas import and export power to each other but are not synchronized electrically.

The electric power industry in the United States is undergoing fundamental changes since the deregulation of the telecommunication, gas, and other industries. The generation business is rapidly becoming market-driven. This is a major change for an industry which, until the last decade, was characterized by large, vertically integrated monopolies. The implementation of open transmission access has resulted in wholesale and retail markets. In the future, utilities may possibly be divided into power generation, transmission, and retail segments. Generating utilities would sell directly to customers instead of to local distributors. This would eliminate the monopoly that distributors currently have. The distributors would sell their services as electricity distributors instead of being a retailer of electricity itself. The retail structure of power distribution would resemble the current structure of the telephone communication industry. The consumer would have a choice as to from which generator they purchase power. If the entire electric power industry were to be deregulated, final consumers could choose from generators across the country. Power brokers and power marketers will assume a major role in this new competitive power industry. Currently, the ability to market electricity to retail end users exists, but only in a limited number of states in pilot programs.

Extensive efforts are being made to create a more competitive environment for electricity markets in order to promote greater efficiency. Thus, the power industry faces many new problems, with one of the highest priority issues being reliability, that is, bringing a steady, uninterruptable power supply to all electricity consumers. The restructuring and deregulation of electric utilities, together with recent progress in technology, introduce unprecedented challenges and opportunities for power systems research and open up new opportunities to young power engineers.

1.3 MODERN POWER SYSTEM

The power system of today is a complex interconnected network as shown in Figure 1.1 (page 7). A power system can be subdivided into four major parts:

- Generation
- Transmission and Subtransmission
- Distribution
- Loads

1.3.1 GENERATION

Generators — One of the essential components of power systems is the three-phase ac generator known as synchronous generator or alternator. Synchronous generators have two synchronously rotating fields: One field is produced by the rotor driven at synchronous speed and excited by dc current. The other field is produced in the stator windings by the three-phase armature currents. The dc current for the rotor windings is provided by excitation systems. In the older units, the exciters are dc generators mounted on the same shaft, providing excitation through slip rings. Today's systems use ac generators with rotating rectifiers, known as *brushless* excitation systems. The generator excitation system maintains generator voltage and controls the reactive power flow. Because they lack the commutator, ac generators can generate high power at high voltage, typically 30 kV. In a power plant, the size of generators can vary from 50 MW to 1500 MW.

The source of the mechanical power, commonly known as the *prime mover*, may be hydraulic turbines at waterfalls, steam turbines whose energy comes from the burning of coal, gas and nuclear fuel, gas turbines, or occasionally internal combustion engines burning oil. The estimated installed generation capacity in 1998 for the United States is presented in Table 1.1.

Steam turbines operate at relatively high speeds of 3600 or 1800 rpm. The generators to which they are coupled are cylindrical rotor, two-pole for 3600 rpm or four-pole for 1800 rpm operation. Hydraulic turbines, particularly those operating with a low pressure, operate at low speed. Their generators are usually a salient type rotor with many poles. In a power station several generators are operated in parallel in the power grid to provide the total power needed. They are connected at a common point called a *bus*.

Today the total installed electric generating capacity is about 760,000 MW. Assuming the United States population to be 270 million,

$$\text{Installed capacity per capita} = \frac{760 \times 10^9}{270 \times 10^6} = 2815 \ \text{W}$$

To realize the significance of this figure, consider the average power of a person to be approximately 50 W. Therefore, the power of 2815 W is equivalent to

$$\frac{2815 \ \text{W}}{50 \ \text{W}} = 56 \ \text{(power slave)}$$

The annual kWh consumption in the United States is about $3,550 \times 10^9$ kWh. The asset of the investment for investor-owned companies is about 200 billion dollars and they employ close to a half million people.

With today's emphasis on environmental consideration and conservation of fossil fuels, many alternate sources are considered for employing the untapped energy sources of the sun and the earth for generation of power. Some of these alternate sources which are being used to some extent are solar power, geothermal power, wind power, tidal power, and biomass. The aspiration for bulk generation of power in the future is the nuclear *fusion*. If nuclear fusion is harnessed economically, it would provide clean energy from an abundant source of fuel, namely water.

Table 1.1 Installed Generation Capacity

Type	**Capacity, MW**	**Percent**	**Fuel**
Steam Plant	478,800	63	Coal, gas, petroleum
Nuclear	106,400	14	Uranium
Hydro and pumped storage	91,200	12	Water
Gas Turbine	60,800	8	Gas, petroleum
Combined cycle	15,200	2	Gas, petroleum
Internal Combustion	4,940	0.65	Gas, petroleum
Others	2,660	0.35	Geothermal, solar, wind
Total	760,000	100.00	

Transformers — Another major component of a power system is the transformer. It transfers power with very high efficiency from one level of voltage to another level. The power transferred to the secondary is almost the same as the primary, except for losses in the transformer, and the product VI on the secondary side is approximately the same as the primary side. Therefore, using a step-up transformer of turns ratio a will reduce the secondary current by a ratio of $1/a$. This will reduce losses in the line, which makes the transmission of power over long distances possible.

The insulation requirements and other practical design problems limit the generated voltage to low values, usually 30 kV. Thus, step-up transformers are used for transmission of power. At the receiving end of the transmission lines step-down transformers are used to reduce the voltage to suitable values for distribution or utilization. In a modern utility system, the power may undergo four or five transformations between generator and ultimate user.

1.3.2 TRANSMISSION AND SUBTRANSMISSION

The purpose of an overhead transmission network is to transfer electric energy from generating units at various locations to the distribution system which ultimately supplies the load. Transmission lines also interconnect neighboring utilities which permits not only economic dispatch of power within regions during normal conditions, but also the transfer of power between regions during emergencies.

Standard transmission voltages are established in the United States by the American National Standards Institute (ANSI). Transmission voltage lines operating at more than 60 kV are standardized at 69 kV, 115 kV, 138 kV, 161 kV, 230 kV, 345 kV, 500 kV, and 765 kV line-to-line. Transmission voltages above 230 kV are usually referred to as extra-high voltage (EHV).

Figure 1.1 shows an elementary diagram of a transmission and distribution system. High voltage transmission lines are terminated in substations, which are called *high-voltage substations*, *receiving substations*, or *primary substations*. The function of some substations is switching circuits in and out of service; they are referred to as *switching stations*. At the primary substations, the voltage is stepped down to a value more suitable for the next part of the journey toward the load. Very large industrial customers may be served from the transmission system.

The portion of the transmission system that connects the high-voltage substations through step-down transformers to the distribution substations are called the *subtransmission* network. There is no clear delineation between transmission and subtransmission voltage levels. Typically, the subtransmission voltage level ranges from 69 to 138 kV. Some large industrial customers may be served from the subtransmission system. Capacitor banks and reactor banks are usually installed in the substations for maintaining the transmission line voltage.

1.3.3 DISTRIBUTION

The distribution system is that part which connects the distribution substations to the consumers' service-entrance equipment. The primary distribution lines are usually in the range of 4 to 34.5 kV and supply the load in a well-defined geographical area. Some small industrial customers are served directly by the primary feeders.

The secondary distribution network reduces the voltage for utilization by commercial and residential consumers. Lines and cables not exceeding a few hun-

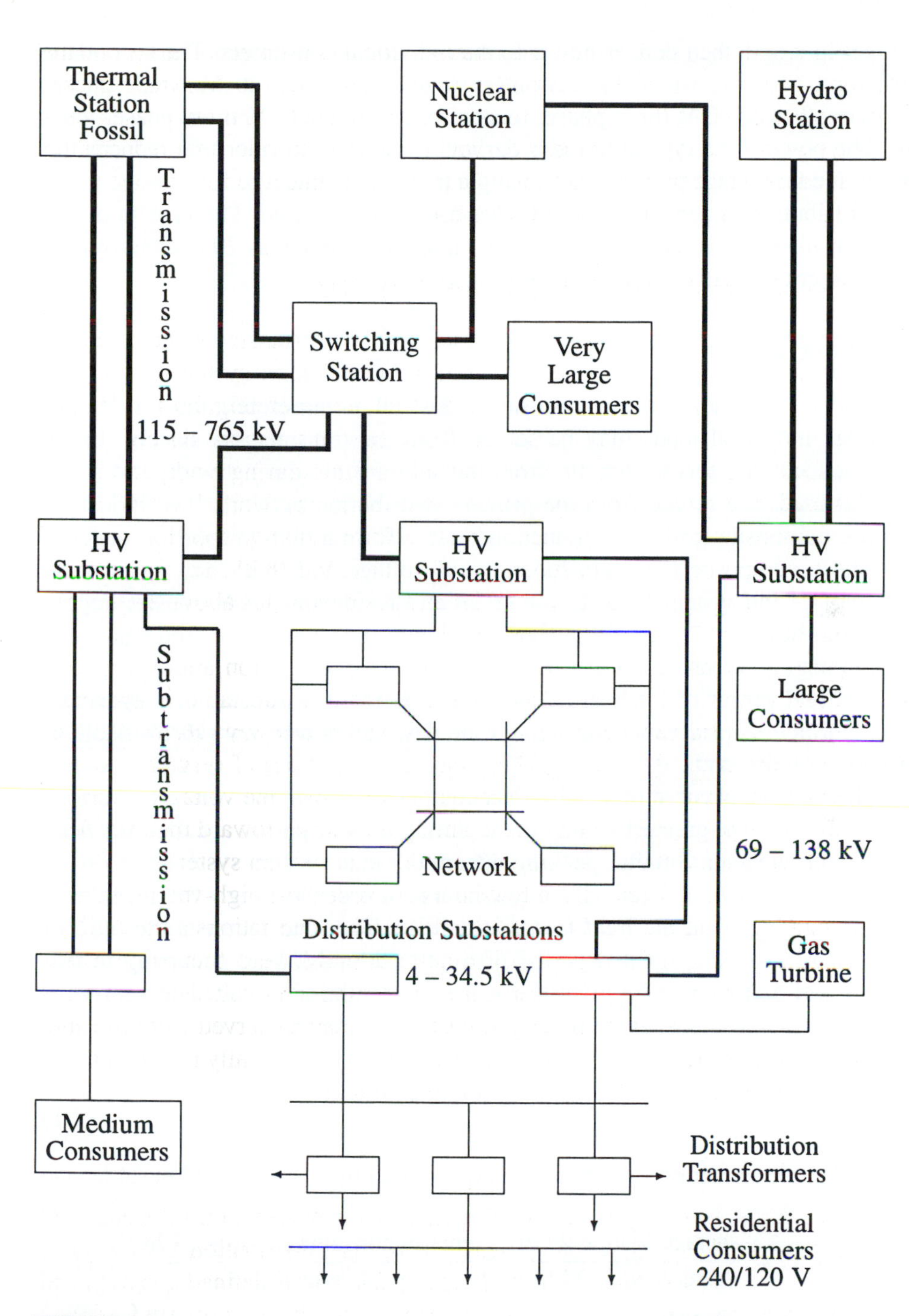

FIGURE 1.1
Basic components of a power system.

dred feet in length then deliver power to the individual consumers. The secondary distribution serves most of the customers at levels of 240/120 V, single-phase, three-wire; 208Y/120 V, three-phase, four-wire; or 480Y/277 V, three-phase, four-wire. The power for a typical home is derived from a transformer that reduces the primary feeder voltage to 240/120 V using a three-wire line.

Distribution systems are both *overhead* and *underground*. The growth of underground distribution has been extremely rapid and as much as 70 percent of new residential construction is served underground.

1.3.4 LOADS

Loads of power systems are divided into industrial, commercial, and residential. Very large industrial loads may be served from the transmission system. Large industrial loads are served directly from the subtransmission network, and small industrial loads are served from the primary distribution network. The industrial loads are composite loads, and induction motors form a high proportion of these load. These composite loads are functions of voltage and frequency and form a major part of the system load. Commercial and residential loads consist largely of lighting, heating, and cooling. These loads are independent of frequency and consume negligibly small reactive power.

The real power of loads are expressed in terms of kilowatts or megawatts. The magnitude of load varies throughout the day, and power must be available to consumers on demand.

The daily-load curve of a utility is a composite of demands made by various classes of users. The greatest value of load during a 24-hr period is called the *peak* or *maximum demand*. Smaller peaking generators may be commissioned to meet the peak load that occurs for only a few hours. In order to assess the usefulness of the generating plant the *load factor* is defined. The load factor is the ratio of average load over a designated period of time to the peak load occurring in that period. Load factors may be given for a day, a month, or a year. The yearly, or annual load factor is the most useful since a year represents a full cycle of time. The daily load factor is

$$\text{Daily L.F.} = \frac{\text{average load}}{\text{peak load}} \tag{1.1}$$

Multiplying the numerator and denominator of (1.1) by a time period of 24 hr, we have

$$\text{Daily L.F.} = \frac{\text{average load} \times 24\text{ hr}}{\text{peak load} \times 24\text{ hr}} = \frac{\text{energy consumed during 24 hr}}{\text{peak load} \times 24\text{ hr}} \tag{1.2}$$

The annual load factor is

$$\text{Annual L.F.} = \frac{\text{total annual energy}}{\text{peak load} \times 8760\text{ hr}} \tag{1.3}$$

Generally there is diversity in the peak load between different classes of loads, which improves the overall system load factor. In order for a power plant to operate economically, it must have a high system load factor. Today's typical system load factors are in the range of 55 to 70 percent.

There are a few other factors used by utilities. *Utilization factor* is the ratio of maximum demand to the installed capacity, and *plant factor* is the ratio of annual energy generation to the plant capacity $\times$ 8760 hr. These factors indicate how well the system capacity is utilized and operated.

A *MATLAB* function **barcycle(data)** is developed which obtains a plot of the load cycle for a given interval. The demand interval and the load must be defined by the variable **data** in a three-column matrix. The first two columns are the demand interval and the third column is the load value. The demand interval may be minutes, hours, or months, in ascending order. Hourly intervals must be expressed in military time.

Example 1.1

The daily load on a power system varies as shown in Table 1.2. Use the **barcycle** function to obtain a plot of the daily load curve. Using the given data compute the average load and the daily load factor (Figure 1.2).

Table 1.2 Daily System Load

Interval, hr			Load, MW
12 A.M.	–	2 A.M.	6
2	–	6	5
6	–	9	10
9	–	12	15
12 P.M.	–	2 P.M.	12
2	–	4	14
4	–	6	16
6	–	8	18
8	–	10	16
10	–	11	12
11	–	12 A.M.	6

The following commands

```
data = [ 0   2    6
         2   6    5
         6   9   10
         9  12   15
        12  14   12
```

```
        14  16   14
        16  18   16
        18  20   18
        20  22   16
        22  23   12
        23  24    6];
P = data(:,3);                          % Column array of load
Dt = data(:, 2) - data(:,1); % Column array of demand interval
W = P'*Dt;                    % Total energy, area under the curve
Pavg = W/sum(Dt)                                  % Average load
Peak = max(P)                                        % Peak load
LF = Pavg/Peak*100                        % Percent load factor
barcycle(data)                            % Plots the load cycle
xlabel('Time, hr'), ylabel('P,  MW')
```

result in

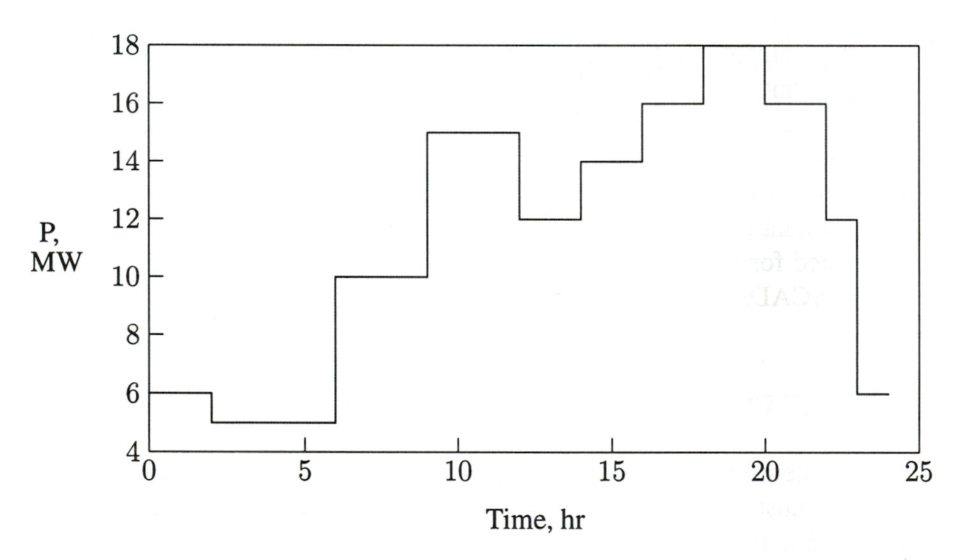

FIGURE 1.2
Daily load cycle for Example 1.1.

```
Pavg = 11.5417
Peak = 18
LF   = 64.12
```

1.4 SYSTEM PROTECTION

In addition to generators, transformers, and transmission lines, other devices are required for the satisfactory operation and protection of a power system. Some of the protective devices directly connected to the circuits are called *switchgear*. They include instrument transformers, circuit breakers, disconnect switches, fuses and lightning arresters. These devices are necessary to deenergize either for normal operation or on the occurrence of faults. The associated control equipment and protective relays are placed on *switchboard* in *control houses*.

1.5 ENERGY CONTROL CENTER

For reliable and economical operation of the power system it is necessary to monitor the entire system in a control center. The modern control center of today is called the *energy control center* (ECC). Energy control centers are equipped with on-line computers performing all signal processing through the remote acquisition system. Computers work in a hierarchical structure to properly coordinate different functional requirements in normal as well as emergency conditions. Every energy control center contains a control console which consists of a visual display unit (VDU), keyboard, and light pen. Computers may give alarms as advance warnings to the operators (dispatchers) when deviation from the normal state occurs. The dispatcher makes judgments and decisions and executes them with the aid of a computer. Simulation tools and software packages written in high-level language are implemented for efficient operation and reliable control of the system. This is referred to as SCADA, an acronym for "supervisory control and data acquisition."

1.6 COMPUTER ANALYSIS

For a power system to be practical it must be safe, reliable, and economical. Thus many analyses must be performed to design and operate an electrical system. However, before going into system analysis we have to model all components of electrical power systems. Therefore, in this text, after reviewing the concepts of power and three-phase circuits, we will calculate the parameters of a multi-circuit transmission line. Then, we will model the transmission line and look at the performance of the transmission line. Since transformers and generators are a part of the system, we will model these devices. Design of a power system, its operation and expansion requires much analysis. This text presents methods of power system analysis with the aid of a personal computer and the use of *MATLAB*. The *MATLAB* environment permits a nearly direct transition from mathematical expression to simulation. Some of the basic analysis covered in this text are:

- Evaluation of transmission line parameters
- Transmission line performance and compensation
- Power flow analysis
- Economic scheduling of generation
- Synchronous machine transient analysis
- Balanced fault
- Symmetrical components and unbalanced fault
- Stability studies
- Power system control

Many *MATLAB* functions are developed for the above studies thus allowing the student to concentrate on analysis and design of practical systems and spend less time on programming.

PROBLEMS

1.1. The demand estimation is the starting point for planning the future electric power supply. The consistency of demand growth over the years has led to numerous attempts to fit mathematical curves to this trend. One of the simplest curves is

$$P = P_0 e^{a(t-t_0)}$$

where a is the average per unit growth rate, P is the demand in year t, and P_0 is the given demand at year t_0.

Assume the peak power demand in the United States in 1984 is 480 GW with an average growth rate of 3.4 percent. Using *MATLAB*, plot the predicated peak demand in GW from 1984 to 1999. Estimate the peak power demand for the year 1999.

1.2. In a certain country, the energy consumption is expected to double in 10 years. Assuming a simple exponential growth given by

$$P = P_0 e^{at}$$

calculate the growth rate a.

1.3. The annual load of a substation is given in the following table. During each month, the power is assumed constant at an average value. Using *MATLAB* and the **barcycle** function, obtain a plot of the annual load curve. Write the necessary statements to find the average load and the annual load factor.

Annual System Load	
Interval, month	Load, MW
January	8
February	6
March	4
April	2
May	6
June	12
July	16
August	14
September	10
October	4
November	6
December	8

CHAPTER 2

BASIC PRINCIPLES

2.1 INTRODUCTION

The concept of power is of central importance in electrical power systems and is the main topic of this chapter. The typical student will already have studied much of this material, and the review here will serve to reinforce the power concepts encountered in the electric circuit theory.

In this chapter, the flow of energy in an ac circuit is investigated. By using various trigonometric identities, the instantaneous power $p(t)$ is resolved into two components. A plot of these components is obtained using *MATLAB* to observe that ac networks not only consume energy at an average rate, but also borrow and return energy to its sources. This leads to the basic definitions of average power P and reactive power Q. The volt-ampere S, which is a mathematical formulation based on the phasor forms of voltage and current, is introduced. Then the complex power balance is demonstrated, and the transmission inefficiencies caused by loads with low power factors are discussed and demonstrated by means of several examples.

Next, the transmission of complex power between two voltage sources is considered, and the dependency of real power on the voltage phase angle and the dependency of reactive power on voltage magnitude is established. *MATLAB* is used conveniently to demonstrate this idea graphically.

Finally, the balanced three-phase circuit is examined. An important property of a balanced three-phase system is that it delivers constant power. That is, the power delivered does not fluctuate with time as in a single-phase system. For the purpose of analysis and modeling, the per-phase equivalent circuit is developed for the three-phase system under balanced condition.

2.2 POWER IN SINGLE-PHASE AC CIRCUITS

Figure 2.1 shows a single-phase sinusoidal voltage supplying a load.

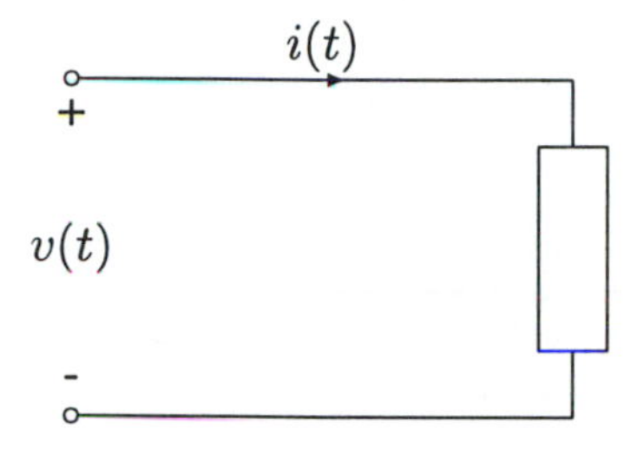

FIGURE 2.1
Sinusoidal source supplying a load.

Let the instantaneous voltage be

$$v(t) = V_m \cos(\omega t + \theta_v) \tag{2.1}$$

and the instantaneous current be given by

$$i(t) = I_m \cos(\omega t + \theta_i) \tag{2.2}$$

The instantaneous power $p(t)$ delivered to the load is the product of voltage $v(t)$ and current $i(t)$ given by

$$p(t) = v(t)\, i(t) = V_m I_m \cos(\omega t + \theta_v) \cos(\omega t + \theta_i) \tag{2.3}$$

In Example 2.1, *MATLAB* is used to plot the instantaneous power $p(t)$, and the result is shown in Figure 2.2. In studying Figure 2.2, we note that the frequency of the instantaneous power is twice the source frequency. Also, note that it is possible for the instantaneous power to be negative for a portion of each cycle. In a passive network, negative power implies that energy that has been stored in inductors or capacitors is now being extracted.

It is informative to write (2.3) in another form using the trigonometric identity

$$\cos A \cos B = \frac{1}{2}\cos(A - B) + \frac{1}{2}\cos(A + B) \tag{2.4}$$

which results in

$$\begin{aligned} p(t) &= \frac{1}{2}V_m I_m[\cos(\theta_v - \theta_i) + \cos(2\omega t + \theta_v + \theta_i)] \\ &= \frac{1}{2}V_m I_m\{\cos(\theta_v - \theta_i) + \cos[2(\omega t + \theta_v) - (\theta_v - \theta_i)]\} \\ &= \frac{1}{2}V_m I_m[\cos(\theta_v - \theta_i) + \cos 2(\omega t + \theta_v)\cos(\theta_v - \theta_i) \\ &\quad + \sin 2(\omega t + \theta_v)\sin(\theta_v - \theta_i)] \end{aligned}$$

The *root-mean-square* (rms) value of $v(t)$ is $|V| = V_m/\sqrt{2}$ and the rms value of $i(t)$ is $|I| = I_m/\sqrt{2}$. Let $\theta = (\theta_v - \theta_i)$. The above equation, in terms of the rms values, is reduced to

$$p(t) = \underbrace{|V||I|\cos\theta[1 + \cos 2(\omega t + \theta_v)]}_{\substack{p_R(t) \\ \text{Energy flow into} \\ \text{the circuit}}} + \underbrace{|V||I|\sin\theta\sin 2(\omega t + \theta_v)}_{\substack{p_X(t) \\ \text{Energy borrowed and} \\ \text{returned by the circuit}}} \tag{2.5}$$

where θ is the angle between voltage and current, or the impedance angle. θ is positive if the load is inductive, (i.e., current is lagging the voltage) and θ is negative if the load is capacitive (i.e., current is leading the voltage).

The instantaneous power has been decomposed into two components. The first component of (2.5) is

$$p_R(t) = |V||I|\cos\theta + |V||I|\cos\theta\cos 2(\omega t + \theta_v)] \tag{2.6}$$

The second term in (2.6), which has a frequency twice that of the source, accounts for the sinusoidal variation in the absorption of power by the resistive portion of the load. Since the average value of this sinusoidal function is zero, the average power delivered to the load is given by

$$P = |V||I|\cos\theta \tag{2.7}$$

This is the power absorbed by the resistive component of the load and is also referred to as the *active power* or *real power*. The product of the rms voltage value and the rms current value $|V||I|$ is called the *apparent power* and is measured in units of volt ampere. The product of the apparent power and the cosine of the angle between voltage and current yields the real power. Because $\cos\theta$ plays a key role in the determination of the average power, it is called *power factor*. When the current lags the voltage, the power factor is considered lagging. When the current leads the voltage, the power factor is considered leading.

The second component of (2.5)

$$p_X(t) = |V||I|\sin\theta\sin 2(\omega t + \theta_v) \tag{2.8}$$

pulsates with twice the frequency and has an average value of zero. This component accounts for power oscillating into and out of the load because of its reactive element (inductive or capacitive). The amplitude of this pulsating power is called *reactive power* and is designated by Q.

$$Q = |V||I| \sin \theta \tag{2.9}$$

Both P and Q have the same dimension. However, in order to distinguish between the real and the reactive power, the term "var" is used for the reactive power (var is an acronym for the phrase "volt-ampere reactive"). For an inductive load, current is lagging the voltage, $\theta = (\theta_v - \theta_i) > 0$ and Q is positive; whereas, for a capacitive load, current is leading the voltage, $\theta = (\theta_v - \theta_i) < 0$ and Q is negative.

A careful study of Equations (2.6) and (2.8) reveals the following characteristics of the instantaneous power.

- For a pure resistor, the impedance angle is zero and the power factor is unity (UPF), so that the apparent and real power are equal. The electric energy is transformed into thermal energy.
- If the circuit is purely inductive, the current lags the voltage by 90° and the average power is zero. Therefore, in a purely inductive circuit, there is no transformation of energy from electrical to nonelectrical form. The instantaneous power at the terminal of a purely inductive circuit oscillates between the circuit and the source. When $p(t)$ is positive, energy is being stored in the magnetic field associated with the inductive elements, and when $p(t)$ is negative, energy is being extracted from the magnetic fields of the inductive elements.
- If the load is purely capacitive, the current leads the voltage by 90°, and the average power is zero, so there is no transformation of energy from electrical to nonelectrical form. In a purely capacitive circuit, the power oscillates between the source and the electric field associated with the capacitive elements.

Example 2.1

The supply voltage in Figure 2.1 is given by $v(t) = 100 \cos \omega t$ and the load is inductive with impedance $Z = 1.25\angle 60^\circ\ \Omega$. Determine the expression for the instantaneous current $i(t)$ and the instantaneous power $p(t)$. Use *MATLAB* to plot $i(t)$, $v(t)$, $p(t)$, $p_R(t)$, and $p_X(t)$ over an interval of 0 to 2π.

$$I_{max} = \frac{100\angle 0^\circ}{1.25\angle 60^\circ} = 80\angle - 60^\circ \ \text{A}$$

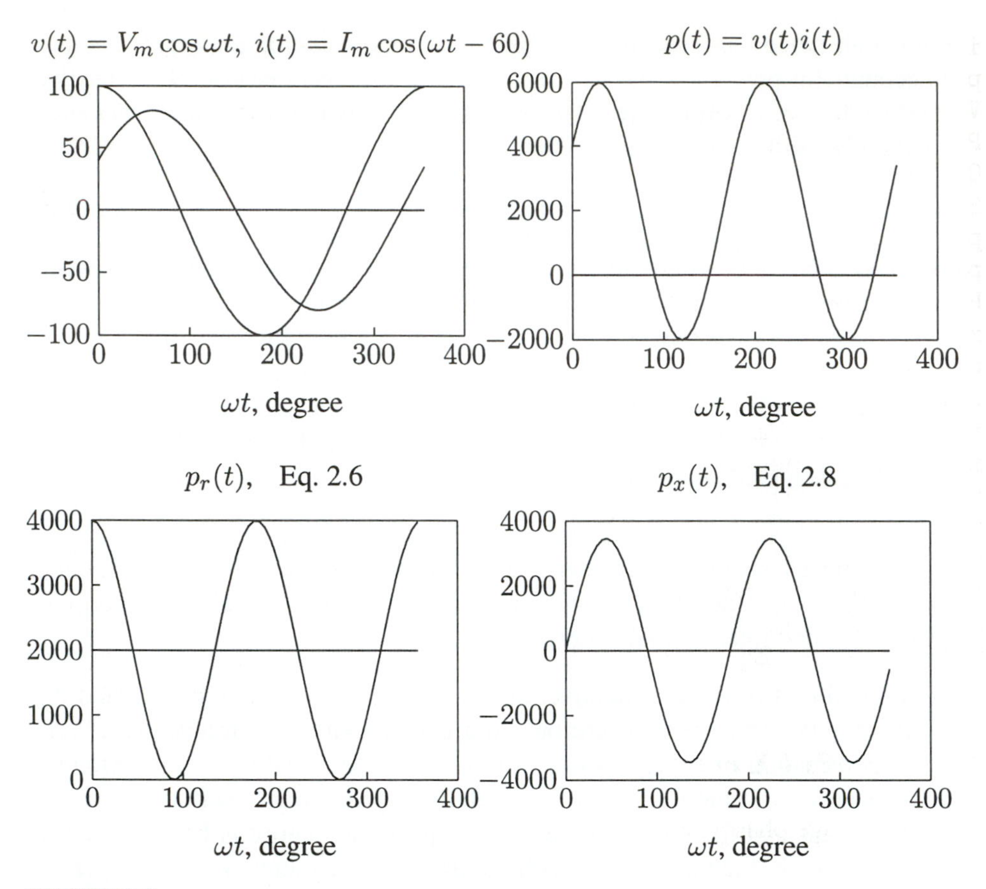

FIGURE 2.2
Instantaneous current, voltage, power, Eqs. 2.6 and 2.8.

therefore

$$i(t) = 80\cos(\omega t - 60^\circ) \text{ A}$$
$$p(t) = v(t)\,i(t) = 8000\cos\omega t\cos(\omega t - 60^\circ) \text{ W}$$

The following statements are used to plot the above instantaneous quantities and the instantaneous terms given by (2.6) and (2.8).

```
Vm = 100; thetav = 0;       % Voltage amplitude and phase angle
Z = 1.25;  gama = 60;     % Impedance magnitude and phase angle
thetai = thetav - gama;        % Current phase angle in degree
theta = (thetav - thetai)*pi/180;              % Degree to radian
Im = Vm/Z;                                    % Current amplitude
wt = 0:.05:2*pi;                              % wt from 0 to 2*pi
v = Vm*cos(wt);                           % Instantaneous voltage
```

```
i = Im*cos(wt + thetai*pi/180);          % Instantaneous current
p = v.*i;                                  % Instantaneous power
V = Vm/sqrt(2); I=Im/sqrt(2);            % rms voltage and current
P = V*I*cos(theta);                              % Average power
Q = V*I*sin(theta);                             % Reactive power
S = P + j*Q                                      % Complex power
pr = P*(1 + cos(2*(wt + thetav)));                  % Eq. (2.6)
px = Q*sin(2*(wt + thetav));                        % Eq. (2.8)
PP = P*ones(1, length(wt));%Average power of length w for plot
xline = zeros(1, length(wt));            %generates a zero vector
wt=180/pi*wt;                        % converting radian to degree
subplot(2,2,1), plot(wt, v, wt,i,wt, xline), grid
title(['v(t)=Vm coswt, i(t)=Im cos(wt+',num2str(thetai), ')'])
xlabel('wt, degree')
subplot(2,2,2), plot(wt, p, wt, xline), grid
title('p(t)=v(t) i(t)'),xlabel('wt, degree')
subplot(2,2,3), plot(wt, pr, wt, PP,wt,xline), grid
title('pr(t)   Eq. 2.6'), xlabel('wt, degree')
subplot(2,2,4), plot(wt, px, wt, xline), grid
title('px(t)  Eq. 2.8'), xlabel('wt, degree'), subplot(111)
```

2.3 COMPLEX POWER

The rms voltage phasor of (2.1) and the rms current phasor of (2.2) shown in Figure 2.3 are

$$V = |V|\angle\theta_v \text{ and } I = |I|\angle\theta_i$$

The term VI^* results in

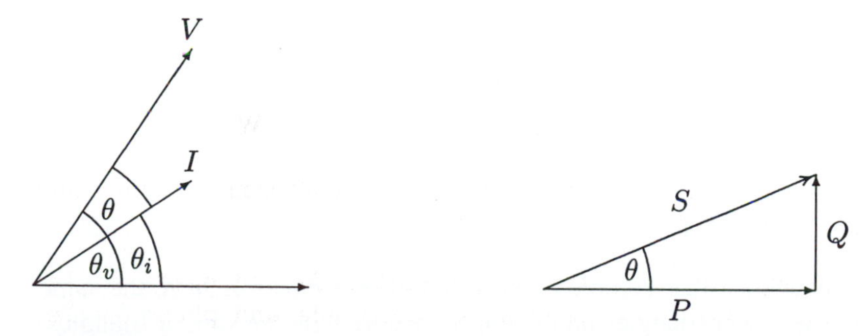

FIGURE 2.3
Phasor diagram and power triangle for an inductive load (lagging PF).

$$VI^* = |V||I|\angle\theta_v - \theta_i = |V||I|\angle\theta$$

$$= |V||I|\cos\theta + j|V||I|\sin\theta$$

The above equation defines a complex quantity where its real part is the average (real) power P and its imaginary part is the reactive power Q. Thus, the complex power designated by S is given by

$$S = VI^* = P + jQ \tag{2.10}$$

The magnitude of S, $|S| = \sqrt{P^2 + Q^2}$, is the apparent power; its unit is volt-amperes and the larger units are kVA or MVA. Apparent power gives a direct indication of heating and is used as a rating unit of power equipment. Apparent power has practical significance for an electric utility company since a utility company must supply both average and apparent power to consumers.

The reactive power Q is positive when the phase angle θ between voltage and current (impedance angle) is positive (i.e., when the load impedance is inductive, and I lags V). Q is negative when θ is negative (i.e., when the load impedance is capacitive and I leads V) as shown in Figure 2.4.

In working with Equation (2.10) it is convenient to think of P, Q, and S as forming the sides of a right triangle as shown in Figures 2.3 and 2.4.

FIGURE 2.4
Phasor diagram and power triangle for a capacitive load (leading PF).

If the load impedance is Z then

$$V = ZI \tag{2.11}$$

substituting for V into (2.10) yields

$$S = VI^* = ZII^* = R|I|^2 + jX|I|^2 \tag{2.12}$$

From (2.12) it is evident that complex power S and impedance Z have the same angle. Because the power triangle and the impedance triangle are similar triangles, the impedance angle is sometimes called the *power angle*.

Similarly, substituting for I from (2.11) into (2.10) yields

$$S = VI^* = \frac{VV^*}{Z^*} = \frac{|V|^2}{Z^*} \tag{2.13}$$

From (2.13), the impedance of the complex power S is given by

$$Z = \frac{|V|^2}{S^*} \tag{2.14}$$

2.4 THE COMPLEX POWER BALANCE

From the conservation of energy, it is clear that real power supplied by the source is equal to the sum of real powers absorbed by the load. At the same time, a balance between the reactive power must be maintained. Thus the total complex power delivered to the loads in parallel is the sum of the complex powers delivered to each. Proof of this is as follows:

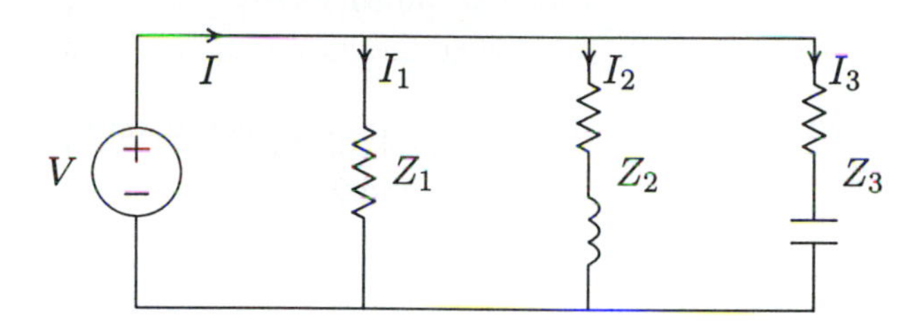

FIGURE 2.5
Three loads in parallel.

For the three loads shown in Figure 2.5, the total complex power is given by

$$S = VI^* = V[I_1 + I_2 + I_3]^* = VI_1^* + VI_2^* + VI_3^* \tag{2.15}$$

Example 2.2

In the above circuit $V = 1200\angle 0°$ V, $Z_1 = 60 + j0\ \Omega$, $Z_2 = 6 + j12\ \Omega$ and $Z_3 = 30 - j30\ \Omega$. Find the power absorbed by each load and the total complex power.

$$I_1 = \frac{1200\angle 0°}{60\angle 0} = 20 + j0 \text{ A}$$
$$I_2 = \frac{1200\angle 0°}{6 + j12} = 40 - j80 \text{ A}$$
$$I_3 = \frac{1200\angle 0°}{30 - j30} = 20 + j20 \text{ A}$$

$$S_1 = VI_1^* = 1200\angle 0^\circ(20 - j0) = 24,000 \text{ W} + j0 \text{ var}$$
$$S_2 = VI_2^* = 1200\angle 0^\circ(40 + j80) = 48,000 \text{ W} + j96,000 \text{ var}$$
$$S_3 = VI_3^* = 1200\angle 0^\circ(20 - j20) = 24,000 \text{ W} - j24,000 \text{ var}$$

The total load complex power adds up to

$$S = S_1 + S_2 + S_3 = 96,000 \text{ W} + j72,000 \text{ var}$$

Alternatively, the sum of complex power delivered to the load can be obtained by first finding the total current.

$$\begin{aligned} I &= I_1 + I_2 + I_3 = (20 + j0) + (40 - j80) + (20 + j20) \\ &= 80 - j60 = 100\angle -36.87^\circ \text{ A} \end{aligned}$$

and

$$\begin{aligned} S = VI^* &= (1200\angle 0^\circ)(100\angle 36.87^\circ) = 120,000\angle 36.87^\circ \text{ VA} \\ &= 96,000 \text{ W} + j72,000 \text{ var} \end{aligned}$$

A final insight is contained in Figure 2.6, which shows the current phasor diagram and the complex power vector representation.

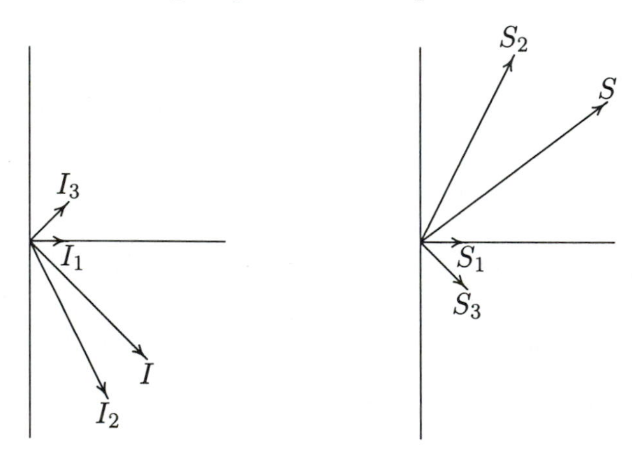

FIGURE 2.6
Current phasor diagram and power plane diagram.

The complex powers may also be obtained directly from (2.14)

$$S_1 = \frac{|V|^2}{Z_1^*} = \frac{(1200)^2}{60} = 24,000 \text{ W} + j\ 0$$
$$S_2 = \frac{|V|^2}{Z_2^*} = \frac{(1200)^2}{6 - j12} = 48,000 \text{ W} + j96,000 \text{ var}$$
$$S_3 = \frac{|V|^2}{Z_3^*} = \frac{(1200)^2}{30 + j30} = 24,000 \text{ W} - j24,000 \text{ var}$$

2.5 POWER FACTOR CORRECTION

It can be seen from (2.7) that the apparent power will be larger than P if the power factor is less than 1. Thus the current I that must be supplied will be larger for $PF < 1$ than it would be for $PF = 1$, even though the average power P supplied is the same in either case. A larger current cannot be supplied without additional cost to the utility company. Thus, it is in the power company's (and its customer's) best interest that major loads on the system have power factors as close to 1 as possible. In order to maintain the power factor close to unity, power companies install banks of capacitors throughout the network as needed. They also impose an additional charge to industrial consumers who operate at low power factors. Since industrial loads are inductive and have low lagging power factors, it is beneficial to install capacitors to improve the power factor. This consideration is not important for residential and small commercial customers because their power factors are close to unity.

Example 2.3

Two loads $Z_1 = 100 + j0\ \Omega$ and $Z_2 = 10 + j20\ \Omega$ are connected across a 200-V rms, 60-Hz source as shown in Figure 2.7.

(a) Find the total real and reactive power, the power factor at the source, and the total current.

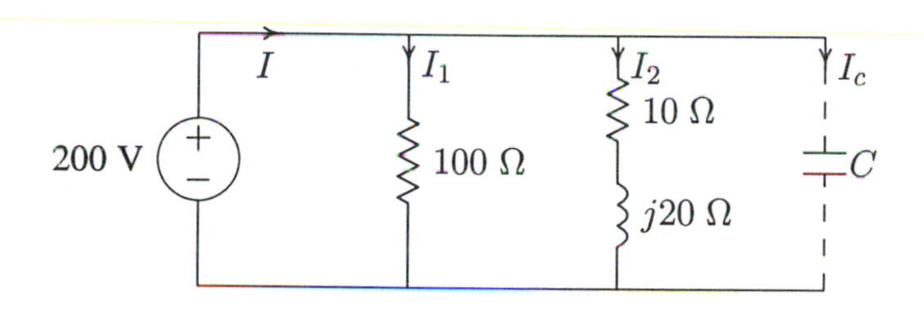

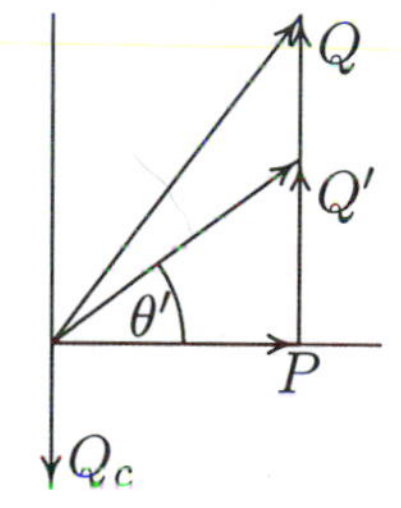

FIGURE 2.7
Circuit for Example 2.3 and the power triangle.

$$I_1 = \frac{200\angle 0°}{100} = 2\angle 0°\ \text{A}$$

$$I_2 = \frac{200\angle 0°}{10 + j20} = 4 - j8\ \text{A}$$

$$S_1 = VI_1^* = 200\angle 0°(2 - j0) = 400\ \text{W} + j0\ \text{var}$$

$$S_2 = VI_2^* = 200\angle 0°(4 + j8) = 800\ \text{W} + j1600\ \text{var}$$

Total apparent power and current are

$$S = P + jQ = 1200 + j1600 = 2000\angle 53.13^\circ \text{ VA}$$
$$I = \frac{S^*}{V^*} = \frac{2000\angle -53.13^\circ}{200\angle 0^\circ} = 10\angle -53.13^\circ \text{ A}$$

Power factor at the source is

$$PF = \cos(53.13) = 0.6 \text{ lagging}$$

(b) Find the capacitance of the capacitor connected across the loads to improve the overall power factor to 0.8 lagging.

Total real power $P = 1200$ W at the new power factor 0.8 lagging. Therefore

$$\theta' = \cos^{-1}(0.8) = 36.87^\circ$$
$$Q' = P \tan\theta' = 1200\tan(36.87^\circ) = 900 \text{ var}$$
$$Q_c = 1600 - 900 = 700 \text{ var}$$
$$Z_c = \frac{|V|^2}{S_c^*} = \frac{(200)^2}{j700} = -j57.14\ \Omega$$
$$C = \frac{10^6}{2\pi(60)(57.14)} = 46.42\ \mu\text{F}$$

The total power and the new current are

$$S' = 1200 + j900 = 1500\angle 36.87^\circ$$
$$I' = \frac{S'^*}{V^*} = \frac{1500\angle -36.87^\circ}{200\angle 0^\circ} = 7.5\angle -36.87^\circ$$

Note the reduction in the supply current from 10 A to 7.5 A.

Example 2.4

Three loads are connected in parallel across a 1400-V rms, 60-Hz single-phase supply as shown in Figure 2.8.

Load 1: Inductive load, 125 kVA at 0.28 power factor.

Load 2: Capacitive load, 10 kW and 40 kvar.

Load 3: Resistive load of 15 kW.

(a) Find the total kW, kvar, kVA, and the supply power factor.

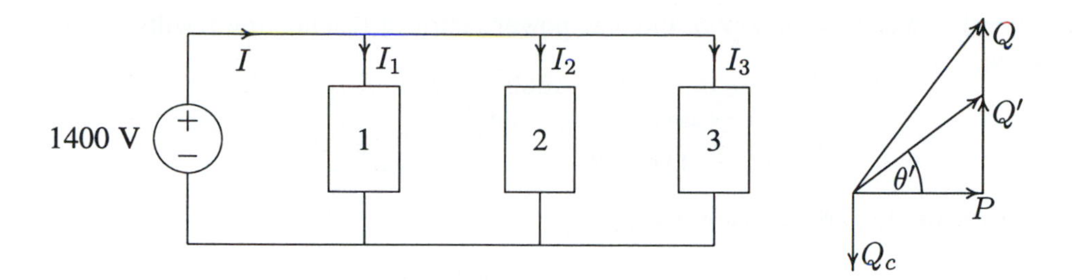

FIGURE 2.8
Circuit for Example 2.4.

An inductive load has a lagging power factor, the capacitive load has a leading power factor, and the resistive load has a unity power factor.

For Load 1:

$$\theta_1 = \cos^{-1}(0.28) = 73.74^\circ \text{ lagging}$$

The load complex powers are

$$\begin{aligned} S_1 &= 125\angle 73.74 \text{ kVA} = 35 \text{ kW} + j120 \text{ kvar} \\ S_2 &= 10 \text{ kW} - j40 \text{ kvar} \\ S_3 &= 15 \text{ kW} + j0 \text{ kvar} \end{aligned}$$

The total apparent power is

$$\begin{aligned} S &= P + jQ = S_1 + S_2 + S_3 \\ &= (35 + j120) + (10 - j40) + (15 + j0) \\ &= 60 \text{ kW} + j80 \text{ kvar} = 100\angle 53.13 \text{ kVA} \end{aligned}$$

The total current is

$$I = \frac{S^*}{V^*} = \frac{100,000\angle -53.13^\circ}{1400\angle 0^\circ} = 71.43\angle -53.13^\circ \text{ A}$$

The supply power factor is

$$PF = \cos(53.13) = 0.6 \text{ lagging}$$

(b) A capacitor of negligible resistance is connected in parallel with the above loads to improve the power factor to 0.8 lagging. Determine the kvar rating of this capacitor and the capacitance in μF.

Total real power $P = 60$ kW at the new power factor of 0.8 lagging results in the new reactive power Q'.

$$\theta' = \cos^{-1}(0.8) = 36.87^\circ$$
$$Q' = 60\tan(36.87^\circ) = 45 \text{ kvar}$$

Therefore, the required capacitor kvar is

$$Q_c = 80 - 45 = 35 \text{ kvar}$$

and

$$X_c = \frac{|V|^2}{S_c^*} = \frac{1400^2}{j35,000} = -j56\ \Omega$$
$$C = \frac{10^6}{2\pi(60)(56)} = 47.37\ \mu\text{F}$$

and the new current is

$$I' = \frac{S'^*}{V^*} = \frac{60,000 - j45,000}{1400\angle 0^\circ} = 53.57\angle{-36.87^\circ} \text{ A}$$

Note the reduction in the supply current from 71.43 A to 53.57 A.

2.6 COMPLEX POWER FLOW

Consider two ideal voltage sources connected by a line of impedance $Z = R + jX\ \Omega$ as shown in Figure 2.9.

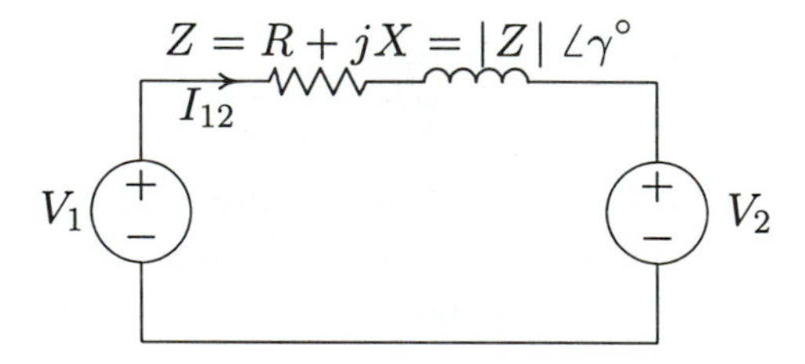

FIGURE 2.9
Two interconnected voltage sources.

Let the phasor voltage be $V_1 = |V_1|\angle\delta_1$ and $V_2 = |V_2|\angle\delta_2$. For the assumed direction of current

$$I_{12} = \frac{|V_1|\angle\delta_1 - |V_2|\angle\delta_2}{|Z|\angle\gamma} = \frac{|V_1|}{|Z|}\angle\delta_1 - \gamma - \frac{|V_2|}{|Z|}\angle\delta_2 - \gamma$$

The complex power S_{12} is given by

$$S_{12} = V_1 I_{12}^* = |V_1|\angle\delta_1 \left[\frac{|V_1|}{|Z|}\angle\gamma - \delta_1 - \frac{|V_2|}{|Z|}\angle\gamma - \delta_2\right]$$
$$= \frac{|V_1|^2}{|Z|}\angle\gamma - \frac{|V_1||V_2|}{|Z|}\angle\gamma + \delta_1 - \delta_2$$

Thus, the real and reactive power at the sending end are

$$P_{12} = \frac{|V_1|^2}{|Z|}\cos\gamma - \frac{|V_1||V_2|}{|Z|}\cos(\gamma + \delta_1 - \delta_2) \tag{2.16}$$
$$Q_{12} = \frac{|V_1|^2}{|Z|}\sin\gamma - \frac{|V_1||V_2|}{|Z|}\sin(\gamma + \delta_1 - \delta_2) \tag{2.17}$$

Power system transmission lines have small resistance compared to the reactance. Assuming $R = 0$ (i.e., $Z = X\angle 90^\circ$), the above equations become

$$P_{12} = \frac{|V_1||V_2|}{X}\sin(\delta_1 - \delta_2) \tag{2.18}$$
$$Q_{12} = \frac{|V_1|}{X}[\,|V_1| - |V_2|\cos(\delta_1 - \delta_2)\,] \tag{2.19}$$

Since $R = 0$, there are no transmission line losses and the real power sent equals the real power received.

From the above results, for a typical power system with small R/X ratio, the following important observations are made :

1. Equation (2.18) shows that small changes in δ_1 or δ_2 will have a significant effect on the real power flow, while small changes in voltage magnitudes will not have appreciable effect on the real power flow. Therefore, the flow of real power on a transmission line is governed mainly by the angle difference of the terminal voltages (i.e., $P_{12} \propto \sin\delta$), where $\delta = \delta_1 - \delta_2$. If V_1 leads V_2, δ is positive and the real power flows from node 1 to node 2. If V_1 lags V_2, δ is negative and power flows from node 2 to node 1.

2. Assuming $R = 0$, the theoretical maximum power (static transmission capacity) occurs when $\delta = 90^\circ$ and the maximum power transfer is given by

$$P_{max} = \frac{|V_1||V_2|}{X} \tag{2.20}$$

In Chapter 3 we learn that increasing δ beyond the static transmission capacity will result in loss of synchronism between the two machines.

3. For maintaining transient stability, the power system is usually operated with small load angle δ. Also, from (2.19) the reactive power flow is determined by the magnitude difference of terminal voltages, (i.e., $Q \propto |V_1| - |V_2|$).

Example 2.5

Two voltage sources $V_1 = 120\angle{-5}$ V and $V_2 = 100\angle 0$ V are connected by a short line of impedance $Z = 1 + j7\,\Omega$ as shown in Figure 2.9. Determine the real and reactive power supplied or received by each source and the power loss in the line.

$$I_{12} = \frac{120\angle{-5^\circ} - 100\angle 0^\circ}{1 + j7} = 3.135\angle{-110.02^\circ} \text{ A}$$
$$I_{21} = \frac{100\angle 0^\circ - 120\angle - 5^\circ}{1 + j7} = 3.135\angle 69.98^\circ \text{ A}$$
$$S_{12} = V_1 I_{12}^* = 376.2\angle 105.02^\circ = -97.5 \text{ W} + j363.3 \text{ var}$$
$$S_{21} = V_2 I_{21}^* = 313.5\angle{-69.98^\circ} = 107.3 \text{ W} - j294.5 \text{ var}$$

Line loss is given by

$$S_L = S_1 + S_2 = 9.8 \text{ W} + j68.8 \text{ var}$$

From the above results, since P_1 is negative and P_2 is positive, source 1 receives 97.5 W, and source 2 generates 107.3 W and the real power loss in the line is 9.8 W. The real power loss in the line can be checked by

$$P_L = R|I_{12}|^2 = (1)(3.135)^2 = 9.8 \text{ W}$$

Also, since Q_1 is positive and Q_2 is negative, source 1 delivers 363.3 var and source 2 receives 294.5 var, and the reactive power loss in the line is 68.6 var. The reactive power loss in the line can be checked by

$$Q_L = X|I_{12}|^2 = (7)(3.135)^2 = 68.8 \text{ var}$$

Example 2.6

This example concerns the direction of power flow between two voltage sources. Write a *MATLAB* program for the system of Example 2.5 such that the phase angle of source 1 is changed from its initial value by $\pm 30^\circ$ in steps of 5°. Voltage magnitudes of the two sources and the voltage phase angle of source 2 is to be kept constant. Compute the complex power for each source and the line loss. Tabulate the real power and plot P_1 , P_2, and P_L versus voltage phase angle δ. The following commands

```
E1 = input('Source # 1 Voltage Mag. = ');
a1 = input('Source# 1 Phase Angle  = ');
E2 = input('Source # 2 Voltage Mag. = ');
a2 = input('Source # 2 Phase Angle  = ');
R  = input('Line Resistance = ');
X  = input('Line Reactance = ');
Z  =  R + j*X;                                   % Line impedance
a1 = (-30+a1:5:30+a1)';         % Change a1 by +/- 30, col. array
a1r = a1*pi/180;                          % Convert degree to radian
k = length(a1);
a2 = ones(k,1)*a2;   % Create col. array of same length for a2
a2r = a2*pi/180;                          % Convert degree to radian
V1 = E1.*cos(a1r) + j*E1.*sin(a1r);
V2 = E2.*cos(a2r) + j*E2.*sin(a2r);
I12 = (V1 - V2)./Z;  I21=-I12;
S1 = V1.*conj(I12);  P1 = real(S1);  Q1 = imag(S1);
S2 = V2.*conj(I21);  P2 = real(S2);  Q2 = imag(S2);
SL = S1+S2;          PL = real(SL);  QL = imag(SL);
Result1 = [a1, P1,  P2, PL];
disp('   Delta 1    P-1        P-2        P-L ')
disp(Result1)
plot(a1, P1,  a1, P2,  a1,PL)
xlabel('Source #1 Voltage Phase Angle')
ylabel(' P, Watts'),
text(-26, -550, 'P1'), text(-26, 600,'P2'),
text(-26, 100, 'PL')
```

result in

```
Source # 1 Voltage Mag. = 120
Source # 1 Phase Angle  = -5
Source # 2 Voltage Mag. = 100
Source # 2 Phase Angle  = 0
Line Resistance = 1
Line Reactance =  7

  Delta 1       P-1         P-2        P-L
-35.0000    -872.2049    967.0119    94.8070
-30.0000    -759.8461    832.1539    72.3078
-25.0000    -639.5125    692.4848    52.9723
-20.0000    -512.1201    549.0676    36.9475
-15.0000    -378.6382    402.9938    24.3556
-10.0000    -240.0828    255.3751    15.2923
 -5.0000     -97.5084    107.3349     9.8265
  0           48.0000    -40.0000     8.0000
```

```
   5.0000       195.3349    -185.5084       9.8265
  10.0000       343.3751    -328.0828      15.2923
  15.0000       490.9938    -466.6382      24.3556
  20.0000       637.0676    -600.1201      36.9475
  25.0000       780.4848    -727.5125      52.9723
```

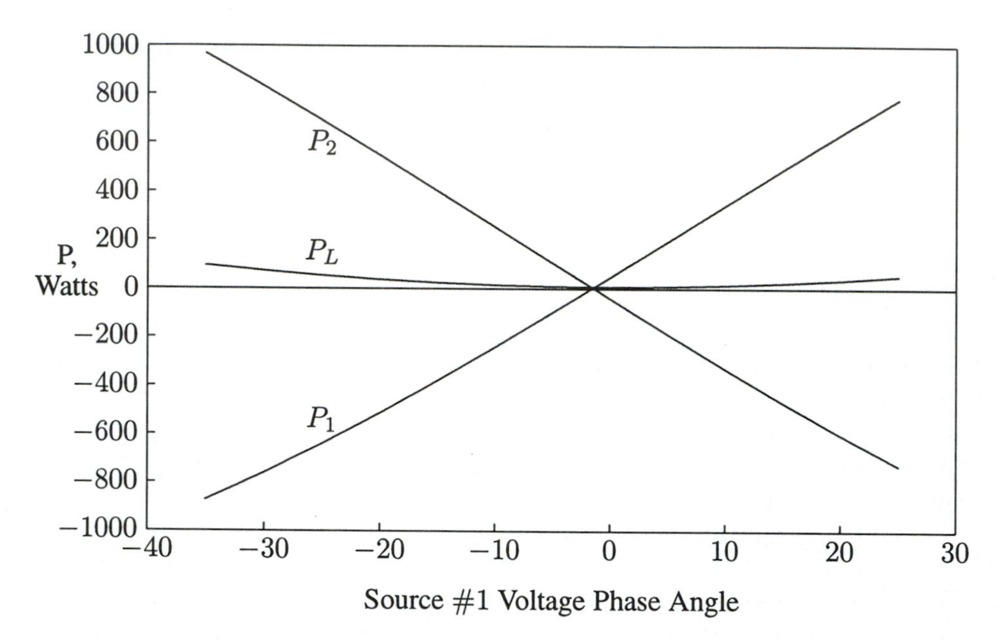

FIGURE 2.10
Real power versus voltage phase angle δ.

Examination of Figure 2.10 shows that the flow of real power along the interconnection is determined by the angle difference of the terminal voltages. Problem 2.9 requires the development of a similar program for demonstrating the dependency of reactive power on the magnitude difference of terminal voltages.

2.7 BALANCED THREE-PHASE CIRCUITS

The generation, transmission and distribution of electric power is accomplished by means of three-phase circuits. At the generating station, three sinusoidal voltages are generated having the same amplitude but displaced in phase by 120°. This is called a *balanced source*. If the generated voltages reach their peak values in the sequential order ABC, the generator is said to have a *positive phase sequence*, shown in Figure 2.11(a). If the phase order is ACB, the generator is said to have a *negative phase sequence*, as shown in Figure 2.11(b).

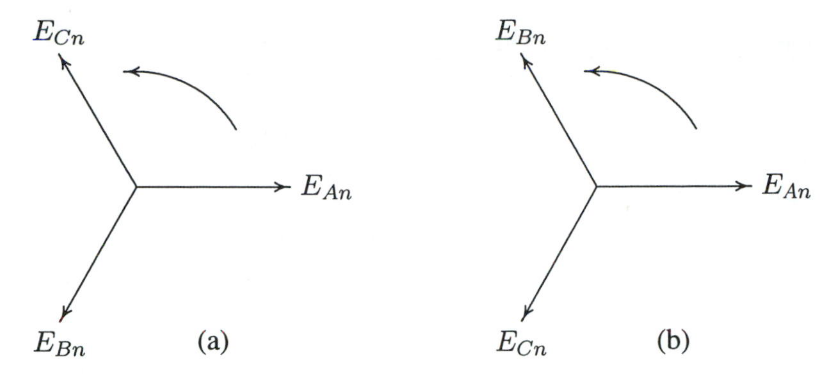

FIGURE 2.11
(a) Positive, or ABC, phase sequence. (b) Negative, or ACB, phase sequence.

In a three-phase system, the instantaneous power delivered to the external loads is constant rather than pulsating as it is in a single-phase circuit. Also, three-phase motors, having constant torque, start and run much better than single-phase motors. This feature of three-phase power, coupled with the inherent efficiency of its transmission compared to single-phase (less wire for the same delivered power), accounts for its universal use.

A power system has Y-connected generators and usually includes both Δ- and Y-connected loads. Generators are rarely Δ-connected, because if the voltages are not perfectly balanced, there will be a net voltage, and consequently a circulating current, around the Δ. Also, the phase voltages are lower in the Y-connected generator, and thus less insulation is required. Figure 2.12 shows a Y-connected generator supplying balanced Y-connected loads through a three-phase line. Assuming a positive phase sequence (phase order ABC) the generated voltages are:

$$\begin{aligned} E_{An} &= |E_p|\angle 0^\circ \\ E_{Bn} &= |E_p|\angle -120^\circ \\ E_{Cn} &= |E_p|\angle -240^\circ \end{aligned} \tag{2.21}$$

In power systems, great care is taken to ensure that the loads of transmission lines are balanced. For balanced loads, the terminal voltages of the generator V_{An}, V_{Bn} and V_{Cn} and the phase voltages V_{an}, V_{bn} and V_{cn} at the load terminals are balanced. For "phase A," these are given by

$$V_{An} = E_{An} - Z_G I_a \tag{2.22}$$
$$V_{an} = V_{An} - Z_L I_a \tag{2.23}$$

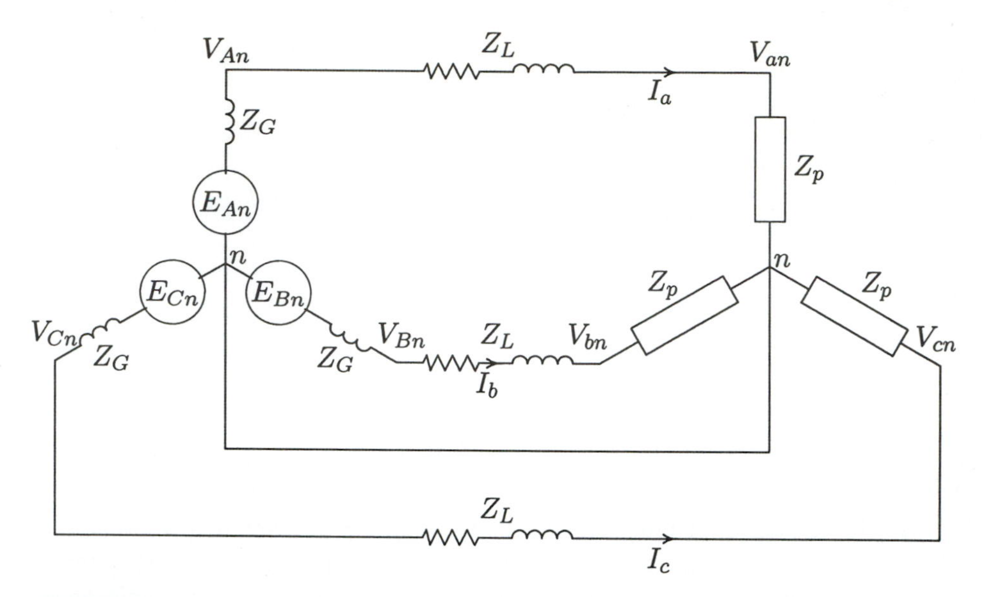

FIGURE 2.12
A Y-connected generator supplying a Y-connected load.

2.8 Y-CONNECTED LOADS

To find the relationship between the line voltages (line-to-line voltages) and the phase voltages (line-to-neutral voltages), we assume a positive, or ABC, sequence. We arbitrarily choose the line-to-neutral voltage of the a-phase as the reference, thus

$$\begin{aligned} V_{an} &= |V_p|\angle 0^\circ \\ V_{bn} &= |V_p|\angle -120^\circ \\ V_{cn} &= |V_p|\angle -240^\circ \end{aligned} \tag{2.24}$$

where $|V_p|$ represents the magnitude of the phase voltage (line-to-neutral voltage).

The line voltages at the load terminals in terms of the phase voltages are found by the application of Kirchhoff's voltage law

$$\begin{aligned} V_{ab} &= V_{an} - V_{bn} = |V_p|(1\angle 0^\circ - 1\angle -120^\circ) = \sqrt{3}|V_p|\angle 30^\circ \\ V_{bc} &= V_{bn} - V_{cn} = |V_p|(1\angle -120^\circ - 1\angle -240^\circ) = \sqrt{3}|V_p|\angle -90^\circ \\ V_{ca} &= V_{cn} - V_{an} = |V_p|(1\angle -240^\circ - 1\angle 0^\circ) = \sqrt{3}|V_p|\angle 150^\circ \end{aligned} \tag{2.25}$$

The voltage phasor diagram of the Y-connected loads of Figure 2.12 is shown in Figure 2.13. The relationship between the line voltages and phase voltages is demonstrated graphically.

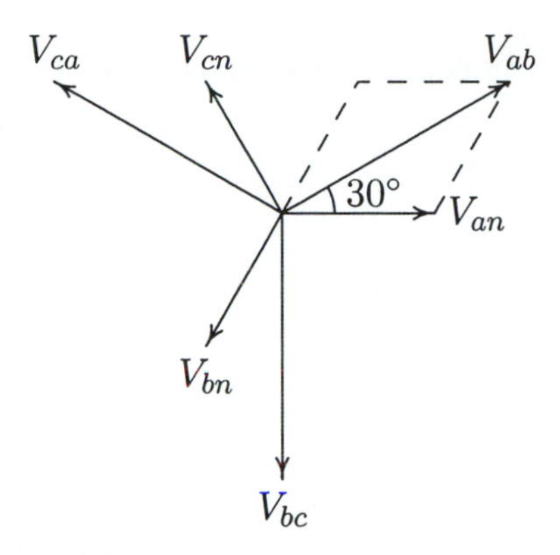

FIGURE 2.13
Phasor diagram showing phase and line voltages.

If the rms value of any of the line voltages is denoted by V_L, then one of the important characteristics of the Y-connected three-phase load may be expressed as

$$V_L = \sqrt{3}\,|V_p|\angle 30^\circ \tag{2.26}$$

Thus in the case of Y-connected loads, the magnitude of the line voltage is $\sqrt{3}$ times the magnitude of the phase voltage, and for a positive phase sequence, the set of line voltages leads the set of phase voltages by 30°.

The three-phase currents in Figure 2.12 also possess three-phase symmetry and are given by

$$\begin{aligned} I_a &= \frac{V_{an}}{Z_p} = |I_p|\angle -\theta \\ I_b &= \frac{V_{bn}}{Z_p} = |I_p|\angle{-120^\circ} - \theta \\ I_c &= \frac{V_{cn}}{Z_p} = |I_p|\angle{-240^\circ} - \theta \end{aligned} \tag{2.27}$$

where θ is the impedance phase angle.

The currents in lines are also the phase currents (the current carried by the phase impedances). Thus

$$I_L = I_p \tag{2.28}$$

2.9 Δ-CONNECTED LOADS

A balanced Δ-connected load (with equal phase impedances) is shown in Figure 2.14.

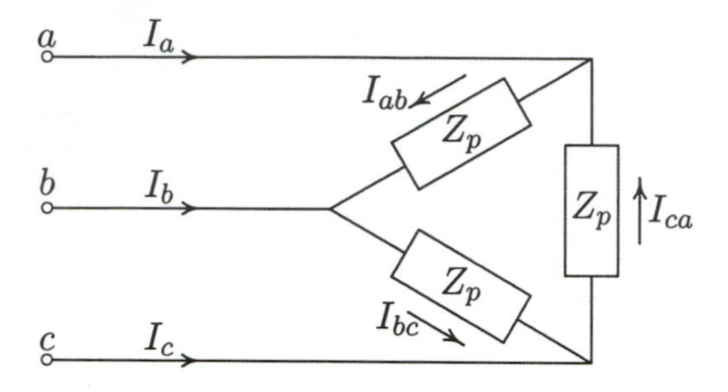

FIGURE 2.14
A Δ-connected load.

It is clear from the inspection of the circuit that the line voltages are the same as phase voltages.

$$V_L = V_p \tag{2.29}$$

Consider the phasor diagram shown in Figure 2.15, where the phase current I_{ab} is arbitrarily chosen as reference. we have

$$\begin{aligned} I_{ab} &= |I_p|\angle 0^\circ \\ I_{bc} &= |I_p|\angle -120^\circ \\ I_{ca} &= |I_p|\angle -240^\circ \end{aligned} \tag{2.30}$$

where $|I_p|$ represents the magnitude of the phase current.

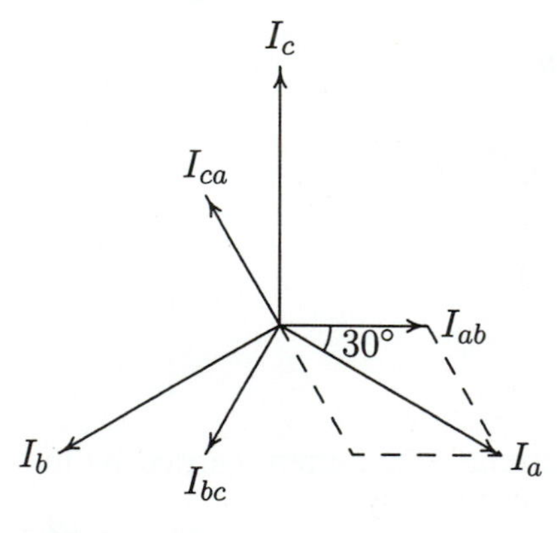

FIGURE 2.15
Phasor diagram showing phase and line currents.

The relationship between phase and line currents can be obtained by applying Kirchhoff's current law at the corners of Δ.

$$\begin{aligned}
I_a &= I_{ab} - I_{ca} = |I_p|(1\angle 0^\circ - 1\angle{-240^\circ}) = \sqrt{3}|I_p|\angle{-30^\circ} \\
I_b &= I_{bc} - I_{ab} = |I_p|(1\angle{-120^\circ} - 1\angle 0^\circ) = \sqrt{3}|I_p|\angle{-150^\circ} \\
I_c &= I_{ca} - I_{bc} = |I_p|(1\angle{-240^\circ} - 1\angle{-120^\circ}) = \sqrt{3}|I_p|\angle 90^\circ
\end{aligned} \tag{2.31}$$

The relationship between the line currents and phase currents is demonstrated graphically in Figure 2.15.

If the rms of any of the line currents is denoted by I_L, then one of the important characteristics of the Δ-connected three-phase load may be expressed as

$$I_L = \sqrt{3}|I_p|\,\angle{-30^\circ} \tag{2.32}$$

Thus in the case of Δ-connected loads, the magnitude of the line current is $\sqrt{3}$ times the magnitude of the phase current, and with positive phase sequence, the set of line currents lags the set of phase currents by 30°.

2.10 Δ-Y TRANSFORMATION

For analyzing network problems, it is convenient to replace the Δ-connected circuit with an equivalent Y-connected circuit. Consider the fictitious Y-connected circuit of Z_Y Ω/phase which is equivalent to a balanced Δ-connected circuit of Z_Δ Ω/phase, as shown in Figure 2.16.

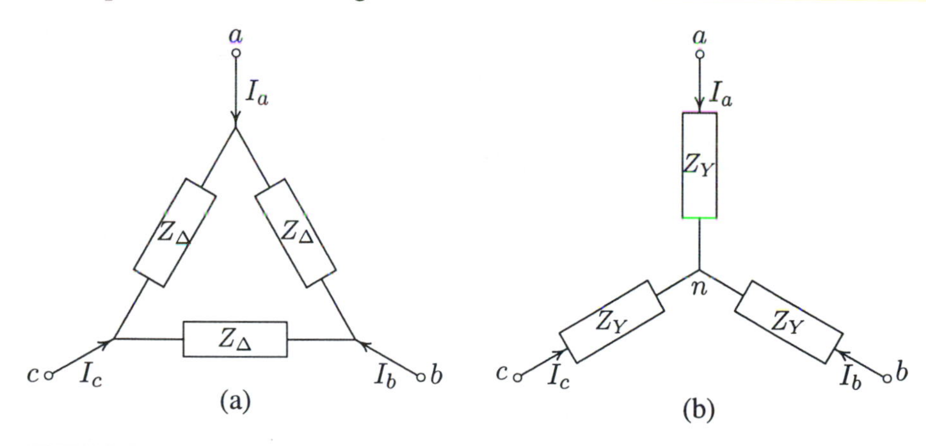

FIGURE 2.16
(a) Δ to (b) Y-connection.

For the Δ-connected circuit, the phase current I_a is given by

$$I_a = \frac{V_{ab}}{Z_\Delta} + \frac{V_{ac}}{Z_\Delta} = \frac{V_{ab} + V_{ac}}{Z_\Delta} \tag{2.33}$$

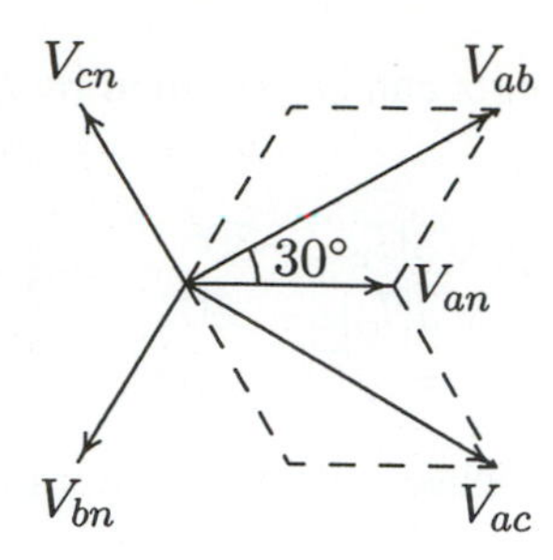

FIGURE 2.17
Phasor diagram showing phase and line voltages.

The phasor diagram in Figure 2.17 shows the relationship between balanced phase and line-to-line voltages. From this phasor diagram, we find

$$V_{ab} + V_{ac} = \sqrt{3}\,|V_{an}|\angle 30^\circ + \sqrt{3}\,|V_{an}|\angle{-30^\circ} \tag{2.34}$$

$$= 3V_{an} \tag{2.35}$$

Substituting in (2.33), we get

$$I_a = \frac{3V_{an}}{Z_\Delta}$$

or

$$V_{an} = \frac{Z_\Delta}{3} I_a \tag{2.36}$$

Now, for the Y-connected circuit, we have

$$V_{an} = Z_Y I_a \tag{2.37}$$

Thus, from (2.36) and (2.37), we find that

$$Z_Y = \frac{Z_\Delta}{3} \tag{2.38}$$

2.11 PER-PHASE ANALYSIS

The current in the neutral of the balanced Y-connected loads shown in Figure 2.12 is given by

$$I_n = I_a + I_b + I_c = 0 \tag{2.39}$$

Since the neutral carries no current, a neutral wire of any impedance may be replaced by any other impedance, including a short circuit and an open circuit. The return line may not actually exist, but regardless, a line of zero impedance is included between the two neutral points. The balanced power system problems are then solved on a "per-phase" basis. It is understood that the other two phases carry identical currents except for the phase shift.

We may then look at only one phase, say "phase A," consisting of the source V_{An} in series with Z_L and Z_p, as shown in Figure 2.18. The neutral is taken as datum and usually a single-subscript notation is used for phase voltages.

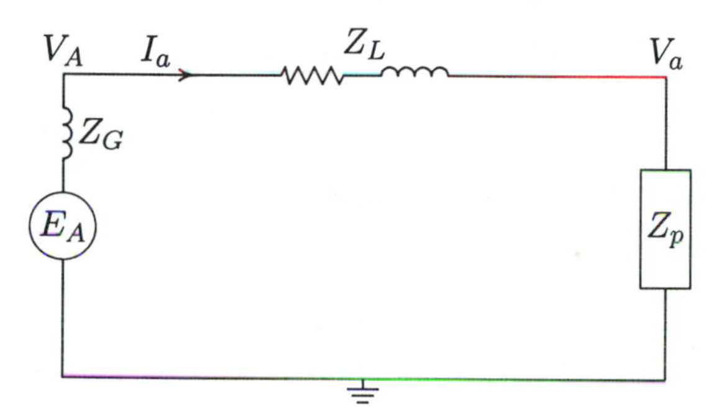

FIGURE 2.18
Single-phase circuit for per-phase analysis.

If the load in a three-phase circuit is connected in a Δ, it can be transformed into a Y by using the Δ-to-Y transformation. When the load is balanced, the impedance of each leg of the Y is one-third the impedance of each leg of the Δ, as given by (2.38), and the circuit is modeled by the single-phase equivalent circuit.

2.12 BALANCED THREE-PHASE POWER

Consider a balanced three-phase source supplying a balanced Y- or Δ- connected load with the following instantaneous voltages

$$\begin{aligned} v_{an} &= \sqrt{2}|V_p|\cos(\omega t + \theta_v) \\ v_{bn} &= \sqrt{2}|V_p|\cos(\omega t + \theta_v - 120^\circ) \\ v_{cn} &= \sqrt{2}|V_p|\cos(\omega t + \theta_v - 240^\circ) \end{aligned} \tag{2.40}$$

For a balanced load the phase currents are

$$\begin{aligned} i_a &= \sqrt{2}|I_p|\cos(\omega t + \theta_i) \\ i_b &= \sqrt{2}|I_p|\cos(\omega t + \theta_i - 120^\circ) \\ i_c &= \sqrt{2}|I_p|\cos(\omega t + \theta_i - 240^\circ) \end{aligned} \tag{2.41}$$

where $|V_p|$ and $|I_P|$ are the magnitudes of the rms phase voltage and current, respectively. The total instantaneous power is the sum of the instantaneous power of each phase, given by

$$p_{3\phi} = v_{an}i_a + v_{bn}i_b + v_{cn}i_c \tag{2.42}$$

Substituting for the instantaneous voltages and currents from (2.40) and (2.41) into (2.42)

$$\begin{aligned} p_{3\phi} = & 2|V_p||I_p|\cos(\omega t + \theta_v)\cos(\omega t + \theta_i) \\ & +2|V_p||I_p|\cos(\omega t + \theta_v - 120°)\cos(\omega t + \theta_i - 120°) \\ & +2|V_p||I_p|\cos(\omega t + \theta_v - 240°)\cos(\omega t + \theta_i - 240°) \end{aligned}$$

Using the trigonometric identity (2.4)

$$\begin{aligned} p_{3\phi} = & |V_p||I_p|[\cos(\theta_v - \theta_i) + \cos(2\omega t + \theta_v + \theta_i)] \\ & +|V_p||I_p|[\cos(\theta_v - \theta_i) + \cos(2\omega t + \theta_v + \theta_i - 240°)] \\ & +|V_p||I_p|[\cos(\theta_v - \theta_i) + \cos(2\omega t + \theta_v + \theta_i - 480°)] \end{aligned} \tag{2.43}$$

The three double frequency cosine terms in (2.43) are out of phase with each other by 120° and add up to zero, and the three-phase instantaneous power is

$$P_{3\phi} = 3|V_p||I_p|\cos\theta \tag{2.44}$$

$\theta = \theta_v - \theta_i$ is the angle between phase voltage and phase current or the impedance angle.

Note that although the power in each phase is pulsating, the total instantaneous power is constant and equal to three times the real power in each phase. Indeed, this constant power is the main advantage of the three-phase system over the single-phase system. Since the power in each phase is pulsating, the power, then, is made up of the real power and the reactive power. In order to obtain formula symmetry between real and reactive powers, the concept of complex or apparent power (S) is extended to three-phase systems by defining the three-phase reactive power as

$$Q_{3\phi} = 3|V_p||I_p|\sin\theta \tag{2.45}$$

Thus, the complex three-phase power is

$$S_{3\phi} = P_{3\phi} + jQ_3\phi \tag{2.46}$$

or

$$S_{3\phi} = 3V_pI_p^* \tag{2.47}$$

Equations (2.44) and (2.45) are sometimes expressed in terms of the rms magnitude of the line voltage and the rms magnitude of the line current. In a Y-connected load the phase voltage $|V_p| = |V_L|/\sqrt{3}$ and the phase current $I_p = I_L$.

In the Δ-connection $V_p = V_L$ and $|I_p| = |I_L|/\sqrt{3}$. Substituting for the phase voltage and phase currents in (2.44) and (2.45), the real and reactive powers for either connection are given by

$$P_{3\phi} = \sqrt{3}|V_L||I_L|\cos\theta \tag{2.48}$$

and

$$Q_{3\phi} = \sqrt{3}|V_L||I_L|\sin\theta \tag{2.49}$$

A comparison of the last two expressions with (2.44) and (2.45) shows that the equation for the power in a three-phase system is the same for either a Y or a Δ connection when the power is expressed in terms of line quantities.

When using (2.48) and (2.49) to calculate the total real and reactive power, remember that θ is the phase angle between the phase voltage and the phase current. As in the case of single-phase systems for the computation of power, it is best to use the complex power expression in terms of phase quantities given by (2.47). The rated power is customarily given for the three-phase and rated voltage is the line-to-line voltage. Thus, in using the per-phase equivalent circuit, care must be taken to use per-phase voltage by dividing the rated voltage by $\sqrt{3}$.

Example 2.7

A three-phase line has an impedance of $2 + j4\ \Omega$ as shown in Figure 2.19.

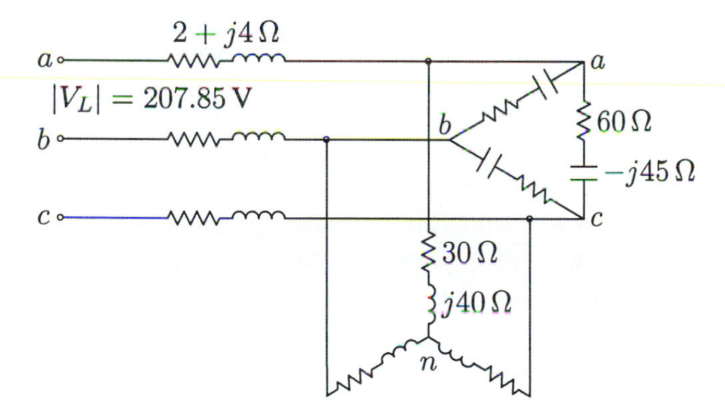

FIGURE 2.19
Three-phase circuit diagram for Example 2.7.

The line feeds two balanced three-phase loads that are connected in parallel. The first load is Y-connected and has an impedance of $30+j40\ \Omega$ per phase. The second load is Δ-connected and has an impedance of $60 - j45\ \Omega$. The line is energized at the sending end from a three-phase balanced supply of line voltage 207.85 V. Taking the phase voltage V_a as reference, determine:

(a) The current, real power, and reactive power drawn from the supply.

(b) The line voltage at the combined loads.
(c) The current per phase in each load.
(d) The total real and reactive powers in each load and the line.

(a) The Δ-connected load is transformed into an equivalent Y. The impedance per phase of the equivalent Y is

$$Z_2 = \frac{60 - j45}{3} = 20 - j15 \ \Omega$$

The phase voltage is

$$V_1 = \frac{207.85}{\sqrt{3}} = 120 \text{ V}$$

The single-phase equivalent circuit is shown in Figure 2.20.

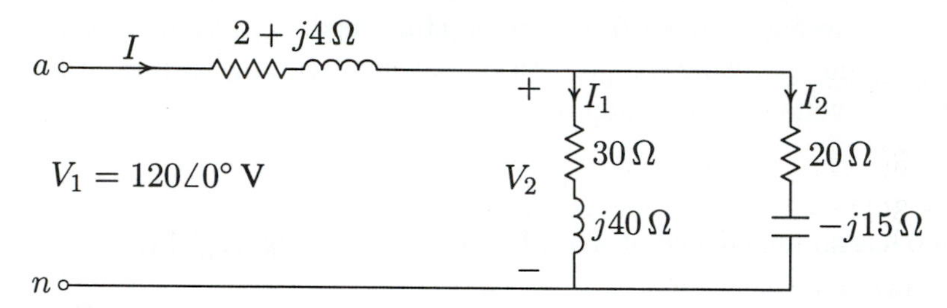

FIGURE 2.20
Single-phase equivalent circuit for Example 2.7.

The total impedance is

$$\begin{aligned} Z &= 2 + j4 + \frac{(30 + j40)(20 - j15)}{(30 + j40) + (20 - j15)} \\ &= 2 + j4 + 22 - j4 = 24 \ \Omega \end{aligned}$$

With the phase voltage V_{an} as reference, the current in phase a is

$$I = \frac{V_1}{Z} = \frac{120\angle 0^\circ}{24} = 5 \text{ A}$$

The three-phase power supplied is

$$S = 3V_1 I^* = 3(120\angle 0^\circ)(5\angle 0^\circ) = 1800 \text{ W}$$

(b) The phase voltage at the load terminal is

$$\begin{aligned} V_2 &= 120\angle 0^\circ - (2 + j4)(5\angle 0^\circ) = 110 - j20 \\ &= 111.8\angle{-10.3^\circ} \text{ V} \end{aligned}$$

The line voltage at the load terminal is

$$V_{2ab} = \sqrt{3}\,\angle 30^\circ\ V_2 = \sqrt{3}\ (111.8)\angle 19.7^\circ = 193.64\angle 19.7^\circ\ \text{V}$$

(c) The current per phase in the Y-connected load and in the equivalent Y of the Δ load is

$$I_1 = \frac{V_2}{Z_1} = \frac{110 - j20}{30 + j40} = 1 - j2 = 2.236\angle - 63.4^\circ\ \text{A}$$
$$I_2 = \frac{V_2}{Z_2} = \frac{110 - j20}{20 - j15} = 4 + j2 = 4.472\angle 26.56^\circ\ \text{A}$$

The phase current in the original Δ-connected load, i.e., I_{ab} is given by

$$I_{ab} = \frac{I_2}{\sqrt{3}\angle{-30^\circ}} = \frac{4.472\angle 26.56^\circ}{\sqrt{3}\angle{-30^\circ}} = 2.582\angle 56.56^\circ\ \text{A}$$

(d) The three-phase power absorbed by each load is

$$S_1 = 3V_2I_1^* = 3(111.8\angle - 10.3^\circ)(2.236\angle 63.4^\circ) = 450\ \text{W} + j600\ \text{var}$$
$$S_2 = 3V_2I_2^* = 3(111.8\angle - 10.3^\circ)(4.472\angle{-26.56^\circ}) = 1200\ \text{W} - j900\ \text{var}$$

The three-phase power absorbed by the line is

$$S_L = 3(R_L + jX_L)|I|^2 = 3(2 + j4)(5)^2 = 150\ \text{W} + j300\ \text{var}$$

It is clear that the sum of load powers and line losses is equal to the power delivered from the supply, i.e.,

$$S_1 + S_2 + S_L = (450 + j600) + (1200 - j900) + (150 + j300)$$
$$= 1800\ \text{W} + j0\ \text{var}$$

Example 2.8

A three-phase line has an impedance of $0.4 + j2.7\ \Omega$ per phase. The line feeds two balanced three-phase loads that are connected in parallel. The first load is absorbing 560.1 kVA at 0.707 power factor lagging. The second load absorbs 132 kW at unity power factor. The line-to-line voltage at the load end of the line is 3810.5 V. Determine:

(a) The magnitude of the line voltage at the source end of the line.
(b) Total real and reactive power loss in the line.
(c) Real power and reactive power supplied at the sending end of the line.

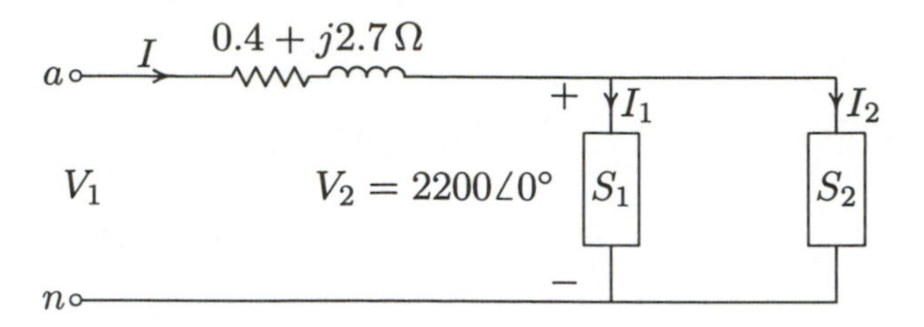

FIGURE 2.21
Single-phase equivalent diagram for Example 2.8.

(a) The phase voltage at the load terminals is

$$V_2 = \frac{3810.5}{\sqrt{3}} = 2200 \text{ V}$$

The single-phase equivalent circuit is shown in Figure 2.21.
The total complex power is

$$\begin{aligned} S_{R(3\phi)} &= 560.1(0.707 + j0.707) + 132 = 528 + j396 \\ &= 660\angle 36.87° \text{ kVA} \end{aligned}$$

With the phase voltage V_2 as reference, the current in the line is

$$I = \frac{S^*_{R(3\phi)}}{3V_2^*} = \frac{660,000\angle -36.87°}{3(2200\angle 0°)} = 100\angle -36.87° \text{ A}$$

The phase voltage at the sending end is

$$V_1 = 2200\angle 0° + (0.4 + j2.7)100\angle -36.87° = 2401.7\angle 4.58° \text{ V}$$

The magnitude of the line voltage at the sending end of the line is

$$|V_{1L}| = \sqrt{3}|V_1| = \sqrt{3}(2401.7) = 4160 \text{ V}$$

(b) The three-phase power loss in the line is

$$\begin{aligned} S_{L(3\phi)} &= 3R|I|^2 + j3X|I|^2 = 3(0.4)(100)^2 + j3(2.7)(100)^2 \\ &= 12 \text{ kW} + j81 \text{ kvar} \end{aligned}$$

(c) The three-phase sending power is

$$S_{S(3\phi)} = 3V_1I^* = 3(2401.7\angle 4.58°)(100\angle 36.87°) = 540 \text{ kW} + j477 \text{ kvar}$$

It is clear that the sum of load powers and the line losses is equal to the power delivered from the supply, i.e.,

$$S_{S(3\phi)} = S_{R(3\phi)} + S_{L(3\phi)} = (528 + j396) + (12 + j81) = 540 \text{ kW}+j477 \text{ kvar}$$

PROBLEMS

2.1. Modify the program in Example 2.1 such that the following quantities can be entered by the user:
The peak amplitude V_m, and the phase angle θ_v of the sinusoidal supply $v(t) = V_m \cos(\omega t + \theta_v)$. The impedance magnitude Z, and the phase angle γ of the load.
The program should produce plots for $i(t)$, $v(t)$, $p(t)$, $p_r(t)$ and $p_x(t)$, similar to Example 2.1. Run the program for $V_m = 100$ V, $\theta_v = 0$ and the following loads:

An inductive load, $Z = 1.25\angle 60°\Omega$
A capacitive load, $Z = 2.0\angle -30°\Omega$
A resistive load, $Z = 2.5\angle 0°\Omega$

(a) From $p_r(t)$ and $p_x(t)$ plots, estimate the real and reactive power for each load. Draw a conclusion regarding the sign of reactive power for inductive and capacitive loads.
(b) Using phasor values of current and voltage, calculate the real and reactive power for each load and compare with the results obtained from the curves.
(c) If the above loads are all connected across the same power supply, determine the total real and reactive power taken from the supply.

2.2. A single-phase load is supplied with a sinusoidal voltage

$$v(t) = 200 \cos(377t)$$

The resulting instantaneous power is

$$p(t) = 800 + 1000 \cos(754t - 36.87°)$$

(a) Find the complex power supplied to the load.
(b) Find the instantaneous current $i(t)$ and the rms value of the current supplied to the load.
(c) Find the load impedance.
(d) Use *MATLAB* to plot $v(t)$, $p(t)$, and $i(t) = p(t)/v(t)$ over a range of 0 to 16.67 ms in steps of 0.1 ms. From the current plot, estimate the peak amplitude, phase angle and the angular frequency of the current, and verify the results obtained in part (b). Note in *MATLAB* the command for array or element-by-element division is ./.

2.3. An inductive load consisting of R and X in series feeding from a 2400-V rms supply absorbs 288 kW at a lagging power factor of 0.8. Determine R and X.

2.4. An inductive load consisting of R and X in parallel feeding from a 2400-V rms supply absorbs 288 kW at a lagging power factor of 0.8. Determine R and X.

2.5. Two loads connected in parallel are supplied from a single-phase 240-V rms source. The two loads draw a total real power of 400 kW at a power factor of 0.8 lagging. One of the loads draws 120 kW at a power factor of 0.96 leading. Find the complex power of the other load.

2.6. The load shown in Figure 2.22 consists of a resistance R in parallel with a capacitor of reactance X. The load is fed from a single-phase supply through a line of impedance $8.4 + j11.2\ \Omega$. The rms voltage at the load terminal is $1200\angle 0°$ V rms, and the load is taking 30 kVA at 0.8 power factor leading.
(a) Find the values of R and X.
(b) Determine the supply voltage V.

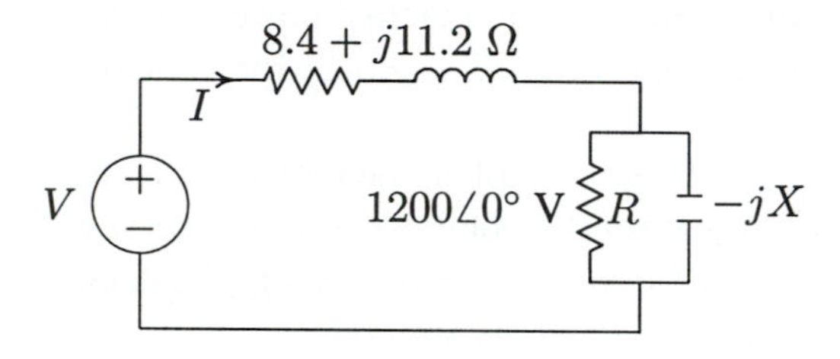

FIGURE 2.22
Circuit for Problem 2.6.

2.7. Two impedances, $Z_1 = 0.8 + j5.6\ \Omega$ and $Z_2 = 8 - j16\ \Omega$, and a single-phase motor are connected in parallel across a 200-V rms, 60-Hz supply as shown in Figure 2.23. The motor draws 5 kVA at 0.8 power factor lagging.

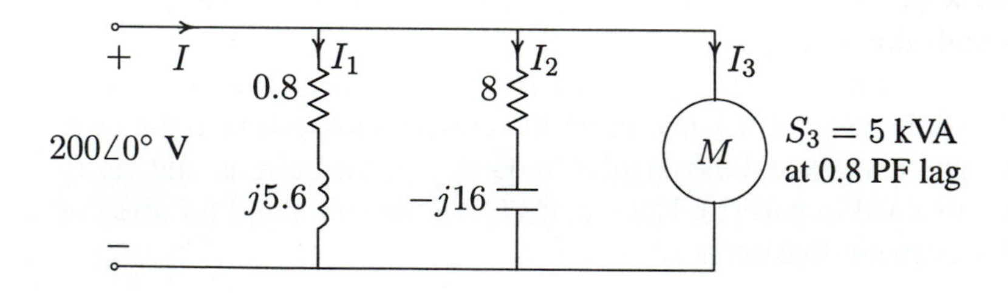

FIGURE 2.23
Circuit for Problem 2.7.

(a) Find the complex powers S_1, S_2 for the two impedances, and S_3 for the motor.
(b) Determine the total power taken from the supply, the supply current, and the overall power factor.
(c) A capacitor is connected in parallel with the loads. Find the kvar and the capacitance in μF to improve the overall power factor to unity. What is the new line current?

2.8. Two single-phase ideal voltage sources are connected by a line of impedance of $0.7 + j2.4\ \Omega$ as shown in Figure 2.24. $V_1 = 500\angle 16.26°$ V and $V_2 = 585\angle 0°$ V. Find the complex power for each machine and determine whether they are delivering or receiving real and reactive power. Also, find the real and the reactive power loss in the line.

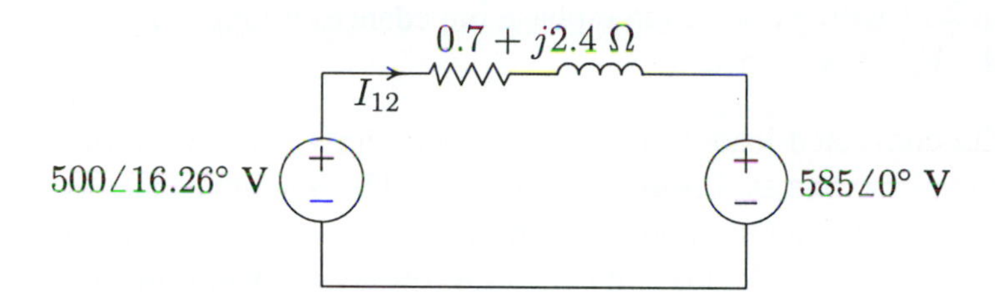

FIGURE 2.24
Circuit for Problem 2.8.

2.9. Write a *MATLAB* program for the system of Example 2.6 such that the voltage magnitude of source 1 is changed from 75 percent to 100 percent of the given value in steps of 1 V. The voltage magnitude of source 2 and the phase angles of the two sources is to be kept constant. Compute the complex power for each source and the line loss. Tabulate the reactive powers and plot Q_1, Q_2, and Q_L versus voltage magnitude $|V_1|$. From the results, show that the flow of reactive power along the interconnection is determined by the magnitude difference of the terminal voltages.

2.10. A balanced three-phase source with the following instantaneous phase voltages

$$v_{an} = 2500 \cos(\omega t)$$
$$v_{bn} = 2500 \cos(\omega t - 120°)$$
$$v_{cn} = 2500 \cos(\omega t - 240°)$$

supplies a balanced Y-connected load of impedance $Z = 250\angle 36.87°\ \Omega$ per phase.
(a) Using *MATLAB*, plot the instantaneous powers p_a, p_b, p_c and their sum versus ωt over a range of 0:0.05: 2π on the same graph. Comment on the nature of the instantaneous power in each phase and the total three-phase real power.
(b) Use (2.44) to verify the total power obtained in part (a).

2.11. A 4157-V rms, three-phase supply is applied to a balanced Y-connected three-phase load consisting of three identical impedances of $48\angle 36.87°\Omega$. Taking the phase to neutral voltage V_{an} as reference, calculate
(a) The phasor currents in each line.
(b) The total active and reactive power supplied to the load.

2.12. Repeat Problem 2.11 with the same three-phase impedances arranged in a Δ connection. Take V_{ab} as reference.

2.13. A balanced delta connected load of $15 + j18\ \Omega$ per phase is connected at the end of a three-phase line as shown in Figure 2.25. The line impedance is $1 + j2\ \Omega$ per phase. The line is supplied from a three-phase source with a line-to-line voltage of 207.85 V rms. Taking V_{an} as reference, determine the following:
(a) Current in phase a.
(b) Total complex power supplied from the source.
(c) Magnitude of the line-to-line voltage at the load terminal.

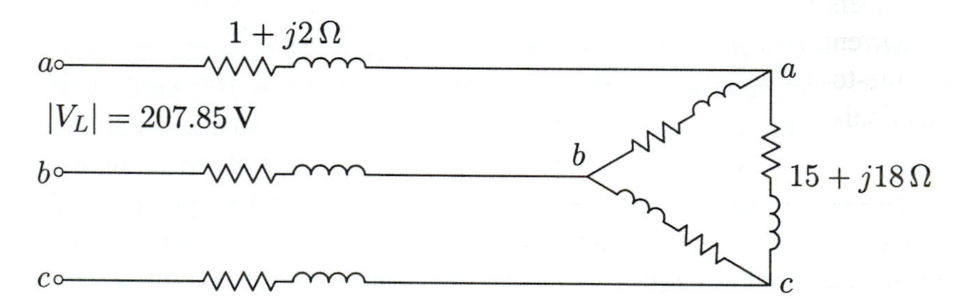

FIGURE 2.25
Circuit for Problem 2.13.

2.14. Three parallel three-phase loads are supplied from a 207.85-V rms, 60-Hz three-phase supply. The loads are as follows:

Load 1: A 15 hp motor operating at full-load, 93.25 percent efficiency, and 0.6 lagging power factor.

Load 2: A balanced resistive load that draws a total of 6 kW.
Load 3: A Y-connected capacitor bank with a total rating of 16 kvar.

(a) What is the total system kW, kvar, power factor, and the supply current per phase?
(b) What is the system power factor and the supply current per phase when the resistive load and induction motor are operating but the capacitor bank is switched off?

2.15. Three loads are connected in parallel across a 12.47 kV three-phase supply.

Load 1: Inductive load, 60 kW and 660 kvar.
Load 2: Capacitive load, 240 kW at 0.8 power factor.
Load 3: Resistive load of 60 kW.

(a) Find the total complex power, power factor, and the supply current.
(b) A Y-connected capacitor bank is connected in parallel with the loads. Find the total kvar and the capacitance per phase in μF to improve the overall power factor to 0.8 lagging. What is the new line current?

2.16. A balanced Δ-connected load consisting of a pure resistances of 18 Ω per phase is in parallel with a purely resistive balanced Y-connected load of 12 Ω per phase as shown in Figure 2.26. The combination is connected to a three-phase balanced supply of 346.41-V rms (line-to-line) via a three-phase line having an inductive reactance of $j3$ Ω per phase. Taking the phase voltage V_{an} as reference, determine
(a) The current, real power, and reactive power drawn from the supply.
(b) The line-to-neutral and the line-to-line voltage of phase a at the combined load terminals.

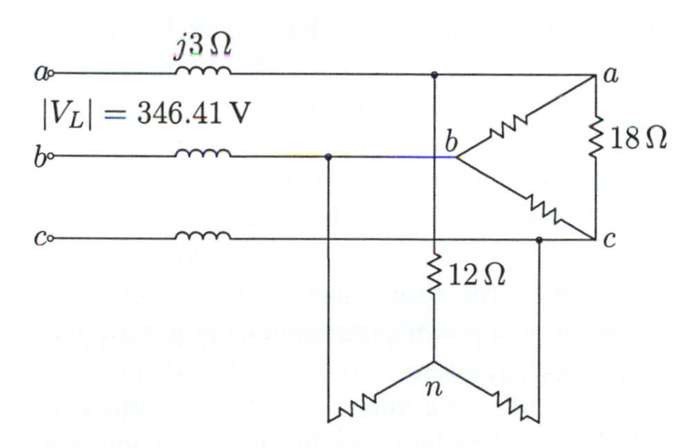

FIGURE 2.26
Circuit for Problem 2.16.

CHAPTER 3

GENERATOR AND TRANSFORMER MODELS; THE PER-UNIT SYSTEM

3.1 INTRODUCTION

Before the power systems network can be solved, it must first be modeled. The three-phase balanced system is represented on a per-phase basis, which was described in Section 2.10. The single-phase representation is also used for unbalanced systems by means of symmetrical components which is treated in a later chapter. In this chapter we deal with the balanced system, where transmission lines are represented by the π model as described in Chapter 4. Other essential components of a power system are generators and transformers; their theory and construction are discussed in standard electric machine textbooks. In this chapter, we represent simple models of generators and transformers for steady-state balanced operation.

Next we review the one-line diagram of a power system showing generators, transformers, transmission lines, capacitors, reactors, and loads. The diagram is usually limited to major transmission systems. As a rule, distribution circuits and small loads are not shown in detail but are taken into account merely as lumped loads on substation busses.

In the analysis of power systems, it is frequently convenient to use the per-unit system. The advantage of this method is the elimination of transformers by simple impedances. The per-unit system is presented, followed by the impedance diagram of the network, expressed to a common MVA base.

3.2 SYNCHRONOUS GENERATORS

Large-scale power is generated by three-phase synchronous generators, known as *alternators*, driven either by steam turbines, hydroturbines, or gas turbines. The armature windings are placed on the stationary part called *stator*. The armature windings are designed for generation of balanced three-phase voltages and are arranged to develop the same number of magnetic poles as the field winding that is on the rotor. The field which requires a relatively small power (0.2–3 percent of the machine rating) for its excitation is placed on the rotor. The rotor is also equipped with one or more short-circuited windings known as *damper windings*. The rotor is driven by a prime mover at constant speed and its field circuit is excited by direct current. The excitation may be provided through slip rings and brushes by means of dc generators (referred to as *exciters*) mounted on the same shaft as the rotor of the synchronous machine. However, modern excitation systems usually use ac generators with rotating rectifiers, and are known as *brushless excitation*. The generator excitation system maintains generator voltage and controls the reactive power flow.

The rotor of the synchronous machine may be of cylindrical or salient construction. The cylindrical type of rotor, also called *round rotor*, has one distributed winding and a uniform air gap. These generators are driven by steam turbines and are designed for high speed 3600 or 1800 rpm (two- and four-pole machines, respectively) operation. The rotor of these generators has a relatively large axial length and small diameter to limit the centrifugal forces. Roughly 70 percent of large synchronous generators are cylindrical rotor type ranging from about 150 to 1500 MVA. The salient type of rotor has concentrated windings on the poles and nonuniform air gaps. It has a relatively large number of poles, short axial length, and large diameter. The generators in hydroelectric power stations are driven by hydraulic turbines, and they have salient-pole rotor construction.

3.2.1 GENERATOR MODEL

An elementary two-pole three-phase generator is illustrated in Figure 3.1. The stator contains three coils, aa', bb', and cc', displaced from each other by 120 electrical degrees. The concentrated full-pitch coils shown here may be considered to represent distributed windings producing sinusoidal mmf waves concentrated on the magnetic axes of the respective phases. When the rotor is excited to produce

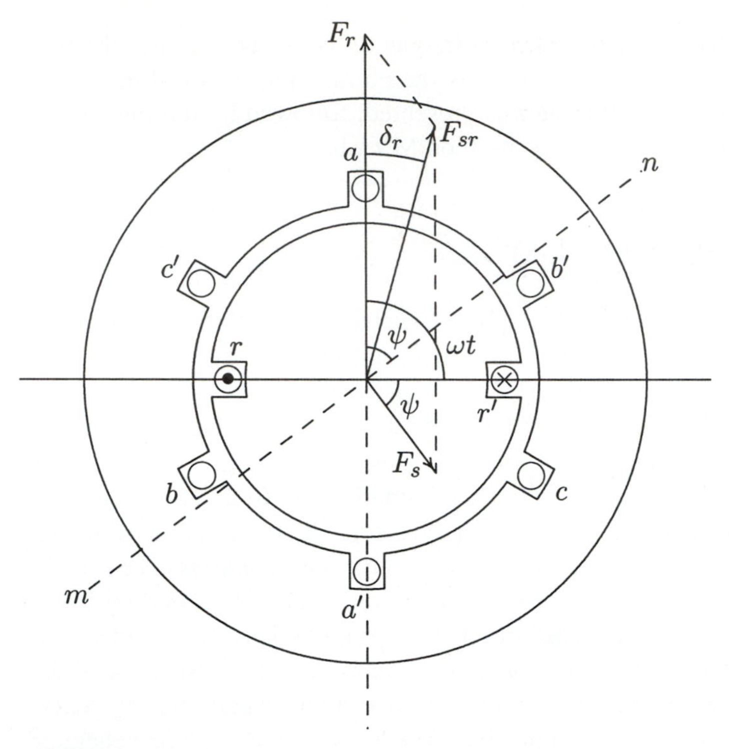

FIGURE 3.1
Elementary two-pole three-phase synchronous generator.

an air gap flux of ϕ per pole and is revolving at constant angular velocity ω, the flux linkage of the coil varies with the position of the rotor mmf axis ωt, where ωt is measured in electrical radians from coil aa' magnetic axis. The flux linkage for an N-turn concentrated coil aa' will be maximum ($N\phi$) at $\omega t = 0$ and zero at $\omega t = \pi/2$. Assuming distributed winding, the flux linkage λ_a will vary as the cosine of the angle ωt. Thus, the flux linkage with coil a is

$$\lambda_a = N\phi \cos \omega t \tag{3.1}$$

The voltage induced in coil aa' is obtained from Faraday's law as

$$\begin{aligned} e_a = -\frac{d\lambda}{dt} &= \omega N\phi \sin \omega t \\ &= E_{max} \sin \omega t \\ &= E_{max} \cos(\omega t - \frac{\pi}{2}) \end{aligned} \tag{3.2}$$

where

$$E_{max} = \omega N\phi = 2\pi f N\phi$$

Therefore, the rms value of the generated voltage is

$$E = 4.44 f N \phi \tag{3.3}$$

where f is the frequency in hertz. In actual ac machine windings, the armature coil of each phase is distributed in a number of slots. Since the emfs induced in different slots are not in phase, their phasor sum is less than their numerical sum. Thus, a reduction factor K_w, called the *winding factor*, must be applied. For most three-phase windings K_w is about 0.85 to 0.95. Therefore, for a distributed phase winding, the rms value of the generated voltage is

$$E = 4.44 K_w f N \phi \tag{3.4}$$

The magnetic field of the rotor revolving at constant speed induces three-phase sinusoidal voltages in the armature, displaced by $2\pi/3$ radians. The frequency of the induced armature voltages depends on the speed at which the rotor runs and on the number of poles for which the machine is wound. The frequency of the armature voltage is given by

$$f = \frac{P}{2}\frac{n}{60} \tag{3.5}$$

where n is the rotor speed in rpm, referred to as *synchronous speed*. During normal conditions, the generator operates synchronously with the power grid. This results in three-phase balanced currents in the armature. Assuming current in phase a is lagging the generated emf e_a by an angle ψ, which is indicated by line mn in Figure 3.1, the instantaneous armature currents are

$$\begin{aligned} i_a &= I_{max} \sin(\omega t - \psi) \\ i_b &= I_{max} \sin(\omega t - \psi - \frac{2\pi}{3}) \\ i_c &= I_{max} \sin(\omega t - \psi - \frac{4\pi}{3}) \end{aligned} \tag{3.6}$$

According to (3.2) the generated emf e_a is maximum when the rotor magnetic axis is under phase a. Since i_a is lagging e_a by an angle ψ, when line mn reaches the axis of coil aa', current in phase a reaches its maximum value. At any instant of time, each phase winding produces a sinusoidally distributed mmf wave with its peak along the axis of the phase winding. These sinusoidally distributed fields can be represented by vectors referred to as *space phasors*. The amplitude of the sinusoidally distributed mmf $f_a(\theta)$ is represented by the vector F_a along the axis of phase a. Similarly, the amplitude of the mmfs $f_b(\theta)$ and $f_c(\theta)$ are shown by vectors F_b and F_c along their respective axis. The mmf amplitudes are proportional to the instantaneous value of the phase current, i.e.,

$$\begin{aligned} F_a &= Ki_a = KI_{max}\sin(\omega t-\psi) = F_m \sin(\omega t-\psi) \\ F_b &= Ki_b = KI_{max}\sin(\omega t-\psi-\frac{2\pi}{3}) = F_m \sin(\omega t-\psi-\frac{2\pi}{3}) \\ F_c &= Ki_c = KI_{max}\sin(\omega t-\psi-\frac{4\pi}{3}) = F_m \sin(\omega t-\psi-\frac{4\pi}{3}) \end{aligned} \tag{3.7}$$

where K is proportional to the number of armature turns per phase and is a function of the winding type. The resultant armature mmf is the vector sum of the above mmfs. A suitable method for finding the resultant mmf is to project these mmfs on line mn and obtain the resultant in-phase and quadrature-phase components. The resultant in-phase components are

$$\begin{aligned} F_1 = F_m \sin(\omega t-\psi)\cos(\omega t-\psi) + F_m \sin(\omega t-\psi-\frac{2\pi}{3}) \\ \cos(\omega t-\psi-\frac{2\pi}{3}) + F_m \sin(\omega t-\psi-\frac{4\pi}{3})\cos(\omega t-\psi-\frac{4\pi}{3}) \end{aligned}$$

Using the trigonometric identity $\sin\alpha\cos\alpha = (1/2)\sin 2\alpha$, the above expression becomes

$$\begin{aligned} F_1 = \frac{F_m}{2}[\sin 2(\omega t-\psi) + \sin 2(\omega t-\psi-\frac{2\pi}{3}) \\ + \sin 2(\omega t-\psi-\frac{4\pi}{3})] \end{aligned}$$

The above expression is the sum of three sinusoidal functions displaced from each other by $2\pi/3$ radians, which adds up to zero, i.e., $F_1 = 0$.

The sum of quadrature components results in

$$\begin{aligned} F_2 = F_m \sin(\omega t-\psi)\sin(\omega t-\psi) + F_m \sin(\omega t-\psi-\frac{2\pi}{3})\sin(\omega t-\psi-\frac{2\pi}{3}) \\ + F_m \sin(\omega t-\psi-\frac{4\pi}{3})\sin(\omega t-\psi-\frac{4\pi}{3}) \end{aligned}$$

Using the trigonometric identity $\sin^2\alpha = (1/2)(1-\cos 2\alpha)$, the above expression becomes

$$\begin{aligned} F_2 = \frac{F_m}{2}[3 - \cos 2(\omega t-\psi) + \cos 2(\omega t-\psi-\frac{2\pi}{3}) \\ + \cos 2(\omega t-\psi-\frac{4\pi}{3})] \end{aligned}$$

The sinusoidal terms of the above expression are displaced from each other by $2\pi/3$ radians and add up to zero, with $F_2 = 3/2F_m$. Thus, the amplitude of the resultant armature mmf or stator mmf becomes

$$F_s = \frac{3}{2}F_m \tag{3.8}$$

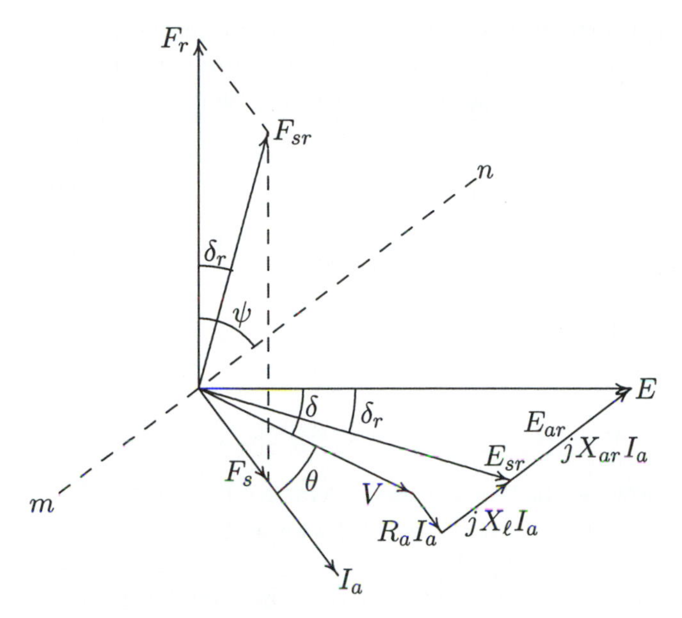

FIGURE 3.2
Combined phasor/vector diagram for one phase of a cylindrical rotor generator.

We thus conclude that the resultant armature mmf has a constant amplitude perpendicular to line mn, and rotates at a constant speed and in synchronism with the field mmf F_r. To see a demonstration of the rotating magnetic field, type **rotfield** at the *MATLAB* prompt.

A typical synchronous machine field alignment for operation as a generator is shown in Figure 3.2, using space vectors to represent the various fields. When the rotor is revolving at synchronous speed and the armature current is zero, the field mmf F_r produces the no-load generated emf E in each phase. The no-load generated voltage which is proportional to the field current is known as the *excitation voltage*. The phasor voltage for phase a, which is lagging F_r by 90°, is combined with the mmf vector diagram as shown in Figure 3.2. This combined phasor/vector diagram leads to a circuit model for the synchronous machine. It must be emphasized that in Figure 3.2 mmfs are space vectors, whereas the emfs are time phasors. When the armature is carrying balanced three-phase currents, F_s is produced perpendicular to line mn. The interaction of armature mmf and the field mmf, known as *armature reaction*, gives rise to the resultant air gap mmf F_{sr}. The resultant mmf F_{sr} is the vector sum of the field mmf F_r and the armature mmf F_s. The resultant mmf is responsible for the resultant air gap flux ϕ_{sr} that induces the generated emf on-load, shown by E_{sr}. The armature mmf F_s induces the emf E_{ar}, known as the *armature reaction voltage*, which is perpendicular to F_s. The voltage E_{ar} leads

I_a by 90° and thus can be represented by a voltage drop across a reactance X_{ar} due to the current I_a. X_{ar} is known as the *reactance of the armature reaction*. The phasor sum of E and E_{ar} is shown by E_{sr} perpendicular to F_{sr}, which represents the on-load generated emf.

$$E = E_{sr} + jX_{ar}I_a \tag{3.9}$$

The terminal voltage V is less than E_{sr} by the amount of resistive voltage drop R_aI_a and leakage reactance voltage drop $X_\ell I_a$. Thus

$$E = V + [R_a + j(X_\ell + X_{ar})]I_a \tag{3.10}$$

or

$$E = V + [R_a + jX_s]I_a \tag{3.11}$$

where $X_s = (X_\ell + X_{ar})$ is known as the *synchronous reactance*. The cosine of the angle between I and V, i.e., $\cos\theta$ represents the power factor at the generator terminals. The angle between E and E_{sr} is equal to the angle between the rotor mmf F_r and the air gap mmf F_{sr}, shown by δ_r. The power developed by the machine is proportional to the product of F_r, F_{sr} and $\sin\delta_r$. The relative positions of these mmfs dictates the action of the synchronous machine. When F_r is ahead of F_{sr} by an angle δ_r, the machine is operating as a generator and when F_r falls behind F_{sr}, the machine will act as a motor. Since E and E_{sr} are proportional to F_r and F_{sr}, respectively, the power developed by the machine is proportional to the products of E, E_{sr}, and $\sin\delta_r$. The angle δ_r is thus known as the *power angle*. This is a very important result because it relates the time angle between the phasor emfs with the space angle between the magnetic fields in the machine. Usually the developed power is expressed in terms of the excitation voltage E, the terminal voltage V, and $\sin\delta$. The angle δ is approximately equal to δ_r because the leakage impedance is very small compared to the magnetization reactance.

Due to the nonlinearity of the machine magnetization curve, the synchronous reactance is not constant. The unsaturated synchronous reactance can be found from the open- and short-circuit data. For operation at or near rated terminal voltage, it is usually assumed that the machine is equivalent to an unsaturated one whose magnetization curve is a straight line through the origin and the rated voltage point on the open-circuit characteristic. For steady-state analysis, a constant value known as the *saturated value of the synchronous reactance* corresponding to the rated voltage is used. A simple per-phase model for a cylindrical rotor generator based on (3.11) is obtained as shown in Figure 3.3. The armature resistance is generally much smaller than the synchronous reactance and is often neglected. The equivalent circuit connected to an infinite bus becomes that shown in Figure 3.4, and (3.11) reduces to

$$E = V + jX_sI_a \tag{3.12}$$

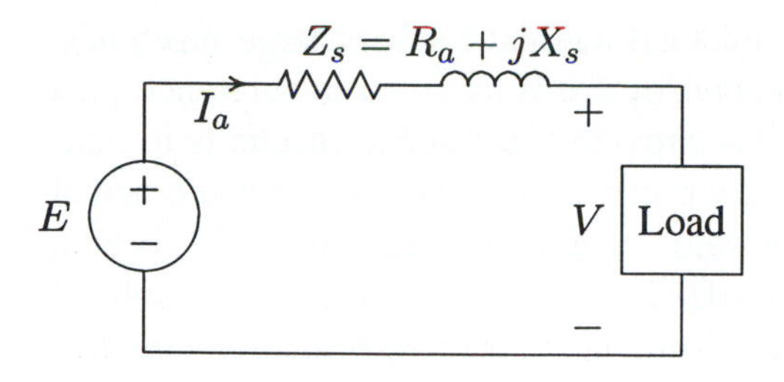

FIGURE 3.3
Synchronous machine equivalent circuit.

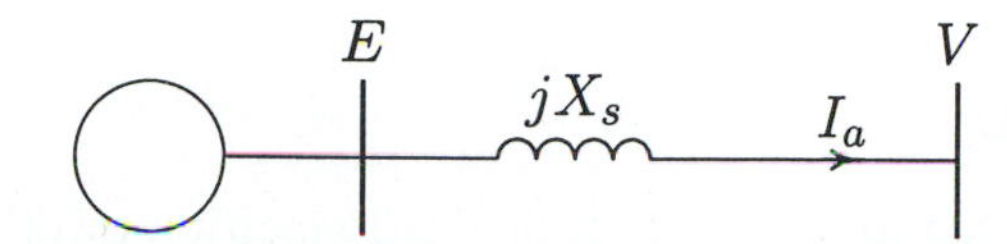

FIGURE 3.4
Synchronous machine connected to an infinite bus.

Figure 3.5 shows the phasor diagram of the generator with terminal voltage as reference for excitations corresponding to lagging, unity, and leading power factors. The voltage regulation of an alternator is a figure of merit used for compari-

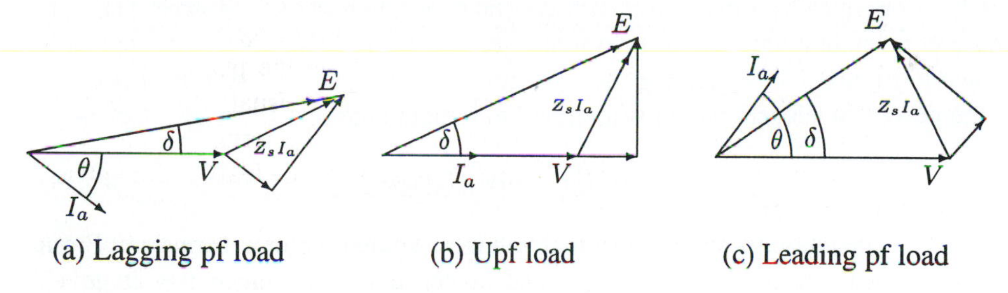

FIGURE 3.5
Synchronous generator phasor diagram.

son with other machines. It is defined as the percentage change in terminal voltage from no-load to rated load. This gives an indication of the change in field current required to maintain system voltage when going from no-load to rated load at some specific power factor.

$$\text{VR} = \frac{|V_{nl}| - |V_{rated}|}{|V_{rated}|} \times 100 = \frac{|E| - |V_{rated}|}{|V_{rated}|} \times 100 \tag{3.13}$$

The no-load voltage for a specific power factor may be determined by operating the machine at rated load conditions and then removing the load and observing the no-load voltage. Since this is not a practical method for very large machines, an accurate analytical method recommended by IEEE as given in reference [43] may be used. An approximate method that provides reasonable results is to consider a hypothetical linearized magnetization curve drawn to intersect the actual magnetization curve at rated voltage. The value of E calculated from (3.12) is then used to find the field current from the linearized curve. Finally, the no-load voltage corresponding to this field current is found from the actual magnetization curve.

3.3 STEADY-STATE CHARACTERISTICS—CYLINDRICAL ROTOR

3.3.1 POWER FACTOR CONTROL

Most synchronous machines are connected to large interconnected electric power networks. These networks have the important characteristic that the system voltage at the point of connection is constant in magnitude, phase angle, and frequency. Such a point in a power system is referred to as an *infinite bus*. That is, the voltage at the generator bus will not be altered by changes in the generator's operating condition.

The ability to vary the rotor excitation is an important feature of the synchronous machine, and we now consider the effect of such a variation when the machine operates as a generator with constant mechanical input power. The per-phase equivalent circuit of a synchronous generator connected to an infinite bus is shown in Figure 3.4. Neglecting the armature resistance, the output power is equal to the power developed, which is assumed to remain constant given by

$$P_{3\phi} = \Re[3VI_a^*] = 3|V||I_a|\cos\theta \tag{3.14}$$

where V is the phase-to-neutral terminal voltage assumed to remain constant. From (3.14) we see that for constant developed power at a fixed terminal voltage V, $I_a \cos\theta$ must be constant. Thus, the tip of the armature current phasor must fall on a vertical line as the power factor is varied by varying the field current as shown in Figure 3.6. From this diagram we have

$$cd = E_1 \sin\delta_1 = X_s I_{a1} \cos\theta_1 \tag{3.15}$$

Thus $E_1 \sin\delta_1$ is a constant, and the locus of E_1 is on the line ef. In Figure 3.6, phasor diagrams are drawn for three armature currents. Application of (3.12) for a lagging power factor armature current I_{a1} results in E_1. If θ is zero, the generator operates at unity power factor and armature current has a minimum value, shown by I_{a2}, which results in E_2. Similarly, E_3 is obtained corresponding to I_{a3} at a leading power factor. Figure 3.6 shows that the generation of reactive power can

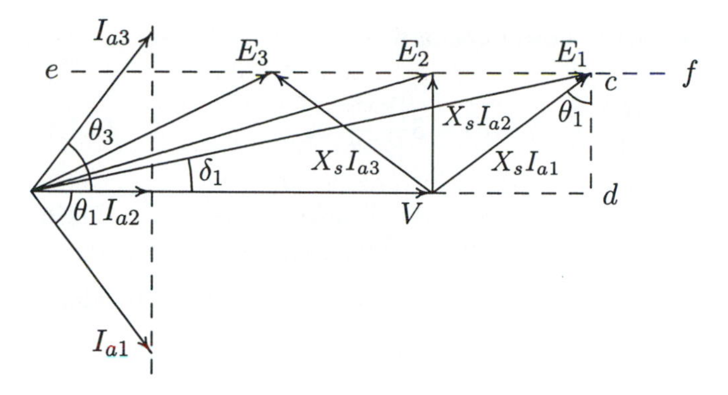

FIGURE 3.6
Variation of field current at constant power.

be controlled by means of the rotor excitation while maintaining a constant real power output. The variation in the magnitude of armature current as the excitation voltage is varied is best shown by a curve. Usually the field current is used as the abscissa instead of excitation voltage because the field current is readily measured. The curve of the armature current as the function of the field current resembles the letter V and is often referred to as the *V curve* of synchronous machines. These curves constitute one of the generator's most important characteristics. There is, of course, a limit beyond which the excitation cannot be reduced. This limit is reached when $\delta = 90°$. Any reduction in excitation below the stability limit for a particular load will cause the rotor to pull out of synchronism. The V curve is illustrated in Figure 3.7 (page 62) for the machine in Example 3.3.

3.3.2 POWER ANGLE CHARACTERISTICS

Consider the per-phase equivalent circuit shown in Figure 3.4. The three-phase complex power at the generator terminal is

$$S_{3\phi} = 3VI_a^* \tag{3.16}$$

Expressing the phasor voltages in polar form, the armature current is

$$I_a = \frac{|E|\angle\delta \ - |V|\angle 0}{|Z_s|\angle\gamma} \tag{3.17}$$

Substituting for I_a^* in (3.16) results in

$$S_{3\phi} = 3\frac{|E||V|}{|Z_s|}\angle\gamma - \delta \ - 3\frac{|V|^2}{|Z_s|}\angle\gamma \tag{3.18}$$

Thus, the real power $P_{3\phi}$ and reactive power $Q_{3\phi}$ are

$$P_{3\phi} = 3\frac{|E||V|}{|Z_s|}\cos(\gamma - \delta) - 3\frac{|V|^2}{|Z_s|}\cos\gamma \tag{3.19}$$

$$Q_{3\phi} = 3\frac{|E||V|}{|Z_s|}\sin(\gamma - \delta) - 3\frac{|V|^2}{|Z_s|}\sin\gamma \tag{3.20}$$

If R_a is neglected, then $Z_s = jX_s$ and $\gamma = 90°$. Equations (3.19) and (3.20) reduce to

$$P_{3\phi} = 3\frac{|E||V|}{X_s}\sin\delta \tag{3.21}$$

$$Q_{3\phi} = 3\frac{|V|}{X_s}(|E|\cos\delta - |V|) \tag{3.22}$$

Equation (3.21) shows that if $|E|$ and $|V|$ are held fixed and the power angle δ is changed by varying the mechanical driving torque, the power transfer varies sinusoidally with the angle δ. From (3.21), the theoretical maximum power occurs when $\delta = 90°$

$$P_{max(3\phi)} = 3\frac{|E||V|}{X_s} \tag{3.23}$$

The behavior of the synchronous machine can be described as follows. If we start with $\delta = 0°$ and increase the driving torque, the machine accelerates, and the rotor mmf F_r advances with respect to the resultant mmf F_{sr}. This results in an increase in δ, causing the machine to deliver electric power. At some value of δ the machine reaches equilibrium where the electric power output balances the increased mechanical power owing to the increased driving torque. It is clear that if an attempt were made to advance δ further than 90° by increasing the driving torque, the electric power output would decrease from the P_{max} point. Therefore, the excess driving torque continues to accelerate the machine, and the mmfs will no longer be magnetically coupled. The machine loses synchronism and automatic equipment disconnects it from the system. The value P_{max} is called the *steady-state stability limit* or *static stability limit*. In general, stability considerations dictate that a synchronous machine achieve steady-state operation for a power angle at considerably less than 90°. The control of real power flow is maintained by the generator governor through the frequency-power control channel.

Equation (3.22) shows that for small δ, $\cos\delta$ is nearly unity and the reactive power can be approximated to

$$Q_{3\phi} \simeq 3\frac{|V|}{X_s}(|E| - |V|) \tag{3.24}$$

From (3.24) we see that when $|E| > |V|$ the generator delivers reactive power to the bus, and the generator is said to be overexcited. If $|E| < |V|$, the reactive power delivered to the bus is negative; that is, the bus is supplying positive reactive power to the generator. Generators are normally operated in the overexcited mode since the generators are the main source of reactive power for inductive load throughout the system. Therefore, we conclude that the flow of reactive power is governed mainly by the difference in the excitation voltage $|E|$ and the bus bar voltage $|V|$. The adjustment in the excitation voltage for the control of reactive power is achieved by the generator excitation system.

Example 3.1

A 50-MVA, 30-kV, three-phase, 60-Hz synchronous generator has a synchronous reactance of 9 Ω per phase and a negligible resistance. The generator is delivering rated power at a 0.8 power factor lagging at the rated terminal voltage to an infinite bus.
(a) Determine the excitation voltage per phase E and the power angle δ.
(b) With the excitation held constant at the value found in (a), the driving torque is reduced until the generator is delivering 25 MW. Determine the armature current and the power factor.
(c) If the generator is operating at the excitation voltage of part (a), what is the steady-state maximum power the machine can deliver before losing synchronism? Also, find the armature current corresponding to this maximum power.

(a) The three-phase apparent power is

$$\begin{aligned} S_{3\phi} = 50\angle\cos^{-1} 0.8 &= 50\angle 36.87^\circ \quad \text{MVA} \\ &= 40 \ \text{MW} + j30 \ \text{Mvar} \end{aligned}$$

The rated voltage per phase is

$$V = \frac{30}{\sqrt{3}} = 17.32\angle 0^\circ \quad \text{kV}$$

The rated current is

$$I_a = \frac{S_{3\phi}^*}{3V^*} = \frac{(50\angle -36.87)10^3}{3(17.32\angle 0^\circ)} = 962.25\angle -36.87^\circ \quad \text{A}$$

The excitation voltage per phase from (3.12) is

$$E = 17320.5 + (j9)(962.25\angle -36.87) = 23558\angle 17.1^\circ \quad \text{V}$$

The excitation voltage per phase (line to neutral) is 23.56 kV and the power angle is 17.1°.

(b) When the generator is delivering 25 MW from (3.21) the power angle is

$$\delta = \sin^{-1}\left[\frac{(25)(9)}{(3)(23.56)(17.32)}\right] = 10.591^\circ$$

The armature current is

$$I_a = \frac{(23,558\angle 10.591^\circ - 17,320\angle 0^\circ)}{j9} = 807.485\angle{-53.43^\circ} \quad \text{A}$$

The power factor is given by $\cos(53.43) = 0.596$ lagging.

(c) The maximum power occurs at $\delta = 90^\circ$

$$P_{max(3\phi)} = 3\frac{|E||V|}{X_s} = 3\frac{(23.56)(17.32)}{9} = 136 \quad \text{MW}$$

The armature current is

$$I_a = \frac{(23,558\angle 90^\circ - 17,320\angle 0^\circ)}{j9} = 3248.85\angle 36.32^\circ \quad \text{A}$$

The power factor is given by $\cos(36.32) = 0.8057$ leading.

Example 3.2

The generator of Example 3.1 is delivering 40 MW at a terminal voltage of 30 kV. Compute the power angle, armature current, and power factor when the field current is adjusted for the following excitations.
(a) The excitation voltage is decreased to 79.2 percent of the value found in Example 3.1.
(b) The excitation voltage is decreased to 59.27 percent of the value found in Example 3.1.
(c) Find the minimum excitation below which the generator will lose synchronism.

(a) The new excitation voltage is

$$E = 0.792 \times 23,558 = 18,657 \quad \text{V}$$

From (3.21) the power angle is

$$\delta = \sin^{-1}\left[\frac{(40)(9)}{(3)(18.657)(17.32)}\right] = 21.8^\circ$$

The armature current is

$$I_a = \frac{(18657\angle 21.8^\circ - 17320\angle 0^\circ)}{j9} = 769.8\angle 0^\circ \quad \text{A}$$

The power factor is given by $\cos(0) = 1$.

(b) The new excitation voltage is

$$E = 0.5927 \times 23,558 = 13,963 \quad \text{V}$$

From (3.21) the power angle is

$$\delta = \sin^{-1}\left[\frac{(40)(9)}{(3)(13.963)(17.32)}\right] = 29.748^\circ$$

The armature current is

$$I_a = \frac{(13,963\angle 29.748^\circ - 17,320\angle 0^\circ)}{j9} = 962.3\angle 36.87^\circ \quad \text{A}$$

From current phase angle, the power factor is $\cos 36.87 = 0.8$ leading. The generator is underexcited and is actually receiving reactive power.

(c) From (3.23), the minimum excitation corresponding to $\delta = 90^\circ$ is

$$E = \frac{(40)(9)}{(3)(17.32)(1)} = 6.928 \quad \text{kV}$$

The armature current is

$$I_a = \frac{(6,928\angle 90^\circ - 17,320\angle 0^\circ)}{j9} = 2073\angle 68.2^\circ \quad \text{A}$$

The current phase angle shows that the power factor is $\cos 68.2 = 0.37$ leading. The generator is underexcited and is receiving reactive power.

Example 3.3

For the generator of Example 3.1, construct the V curve for the rated power of 40 MW with varying field excitation from 0.4 power factor leading to 0.4 power factor lagging. Assume the open-circuit characteristic in the operating region is given by $E = 2000I_f$ V.

The following *MATLAB* command results in the V curve shown in Figure 3.7.

```
P = 40;                                          % real power, MW
V = 30/sqrt(3)+ j*0;                          % phase voltage, kV
Zs = j*9;                                 % synchronous impedance
ang = acos(0.4);
theta=ang:-0.01:-ang;%Angle 0.4 leading to 0.4 lagging pf
P = P*ones(1,length(theta));%generates array of same size
Iam = P./(3*abs(V)*cos(theta));    % current magnitude kA
Ia = Iam.*(cos(theta) + j*sin(theta));   % current phasor
E = V + Zs.*Ia;                    % excitation voltage phasor
Em = abs(E);            % excitation voltage magnitude, kV
If = Em*1000/2000;                          % field current, A
plot(If, Iam), grid, xlabel('If - A')
ylabel('Ia - kA'), text(3.4, 1, 'Leading pf')
text(13, 1, 'Lagging pf'), text(9, .71, 'Upf')
```

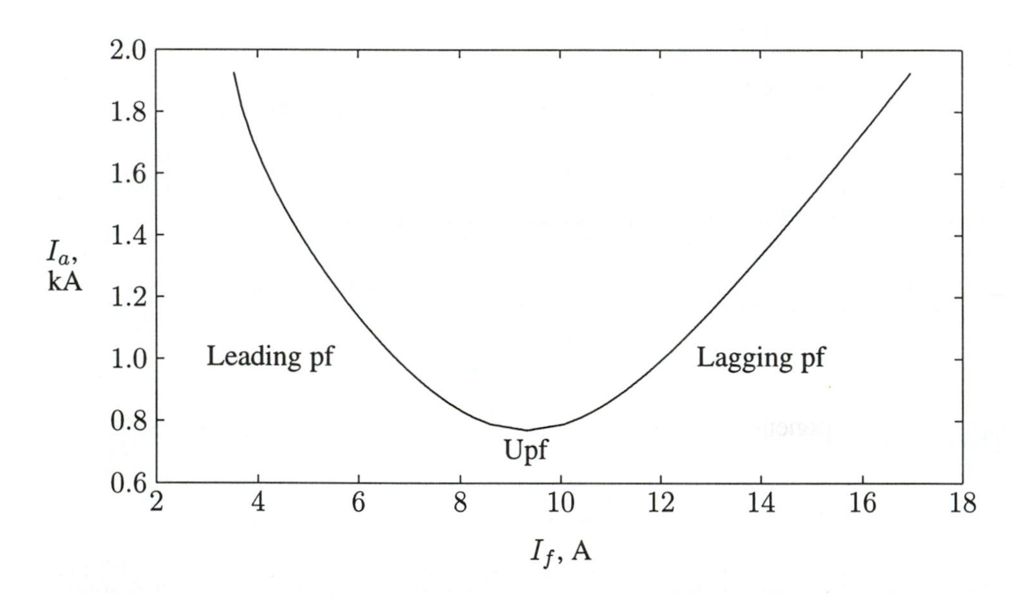

FIGURE 3.7
V curve for generator of Example 3.3.

3.4 SALIENT-POLE SYNCHRONOUS GENERATORS

The model developed in Section 3.2 is only valid for cylindrical rotor generators with uniform air gaps. The salient-pole rotor results in nonuniformity of the magnetic reluctance of the air gap. The reluctance along the polar axis, commonly referred to as the rotor *direct axis*, is appreciably less than that along the interpolar axis, commonly referred to as the *quadrature axis*. Therefore, the reactance has a high value X_d along the direct axis, and a low value X_q along the quadrature axis. These reactances produce voltage drop in the armature and can be taken into account by resolving the armature current I_a into two components I_q, in phase, and I_d in time quadrature, with the excitation voltage. The phasor diagram with the armature resistance neglected is shown in Figure 3.8. It is no longer possible to rep-

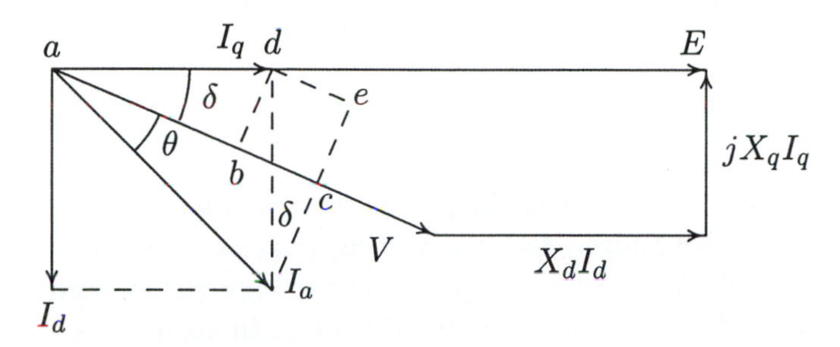

FIGURE 3.8
Phasor diagram for a salient-pole generator.

resent the machine by a simple equivalent circuit. The excitation voltage magnitude is

$$|E| = |V|\cos\delta + X_d I_d \tag{3.25}$$

The three-phase real power at the generator terminal is

$$P = 3|V||I_a|\cos\theta \tag{3.26}$$

The power component of the armature current can be expressed in terms of I_d and I_q as follows.

$$\begin{aligned} |I_a|\cos\theta &= ab + de \\ &= I_q\cos\delta + I_d\sin\delta \end{aligned} \tag{3.27}$$

Substituting from (3.27) into (3.26), we have

$$P = 3|V|(I_q\cos\delta + I_d\sin\delta) \tag{3.28}$$

Now from the phasor diagram given in Figure 3.8,

$$|V|\sin\delta = X_q I_q \tag{3.29}$$

or

$$I_q = \frac{|V|\sin\delta}{X_q} \tag{3.30}$$

Also, from (3.25), I_d is given by

$$I_d = \frac{|E| - |V|\cos\delta}{X_d} \tag{3.31}$$

Substituting for I_d and I_q from (3.31) and (3.30) into (3.28), the real power with armature current neglected becomes

$$P_{3\phi} = 3\frac{|E||V|}{X_d}\sin\delta + 3|V|^2\frac{X_d - X_q}{2X_dX_q}\sin 2\delta \tag{3.32}$$

The power equation contains an additional term known as the *reluctance power*. Equations (3.25) and (3.32) can be utilized for steady-state analysis. For short-circuit analysis, assuming a high X/R ratio, the power factor approaches zero and the quadrature component of current can often be neglected. In such a case, X_d merely replaces the X_s used for the cylindrical rotor machine. Generators are thus modeled by their direct axis reactance in series with a constant-voltage power source. Later in the text it will be shown that X_d takes on different values, depending upon the transient time following the short circuit. These reactances are usually expressed in per-unit and are available from the manufacturer's data.

3.5 POWER TRANSFORMER

Transformers are essential elements in any power system. They allow the relatively low voltages from generators to be raised to a very high level for efficient power transmission. At the user end of the system, transformers reduce the voltage to values most suitable for utilization. In modern utility systems, the energy may undergo four or five transformations between generator and ultimate user. As a result, a given system is likely to have about five times more kVA of installed capacity of transformers than of generators.

3.6 EQUIVALENT CIRCUIT OF A TRANSFORMER

The equivalent circuit model of a single-phase transformer is shown in Figure 3.9. The equivalent circuit consists of an ideal transformer of ratio $N_1{:}N_2$ together with elements which represent the imperfections of the real transformer. An ideal transformer would have windings with zero resistance and a lossless, infinite permeability core. The voltage E_1 across the primary of the ideal transformer represents the rms voltage induced in the primary winding by the mutual flux ϕ. This is the portion of the core flux which links both primary and secondary coils. Assuming

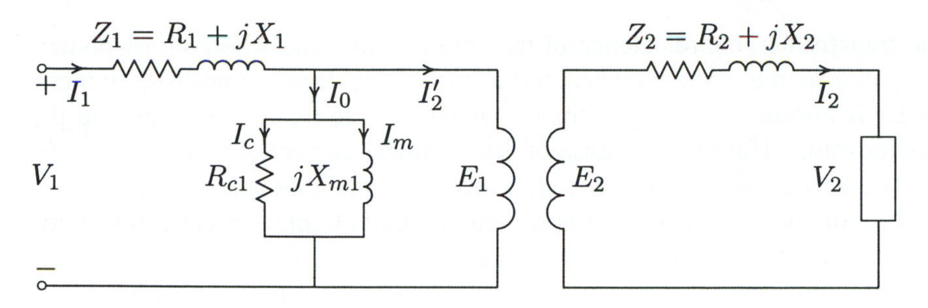

FIGURE 3.9
Equivalent circuit of a transformer.

sinusoidal flux $\phi = \Phi_{max} \cos \omega t$, the instantaneous voltage e_1 is

$$\begin{aligned} e_1 &= N_1 \frac{d\phi}{dt} \\ &= -\omega N_1 \Phi_{max} \sin \omega t \\ &= E_{1max} \cos(\omega t + 90°) \end{aligned} \tag{3.33}$$

where

$$E_{1max} = 2\pi f N_1 \Phi_{max} \tag{3.34}$$

or the rms voltage magnitude E_1 is

$$E_1 = 4.44 f N_1 \Phi_{max} \tag{3.35}$$

It is important to note that the phasor flux is lagging the induced voltage E_1 by 90°. Similarly the rms voltage E_2 across the secondary of the ideal transformer represents the voltage induced in the secondary winding by the mutual flux ϕ, given by

$$E_2 = 4.44 f N_2 \Phi_{max} \tag{3.36}$$

In the ideal transformer, the core is assumed to have a zero reluctance and there is an exact mmf balanced between the primary and secondary. If I_2' represents the component of current to neutralize the secondary mmf, then

$$I_2' N_1 = I_2 N_2 \tag{3.37}$$

Therefore, for an ideal transformer, from (3.35) through (3.37) we have

$$\frac{E_1}{E_2} = \frac{I_2}{I_2'} = \frac{N_1}{N_2} \tag{3.38}$$

In a real transformer, the reluctance of the core is finite, and when the secondary current I_2 is zero, the primary current has a finite value. Since at no-load, induced voltage E_1 is almost equal to the supply voltage V_1, the induced voltage and the flux are sinusoidal. However, because of the nonlinear characteristics of the ferromagnetic core, the no-load current is not sinusoidal and contains odd harmonics. The third harmonic is particularly troublesome in certain three-phase connections of transformers. For the purpose of modeling, we assume a sinusoidal no-load current with the rms value of I_0, known as the *no-load current*. This current has a component I_m, in phase with flux, known as the *magnetizing current*, to set up the core flux. Since flux is lagging the induced voltage E_1 by 90°, I_m is also lagging the induced voltage E_1 by 90°. Thus, this component can be represented in the circuit by the magnetizing reactance jX_{m1}. The other component of I_0 is I_c, which supplies the eddy-current and hysteresis losses in the core. Since this is a power component, it is in phase with E_1 and is represented by the resistance R_{c1} as shown in Figure 3.9.

In a real transformer with finite reluctance, all of the flux is not common to both primary and secondary windings. The flux has three components: mutual flux, primary leakage flux, and secondary leakage flux. The leakage flux associated with one winding does not link the other, and the voltage drops caused by the leakage flux are expressed in terms of leakage reactances X_1 and X_2. Finally, R_1 and R_2 are included to represent the primary and secondary winding resistances.

To obtain the performance characteristics of a transformer, it is convenient to use an equivalent circuit model referred to one side of the transformer. From Kirchhoff's voltage law (KVL), the voltage equation of the secondary side is

$$E_2 = V_2 + Z_2 I_2 \tag{3.39}$$

From the relationship (3.38) developed for the ideal transformer, the secondary induced voltage and current are $E_2 = (N_2/N_1)E_1$ and $I_2 = (N_1/N_2)I_2'$, respectively. Upon substitution, (3.39) reduces to

$$\begin{aligned} E_1 &= \frac{N_1}{N_2}V_2 + \left(\frac{N_1}{N_2}\right)^2 Z_2 I_2' \\ &= V_2' + Z_2' I_2' \end{aligned} \tag{3.40}$$

where

$$Z_2' = R_2' + jX_2' = \left(\frac{N_1}{N_2}\right)^2 R_2 + j\left(\frac{N_1}{N_2}\right)^2 X_2$$

Relation (3.40) is the KVL equation of the secondary side referred to the primary, and the equivalent circuit of Figure 3.9 can be redrawn as shown in Figure 3.10, so the same effects are produced in the primary as would be produced in the secondary.

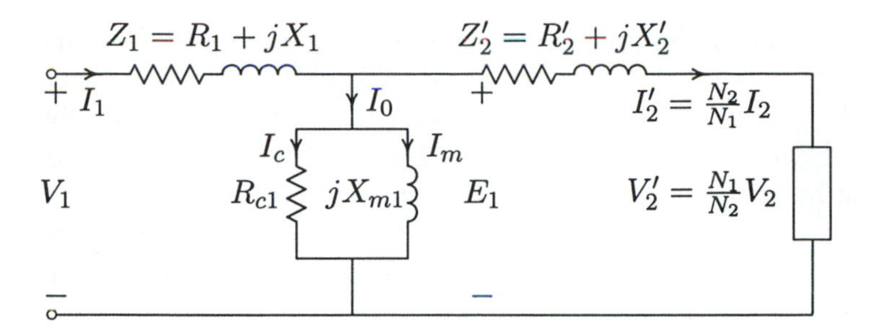

FIGURE 3.10
Exact equivalent circuit referred to the primary side.

On no-load, the primary voltage drop is very small, and V_1 can be used in place of E_1 for computing the no-load current I_0. Thus, the shunt branch can be moved to the left of the primary series impedance with very little loss of accuracy. In this manner, the primary quantities R_1 and X_1 can be combined with the referred secondary quantities R_2' and X_2' to obtain the equivalent primary quantities R_{e1} and X_{e1}. The equivalent circuit is shown in Figure 3.11 where we have dispensed with the coils of the ideal transformer. From Figure 3.11

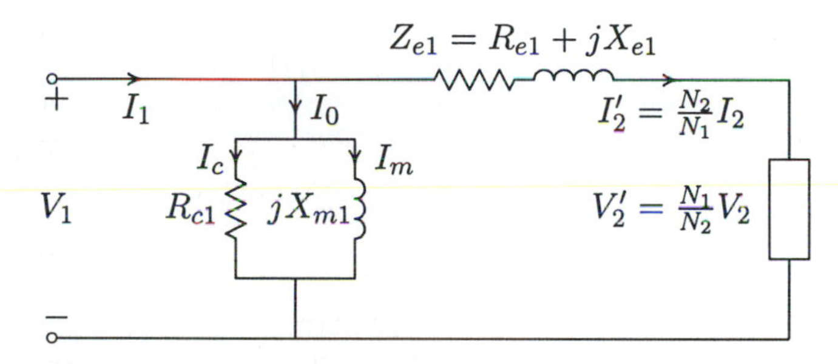

FIGURE 3.11
Approximate equivalent circuit referred to the primary.

$$V_1 = V_2' + (R_{e1} + jX_{e1})I_2' \tag{3.41}$$

where

$$R_{e1} = R_1 + \left(\frac{N_1}{N_2}\right)^2 R_2 \quad X_{e1} = X_1 + \left(\frac{N_1}{N_2}\right)^2 X_2 \quad \text{and} \quad I_2' = \frac{S_L^*}{3V_2'^*}$$

The equivalent circuit referred to the secondary is also shown in Figure 3.12. From Figure 3.12 the referred primary voltage V_1' is given by

$$V_1' = V_2 + (R_{e2} + jX_{e2})I_2 \tag{3.42}$$

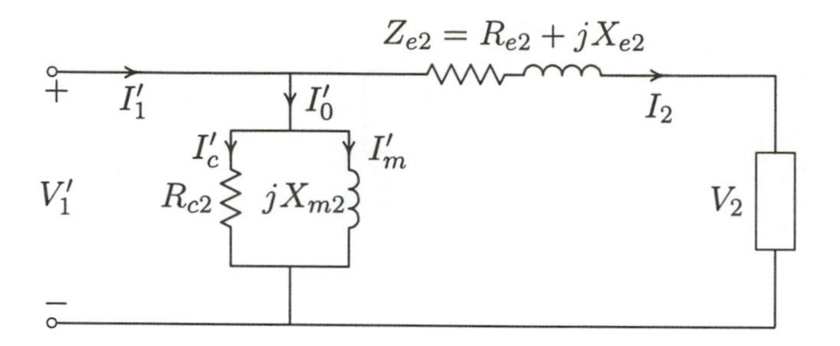

FIGURE 3.12
Approximate equivalent circuit referred to the secondary.

Power transformers are generally designed with very high permeability core and very small core loss. Consequently, a further approximation of the equivalent circuit can be made by omitting the shunt branch, as shown in Figure 3.13. The equivalent circuit referred to the secondary is also shown in Figure 3.13.

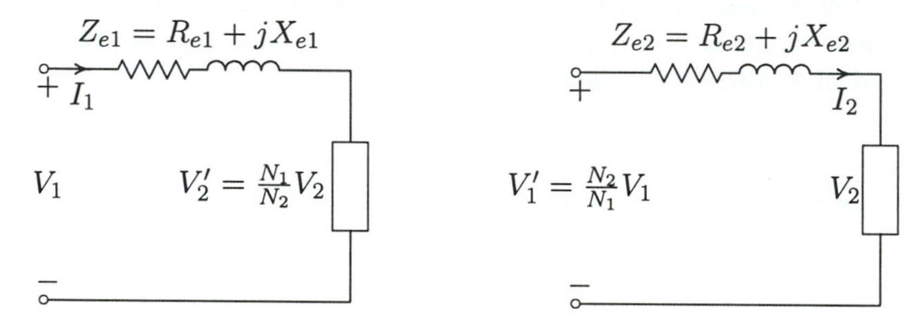

FIGURE 3.13
Simplified circuits referred to one side.

3.7 DETERMINATION OF EQUIVALENT CIRCUIT PARAMETERS

The parameters of the approximate equivalent circuit are readily obtained from open-circuit and short-circuit tests. In the open-circuit test, rated voltage is applied at the terminals of one winding while the other winding terminals are open-circuited. Instruments are connected to measure the input voltage V_1, the no-load input current I_0, and the input power P_0. If the secondary is open-circuited, the referred secondary current I_2' will be zero, and only a small no-load current will be drawn from the supply. Also, the primary voltage drop $(R_1 + jX_1)I_0$ can be neglected, and the equivalent circuit reduces to the form shown in Figure 3.14.

Since the secondary winding copper loss (resistive power loss) is zero and the

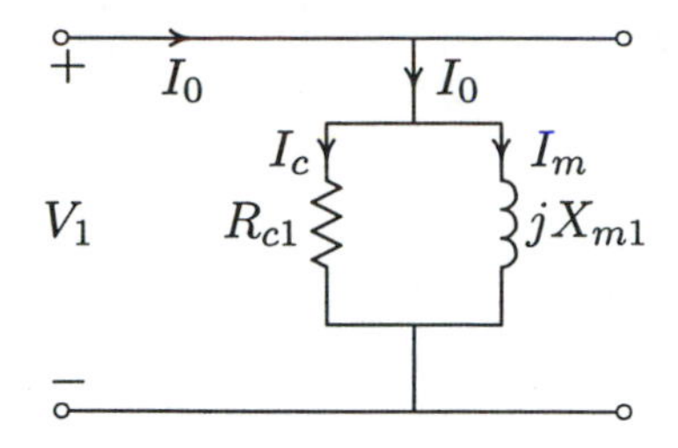

FIGURE 3.14
Equivalent circuit for the open-circuit test.

primary copper loss $R_1 {I_0}^2$ is negligible, the no-load input power P_0 represents the transformer core loss commonly referred to as *iron loss*. The shunt elements R_c and X_m may then be determined from the relations

$$R_{c1} = \frac{{V_1}^2}{P_0} \tag{3.43}$$

The two components of the no-load current are

$$I_c = \frac{V_1}{R_{c1}} \tag{3.44}$$

and

$$I_m = \sqrt{{I_0}^2 - {I_c}^2} \tag{3.45}$$

Therefore, the magnetizing reactance is

$$X_{m1} = \frac{V_1}{I_m} \tag{3.46}$$

In the short-circuit test, a reduced voltage V_{sc} is applied at the terminals of one winding while the other winding terminals are short-circuited. Instruments are connected to measure the input voltage V_{sc}, the input current I_{sc}, and the input power P_{sc}. The applied voltage is adjusted until rated currents are flowing in the windings. The primary voltage required to produce rated current is only a few percent of the rated voltage. At the correspondingly low value of core flux, the exciting current and core losses are entirely negligible, and the shunt branch can be omitted. Thus, the power input can be taken to represent the winding copper loss. The transformer appears as a short when viewed from the primary with the equivalent leakage impedance Z_{e1} consisting of the primary leakage impedance and the referred secondary leakage impedance as shown in Figure 3.15. The series elements

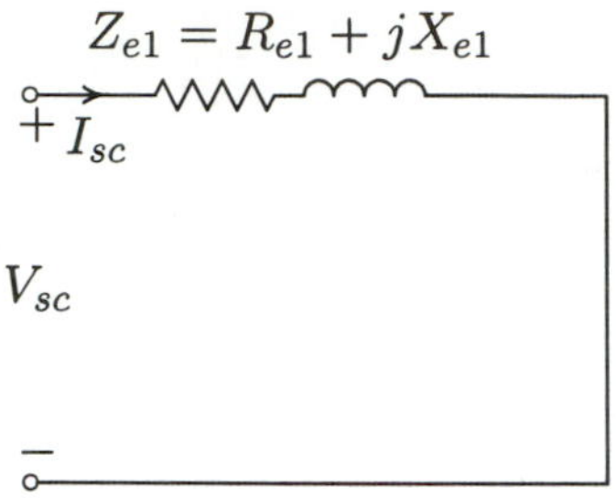

$Z_{e1} = R_{e1} + jX_{e1}$

$+$ I_{sc}

V_{sc}

$-$

FIGURE 3.15
Equivalent circuit for the short-circuit test.

R_{e1} and X_{e1} may then be determined from the relations

$$Z_{e1} = \frac{V_{sc}}{I_{sc}}$$

and

$$R_{e1} = \frac{P_{sc}}{(I_{sc})^2} \tag{3.47}$$

Therefore, the equivalent leakage reactance is

$$X_{e1} = \sqrt{{Z_{e1}}^2 - {R_{e1}}^2} \tag{3.48}$$

3.8 TRANSFORMER PERFORMANCE

The equivalent circuit can now be used to predict the performance characteristics of the transformer. An important aspect is the transformer efficiency. Power transformer efficiencies very from 95 percent to 99 percent, the higher efficiencies being obtained from transformers with the greater ratings. The actual efficiency of a transformer in percent is given by

$$\eta = \frac{\text{output power}}{\text{input power}} \tag{3.49}$$

and the conventional efficiency of a transformer at n fraction of the full-load power is given by

$$\eta = \frac{n \times S \times PF}{(n \times S \times PF) + n^2 \times P_{cu} + P_c} \tag{3.50}$$

where S is the full-load rated volt-ampere, P_{cu} is the full-load copper loss, and for a three-phase transformer, they are given by

$$\begin{aligned} S &= 3|V_2||I_2| \\ P_{cu} &= 3R_{e2}|I_2|^2 \end{aligned}$$

and P_c is the iron loss at rated voltage. For varying I_2 at constant power factor, maximum efficiency occurs when

$$\frac{d\eta}{d|I_2|} = 0$$

For the above condition, it can be easily shown that maximum efficiency occurs when copper loss equals core loss at n per-unit loading given by

$$n = \sqrt{\frac{P_c}{P_{cu}}} \tag{3.51}$$

Another important performance characteristic of a transformer is change in the secondary voltage from no-load to full-load. A figure of merit used to compare the relative performance of different transformers is the voltage regulation. Voltage regulation is defined as the change in the magnitude of the secondary terminal voltage from no-load to full-load expressed as a percentage of the full-load value.

$$\text{Regulation} = \frac{|V_{2nl}| - |V_2|}{|V_2|} \times 100 \tag{3.52}$$

where V_2 is the full-load rated voltage. V_{2nl} in (3.52) can be calculated by using equivalent circuits referred to either primary or secondary. When the equivalent circuit is referred to the primary side, the primary no-load voltage is found from (3.41), and the voltage regulation becomes

$$\text{Regulation} = \frac{|V_1| - |V_2'|}{|V_2'|} \times 100 \tag{3.53}$$

When the equivalent circuit is referred to the secondary side, the secondary no-load voltage is found from (3.42), and the voltage regulation becomes

$$\text{Regulation} = \frac{|V_1'| - |V_2|}{|V_2|} \times 100 \tag{3.54}$$

An interesting feature arises with a capacitive load. Because partial resonance is set up between the capacitance and the reactance, the secondary voltage may actually tend to rise as the capacitive load value increases.

A program called **trans** is developed for obtaining the transformer performance characteristics. The command **trans** displays a menu with three options:

Option 1 calls upon the function **[Rc, Xm] = troct(Vo, Io, Po)** which prompts the user to enter the no-load test data and returns the shunt branch parameters. Then **Ze = trsct(Vsc, Isc, Psc)** is loaded which prompts the user to enter the short-circuit test data and returns the equivalent leakage impedance.

Option 2 calls upon the function **[Zelv, Zehv] = wz2eqz(Elv, Ehv, Zlv, Zhv)** which prompts the user to enter the individual winding impedances and the shunt branch. This function returns the referred equivalent circuit for both sides.

Option 3 prompts the user to enter the parameters of the equivalent circuit.

The above functions can be used independently when the arguments of the functions are defined in the *MATLAB* environment. If the above functions are typed without the parenthesis and the arguments, the user will be prompted to enter the required data.

After the selection of any of the above options, the program prompts the user to enter the load specifications and proceeds to obtain the transformer performance characteristics including an efficiency curve from 25 to 125 percent of full-load.

Example 3.4

Data obtained from short-circuit and open-circuit tests of a 240-kVA, 4800/240-V, 60-Hz transformer are:

Open-circuit test, low-side data	Short-circuit test, high-side data
$V_1 = 240$ V	$V_{sc} = 187.5$ V
$I_0 = 10$ A	$I_{sc} = 50$ A
$P_0 = 1440$ W	$P_{sc} = 2625$ W

Determine the parameters of the equivalent circuit

The commands

```
trans
```

display the following menu

```
Type of parameters for input                          Select
To obtain equivalent circuit from tests                  1
To input individual winding impedances                   2
To input transformer equivalent impedance                3
To quit                                                  0

Select number of menu → 1
Enter Transformer rated power in kVA, S = 240
Enter rated low voltage in volts = 240
Enter rated high voltage in volts = 4800

Open circuit test data
Enter 'lv' within quotes for data ref. to low side or
enter 'hv' within quotes for data ref. to high side → 'lv'
Enter input voltage, in volts, Vo = 240
Enter no-load current in Amp, Io = 10
Enter no-load input power in Watt, Po = 1440

Short circuit test data
Enter 'lv' within quotes for data ref. to low side or
enter 'hv' within quotes for data ref. to high side → 'hv'
Enter reduced input voltage in volts, Vsc = 187.5
Enter input current in Amp, Isc = 50
Enter input power in Watt, Psc = 2625

Shunt branch ref. to LV side      Shunt branch ref. to HV side
Rc = 40.000 ohm                   Rc = 16000.000 ohm
Xm = 30.000 ohm                   Xm = 12000.000 ohm

Series branch ref. to LV side     Series branch ref. to HV side
Ze = 0.002625 + j 0.0090 ohm      Ze = 1.0500 + j 3.6000 ohm

Hit return to continue
```

At this point the user is prompted to enter the load apparent power, power factor, and voltage. The program then obtains the performance characteristics of the transformer including the efficiency curve from 25 to 125 percent of full load as shown in Figure 3.16.

```
Enter load kVA, S2 = 240
Enter load power factor, pf = 0.8
Enter 'lg' within quotes for lagging pf
or 'ld' within quotes for leading pf -> 'lg'
Enter load terminal voltage in volt, V2 = 240
```

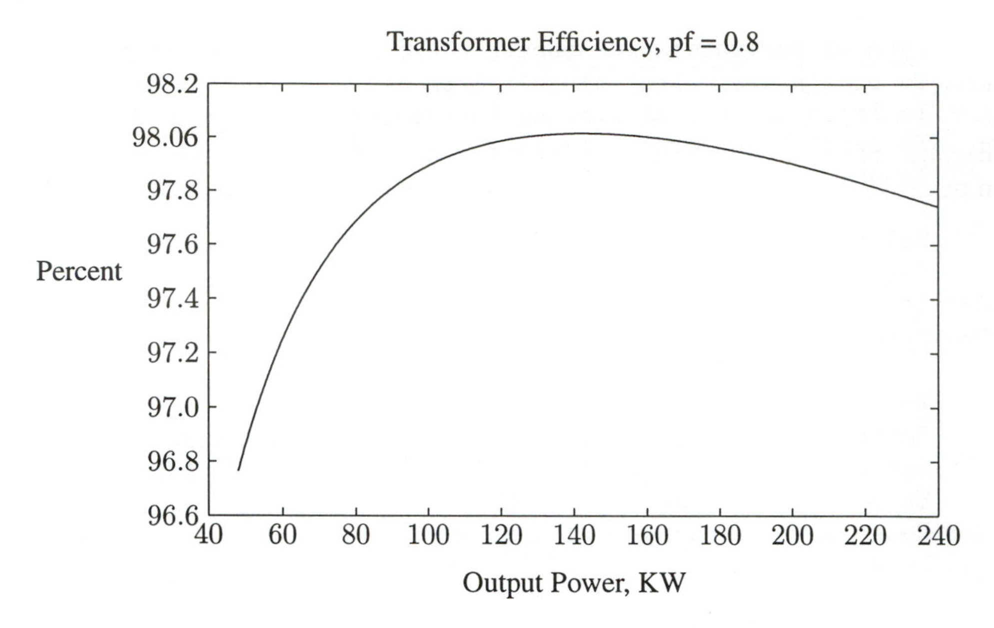

FIGURE 3.16
Efficiency curve of Example 3.4.

```
Secondary load voltage    =   240.000 V
Secondary load current    =  1000.000 A  at  -36.87 degrees
Current ref. to primary   =    50.000 A  at  -36.87 degrees
Primary no-load current   =     0.516 A  at  -53.13 degrees
Primary input current     =    50.495 A  at  -37.03 degrees
Primary input voltage     =  4951.278 V  at    1.30 degrees
Voltage regulation        =     3.152 %
Transformer efficiency    =    97.927 %

Maximum efficiency is 98.015 percent, occurs at 177.757 kVA
with 0.80 pf.
```

At the end of this analysis the program menu is displayed.

3.9 THREE-PHASE TRANSFORMER CONNECTIONS

Three-phase power is transformed by use of three-phase units. However, in large extra high voltage (EHV) units, the insulation clearances and shipping limitations may require a bank of three single-phase transformers connected in three-phase arrangements.

The primary and secondary windings can be connected in either wye (Y) or delta (Δ) configurations. This results in four possible combinations of connections: Y–Y, Δ–Δ, Y–Δ and Δ–Y shown by the simple schematic in Figure 3.17. In this diagram, transformer windings are indicated by heavy lines. The windings shown in parallel are located on the same core and their voltages are in phase. The Y–Y

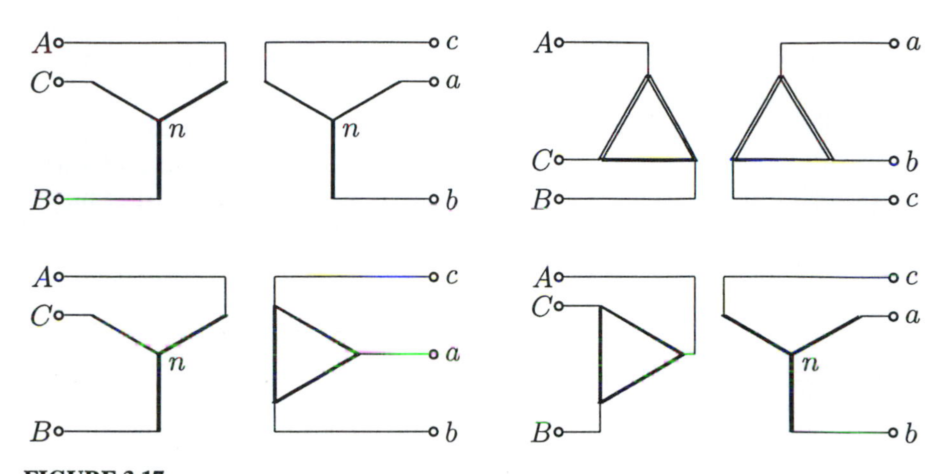

FIGURE 3.17
Three-phase transformer connections.

connection offers advantages of decreased insulation costs and the availability of the neutral for grounding purposes. However, because of problems associated with third harmonics and unbalanced operation, this connection is rarely used. To eliminate the harmonics, a third set of windings, called a *tertiary* winding, connected in Δ is normally fitted on the core to provide a path for the third harmonic currents. This is known as the *three-winding* transformer. The tertiary winding can be loaded with switched reactors or capacitors for reactive power compensation. The Δ–Δ provides no neutral connection and each transformer must withstand full line-to-line voltage. The Δ connection does, however, provide a path for third harmonic currents to flow. This connection has the advantage that one transformer can be removed for repair and the remaining two can continue to deliver three-phase power at a reduced rating of 58 percent of the original bank. This is known as the V connection. The most common connection is the Y–Δ or Δ–Y. This connection is more stable with respect to unbalanced loads, and if the Y connection is used on the high voltage side, insulation costs are reduced. The Y–Δ connection is commonly used to step down a high voltage to a lower voltage. The neutral point on the high voltage side can be grounded. This is desirable in most cases. The Δ–Y connection is commonly used for stepping up to a high voltage.

3.9.1 THE PER-PHASE MODEL OF A THREE-PHASE TRANSFORMER

In Y–Y and Δ–Δ connections, the ratio of the line voltages on HV and LV sides are the same as the ratio of the phase voltages on the HV and LV sides. Furthermore, there is no phase shift between the corresponding line voltages on the HV and LV sides. However, the Y–Δ and the Δ–Y connections will result in a phase shift of 30° between the primary and secondary line-to-line voltages. The windings are arranged in accordance to the ASA (American Standards Association) such that the line voltage on the HV side leads the corresponding line voltage on the LV side by 30° regardless of which side is Y or Δ. Consider the Y–Δ schematic diagram shown in Figure 3.17. The positive phase sequence voltage phasor diagram for this connection is shown in Figure 3.18, where V_{An} is taken as reference. Let the Y

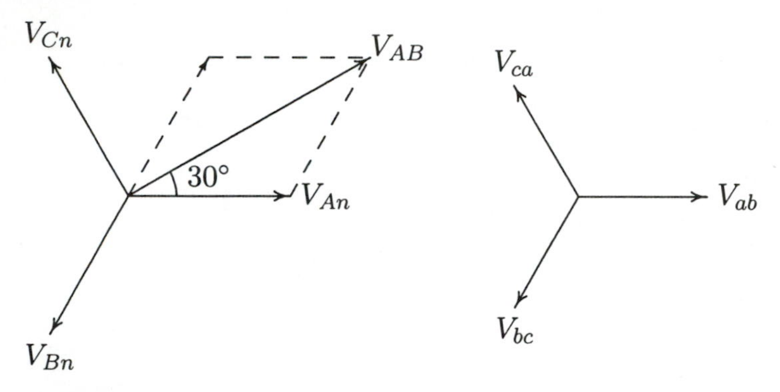

FIGURE 3.18
30° phase shift in line-to-line voltages of Y–Δ connection.

connection be the high voltage side shown by letter H and the Δ connection the low voltage side shown by X. We consider phase a only and use subscript L for line and P for phase quantities. If N_H is the number of turns on one phase of the high voltage winding and N_X is the number of turns on one phase of the low voltage winding, the transformer turns ratio is $a = N_H/N_X = V_{HP}/V_{XP}$. The relationship between the line voltage and phase voltage magnitudes is

$$V_{HL} = \sqrt{3}\, V_{HP}$$
$$V_{XL} = V_{XP}$$

Therefore, the ratio of the line voltage magnitudes for Y–Δ transformer is

$$\frac{V_{HL}}{V_{XL}} = \sqrt{3}\, a \tag{3.55}$$

Because the core losses and magnetization current for power transformers are on the order of 1 percent of the maximum ratings, the shunt impedance is neglected and only the winding resistance and leakage reactance are used to model the transformer. In dealing with Y–Δ or Δ–Y banks, it is convenient to replace the Δ connection by an equivalent Y connection and then work with only one phase. Since for balanced operations, the Y neutral and the neutral of the equivalent Y of the Δ connection are at the same potential, they can be connected together and represented by a neutral conductor. When the equivalent series impedance of one transformer is referred to the delta side, the Δ connected impedances of the transformer are replaced by equivalent Y-connected impedances, given by $Z_Y = Z_\Delta/3$. The per phase equivalent model with the shunt branch neglected is shown in Figure 3.19. Z_{e1} and Z_{e2} are the equivalent impedances based on the line-to-neutral connections, and the voltages are the line-to-neutral values.

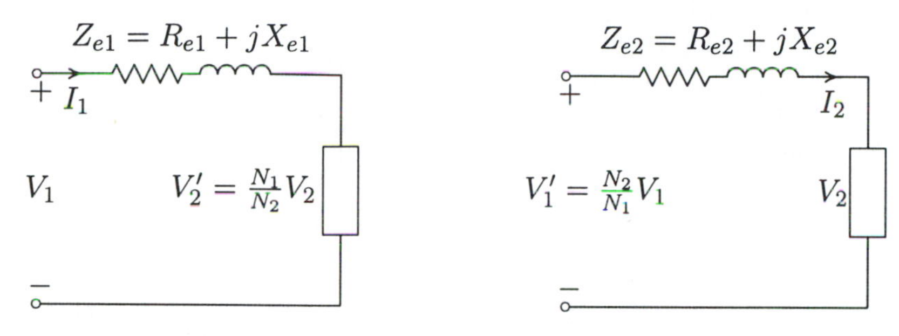

FIGURE 3.19
The per phase equivalent circuit.

3.10 AUTOTRANSFORMERS

Transformers can be constructed so that the primary and secondary coils are electrically connected. This type of transformer is called an autotransformer. A conventional two-winding transformer can be changed into an autotransformer by connecting the primary and secondary windings in series. Consider the two-winding transformer shown in Figure 3.20(a). The two-winding transformer is converted to an autotransformer arrangement as shown in Figure 3.20(b) by connecting the two windings electrically in series so that the polarities are additive. The winding from X_1 to X_2 is called the series winding, and the winding from H_1 to H_2 is called the common winding. From an inspection of this figure it follows that an autotransformer can operate as a step-up as well as a step-down transformer. In both cases, winding part H_1H_2 is common to the primary as well as the secondary side of the transformer. The performance of an autotransformer is governed by the fundamental considerations already discussed for transformers having two separate windings. For determining the power rating as an autotransformer, the ideal transformer relations are ordinarily used, which provides an adequate approximation to the actual transformer values.

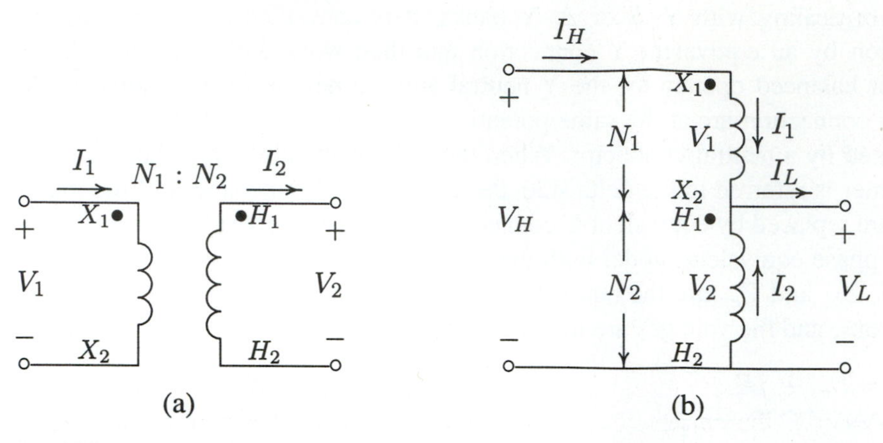

FIGURE 3.20
(a) Two-winding transformer, (b) reconnected as an autotransformer.

From Figure 3.20(a), the two-winding voltages and currents are related by

$$\frac{V_1}{V_2} = \frac{N_1}{N_2} = a \tag{3.56}$$

and

$$\frac{I_2}{I_1} = \frac{N_1}{N_2} = a \tag{3.57}$$

where a is the turns ratio of the two-winding transformer. From Figure 3.20(b), we have

$$V_H = V_2 + V_1 \tag{3.58}$$

Substituting for V_1 from (3.56) into (3.58) yields

$$V_H = V_2 + \frac{N_1}{N_2}V_2 \tag{3.59}$$

Since $V_2 = V_L$, the voltage relationship between the two sides of an autotransformer becomes

$$\begin{aligned} V_H &= V_L + \frac{N_1}{N_2}V_L \\ &= (1+a)V_L \end{aligned} \tag{3.60}$$

or

$$\frac{V_H}{V_L} = 1 + a \tag{3.61}$$

Since the transformer is ideal, the mmf due to I_1 must be equal and opposite to the mmf produced by I_2. As a result, we have

$$N_2 I_2 = N_1 I_1 \tag{3.62}$$

From Kirchhoff's law, $I_2 = I_L - I_1$, and the above equation becomes

$$N_2(I_L - I_1) = N_1 I_1 \tag{3.63}$$

or

$$I_L = \frac{N_1 + N_2}{N_2} I_1 \tag{3.64}$$

Since $I_1 = I_H$, the current relationship between the two sides of an autotransformer becomes

$$\frac{I_L}{I_H} = 1 + a \tag{3.65}$$

The ratio of the apparent power rating of an autotransformer to a two-winding transformer, known as the *power rating advantage*, is found from

$$\frac{S_{auto}}{S_{2-w}} = \frac{(V_1 + V_2)I_1}{V_1 I_1} = 1 + \frac{N_2}{N_1} = 1 + \frac{1}{a} \tag{3.66}$$

From (3.66), we can see that a higher rating is obtained as an autotransformer with a higher number of turns of the common winding (N_2). The higher rating as an autotransformer is a consequence of the fact that only S_{2-w} is transformed by the electromagnetic induction. The rest passes from the primary to secondary without being coupled through the transformer's windings. This is known as the *conducted power*. Compared with a two-winding transformer of the same rating, autotransformers are smaller, more efficient, and have lower internal impedance. Three-phase autotransformers are used extensively in power systems where the voltages of the two systems coupled by the transformers do not differ by a factor greater than about three.

Example 3.5

A two-winding transformer is rated at 60 kVA, 240/1200 V, 60 Hz. When operated as a conventional two-winding transformer at rated load, 0.8 power factor, its efficiency is 0.96. This transformer is to be used as a 1440/1200-V step-down autotransformer in a power distribution system.
(a) Assuming ideal transformer, find the transformer kVA rating when used as an autotransformer.

(b) Find the efficiency with the kVA loading of part (a) and 0.8 power factor.

The two-winding transformer rated currents are:

$$I_1 = \frac{60,000}{240} = 250 \text{ A}$$
$$I_2 = \frac{60,000}{1200} = 50 \text{ A}$$

The autotransformer connection is as shown in Figure 3.21.

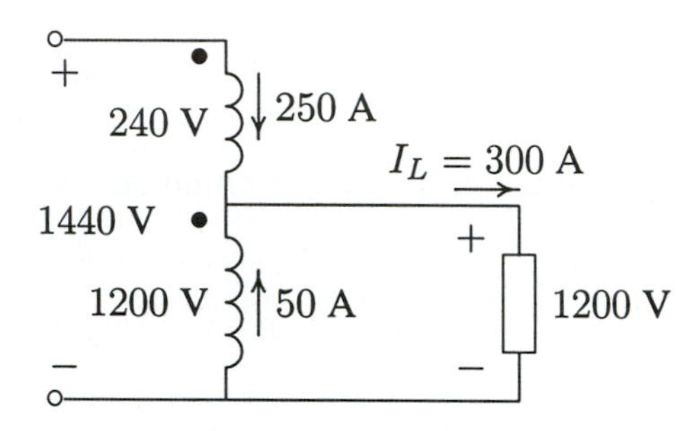

FIGURE 3.21
Auto transformer connection for Example 3.5.

(a) The autotransformer secondary current is

$$I_L = 250 + 50 = 300 \text{ A}$$

With windings carrying rated currents, the autotransformer rating is

$$S = (1200)(300)(10^{-3}) = 360 \text{ kVA}$$

Therefore, the power advantage of the autotransformer is

$$\frac{S_{auto}}{S_{2-w}} = \frac{360}{60} = 6$$

(b) When operated as a two-winding transformer at full-load, 0.8 power factor, the losses are found from the efficiency formula

$$\frac{(60)(0.8)}{(60)(0.8) + P_{loss}} = 0.96$$

Solving the above equation, the total transformer loss is

$$P_{loss} = \frac{48(1 - 0.96)}{0.96} = 2.0 \text{ kW}$$

Since the windings are subjected to the same rated voltages and currents as the two-winding transformer, the autotransformer copper loss and the core loss at the rated values are the same as the two-winding transformer. Therefore, the autotransformer efficiency at rated load, 0.8 power factor, is

$$\eta = \frac{(360)(0.8)}{(360)(0.8) + 2} \times 100 = 99.31 \text{ percent}$$

3.10.1 AUTOTRANSFORMER MODEL

When a two-winding transformer is connected as an autotransformer, its equivalent impedance expressed in per-unit is much smaller compared to the equivalent value of the two-winding connection. It can be shown that the effective per-unit impedance of an autotransformer is smaller by a factor equal to the reciprocal of the power advantage of the autotransformer connection. It is common practice to consider an autotransformer as a two-winding transformer with its two winding connected in series as shown in Figure 3.22, where the equivalent impedance is referred to the N_1-turn side.

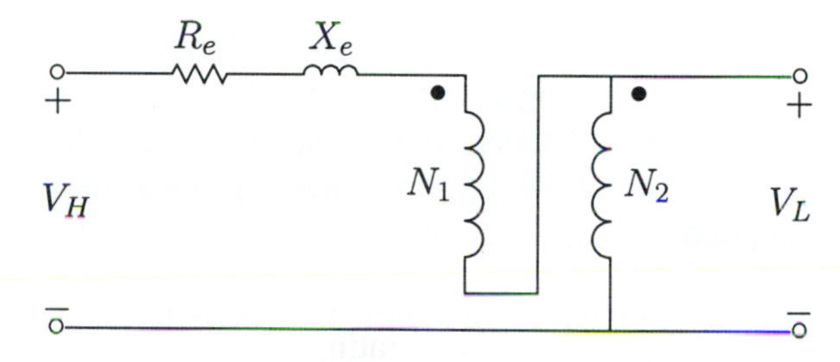

FIGURE 3.22
Autotransformer equivalent circuit.

3.11 THREE-WINDING TRANSFORMERS

Transformers having three windings are often used to interconnect three circuits which may have different voltages. These windings are called primary, secondary, and tertiary windings. Typical applications of three-winding transformers in power systems are for the supply of two independent loads at different voltages from the same source and interconnection of two transmission systems of different voltages. Usually the tertiary windings are used to provide voltage for auxiliary power purposes in the substation or to supply a local distribution system. In addition, the switched reactor or capacitors are connected to the tertiary bus for the purpose of reactive power compensation. Sometimes three-phase Y-Y transformers and Y–

connected autotransformers are provided with Δ–connected tertiary windings for harmonic suppression.

3.11.1 THREE-WINDING TRANSFORMER MODEL

If the exciting current of a three-winding transformer is neglected, it is possible to draw a simple single-phase equivalent T-circuit as shown in Figure 3.23.

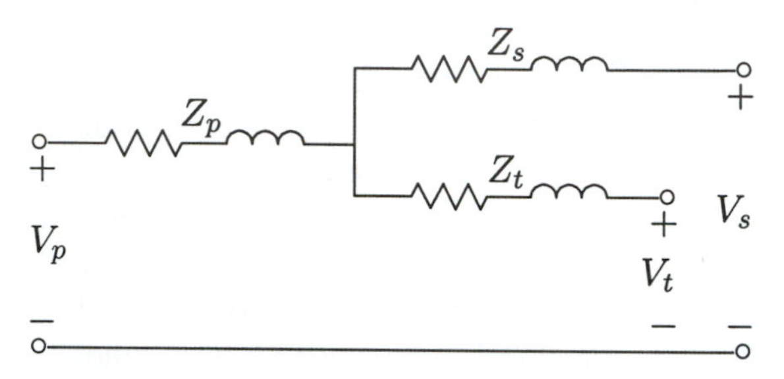

FIGURE 3.23
Equivalent circuit of three-winding transformer.

Three short-circuit tests are carried out on a three-winding transformer with N_p, N_s, and N_t turns per phase on the three windings, respectively. The three tests are similar in that in each case one winding is open, one shorted, and reduced voltage is applied to the remaining winding. The following impedances are measured on the side to which the voltage is applied.

Z_{ps} = impedance measured in the primary circuit with the secondary short-circuited and the tertiary open.
Z_{pt} = impedance measured in the primary circuit with the tertiary short-circuited and the secondary open.
Z'_{st} = impedance measured in the secondary circuit with the tertiary short-circuited and the primary open.

Referring Z'_{st} to the primary side, we obtain

$$Z_{st} = \left(\frac{N_p}{N_s}\right)^2 Z'_{st} \tag{3.67}$$

If Z_p, Z_s, and Z_t are the impedances of the three separate windings referred to the primary side, then

$$\begin{aligned} Z_{ps} &= Z_p + Z_s \\ Z_{pt} &= Z_p + Z_t \\ Z_{st} &= Z_s + Z_t \end{aligned} \tag{3.68}$$

Solving the above equations, we have

$$
\begin{aligned}
Z_p &= \frac{1}{2}(Z_{ps} + Z_{pt} - Z_{st}) \\
Z_s &= \frac{1}{2}(Z_{ps} + Z_{st} - Z_{pt}) \\
Z_t &= \frac{1}{2}(Z_{pt} + Z_{st} - Z_{ps})
\end{aligned}
\tag{3.69}
$$

3.12 VOLTAGE CONTROL OF TRANSFORMERS

Voltage control in transformers are required to compensate for varying voltage drops in the system and to control reactive power flow over transmission lines. Transformers may also be used to control phase angle and, therefore, active power flow. The two commonly used methods are tap changing transformers and regulating transformers.

3.12.1 TAP CHANGING TRANSFORMERS

Practically all power transformers and many distribution transformers have taps in one or more windings for changing the turns ratio. This method is the most popular since it can be used for controlling voltages at all levels. Tap changing, by altering the voltage magnitude, affects the distribution of vars and may therefore be used to control the flow of reactive power. There are two types of tap changing transformers

(i) Off-load tap changing transformers.
(ii) Tap changing under load (TCUL) transformers.

The off-load tap changing transformer requires the disconnection of the transformer when the tap setting is to be changed. Off-load tap changers are used when it is expected that the ratio will need to be changed only infrequently, because of load growth or some seasonal change. A typical transformer might have four taps in addition to the nominal setting, with spacing of 2.5 percent of full-load voltage between them. Such an arrangement provides for adjustments of up to 5 percent above or below the nominal voltage rating of the transformer.

Tap changing under load (TCUL) is used when changes in ratio may be frequent or when it is undesirable to de-energize the transformer to change a tap. A large number of units are now being built with load tap changing equipment. It is used on transformers and autotransformers for transmission tie, for bulk distribution units, and at other points of load service. Basically, a TCUL transformer is a transformer with the ability to change taps while power is connected. A TCUL transformer may have built-in voltage sensing circuitry that automatically changes taps to keep the system voltage constant. Such special transformers are very common in modern power systems. Special tap changing gear are required for TCUL transformers, and the position of taps depends on a number of factors and requires special consideration to arrive at an optimum location for the TCUL equipment. Step-down units usually have TCUL in the low voltage winding and de-energized taps in the high voltage winding. For example, the high voltage winding might be equipped with a nominal voltage turns ratio plus four 2.5 percent fixed tap settings to yield ± 5 percent buck or boost voltage. In addition to this, there could be provision, on the low voltage windings, for 32 incremental steps of $\frac{5}{8}$ each, giving an automatic range of ± 10 percent.

Tapping on both ends of a radial transmission line can be adjusted to compensate for the voltage drop in the line. Consider one phase of a three-phase transmission line with a step-up transformer at the sending end and a step-down transformer at the receiving end of the line. A single-line representation is shown in Figure 3.24, where t_S and t_R are the tap setting in per-unit. In this diagram, V_1' is the supply phase voltage referred to the high voltage side, and V_2' is the load phase voltage, also referred to the high voltage side. The impedance shown includes the

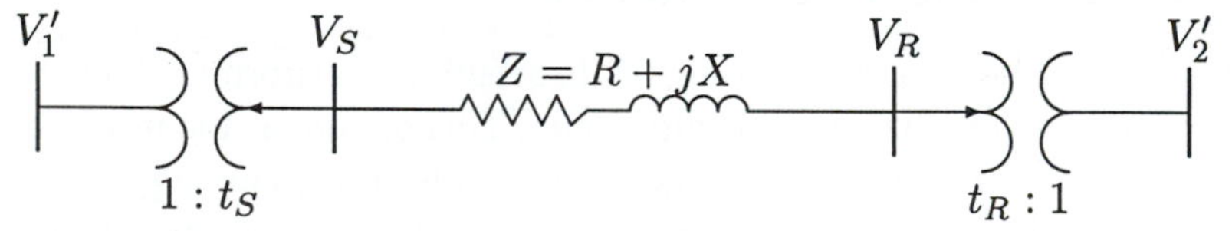

FIGURE 3.24
A radial line with tap changing transformers at both ends.

line impedance plus the referred impedances of the sending end and the receiving end transformers to the high voltage side. If V_S and V_R are the phase voltages at both ends of the line, we have

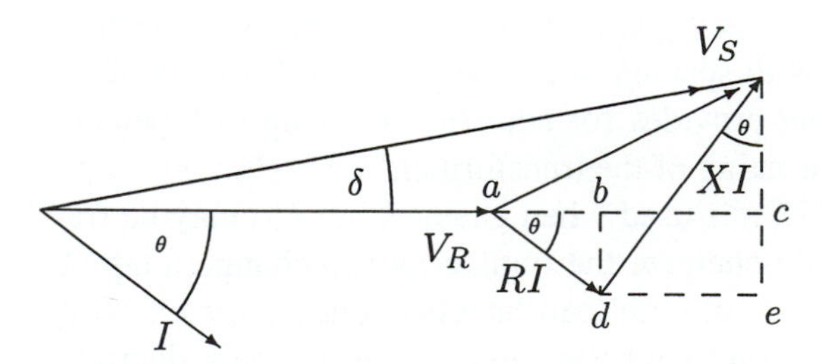

FIGURE 3.25
Voltage phasor diagram.

$$V_R = V_S + (R + jX)I \tag{3.70}$$

The phasor diagram for the above equation is shown in Figure 3.25.

The phase shift δ between the two ends of the line is usually small, and we can neglect the vertical component of V_S. Approximating V_S by its horizontal component results in

$$\begin{aligned}|V_S| &= |V_R| + ab + de \\ &= |V_R| + |I|R\cos\theta + |I|X\sin\theta\end{aligned} \tag{3.71}$$

Substituting for $|I|$ from $P_\phi = |V_R||I|\cos\theta$ and $Q_\phi = |V_R||I|\sin\theta$ will result in

$$|V_S| = |V_R| + \frac{RP_\phi + XQ_\phi}{|V_R|} \tag{3.72}$$

Since $V_S = t_S V_1'$ and $V_R = t_R V_2'$, the above relation in terms of V_1', and V_2' becomes

$$t_S|V_1'| = t_R|V_2'| + \frac{RP_\phi + XQ_\phi}{t_R|V_2'|} \tag{3.73}$$

or

$$t_S = \frac{1}{|V_1'|}\left(t_R|V_2'| + \frac{RP_\phi + XQ_\phi}{t_R|V_2'|}\right) \tag{3.74}$$

Assuming the product of t_S and t_R is unity, i.e., $t_S t_R = 1$, and substituting for t_R in (3.74), the following expression is found for t_S.

$$t_S = \sqrt{\frac{\frac{|V_2'|}{|V_1'|}}{1 - \frac{RP_\phi + XQ_\phi}{|V_1'||V_2'|}}} \tag{3.75}$$

Example 3.6

A three-phase transmission line is feeding from a 23/230-kV transformer at its sending end. The line is supplying a 150-MVA, 0.8 power factor load through a step-down transformer of 230/23 kV. The impedance of the line and transformers at 230 kV is $18 + j60\ \Omega$. The sending end transformer is energized from a 23-kV supply. Determine the tap setting for each transformer to maintain the voltage at the load at 23 kV.

The load real and reactive power per phase are

$$P_\phi = \frac{1}{3}(150)(0.8) = 40 \text{ MW}$$
$$Q_\phi = \frac{1}{3}(150)(0.6) = 30 \text{ Mvar}$$

The source and the load phase voltages referred to the high voltage side are

$$|V_1'| = |V_2'| = \left(\frac{230}{23}\right)\left(\frac{23}{\sqrt{3}}\right) = \frac{230}{\sqrt{3}}$$

From (3.75), we have

$$t_S = \sqrt{\frac{1}{1 - \frac{(18)(40)+(60)(30)}{(230/\sqrt{3})^2}}} = 1.08 \text{ pu}$$

and

$$t_R = \frac{1}{1.08} = 0.926 \text{ pu}$$

3.12.2 REGULATING TRANSFORMERS OR BOOSTERS

Regulating transformers, also known as *boosters*, are used to change the voltage magnitude and phase angle at a certain point in the system by a small amount. A booster consists of an exciting transformer and a series transformer.

VOLTAGE MAGNITUDE CONTROL

Figure 3.26 shows the connection of a regulating transformer for phase a of a three-phase system for voltage magnitude control. Other phases have identical arrangement. The secondary of the exciting transformer is tapped, and the voltage obtained from it is applied to the primary of the series transformer. The corresponding voltage on the secondary of the series transformer is added to the input voltage. Thus, the output voltage is

$$V'_{an} = V_{an} + \Delta V_{an} \tag{3.76}$$

Since the voltages are in phase, a booster of this type is called an *in-phase booster*. The output voltage can be adjusted by changing the excitation transformer taps. By changing the switch from position 1 to 2, the polarity of the voltage across the series transformer is reversed, so that the output voltage is now less than the input voltage.

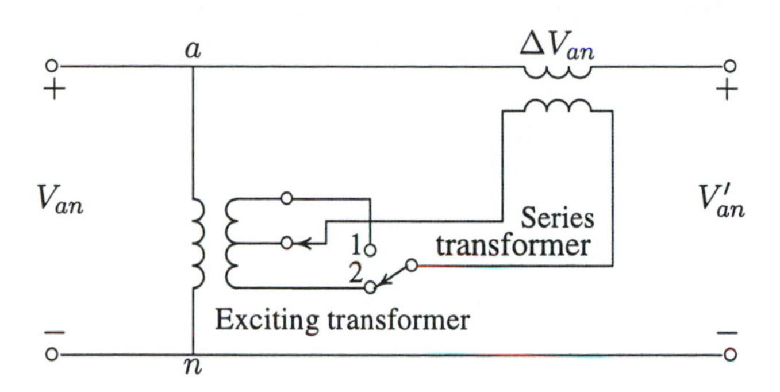

FIGURE 3.26
Regulating transformer for voltage magnitude control.

PHASE ANGLE CONTROL

Regulating transformers are also used to control the voltage phase angle. If the injected voltage is out of phase with the input voltage, the resultant voltage will have a phase shift with respect to the input voltage. Phase shifting is used to control active power flow at major intertie buses. A typical arrangement for phase a of a three-phase system is shown in Figure 3.27.

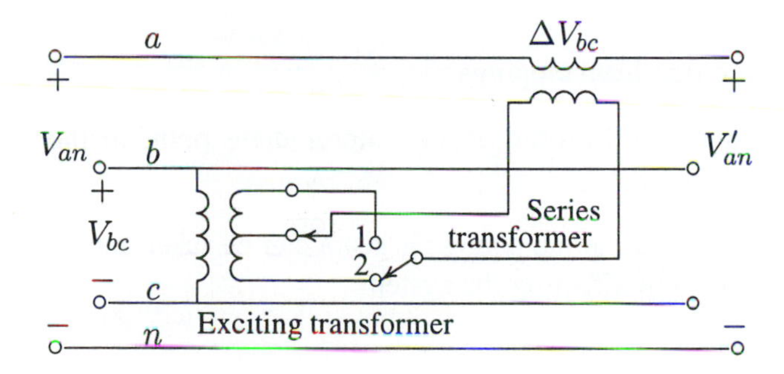

FIGURE 3.27
Regulating transformer for voltage phase angle control.

The series transformer of phase a is supplied from the secondary of the exciting transformer bc. The injected voltage ΔV_{bc} is in quadrature with the voltage V_{an}, thus the resultant voltage V'_{an} goes through a phase shift α, as shown in Figure 3.28. The output voltage is

$$V'_{an} = V_{an} + \Delta V_{bc} \tag{3.77}$$

Similar connections are made for the remaining phases, resulting in a balanced three phase output voltage. The amount of phase shift can be adjusted by changing the excitation transformer taps. By changing the switch from position 1 to 2, the

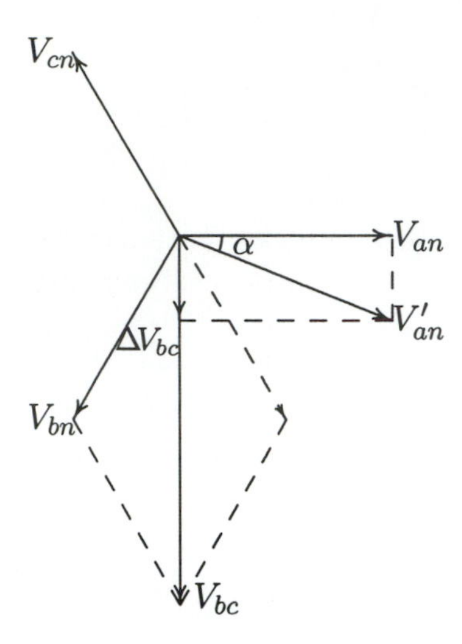

FIGURE 3.28
Voltage phasor diagram showing phase shifting effect for phase a.

output voltage can be made to lag or lead the input voltage. The advantages of the regulating transformers are

1. The main transformers are free from tappings.
2. The regulating transformers can be used at any intermediate point in the system.
3. The regulating transformers and the tap changing gears can be taken out of service for maintenance without affecting the system.

3.13 THE PER-UNIT SYSTEM

The solution of an interconnected power system having several different voltage levels requires the cumbersome transformation of all impedances to a single voltage level. However, power system engineers have devised the *per-unit system* such that the various physical quantities such as power, voltage, current and impedance are expressed as a decimal fraction or multiples of base quantities. In this system, the different voltage levels disappear, and a power network involving generators, transformers, and lines (of different voltage levels) reduces to a system of simple impedances. The per-unit value of any quantity is defined as

$$\text{Quantity in per-unit} = \frac{\text{actual quantity}}{\text{base value of quantity}} \tag{3.78}$$

For example,

$$S_{pu} = \frac{S}{S_B} \quad V_{pu} = \frac{V}{V_B} \quad I_{pu} = \frac{I}{I_B} \quad \text{and} \quad Z_{pu} = \frac{Z}{Z_B}$$

where the numerators (actual values) are phasor quantities or complex values and the denominators (base values) are always real numbers. A minimum of four base quantities are required to completely define a per-unit system: volt-ampere, voltage, current, and impedance. Usually, the three-phase base volt-ampere S_B or MVA_B and the line-to-line base voltage V_B or kV_B are selected. Base current and base impedance are then dependent on S_B and V_B and must obey the circuit laws. These are given by

$$I_B = \frac{S_B}{\sqrt{3}\,V_B} \tag{3.79}$$

and

$$Z_B = \frac{V_B/\sqrt{3}}{I_B} \tag{3.80}$$

Substituting for I_B from (3.79), the base impedance becomes

$$Z_B = \frac{(V_B)^2}{S_B}$$

$$Z_B = \frac{(kV_B)^2}{MVA_B} \tag{3.81}$$

The phase and line quantities expressed in per-unit are the same, and the circuit laws are valid, i.e.,

$$S_{pu} = V_{pu} I_{pu}^* \tag{3.82}$$

and

$$V_{pu} = Z_{pu} I_{pu} \tag{3.83}$$

The load power at its rated voltage can also be expressed by a per-unit impedance. If $S_{L(3\phi)}$ is the complex load power, the load current per phase at the phase voltage V_P is given by

$$S_{L(3\phi)} = 3 V_P I_P^* \tag{3.84}$$

The phase current in terms of the ohmic load impedance is

$$I_P = \frac{V_P}{Z_P} \tag{3.85}$$

Substituting for I_P from (3.85) into (3.84) results in the ohmic value of the load impedance

$$Z_P = \frac{3|V_P|^2}{S^*_{L(3\phi)}} = \frac{|V_{L-L}|^2}{S^*_{L(3\phi)}} \tag{3.86}$$

From (3.81) the load impedance in per-unit is

$$Z_{pu} = \frac{Z_P}{Z_B} = \left|\frac{V_{L-L}}{V_B}\right|^2 \frac{S_B}{S^*_{L(3\phi)}} \tag{3.87}$$

or

$$Z_{pu} = \frac{|V_{pu}|^2}{S^*_{L(pu)}} \tag{3.88}$$

3.14 CHANGE OF BASE

The impedance of individual generators and transformers, as supplied by the manufacturer, are generally in terms of percent or per-unit quantities based on their own ratings. The impedance of transmission lines are usually expressed by their ohmic values. For power system analysis, all impedances must be expressed in per unit on a common system base. To accomplish this, an arbitrary base for apparent power is selected; for example, 100 MVA. Then, the voltage bases must be selected. Once a voltage base has been selected for a point in a system, the remaining voltage bases are no longer independent; they are determined by the various transformer turns ratios. For example, if on a low-voltage side of a 34.5/115-kV transformer the base voltage of 36 kV is selected, the base voltage on the high-voltage side must be $36(115/34.5) = 120$ kV. Normally, we try to select the voltage bases that are the same as the nominal values.

Let Z_{pu}^{old} be the per-unit impedance on the power base S_B^{old} and the voltage base V_B^{old}, which is expressed by

$$Z_{pu}^{old} = \frac{Z_\Omega}{Z_B^{old}} = Z_\Omega \frac{S_B^{old}}{(V_B^{old})^2} \tag{3.89}$$

Expressing Z_Ω to a new power base and a new voltage base, results in the new per-unit impedance

$$Z_{pu}^{new} = \frac{Z_\Omega}{Z_B^{new}} = Z_\Omega \frac{S_B^{new}}{(V_B^{new})^2} \tag{3.90}$$

From (3.89) and (3.90), the relationship between the old and the new per-unit values is

$$Z_{pu}^{new} = Z_{pu}^{old} \frac{S_B^{new}}{S_B^{old}} \left(\frac{V_B^{old}}{V_B^{new}} \right)^2 \tag{3.91}$$

If the voltage bases are the same, (3.91) reduces to

$$Z_{pu}^{new} = Z_{pu}^{old} \frac{S_B^{new}}{S_B^{old}} \tag{3.92}$$

The advantages of the per-unit system for analysis are described below.

- The per-unit system gives us a clear idea of relative magnitudes of various quantities, such as voltage, current, power and impedance.
- The per-unit impedance of equipment of the same general type based on their own ratings fall in a narrow range regardless of the rating of the equipment. Whereas their impedance in ohms vary greatly with the rating.
- The per-unit values of impedance, voltage and current of a transformer are the same regardless of whether they are referred to the primary or the secondary side. This is a great advantage since the different voltage levels disappear and the entire system reduces to a system of simple impedance.
- The per-unit systems are ideal for the computerized analysis and simulation of complex power system problems.
- The circuit laws are valid in per-unit systems, and the power and voltage equations as given by (3.82) and (3.83) are simplified since the factors of $\sqrt{3}$ and 3 are eliminated in the per-unit system.

Example 3.7 demonstrates how a per-unit impedance diagram is obtained for a simple power system network.

Example 3.7

The one-line diagram of a three-phase power system is shown in Figure 3.29. Select a common base of 100 MVA and 22 kV on the generator side. Draw an impedance diagram with all impedances including the load impedance marked in per-unit. The manufacturer's data for each device is given as follow:

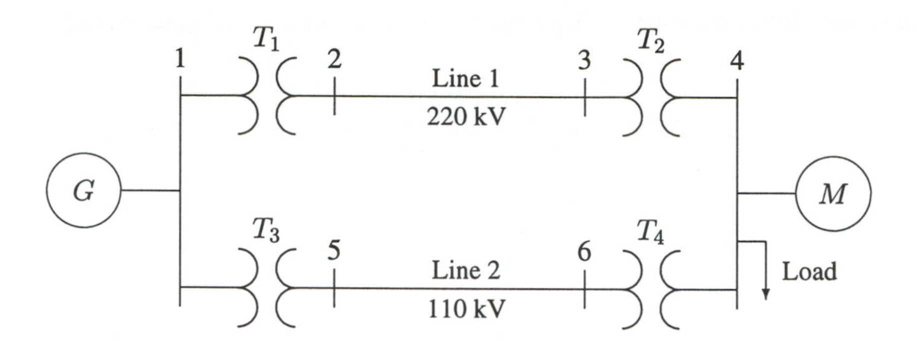

FIGURE 3.29
One-line diagram for Example 3.7.

G:	90 MVA	22 kV	$X = 18\ \%$
T_1:	50 MVA	22/220 kV	$X = 10\ \%$
T_2:	40 MVA	220/11 kV	$X = 6.0\ \%$
T_3:	40 MVA	22/110 kV	$X = 6.4\ \%$
T_4:	40 MVA	110/11 kV	$X = 8.0\ \%$
M:	66.5 MVA	10.45 kV	$X = 18.5\ \%$

The three-phase load at bus 4 absorbs 57 MVA, 0.6 power factor lagging at 10.45 kV. Line 1 and line 2 have reactances of 48.4 and 65.43 Ω, respectively.

First, the voltage bases must be determined for all sections of the network. The generator rated voltage is given as the base voltage at bus 1. This fixes the voltage bases for the remaining buses in accordance to the transformer turns ratios. The base voltage V_{B1} on the LV side of T_1 is 22 kV. Hence the base on its HV side is

$$V_{B2} = 22\left(\frac{220}{22}\right) = 220 \quad \text{kV}$$

This fixes the base on the HV side of T_2 at $V_{B3} = 220$ kV, and on its LV side at

$$V_{B4} = 220\left(\frac{11}{220}\right) = 11 \quad \text{kV}$$

Similarly, the voltage base at buses 5 and 6 are

$$V_{B5} = V_{B6} = 22\left(\frac{110}{22}\right) = 110 \quad \text{kV}$$

Since generator and transformer voltage bases are the same as their rated values, their per-unit reactances on a 100 MVA base, from (3.92) are

$$G\colon\ X = 0.18\left(\frac{100}{90}\right) = 0.20 \quad \text{pu}$$

$$T_1\colon\ X = 0.10\left(\frac{100}{50}\right) = 0.20 \quad \text{pu}$$

$$T_2\colon\ X = 0.06\left(\frac{100}{40}\right) = 0.15 \quad \text{pu}$$

$$T_3\colon\ X = 0.064\left(\frac{100}{40}\right) = 0.16 \quad \text{pu}$$

$$T_4\colon\ X = 0.08\left(\frac{100}{40}\right) = 0.2 \quad \text{pu}$$

The motor reactance is expressed on its nameplate rating of 66.5 MVA and 10.45 kV. However, the base voltage at bus 4 for the motor is 11 kV. From (3.91) the motor reactance on a 100 MVA, 11-kV base is

$$M\colon\ X = 0.185\left(\frac{100}{66.5}\right)\left(\frac{10.45}{11}\right)^2 = 0.25 \quad \text{pu}$$

Impedance bases for lines 1 and 2, from (3.81) are

$$Z_{B2} = \frac{(220)^2}{100} = 484 \quad \Omega$$

$$Z_{B5} = \frac{(110)^2}{100} = 121 \quad \Omega$$

Line 1 and 2 per-unit reactances are

$$\text{Line 1:}\ \ X = \left(\frac{48.4}{484}\right) = 0.10 \quad \text{pu}$$

$$\text{Line 2:}\ \ X = \left(\frac{65.43}{121}\right) = 0.54 \quad \text{pu}$$

The load apparent power at 0.6 power factor lagging is given by

$$S_{L(3\phi)} = 57\angle 53.13^\circ \quad \text{MVA}$$

Hence, the load impedance in ohms is

$$Z_L = \frac{(V_{L-L})^2}{S^*_{L(3\phi)}} = \frac{(10.45)^2}{57\angle -53.13^\circ} = 1.1495 + j1.53267 \quad \Omega$$

The base impedance for the load is

$$Z_{B4} = \frac{(11)^2}{100} = 1.21 \quad \Omega$$

Therefore, the load impedance in per-unit is

$$Z_{L(pu)} = \frac{1.1495 + j1.53267}{1.21} = 0.95 + j1.2667 \quad \text{pu}$$

The per-unit equivalent circuit is shown in Figure 3.30.

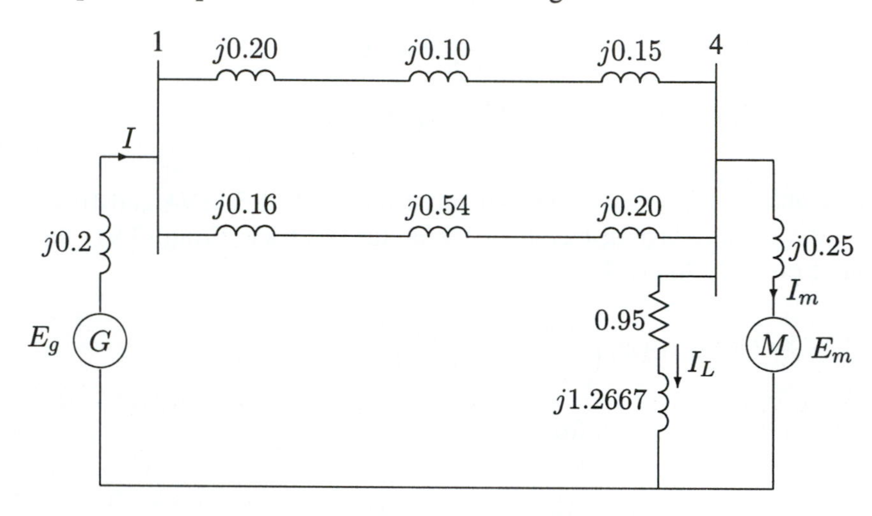

FIGURE 3.30
Per-unit impedance diagram for Example 3.7.

Example 3.8

The motor of Example 3.7 operates at full-load 0.8 power factor leading at a terminal voltage of 10.45 kV.

(a) Determine the voltage at the generator bus bar (bus 1).
(b) Determine the generator and the motor internal emfs.

(a) The per-unit voltage at bus 4, taken as reference is

$$V_4 = \frac{10.45}{11} = 0.95\angle 0^\circ \quad \text{pu}$$

The motor apparent power at 0.8 power factor leading is given by

$$S_m = \frac{66.5}{100}\angle{-36.87^\circ} \quad \text{pu}$$

Therefore, current drawn by the motor is

$$I_m = \frac{S_m^*}{V_4^*} = \frac{0.665\angle 36.87}{0.95\angle 0^\circ} = 0.56 + j0.42 \quad \text{pu}$$

and current drawn by the load is

$$I_L = \frac{V_4}{Z_L} = \frac{0.95\angle 0^\circ}{0.95 + j1.2667} = 0.36 - j0.48 \quad \text{pu}$$

Total current drawn from bus 4 is

$$I = I_m + I_L = (0.56 + j0.42) + (0.36 - j0.48) = 0.92 - j0.06 \quad \text{pu}$$

The equivalent reactance of the parallel branches is

$$X_{\|} = \frac{0.45 \times 0.9}{0.45 + 0.9} = 0.3 \quad \text{pu}$$

The generator terminal voltage is

$$\begin{aligned} V_1 = V_4 + Z_{\|}I = 0.95\angle 0^\circ + j0.3(0.92 - j0.06) &= 0.968 + j0.276 \\ &= 1.0\angle 15.91^\circ \quad \text{pu} \\ &= 22\angle 15.91^\circ \quad \text{kV} \end{aligned}$$

(b) The generator internal emf is

$$\begin{aligned} E_g = V_1 + Z_g I = 0.968 + j0.276 + j0.20(0.92 - j0.06) &= 1.0826\angle 25.14^\circ \ \text{pu} \\ &= 23.82\angle 25.14^\circ \quad \text{kV} \end{aligned}$$

and the motor internal emf is

$$\begin{aligned} E_m = V_4 - Z_m I_m = 0.95 + j0 - j0.25(0.56 + j0.42) &= 1.064\angle -7.56^\circ \quad \text{pu} \\ &= 11.71\angle -7.56^\circ \quad \text{kV} \end{aligned}$$

PROBLEMS

3.1. A three-phase, 318.75-kVA, 2300-V alternator has an armature resistance of 0.35 Ω/phase and a synchronous reactance of 1.2 Ω/phase. Determine the no-load line-to-line generated voltage and the voltage regulation at

(a) Full-load kVA, 0.8 power factor lagging, and rated voltage.

(b) Full-load kVA, 0.6 power factor leading, and rated voltage.

3.2. A 60-MVA, 69.3-kV, three-phase synchronous generator has a synchronous reactance of 15 Ω/phase and negligible armature resistance.

(a) The generator is delivering rated power at 0.8 power factor lagging at the rated terminal voltage to an infinite bus bar. Determine the magnitude of the generated emf per phase and the power angle δ.

(b) If the generated emf is 36 kV per phase, what is the maximum three-phase power that the generator can deliver before losing its synchronism?

(c) The generator is delivering 48 MW to the bus bar at the rated voltage with its field current adjusted for a generated emf of 46 kV per phase. Determine the armature current and the power factor. State whether power factor is lagging or leading?

3.3. A 24,000-kVA, 17.32-kV, 60-Hz, three-phase synchronous generator has a synchronous reactance of 5 Ω/phase and negligible armature resistance.

(a) At a certain excitation, the generator delivers rated load, 0.8 power factor lagging to an infinite bus bar at a line-to-line voltage of 17.32 kV. Determine the excitation voltage per phase.

(b) The excitation voltage is maintained at 13.4 kV/phase and the terminal voltage at 10 kV/phase. What is the maximum three-phase real power that the generator can develop before pulling out of synchronism?

(c) Determine the armature current for the condition of part (b).

3.4. A 34.64-kV, 60-MVA, three-phase salient-pole synchronous generator has a direct axis reactance of 13.5 Ω and a quadrature-axis reactance of 9.333 Ω. The armature resistance is negligible.

(a) Referring to the phasor diagram of a salient-pole generator shown in Figure 3.8, show that the power angle δ is given by

$$\delta = \tan^{-1}\left(\frac{X_q|I_a|\cos\theta}{V + X_q|I_a|\sin\theta}\right)$$

(b) Compute the load angle δ and the per phase excitation voltage E when the generator delivers rated MVA, 0.8 power factor lagging to an infinite bus bar of 34.64-kV line-to-line voltage.

(c) The generator excitation voltage is kept constant at the value found in part (b). Use *MATLAB* to obtain a plot of the power angle curve, i.e., equation (3.32) over a range of δ = 0:0.05:180°. Use the command **Pmax, k = max(P); dmax = d(k)**, to obtain the steady-state maximum power **Pmax** and the corresponding power angle dmax.

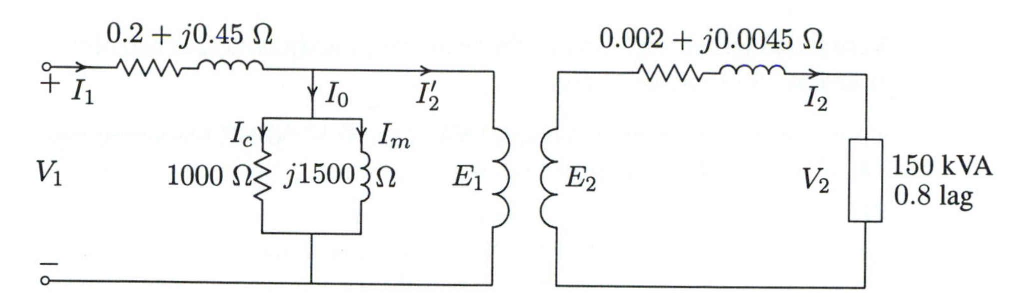

FIGURE 3.31
Transformer circuit for Problem 3.5

3.5. A 150-kVA, 2400/240-V single-phase transformer has the parameters as shown in Figure 3.31.

(a) Determine the equivalent circuit referred to the high-voltage side.

(b) Find the primary voltage and voltage regulation when transformer is operating at full load 0.8 power factor lagging and 240 V.

(c) Find the primary voltage and voltage regulation when the transformer is operating at full-load 0.8 power factor leading.

(d) Verify your answers by running the **trans** program in *MATLAB* and obtain the transformer efficiency curve.

3.6. A 60-kVA, 4800/2400-V single-phase transformer gave the following test results:

1. Rated voltage is applied to the low voltage winding and the high voltage winding is open-circuited. Under this condition, the current into the low voltage winding is 2.4 A and the power taken from the 2400 V source is 3456 W.

2. A reduced voltage of 1250 V is applied to the high voltage winding and the low voltage winding is short-circuited. Under this condition, the current flowing into the high voltage winding is 12.5 A and the power taken from the 1250 V source is 4375 W.

(a) Determine parameters of the equivalent circuit referred to the high voltage side.

(b) Determine voltage regulation and efficiency when transformer is operating at full-load, 0.8 power factor lagging, and a terminal voltage of 2400 V.

(c) What is the load kVA for maximum efficiency and the maximum efficiency at 0.8 power factor?

(d) Determine the efficiency when transformer is operating at 3/4 full-load, 0.8 power factor lagging, and a terminal voltage of 2400 V.

(e) Verify your answers by running the **trans** program in *MATLAB* and obtain the transformer efficiency curve.

3.7. A two-winding transformer rated at 9-kVA, 120/90-V, 60-HZ has a core loss of 200 W and a full-load copper loss of 500 W.

(a) The above transformer is to be connected as an auto transformer to supply a load at 120 V from a 210-V source. What kVA load can be supplied without exceeding the current rating of the windings? (For this part assume an ideal transformer.)

(b) Find the efficiency with the kVA loading of part (a) and 0.8 power factor.

3.8. Three identical 9-MVA, 7.2-kV/4.16-kV, single-phase transformers are connected in wye on the high-voltage side and delta on the low voltage side. The equivalent series impedance of each transformer referred to the high-voltage side is $0.12 + j0.82\ \Omega$ per phase. The transformer supplies a balanced three-phase load of 18 MVA, 0.8 power factor lagging at 4.16 kV. Determine the line-to-line voltage at the high-voltage terminals of the transformer.

3.9. A 400-MVA, 240-kV/24-kV, three-phase Y-Δ transformer has an equivalent series impedance of $1.2 + j6\ \Omega$ per phase referred to the high-voltage side. The transformer is supplying a three-phase load of 400-MVA, 0.8 power factor lagging at a terminal voltage of 24 kV (line to line) on its low-voltage side. The primary is supplied from a feeder with an impedance of $0.6 + j1.2\ \Omega$ per phase. Determine the line-to-line voltage at the high-voltage terminals of the transformer and the sending-end of the feeder.

3.10. In Problem 3.9, with transformer rated values as base quantities, express all impedances in per-unit. Working with per-unit values, determine the line-to-line voltage at the high-voltage terminals of the transformer and the sending-end of the feeder.

3.11. A three-phase, Y-connected, 75-MVA, 27-kV synchronous generator has a synchronous reactance of $9.0\ \Omega$ per phase. Using rated MVA and voltage as base values, determine the per-unit reactance. Then refer this per-unit value to a 100-MVA, 30-kV base.

3.12. A 40-MVA, 20-kV/400-kV, single-phase transformer has the following series impedances:

$Z_1 = 0.9 + j1.8\ \Omega$ and $Z_2 = 128 + j288\ \Omega$

Using the transformer rating as base, determine the per-unit impedance of the transformer from the ohmic value referred to the low-voltage side. Compute the per-unit impedance using the ohmic value referred to the high-voltage side.

3.13. Draw an impedance diagram for the electric power system shown in Figure 3.32 showing all impedances in per unit on a 100-MVA base. Choose 20-kV as the voltage base for generator. The three-phase power and line-line ratings are given below.

G_1 :	90 MVA	20 kV	$X = 9\%$
T_1 :	80 MVA	20/200 kV	$X = 16\%$
T_2 :	80 MVA	200/20 kV	$X = 20\%$
G_2 :	90 MVA	18 kV	$X = 9\%$
Line:		200 kV	$X = 120\ \Omega$
Load:		200 kV	$S = 48$ MW $+j64$ Mvar

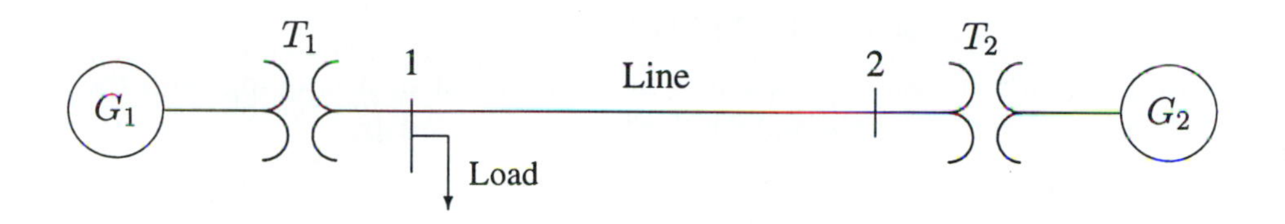

FIGURE 3.32
One-line diagram for Problem 3.13

3.14. The one-line diagram of a power system is shown in Figure 3.33.

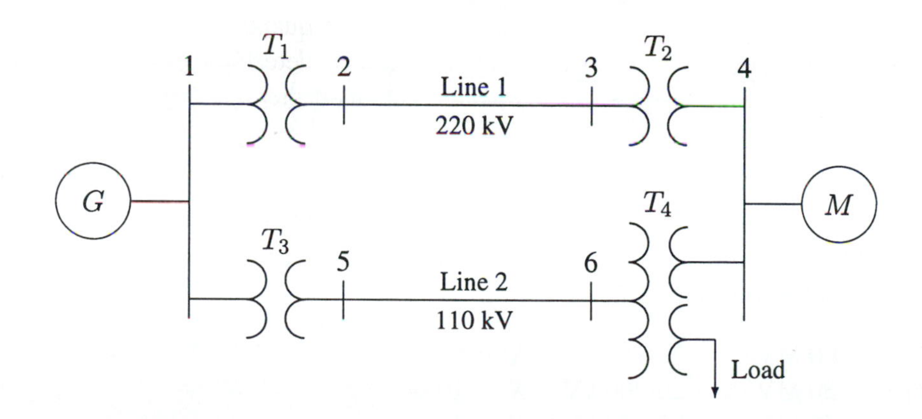

FIGURE 3.33
One-line diagram for Problem 3.14

The three-phase power and line-line ratings are given below.

G:	80 MVA	22 kV	$X = 24\%$
T_1:	50 MVA	22/220 kV	$X = 10\%$
T_2:	40 MVA	220/22 kV	$X = 6.0\%$
T_3:	40 MVA	22/110 kV	$X = 6.4\%$
Line 1:		220 kV	$X = 121\ \Omega$
Line 2:		110 kV	$X = 42.35\ \Omega$
M:	68.85 MVA	20 kV	$X = 22.5\%$
Load:	10 Mvar	4 kV	Δ-connected capacitors

The three-phase ratings of the three-phase transformer are

Primary:	Y-connected	40MVA, 110 kV
Secondary:	Y-connected	40 MVA, 22 kV
Tertiary:	Δ-connected	15 MVA, 4 kV

The per phase measured reactances at the terminal of a winding with the second one short-circuited and the third open-circuited are

$Z_{ps} = 9.6\%$	40 MVA, 110 kV/22 kV
$Z_{pt} = 7.2\%$	40 MVA, 110 kV/4 kV
$Z_{st} = 12\%$	40 MVA, 22kV/4 kV

Obtain the T-circuit equivalent impedances of the three-winding transformer to the common 100-MVA base. Draw an impedance diagram showing all impedances in per-unit on a 100-MVA base. Choose 22 kV as the voltage base for generator.

3.15. The three-phase power and line-line ratings of the electric power system shown in Figure 3.34 are given below.

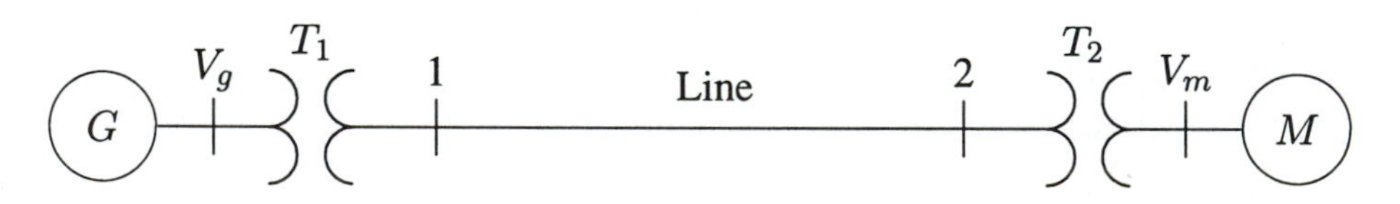

FIGURE 3.34
One-line diagram for Problem 3.15

G_1 :	60 MVA	20 kV	$X = 9\%$
T_1 :	50 MVA	20/200 kV	$X = 10\%$
T_2 :	50 MVA	200/20 kV	$X = 10\%$
M :	43.2 MVA	18 kV	$X = 8\%$
Line:		200 kV	$Z = 120 + j200\ \Omega$

(a) Draw an impedance diagram showing all impedances in per-unit on a 100-MVA base. Choose 20 kV as the voltage base for generator.

(b) The motor is drawing 45 MVA, 0.80 power factor lagging at a line-to-line terminal voltage of 18 kV. Determine the terminal voltage and the internal emf of the generator in per-unit and in kV.

3.16. The one-line diagram of a three-phase power system is as shown in Figure 3.35. Impedances are marked in per-unit on a 100-MVA, 400-kV base. The load at bus 2 is $S_2 = 15.93$ MW $-j33.4$ Mvar, and at bus 3 is $S_3 = 77$ MW $+j14$ Mvar. It is required to hold the voltage at bus 3 at $400\angle 0°$ kV. Working in per-unit, determine the voltage at buses 2 and 1.

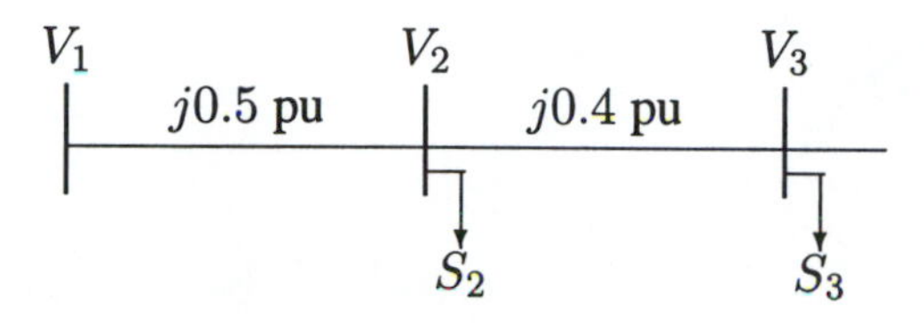

FIGURE 3.35
One-line diagram for Problem 3.16

3.17. The one-line diagram of a three-phase power system is as shown in Figure 3.36. The transformer reactance is 20 percent on a base of 100 MVA, 23/115 kV and the line impedance is $Z = j66.125\Omega$. The load at bus 2 is $S_2 = 184.8$ MW $+j6.6$ Mvar, and at bus 3 is $S_3 = 0$ MW $+j20$ Mvar. It is required to hold the voltage at bus 3 at $115\angle 0°$ kV. Working in per-unit, determine the voltage at buses 2 and 1.

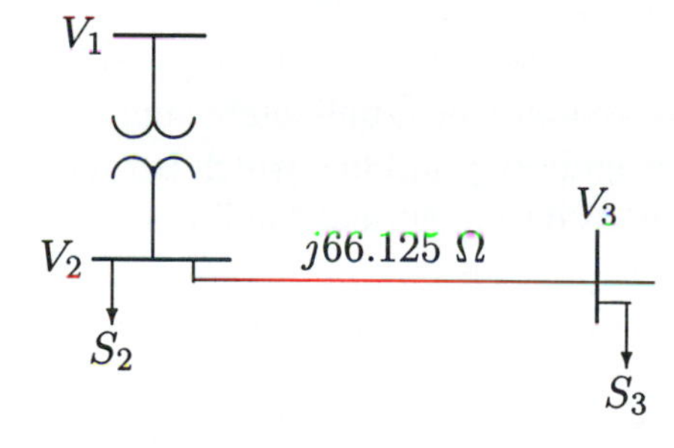

FIGURE 3.36
One-line diagram for Problem 3.17

CHAPTER
4

TRANSMISSION LINE PARAMETERS

4.1 INTRODUCTION

The purpose of a transmission network is to transfer electric energy from generating units at various locations to the distribution system which ultimately supplies the load. Transmission lines also interconnect neighboring utilities which permits not only economic dispatch of power within regions during normal conditions, but also transfer of power between regions during emergencies.

All transmission lines in a power system exhibit the electrical properties of resistance, inductance, capacitance, and conductance. The inductance and capacitance are due to the effects of magnetic and electric fields around the conductor. These parameters are essential for the development of the transmission line models used in power system analysis. The shunt conductance accounts for leakage currents flowing across insulators and ionized pathways in the air. The leakage currents are negligible compared to the current flowing in the transmission lines and may be neglected.

The first part of this chapter deals with the determination of inductance and capacitance of overhead lines. The concept of *geometric mean radius*, *GMR* and *geometric mean distance GMD* are discussed, and the function **[GMD, GMRL,**

GMRC] = gmd is developed for the evaluation of GMR and GMD. This function is very useful for computing the inductance and capacitance of single-circuit or double-circuit transmission lines with bundled conductors. Alternatively, the function **[L, C] = gmd2LC** returns the line inductance in mH per km and the shunt capacitance in μF per km. Finally the effects of electromagnetic and electrostatic induction are discussed.

4.2 OVERHEAD TRANSMISSION LINES

A transmission circuit consists of conductors, insulators, and usually shield wires, as shown in Figure 4.1. Transmission lines are hung overhead from a tower usually made of steel, wood or reinforced concrete with its own right-of-way. Steel towers may be single-circuit or double-circuit designs. Multicircuit steel towers have been built, where the tower supports three to ten 69-kV lines over a given width of right-of-way. Less than 1 percent of the nation's total transmission lines are placed underground. Although underground ac transmission would present a solution to some of the environmental and aesthetic problems involved with overhead transmission lines, there are technical and economic reasons that make the use of underground ac transmission prohibitive.

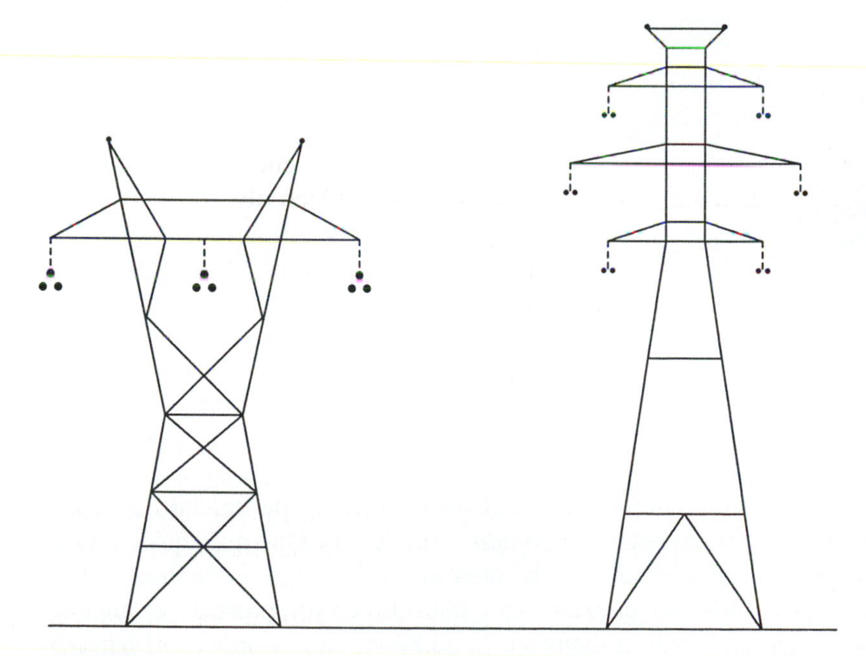

FIGURE 4.1
Typical lattice-type structure for 345-kV transmission line.

The selection of an economical voltage level for the transmission line is based on the amount of power and the distance of transmission. The voltage choice together with the selection of conductor size is mainly a process of weighing RI^2 losses, audible noise, and radio interference level against fixed charges on the investment. Standard transmission voltages are established in the United States by the American National Standards Institute (ANSI). Transmission voltage lines operating at more than 60 kV are standardized at 69 kV, 115 kV, 138 kV, 161 kV, 230 kV, 345 kV, 500 kV, 765 kV line-to-line. Transmission voltages above 230 kV are usually referred to as *extra-high voltage* (EHV) and those at 765 kV and above are referred to as *ultra-high voltage* (UHV). The most commonly used conductor materials for high voltage transmission lines are ACSR (aluminum conductor steel-reinforced), AAC (all-aluminum conductor), AAAC (all-aluminum alloy conductor), and ACAR (aluminum conductor alloy-reinforced). The reason for their popularity is their low relative cost and high strength-to-weight ratio as compared to copper conductors. Also, aluminum is in abundant supply, while copper is limited in quantity. A table of the most commonly used ACSR conductors is stored in file **acsr.m** Characteristics of other conductors can be found in conductor handbooks or manufacturer's literature. The conductors are stranded to have flexibility. The ACSR conductor consists of a center core of steel strands surrounded by layers of aluminum as shown in Figure 4.2. Each layer of strands is spiraled in the opposite direction of its adjacent layer. This spiraling holds the strands in place.

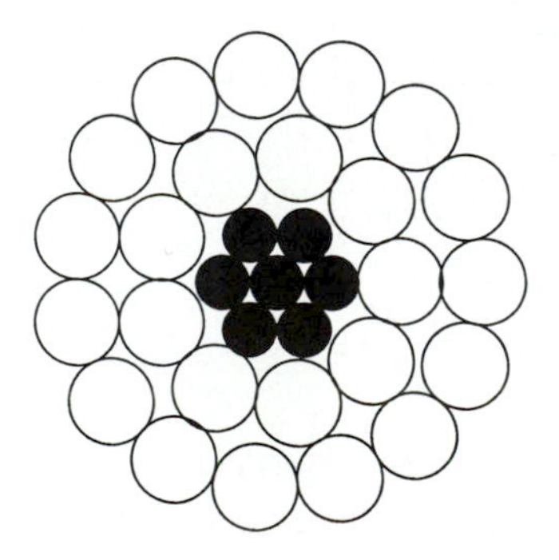

FIGURE 4.2
Cross-sectional view of a 24/7 ACSR conductor.

Conductor manufacturers provide the characteristics of the standard conductors with conductor sizes expressed in *circular mils* (cmil). One mil equals 0.001 inch, and for a solid round conductor the area in circular mils is defined as the square of diameter in mils. As an example, $1,000,000$ cmil represents an area of a solid round conductor 1 inch in diameter. In addition, code words (bird names) have been assigned to each conductor for easy reference.

At voltages above 230 kV, it is preferable to use more than one conductor per phase, which is known as *bundling* of conductors. The bundle consists of two, three, or four conductors. Bundling increases the effective radius of the line's conductor and reduces the electric field strength near the conductors, which reduces corona power loss, audible noise, and radio interference. Another important advantage of bundling is reduced line reactance.

4.3 LINE RESISTANCE

The resistance of the conductor is very important in transmission efficiency evaluation and economic studies. The dc resistance of a solid round conductor at a specified temperature is given by

$$R_{dc} = \frac{\rho l}{A} \tag{4.1}$$

where ρ = conductor resistivity
l = conductor length
A = conductor cross-sectional area

The conductor resistance is affected by three factors: frequency, spiraling, and temperature.

When ac flows in a conductor, the current distribution is not uniform over the conductor cross-sectional area and the current density is greatest at the surface of the conductor. This causes the ac resistance to be somewhat higher than the dc resistance. This behavior is known as *skin effect*. At 60 Hz, the ac resistance is about 2 percent higher than the dc resistance.

Since a stranded conductor is spiraled, each strand is longer than the finished conductor. This results in a slightly higher resistance than the value calculated from 4.1.

The conductor resistance increases as temperature increases. This change can be considered linear over the range of temperature normally encountered and may be calculated from

$$R_2 = R_1 \frac{T + t_2}{T + t_1} \tag{4.2}$$

where R_2 and R_1 are conductor resistances at t_2 and t_1-C°, respectively. T is a temperature constant that depends on the conductor material. For aluminum $T \simeq 228$.

Because of the above effects, the conductor resistance is best determined from manufacturers' data.

4.4 INDUCTANCE OF A SINGLE CONDUCTOR

A current-carrying conductor produces a magnetic field around the conductor. The magnetic flux lines are concentric closed circles with direction given by the right-hand rule. With the thumb pointing in the direction of the current, the fingers of the right hand encircled the wire point in the direction of the magnetic field. When the current changes, the flux changes and a voltage is induced in the circuit. By definition, for nonmagnetic material, the inductance L is the ratio of its total magnetic flux linkage to the current I, given by

$$L = \frac{\lambda}{I} \tag{4.3}$$

where λ = flux linkages, in Weber turns.

Consider a long round conductor with radius r, carrying a current I as shown in Figure 4.3.

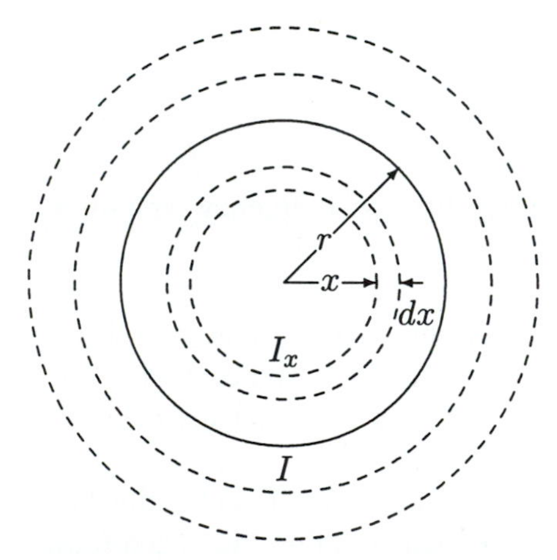

FIGURE 4.3
Flux linkage of a long round conductor.

The magnetic field intensity H_x, around a circle of radius x, is constant and tangent to the circle. The Ampere's law relating H_x to the current I_x is given by

$$\int_0^{2\pi x} H_x \cdot dl = I_x \tag{4.4}$$

or

$$H_x = \frac{I_x}{2\pi x} \tag{4.5}$$

where I_x is the current enclosed at radius x. As shown in Figure 4.3, Equation (4.5) is all that is required for evaluating the flux linkage λ of a conductor. The inductance of the conductor can be defined as the sum of contributions from flux linkages internal and external to the conductor.

4.4.1 INTERNAL INDUCTANCE

A simple expression can be obtained for the internal flux linkage by neglecting the skin effect and assuming uniform current density throughout the conductor cross section, i.e.,

$$\frac{I}{\pi r^2} = \frac{I_x}{\pi x^2} \tag{4.6}$$

Substituting for I_x in (4.5) yields

$$H_x = \frac{I}{2\pi r^2} x \tag{4.7}$$

For a nonmagnetic conductor with constant permeability μ_0, the magnetic flux density is given by $B_x = \mu_0 H_x$, or

$$B_x = \frac{\mu_0 I}{2\pi r^2} x \tag{4.8}$$

where μ_0 is the permeability of free space (or air) and is equal to $4\pi \times 10^{-7}$H/m. The differential flux $d\phi$ for a small region of thickness dx and one meter length of the conductor is

$$d\phi_x = B_x dx \cdot 1 = \frac{\mu_0 I}{2\pi r^2} x dx \tag{4.9}$$

The flux $d\phi_x$ links only the fraction of the conductor from the center to radius x. Thus, on the assumption of uniform current density, only the fraction $\pi x^2/\pi r^2$ of the total current is linked by the flux, i.e.,

$$d\lambda_x = (\frac{x^2}{r^2}) d\phi_x = \frac{\mu_0 I}{2\pi r^4} x^3 dx \tag{4.10}$$

The total flux linkage is found by integrating $d\lambda_x$ from 0 to r.

$$\begin{aligned} \lambda_{int} &= \frac{\mu_0 I}{2\pi r^4} \int_0^r x^3 dx \\ &= \frac{\mu_0 I}{8\pi} \text{ Wb/m} \end{aligned} \tag{4.11}$$

From (4.3), the inductance due to the internal flux linkage is

$$L_{int} = \frac{\mu_0}{8\pi} = \frac{1}{2} \times 10^{-7} \text{ H/m} \tag{4.12}$$

Note that L_{int} is independent of the conductor radius r.

4.4.2 INDUCTANCE DUE TO EXTERNAL FLUX LINKAGE

Consider H_x external to the conductor at radius $x > r$ as shown in Figure 4.4. Since the circle at radius x encloses the entire current, $I_x = I$ and in (4.5) I_x is replaced by I and the flux density at radius x becomes

$$B_x = \mu_0 H_x = \frac{\mu_0 I}{2\pi x} \tag{4.13}$$

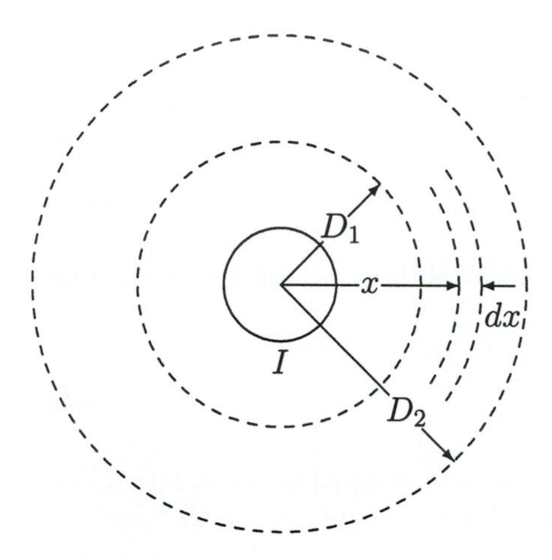

FIGURE 4.4
Flux linkage between D_1 and D_2.

Since the entire current I is linked by the flux outside the conductor, the flux linkage $d\lambda_x$ is numerically equal to the flux $d\phi_x$. The differential flux $d\phi_x$ for a small region of thickness dx and one meter length of the conductor is then given by

$$d\lambda_x = d\phi_x = B_x dx \cdot 1 = \frac{\mu_0 I}{2\pi x} dx \tag{4.14}$$

The external flux linkage between two points D_1 and D_2 is found by integrating $d\lambda_x$ from D_1 to D_2.

$$\begin{aligned} \lambda_{ext} &= \frac{\mu_0 I}{2\pi} \int_{D_1}^{D_2} \frac{1}{x} dx \\ &= 2 \times 10^{-7} I \ln \frac{D_2}{D_1} \quad \text{Wb/m} \end{aligned} \tag{4.15}$$

The inductance between two points external to a conductor is then

$$L_{ext} = 2 \times 10^{-7} \ln \frac{D_2}{D_1} \quad \text{H/m} \tag{4.16}$$

4.5 INDUCTANCE OF SINGLE-PHASE LINES

Consider one meter length of a single-phase line consisting of two solid round conductors of radius r_1 and r_2 as shown in Figure 4.5. The two conductors are separated by a distance D. Conductor 1 carries the phasor current I_1 referenced into the page and conductor 2 carries return current $I_2 = -I_1$. These currents set up magnetic field lines that links between the conductors as shown.

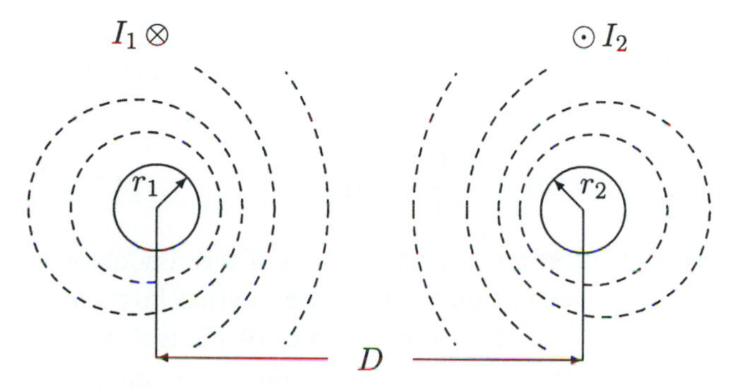

FIGURE 4.5
Single-phase two-wire line.

Inductance of conductor 1 due to internal flux is given by (4.12). The flux beyond D links a net current of zero and does not contribute to the net magnetic flux linkages in the circuit. Thus, to obtain the inductance of conductor 1 due to the net external flux linkage, it is necessary to evaluate (4.16) from $D_1 = r_1$ to $D_2 = D$.

$$L_{1(ext)} = 2 \times 10^{-7} \ln \frac{D}{r_1} \quad \text{H/m} \tag{4.17}$$

The total inductance of conductor 1 is then

$$L_1 = \frac{1}{2} \times 10^{-7} + 2 \times 10^{-7} \ln \frac{D}{r_1} \quad \text{H/m} \tag{4.18}$$

Equation (4.18) is often rearranged as follows:

$$\begin{aligned} L_1 &= 2 \times 10^{-7} \left(\frac{1}{4} + \ln \frac{D}{r_1} \right) \\ &= 2 \times 10^{-7} \left(\ln e^{\frac{1}{4}} + \ln \frac{1}{r_1} + \ln \frac{D}{1} \right) \\ &= 2 \times 10^{-7} \left(\ln \frac{1}{r_1 e^{-1/4}} + \ln \frac{D}{1} \right) \end{aligned} \tag{4.19}$$

Let $r_1' = r_1 e^{-\frac{1}{4}}$, the inductance of conductor 1 becomes

$$L_1 = 2 \times 10^{-7} \ln \frac{1}{r_1'} + 2 \times 10^{-7} \ln \frac{D}{1} \quad \text{H/m} \tag{4.20}$$

Similarly, the inductance of conductor 2 is

$$L_2 = 2 \times 10^{-7} \ln \frac{1}{r_2'} + 2 \times 10^{-7} \ln \frac{D}{1} \quad \text{H/m} \tag{4.21}$$

If the two conductors are identical, $r_1 = r_2 = r$ and $L_1 = L_2 = L$, and the inductance per phase per meter length of the line is given by

$$L = 2 \times 10^{-7} \ln \frac{1}{r'} + 2 \times 10^{-7} \ln \frac{D}{1} \quad \text{H/m} \tag{4.22}$$

Examination of (4.22) reveals that the first term is only a function of the conductor radius. This term is considered as the inductance due to both the internal flux and that external to conductor 1 to a radius of 1 m. The second term of (4.22) is dependent only upon conductor spacing. This term is known as the *inductance spacing factor*. The above terms are usually expressed as inductive reactances at 60 Hz and are available in the manufacturers table in English units.

The term $r' = re^{-\frac{1}{4}}$ is known mathematically as the *self-geometric mean distance* of a circle with radius r and is abbreviated by GMR. r' can be considered as the radius of a fictitious conductor assumed to have no internal flux but with the same inductance as the actual conductor with radius r. GMR is commonly referred to as *geometric mean radius* and will be designated by D_s. Thus, the inductance per phase in millihenries per kilometer becomes

$$L = 0.2 \ln \frac{D}{D_s} \quad \text{mH/km} \tag{4.23}$$

4.6 FLUX LINKAGE IN TERMS OF SELF- AND MUTUAL INDUCTANCES

The series inductance per phase for the above single-phase two-wire line can be expressed in terms of self-inductance of each conductor and their mutual inductance. Consider one meter length of the single-phase circuit represented by two coils characterized by the self-inductances L_{11} and L_{22} and the mutual inductance L_{12}. The magnetic polarity is indicated by dot symbols as shown in Figure 4.6.

The flux linkages λ_1 and λ_2 are given by

$$\begin{aligned} \lambda_1 &= L_{11} I_1 + L_{12} I_2 \\ \lambda_2 &= L_{21} I_1 + L_{22} I_2 \end{aligned} \tag{4.24}$$

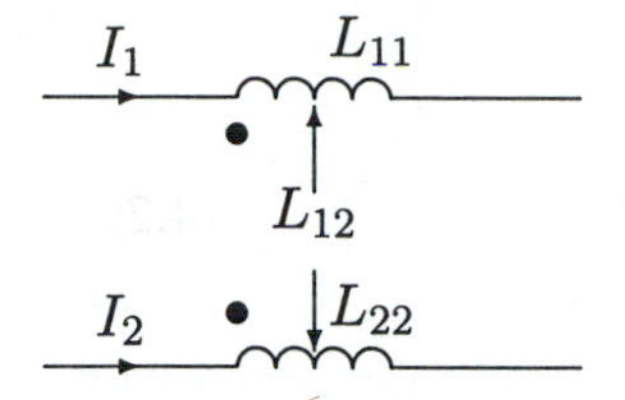

FIGURE 4.6
The single-phase line viewed as two magnetically coupled coils.

Since $I_2 = -I_1$, we have

$$\begin{aligned} \lambda_1 &= (L_{11} - L_{12})I_1 \\ \lambda_2 &= (-L_{21} + L_{22})I_2 \end{aligned} \tag{4.25}$$

Comparing (4.25) with (4.20) and (4.21), we conclude the following equivalent expressions for the self- and mutual inductances:

$$\begin{aligned} L_{11} &= 2 \times 10^{-7} \ln \frac{1}{r'_1} \\ L_{22} &= 2 \times 10^{-7} \ln \frac{1}{r'_2} \\ L_{12} &= L_{21} = 2 \times 10^{-7} \ln \frac{1}{D} \end{aligned} \tag{4.26}$$

The concept of self- and mutual inductance can be extended to a group of n conductors. Consider n conductors carrying phasor currents I_1, I_2, ..., I_n, such that

$$I_1 + I_2 + \cdots + I_i + \cdots + I_n = 0 \tag{4.27}$$

Generalizing (4.24), the flux linkages of conductor i are

$$\lambda_i = L_{ii}I_i + \sum_{j=1}^{n} L_{ij}I_j \qquad j \neq i \tag{4.28}$$

or

$$\lambda_i = 2 \times 10^{-7} \left(I_i \ln \frac{1}{r'_i} + \sum_{j=1}^{n} I_j \ln \frac{1}{D_{ij}} \right) \qquad j \neq i \tag{4.29}$$

4.7 INDUCTANCE OF THREE-PHASE TRANSMISSION LINES

4.7.1 SYMMETRICAL SPACING

Consider one meter length of a three-phase line with three conductors, each with radius r, symmetrically spaced in a triangular configuration as shown in Figure 4.7.

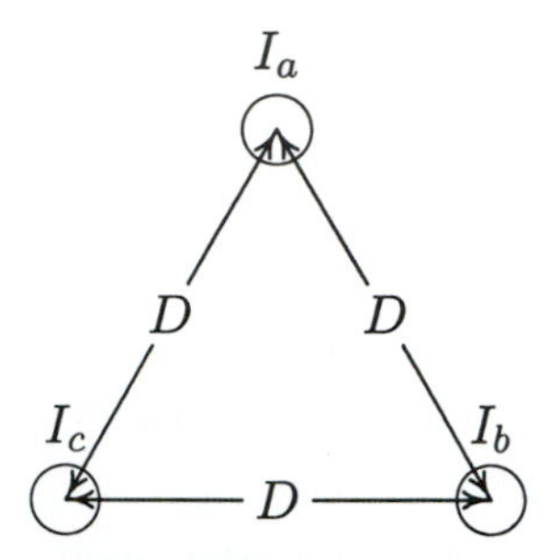

FIGURE 4.7
Three-phase line with symmetrical spacing.

Assuming balanced three-phase currents, we have

$$I_a + I_b + I_c = 0 \tag{4.30}$$

From (4.29) the total flux linkage of phase a conductor is

$$\lambda_a = 2 \times 10^{-7} \left(I_a \ln \frac{1}{r'} + I_b \ln \frac{1}{D} + I_c \ln \frac{1}{D} \right) \tag{4.31}$$

Substituting for $I_b + I_c = -I_a$

$$\begin{aligned} \lambda_a &= 2 \times 10^{-7} \left(I_a \ln \frac{1}{r'} - I_a \ln \frac{1}{D} \right) \\ &= 2 \times 10^{-7} I_a \ln \frac{D}{r'} \end{aligned} \tag{4.32}$$

Because of symmetry, $\lambda_b = \lambda_c = \lambda_a$, and the three inductances are identical. Therefore, the inductance per phase per kilometer length is

$$L = 0.2 \ln \frac{D}{D_s} \quad \text{mH/km} \tag{4.33}$$

where r' is the geometric mean radius, GMR, and is shown by D_s. For a solid round conductor, $D_s = re^{-\frac{1}{4}}$ for stranded conductor D_s can be evaluated from (4.50). Comparison of (4.33) with (4.23) shows that inductance per phase for a three-phase circuit with equilateral spacing is the same as for one conductor of a single-phase circuit.

4.7.2 ASYMMETRICAL SPACING

Practical transmission lines cannot maintain symmetrical spacing of conductors because of construction considerations. With asymmetrical spacing, even with balanced currents, the voltage drop due to line inductance will be unbalanced. Consider one meter length of a three-phase line with three conductors, each with radius r. The Conductors are asymmetrically spaced with distances shown in Figure 4.8.

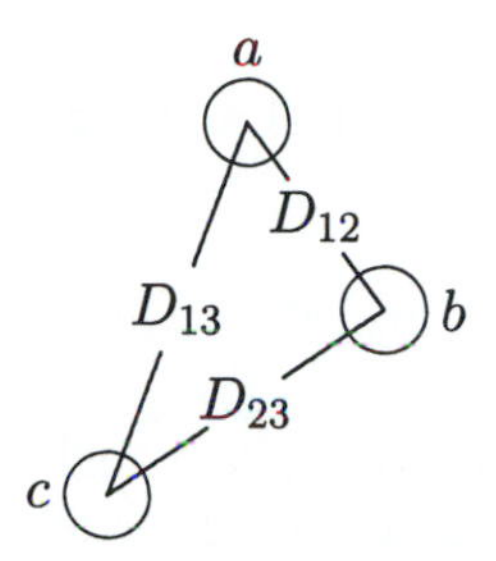

FIGURE 4.8
Three-phase line with asymmetrical spacing.

The application of (4.29) will result in the following flux linkages.

$$
\begin{aligned}
\lambda_a &= 2 \times 10^{-7} \left(I_a \ln \frac{1}{r'} + I_b \ln \frac{1}{D_{12}} + I_c \ln \frac{1}{D_{13}} \right) \\
\lambda_b &= 2 \times 10^{-7} \left(I_a \ln \frac{1}{D_{12}} + I_b \ln \frac{1}{r'} + I_c \ln \frac{1}{D_{23}} \right) \\
\lambda_c &= 2 \times 10^{-7} \left(I_a \ln \frac{1}{D_{13}} + I_b \ln \frac{1}{D_{23}} + I_c \ln \frac{1}{r'} \right)
\end{aligned} \tag{4.34}
$$

or in matrix form

$$\boldsymbol{\lambda} = \boldsymbol{LI} \tag{4.35}$$

where the symmetrical inductance matrix L is given by

$$L = 2 \times 10^{-7} \begin{bmatrix} \ln \frac{1}{r'} & \ln \frac{1}{D_{12}} & \ln \frac{1}{D_{13}} \\ \ln \frac{1}{D_{12}} & \ln \frac{1}{r'} & \ln \frac{1}{D_{23}} \\ \ln \frac{1}{D_{13}} & \ln \frac{1}{D_{23}} & \ln \frac{1}{r'} \end{bmatrix} \tag{4.36}$$

For balanced three-phase currents with I_a as reference, we have

$$
\begin{aligned}
I_b &= I_a \angle 240^\circ = a^2 I_a \\
I_c &= I_a \angle 120^\circ = a I_a
\end{aligned} \tag{4.37}
$$

where the operator $a = 1\angle 120°$ and $a^2 = 1\angle 240°$. Substituting in (4.34) results in

$$
\begin{aligned}
L_a &= \frac{\lambda_a}{I_a} = 2 \times 10^{-7}\left(\ln\frac{1}{r'} + a^2 \ln\frac{1}{D_{12}} + a\ln\frac{1}{D_{13}}\right)\\
L_b &= \frac{\lambda_b}{I_b} = 2 \times 10^{-7}\left(a\ln\frac{1}{D_{12}} + \ln\frac{1}{r'} + a^2 \ln\frac{1}{D_{23}}\right)\\
L_c &= \frac{\lambda_c}{I_c} = 2 \times 10^{-7}\left(a^2\ln\frac{1}{D_{13}} + a\ln\frac{1}{D_{23}} + \ln\frac{1}{r'}\right)
\end{aligned} \tag{4.38}
$$

Examination of (4.38) shows that the phase inductances are not equal and they contain an imaginary term due to the mutual inductance.

4.7.3 TRANSPOSE LINE

A per-phase model of the transmission line is required in most power system analysis. One way to regain symmetry in good measure and obtain a per-phase model is to consider transposition. This consists of interchanging the phase configuration every one-third the length so that each conductor is moved to occupy the next physical position in a regular sequence. Such a transposition arrangement is shown in Figure 4.9.

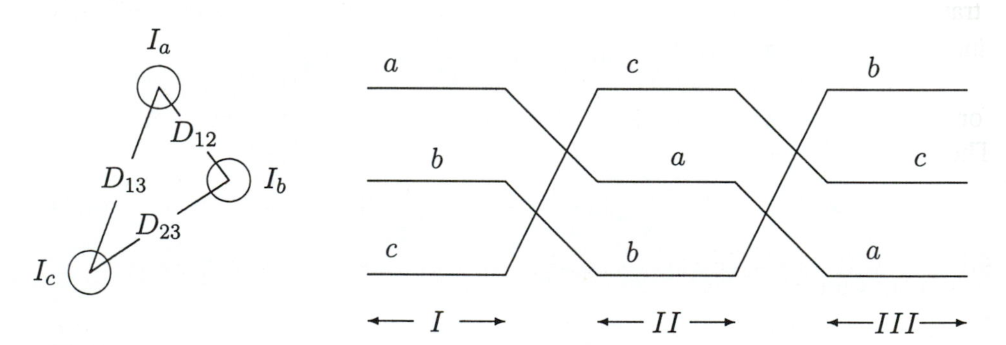

FIGURE 4.9
A transposed three-phase line.

Since in a transposed line each phase takes all three positions, the inductance per phase can be obtained by finding the average value of (4.38).

$$
L = \frac{L_a + L_b + L_c}{3} \tag{4.39}
$$

Noting $a + a^2 = 1\angle 120° + 1\angle 240° = -1$, the average of (4.38) becomes

$$
L = \frac{2 \times 10^{-7}}{3}\left(3\ln\frac{1}{r'} - \ln\frac{1}{D_{12}} - \ln\frac{1}{D_{23}} - \ln\frac{1}{D_{13}}\right)
$$

or

$$L = 2 \times 10^{-7} \left(\ln \frac{1}{r'} - \ln \frac{1}{(D_{12}D_{23}D_{13})^{\frac{1}{3}}} \right)$$

$$= 2 \times 10^{-7} \ln \frac{(D_{12}D_{23}D_{13})^{\frac{1}{3}}}{r'} \tag{4.40}$$

or the inductance per phase per kilometer length is

$$L = 0.2 \ln \frac{GMD}{D_S} \quad \text{mH/km} \tag{4.41}$$

where

$$GMD = \sqrt[3]{D_{12}D_{23}D_{13}} \tag{4.42}$$

This again is of the same form as the expression for the inductance of one phase of a single-phase line. GMD (geometric mean distance) is the equivalent conductor spacing. For the above three-phase line this is the cube root of the product of the three-phase spacings. D_s is the geometric mean radius, GMR. For stranded conductor D_s is obtained from the manufacturer's data. For solid conductor, $D_s = r' = re^{-\frac{1}{4}}$.

In modern transmission lines, transposition is not generally used. However, for the purpose of modeling, it is most practical to treat the circuit as transposed. The error introduced as a result of this assumption is very small.

4.8 INDUCTANCE OF COMPOSITE CONDUCTORS

In the evaluation of inductance, solid round conductors were considered. However, in practical transmission lines, stranded conductors are used. Also, for reasons of economy, most EHV lines are constructed with bundled conductors. In this section an expression is found for the inductance of composite conductors. The result can be used for evaluating the GMR of stranded or bundled conductors. It is also useful in finding the equivalent GMR and GMD of parallel circuits. Consider a single-phase line consisting of two composite conductors x and y as shown in Figure 4.10. The current in x is I referenced into the page, and the return current in y is $-I$. Conductor x consists of n identical strands or subconductors, each with radius r_x. Conductor y consists of m identical strands or subconductors, each with radius r_y. The current is assumed to be equally divided among the subconductors. The current per strand is I/n in x and I/m in y. The application of (4.29) will result in the following expression for the total flux linkage of conductor a

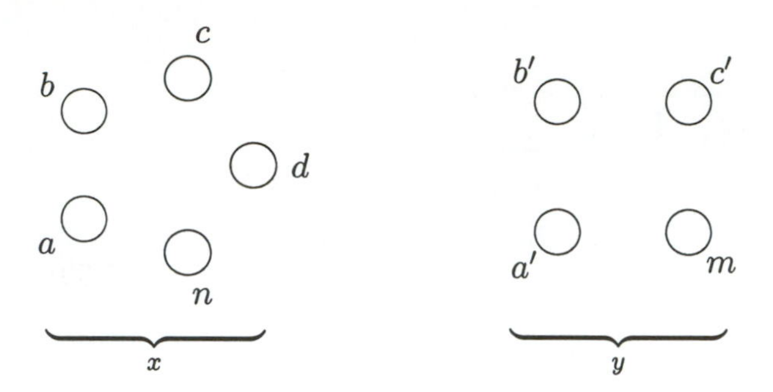

FIGURE 4.10
Single-phase line with two composite conductors.

$$\lambda_a = 2 \times 10^{-7}\frac{I}{n}\left(\ln\frac{1}{r'_x} + \ln\frac{1}{D_{ab}} + \ln\frac{1}{D_{ac}} + \cdots + \ln\frac{1}{D_{an}}\right)$$
$$-2 \times 10^{-7}\frac{I}{m}\left(\ln\frac{1}{D_{aa'}} + \ln\frac{1}{D_{ab'}} + \ln\frac{1}{D_{ac'}} + \cdots + \ln\frac{1}{D_{am}}\right)$$

or

$$\lambda_a = 2 \times 10^{-7} I \ln\frac{\sqrt[m]{D_{aa'}D_{ab'}D_{ac'}\cdots D_{am}}}{\sqrt[n]{r'_x D_{ab}D_{ac}\cdots D_{an}}} \tag{4.43}$$

The inductance of subconductor a is

$$L_a = \frac{\lambda_a}{I/n} = 2n \times 10^{-7}\ln\frac{\sqrt[m]{D_{aa'}D_{ab'}D_{ac'}\cdots D_{am}}}{\sqrt[n]{r'_x D_{ab}D_{ac}\cdots D_{an}}} \tag{4.44}$$

Using (4.29), the inductance of other subconductors in x are similarly obtained. For example, the inductance of the subconductor n is

$$L_n = \frac{\lambda_n}{I/n} = 2n \times 10^{-7}\ln\frac{\sqrt[m]{D_{na'}D_{nb'}D_{nc'}\cdots D_{nm}}}{\sqrt[n]{r'_x D_{na}D_{nb}\cdots D_{nc}}} \tag{4.45}$$

The average inductance of any one subconductor in group x is

$$L_{av} = \frac{L_a + L_b + L_c + \cdots + L_n}{n} \tag{4.46}$$

Since all the subconductors of conductor x are electrically parallel, the inductance of x will be

$$L_x = \frac{L_{av}}{n} = \frac{L_a + L_b + L_c + \cdots + L_n}{n^2} \tag{4.47}$$

substituting the values of $L_a,\ L_b,\ L_c, \cdots, L_n$ in (4.47) results in

$$L_x = 2 \times 10^{-7}\ln\frac{GMD}{GMR_x}\ \text{H/meter} \tag{4.48}$$

where

$$GMD = \sqrt[mn]{(D_{aa'}D_{ab'}\cdots D_{am})\cdots(D_{na'}D_{nb'}\cdots D_{nm})} \tag{4.49}$$

and

$$GMR_x = \sqrt[n^2]{(D_{aa}D_{ab}\cdots D_{an})\cdots(D_{na}D_{nb}\cdots D_{nn})} \tag{4.50}$$

where $D_{aa} = D_{bb}\cdots = D_{nn} = r'_x$

GMD is the mnth root of the product of the mnth distances between n strands of conductor x and m strands of conductor y. GMR_x is the n^2 root of the product of n^2 terms consisting of r' of every strand times the distance from each strand to all other strands within group x.

The inductance of conductor y can also be similarly obtained. The geometric mean radius GMR_y will be different. The geometric mean distance GMD, however, is the same.

Example 4.1

A stranded conductor consists of seven identical strands each having a radius r as shown in Figure 4.11. Determine the GMR of the conductor in terms of r.

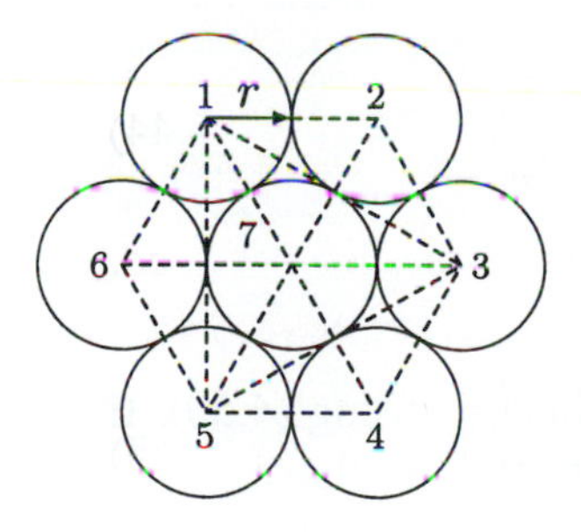

FIGURE 4.11
Cross section of a stranded conductor.

From Figure 4.11, the distance from strand 1 to all other strands is:

$$D_{12} = D_{16} = D_{17} = 2r$$
$$D_{14} = 4r$$
$$D_{13} = D_{15} = \sqrt{D_{14}^2 - D_{45}^2} = 2\sqrt{3}\,r$$

From (4.50) the GMR of the above conductor is

$$GMR = \sqrt[49]{(r' \cdot 2r \cdot 2\sqrt{3}\,r \cdot 4r \cdot 2\sqrt{3}\,r \cdot 2r \cdot 2r)^6 \cdot r'(2r)^6}$$

$$= r\sqrt[7]{(e)^{-\frac{1}{4}}\,(2)^6\,(3)^{\frac{6}{7}}\,(2)^{\frac{6}{7}}}$$
$$= 2.1767r$$

With a large number of strands the calculation of GMR can become very tedious. Usually these are available in the manufacturer's data.

4.8.1 *GMR* OF BUNDLED CONDUCTORS

Extra-high voltage transmission lines are usually constructed with bundled conductors. Bundling reduces the line reactance, which improves the line performance and increases the power capability of the line. Bundling also reduces the voltage surface gradient, which in turn reduces corona loss, radio interference, and surge impedance. Typically, bundled conductors consist of two, three, or four subconductors symmetrically arranged in configuration as shown in Figure 4.12. The subconductors within a bundle are separated at frequent intervals by spacer-dampers. Spacer-dampers prevent clashing, provide damping, and connect the subconductors in parallel.

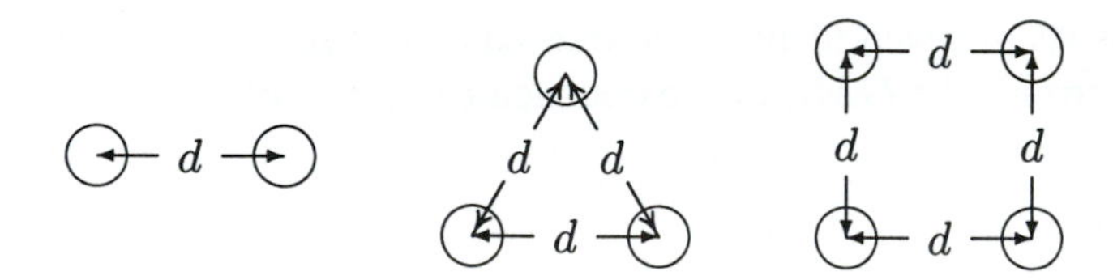

FIGURE 4.12
Examples of bundled arrangements.

The GMR of the equivalent single conductor is obtained by using (4.50). If D_s is the GMR of each subconductor and d is the bundle spacing, we have

for the two-subconductor bundle

$$D_s^b = \sqrt[4]{(D_s \times d)^2} = \sqrt{D_s \times d} \tag{4.51}$$

for the three-subconductor bundle

$$D_s^b = \sqrt[9]{(D_s \times d \times d)^3} = \sqrt[3]{D_s \times d^2} \tag{4.52}$$

for the four-subconductor bundle

$$D_s^b = \sqrt[16]{(D_s \times d \times d \times d \times 2^{\frac{1}{2}})^4} = 1.09\sqrt[4]{D_s \times d^3} \tag{4.53}$$

4.9 INDUCTANCE OF THREE-PHASE DOUBLE-CIRCUIT LINES

A three-phase double-circuit line consists of two identical three-phase circuits. The circuits are operated with a_1–a_2, b_1–b_2, and c_1–c_2 in parallel. Because of geometrical differences between conductors, voltage drop due to line inductance will be unbalanced. To achieve balance, each phase conductor must be transposed within its group and with respect to the parallel three-phase line. Consider a three-phase double-circuit line with relative phase positions $a_1b_1c_1$–$c_2b_2a_2$, as shown in Figure 4.13.

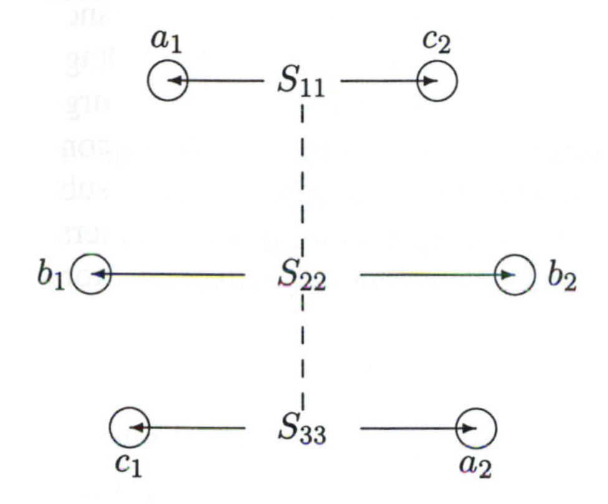

FIGURE 4.13
Transposed double-circuit line.

The method of *GMD* can be used to find the inductance per phase. To do this, we group identical phases together and use (4.49) to find the *GMD* between each phase group

$$D_{AB} = \sqrt[4]{D_{a_1b_1}D_{a_1b_2}D_{a_2b_1}D_{a_2b_2}}$$
$$D_{BC} = \sqrt[4]{D_{b_1c_1}D_{b_1c_2}D_{b_2c_1}D_{b_2c_2}}$$
$$D_{AC} = \sqrt[4]{D_{a_1c_1}D_{a_1c_2}D_{a_2c_1}D_{a_2c_2}} \tag{4.54}$$

The equivalent *GMD* per phase is then

$$GMD = \sqrt[3]{D_{AB}D_{BC}D_{AC}} \tag{4.55}$$

Similarly, from (4.50), the *GMR* of each phase group is

$$D_{SA} = \sqrt[4]{(D_s^b D_{a_1a_2})^2} = \sqrt{D_s^b D_{a_1a_2}}$$
$$D_{SB} = \sqrt[4]{(D_s^b D_{b_1b_2})^2} = \sqrt{D_s^b D_{b_1b_2}}$$
$$D_{SC} = \sqrt[4]{(D_s^b D_{c_1c_2})^2} = \sqrt{D_s^b D_{c_1c_2}} \tag{4.56}$$

where D_s^b is the geometric mean radius of the bundled conductors given by (4.51)–(4.53). The equivalent geometric mean radius for calculating the per-phase inductance to neutral is

$$GMR_L = \sqrt[3]{D_{SA} D_{SB} D_{SC}} \tag{4.57}$$

The inductance per phase in millihenries per kilometer is

$$L = 0.2 \ln \frac{GMD}{GMR_L} \quad \text{mH/km} \tag{4.58}$$

4.10 LINE CAPACITANCE

Transmission line conductors exhibit capacitance with respect to each other due to the potential difference between them. The amount of capacitance between conductors is a function of conductor size, spacing, and height above ground. By definition, the capacitance C is the ratio of charge q to the voltage V, given by

$$C = \frac{q}{V} \tag{4.59}$$

Consider a long round conductor with radius r, carrying a charge of q coulombs per meter length as shown in Figure 4.14.

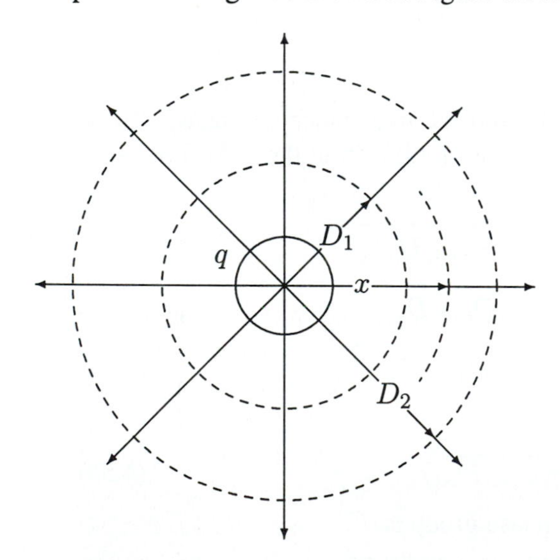

FIGURE 4.14
Electric field around a long round conductor.

The charge on the conductor gives rise to an electric field with radial flux lines. The total electric flux is numerically equal to the value of charge on the conductor. The intensity of the field at any point is defined as the force per unit charge and is termed *electric field intensity* designated as E. Concentric cylinders surrounding the conductor are equipotential surfaces and have the same electric flux density. From Gauss's law, for one meter length of the conductor, the electric flux density at a cylinder of radius x is given by

$$D = \frac{q}{A} = \frac{q}{2\pi x(1)} \tag{4.60}$$

The electric field intensity E may be found from the relation

$$E = \frac{D}{\varepsilon_0} \tag{4.61}$$

where ε_0 is the permittivity of free space and is equal to 8.85×10^{-12} F/m. Substituting (4.60) in (4.61) results in

$$E = \frac{q}{2\pi\varepsilon_0 x} \tag{4.62}$$

The potential difference between cylinders from position D_1 to D_2 is defined as the work done in moving a unit charge of one coulomb from D_2 to D_1 through the electric field produced by the charge on the conductor. This is given by

$$V_{12} = \int_{D_1}^{D_2} E dx = \int_{D_1}^{D_2} \frac{q}{2\pi\varepsilon_0 x} dx = \frac{q}{2\pi\varepsilon_0} \ln \frac{D_2}{D_1} \tag{4.63}$$

The notation V_{12} implies the voltage drop from 1 relative to 2, that is, 1 is understood to be positive relative to 2. The charge q carries its own sign.

4.11 CAPACITANCE OF SINGLE-PHASE LINES

Consider one meter length of a single-phase line consisting of two long solid round conductors each having a radius r as shown in Figure 4.15. The two conductors are separated by a distance D. Conductor 1 carries a charge of q_1 coulombs/meter and conductor 2 carries a charge of q_2 coulombs/meter. The presence of the second conductor and ground disturbs the field of the first conductor. The distance of separation of the wires D is great with respect to r and the height of conductors is much larger compared with D. Therefore, the distortion effect is small and the charge is assumed to be uniformly distributed on the surface of the conductors.

Assuming conductor 1 alone to have a charge of q_1, the voltage between conductor 1 and 2 is

$$V_{12(q_1)} = \frac{q_1}{2\pi\varepsilon_0} \ln \frac{D}{r} \tag{4.64}$$

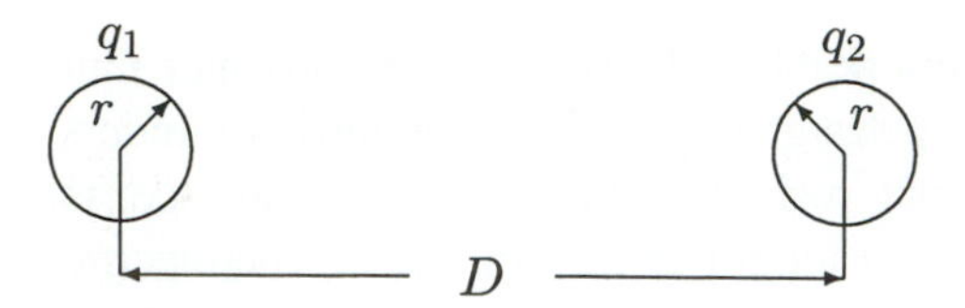

FIGURE 4.15
Single-phase two-wire line.

Now assuming only conductor 2, having a charge of q_2, the voltage between conductors 2 and 1 is

$$V_{21(q_2)} = \frac{q_2}{2\pi\varepsilon_0} \ln \frac{D}{r}$$

Since $V_{12(q_2)} = -V_{21(q_2)}$, we have

$$V_{12(q_2)} = \frac{q_2}{2\pi\varepsilon_0} \ln \frac{r}{D} \tag{4.65}$$

From the principle of superposition, the potential difference due to presence of both charges is

$$V_{12} = V_{12(q_1)} + V_{12(q_2)} = \frac{q_1}{2\pi\varepsilon_0} \ln \frac{D}{r} + \frac{q_2}{2\pi\varepsilon_0} \ln \frac{r}{D} \tag{4.66}$$

For a single-phase line $q_2 = -q_1 = -q$, and (4.66) reduces to

$$V_{12} = \frac{q}{\pi\varepsilon_0} \ln \frac{D}{r} \quad \text{F/m} \tag{4.67}$$

From (4.59), the capacitance between conductors is

$$C_{12} = \frac{\pi\varepsilon_0}{\ln \frac{D}{r}} \quad \text{F/m} \tag{4.68}$$

Equation (4.68) gives the line-to-line capacitance between the conductors. For the purpose of transmission line modeling, we find it convenient to define a capacitance C between each conductor and a neutral as illustrated in Figure 4.16. Since the

FIGURE 4.16
Illustration of capacitance to neutral.

voltage to neutral is half of V_{12}, the capacitance to neutral $C = 2C_{12}$, or

$$C = \frac{2\pi\varepsilon_0}{\ln \frac{D}{r}} \quad \text{F/m} \tag{4.69}$$

Recalling $\varepsilon_0 = 8.85 \times 10^{-12}$ F/m and converting to μF per kilometer, we have

$$C = \frac{0.0556}{\ln \frac{D}{r}} \quad \mu\text{F/km} \tag{4.70}$$

The capacitance per phase contains terms analogous to those derived for inductance per phase. However, unlike inductance where the conductor geometric mean radius (GMR) is used, in capacitance formula the actual conductor radius r is used.

4.12 POTENTIAL DIFFERENCE IN A MULTICONDUCTOR CONFIGURATION

Consider n parallel long conductors with charges $q_1, q_2, \ldots, q_n$ coulombs/meter as shown in Figure 4.17.

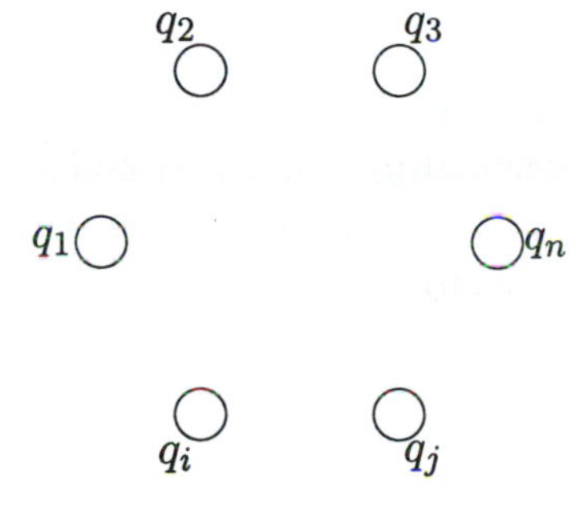

FIGURE 4.17
Multiconductor configuration.

Assume that the distortion effect is negligible and the charge is uniformly distributed around the conductor, with the following constraint

$$q_1 + q_2 + \cdots + q_n = 0 \tag{4.71}$$

Using superposition and (4.63), potential difference between conductors i and j due to the presence of all charges is

$$V_{ij} = \frac{1}{2\pi\varepsilon_0} \sum_{k=1}^{n} q_k \ln \frac{D_{kj}}{D_{ki}} \tag{4.72}$$

When $k = i$, D_{ii} is the distance between the surface of the conductor and its center, namely its radius r.

4.13 CAPACITANCE OF THREE-PHASE LINES

Consider one meter length of a three-phase line with three long conductors, each with radius r, with conductor spacing as shown Figure 4.18.

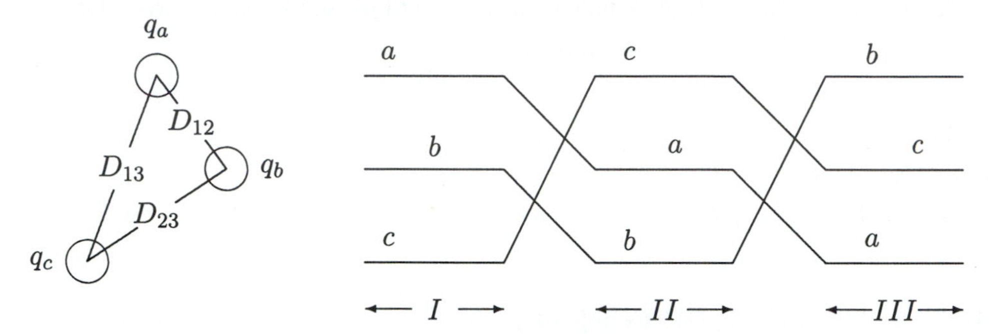

FIGURE 4.18
Three-phase transmission line.

Since we have a balanced three-phase system

$$q_a + q_b + q_c = 0 \tag{4.73}$$

We shall neglect the effect of ground and the shield wires. Assume that the line is transposed. We proceed with the calculation of the potential difference between a and b for each section of transposition. Applying (4.72) to the first section of the transposition, V_{ab} is

$$V_{ab(I)} = \frac{1}{2\pi\varepsilon_0}\left(q_a \ln\frac{D_{12}}{r} + q_b \ln\frac{r}{D_{12}} + q_c \ln\frac{D_{23}}{D_{13}}\right) \tag{4.74}$$

Similarly, for the second section of the transposition, we have

$$V_{ab(II)} = \frac{1}{2\pi\varepsilon_0}\left(q_a \ln\frac{D_{23}}{r} + q_b \ln\frac{r}{D_{23}} + q_c \ln\frac{D_{13}}{D_{12}}\right) \tag{4.75}$$

and for the last section

$$V_{ab(III)} = \frac{1}{2\pi\varepsilon_0}\left(q_a \ln\frac{D_{13}}{r} + q_b \ln\frac{r}{D_{13}} + q_c \ln\frac{D_{12}}{D_{23}}\right) \tag{4.76}$$

The average value of V_{ab} is

$$V_{ab} = \frac{1}{(3)2\pi\varepsilon_0}\left(q_a \ln\frac{D_{12}D_{23}D_{13}}{r^3} + q_b \ln\frac{r^3}{D_{12}D_{23}D_{13}} + q_c \ln\frac{D_{12}D_{23}D_{13}}{D_{12}D_{23}D_{13}}\right) \tag{4.77}$$

or

$$V_{ab} = \frac{1}{2\pi\varepsilon_0}\left(q_a \ln \frac{(D_{12}D_{23}D_{13})^{\frac{1}{3}}}{r} + q_b \ln \frac{r}{(D_{12}D_{23}D_{13})^{\frac{1}{3}}}\right) \tag{4.78}$$

Note that the GMD of the conductor appears in the logarithm arguments and is given by

$$GMD = \sqrt[3]{D_{12}D_{23}D_{13}} \tag{4.79}$$

Therefore, V_{ab} is

$$V_{ab} = \frac{1}{2\pi\varepsilon_0}\left(q_a \ln \frac{GMD}{r} + q_b \ln \frac{r}{GMD}\right) \tag{4.80}$$

Similarly, we find the average voltage V_{ac} as

$$V_{ac} = \frac{1}{2\pi\varepsilon_0}\left(q_a \ln \frac{GMD}{r} + q_c \ln \frac{r}{GMD}\right) \tag{4.81}$$

Adding (4.80) and (4.81) and substituting for $q_b + q_c = -q_a$, we have

$$V_{ab} + V_{ac} = \frac{1}{2\pi\varepsilon}\left(2q_a \ln \frac{GMD}{r} - q_a \ln \frac{r}{GMD}\right) = \frac{3q_a}{2\pi\varepsilon_0} \ln \frac{GMD}{r} \tag{4.82}$$

For balanced three-phase voltages,

$$\begin{aligned} V_{ab} &= V_{an}\angle 0^\circ - V_{an}\angle -120^\circ \\ V_{ac} &= V_{an}\angle 0^\circ - V_{an}\angle -240^\circ \end{aligned} \tag{4.83}$$

Therefore,

$$V_{ab} + V_{ac} = 3V_{an} \tag{4.84}$$

Substituting in (4.82) the capacitance per phase to neutral is

$$C = \frac{q_a}{V_{an}} = \frac{2\pi\varepsilon_0}{\ln \frac{GMD}{r}} \text{ F/m} \tag{4.85}$$

or capacitance to neutral in μF per kilometer is

$$C = \frac{0.0556}{\ln \frac{GMD}{r}} \ \mu\text{F/km} \tag{4.86}$$

This is of the same form as the expression for the capacitance of one phase of a single-phase line. GMD (geometric mean distance) is the equivalent conductor spacing. For the above three-phase line this is the cube root of the product of the three-phase spacings.

4.14 EFFECT OF BUNDLING

The procedure for finding the capacitance per phase for a three-phase transposed line with bundle conductors follows the same steps as the procedure in Section 3.13. The capacitance per phase is found to be

$$C = \frac{2\pi\varepsilon_0}{\ln\frac{GMD}{r^b}} \quad \text{F/m} \tag{4.87}$$

The effect of bundling is to introduce an equivalent radius r^b. The equivalent radius r^b is similar to the GMR (geometric mean radius) calculated earlier for the inductance with the exception that radius r of each subconductor is used instead of D_s. If d is the bundle spacing, we obtain for the two-subconductor bundle

$$r^b = \sqrt{r \times d} \tag{4.88}$$

for the three-subconductor bundle

$$r^b = \sqrt[3]{r \times d^2} \tag{4.89}$$

for the four-subconductor bundle

$$r^b = 1.09\sqrt[4]{r \times d^3} \tag{4.90}$$

4.15 CAPACITANCE OF THREE-PHASE DOUBLE-CIRCUIT LINES

Consider a three-phase double-circuit line with relative phase positions $a_1b_1c_1$ - $c_2b_2a_2$, as shown in Figure 4.13. Each phase conductor is transposed within its group and with respect to the parallel three-phase line. The effect of shield wires and the ground are considered to be negligible for this balanced condition. Following the procedure of section 4.13, the average voltages V_{ab}, V_{ac} and V_{an} are calculated and the per-phase equivalent capacitance to neutral is obtained to be

$$C = \frac{2\pi\varepsilon_0}{\ln\frac{GMD}{GMR_c}} \quad \text{F/m} \tag{4.91}$$

or capacitance to neutral in μF per kilometer is

$$C = \frac{0.0556}{\ln\frac{GMD}{GMR_c}} \quad \mu\text{F/km} \tag{4.92}$$

The expression for GMD is the same as was found for inductance calculation and is given by (4.55). The GMR_c of each phase group is similar to the GMR_L, with the exception that in (4.56) r^b is used instead of D_s^b. This will result in the following equations

$$r_A = \sqrt{r^b \, D_{a_1 a_2}}$$
$$r_B = \sqrt{r^b \, D_{b_1 b_2}}$$
$$r_C = \sqrt{r^b \, D_{c_1 c_2}} \tag{4.93}$$

where r^b is the geometric mean radius of the bundled conductors given by (4.88) – (4.90). The equivalent geometric mean radius for calculating the per-phase capacitance to neutral is

$$GMR_C = \sqrt[3]{r_A \, r_B \, r_C} \tag{4.94}$$

4.16 EFFECT OF EARTH ON THE CAPACITANCE

For an isolated charged conductor the electric flux lines are radial and are orthogonal to the cylindrical equipotential surfaces. The presence of earth will alter the distribution of electric flux lines and equipotential surfaces, which will change the effective capacitance of the line.

The earth level is an equipotential surface, therefore the flux lines are forced to cut the surface of the earth orthogonally. The effect of the presence of earth can be accounted for by the method of *image charges* introduced by Kelvin. To illustrate this method, consider a conductor with a charge q coulombs/meter at a height H above ground. Also, imagine a charge $-q$ placed at a depth H below the surface of earth. This configuration without the presence of the earth surface will produce the same field distribution as a single charge and the earth surface. Thus, the earth can be replaced for the calculation of electric field potential by a fictitious charged conductor with charge equal and opposite to the charge on the actual conductor and at a depth below the surface of the earth the same as the height of the actual conductor above earth. This imaginary conductor is called the image of the actual conductor. The procedure of Section 4.13 can now be used for the computation of the capacitance.

The effect of the earth is to increase the capacitance. But normally the height of the conductor is large as compared to the distance between the conductors, and the earth effect is negligible. Therefore, for all line models used for balanced steady-state analysis, the effect of earth on the capacitance can be neglected. However, for unbalanced analysis such as unbalanced faults, the earth's effect as well as the shield wires should be considered.

Example 4.2

A 500-kV three-phase transposed line is composed of one $ACSR$ $1,272,000$–cmil, $45/7$ Bittern conductor per phase with horizontal conductor configuration as shown in Figure 4.19. The conductors have a diameter of 1.345 in and a GMR of 0.5328 in. Find the inductance and capacitance per phase per kilometer of the line.

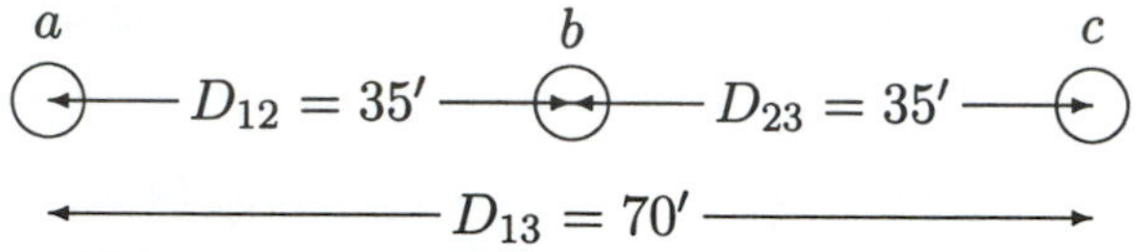

FIGURE 4.19
Conductor layout for Example 4.2.

Conductor radius is $r = \frac{1.345}{2\times 12} = 0.056$ ft, and $GMR_L = 0.5328/12 = 0.0444$ ft. GMD is obtained using (4.42)

$$GMD = \sqrt[3]{35 \times 35 \times 70} = 44.097 \quad \text{ft}$$

From (4.58) the inductance per phase is

$$L = 0.2 \ln \frac{44.097}{0.0444} = 1.38 \quad \text{mH/km}$$

and from (4.92) the capacitance per phase is

$$C = \frac{0.0556}{\ln \frac{44.097}{0.056}} = 0.0083 \quad \mu\text{F/km}$$

Example 4.3

The line in Example 4.2 is replaced by two $ACSR$ $636,000$-cmil, $24/7$ Rook conductors which have the same total cross-sectional area of aluminum as one Bittern conductor. The line spacing as measured from the center of the bundle is the same as before and is shown in Figure 4.20.

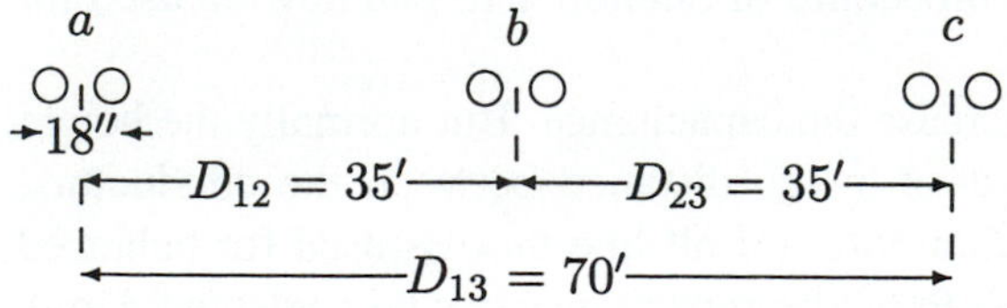

FIGURE 4.20
Conductor layout for Example 4.3.

The conductors have a diameter of 0.977 in and a GMR of 0.3924 in. Bundle spacing is 18 in. Find the inductance and capacitance per phase per kilometer of the line and compare it with that of Example 4.2.

Conductor radius is $r = \frac{0.977}{2} = 0.4885$ in, and from Example 4.2 $GMD =$ 44.097 ft. The equivalent geometric mean radius with two conductors per bundle, for calculating inductance and capacitance, are given by (4.51) and (4.88)

$$GMR_L = \frac{\sqrt{d \times D_s}}{12} = \frac{\sqrt{18 \times 0.3924}}{12} = 0.22147 \quad \text{ft}$$

and

$$GMR_c = \frac{\sqrt{d \times r}}{12} = \frac{\sqrt{18 \times 0.4885}}{12} = 0.2471 \quad \text{ft}$$

From (4.58) the inductance per phase is

$$L = 0.2 \ln \frac{44.097}{0.22147} = 1.0588 \quad \text{mH/km}$$

and from (4.92) the capacitance per phase is

$$C = \frac{0.0556}{\ln \frac{44.097}{0.2471}} = 0.0107 \quad \mu\text{F/km}$$

Comparing with the results of Example 4.2, there is a 23.3 percent reduction in the inductance and a 28.9 percent increase in the capacitance.

The function **[GMD, GMRL, GMRC] = gmd** is developed for the computation of GMD, GMR_L, and GMR_C for single-circuit, double-circuit vertical, and horizontal transposed lines with up to four bundled conductors. A menu is displayed for the selection of any of the above three circuits. The user is prompted to input the phase spacing, number of bundled conductors and their spacing, conductor diameter, and the GMR of the individual conductor. The specifications for some common $ACSR$ conductors are contained in a file named **acsr.m**. The command **acsr** will display the characteristics of $ACSR$ conductors. Also, the function **[L, C] = gmd2lc** in addition to the geometric mean values returns the inductance in mH per km and the capacitance in μF per km.

Example 4.4

A 735-kV three-phase transposed line is composed of four $ACSR$, 954,000-cmil, 45/7 Rail conductors per phase with horizontal conductor configuration as shown in Figure 4.21. Bundle spacing is 46 cm. Use **acsr** in *MATLAB* to obtain the conductor size and the electrical characteristics for the Rail conductor. Find the inductance and capacitance per phase per kilometer of the line.

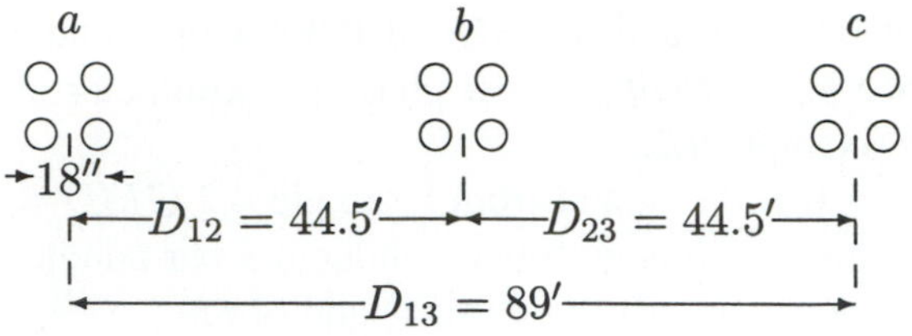

FIGURE 4.21
Conductor layout for Example 4.4.

The command **acsr** displays the conductor code name and the area in cmils for the ACSR conductors. The user is then prompted to enter the conductor code name within single quotes.

```
Enter ACSR code name within single quotes -> 'rail'
Al Area Strand  Diameter GMR  Resistance Ohm/km  Ampacity
 cmil   Al/St     cm      cm   60Hz 25C  60Hz 50C  Ampere
954000   45/7    2.959   1.173  0.0624   0.0683     1000
```

The following commands

```
[GMD, GMRL, GMRC] = gmd;
L=0.2*log(GMD/GMRL)           % mH/km        Eq. (4.58)
C = 0.0556/log(GMD/GMRC)      % micro F/km   Eq. (4.92)
```

result in

Number of three-phase circuits	Enter
Single-circuit	1
Double-circuit vertical configuration	2
Double-circuit horizontal configuration	3
To quit	0

```
Select number of menu → 1
Enter spacing unit within quotes 'm' or 'ft' → 'ft'
Enter row vector [D12, D23, D13] = [44.5  44.5  89]
Cond. size, bundle spacing unit: 'cm' or 'in' → 'cm'
Conductor diameter in cm = 2.959
Geometric Mean Radius in cm = 1.173
No. of bundled cond. (enter 1 for single cond.) = 4
Bundle spacing in cm = 46
GMD = 56.06649 ft
GMRL = 0.65767 ft     GMRC = 0.69696 ft
L = 0.8891
C = 0.0127
```

Example 4.5

A 345-kV double-circuit three-phase transposed line is composed of two $ACSR$, 1, 431, 000-cmil, 45/7 Bobolink conductors per phase with vertical conductor configuration as shown in Figure 4.22. The conductors have a diameter of 1.427 in and a GMR of 0.564 in. The bundle spacing in 18 in. Find the inductance and capacitance per phase per kilometer of the line. The following commands

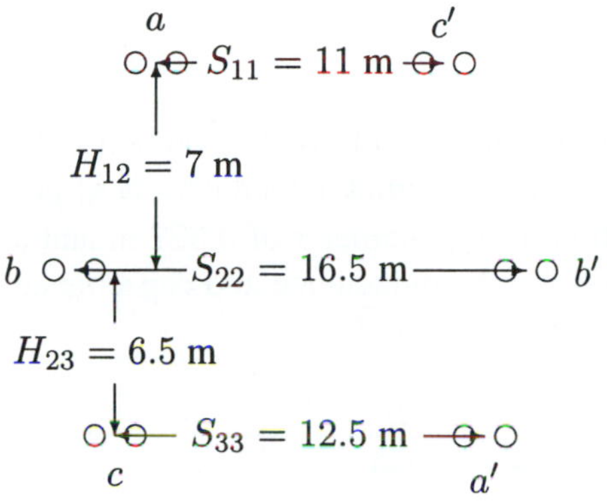

FIGURE 4.22
Conductor layout for Example 4.5.

```
[GMD, GMRL, GMRC] = gmd;
L=0.2*log(GMD/GMRL)          % mH/km        Eq. (4.58)
C = 0.0556/log(GMD/GMRC)     % micro F/km   Eq. (4.92)
```

result in

```
  Number of three-phase circuits              Enter
  Single-circuit                                1
  Double-circuit vertical configuration         2
  Double-circuit horizontal configuration       3
  To quit                                       0
Select number of menu → 2

Circuit Arrangements
(1) abc-c'b'a'
(2) abc-a'b'c'

Enter (1 or 2) → 1
Enter spacing unit within quotes 'm' or 'ft' → 'm'
Enter row vector [S11, S22, S33] = [11  16.5  12.5]
Enter row vector [H12, H23] = [7  6.5]
Cond. size, bundle spacing unit: 'cm' or 'in' → 'in'
```

```
Conductor diameter in inch = 1.427
Geometric Mean Radius in inch = 0.564
No. of bundled cond. (enter 1 for single cond.) = 2
Bundle spacing in inch = 18
GMD = 11.21352 m
GMRL = 1.18731 m      GMRC = 1.25920 m
L = 0.4491
C = 0.0254
```

Example 4.6

A 345-kV double-circuit three-phase transposed line is composed of one $ACSR$, 556, 500-cmil, 26/7 Dove conductor per phase with horizontal conductor configuration as shown in Figure 4.23. The conductors have a diameter of 0.927 in and a GMR of 0.3768 in. Bundle spacing is 18 in. Find the inductance and capacitance per phase per kilometer of the line. The following commands

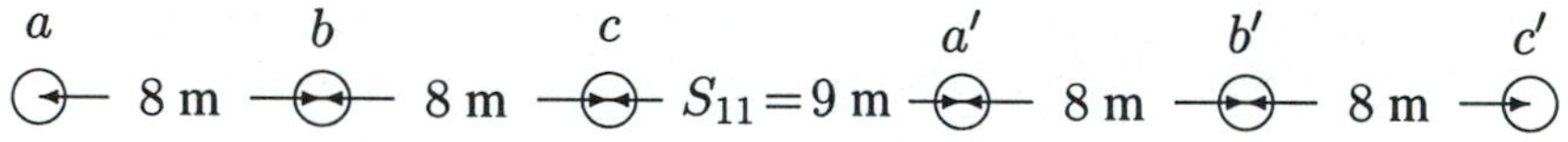

a — 8 m — b — 8 m — c — $S_{11}=9$ m — a' — 8 m — b' — 8 m — c'

FIGURE 4.23
Conductor layout for Example 4.6.

```
[GMD, GMRL, GMRC] = gmd;
L=0.2*log(GMD/GMRL)          % mH/km        Eq. (4.58)
C = 0.0556/log(GMD/GMRC)     % micro F/km   Eq. (4.92)
```

result in

```
Number of three-phase circuits                 Enter
Single-circuit                                   1
Double-circuit vertical configuration            2
Double-circuit horizontal configuration          3
To quit                                          0

Select number of menu → 3

Circuit Arrangements
(1) abc-a'b'c'
(2) abc-c'b'a'

Enter (1 or 2) → 1
Enter spacing unit within quotes 'm' or 'ft' → 'm'
Enter row vector [D12, D23, S13] = [8  8  16]
Enter distance between two circuits, S11 = 9
```

```
Cond. size, bundle spacing unit: 'cm' or 'in' → 'in'
Conductor diameter in inch = 0.927
Geometric Mean Radius in inch = 0.3768
No. of bundled cond. (enter 1 for single cond.) = 1
GMD = 14.92093 m
GMRL = 0.48915 m     GMRC = 0.54251 m
L = 0.6836
C = 0.0168
```

4.17 MAGNETIC FIELD INDUCTION

Transmission line magnetic fields affect objects in the proximity of the line. The magnetic fields, related to the currents in the line, induces voltage in objects that have a considerable length parallel to the line, such as fences, pipelines, and telephone wires.

The magnetic field is affected by the presence of earth return currents. Carson [14] presents an equation for computation of mutual resistance and inductance which are functions of the earth's resistivity. For balanced three-phase systems the total earth return current is zero. Under normal operating conditions, the magnetic field in proximity to balanced three-phase lines may be calculated considering the currents in the conductors and neglecting earth currents.

Magnetic fields have been reported to affect blood composition, growth, behavior, immune systems, and neural functions. There are general concerns regarding the biological effects of electromagnetic and electrostatic fields on people. Long-term effects are the subject of several worldwide research efforts.

Example 4.7

A three-phase untransposed transmission line and a telephone line are supported on the same towers as shown in Figure 4.24. The power line carries a 60-Hz balanced current of 200 A per phase. The telephone line is located directly below phase b. Assuming balanced three-phase currents in the power line, find the voltage per kilometer induced in the telephone line.

From (4.15) the flux linkage between conductors 1 and 2 due to current I_a is

$$\lambda_{12(I_a)} = 0.2 I_a \ln \frac{D_{a2}}{D_{a1}} \quad \text{mWb/km}$$

Since $D_{b1} = D_{b2}$, λ_{12} due to I_b is zero. The flux linkage between conductors 1 and 2 due to current I_c is

$$\lambda_{12(I_c)} = 0.2 I_c \ln \frac{D_{c2}}{D_{c1}} \quad \text{mWb/km}$$

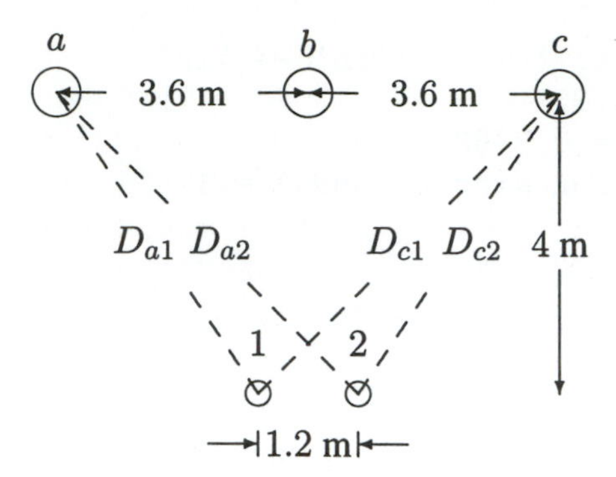

FIGURE 4.24
Conductor layout for Example 4.6.

Total flux linkage between conductors 1 and 2 due to all currents is

$$\lambda_{12} = 0.2 I_a \ln \frac{D_{a2}}{D_{a1}} + 0.2 I_c \ln \frac{D_{c2}}{D_{c1}} \quad \text{mWb/km}$$

For positive phase sequence, with I_a as reference, $I_c = I_a \angle -240^\circ$ and we have

$$\lambda_{12} = 0.2 I_a \left(\ln \frac{D_{a2}}{D_{a1}} + 1\angle -240^\circ \ln \frac{D_{c2}}{D_{c1}} \right) \quad \text{mH/km}$$

With I_a as reference, the instantaneous flux linkage is

$$\lambda_{12}(t) = \sqrt{2}\, |\lambda_{12}| \cos(\omega t + \alpha)$$

Thus, the induced voltage in the telephone line per kilometer length is

$$v = \frac{d\lambda_{12}(t)}{dt} = \sqrt{2}\, \omega |\lambda_{12}| \cos(\omega t + \alpha + 90^\circ)$$

The rms voltage induced in the telephone line per kilometer is

$$V = \omega |\lambda_{12}| \angle \alpha + 90^\circ = j\omega \lambda_{12}$$

From the circuits geometry

$$\begin{aligned} D_{a1} &= D_{c2} = (3^2 + 4^2)^{\frac{1}{2}} = 5 \ \text{m} \\ D_{a2} &= D_{c1} = (4.2^2 + 4^2)^{\frac{1}{2}} = 5.8 \ \text{m} \end{aligned}$$

The total flux linkage is

$$\begin{aligned} \lambda_{12} &= 0.2 \times 200\angle 0^\circ \ln \frac{5.8}{5} + 0.2 \times 200\angle -240^\circ \ln \frac{5}{5.8} \\ &= 10.283\angle -30^\circ \ \text{mWb/km} \end{aligned}$$

The voltage induced in the telephone line per kilometer is

$$V = j\omega \lambda_{12} = j2\pi 60 (10.283\angle -30^\circ)(10^{-3}) = 3.88\angle 60^\circ \quad \text{V/km}$$

4.18 ELECTROSTATIC INDUCTION

Transmission line electric fields affect objects in the proximity of the line. The electric field produced by high voltage lines induces current in objects which are in the area of the electric fields. The effects of electric fields becomes of increasing concern at higher voltages. Electric fields, related to the voltage of the line, are the primary cause of induction to vehicles, buildings, and objects of comparable size. The human body is affected with exposure to electric discharges from charged objects in the field of the line. These may be steady current or spark discharges. The current densities in humans induced by electric fields of transmission lines are known to be much higher than those induced by magnetic fields.

The resultant electric field in proximity to a transmission line can be obtained by representing the earth effect by image charges located below the conductors at a depth equal to the conductor height.

4.19 CORONA

When the surface potential gradient of a conductor exceeds the dielectric strength of the surrounding air, ionization occurs in the area close to the conductor surface. This partial ionization is known as *corona*. The dielectric strength of air during fair weather and at NTP (25°C and 76 cm of Hg) is about 30 kV/cm.

Corona produces power loss, audible hissing sound in the vicinity of the line, ozone and radio and television interference. The audible noise is an environmental concern and occurs in foul weather. Radio interference occurs in the AM band. Rain and snow may produce moderate TVI in a low signal area. Corona is a function of conductor diameter, line configuration, type of conductor, and condition of its surface. Atmospheric conditions such as air density, humidity, and wind influence the generation of corona. Corona losses in rain or snow are many times the losses during fair weather. On a conductor surface, an irregularity such as a contaminating particle causes a voltage gradient that may become the point source of a discharge. Also, insulators are contaminated by dust or chemical deposits which will lower the disruptive voltage and increase the corona loss. The insulators are cleaned periodically to reduce the extent of the problem. Corona can be reduced by increasing the conductor size and the use of conductor bundling.

The power loss associated with corona can be represented by shunt conductance. However, under normal operating conditions g, which represents the resistive leakage between a phase and ground, has negligible effect on performance and is customarily neglected. (i.e., $g = 0$).

PROBLEMS

4.1. A solid cylindrical aluminum conductor 25 km long has an area of 336,400 circular mils. Obtain the conductor resistance at (a) 20°C and (b) 50°C. The resistivity of aluminum at 20°C is 2.8×10^{-8} Ω-m.

4.2. A transmission-line cable consists of 12 identical strands of aluminum, each 3 mm in diameter. The resistivity of aluminum strand at 20°C is 2.8×10^{-8} Ω-m. Find the 50°C ac resistance per km of the cable. Assume a skin-effect correction factor of 1.02 at 60 Hz.

4.3. A three-phase transmission line is designed to deliver 190.5 MVA at 220 kV over a distance of 63 km. The total transmission line loss is not to exceed 2.5 percent of the rated line MVA. If the resistivity of the conductor material is 2.84×10^{-8} Ω-m, determine the required conductor diameter and the conductor size in circular mils.

4.4. A single-phase transmission line 35 km long consists of two solid round conductors, each having a diameter of 0.9 cm. The conductor spacing is 2.5 m. Calculate the equivalent diameter of a fictitious hollow, thin-walled conductor having the same equivalent inductance as the original line. What is the value of the inductance per conductor?

4.5. Find the geometric mean radius of a conductor in terms of the radius r of an individual strand for

(a) Three equal strands as shown in Figure 4.25(a)
(b) Four equal strands as shown in Figure 4.25(b)

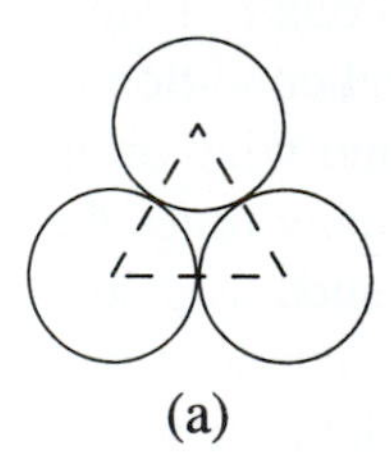

(a)

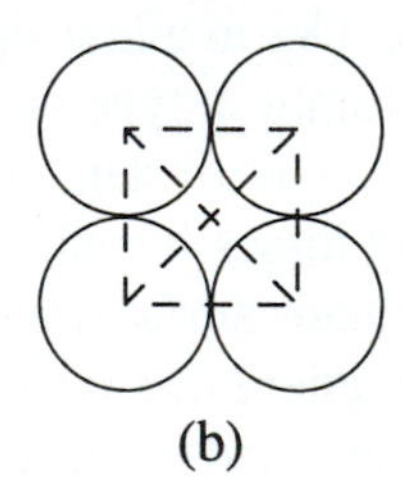

(b)

FIGURE 4.25
Cross section of the stranded conductor for Problem 4.5.

4.6. One circuit of a single-phase transmission line is composed of three solid 0.5-cm radius wires. The return circuit is composed of two solid 2.5-cm radius wires. The arrangement of conductors is as shown in Figure 4.26. Applying the concept of the GMD and GMR, find the inductance of the complete line in millihenry per kilometer.

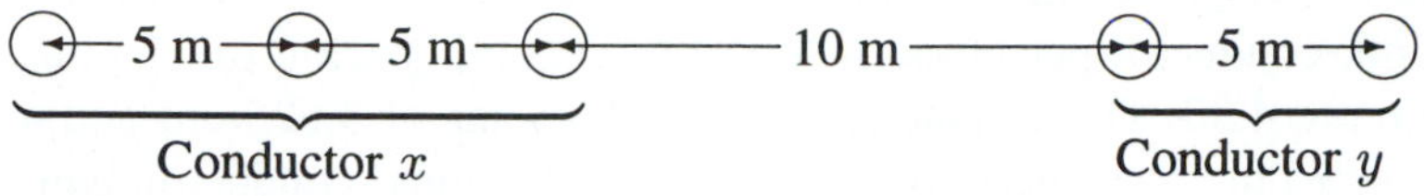

5 m — 5 m — 10 m — 5 m

Conductor x Conductor y

FIGURE 4.26
Conductor layout for Problem 4.6.

4.7. A three-phase, 60-Hz transposed transmission line has a flat horizontal configuration as shown in Figure 4.27. The line reactance is 0.486 Ω per kilometer. The conductor geometric mean radius is 2.0 cm. Determine the phase spacing D in meters.

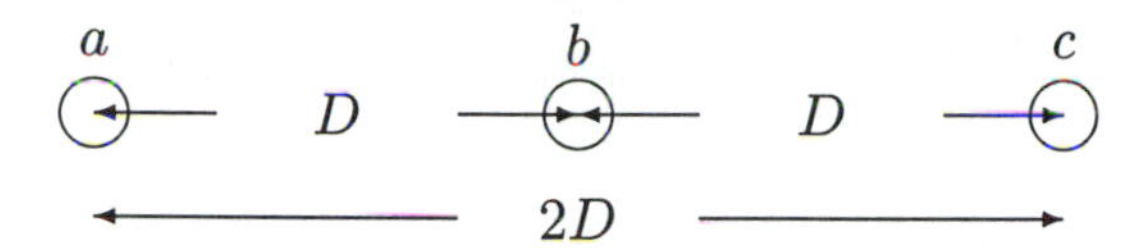

FIGURE 4.27
Conductor layout for Problem 4.7.

4.8. A three-phase transposed line is composed of one ACSR 159,000-cmil, 54/19 Lapwing conductor per phase with flat horizontal spacing of 8 m as shown in Figure 4.28. The GMR of each conductor is 1.515 cm.
(a) Determine the inductance per phase per kilometer of the line.
(b) This line is to be replaced by a two-conductor bundle with 8 m spacing measured from the center of the bundles as shown in Figure 4.29. The spacing between the conductors in the bundle is 40 cm. If the line inductance per phase is to be 77 percent of the inductance in part (a), what would be the GMR of each new conductor in the bundle?

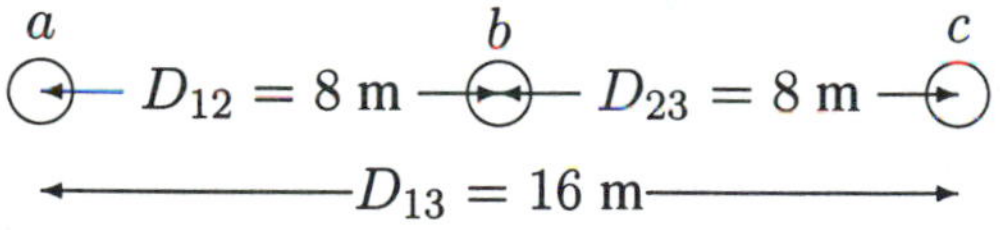

a b c

$D_{12} = 8$ m $D_{23} = 8$ m

$D_{13} = 16$ m

FIGURE 4.28
Conductor layout for Problem 4.8 (a).

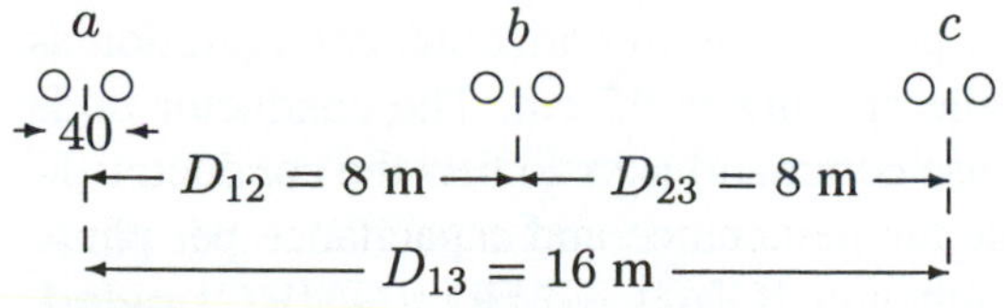

a b c

40

$D_{12} = 8$ m $D_{23} = 8$ m

$D_{13} = 16$ m

FIGURE 4.29
Conductor layout for Problem 4.8 (b).

4.9. A three-phase transposed line is composed of one ACSR, 1,431,000-cmil, 47/7 Bobolink conductor per phase with flat horizontal spacing of 11 m as shown in Figure 4.30. The conductors have a diameter of 3.625 cm and a *GMR* of 1.439 cm. The line is to be replaced by a three-conductor bundle of ACSR, 477,000-cmil, 26/7 Hawk conductors having the same cross-sectional area of aluminum as the single-conductor line. The conductors have a diameter of 2.1793 cm and a *GMR* of 0.8839 cm. The new line will also have a flat horizontal configuration, but it is to be operated at a higher voltage and therefore the phase spacing is increased to 14 m as measured from the center of the bundles as shown in Figure 4.31. The spacing between the conductors in the bundle is 45 cm. Determine

(a) The percentage change in the inductance.
(b) The percentage change in the capacitance.

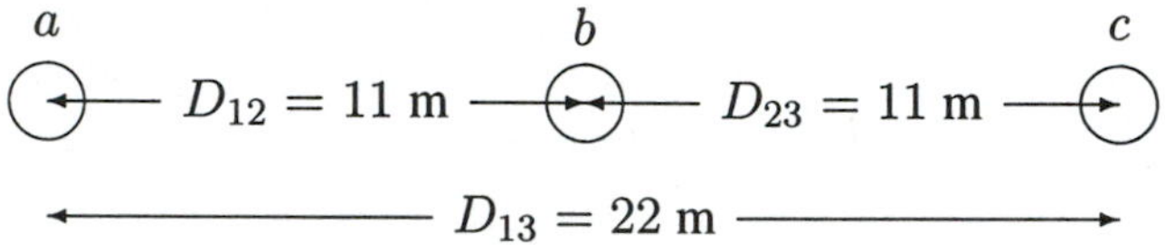

a b c

$D_{12} = 11$ m $D_{23} = 11$ m

$D_{13} = 22$ m

FIGURE 4.30
Conductor layout for Problem 4.9 (a).

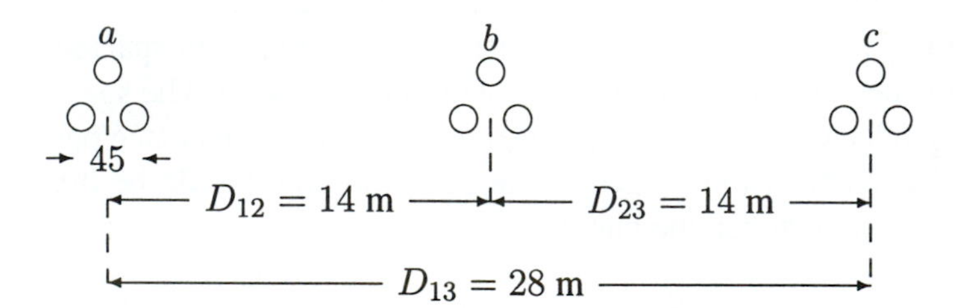

FIGURE 4.31
Conductor layout for Problem 4.9 (b).

4.10. A single-circuit three-phase transposed transmission line is composed of four ACSR, 1,272,000-cmil conductor per phase with horizontal configuration as shown in Figure 4.32. The bundle spacing is 45 cm. The conductor code name is *pheasant*. In *MATLAB*, use command **acsr** to find the conductor diameter and its *GMR*. Determine the inductance and capacitance per phase per kilometer of the line. Use function **[GMD, GMRL, GMRC] =gmd**, (4.58) and (4.92) in *MATLAB* to verify your results.

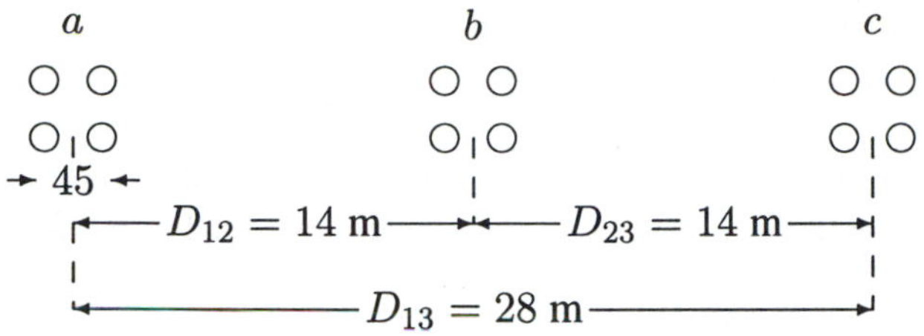

a b c

45

$D_{12} = 14$ m $D_{23} = 14$ m

$D_{13} = 28$ m

FIGURE 4.32
Conductor layout for Problem 4.10.

4.11. A double circuit three-phase transposed line is composed of two ACSR, 2,16,7000-cmil, 72/7 Kiwi conductor per phase with vertical configuration as shown in Figure 4.33. The conductors have a diameter of 4.4069 cm and a *GMR* of 1.7374 cm. The bundle spacing is 45 cm. The circuit arrangement is $a_1b_1c_1$, $c_2b_2a_2$. Find the inductance and capacitance per phase per kilometer of the line. Find these values when the circuit arrangement is $a_1b_1c_1$, $a_2b_2c_2$. Use function **[GMD, GMRL, GMRC] =gmd**, (4.58) and (4.92) in *MATLAB* to verify your results.

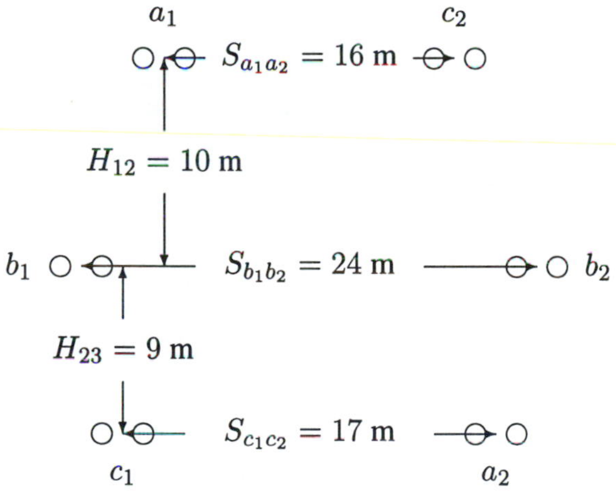

a_1 c_2

$S_{a_1a_2} = 16$ m

$H_{12} = 10$ m

b_1 $S_{b_1b_2} = 24$ m b_2

$H_{23} = 9$ m

$S_{c_1c_2} = 17$ m

c_1 a_2

FIGURE 4.33
Conductor layout for Problem 4.11.

4.12. The conductors of a double-circuit three-phase transmission line are placed on the corner of a hexagon as shown in Figure 4.34. The two circuits are in parallel and are sharing the balanced load equally. The conductors of the circuits are identical, each having a radius r. Assume that the line is symmetrically transposed. Using the method of GMD, determine an expression for the capacitance per phase per meter of the line.

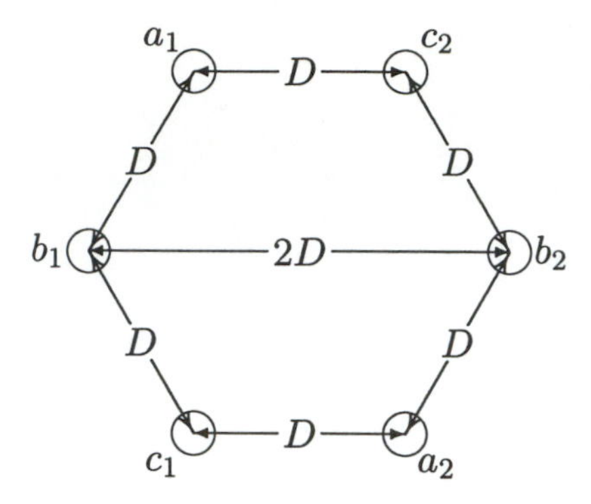

FIGURE 4.34
Conductor layout for Problem 4.12.

4.13. A 60-Hz, single-phase power line and a telephone line are parallel to each other as shown in Figure 4.35. The telephone line is symmetrically positioned directly below phase b. The power line carries an rms current of 226 A. Assume zero current flows in the ungrounded telephone wires. Find the magnitude of the voltage per km induced in the telephone line.

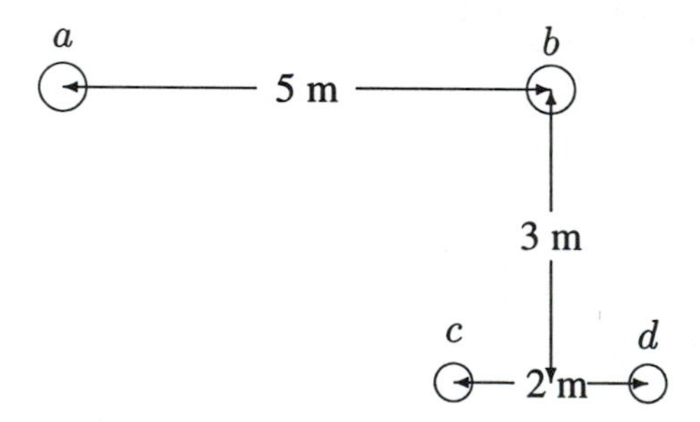

FIGURE 4.35
Conductor layout for Problem 4.13.

4.14. A three-phase, 60-Hz untransposed transmission line runs in parallel with a telephone line for 20 km. The power line carries a balanced three-phase rms current of $I_a = 320\angle 0°$ A, $I_b = 320\angle -120°$ A, and $I_c = 320\angle -240°$ A. The line configuration is as shown in Figure 4.36. Assume zero current flows in the ungrounded telephone wires. Find the magnitude of the voltage induced in the telephone line.

4.15. Since earth is an equipotential plane, the electric flux lines are forced to cut the surface of the earth orthogonally. The earth effect can be represented by placing an oppositely charged conductor a depth H below the surface of the earth as shown in Figure 4.37(a). This configuration without the presence

FIGURE 4.36
Conductor layout for Problem 4.14.

of the earth will produce the same field as a single charge and the earth surface. This imaginary conductor is called the image conductor. Figure 4.37(b) shows a single-phase line with its image conductors. Find the potential difference V_{ab} and show that the equivalent capacitance to neutral is given by

$$C_{an} = C_{bn} = \frac{2\pi\varepsilon}{\ln(\frac{D}{r}\frac{2H}{H_{12}})}$$

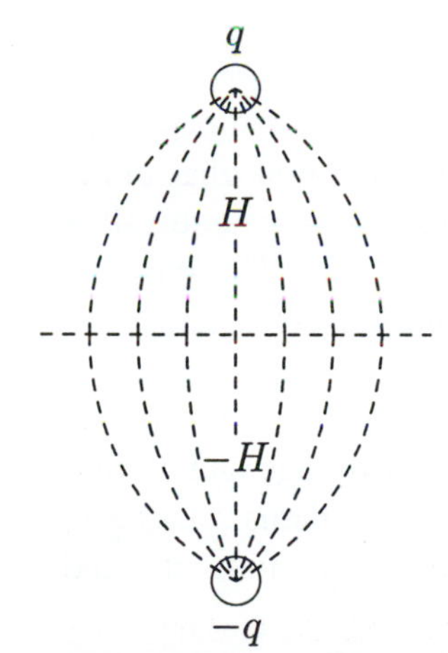

(a) Earth plane replaced by image conductor

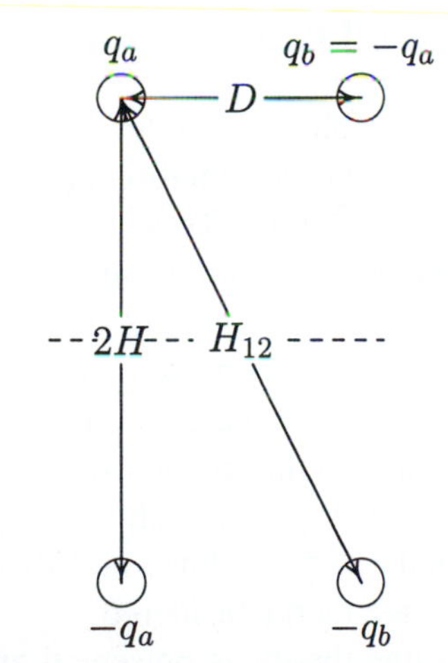

(b) Single-phase line and its image

FIGURE 4.37
Conductor layout for Problem 4.15.

CHAPTER 5

LINE MODEL AND PERFORMANCE

5.1 INTRODUCTION

In Chapter 4 the per-phase parameters of transmission lines were obtained. This chapter deals with the representation and performance of transmission lines under normal operating conditions. Transmission lines are represented by an equivalent model with appropriate circuit parameters on a "per-phase" basis. The terminal voltages are expressed from one line to neutral, the current for one phase and, thus, the three-phase system is reduced to an equivalent single-phase system.

The model used to calculate voltages, currents, and power flows depends on the length of the line. In this chapter the circuit parameters and voltage and current relations are first developed for "short" and "medium" lines. Problems relating to the regulation and losses of lines and their operation under conditions of fixed terminal voltages are then considered.

Next, long line theory is presented and expressions for voltage and current along the distributed line model are obtained. Propagation constant and characteristic impedance are defined, and it is demonstrated that the electrical power is being transmitted over the lines at approximately the speed of light. Since the terminal conditions at the two ends of the line are of primary importance, an equivalent

π model is developed for the long lines. Several *MATLAB* functions are developed for calculation of line parameters and performance. Finally, line compensations are discussed for improving the line performance for unloaded and loaded transmission lines.

5.2 SHORT LINE MODEL

Capacitance may often be ignored without much error if the lines are less than about 80 km (50 miles) long, or if the voltage is not over 69 kV. The short line model is obtained by multiplying the series impedance per unit length by the line length.

$$\begin{aligned} Z &= (r + j\omega L)\ell \\ &= R + jX \end{aligned} \tag{5.1}$$

where r and L are the per-phase resistance and inductance per unit length, respectively, and ℓ is the line length. The short line model on a per-phase basis is shown in Figure 5.1. V_S and I_S are the phase voltage and current at the sending end of the line, and V_R and I_R are the phase voltage and current at the receiving end of the line.

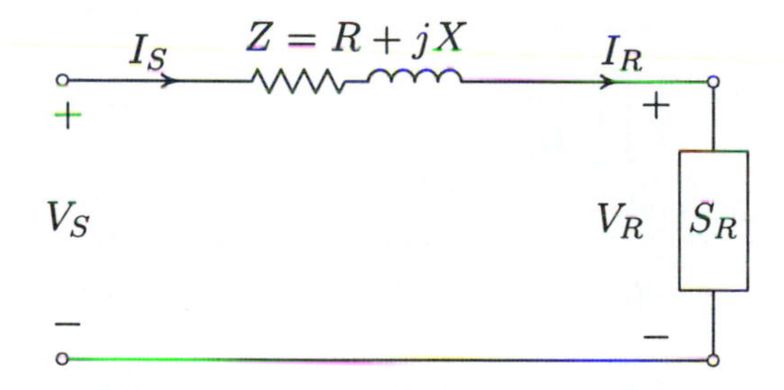

FIGURE 5.1
Short line model.

If a three-phase load with apparent power $S_{R(3\phi)}$ is connected at the end of the transmission line, the receiving end current is obtained by

$$I_R = \frac{S^*_{R(3\phi)}}{3V^*_R} \tag{5.2}$$

The phase voltage at the sending end is

$$V_S = V_R + ZI_R \tag{5.3}$$

and since the shunt capacitance is neglected, the sending end and the receiving end current are equal, i.e.,

$$I_S = I_R \tag{5.4}$$

The transmission line may be represented by a two-port network as shown in Figure 5.2, and the above equations can be written in terms of the generalized circuit constants commonly known as the $ABCD$ constants

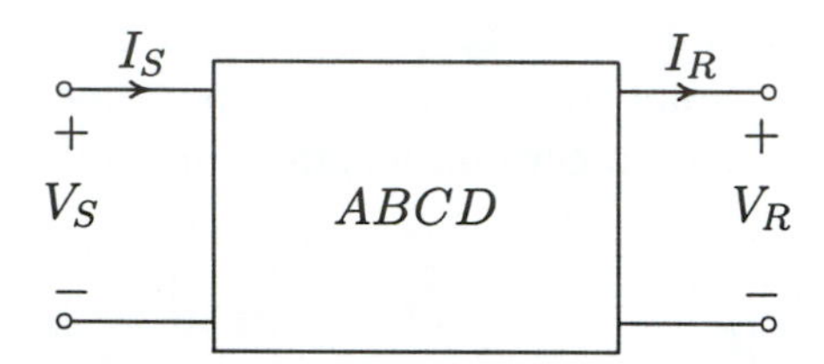

FIGURE 5.2
Two-port representation of a transmission line.

$$V_S = AV_R + BI_R \tag{5.5}$$

$$I_S = CV_R + DI_R \tag{5.6}$$

or in matrix form

$$\begin{bmatrix} V_S \\ I_S \end{bmatrix} = \begin{bmatrix} A & B \\ C & D \end{bmatrix} \begin{bmatrix} V_R \\ I_R \end{bmatrix} \tag{5.7}$$

According to (5.3) and (5.4), for short line model

$$A = 1 \qquad B = Z \qquad C = 0 \qquad D = 1 \tag{5.8}$$

Voltage regulation of the line may be defined as the percentage change in voltage at the receiving end of the line (expressed as percent of full-load voltage) in going from no-load to full-load.

$$\text{Percent } VR = \frac{|V_{R(NL)}| - |V_{R(FL)}|}{|V_{R(FL)}|} \times 100 \tag{5.9}$$

At no-load $I_R = 0$ and from (5.5)

$$V_{R(NL)} = \frac{V_S}{A} \tag{5.10}$$

For a short line, $A = 1$ and $V_{R(NL)} = V_S$. Voltage regulation is a measure of line voltage drop and depends on the load power factor. Voltage regulation will be

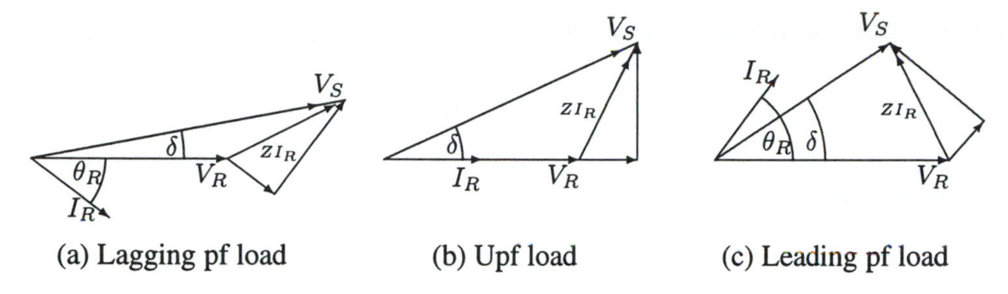

FIGURE 5.3
Phasor diagram for short line.

poorer at low lagging power factor loads. With capacitive loads, i.e., leading power factor loads, regulation may become negative. This is demonstrated by the phasor diagram of Figure 5.3.

Once the sending end voltage is calculated the sending-end power is obtained by

$$S_{S(3\phi)} = 3V_S I_S^* \tag{5.11}$$

The total line loss is then given by

$$S_{L(3\phi)} = S_{S(3\phi)} - S_{R(3\phi)} \tag{5.12}$$

and the transmission line efficiency is given by

$$\eta = \frac{P_{R(3\phi)}}{P_{S(3\phi)}} \tag{5.13}$$

where $P_{R(3\phi)}$ and $P_{S(3\phi)}$ are the total real power at the receiving end and sending end of the line, respectively.

Example 5.1

A 220-kV, three-phase transmission line is 40 km long. The resistance per phase is 0.15 Ω per km and the inductance per phase is 1.3263 mH per km. The shunt capacitance is negligible. Use the short line model to find the voltage and power at the sending end and the voltage regulation and efficiency when the line is supplying a three-phase load of

(a) 381 MVA at 0.8 power factor lagging at 220 kV.
(b) 381 MVA at 0.8 power factor leading at 220 kV.

(a) The series impedance per phase is

$$Z = (r + j\omega L)\ell = (0.15 + j2\pi \times 60 \times 1.3263 \times 10^{-3})40 = 6 + j20 \ \Omega$$

The receiving end voltage per phase is

$$V_R = \frac{220\angle 0^\circ}{\sqrt{3}} = 127\angle 0^\circ \ \ \text{kV}$$

The apparent power is

$$S_{R(3\phi)} = 381\angle \cos^{-1} 0.8 = 381\angle 36.87^\circ = 304.8 + j228.6 \ \ \text{MVA}$$

The current per phase is given by

$$I_R = \frac{S^*_{R(3\phi)}}{3\,V^*_R} = \frac{381\angle -36.87^\circ \times 10^3}{3 \times 127\angle 0^\circ} = 1000\angle -36.87^\circ \ \ \text{A}$$

From (5.3) the sending end voltage is

$$\begin{aligned} V_S = V_R + ZI_R &= 127\angle 0^\circ + (6 + j20)(1000\angle -36.87^\circ)(10^{-3}) \\ &= 144.33\angle 4.93^\circ \ \ \text{kV} \end{aligned}$$

The sending end line-to-line voltage magnitude is

$$|V_{S(L-L)}| = \sqrt{3}\,|V_S| = 250 \ \ \text{kV}$$

The sending end power is

$$\begin{aligned} S_{S(3\phi)} = 3V_S I^*_S &= 3 \times 144.33\angle 4.93 \times 1000\angle 36.87^\circ \times 10^{-3} \\ &= 322.8 \ \ \text{MW} + j288.6 \ \ \text{Mvar} \\ &= 433\angle 41.8^\circ \ \ \text{MVA} \end{aligned}$$

Voltage regulation is

$$\text{Percent } VR = \frac{250 - 220}{220} \times 100 = 13.6\%$$

Transmission line efficiency is

$$\eta = \frac{P_{R(3\phi)}}{P_{S(3\phi)}} = \frac{304.8}{322.8} \times 100 = 94.4\%$$

(b) The current for 381 MVA with 0.8 leading power factor is

$$I_R = \frac{S^*_{R(3\phi)}}{3\,V^*_R} = \frac{381\angle 36.87^\circ \times 10^3}{3 \times 127\angle 0^\circ} = 1000\angle 36.87^\circ \ \ \text{A}$$

The sending end voltage is

$$\begin{aligned} V_S = V_R + ZI_R &= 127\angle 0^\circ + (6 + j20)(1000\angle 36.87^\circ)(10^{-3}) \\ &= 121.39\angle 9.29^\circ \ \ \text{kV} \end{aligned}$$

The sending end line-to-line voltage magnitude is

$$|V_{S(L-L)}| = \sqrt{3}\, V_S = 210.26 \ \ \text{kV}$$

The sending end power is

$$\begin{aligned} S_{S(3\phi)} = 3V_S I_S^* &= 3 \times 121.39\angle 9.29 \times 1000\angle -36.87^\circ \times 10^{-3} \\ &= 322.8 \ \ \text{MW} - j168.6 \ \ \text{Mvar} \\ &= 364.18\angle -27.58^\circ \ \ \text{MVA} \end{aligned}$$

Voltage regulation is

$$\text{Percent } VR = \frac{210.26 - 220}{220} \times 100 = -4.43\%$$

Transmission line efficiency is

$$\eta = \frac{P_{R(3\phi)}}{P_{S(3\phi)}} = \frac{304.8}{322.8} \times 100 = 94.4\%$$

5.3 MEDIUM LINE MODEL

As the length of line increases, the line charging current becomes appreciable and the shunt capacitance must be considered. Lines above 80 km (50 miles) and below 250 km (150 miles) in length are termed as *medium length lines*. For medium length lines, half of the shunt capacitance may be considered to be lumped at each end of the line. This is referred to as the *nominal* π model and is shown in Figure 5.4. Z is the total series impedance of the line given by (5.1), and Y is the total shunt admittance of the line given by

$$Y = (g + j\omega C)\ell \tag{5.14}$$

Under normal conditions, the shunt conductance per unit length, which represents the leakage current over the insulators and due to corona, is negligible and g is assumed to be zero. C is the line to neutral capacitance per km, and ℓ is the line length. The sending end voltage and current for the nominal π model are obtained as follows:

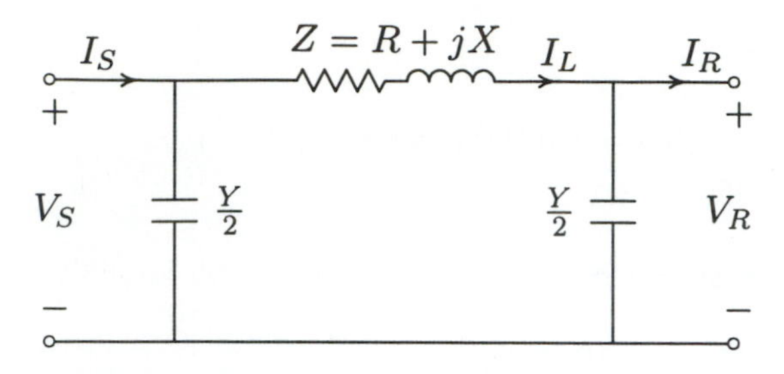

FIGURE 5.4
Nominal π model for medium length line.

From KCL the current in the series impedance designated by I_L is

$$I_L = I_R + \frac{Y}{2} V_R \tag{5.15}$$

From KVL the sending end voltage is

$$V_S = V_R + Z I_L \tag{5.16}$$

Substituting for I_L from (5.15), we obtain

$$V_S = \left(1 + \frac{ZY}{2}\right) V_R + Z I_R \tag{5.17}$$

The sending end current is

$$I_S = I_L + \frac{Y}{2} V_S \tag{5.18}$$

Substituting for I_L and V_S

$$I_S = Y\left(1 + \frac{ZY}{4}\right) V_R + \left(1 + \frac{ZY}{2}\right) I_R \tag{5.19}$$

Comparing (5.17) and (5.19) with (5.5) and (5.6), the $ABCD$ constants for the nominal π model are given by

$$A = \left(1 + \frac{ZY}{2}\right) \qquad B = Z \tag{5.20}$$

$$C = Y\left(1 + \frac{ZY}{4}\right) \qquad D = \left(1 + \frac{ZY}{2}\right) \tag{5.21}$$

In general, the $ABCD$ constants are complex and since the π model is a symmetrical two-port network, $A = D$. Furthermore, since we are dealing with a linear passive, bilateral two-port network, the determinant of the transmission matrix in (5.7) is unity, i.e.,

$$AD - BC = 1 \tag{5.22}$$

Solving (5.7), the receiving end quantities can be expressed in terms of the sending end quantities by

$$\begin{bmatrix} V_R \\ I_R \end{bmatrix} = \begin{bmatrix} D & -B \\ -C & A \end{bmatrix} \begin{bmatrix} V_S \\ I_S \end{bmatrix} \tag{5.23}$$

Two *MATLAB* functions are written for computation of the transmission matrix. Function **[Z, Y, ABCD] = rlc2abcd(r, L, C, g, f, Length)** is used when resistance in ohm, inductance in mH and capacitance in μF per unit length are specified, and function **[Z, Y, ABCD] = zy2abcd(z, y, Length)** is used when series impedance in ohm and shunt admittance in siemens per unit length are specified. The above functions provide options for the nominal π model and the equivalent π model discussed in Section 5.4.

Example 5.2

A 345-kV, three-phase transmission line is 130 km long. The resistance per phase is 0.036 Ω per km and the inductance per phase is 0.8 mH per km. The shunt capacitance is 0.0112 μF per km. The receiving end load is 270 MVA with 0.8 power factor lagging at 325 kV. Use the medium line model to find the voltage and power at the sending end and the voltage regulation.

The function **[Z, Y, ABCD] = rlc2abcd(r, L, C, g, f, Length)** is used to obtain the transmission matrix of the line. The following commands

```
r = .036; g = 0; f = 60;
L = 0.8;          % milli-Henry
C = 0.0112;       % micro-Farad
Length = 130;  VR3ph = 325;
VR = VR3ph/sqrt(3) + j*0;     % kV (receiving end phase voltage)
[Z, Y, ABCD] = rlc2abcd(r, L, C, g, f, Length);
AR = acos(0.8);
SR = 270*(cos(AR) + j*sin(AR));  %   MVA (receiving end power)
IR = conj(SR)/(3*conj(VR));      %  kA (receiving end current)
VsIs = ABCD* [VR; IR];           %      column vector [Vs; Is]
Vs = VsIs(1);
Vs3ph = sqrt(3)*abs(Vs);         % kV(sending end L-L voltage)
Is = VsIs(2); Ism = 1000*abs(Is);%     A (sending end current)
pfs= cos(angle(Vs)- angle(Is));  %  (sending end power factor)
Ss = 3*Vs*conj(Is);              %     MVA (sending end power)
REG = (Vs3ph/abs(ABCD(1,1)) - VR3ph)/VR3ph *100;
```

```
fprintf(' Is = %g A', Ism), fprintf('  pf = %g', pfs)
fprintf(' Vs = %g L-L kV', Vs3ph)
fprintf(' Ps = %g MW', real(Ss)),
fprintf('  Qs = %g Mvar', imag(Ss))
fprintf(' Percent voltage Reg. = %g', REG)
```

result in

```
Enter 1 for Medium line or 2 for long line → 1
Nominal π model
Z = 4.68 + j 39.2071 ohms
Y = 0 + j 0.000548899 siemens
```

$$\text{ABCD} = \begin{bmatrix} \texttt{0.98924} & + \ \texttt{j 0.0012844} & \texttt{4.68} & + \ \texttt{j 39.207} \\ \texttt{-3.5251e-07} & + \ \texttt{j 0.00054595} & \texttt{0.98924} & + \ \texttt{j 0.0012844} \end{bmatrix}$$

```
Is = 421.132 A        pf = 0.869657
Vs = 345.002 L-L kV
Ps = 218.851 MW     Qs = 124.23 Mvar
Percent voltage Reg. = 7.30913
```

Example 5.3

A 345-kV, three-phase transmission line is 130 km long. The series impedance is $z = 0.036 + j0.3$ Ω per phase per km, and the shunt admittance is $y = j4.22 \times 10^{-6}$ siemens per phase per km. The sending end voltage is 345 kV, and the sending end current is 400 A at 0.95 power factor lagging. Use the medium line model to find the voltage, current and power at the receiving end and the voltage regulation.

The function **[Z, Y, ABCD] = zy2abcd(z, y, Length)** is used to obtain the transmission matrix of the line. The following commands

```
z = .036 + j* 0.3;  y = j*4.22/1000000;  Length = 130;
Vs3ph = 345;  Ism = 0.4;  %kA;
As = -acos(0.95);
Vs = Vs3ph/sqrt(3) + j*0;      % kV (sending end phase voltage)
Is = Ism*(cos(As) + j*sin(As));
[Z,Y, ABCD] = zy2abcd(z, y, Length);
VrIr = inv(ABCD)* [Vs; Is];     %       column vector [Vr; Ir]
Vr = VrIr(1);
Vr3ph = sqrt(3)*abs(Vr);        % kV(receiving end L-L voltage)
Ir = VrIr(2); Irm = 1000*abs(Ir);  % A (receiving end current)
pfr= cos(angle(Vr)- angle(Ir));  %(receiving end power factor)
Sr = 3*Vr*conj(Ir);                    % MVA (receiving end power)
```

```
REG = (Vs3ph/abs(ABCD(1,1)) - Vr3ph)/Vr3ph *100;
fprintf(' Ir = %g A', Irm), fprintf('  pf = %g', pfr)
fprintf(' Vr = %g L-L kV', Vr3ph)
fprintf(' Pr = %g MW', real(Sr))
fprintf('  Qr = %g Mvar', imag(Sr))
fprintf(' Percent voltage Reg. = %g', REG)
```

result in

```
Enter 1 for Medium line or 2 for long line → 1
Nominal π model
Z = 4.68 + j 39 ohms
Y = 0 + j 0.0005486 siemens
```

$$\text{ABCD} = \begin{bmatrix} 0.9893 + j\,0.0012837 & 4.68 + j\,39 \\ -3.5213\text{e-}07 + j\,0.00054565 & 0.9893 + j\,0.0012837 \end{bmatrix}$$

```
Ir = 441.832 A       pf = 0.88750
Vr = 330.68 L-L kV
Pr = 224.592 MW     Qr = 116.612 Mvar
Percent voltage Reg. = 5.45863
```

5.4 LONG LINE MODEL

For the short and medium length lines reasonably accurate models were obtained by assuming the line parameters to be lumped. For lines 250 km (150 miles) and longer and for a more accurate solution the exact effect of the distributed parameters must be considered. In this section expressions for voltage and current at any point on the line are derived. Then, based on these equations an equivalent π model is obtained for the long line. Figure 5.5 shows one phase of a distributed line of length ℓ km.

The series impedance per unit length is shown by the lowercase letter z, and the shunt admittance per phase is shown by the lowercase letter y, where $z = r + j\omega L$ and $y = g + j\omega C$. Consider a small segment of line Δx at a distance x from the receiving end of the line. The phasor voltages and currents on both sides of this segment are shown as a function of distance. From Kirchhoff's voltage law

$$V(x + \Delta x) = V(x) + z\,\Delta x\,I(x) \tag{5.24}$$

or

$$\frac{V(x + \Delta x) - V(x)}{\Delta x} = z\,I(x) \tag{5.25}$$

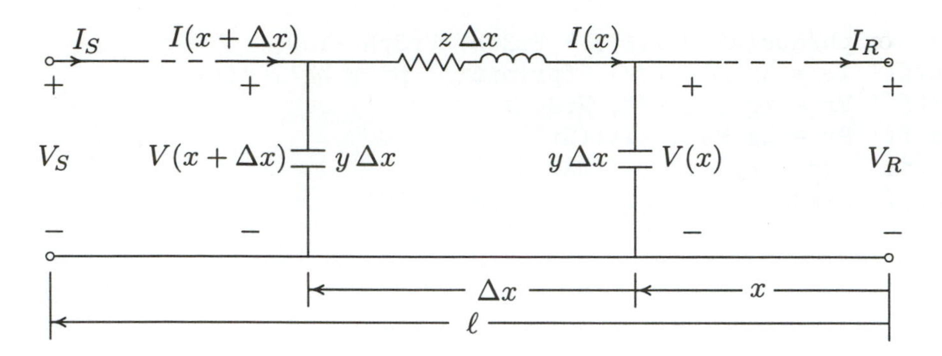

FIGURE 5.5
Long line with distributed parameters.

Taking the limit as $\Delta x \to 0$, we have

$$\frac{dV(x)}{dx} = z\,I(x) \tag{5.26}$$

Also, from Kirchhoff's current law

$$I(x+\Delta x) = I(x) + y\,\Delta x\,V(x+\Delta x) \tag{5.27}$$

or

$$\frac{I(x+\Delta x) - I(x)}{\Delta x} = y\,V(x+\Delta x) \tag{5.28}$$

Taking the limit as $\Delta x \to 0$, we have

$$\frac{dI(x)}{dx} = y\,V(x) \tag{5.29}$$

Differentiating (5.26) and substituting from (5.29), we get

$$\begin{aligned} \frac{d^2V(x)}{dx^2} &= z\,\frac{dI(x)}{dx} \\ &= zy\,V(x) \end{aligned} \tag{5.30}$$

Let

$$\gamma^2 = zy \tag{5.31}$$

The following second-order differential equation will result.

$$\frac{d^2V(x)}{dx^2} - \gamma^2\,V(x) = 0 \tag{5.32}$$

The solution of the above equation is

$$V(x) = A_1 e^{\gamma x} + A_2 e^{-\gamma x} \tag{5.33}$$

where γ, known as the *propagation constant*, is a complex expression given by (5.31) or

$$\gamma = \alpha + j\beta = \sqrt{zy} = \sqrt{(r + j\omega L)(g + j\omega C)} \tag{5.34}$$

The real part α is known as the *attenuation constant*, and the imaginary component β is known as the *phase constant*. β is measured in radian per unit length.

From (5.26), the current is

$$\begin{aligned} I(x) = \frac{1}{z}\frac{dV(x)}{dx} &= \frac{\gamma}{z}(A_1 e^{\gamma x} - A_2 e^{-\gamma x}) \\ &= \sqrt{\frac{y}{z}}(A_1 e^{\gamma x} - A_2 e^{-\gamma x}) \end{aligned} \tag{5.35}$$

or

$$I(x) = \frac{1}{Z_c}(A_1 e^{\gamma x} - A_2 e^{-\gamma x}) \tag{5.36}$$

where Z_c is known as the *characteristic impedance*, given by

$$Z_c = \sqrt{\frac{z}{y}} \tag{5.37}$$

To find the constants A_1 and A_2 we note that when $x = 0$, $V(x) = V_R$, and $I(x) = I_R$. From (5.33) and (5.36) these constants are found to be

$$\begin{aligned} A_1 &= \frac{V_R + Z_c I_R}{2} \\ A_2 &= \frac{V_R - Z_c I_R}{2} \end{aligned} \tag{5.38}$$

Upon substitution in (5.33) and (5.36), the general expressions for voltage and current along a long transmission line become

$$V(x) = \frac{V_R + Z_c I_R}{2} e^{\gamma x} + \frac{V_R - Z_c I_R}{2} e^{-\gamma x} \tag{5.39}$$

$$I(x) = \frac{\frac{V_R}{Z_c} + I_R}{2} e^{\gamma x} - \frac{\frac{V_R}{Z_c} - I_R}{2} e^{-\gamma x} \tag{5.40}$$

The equations for voltage and currents can be rearranged as follows:

$$V(x) = \frac{e^{\gamma x} + e^{-\gamma x}}{2} V_R + Z_c \frac{e^{\gamma x} - e^{-\gamma x}}{2} I_R \tag{5.41}$$

$$I(x) = \frac{1}{Z_c} \frac{e^{\gamma x} - e^{-\gamma x}}{2} V_R + \frac{e^{\gamma x} + e^{-\gamma x}}{2} I_R \tag{5.42}$$

Recognizing the hyperbolic functions sinh, and cosh, the above equations are written as follows:

$$V(x) = \cosh \gamma x \, V_R + Z_c \sinh \gamma x \, I_R \tag{5.43}$$

$$I(x) = \frac{1}{Z_c} \sinh \gamma x \, V_R + \cosh \gamma x \, I_R \tag{5.44}$$

We are particularly interested in the relation between the sending end and the receiving end of the line. Setting $x = \ell$, $V(\ell) = V_s$ and $I(\ell) = I_s$, the result is

$$V_s = \cosh \gamma \ell \, V_R + Z_c \sinh \gamma \ell \, I_R \tag{5.45}$$

$$I_s = \frac{1}{Z_c} \sinh \gamma \ell \, V_R + \cosh \gamma \ell \, I_R \tag{5.46}$$

Rewriting the above equations in terms of the $ABCD$ constants as before, we have

$$\begin{bmatrix} V_S \\ I_S \end{bmatrix} = \begin{bmatrix} A & B \\ C & D \end{bmatrix} \begin{bmatrix} V_R \\ I_R \end{bmatrix} \tag{5.47}$$

where

$$A = \cosh \gamma \ell \qquad B = Z_c \sinh \gamma \ell \tag{5.48}$$

$$C = \frac{1}{Z_c} \sinh \gamma \ell \qquad D = \cosh \gamma \ell \tag{5.49}$$

Note that, as before, $A = D$ and $AD - BC = 1$.

It is now possible to find an accurate equivalent π model, shown in Figure 5.6, to replace the $ABCD$ constants of the two-port network. Similar to the expressions (5.17) and (5.19) obtained for the nominal π, for the equivalent π model we have

$$V_S = \left(1 + \frac{Z'Y'}{2}\right) V_R + Z' I_R \tag{5.50}$$

$$I_S = Y' \left(1 + \frac{Z'Y'}{4}\right) V_R + \left(1 + \frac{Z'Y'}{2}\right) I_R \tag{5.51}$$

Comparing (5.50) and (5.51) with (5.45) and (5.46), respectively, and making use of the identity

$$\tanh \frac{\gamma \ell}{2} = \frac{\cosh \gamma \ell - 1}{\sinh \gamma \ell} \tag{5.52}$$

the parameters of the equivalent π model are obtained.

$$Z' = Z_c \sinh \gamma\ell = Z\frac{\sinh \gamma\ell}{\gamma\ell} \tag{5.53}$$

$$\frac{Y'}{2} = \frac{1}{Z_c}\tanh\frac{\gamma\ell}{2} = \frac{Y}{2}\frac{\tanh \gamma\ell/2}{\gamma\ell/2} \tag{5.54}$$

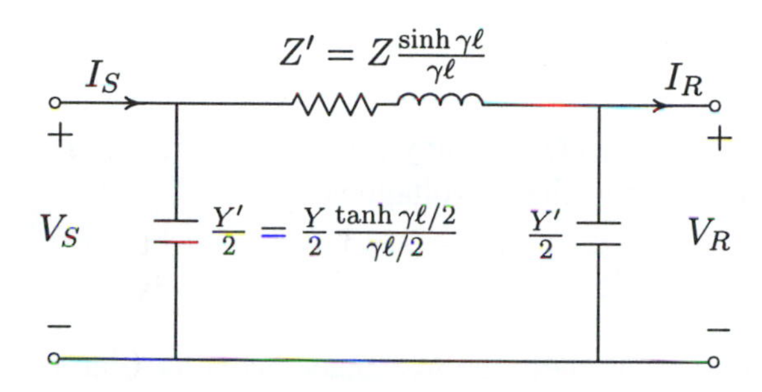

FIGURE 5.6
Equivalent π model for long length line.

The functions **[Z, Y, ABCD] = rlc2abcd(r, L, C, g, f, Length)** and **[Z, Y, ABCD] = zy2abcd(z, y, Length)** with option 2 can be used for the evaluation of the transmission matrix and the equivalent π parameters. However, Example 5.4 shows how these hyperbolic functions can be evaluated easily with simple *MATLAB* commands.

Example 5.4

A 500-kV, three-phase transmission line is 250 km long. The series impedance is $z = 0.045 + j0.4\ \Omega$ per phase per km and the shunt admittance is $y = j4 \times 10^{-6}$ siemens per phase per km. Evaluate the equivalent π model and the transmission matrix

The following commands

```
z = 0.045 + j*.4;     y = j*4.0/1000000; Length = 250;
gamma = sqrt(z*y);  Zc = sqrt(z/y);
A = cosh(gamma*Length);  B = Zc*sinh(gamma*Length);
C = 1/Zc * sinh(gamma*Length);  D = A;
ABCD = [A  B; C  D]
Z = B; Y = 2/Zc * tanh(gamma*Length/2)
```

result in

```
ABCD =
        0.9504 + 0.0055i  10.8778 +98.3624i
       -0.0000 + 0.0010i   0.9504 + 0.0055i
Z =
       10.8778 +98.3624i
Y =
        0.0000 + 0.0010i
```

5.5 VOLTAGE AND CURRENT WAVES

The rms expression for the phasor value of voltage at any point along the line is given by (5.33). Substituting $\alpha + j\beta$ for γ, the phasor voltage is

$$V(x) = A_1 e^{\alpha x} e^{j\beta x} + A_2 e^{-\alpha x} e^{-j\beta x}$$

Transforming from phasor domain to time domain, the instantaneous voltage as a function of t and x becomes

$$v(t,x) = \sqrt{2}\,\Re\, A_1 e^{\alpha x} e^{j(\omega t + \beta x)} + \sqrt{2}\,\Re\, A_2 e^{-\alpha x} e^{j(\omega t - \beta x)} \tag{5.55}$$

As x increases (moving away from the receiving end), the first term becomes larger because of $e^{\alpha x}$ and is called the *incident wave*. The second term becomes smaller because of $e^{-\alpha x}$ and is called the *reflected wave*. At any point along the line, voltage is the sum of these two components.

$$v(t,x) = v_1(t,x) + v_2(t,x) \tag{5.56}$$

where

$$v_1(t,x) = \sqrt{2}\, A_1 e^{\alpha x} \cos(\omega t + \beta x) \tag{5.57}$$

$$v_2(t,x) = \sqrt{2}\, A_2 e^{-\alpha x} \cos(\omega t - \beta x) \tag{5.58}$$

As the current expression is similar to the voltage, the current can also be considered as the sum of incident and reflected current waves.

Equations (5.57) and (5.58) behave like traveling waves as we move along the line. This is similar to the disturbance in the water at some sending point. To see this, consider the reflected wave $v_2(t,x)$ and imagine that we ride along with the wave. To observe the instantaneous value, for example the peak amplitude requires that

$$\omega t - \beta x = 2K\pi \quad \text{or} \quad x = \frac{\omega}{\beta}t - \frac{2K\pi}{\beta}$$

Thus, to keep up with the wave and observe the peak amplitude we must travel with the speed

$$\frac{dx}{dt} = \frac{\omega}{\beta} \tag{5.59}$$

Thus, the velocity of propagation is given by

$$v = \frac{\omega}{\beta} = \frac{2\pi f}{\beta} \tag{5.60}$$

The wavelength λ or distance x on the wave which results in a phase shift of 2π radian is

$$\beta\lambda = 2\pi$$

or

$$\lambda = \frac{2\pi}{\beta} \tag{5.61}$$

When line losses are neglected, i.e., when $g = 0$ and $r = 0$, the real part of the propagation constant $\alpha = 0$, and from (5.34) the phase constant becomes

$$\beta = \omega\sqrt{LC} \tag{5.62}$$

Also, the characteristic impedance is purely resistive and (5.37) becomes

$$Z_c = \sqrt{\frac{L}{C}} \tag{5.63}$$

which is commonly referred to as the *surge impedance*. Substituting for β in (5.60) and (5.61), for a lossless line the velocity of propagation and the wavelength become

$$v = \frac{1}{\sqrt{LC}} \tag{5.64}$$

$$\lambda = \frac{1}{f\sqrt{LC}} \tag{5.65}$$

The expressions for the inductance per unit length L and capacitance per unit length C of a transmission line were derived in Chapter 4, given by (4.58) and (4.91). When the internal flux linkage of a conductor is neglected $GMR_L = GMR_C$, and upon substitution (5.64) and (5.65) become

$$v \simeq \frac{1}{\sqrt{\mu_0\varepsilon_0}} \tag{5.66}$$

$$\lambda \simeq \frac{1}{f\sqrt{\mu_0\varepsilon_0}} \tag{5.67}$$

Substituting for $\mu_0 = 4\pi \times 10^{-7}$ and $\varepsilon_0 = 8.85 \times 10^{-12}$, the velocity of the wave is obtained to be approximately 3×10^8 m/sec, i.e., the velocity of light. At 60 Hz, the wavelength is 5000 km. Similarly, substituting for L and C in (5.63), we have

$$\begin{aligned} Z_c &\simeq \frac{1}{2\pi}\sqrt{\frac{\mu_0}{\varepsilon_0}} \ln \frac{GMD}{GMR_c} \\ &\simeq 60 \ln \frac{GMD}{GMR_c} \end{aligned} \tag{5.68}$$

For typical transmission lines the surge impedance varies from approximately 400Ω for 69-kV lines down to around 250 Ω for double-circuit 765-kV transmission lines.

For a lossless line $\gamma = j\beta$ and the hyperbolic functions $\cosh \gamma x = \cosh j\beta x = \cos \beta x$ and $\sinh \gamma x = \sinh j\beta x = j \sin \beta x$, the equations for the rms voltage and current along the line, given by (5.43) and (5.44), become

$$V(x) = \cos \beta x \, V_R + jZ_c \sin \beta x \, I_R \tag{5.69}$$

$$I(x) = j\frac{1}{Z_c} \sin \beta x \, V_R + \cos \beta x \, I_R \tag{5.70}$$

At the sending end $x = \ell$

$$V_S = \cos \beta\ell \, V_R + jZ_c \sin \beta\ell \, I_R \tag{5.71}$$

$$I_S = j\frac{1}{Z_c} \sin \beta\ell \, V_R + \cos \beta\ell \, I_R \tag{5.72}$$

For hand calculation it is easier to use (5.71) and (5.72), and for more accurate calculations (5.47) through (5.49) can be used in *MATLAB*. The terminal conditions are readily obtained from the above equations. For example, for the open-circuited line $I_R = 0$, and from (5.71) the no-load receiving end voltage is

$$V_{R(nl)} = \frac{V_S}{\cos \beta\ell} \tag{5.73}$$

At no-load, the line current is entirely due to the line charging capacitive current and the receiving end voltage is higher than the sending end voltage. This is evident from (5.73), which shows that as the line length increases $\beta\ell$ increases and $\cos \beta\ell$ decreases, resulting in a higher no-load receiving end voltage.

For a solid short circuit at the receiving end, $V_R = 0$ and (5.71) and (5.72) reduce to

$$V_S = jZ_c \sin \beta\ell \, I_R \tag{5.74}$$

$$I_S = \cos \beta\ell \, I_R \tag{5.75}$$

The above equations can be used to find the short circuit currents at both ends of the line.

5.6 SURGE IMPEDANCE LOADING

When the line is loaded by being terminated with an impedance equal to its characteristic impedance, the receiving end current is

$$I_R = \frac{V_R}{Z_c} \tag{5.76}$$

For a lossless line Z_c is purely resistive. The load corresponding to the surge impedance at rated voltage is known as the *surge impedance loading* (SIL), given by

$$SIL = 3V_R I_R^* = \frac{3|V_R|^2}{Z_c} \tag{5.77}$$

Since $V_R = V_{Lrated}/\sqrt{3}$, SIL in MW becomes

$$SIL = \frac{(kV_{Lrated})^2}{Z_c} \text{ MW} \tag{5.78}$$

Substituting for I_R in (5.69) and V_R in (5.70) will result in

$$V(x) = (\cos\beta x + j\sin\beta x)V_R \quad \text{or} \quad V(x) = V_R\angle\beta x \tag{5.79}$$
$$I(x) = (\cos\beta x + j\sin\beta x)I_R \quad \text{or} \quad I(x) = I_R\angle\beta x \tag{5.80}$$

Equations (5.79) and (5.80) show that in a lossless line under surge impedance loading the voltage and current at any point along the line are constant in magnitude and are equal to their sending end values. Since Z_c has no reactive component, there is no reactive power in the line, $Q_S = Q_R = 0$. This indicates that for SIL, the reactive losses in the line inductance are exactly offset by reactive power supplied by the shunt capacitance or $\omega L|I_R|^2 = \omega C|V_R|^2$. From this relation, we find that $Z_c = V_R/I_R = \sqrt{L/C}$, which verifies the result in (5.63). SIL for typical transmission lines varies from approximately 150 MW for 230-kV lines to about 2000 MW for 765-kV lines. SIL is a useful measure of transmission line capacity as it indicates a loading where the line's reactive requirements are small. For loads significantly above SIL, shunt capacitors may be needed to minimize voltage drop along the line, while for light loads significantly below SIL , shunt inductors may be needed. Generally the transmission line full-load is much higher than SIL. The voltage profile for various loading conditions is illustrated in Figure 5.11 (page 182) in Example 5.9(h).

Example 5.5

A three-phase, 60-Hz, 500-kV transmission line is 300 km long. The line inductance is 0.97 mH/km per phase and its capacitance is 0.0115 μF/km per phase. Assume a lossless line.

(a) Determine the line phase constant β, the surge impedance Z_C, velocity of propagation v and the line wavelength λ.
(b) The receiving end rated load is 800 MW, 0.8 power factor lagging at 500 kV. Determine the sending end quantities and the voltage regulation.

(a) For a lossless line, from (5.62) we have

$$\beta = \omega\sqrt{LC} = 2\pi \times 60\sqrt{0.97 \times 0.0115 \times 10^{-9}} = 0.001259 \quad \text{rad/km}$$

and from (5.63)

$$Z_c = \sqrt{\frac{L}{C}} = \sqrt{\frac{0.97 \times 10^{-3}}{0.0115 \times 10^{-6}}} = 290.43 \ \ \Omega$$

Velocity of propagation is

$$v = \frac{1}{\sqrt{LC}} = \frac{1}{\sqrt{0.97 \times 0.0115 \times 10^{-9}}} = 2.994 \times 10^5 \ \ \text{km/s}$$

and the line wavelength is

$$\lambda = \frac{v}{f} = \frac{1}{60}(2.994 \times 10^5) = 4990 \ \ \text{km}$$

(b) $\beta\ell = 0.001259 \times 300 = 0.3777 \ \ \text{rad} = 21.641°$

The receiving end voltage per phase is

$$V_R = \frac{500\angle 0°}{\sqrt{3}} = 288.675\angle 0° \ \ \text{kV}$$

The receiving end apparent power is

$$S_{R(3\phi)} = \frac{800}{0.8}\angle\cos^{-1} 0.8 = 1000\angle 36.87° = 800 + j600 \ \ \text{MVA}$$

The receiving end current per phase is given by

$$I_R = \frac{S^*_{R(3\phi)}}{3\,V^*_R} = \frac{1000\angle - 36.87° \times 10^3}{3 \times 288.675\angle 0°} = 1154.7\angle - 36.87° \ \ \text{A}$$

From (5.71) the sending end voltage is

$$\begin{aligned} V_S &= \cos\beta\ell\, V_R + jZ_c \sin\beta\ell\, I_R \\ &= (0.9295)288.675\angle 0^\circ + j(290.43)(0.3688)(1154.7\angle -36.87^\circ)(10^{-3}) \\ &= 356.53\angle 16.1^\circ \text{ kV} \end{aligned}$$

The sending end line-to-line voltage magnitude is

$$|V_{S(L-L)}| = \sqrt{3}\,|V_S| = 617.53 \text{ kV}$$

From (5.72) the sending end current is

$$\begin{aligned} I_S &= j\frac{1}{Z_c}\sin\beta\ell\, V_R + \cos\beta\ell\, I_R \\ &= j\frac{1}{290.43}(0.3688)(288.675\angle 0^\circ)(10^3) + (0.9295)(1154.7\angle -36.87^\circ) \\ &= 902.3\angle -17.9^\circ \text{ A} \end{aligned}$$

The sending end power is

$$\begin{aligned} S_{S(3\phi)} = 3V_S I_S^* &= 3 \times 356.53\angle 16.1 \times 902.3\angle{-17.9^\circ} \times 10^{-3} \\ &= 800 \text{ MW} + j539.672 \text{ Mvar} \\ &= 965.1\angle 34^\circ \text{ MVA} \end{aligned}$$

Voltage regulation is

$$\text{Percent } VR = \frac{356.53/0.9295 - 288.675}{288.675} \times 100 = 32.87\%$$

The line performance of the above transmission line including the line resistance is obtained in Example 5.9 using the **lineperf** program. When a line is operating at the rated load, the exact solution results in $V_{S(L-L)} = 623.5\angle 15.57^\circ$ kV, and $I_s = 903.1\angle{-17.7^\circ}$ A. This shows that the lossless assumption yields acceptable results and is suitable for hand calculation.

5.7 COMPLEX POWER FLOW THROUGH TRANSMISSION LINES

Specific expressions for the complex power flow on a line may be obtained in terms of the sending end and receiving end voltage magnitudes and phase angles and the $ABCD$ constants. Consider Figure 5.2 where the terminal relations are given by (5.5) and (5.6). Expressing the $ABCD$ constants in polar form as $A = |A|\angle\theta_A$,

$B = |B|\angle\theta_B$, the sending end voltage as $V_S = |V_S|\angle\delta$, and the receiving end voltage as reference $V_R = |V_R|\angle 0$, from (5.5) I_R can be written as

$$\begin{aligned} I_R &= \frac{|V_S|\angle\delta - |A|\angle\theta_A|V_R|\angle 0}{|B|\angle\theta_B} \\ &= \frac{|V_S|}{|B|}\angle\delta - \theta_B - \frac{|A||V_R|}{|B|}\angle\theta_A - \theta_B \end{aligned} \tag{5.81}$$

The receiving end complex power is

$$S_{R(3\phi)} = P_{R(3\phi)} + jQ_{R(3\phi)} = 3V_R I_R^* \tag{5.82}$$

Substituting for I_R from (5.81), we have

$$S_{R(3\phi)} = 3\frac{|V_S||V_R|}{|B|}\angle\theta_B - \delta - 3\frac{|A||V_R|^2}{|B|}\angle\theta_B - \theta_A \tag{5.83}$$

or in terms of the line-to-line voltages, we have

$$S_{R(3\phi)} = \frac{|V_{S(L-L)}||V_{R(L-L)}|}{|B|}\angle\theta_B - \delta - \frac{|A||V_{R(L-L)}|^2}{|B|}\angle\theta_B - \theta_A \tag{5.84}$$

The real and reactive power at the receiving end of the line are

$$P_{R(3\phi)} = \frac{|V_{S(L-L)}||V_{R(L-L)}|}{|B|}\cos(\theta_B - \delta) - \frac{|A||V_{R(L-L)}|^2}{|B|}\cos(\theta_B - \theta_A) \tag{5.85}$$

$$Q_{R(3\phi)} = \frac{|V_{S(L-L)}||V_{R(L-L)}|}{|B|}\sin(\theta_B - \delta) - \frac{|A||V_{R(L-L)}|^2}{|B|}\sin(\theta_B - \theta_A) \tag{5.86}$$

The sending end power is

$$S_{S(3\phi)} = P_{S(3\phi)} + jQ_{S(3\phi)} = 3V_S I_S^* \tag{5.87}$$

From (5.23), I_S can be written as

$$I_S = \frac{|A|\angle\theta_A|V_S|\angle\delta\ - |V_R|\angle 0}{|B|\angle\theta_B} \tag{5.88}$$

Substituting for I_S in (5.87) yields

$$P_{S(3\phi)} = \frac{|A||V_{S(L-L)}|^2}{|B|}\cos(\theta_B - \theta_A) - \frac{|V_{S(L-L)}||V_{R(L-L)}|}{|B|}\cos(\theta_B + \delta) \tag{5.89}$$

$$Q_{S(3\phi)} = \frac{|A||V_{S(L-L)}|^2}{|B|}\sin(\theta_B - \theta_A) - \frac{|V_{S(L-L)}||V_{R(L-L)}|}{|B|}\sin(\theta_B + \delta) \tag{5.90}$$

The real and reactive transmission line losses are

$$P_{L(3\phi)} = P_{S(3\phi)} - P_{R(3\phi)} \tag{5.91}$$

$$Q_{L(3\phi)} = Q_{S(3\phi)} - Q_{R(3\phi)} \tag{5.92}$$

The locus of all points obtained by plotting $Q_{R(3\phi)}$ versus $P_{R(3\phi)}$ for fixed line voltages and varying load angle δ is a circle known as the *receiving end power circle diagram.* A family of such circles with fixed receiving end voltage and varying sending end voltage is extremely useful in assessing the performance characteristics of the transmission line. A function called **pwrcirc(ABCD)** is developed for the construction of the receiving end power circle diagram, and its use is demonstrated in Example 5.9(g).

For a lossless line $B = jX'$, $\theta_A = 0$, $\theta_B = 90°$, and $A = \cos\beta\ell$, and the real power transferred over the line is given by

$$P_{3\phi} = \frac{|V_{S(L-L)}||V_{R(L-L)}|}{X'}\sin\delta \tag{5.93}$$

and the receiving end reactive power is

$$Q_{R3\phi} = \frac{|V_{S(L-L)}||V_{R(L-L)}|}{X'}\cos\delta - \frac{|V_{R(L-L)}|^2}{X'}\cos\beta\ell \tag{5.94}$$

For a given system operating at constant voltage, the power transferred is proportional to the sine of the power angle δ. As the load increases, δ increases. For a lossless line, the maximum power that can be transmitted under stable steady-state condition occurs for an angle of 90°. However, a transmission system with its connected synchronous machines must also be able to withstand, without loss of stability, sudden changes in generation, load, and faults. To assure an adequate margin of stability, the practical operating load angle is usually limited to 35 to 45°.

5.8 POWER TRANSMISSION CAPABILITY

The power handling ability of a line is limited by the thermal loading limit and the stability limit. The increase in the conductor temperature, due to the real power loss, stretches the conductors. This will increase the sag between transmission towers. At higher temperatures this may result in irreversible stretching. The thermal limit is specified by the current-carrying capacity of the conductor and is available in the manufacturer's data. If the current-carrying capacity is denoted by $I_{thermal}$, the thermal loading limit of a line is

$$S_{thermal} = 3V_{\phi rated}I_{thermal} \tag{5.95}$$

The expression for real power transfer over the line for a lossless line is given by (5.93). The theoretical maximum power transfer is when $\delta = 90°$. The practical operating load angle for the line alone is limited to no more than 30 to 45°. This is because of the generator and transformer reactances which, when added to the line, will result in a larger δ for a given load. For planning and other purposes, it is very useful to express the power transfer formula in terms of SIL, and construct the line loadability curve. For a lossless line $X' = Z_c \sin \beta\ell$, and (5.93) may be written as

$$P_{3\phi} = \left(\frac{|V_{S(L-L)}|}{V_{rated}}\right)\left(\frac{|V_{R(L-L)}|}{V_{rated}}\right)\left(\frac{V_{rated}^2}{Z_c}\right)\frac{\sin\delta}{\sin\beta\ell} \tag{5.96}$$

The first two terms within parenthesis are the per-unit voltages denoted by V_{Spu} and V_{Rpu}, and the third term is recognized as SIL. Equation (5.96) may be written as

$$\begin{aligned} P_{3\phi} &= \frac{|V_{Spu}||V_{Rpu}|SIL}{\sin\beta\ell}\sin\delta \\ &= \frac{|V_{Spu}||V_{Rpu}|SIL}{\sin(\frac{2\pi\ell}{\lambda})}\sin\delta \end{aligned} \tag{5.97}$$

The function **loadabil(L, C, f)** obtains the loadability curve and thermal limit curve of the line. The loadability curve as obtained in Figure 5.12 (page 182) for Example 5.9(i) shows that for short and medium lines the thermal limit dictates the maximum power transfer. Whereas, for longer lines the limit is set by the practical line loadability curve. As we see in the next section, for longer lines it may be necessary to use series capacitors in order to increase the power transfer over the line.

Example 5.6

A three-phase power of 700-MW is to be transmitted to a substation located 315 km from the source of power. For a preliminary line design assume the following parameters:

$V_S = 1.0$ per unit, $V_R = 0.9$ per unit, $\lambda = 5000$ km, $Z_c = 320\ \Omega$, and $\delta = 36.87°$

(a) Based on the practical line loadability equation determine a nominal voltage level for the transmission line.

(b) For the transmission voltage level obtained in (a) calculate the theoretical maximum power that can be transferred by the transmission line.

(a) From (5.61), the line phase constant is

$$\begin{aligned} \beta\ell &= \frac{2\pi}{\lambda}\ell \text{ rad} \\ &= \frac{360}{\lambda}\ell = \frac{360}{5000}(315) = 22.68° \end{aligned}$$

From the practical line loadability given by (5.97), we have

$$700 = \frac{(1.0)(0.9)(SIL)}{\sin(22.68°)}\sin(36.87°)$$

Thus

$$SIL = 499.83 \text{ MW}$$

From (5.78)

$$kV_L = \sqrt{(Z_c)(SIL)} = \sqrt{(320)(499.83)} = 400 \text{ kV}$$

(b) The equivalent line reactance for a lossless line is given by

$$X' = Z_c \sin\beta\ell = 320\sin(22.68) = 123.39 \ \Omega$$

For a lossless line, the maximum power that can be transmitted under steady state condition occurs for a load angle of 90°. Thus, from (5.93), assuming $|V_S| = 1.0$ pu and $|V_R| = 0.9$ pu, the theoretical maximum power is

$$P_{3\phi(max)} = \frac{(400)(0.9)(400)}{123.39}(1) = 1167 \text{ MW}$$

5.9 LINE COMPENSATION

We have noted that a transmission line loaded to its surge impedance loading has no net reactive power flow into or out of the line and will have approximately a flat voltage profile along its length. On long transmission lines, light loads appreciably less than SIL result in a rise of voltage at the receiving end, and heavy loads appreciably greater than SIL will produce a large dip in voltage. The voltage profile of a long line for various loading conditions is shown in Figure 5.11 (page 182). Shunt reactors are widely used to reduce high voltages under light load or open line conditions. If the transmission system is heavily loaded, shunt capacitors, static var control, and synchronous condensers are used to improve voltage, increase power transfer, and improve the system stability.

5.9.1 SHUNT REACTORS

Shunt reactors are applied to compensate for the undesirable voltage effects associated with line capacitance. The amount of reactor compensation required on a transmission line to maintain the receiving end voltage at a specified value can be obtained as follows.

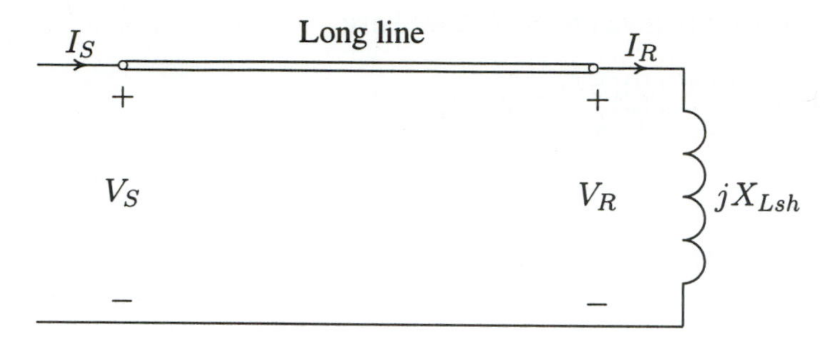

FIGURE 5.7
Shunt reactor compensation.

Consider a reactor of reactance X_{Lsh}, connected at the receiving end of a long transmission line as shown in Figure 5.7. The receiving end current is

$$I_R = \frac{V_R}{jX_{Lsh}} \tag{5.98}$$

Substituting I_R into (5.71) results in

$$V_S = V_R(\cos\beta\ell + \frac{Z_c}{X_{Lsh}}\sin\beta\ell)$$

Note that V_S and V_R are in phase, which is consistent with the fact that no real power is being transmitted over the line. Solving for X_{Lsh} yields

$$X_{Lsh} = \frac{\sin\beta\ell}{\frac{V_S}{V_R} - \cos\beta\ell} Z_c \tag{5.99}$$

For $V_S = V_R$, the required inductor reactance is

$$X_{Lsh} = \frac{\sin\beta\ell}{1 - \cos\beta\ell} Z_c \tag{5.100}$$

To find the relation between I_S and I_R, we substitute for V_R from (5.98) into (5.72)

$$I_S = \left(-\frac{1}{Z_c}\sin\beta\ell\, X_{Lsh} + \cos\beta\ell\right) I_R$$

Substituting for X_{Lsh} from (5.100) for the case when $V_S = V_R$ results in

$$I_S = -I_R \tag{5.101}$$

With one reactor only at the receiving end, the voltage profile will not be uniform, and the maximum rise occurs at the midspan. It is left as an exercise to show that for $V_S = V_R$, the voltage at the midspan is given by

$$V_m = \frac{V_R}{\cos\frac{\beta\ell}{2}} \tag{5.102}$$

Also, the current at the midspan is zero. The function **openline(ABCD)** is used to find the receiving end voltage of an open line and to determine the Mvar of the reactor required to maintain the no-load receiving end voltage at a specified value. Example 5.9(d) illustrates the reactor compensation. Installing reactors at both ends of the line will improve the voltage profile and reduce the tension at midspan.

Example 5.7

For the transmission line of Example 5.5:
(a) Calculate the receiving end voltage when line is terminated in an open circuit and is energized with 500 kV at the sending end.
(b) Determine the reactance and the Mvar of a three-phase shunt reactor to be installed at the receiving end to keep the no-load receiving end voltage at the rated value.

(a) The line is energized with 500 kV at the sending end. The sending end voltage per phase is

$$V_S = \frac{500\angle 0^\circ}{\sqrt{3}} = 288.675 \text{ kV}$$

From Example 5.5, $Z_c = 290.43$ and $\beta\ell = 21.641°$.

When the line is open $I_R = 0$ and from (5.71) the no-load receiving end voltage is given by

$$V_{R(nl)} = \frac{V_S}{\cos\beta\ell} = \frac{288.675}{0.9295} = 310.57 \text{ kV}$$

The no-load receiving end line-to-line voltage is

$$V_{R(L-L)(nl)} = \sqrt{3}\,V_{R(nl)} = 537.9 \text{ kV}$$

(b) For $V_S = V_R$, the required inductor reactance given by (5.100) is

$$X_{Lsh} = \frac{\sin(21.641^\circ)}{1 - \cos(21.641^\circ)}(290.43) = 1519.5 \ \Omega$$

The three-phase shunt reactor rating is

$$Q_{3\phi} = \frac{(kV_{Lrated})^2}{X_{Lsh}} = \frac{(500)^2}{1519.5} = 164.53 \text{ Mvar}$$

5.9.2 SHUNT CAPACITOR COMPENSATION

Shunt capacitors are used for lagging power factor circuits created by heavy loads. The effect is to supply the requisite reactive power to maintain the receiving end voltage at a satisfactory level. Capacitors are connected either directly to a bus bar or to the tertiary winding of a main transformer and are disposed along the route to minimize the losses and voltage drops. Given V_S and V_R, (5.85) and (5.86) can be used conveniently to compute the required capacitor Mvar at the receiving end for a specified load. A function called **shntcomp(ABCD)** is developed for this purpose, and its use is demonstrated in Example 5.9(f).

5.9.3 SERIES CAPACITOR COMPENSATION

Series capacitors are connected in series with the line, usually located at the mid-point, and are used to reduce the series reactance between the load and the supply point. This results in improved transient and steady-state stability, more economical loading, and minimum voltage dip on load buses. Series capacitors have the good characteristics that their reactive power production varies concurrently with the line loading. Studies have shown that the addition of series capacitors on EHV transmission lines can more than double the transient stability load limit of long lines at a fraction of the cost of a new transmission line.

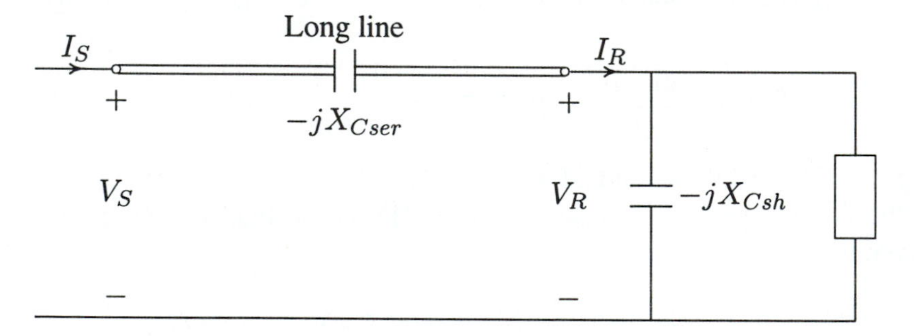

FIGURE 5.8
Shunt and series capacitor compensation.

With the series capacitor switched on as shown in Figure 5.8, from (5.93), the power transfer over the line for a lossless line becomes

$$P_{3\phi} = \frac{|V_{S(L-L)}||V_{R(L-L)}|}{X' - X_{Cser}} \sin \delta \qquad (5.103)$$

Where X_{Cser} is the series capacitor reactance. The ratio X_{Cser}/X' expressed as a percentage is usually referred to as the *percentage compensation*. The percentage compensation is in the range of 25 to 70 percent.

One major drawback with series capacitor compensation is that special protective devices are required to protect the capacitors and bypass the high current produced when a short circuit occurs. Also, inclusion of series capacitors establishes a resonant circuit that can oscillate at a frequency below the normal synchronous frequency when stimulated by a disturbance. This phenomenon is referred to as *subsynchronous resonance* (SSR). If the synchronous frequency minus the electrical resonant frequency approaches the frequency of one of the turbine-generator natural torsional modes, considerable damage to the turbine-generator may result. If L' is the lumped line inductance corrected for the effect of distribution and C_{ser} is the capacitance of the series capacitor, the subsynchronous resonant frequency is

$$f_r = f_s \sqrt{\frac{1}{L'C_{ser}}} \tag{5.104}$$

where f_s is the synchronous frequency. The function **sercomp(ABCD)** can be used to obtain the line performance for a specified percentage compensation. Finally, when line is compensated with both series and shunt capacitors, for the specified terminal voltages, the function **srshcomp(ABCD)** is used to obtain the line performance and the required shunt capacitor. These compensations are also demonstrated in Example 5.9(f).

Example 5.8

The transmission line in Example 5.5 supplies a load of 1000 MVA, 0.8 power factor lagging at 500 kV.

(a) Determine the Mvar and the capacitance of the shunt capacitors to be installed at the receiving end to keep the receiving end voltage at 500 kV when the line is energized with 500 kV at the sending end.

(b) Only series capacitors are installed at the midpoint of the line providing 40 percent compensation. Find the sending end voltage and voltage regulation.

(a) From Example 5.5, $Z_c = 290.43$ and $\beta\ell = 21.641°$. Thus, the equivalent line reactance for a lossless line is given by

$$X' = Z_c \sin\beta\ell = 290.43 \sin(21.641°) = 107.11 \ \Omega$$

The receiving end power is

$$S_{R(3\phi)} = 1000\angle\cos^{-1}(0.8) = 800 + j600 \ \text{MVA}$$

For the above operating condition, the power angle δ is obtained from (5.93)

$$800 = \frac{(500)(500)}{107.11} \sin\delta$$

which results in $\delta = 20.044°$. Using the approximate relation given by (5.94), the net reactive power at the receiving end is

$$Q_{R(3\phi)} = \frac{(500)(500)}{107.11}\cos(20.044°) - \frac{(500)^2}{107.11}\cos(21.641°) = 23.15 \text{ Mvar}$$

Thus, the required capacitor Mvar is $S_C = j23.15 - j600 = -j576.85$

The capacitive reactance is given by

$$X_C = \frac{|V_L|^2}{S_C^*} = \frac{(500)^2}{j576.85} = -j433.38 \ \Omega$$

or

$$C = \frac{10^6}{2\pi(60)(433.38)} = 6.1 \ \mu\text{F}$$

The shunt compensation for the above transmission line including the line resistance is obtained in Example 5.9(f) using the **lineperf** program. The exact solution results in 613.8 Mvar for capacitor reactive power as compared to 576.85 Mvar obtained from the approximate formula for the lossless line. This represents approximately an error of 6 percent.

(b) For 40 percent compensation, the series capacitor reactance per phase is

$$X_{ser} = 0.4X' = 0.4(107.1) = 42.84 \ \Omega$$

The new equivalent π circuit parameters are given by

$$Z' = j(X' - X_{ser}) = j(107.1 - 42.84) = j64.26 \ \Omega$$

$$Y' = = j\frac{2}{Z_c}\tan(\beta\ell/2) = j\frac{2}{290.43}\tan(21.641°/2) = j0.001316 \text{ siemens}$$

The new B constant is $B = j64.26$ and the new A constant is given by

$$A = 1 + \frac{Z'Y'}{2} = 1 + \frac{(j64.26)(j0.001316)}{2} = 0.9577$$

The receiving end voltage per phase is

$$V_R = \frac{500}{\sqrt{3}} = 288.675 \text{ kV}$$

and the receiving end current is

$$I_R = \frac{S^*_{R(3\phi)}}{3V_R^*} = \frac{1000\angle{-36.87°}}{3 \times 288.675\angle 0°} = 1.1547\angle{-36.87°} \text{ kA}$$

Thus, the sending end voltage is

$$\begin{aligned} V_S = AV_R + BI_R &= 0.9577 \times 288.675 + j64.26 \times 1.1547\angle{-36.87^\circ} \\ &= 326.4\angle 10.47^\circ \text{ kV} \end{aligned}$$

and the line-to-line voltage magnitude is $|V_{S(L-L)}| = \sqrt{3}\,V_S = 565.4$ kV. Voltage regulation is

$$\text{Percent } VR = \frac{565.4/0.958 - 500}{500} \times 100 = 18\%$$

The exact solution obtained in Example 5.9(f) results in $V_{S(L-L)} = 571.9$ kV. This represents an error of 1.0 percent.

5.10 LINE PERFORMANCE PROGRAM

A program called **lineperf** is developed for the complete analysis and compensation of a transmission line. The command **lineperf** displays a menu with five options for the computation of the parameters of the π models and the transmission constants. Selection of these options will call upon the following functions.

[Z, Y, ABCD] = rlc2abcd(r, L, C, g, f, Length) computes and returns the π model parameters and the transmission constants when r in ohm, L in mH, and C in μF per unit length, frequency, and line length are specified.

[Z, Y, ABCD] = zy2abcd(z, y, Length) computes and returns the π model parameters and the transmission constants when impedance and admittance per unit length are specified.

[Z, Y, ABCD] = pi2abcd(Z, Y) returns the ABCD constants when the π model parameters are specified.

[Z, Y, ABCD] = abcd2pi(A, B, C) returns the π model parameters when the transmission constants are specified.

[L , C] = gmd2lc computes and returns the inductance and capacitance per phase when the line configuration and conductor dimensions are specified.

[r, L, C, f] = abcd2rlc(ABCD) returns the line parameters per unit length and frequency when the transmission constants are specified.

Any of the above functions can be used independently when the arguments of the functions are defined in the *MATLAB* environment. If the above functions are typed without the parenthesis and the arguments, the user will be prompted to enter the required data. Next the **lineperf** loads the program **listmenu** which displays a list of eight options for transmission line analysis and compensation. Selection of these options will call upon the following functions.

givensr(ABCD) prompts the user to enter V_R, P_R and Q_R. This function computes V_S, P_S, Q_S, line losses, voltage regulation, and transmission efficiency.

givenss(ABCD) prompts the user to enter V_S, P_S and Q_S. This function computes V_R, P_R, Q_R, line losses, voltage regulation, and transmission efficiency.

givenzl(ABCD) prompts the user to enter V_R and the load impedance. This function computes V_S, P_S, Q_S, line losses, voltage regulation, and transmission efficiency.

openline(ABCD) prompts the user to enter V_S. This function computes V_R for the open-ended line. Also, the reactance and the Mvar of the necessary reactor to maintain the receiving end voltage at a specified value are obtained. In addition, the function plots the voltage profile of the line.

shcktlin(ABCD) prompts the user to enter V_S. This function computes the current at both ends of the line for a solid short circuit at the receiving end.

Option 6 is for capacitive compensation and calls upon **compmenu** which displays three options. Selection of these options will call upon the following functions.

shntcomp(ABCD) prompts the user to enter V_S, P_R, Q_R and the desired V_R. This function computes the capacitance and the Mvar of the shunt capacitor bank to be installed at the receiving end in order to maintain the specified V_R. Then, V_S, P_S, Q_S, line losses, voltage regulation, and transmission efficiency are found.

sercomp(ABCD) prompts the user to enter V_R, P_R, Q_R, power, and the percentage compensation (i.e., $X_{Cser}/X_{line} \times 100$). This function computes the Mvar of the specified series capacitor and V_S, P_S, Q_S, line losses, voltage regulation, and transmission efficiency for the compensated line.

srshcomp(ABCD) prompts the user to enter V_S, P_R, Q_R, the desired V_R and the percentage series capacitor compensation. This function computes the capacitance and the Mvar of a shunt capacitor to be installed at the receiving end in order to maintain the specified V_R. Also, V_S, P_S, Q_S, line losses, voltage regulation, and transmission efficiency are obtained for the compensated line.

Option 7 loads the **pwrcirc(ABCD)** which prompts for the receiving end voltage. This function constructs the receiving end power circle diagram for various values of V_S from V_R up to $1.3V_R$.

Option 8 calls upon **profmenu** which displays two options. Selection of these options will call upon the following functions:

vprofile(r, L, C, f) prompts the user to enter V_S, rated MVA, power factor, V_R, P_R, and Q_R. This function displays a graph consisting of voltage profiles for line length up to $1/8$ of the line wavelength for the following cases: open-ended line, line terminated in SIL, short-circuited line, and full-load.

loadabil(L, C, f) prompts the user for V_S, V_R, rated line voltage, and current-carrying capacity of the line. This function displays a graph consisting of the practical line loadability curve for $\delta = 30°$, the theoretical stability limit curve, and the thermal limit. This function assumes a lossless line and the plots are obtained for a line length up to $1/4$ of the line wavelength.

Any of the above functions can be used independently when the arguments of the functions are defined in the *MATLAB* environment. The $ABCD$ constant is entered as a matrix. If the above functions are typed without the parenthesis and the arguments, the user will be prompted to enter the required data.

Example 5.9

A three-phase, 60-Hz, 550-kV transmission line is 300 km long. The line parameters per phase per unit length are found to be

$$r = 0.016\ \Omega/\text{km} \quad L = 0.97\ \text{mH/km} \quad C = 0.0115\ \mu\text{F/km}$$

(a) Determine the line performance when load at the receiving end is 800 MW, 0.8 power factor lagging at 500 kV.

The command:

```
lineperf
```

displays the following menu

```
Type of parameters for input                           Select
Parameters per unit length
r (Ω), g (siemens), L (mH), C (μF)                       1

Complex z and y per unit length
r + j*x (Ω), g + j*b (siemens)                           2

Nominal π or Eq. π model                                 3

A, B, C, D constants                                     4

Conductor configuration and dimension                    5

To quit                                                  0
Select number of menu → 1
Enter line length = 300
Enter frequency in Hz = 60
Enter line resistance/phase in Ω/unit length, r = 0.016
Enter line inductance in mH per unit length, L = 0.97
Enter line capacitance in μF per unit length, C = .0115
Enter line conductance in siemens per unit length, g = 0
Enter 1 for medium line or 2 for long line → 2

Equivalent π model
Z' = 4.57414 + j 107.119 ohms
Y' = 6.9638e-07 + j 0.00131631 siemens
Zc = 290.496 + j -6.35214 ohms
αℓ = 0.00826172 neper βℓ = 0.377825 radian = 21.6478°
```

$$\text{ABCD} = \begin{bmatrix} 0.9295 & + & j0.0030478 & 4.5741 & + & j107.12 \\ -1.3341e-06 & + & j0.0012699 & 0.9295 & + & j0.0030478 \end{bmatrix}$$

At this point the program **listmenu** is automatically loaded and displays the following menu.

```
         Transmission line performance
                 Analysis                              Select

To calculate sending end quantities
for specified receiving end MW, Mvar                     1
```

```
To calculate receiving end quantities
for specified sending end MW, Mvar                        2

To calculate sending end quantities
when load impedance is specified                          3

Open-end line and reactive compensation                   4

Short-circuited line                                      5

Capacitive compensation                                   6

Receiving end circle diagram                              7

Loadability curve and voltage profile                     8


To quit                                                   0

Select number of menu → 1

Enter receiving end line-line voltage kV = 500
Enter receiving end voltage phase angle° = 0
Enter receiving end 3-phase power MW = 800
Enter receiving end 3-phase reactive power
(+ for lagging and - for leading power factor) Mvar = 600
```

<u>Line performance for specified receiving end quantities</u>

```
Vr = 500 kV (L-L) at 0°
Pr = 800 MW Qr = 600 Mvar
Ir = 1154.7 A at -36.8699° PFr = 0.8 lagging
Vs = 623.511 kV (L-L) at 15.5762°
Is = 903.113 A at -17.6996°, PFs = 0.836039 lagging
Ps = 815.404 MW,      Qs = 535.129 Mvar
PL = 15.4040 MW,      QL = -64.871 Mvar
Percent Voltage Regulation = 34.1597
Transmission line efficiency = 98.1108
```

At the end of this analysis the **listmenu** (Analysis Menu) is displayed.

(b) Determine the receiving end quantities and the line performance when 600 MW and 400 Mvar are being transmitted at 525 kV from the sending end.

Selecting option 2 of the **listmenu** results in

```
Enter sending end line-line voltage kV = 525
Enter sending end voltage phase angle° = 0
Enter sending end 3-phase power MW = 600
Enter sending end 3-phase reactive power
(+ for lagging and - for leading power factor) Mvar = 400

Line performance for specified sending end quantities

Vs = 525 kV (L-L) at 0°
Ps = 600 MW,      Qs = 400 Mvar
Is = 793.016 A at -33.6901°, PFs = 0.83205 lagging
Vr = 417.954 kV (L-L) at -16.3044°
Ir = 1002.6 A at -52.16° PFr = 0.810496 lagging
Pr = 588.261 MW,     Qr = 425.136 Mvar
PL = 11.7390 MW,     QL = -25.136 Mvar
Percent Voltage Regulation = 35.1383
Transmission line efficiency = 98.0435
```

(c) Determine the sending end quantities and the line performance when the receiving end load impedance is 290 Ω at 500 kV.

Selecting option 3 of the listmenu results in

```
Enter receiving end line-line voltage kV = 500
Enter receiving end voltage phase angle° = 0
Enter sending end complex load impedance 290 + j * 0

Line performance for specified load impedance

Vr = 500 kV (L-L) at 0°
Ir = 995.431 A at 0° PFr = 1
Pr = 862.069 MW,     Qr = 0 Mvar
Vs = 507.996 kV (L-L) at 21.5037°
Is = 995.995 A at 21.7842°, PFs = 0.999988 leading
Ps = 876.341 MW Qs = -4.290 Mvar
PL = 14.272 MW QL = -4.290 Mvar
Percent Voltage Regulation = 9.30464
Transmission line efficiency = 98.3714
```

(d) Find the receiving end voltage when the line is terminated in an open circuit and is energized with 500 kV at the sending end. Also, determine the reactance and the Mvar of a three-phase shunt reactor to be installed at the receiving end in order to limit the no-load receiving end voltage to 500 kV.

Selecting option 4 of the **listmenu** results in

```
Enter sending end line-line voltage kV = 500
Enter sending end voltage phase angle° = 0

Open line and shunt reactor compensation

Vs = 500 kV (L-L) at 0°
Vr = 537.92 kV (L-L) at -0.00327893°
Is = 394.394 A at 89.8723°, PFs = 0.0022284 leading
Desired no load receiving end voltage = 500 kV
Shunt reactor reactance = 1519.4 Ω
Shunt reactor rating = 164.538 Mvar
```

The voltage profile for the uncompensated and the compensated line is also found as shown in Figure 5.9.

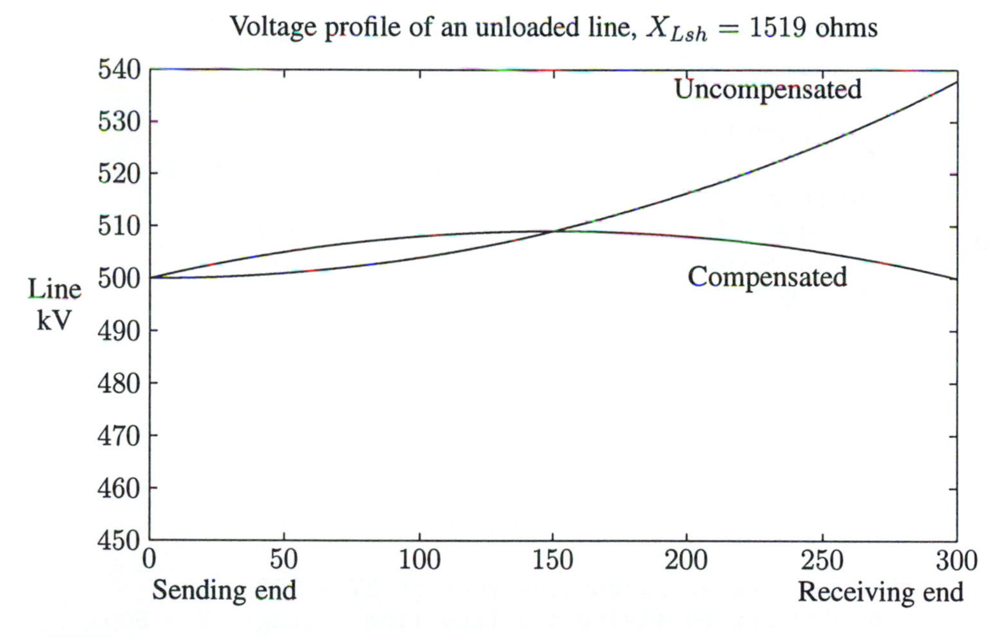

FIGURE 5.9
Compensated and uncompensated voltage profile of open-ended line.

(e) Find the receiving end and the sending end currents when the line is terminated in a short circuit.

Selecting option 5 of the **listmenu** results in

```
Enter sending end line-line voltage kV = 500
Enter sending end voltage phase angle° = 0

Line short-circuited at the receiving end

Vs = 500 kV (L-L) at 0°
Ir = 2692.45 A at -87.5549°
Is = 2502.65 A at -87.367°
```

(f) The line loading in part (a) resulted in a voltage regulation of 34.16 percent, which is unacceptably high. To improve the line performance, the line is compensated with series and shunt capacitors. For the loading condition in (a):

(1) Determine the Mvar and the capacitance of the shunt capacitors to be installed at the receiving end to keep the receiving end voltage at 500 kV when the line is energized with 500 kV at the sending end.

Selecting option 6 will display the **compmenu** as follows:

```
           Capacitive compensation
                 Analysis                      Select

Shunt capacitive compensation                    1

Series capacitive compensation                   2

Series and shunt capacitive compensation         3

To quit                                          0
```

Selecting option 1 of the **compmenu** results in

```
Enter sending end line-line voltage kV = 500
Enter desired receiving end line-line voltage kV = 500
Enter receiving end voltage phase angle° = 0
Enter receiving end 3-phase power MW = 800
Enter receiving end 3-phase reactive power
(+ for lagging and - for leading power factor) Mvar = 600
```

```
Shunt capacitive compensation

Vs = 500 kV (L-L) at 20.2479°
Vr = 500 kV (L-L) at 0°
Pload = 800 MW,     Qload = 600 Mvar
Load current = 1154.7 A at -36.8699°, PFl = 0.8 lagging
Required shunt capacitor: 407.267 Ω, 6.51314 μF,613.849 Mvar
Shunt capacitor current = 708.811 A at 90°
Pr = 800.000 MW,     Qr = -13.849 Mvar
Ir = 923.899 A at 0.991732°, PFr = 0.99985 leading
Is = 940.306 A at 24.121° PFs = 0.997716 leading
Ps = 812.469 MW,     Qs = -55.006 Mvar
PL = 12.469 MW,     QL = -41.158 Mvar
Percent Voltage Regulation = 7.58405
Transmission line efficiency = 98.4653
```

(2) Determine the line performance when the line is compensated by series capacitors for 40 percent compensation with the load condition in (a) at 500 kV.

Selecting option 2 of the **compmenu** results in

```
Enter receiving end line-line voltage kV = 500
Enter receiving end voltage phase angle° = 0
Enter receiving end 3-phase power MW = 800
Enter receiving end 3-phase reactive power
(+ for lagging and - for leading power factor) Mvar = 600
Enter percent compensation for series capacitor
(Recommended range 25 to 75% of the line reactance) = 40

Series capacitor compensation

Vr = 500 kV (L-L) at 0°
Pr = 800 MW,     Qr = 600 Mvar
Required series capacitor: 42.8476 Ω, 61.9074 μF, 47.4047 Mvar
Subsynchronous resonant frequency = 37.9473 Hz
Ir = 1154.7 A at -36.8699°, PFr = 0.8 lagging
Vs = 571.904 kV (L-L) at 9.95438°
Is = 932.258 A at -18.044°, PFs = 0.882961 lagging
Ps = 815.383 MW,     Qs = 433.517 Mvar
PL = 15.383 MW,     QL = -166.483 Mvar
Percent Voltage Regulation = 19.4322
Transmission line efficiency = 98.1134
```

(3) The line has 40 percent series capacitor compensation and supplies the load in (a). Determine the Mvar and the capacitance of the shunt capacitors to be installed at the receiving end to keep the receiving end voltage at 500 kV when line is energized with 500 kV at the sending end.

Selecting option 3 of the **compmenu** results in

```
Enter sending end line-line voltage kV = 500
Enter desired receiving end line-line voltage kV = 500
Enter receiving end voltage phase angle° = 0
Enter receiving end 3-phase power MW = 800
Enter receiving end 3-phase reactive power
(+ for lagging and - for leading power factor) Mvar = 600
Enter percent compensation for series capacitor
(Recommended range 25 to 75% of the line reactance) = 40
```

```
Series and shunt capacitor compensation

Vs = 500 kV (L-L) at 12.0224°
Vr = 500 kV (L-L) at 0°
Pload = 800 MW,      Qload = 600 Mvar
Load current = 1154.7 A at -36.8699°, PFl = 0.8 lagging
Required shunt capacitor: 432.736 Ω, 6.1298 μF, 577.72 Mvar
Shunt capacitor current = 667.093 A at 90°
Required series capacitor: 42.8476 Ω, 61.9074 μF,37.7274 Mvar
Subsynchronous resonant frequency = 37.9473 Hz
Pr = 800 MW,     Qr = 22.2804 Mvar
Ir = 924.119 A at -1.5953°, PFr = 0.999612 lagging
Is = 951.165 A at 21.5977°, PFs = 0.986068 leading
Ps = 812.257 MW,     Qs = -137.023 Mvar
PL = 12.257 MW,      QL = -159.304 Mvar
Percent Voltage Regulation = 4.41619
Transmission line efficiency = 98.491
```

(g) Construct the receiving end circle diagram.

Selecting option 7 of the **listmenu** results in

```
Enter receiving end line-line voltage kV = 500
```

A plot of the receiving end circle diagram is obtained as shown in Figure 5.10.

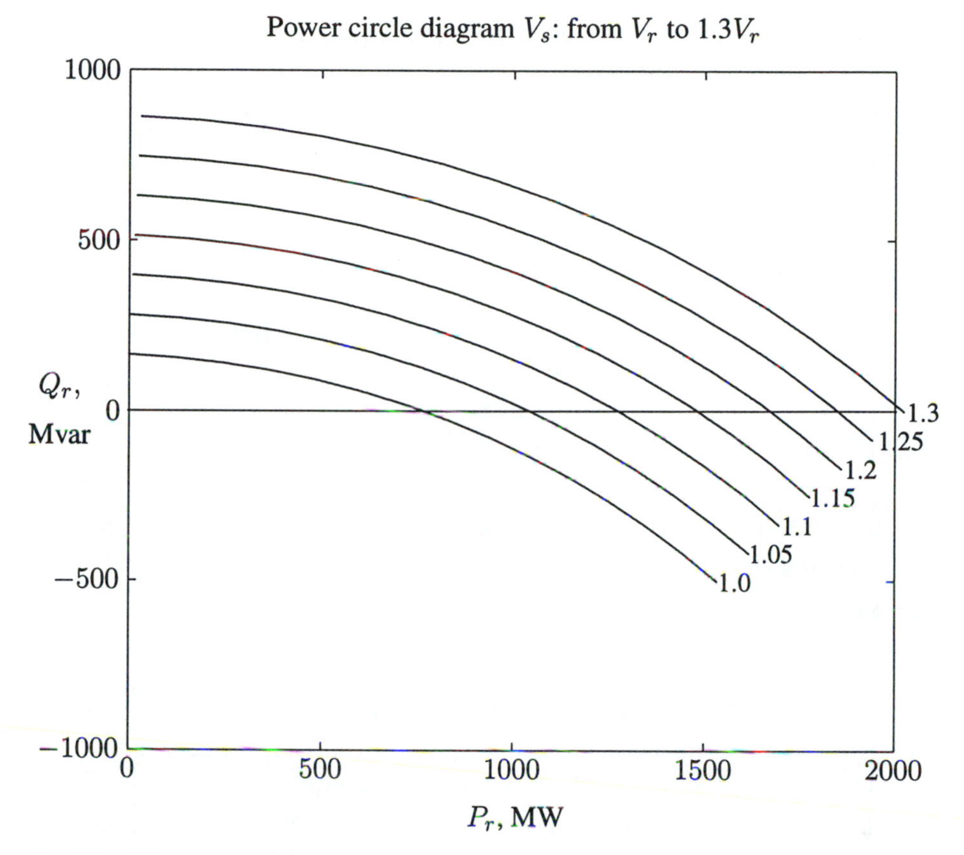

FIGURE 5.10
Receiving end circle diagram.

(h) Determine the line voltage profile for the following cases: no-load, rated load, line terminated in the SIL, and short-circuited line.

Selecting option 8 of the **listmenu** results in

```
         Voltage profile and line loadability
                Analysis                          Select

     Voltage profile curves                         1
     Line loadability curve                         2
     To quit                                        0
```

Selecting option 1 of the **profmenu** results in

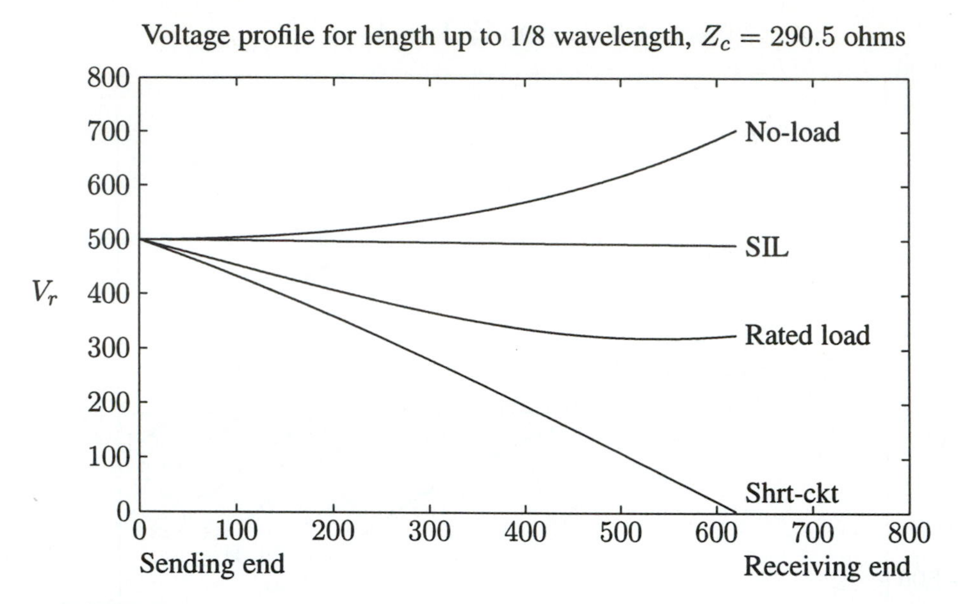

FIGURE 5.11
Voltage profile for length up to 1/8 wavelength.

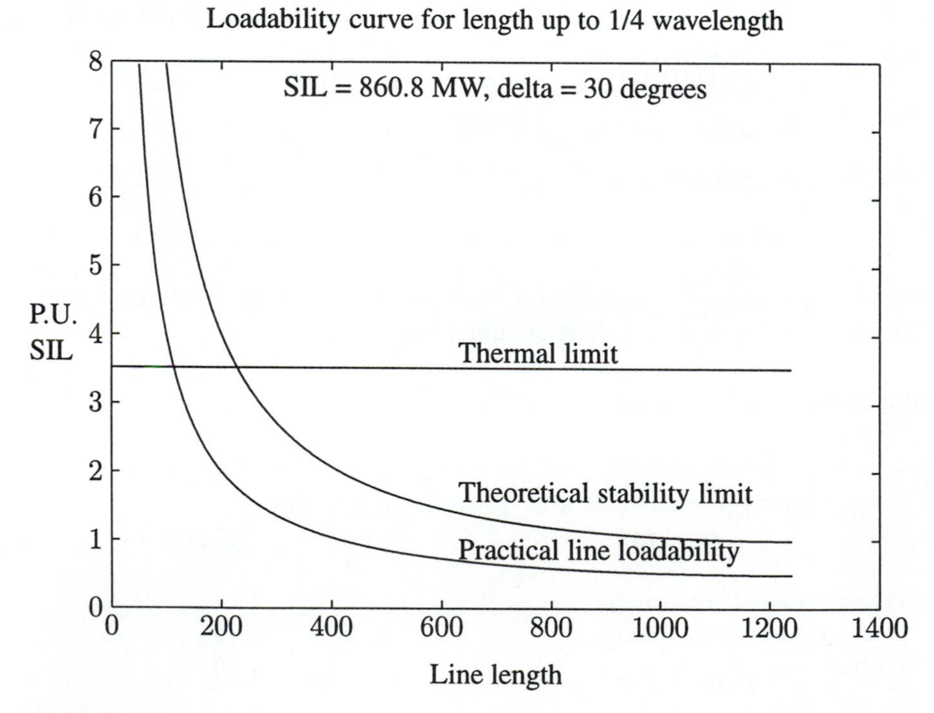

FIGURE 5.12
Line loadability curve for length up to 1/4 wavelength.

```
Enter sending end line-line voltage kV = 500
Enter rated sending end power, MVA = 1000
Enter power factor = 0.8
```

A plot of the voltage profile is obtained as shown in Figure 5.11 (page 182).

(i) Obtain the line loadability curves.
Selecting option 2 of the **profmenu** results in

```
Enter sending end line-line voltage kV = 500
Enter receiving end line-line voltage kV = 500
Enter rated line-line voltage kV = 500
Enter line current-carrying capacity, Amp/phase = 3500
```

The line loadability curve is obtained as shown in Figure 5.12 (page 182).

PROBLEMS

5.1. A 69-kV, three-phase short transmission line is 16 km long. The line has a per phase series impedance of $0.125 + j0.4375\ \Omega$ per km. Determine the sending end voltage, voltage regulation, the sending end power, and the transmission efficiency when the line delivers

(a) 70 MVA, 0.8 lagging power factor at 64 kV.
(b) 120 MW, unity power factor at 64 kV.

Use **lineperf** program to verify your results.

5.2. Shunt capacitors are installed at the receiving end to improve the line performance of Problem 5.1. The line delivers 70 MVA, 0.8 lagging power factor at 64 kV. Determine the total Mvar and the capacitance per phase of the Y-connected capacitors when the sending end voltage is

(a) 69 kV.
(b) 64 kV.

Hint: Use (5.85) and (5.86) to compute the power angle δ and the receiving end reactive power.

(c) Use **lineperf** to obtain the compensated line performance.

5.3. A 230-kV, three-phase transmission line has a per phase series impedance of $z = 0.05 + j0.45\ \Omega$ per km and a per phase shunt admittance of $y = j3.4 \times 10^{-6}$ siemens per km. The line is 80 km long. Using the nominal π model, determine

(a) The transmission line ABCD constants.

Find the sending end voltage and current, voltage regulation, the sending end power and the transmission efficiency when the line delivers

(b) 200 MVA, 0.8 lagging power factor at 220 kV.
(c) 306 MW, unity power factor at 220 kV.

Use **lineperf** program to verify your results.

5.4. Shunt capacitors are installed at the receiving end to improve the line performance of Problem 5.3. The line delivers 200 MVA, 0.8 lagging power factor at 220 kV.

(a)Determine the total Mvar and the capacitance per phase of the Y-connected capacitors when the sending end voltage is 220 kV. *Hint*: Use (5.85) and (5.86) to compute the power angle δ and the receiving end reactive power.
(b) Use **lineperf** to obtain the compensated line performance.

5.5. A three-phase, 345-kV, 60-Hz transposed line is composed of two ACSR, 1,113,000-cmil, 45/7 Bluejay conductors per phase with flat horizontal spacing of 11 m. The conductors have a diameter of 3.195 cm and a *GMR* of 1.268 cm. The bundle spacing is 45 cm. The resistance of each conductor in the bundle is 0.0538 Ω per km and the line conductance is negligible. The line is 150 km long. Using the nominal π model, determine the ABCD constant of the line. Use **lineperf** and option 5 to verify your results.

5.6. The ABCD constants of a three-phase, 345-kV transmission line are

$$A = D = 0.98182 + j0.0012447$$
$$B = 4.035 + j58.947$$
$$C = j0.00061137$$

The line delivers 400 MVA at 0.8 lagging power factor at 345 kV. Determine the sending end quantities, voltage regulation, and transmission efficiency.

5.7. Write a *MATLAB* function named **[ABCD] = abcdm(z, y, Lngt)** to evaluate and return the ABCD transmission matrix for a medium-length transmission line where **z** is the per phase series impedance per unit length, **y** is the shunt admittance per unit length, and **Lngt** is the line length. Then, write a program that uses the above function and computes the receiving end quantities, voltage regulation, and the line efficiency when sending end quantities are specified. The program should prompt for the following quantities:

The sending end line-to-line voltage magnitude in kV
The sending end voltage phase angle in degrees

The three-phase sending end real power in MW
The three-phase sending end reactive power in Mvar

Use your program to obtain the solution for the following case.

A three-phase transmission line has a per phase series impedance of $z = 0.03 + j0.4\ \Omega$ per km and a per phase shunt admittance of $y = j4.0 \times 10^{-6}$ siemens per km. The line is 125 km long. Obtain the ABCD transmission matrix. Determine the receiving end quantities, voltage regulation, and the line efficiency when the line is sending 407 MW, 7.833 Mvar at 350 kV.

5.8. Obtain the solution for Problems 5.8 through 5.13 using the **lineperf** program. Then, solve each problem using hand calculations.

A three-phase, 765-kV, 60-Hz transposed line is composed of four ACSR, 1,431,000-cmil, 45/7 Bobolink conductors per phase with flat horizontal spacing of 14 m. The conductors have a diameter of 3.625 cm and a GMR of 1.439 cm. The bundle spacing is 45 cm. The line is 400 km long, and for the purpose of this problem, a lossless line is assumed.

(a) Determine the transmission line surge impedance Z_c, phase constant β, wavelength λ, the surge impedance loading SIL, and the ABCD constant.
(b) The line delivers 2000 MVA at 0.8 lagging power factor at 735 kV. Determine the sending end quantities and voltage regulation.
(c) Determine the receiving end quantities when 1920 MW and 600 Mvar are being transmitted at 765 kV at the sending end.
(d) The line is terminated in a purely resistive load. Determine the sending end quantities and voltage regulation when the receiving end load resistance is 264.5 Ω at 735 kV.

5.9. The transmission line in Problem 5.8 is energized with 765 kV at the sending end when the load at the receiving end is removed.

(a) Find the receiving end voltage.
(b) Determine the reactance and the Mvar of a three-phase shunt reactor to be installed at the receiving end in order to limit the no-load receiving end voltage to 735 kV.

5.10. The transmission line in Problem 5.8 is energized with 765 kV at the sending end when a three-phase short-circuit occurs at the receiving end. Determine the receiving end current and the sending end current.

5.11. Shunt capacitors are installed at the receiving end to improve the line performance of Problem 5.8. The line delivers 2000 MVA, 0.8 lagging power factor. Determine the total Mvar and the capacitance per phase of the Y-connected capacitors to keep the receiving end voltage at 735 kV when the sending end voltage is 765 kV. *Hint*: Use (5.93) and (5.94) to compute the power angle δ and the receiving end reactive power. Find the sending end quantities and voltage regulation for the compensated line.

5.12. Series capacitors are installed at the midpoint of the line in Problem 5.8, providing 40 percent compensation. Determine the sending end quantities and the voltage regulation when the line delivers 2000 MVA at 0.8 lagging power factor at 735 kV.

5.13. Series capacitors are installed at the midpoint of the line in Problem 5.8, providing 40 percent compensation. In addition, shunt capacitors are installed at the receiving end. The line delivers 2000 MVA, 0.8 lagging power factor. Determine the total Mvar and the capacitance per phase of the series and shunt capacitors to keep the receiving end voltage at 735 kV when the sending end voltage is 765 kV. Find the sending end quantities and voltage regulation for the compensated line.

5.14. The transmission line in Problem 5.8 has a per phase resistance of 0.011 Ω per km. Using the **lineperf** program, perform the following analysis and present a summary of the calculation along with your conclusions and recommendations.

(a) Determine the sending end quantities for the specified receiving end quantities of $735\angle0°$, 1600 MW, 1200 Mvar.
(b) Determine the receiving end quantities for the specified sending end quantities of $765\angle0°$, 1920 MW, 600 Mvar.
(c) Determine the sending end quantities for a load impedance of $282.38 + j0\ \Omega$ at 735 kV.
(d) Find the receiving end voltage when the line is terminated in an open circuit and is energized with 765 kV at the sending end. Also, determine the reactance and the Mvar of a three-phase shunt reactor to be installed at the receiving end in order to limit the no-load receiving end voltage to 765 kV. Obtain the voltage profile for the uncompensated and the compensated line.
(e) Find the receiving end and the sending end current when the line is terminated in a three-phase short circuit.
(f) For the line loading of part (a), determine the Mvar and the capacitance of the shunt capacitors to be installed at the receiving end to keep the receiving end voltage at 735 kV when line is energized with 765 kV. Obtain the line performance of the compensated line.
(g) Determine the line performance when the line is compensated by series capacitor for 40 percent compensation with the load condition in part (a) at 735 kV.

(h) The line has 40 percent series capacitor compensation and supplies the load in part (a). Determine the Mvar and the capacitance of the shunt capacitors to be installed at the receiving end to keep the receiving end voltage at 735 kV when line is energized with 765 kV at the sending end.
(i) Obtain the receiving end circle diagram.
(j) Obtain the line voltage profile for a sending end voltage of 765 kV.
(k) Obtain the line loadability curves when the sending end voltage is 765 kV, and the receiving end voltage is 735 kV. The current-carrying capacity of the line is 5000 A per phase.

5.15. The ABCD constants of a lossless three-phase, 500-kV transmission line are

$$A = D = 0.86 + j0$$
$$B = 0 + j130.2$$
$$C = j0.002$$

(a) Obtain the sending end quantities and the voltage regulation when line delivers 1000 MVA at 0.8 lagging power factor at 500 kV.

To improve the line performance, series capacitors are installed at both ends in each phase of the transmission line. As a result of this, the compensated ABCD constants become

$$\begin{bmatrix} A' & B' \\ C' & D' \end{bmatrix} = \begin{bmatrix} 1 & -\frac{1}{2}jX_c \\ 0 & 1 \end{bmatrix} \begin{bmatrix} A & B \\ C & D \end{bmatrix} \begin{bmatrix} 1 & -\frac{1}{2}jX_c \\ 0 & 1 \end{bmatrix}$$

where X_c is the total reactance of the series capacitor. If $X_c = 100\ \Omega$

(b) Determine the compensated ABCD constants.
(c) Determine the sending end quantities and the voltage regulation when line delivers 1000 MVA at 0.8 lagging power factor at 500 kV.

5.16. A three-phase 420-kV, 60-HZ transmission line is 463 km long and may be assumed lossless. The line is energized with 420 kV at the sending end. When the load at the receiving end is removed, the voltage at the receiving end is 700 kV, and the per phase sending end current is $646.6\angle 90°$ A.

(a) Find the phase constant β in radians per km and the surge impedance Z_c in Ω.
(b) Ideal reactors are to be installed at the receiving end to keep $|V_S| = |V_R| = 420$ kV when load is removed. Determine the reactance per phase and the required three-phase kvar.

5.17. A three-phase power of 3600 MW is to be transmitted via four identical 60-Hz transmission lines for a distance of 300 km. From a preliminary line design, the line phase constant and surge impedance are given by $\beta = 9.46 \times 10^{-4}$ radian/km and $Z_c = 343\ \Omega$, respectively.

Based on the practical line loadability criteria determine the suitable nominal voltage level in kV for each transmission line. Assume $V_S = 1.0$ per unit, $V_R = 0.9$ per unit, and the power angle $\delta = 36.87°$.

5.18. Power system studies on an existing system have indicated that 2400 MW are to be transmitted for a distance of 400 km. The voltage levels being considered include 345 kV, 500 kV, and 765 kV. For a preliminary design based on the practical line loadability, you may assume the following surge impedances

345 kV $Z_C = 320\ \Omega$
500 kV $Z_C = 290\ \Omega$
765 kV $Z_C = 265\ \Omega$

The line wavelength may be assumed to be 5000 km. The practical line loadability may be based on a load angle δ of 35°. Assume $|V_S| = 1.0$ pu and $|V_R| = 0.9$ pu. Determine the number of three-phase transmission circuits required for each voltage level. Each transmission tower may have up to two circuits. To limit the corona loss, all 500-kV lines must have at least two conductors per phase, and all 765-kV lines must have at least four conductors per phase. The bundle spacing is 45 cm. The conductor size should be such that the line would be capable of carrying current corresponding to at least 5000 MVA. Use **acsr** command in *MATLAB* to find a suitable conductor size. Following are the minimum recommended spacings between adjacent phase conductors at various voltage levels.

Voltage level, kV	Spacing meter
345	7.0
500	9.0
765	12.5

(a) Select a suitable voltage level, and conductor size, and tower structure. Use **lineperf** program and option 1 to obtain the voltage regulation and transmission efficiency based on a receiving end power of 3000 MVA at 0.8 power factor lagging at the selected rated voltage. Modify your design and select a conductor size for a line efficiency of at least 94 percent for the above specified load.

(b) Obtain the line performance including options 4–8 of the **lineperf** program for your final selection. Summarize the line characteristics and the required line compensation.

CHAPTER

6

POWER FLOW ANALYSIS

6.1 INTRODUCTION

In the previous chapters, modeling of the major components of an electric power system was discussed. This chapter deals with the steady-state analysis of an interconnected power system during normal operation. The system is assumed to be operating under balanced condition and is represented by a single-phase network. The network contains hundreds of nodes and branches with impedances specified in per unit on a common MVA base.

Network equations can be formulated systematically in a variety of forms. However, the node-voltage method, which is the most suitable form for many power system analyses, is commonly used. The formulation of the network equations in the nodal admittance form results in complex linear simultaneous algebraic equations in terms of node currents. When node currents are specified, the set of linear equations can be solved for the node voltages. However, in a power system, powers are known rather than currents. Thus, the resulting equations in terms of power, known as the *power flow equation*, become nonlinear and must be solved by iterative techniques. Power flow studies, commonly referred to as *load flow*, are the backbone of power system analysis and design. They are necessary for planning, operation, economic scheduling and exchange of power between utilities. In addition, power flow analysis is required for many other analyses such as transient stability and contingency studies.

In this chapter, the bus admittance matrix of the node-voltage equation is formulated, and a *MATLAB* function named **ybus** is developed for the systematic formation of the bus admittance matrix. Next, two commonly used iterative techniques, namely Gauss-Seidel and Newton-Raphson methods for the solution of nonlinear algebraic equations, are discussed. These techniques are employed in the solution of power flow problems. Three programs **lfgauss**, **lfnewton**, and **decouple** are developed for the solution of power flow problems by Gauss-Seidel, Newton-Raphson, and the fast decoupled power flow, respectively.

6.2 BUS ADMITTANCE MATRIX

In order to obtain the node-voltage equations, consider the simple power system shown in Figure 6.1 where impedances are expressed in per unit on a common MVA base and for simplicity resistances are neglected. Since the nodal solution is based upon Kirchhoff's current law, impedances are converted to admittance, i.e.,

$$y_{ij} = \frac{1}{z_{ij}} = \frac{1}{r_{ij} + jx_{ij}}$$

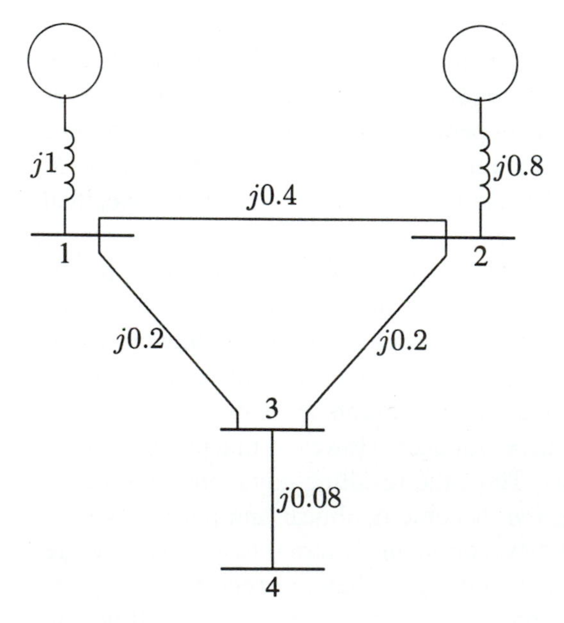

FIGURE 6.1
The impedance diagram of a simple system.

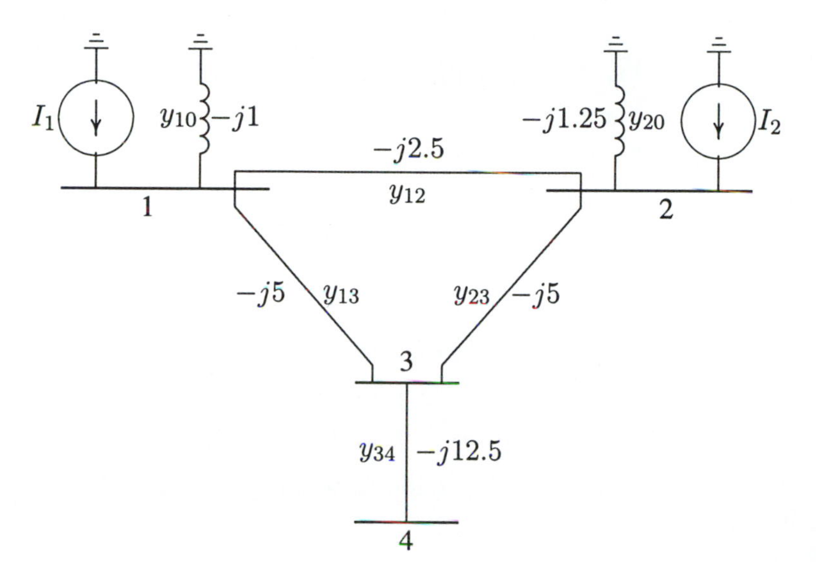

FIGURE 6.2
The admittance diagram for system of Figure 6.1.

The circuit has been redrawn in Figure 6.2 in terms of admittances and transformation to current sources. Node 0 (which is normally ground) is taken as reference. Applying KCL to the independent nodes 1 through 4 results in

$$
\begin{aligned}
I_1 &= y_{10}V_1 + y_{12}(V_1 - V_2) + y_{13}(V_1 - V_3) \\
I_2 &= y_{20}V_2 + y_{12}(V_2 - V_1) + y_{23}(V_2 - V_3) \\
0 &= y_{23}(V_3 - V_2) + y_{13}(V_3 - V_1) + y_{34}(V_3 - V_4) \\
0 &= y_{34}(V_4 - V_3)
\end{aligned}
$$

Rearranging these equations yields

$$
\begin{aligned}
I_1 &= (y_{10} + y_{12} + y_{13})V_1 - y_{12}V_2 - y_{13}V_3 \\
I_2 &= -y_{12}V_1 + (y_{20} + y_{12} + y_{23})V_2 - y_{23}V_3 \\
0 &= -y_{13}V_1 - y_{23}V_2 + (y_{13} + y_{23} + y_{34})V_3 - y_{34}V_4 \\
0 &= -y_{34}V_3 + y_{34}V_4
\end{aligned}
$$

We introduce the following admittances

$$
\begin{aligned}
Y_{11} &= y_{10} + y_{12} + y_{13} \\
Y_{22} &= y_{20} + y_{12} + y_{23}
\end{aligned}
$$

$$Y_{33} = y_{13} + y_{23} + y_{34}$$
$$Y_{44} = y_{34}$$
$$Y_{12} = Y_{21} = -y_{12}$$
$$Y_{13} = Y_{31} = -y_{13}$$
$$Y_{23} = Y_{32} = -y_{23}$$
$$Y_{34} = Y_{43} = -y_{34}$$

The node equation reduces to

$$I_1 = Y_{11}V_1 + Y_{12}V_2 + Y_{13}V_3 + Y_{14}V_4$$
$$I_2 = Y_{21}V_1 + Y_{22}V_2 + Y_{23}V_3 + Y_{24}V_4$$
$$I_3 = Y_{31}V_1 + Y_{32}V_2 + Y_{33}V_3 + Y_{34}V_4$$
$$I_4 = Y_{41}V_1 + Y_{42}V_2 + Y_{43}V_3 + Y_{44}V_4$$

In the above network, since there is no connection between bus 1 and 4, $Y_{14} = Y_{41} = 0$; similarly $Y_{24} = Y_{42} = 0$.

Extending the above relation to an n bus system, the node-voltage equation in matrix form is

$$\begin{bmatrix} I_1 \\ I_2 \\ \vdots \\ I_i \\ \vdots \\ I_n \end{bmatrix} = \begin{bmatrix} Y_{11} & Y_{12} & \cdots & Y_{1i} & \cdots & Y_{1n} \\ Y_{21} & Y_{22} & \cdots & Y_{2i} & \cdots & Y_{2n} \\ \vdots & \vdots & & \vdots & & \vdots \\ Y_{i1} & Y_{i2} & \cdots & Y_{ii} & \cdots & Y_{in} \\ \vdots & \vdots & & \vdots & & \vdots \\ Y_{n1} & Y_{n2} & \cdots & Y_{ni} & \cdots & Y_{nn} \end{bmatrix} \begin{bmatrix} V_1 \\ V_2 \\ \vdots \\ V_i \\ \vdots \\ V_n \end{bmatrix} \tag{6.1}$$

or

$$\mathbf{I}_{bus} = \mathbf{Y}_{bus}\,\mathbf{V}_{bus} \tag{6.2}$$

where $\mathbf{I}_{bus}$ is the vector of the injected bus currents (i.e., external current sources). The current is positive when flowing towards the bus, and it is negative if flowing away from the bus. $\mathbf{V}_{bus}$ is the vector of bus voltages measured from the reference node (i.e., node voltages). $\mathbf{Y}_{bus}$ is known as the *bus admittance matrix*. The diagonal element of each node is the sum of admittances connected to it. It is known as the *self-admittance* or *driving point admittance*, i.e.,

$$Y_{ii} = \sum_{j=0}^{n} y_{ij} \qquad j \neq i \tag{6.3}$$

The off-diagonal element is equal to the negative of the admittance between the nodes. It is known as the *mutual admittance* or *transfer admittance*, i.e.,

$$Y_{ij} = Y_{ji} = -y_{ij} \tag{6.4}$$

When the bus currents are known, (6.2) can be solved for the n bus voltages.

$$\mathbf{V}_{bus} = \mathbf{Y}_{bus}^{-1}\,\mathbf{I}_{bus} \tag{6.5}$$

The inverse of the bus admittance matrix is known as the *bus impedance matrix* Z_{bus}. The admittance matrix obtained with one of the buses as reference is nonsingular. Otherwise the nodal matrix is singular.

Inspection of the bus admittance matrix reveals that the matrix is symmetric along the leading diagonal, and we need to store the upper triangular nodal admittance matrix only. In a typical power system network, each bus is connected to only a few nearby buses. Consequently, many off-diagonal elements are zero. Such a matrix is called *sparse*, and efficient numerical techniques can be applied to compute its inverse. By means of an appropriately ordered triangular decomposition, the inverse of a sparse matrix can be expressed as a product of sparse matrix factors, thereby giving an advantage in computational speed, storage and reduction of round-off errors. However, Z_{bus}, which is required for short-circuit analysis, can be obtained directly by the method of *building algorithm* without the need for matrix inversion. This technique is discussed in Chapter 9.

Based on (6.3) and (6.4), the bus admittance matrix for the network in Figure 6.2 obtained by inspection is

$$Y_{bus} = \begin{bmatrix} -j8.50 & j2.50 & j5.00 & 0 \\ j2.50 & -j8.75 & j5.00 & 0 \\ j5.00 & j5.00 & -j22.50 & j12.50 \\ 0 & 0 & j12.50 & -j12.50 \end{bmatrix}$$

A function called **Y = ybus(zdata)** is written for the formation of the bus admittance matrix. **zdata** is the line data input and contains four columns. The first two columns are the line bus numbers and the remaining columns contain the line resistance and reactance in per unit. The function returns the bus admittance matrix. The algorithm for the bus admittance program is very simple and basic to power system programming. Therefore, it is presented here for the reader to study and understand the method of solution. In the program, the line impedances are first converted to admittances. Y is then initialized to zero. In the first loop, the line data is searched, and the off-diagonal elements are entered. Finally, in a nested loop, line data is searched to find the elements connected to a bus, and the diagonal elements are thus formed.

The following is a program for building the bus admittance matrix:

```
function[Y] = ybus(zdata)
nl=zdata(:,1); nr=zdata(:,2); R=zdata(:,3); X=zdata(:,4);
nbr=length(zdata(:,1)); nbus = max(max(nl), max(nr));
Z = R + j*X;                                  %branch impedance
```

```
y= ones(nbr,1)./Z;                               %branch admittance
Y = zeros(nbus,nbus);                       % initialize Y to zero
for k = 1:nbr;    % formation of the off diagonal elements
    if nl(k) > 0 & nr(k) > 0
    Y(nl(k),nr(k)) = Y(nl(k),nr(k)) - y(k);
    Y(nr(k),nl(k)) = Y(nl(k),nr(k));
    end
end
for n = 1:nbus         % formation of the diagonal elements
    for k = 1:nbr
    if nl(k) == n | nr(k) == n
    Y(n,n) = Y(n,n) + y(k);
    else, end
    end
end
```

Example 6.1

The emfs shown in Figure 6.1 are $E_1 = 1.1\angle 0°$ and $E_2 = 1.0\angle 0°$. Use the function **Y = ybus(zdata)** to obtain the bus admittance matrix. Find the bus impedance matrix by inversion, and solve for the bus voltages.

With source transformation, the equivalent current sources are

$$I_1 = \frac{1.1}{j1.0} = -j1.1 \text{ pu}$$

$$I_2 = \frac{1.0}{j0.8} = -j1.25 \text{ pu}$$

The following commands

```
%     From  To   R    X
z = [ 0     1    0   1.0
      0     2    0   0.8
      1     2    0   0.4
      1     3    0   0.2
      2     3    0   0.2
      3     4    0   0.08];
Y = ybus(z)                               % bus admittance matrix
Ibus = [-j*1.1; -j*1.25; 0; 0];  % vector of bus currents
Zbus = inv(Y)                              % bus impedance matrix
Vbus = Zbus*Ibus
```

result in

```
  Y  =
        0 - 8.50i     0 + 2.50i     0 +  5.00i     0 +  0.00i
        0 + 2.50i     0 - 8.75i     0 +  5.00i     0 +  0.00i
        0 + 5.00i     0 + 5.00i     0 - 22.50i     0 + 12.50i
        0 + 0.00i     0 + 0.00i     0 + 12.50i     0 - 12.50i
Zbus =
        0 + 0.50i     0 + 0.40i     0 + 0.450i     0 + 0.450i
        0 + 0.40i     0 + 0.48i     0 + 0.440i     0 + 0.440i
        0 + 0.45i     0 + 0.44i     0 + 0.545i     0 + 0.545i
        0 + 0.45i     0 + 0.44i     0 + 0.545i     0 + 0.625i
Vbus =
        1.0500
        1.0400
        1.0450
        1.0450
```

The solution of equation $\mathbf{I}_{bus} = \mathbf{Y}_{bus}\mathbf{V}_{bus}$ by inversion is very inefficient. It is not necessary to obtain the inverse of $\mathbf{Y}_{bus}$. Instead, direct solution is obtained by optimally ordered triangular factorization. In *MATLAB*, the solution of linear simultaneous equations $AX = B$ is obtained by using the matrix division operator $\backslash$ (i.e., $X = A \backslash B$), which is based on the triangular factorization and Gaussian elimination. This technique is superior in both execution time and numerical accuracy. It is two to three times as fast and produces residuals on the order of machine accuracy.

In Example 6.1, obtain the direct solution by replacing the statements `Zbus = inv(Y)` and `Vbus = Zbus*Ibus` with `Vbus = Y\ Ibus`.

6.3 SOLUTION OF NONLINEAR ALGEBRAIC EQUATIONS

The most common techniques used for the iterative solution of nonlinear algebraic equations are Gauss-Seidel, Newton-Raphson, and Quasi-Newton methods. The Gauss-Seidel and Newton-Raphson methods are discussed for one-dimensional equation, and are then extended to n-dimensional equations.

6.3.1 GAUSS-SEIDEL METHOD

The Gauss-Seidel method is also known as the method of successive displacements. To illustrate the technique, consider the solution of the nonlinear equation given by

$$f(x) = 0 \tag{6.6}$$

The above function is rearranged and written as

$$x = g(x) \tag{6.7}$$

If $x^{(k)}$ is an initial estimate of the variable x, the following iterative sequence is formed.

$$x^{(k+1)} = g(x^{(k)}) \tag{6.8}$$

A solution is obtained when the difference between the absolute value of the successive iteration is less than a specified accuracy, i.e.,

$$|x^{(k+1)} - x^{(k)}| \leq \epsilon \tag{6.9}$$

where ϵ is the desired accuracy.

Example 6.2

Use the Gauss-Seidel method to find a root of the following equation

$$f(x) = x^3 - 6x^2 + 9x - 4 = 0$$

Solving for x, the above expression is written as

$$\begin{aligned} x &= -\frac{1}{9}x^3 + \frac{6}{9}x^2 + \frac{4}{9} \\ &= g(x) \end{aligned}$$

The *MATLAB* **plot** command is used to plot $g(x)$ and x over a range of 0 to 4.5, as shown in Figure 6.3. The intersections of $g(x)$ and x results in the roots of $f(x)$. From Figure 6.3 two of the roots are found to be 1 and 4. Actually, there is a repeated root at $x = 1$. Apply the Gauss-Seidel algorithm, and use an initial estimate of

$$x^{(0)} = 2$$

From (6.8), the first iteration is

$$x^{(1)} = g(2) = -\frac{1}{9}(2)^3 + \frac{6}{9}(2)^2 + \frac{4}{9} = 2.2222$$

The second iteration is

$$x^{(2)} = g(2.2222) = -\frac{1}{9}(2.2222)^3 + \frac{6}{9}(2.2222)^2 + \frac{4}{9} = 2.5173$$

The subsequent iterations result in 2.8966, 3.3376, 3.7398, 3.9568, 3.9988 and 4.0000. The process is repeated until the change in variable is within the desired

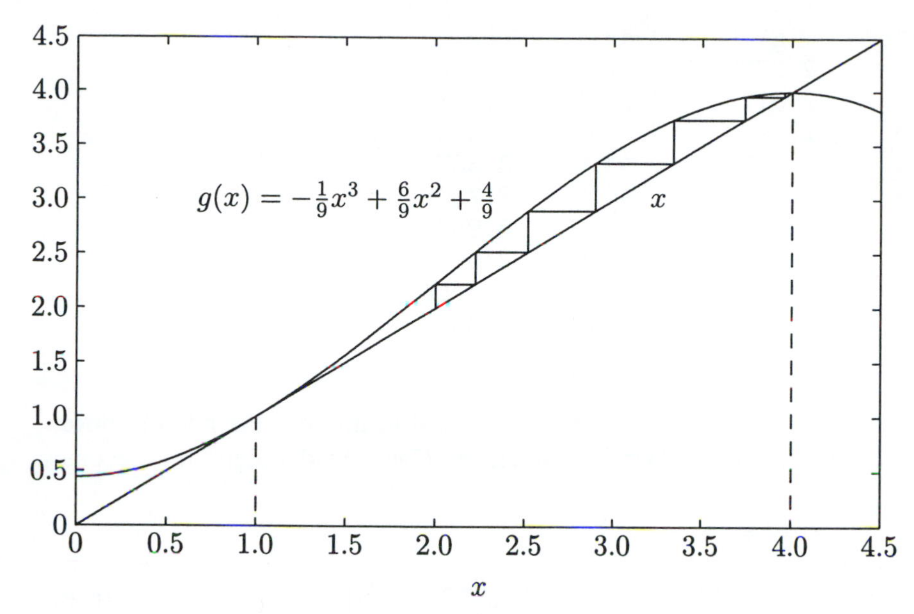

FIGURE 6.3
Graphical illustration of the Gauss-Seidel method.

accuracy. It can be seen that the Gauss-Seidel method needs many iterations to achieve the desired accuracy, and there is no guarantee for the convergence. In this example, since the initial estimate was within a "boxed in" region, the solution converged in a zigzag fashion to one of the roots. In fact, if the initial estimate was outside this region, say $x^{(0)} = 6$, the process would diverge. A test of convergence, especially for the n-dimensional case, is difficult, and no general methods are known.

The following commands show the procedure for the solution of the given equation starting with an initial estimate of $x^{(0)} = 2$.

```
dx=1;           % Change in variable is set to a high value
x=2;                                        % Initial estimate
iter = 0;                                   % Iteration counter
disp('Iter    g         dx        x')%Heading for results
while abs(dx) >= 0.001 & iter < 100 %Test for convergence
iter = iter + 1;                            % No. of iterations
g = -1/9*x^3+6/9*x^2+4/9 ;
dx = g-x;                                   % Change in variable
x = x + dx;                          % Successive approximation
fprintf('%g', iter), disp([g, dx, x])
end
```

The result is

```
Iter      g         dx        x
1      2.2222    0.2222    2.2222
2      2.5173    0.2951    2.5173
3      2.8966    0.3793    2.8966
4      3.3376    0.4410    3.3376
5      3.7398    0.4022    3.7398
6      3.9568    0.2170    3.9568
7      3.9988    0.0420    3.9988
8      4.0000    0.0012    4.0000
9      4.0000    0.0000    4.0000
```

In some cases, an acceleration factor can be used to improve the rate of convergence. If $\alpha > 1$ is the acceleration factor, the Gauss-Seidel algorithm becomes

$$x^{(k+1)} = x^{(k)} + \alpha[g(x^{(k)}) - x^{(k)}] \tag{6.10}$$

Example 6.3

Find a root of the equation in Example 6.2, using the Gauss-Seidel method with an acceleration factor of $\alpha = 1.25$:

Starting with an initial estimate of $x^{(0)} = 2$ and using (6.10), the first iteration is

$$\begin{aligned} g(2) &= -\frac{1}{9}(2)^3 + \frac{6}{9}(2)^2 + \frac{4}{9} = 2.2222 \\ x^{(1)} &= 2 + 1.25[2.2222 - 2] = 2.2778 \end{aligned}$$

The second iteration is

$$\begin{aligned} g(2.2778) &= -\frac{1}{9}(2.2778)^3 + \frac{6}{9}(2.2778)^2 + \frac{4}{9} = 2.5902 \\ x^{(2)} &= 2.2778 + 1.25[2.5902 - 2.2778] = 2.6683 \end{aligned}$$

The subsequent iterations result in 3.0801, 3.1831, 3.7238, 4.0084, 3.9978 and 4.0005. The effect of acceleration is shown graphically in Figure 6.4. Care must be taken not to use a very large acceleration factor since the larger step size may result in an overshoot. This can cause an increase in the number of iterations or even result in divergence. In the *MATLAB* command of Example 6.2, replace the command before the end statement by $x = x + 1.25 * dx$ to reflect the effect of the acceleration factor and run the program.

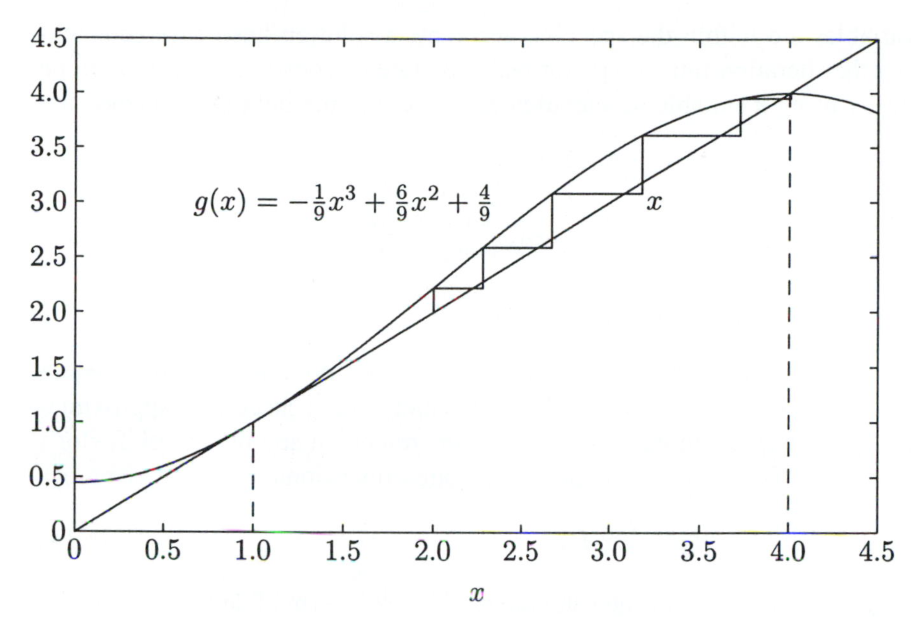

FIGURE 6.4
Graphical illustration of the Gauss-Seidel method using acceleration factor.

We now consider the system of n equations in n variables

$$\begin{aligned} f_1(x_1, x_2, \cdots, x_n) &= c_1 \\ f_2(x_1, x_2, \cdots, x_n) &= c_2 \\ \cdots\cdots&\cdots\cdots \\ f_n(x_1, x_2, \cdots, x_n) &= c_n \end{aligned} \tag{6.11}$$

Solving for one variable from each equation, the above functions are rearranged and written as

$$\begin{aligned} x_1 &= c_1 + g_1(x_1, x_2, \cdots, x_n) \\ x_2 &= c_2 + g_2(x_1, x_2, \cdots, x_n) \\ \cdots\cdots&\cdots\cdots \\ x_n &= c_n + g_n(x_1, x_2, \cdots, x_n) \end{aligned} \tag{6.12}$$

The iteration procedure is initiated by assuming an approximate solution for each of the independent variables $(x_1^{(0)}, x_2^{(0)} \cdots, x_n^{(0)})$. Equation (6.12) results in a new approximate solution $(x_1^{(1)}, x_2^{(1)} \cdots, x_n^{(1)})$. In the Gauss-Seidel method, the updated values of the variables calculated in the preceding equations are immediately used in the solution of the subsequent equations. At the end of each iteration, the calculated values of all variables are tested against the previous values. If all changes in the variables are within the specified accuracy, a solution has converged, otherwise another iteration must be performed. The rate of convergence can often be increased by using a suitable acceleration factor α, and the iterative sequence becomes

$$x_i^{(k+1)} = x_i^{(k)} + \alpha(x_{i\,cal}^{(k+1)} - x_i^{(k)}) \tag{6.13}$$

6.3.2 NEWTON-RAPHSON METHOD

The most widely used method for solving simultaneous nonlinear algebraic equations is the Newton-Raphson method. Newton's method is a successive approximation procedure based on an initial estimate of the unknown and the use of Taylor's series expansion. Consider the solution of the one-dimensional equation given by

$$f(x) = c \tag{6.14}$$

If $x^{(0)}$ is an initial estimate of the solution, and $\Delta x^{(0)}$ is a small deviation from the correct solution, we must have

$$f(x^{(0)} + \Delta x^{(0)}) = c$$

Expanding the left-hand side of the above equation in Taylor's series about $x^{(0)}$ yields

$$f(x^{(0)}) + \left(\frac{df}{dx}\right)^{(0)} \Delta x^{(0)} + \frac{1}{2!}\left(\frac{d^2 f}{dx^2}\right)^{(0)} (\Delta x^{(0)})^2 + \cdots = c$$

Assuming the error $\Delta x^{(0)}$ is very small, the higher-order terms can be neglected, which results in

$$\Delta c^{(0)} \simeq \left(\frac{df}{dx}\right)^{(0)} \Delta x^{(0)}$$

where

$$\Delta c^{(0)} = c - f(x^{(0)})$$

Adding $\Delta x^{(0)}$ to the initial estimate will result in the second approximation

$$x^{(1)} = x^{(0)} + \frac{\Delta c^{(0)}}{\left(\frac{df}{dx}\right)^{(0)}}$$

Successive use of this procedure yields the Newton-Raphson algorithm

$$\Delta c^{(k)} = c - f(x^{(k)}) \tag{6.15}$$

$$\Delta x^{(k)} = \frac{\Delta c^{(k)}}{\left(\frac{df}{dx}\right)^{(k)}} \tag{6.16}$$

$$x^{(k+1)} = x^{(k)} + \Delta x^{(k)} \tag{6.17}$$

(6.16) can be rearranged as

$$\Delta c^{(k)} = j^{(k)} \Delta x^{(k)} \tag{6.18}$$

where

$$j^{(k)} = \left(\frac{df}{dx}\right)^{(k)}$$

The relation in (6.18) demonstrates that the nonlinear equation $f(x) - c = 0$ is approximated by the tangent line on the curve at $x^{(k)}$. Therefore, a linear equation is obtained in terms of the small changes in the variable. The intersection of the tangent line with the x-axis results in $x^{(k+1)}$. This idea is demonstrated graphically in Example 6.4.

Example 6.4

Use the Newton-Raphson method to find a root of the equation given in Example 6.2. Assume an initial estimate of $x^{(0)} = 6$.

The *MATLAB* **plot** command is used to plot $f(x) = x^3 - 6x^2 + 9x - 4$ over a range of 0 to 6 as shown in Figure 6.5. The intersections of $f(x)$ with the x-axis results in the roots of $f(x)$. From Figure 6.5, two of the roots are found to be 1 and 4. Actually, there is a repeated root at $x = 1$.

Also, Figure 6.5 gives a graphical description of the Newton-Raphson method. Starting with an initial estimate of $x^{(0)} = 6$, we extrapolate along the tangent to its intersection with the x-axis and take that as the next approximation. This is continued until successive x-values are sufficiently close.

The analytical solution given by the Newton-Raphson algorithm is

$$\frac{df(x)}{dx} = 3x^2 - 12x + 9$$

$$\Delta c^{(0)} = c - f(x^{(0)}) = 0 - [(6)^3 - 6(6)^2 + 9(6) - 4] = -50$$

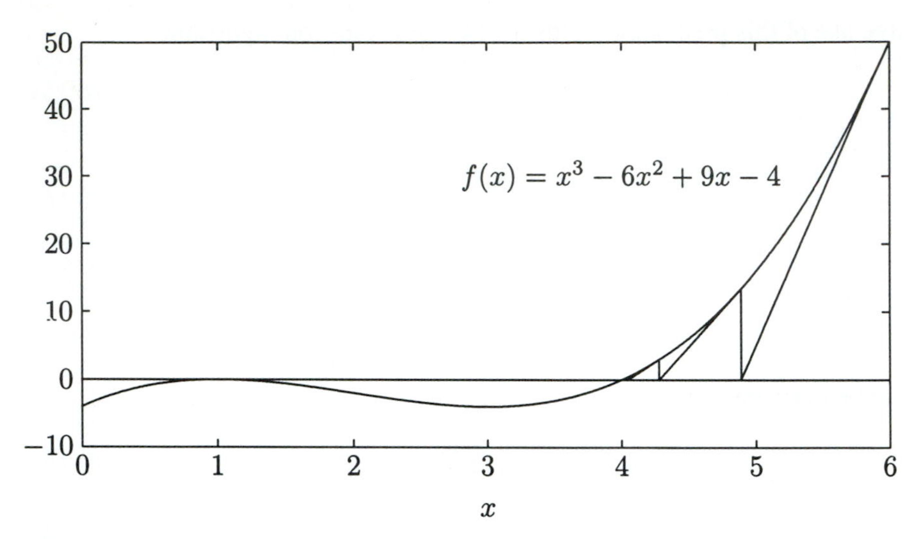

FIGURE 6.5
Graphical illustration of the Newton-Raphson algorithm.

$$\left(\frac{df}{dx}\right)^{(0)} = 3(6)^2 - 12(6) + 9 = 45$$

$$\Delta x^{(0)} = \frac{\Delta c^{(0)}}{\left(\frac{df}{dx}\right)^{(0)}} = \frac{-50}{45} = -1.1111$$

Therefore, the result at the end of the first iteration is

$$x^{(1)} = x^{(0)} + \Delta x^{(0)} = 6 - 1.1111 = 4.8889$$

The subsequent iterations result in

$$x^{(2)} = x^{(1)} + \Delta x^{(1)} = 4.8889 - \frac{13.4431}{22.037} = 4.2789$$

$$x^{(3)} = x^{(2)} + \Delta x^{(2)} = 4.2789 - \frac{2.9981}{12.5797} = 4.0405$$

$$x^{(4)} = x^{(3)} + \Delta x^{(3)} = 4.0405 - \frac{0.3748}{9.4914} = 4.0011$$

$$x^{(5)} = x^{(4)} + \Delta x^{(4)} = 4.0011 - \frac{0.0095}{9.0126} = 4.0000$$

We see that Newton's method converges considerably more rapidly than the Gauss-Seidel method. The method may converge to a root different from the expected one or diverge if the starting value is not close enough to the root.

The following commands show the procedure for the solution of the given equation by the Newton-Raphson method.

```
dx=1;           % Change in variable is set to a high value
x=input('Enter initial estimate -> '); % Initial estimate
iter = 0;                                  % Iteration counter
disp('iter    Dc         J         dx          x') % Heading
while abs(dx) >= 0.001 & iter < 100  Test for convergence
iter = iter + 1;                           % No. of iterations
Dc = 0 - (x^3 - 6*x^2 + 9*x - 4);                 % Residual
J = 3*x^2 - 12*x + 9;                           % Derivative
dx= Dc/J;                                  %Change in variable
x=x + dx;                                % Successive solution
fprintf('%g', iter), disp([Dc, J, dx, x])
end
```

The result is

```
Enter the initial estimate -> 6
iter    Dc           J          dx          x
 1  -50.0000   45.0000   -1.1111    4.8889
 2  -13.4431   22.0370   -0.6100    4.2789
 3   -2.9981   12.5797   -0.2383    4.0405
 4   -0.3748    9.4914   -0.0395    4.0011
 5   -0.0095    9.0126   -0.0011    4.0000
 6   -0.0000    9.0000   -0.0000    4.0000
```

Now consider the n-dimensional equations given by (6.11). Expanding the left-hand side of the equations (6.11) in the Taylor's series about the initial estimates and neglecting all higher order terms, leads to the expression

$$(f_1)^{(0)} + \left(\frac{\partial f_1}{\partial x_1}\right)^{(0)} \Delta x_1^{(0)} + \left(\frac{\partial f_1}{\partial x_2}\right)^{(0)} \Delta x_2^{(0)} + \cdots + \left(\frac{\partial f_1}{\partial x_n}\right)^{(0)} \Delta x_n^{(0)} = c_1$$

$$(f_2)^{(0)} + \left(\frac{\partial f_2}{\partial x_1}\right)^{(0)} \Delta x_1^{(0)} + \left(\frac{\partial f_2}{\partial x_2}\right)^{(0)} \Delta x_2^{(0)} + \cdots + \left(\frac{\partial f_2}{\partial x_n}\right)^{(0)} \Delta x_n^{(0)} = c_2$$

$$\vdots$$

$$(f_n)^{(0)} + \left(\frac{\partial f_n}{\partial x_1}\right)^{(0)} \Delta x_1^{(0)} + \left(\frac{\partial f_n}{\partial x_2}\right)^{(0)} \Delta x_2^{(0)} + \cdots + \left(\frac{\partial f_n}{\partial x_n}\right)^{(0)} \Delta x_n^{(0)} = c_n$$

or in matrix form

$$\begin{bmatrix} c_1 - (f_1)^{(0)} \\ c_2 - (f_2)^{(0)} \\ \vdots \\ c_n - (f_n)^{(0)} \end{bmatrix} = \begin{bmatrix} \left(\frac{\partial f_1}{\partial x_1}\right)^{(0)} & \left(\frac{\partial f_1}{\partial x_2}\right)^{(0)} & \cdots & \left(\frac{\partial f_1}{\partial x_n}\right)^{(0)} \\ \left(\frac{\partial f_2}{\partial x_1}\right)^{(0)} & \left(\frac{\partial f_2}{\partial x_2}\right)^{(0)} & \cdots & \left(\frac{\partial f_2}{\partial x_n}\right)^{(0)} \\ \vdots & \vdots & \ddots & \cdots \\ \left(\frac{\partial f_n}{\partial x_1}\right)^{(0)} & \left(\frac{\partial f_n}{\partial x_2}\right)^{(0)} & \cdots & \left(\frac{\partial f_n}{\partial x_n}\right)^{(0)} \end{bmatrix} \begin{bmatrix} \Delta x_1^{(0)} \\ \Delta x_2^{(0)} \\ \vdots \\ \Delta x_n^{(0)} \end{bmatrix}$$

In short form, it can be written as

$$\Delta C^{(k)} = J^{(k)} \Delta X^{(k)}$$

or

$$\Delta X^{(k)} = [J^{(k)}]^{-1} \Delta C^{(k)} \tag{6.19}$$

and the Newton-Raphson algorithm for the n-dimensional case becomes

$$X^{(k+1)} = X^{(k)} + \Delta X^{(k)} \tag{6.20}$$

where

$$\Delta X^{(k)} = \begin{bmatrix} \Delta x_1^{(k)} \\ \Delta x_2^{(k)} \\ \vdots \\ \Delta x_n^{(k)} \end{bmatrix} \quad \text{and} \quad \Delta C^{(k)} = \begin{bmatrix} c_1 - (f_1)^{(k)} \\ c_2 - (f_2)^{(k)} \\ \vdots \\ c_n - (f_n)^{(k)} \end{bmatrix} \tag{6.21}$$

$$J^{(k)} = \begin{bmatrix} \left(\frac{\partial f_1}{\partial x_1}\right)^{(k)} & \left(\frac{\partial f_1}{\partial x_2}\right)^{(k)} & \cdots & \left(\frac{\partial f_1}{\partial x_n}\right)^{(k)} \\ \left(\frac{\partial f_2}{\partial x_1}\right)^{(k)} & \left(\frac{\partial f_2}{\partial x_2}\right)^{(k)} & \cdots & \left(\frac{\partial f_2}{\partial x_n}\right)^{(k)} \\ \vdots & \vdots & \ddots & \cdots \\ \left(\frac{\partial f_n}{\partial x_1}\right)^{(k)} & \left(\frac{\partial f_n}{\partial x_2}\right)^{(k)} & \cdots & \left(\frac{\partial f_n}{\partial x_n}\right)^{(k)} \end{bmatrix} \tag{6.22}$$

$J^{(k)}$ is called the *Jacobian matrix*. Elements of this matrix are the partial derivatives evaluated at $X^{(k)}$. It is assumed that $J^{(k)}$ has an inverse during each iteration. Newton's method, as applied to a set of nonlinear equations, reduces the problem to solving a set of linear equations in order to determine the values that improve the accuracy of the estimates.

The solution of (6.19) by inversion is very inefficient. It is not necessary to obtain the inverse of $J^{(k)}$. Instead, a direct solution is obtained by optimally ordered triangular factorization. In *MATLAB*, the solution of linear simultaneous equations $\Delta C = J \Delta X$ is obtained by using the matrix division operator \ (i.e., $\Delta X = J \setminus \Delta C$) which is based on the triangular factorization and Gaussian elimination.

Example 6.5

Use the Newton-Raphson method to find the intersections of the curves

$$\begin{aligned} x_1^2 + x_2^2 &= 4 \\ e^{x_1} + x_2 &= 1 \end{aligned}$$

Graphically, the solution to this system is represented by the intersections of the circle $x_1^2 + x_2^2 = 4$ with the curve $e^{x_1} + x_2 = 1$. Figure 6.6 shows that these are near (1, −1.7) and (−1.8, 0.8).

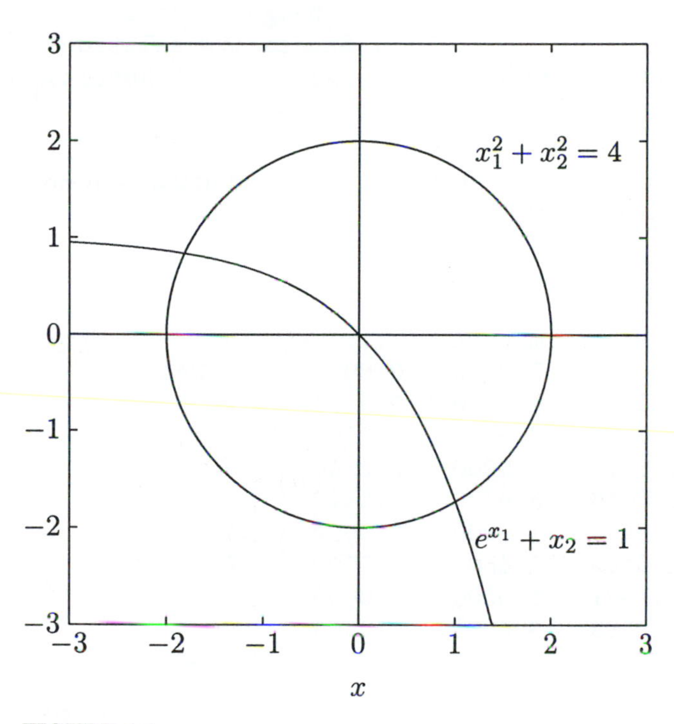

FIGURE 6.6
Graphs of Example 6.5.

Taking partial derivatives of the above functions results in the Jacobian matrix

$$J = \begin{bmatrix} 2x_1 & 2x_2 \\ e^{x_1} & 1 \end{bmatrix}$$

The Newton-Raphson algorithm for the above system is presented in the following statements.

```
iter = 0;                                % Iteration counter
```

```
x=input('Enter initial estimates, col. vector[x1;x2]->');
Dx = [1; 1];  % Change in variable is set to a high value
C=[4; 1];
disp('Iter   DC      Jacobian matrix        Dx          x');
                                            % Heading for results
while max(abs(Dx)) >= 0.0001 & iter <10 %Convergence test
iter=iter+1;                                 % Iteration counter
f = [x(1)^2+x(2)^2;  exp(x(1))+x(2)];              % Functions
DC = C - f;                                          % Residuals
J = [2*x(1)      2*x(2)                        % Jacobian matrix
     exp(x(1))        1];
Dx=J\DC;                                   % Change in variables
x=x+Dx;                                   % Successive solutions
fprintf('%g', iter), disp([DC, J, Dx, x])            % Results
end
```

When the program is run, the user is prompted to enter the initial estimate. Let us try an initial estimate given by [0.5; -1].

Enter Initial estimates, col. vector $[x_1;\ x_2] \rightarrow [0.5;\ -1]$

Iter	ΔC	Jacobian matrix		Δx	x
1	2.7500	1.0000	-2.0000	0.8034	1.3034
	0.3513	1.6487	1.0000	-0.9733	-1.9733
2	-1.5928	2.6068	-3.9466	-0.2561	1.0473
	-0.7085	3.6818	1.0000	0.2344	-1.7389
3	-0.1205	2.0946	-3.4778	-0.0422	1.0051
	-0.1111	2.8499	1.0000	0.0092	-1.7296
4	-0.0019	2.0102	-3.4593	-0.0009	1.0042
	-0.0025	2.7321	1.0000	0.0000	-1.7296
5	-0.0000	2.0083	-3.4593	-0.0000	1.0042
	-0.0000	2.7296	1.0000	-0.0000	-1.7296

After five iterations, the solution converges to $x_1 = 1.0042$ and $x_2 = -1.7296$ accurate to four decimal places. Starting with an initial value of $[-0.5; 1]$, which is closer to the other intersection, results in $x_1 = -1.8163$ and $x_2 = 0.8374$.

Example 6.6

Starting with the initial values, $x_1 = 1$, $x_2 = 1$, and $x_3 = 1$, solve the following system of equations by the Newton-Raphson method.

$$\begin{aligned} x_1^2 - x_2^2 + x_3^2 &= 11 \\ x_1 x_2 + x_2^2 - 3x_3 &= 3 \\ x_1 - x_1 x_3 + x_2 x_3 &= 6 \end{aligned}$$

Taking partial derivatives of the above functions results in the Jacobian matrix

$$J = \begin{bmatrix} 2x_1 & -2x_2 & 2x_3 \\ x_2 & x_1 + 2x_2 & -3 \\ 1 - x_3 & x_3 & -x_1 + x_2 \end{bmatrix}$$

The following statements solve the given system of equations by the Newton-Raphson algorithm

```
Dx=[10;10;10]; %Change in variable is set to a high value
x=[1; 1; 1];                              % Initial estimate
C=[11; 3; 6];
iter = 0;                                 % Iteration counter
while max(abs(Dx))>=.0001 & iter<10;%Test for convergence
iter = iter + 1                           % No. of iterations
F = [x(1)^2-x(2)^2+x(3)^2                         % Functions
     x(1)*x(2)+x(2)^2-3*x(3)
     x(1)-x(1)*x(3)+x(2)*x(3)];
DC =C - F                                         % Residuals
J = [2*x(1)   -2*x(2)       2*x(3)        % Jacobian matrix
     x(2)     x(1)+2*x(2)   -3
     1-x(3)   x(3)          -x(1)+x(2)]
Dx=J\DC                                   %Change in variable
x=x+Dx                                  % Successive solution
end
```

The program results for the first iteration are

```
DC =                J =
      10                 2    -2     2
       4                 1     3    -3
       5                 0     1     0
Dx =                x =
      4.750              5.750
      5.000              6.000
      5.250              6.250
```

After six iterations, the solution converges to $x_1 = 2.0000$, $x_2 = 3.0000$, and $x_3 = 4.0000$.

Newton's method has the advantage of converging quadratically when we are near a root. However, more functional evaluations are required during each iteration. A very important limitation is that it does not generally converge to a solution from an arbitrary starting point.

6.4 POWER FLOW SOLUTION

Power flow studies, commonly known as *load flow*, form an important part of power system analysis. They are necessary for planning, economic scheduling, and control of an existing system as well as planning its future expansion. The problem consists of determining the magnitudes and phase angle of voltages at each bus and active and reactive power flow in each line.

In solving a power flow problem, the system is assumed to be operating under balanced conditions and a single-phase model is used. Four quantities are associated with each bus. These are voltage magnitude $|V|$, phase angle δ, real power P, and reactive power Q. The system buses are generally classified into three types.

Slack bus One bus, known as *slack* or *swing bus*, is taken as reference where the magnitude and phase angle of the voltage are specified. This bus makes up the difference between the scheduled loads and generated power that are caused by the losses in the network.

Load buses At these buses the active and reactive powers are specified. The magnitude and the phase angle of the bus voltages are unknown. These buses are called P-Q buses.

Regulated buses These buses are the *generator buses*. They are also known as *voltage-controlled buses*. At these buses, the real power and voltage magnitude are specified. The phase angles of the voltages and the reactive power are to be determined. The limits on the value of the reactive power are also specified. These buses are called P-V buses.

6.4.1 POWER FLOW EQUATION

Consider a typical bus of a power system network as shown in Figure 6.7. Transmission lines are represented by their equivalent π models where impedances have been converted to per unit admittances on a common MVA base.

Application of KCL to this bus results in

$$\begin{aligned} I_i &= y_{i0}V_i + y_{i1}(V_i - V_1) + y_{i2}(V_i - V_2) + \cdots + y_{in}(V_i - V_n) \\ &= (y_{i0} + y_{i1} + y_{i2} + \cdots + y_{in})V_i - y_{i1}V_1 - y_{i2}V_2 - \cdots - y_{in}V_n \quad (6.23) \end{aligned}$$

or

$$I_i = V_i \sum_{j=0}^{n} y_{ij} - \sum_{j=1}^{n} y_{ij}V_j \qquad j \neq i \qquad (6.24)$$

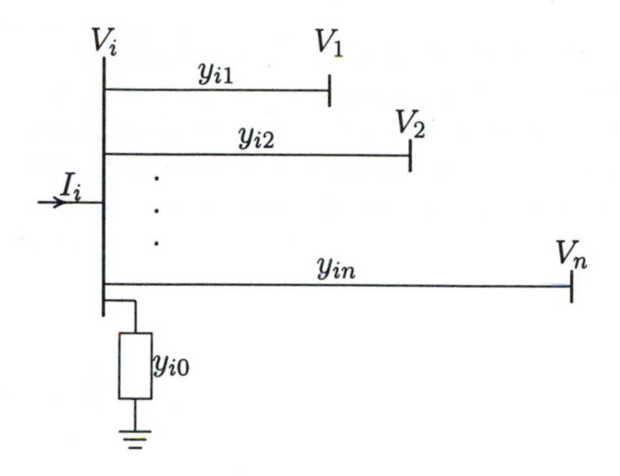

FIGURE 6.7
A typical bus of the power system.

The real and reactive power at bus i is

$$P_i + jQ_i = V_i I_i^* \tag{6.25}$$

or

$$I_i = \frac{P_i - jQ_i}{V_i^*} \tag{6.26}$$

Substituting for I_i in (6.24) yields

$$\frac{P_i - jQ_i}{V_i^*} = V_i \sum_{j=0}^{n} y_{ij} - \sum_{j=1}^{n} y_{ij} V_j \qquad j \neq i \tag{6.27}$$

From the above relation, the mathematical formulation of the power flow problem results in a system of algebraic nonlinear equations which must be solved by iterative techniques.

6.5 GAUSS-SEIDEL POWER FLOW SOLUTION

In the power flow study, it is necessary to solve the set of nonlinear equations represented by (6.27) for two unknown variables at each node. In the Gauss-Seidel method (6.27) is solved for V_i, and the iterative sequence becomes

$$V_i^{(k+1)} = \frac{\frac{P_i^{sch} - jQ_i^{sch}}{V_i^{*(k)}} + \sum y_{ij} V_j^{(k)}}{\sum y_{ij}} \qquad j \neq i \tag{6.28}$$

where y_{ij} shown in lowercase letters is the actual admittance in per unit. P_i^{sch} and Q_i^{sch} are the net real and reactive powers expressed in per unit. In writing the KCL, current entering bus i was assumed positive. Thus, for buses where real and reactive powers are injected into the bus, such as generator buses, P_i^{sch} and Q_i^{sch} have positive values. For load buses where real and reactive powers are flowing away from the bus, P_i^{sch} and Q_i^{sch} have negative values. If (6.27) is solved for P_i and Q_i, we have

$$P_i^{(k+1)} = \Re\{V_i^{*(k)}[V_i^{(k)} \sum_{j=0}^{n} y_{ij} - \sum_{j=1}^{n} y_{ij} V_j^{(k)}]\} \qquad j \neq i \tag{6.29}$$

$$Q_i^{(k+1)} = -\Im\{V_i^{*(k)}[V_i^{(k)} \sum_{j=0}^{n} y_{ij} - \sum_{j=1}^{n} y_{ij} V_j^{(k)}]\} \qquad j \neq i \tag{6.30}$$

The power flow equation is usually expressed in terms of the elements of the bus admittance matrix. Since the off-diagonal elements of the bus admittance matrix Y_{bus}, shown by uppercase letters, are $Y_{ij} = -y_{ij}$, and the diagonal elements are $Y_{ii} = \sum y_{ij}$, (6.28) becomes

$$V_i^{(k+1)} = \frac{\frac{P_i^{sch} - jQ_i^{sch}}{V_i^{*(k)}} - \sum_{j \neq i} Y_{ij} V_j^{(k)}}{Y_{ii}} \tag{6.31}$$

and

$$P_i^{(k+1)} = \Re\{V_i^{*(k)}[V_i^{(k)} Y_{ii} + \sum_{\substack{j=1 \\ j \neq i}}^{n} Y_{ij} V_j^{(k)}]\} \quad j \neq i \tag{6.32}$$

$$Q_i^{(k+1)} = -\Im\{V_i^{*(k)}[V_i^{(k)} Y_{ii} + \sum_{\substack{j=1 \\ j \neq i}}^{n} Y_{ij} V_j^{(k)}]\} \quad j \neq i \tag{6.33}$$

Y_{ii} includes the admittance to ground of line charging susceptance and any other fixed admittance to ground. In Section 6.7, a model is presented for transformers containing off-nominal ratio, which includes the effect of transformer tap setting.

Since both components of voltage are specified for the slack bus, there are $2(n-1)$ equations which must be solved by an iterative method. Under normal operating conditions, the voltage magnitude of buses are in the neighborhood of 1.0 per unit or close to the voltage magnitude of the slack bus. Voltage magnitude at load buses are somewhat lower than the slack bus value, depending on the reactive power demand, whereas the scheduled voltage at the generator buses are somewhat higher. Also, the phase angle of the load buses are below the reference angle in accordance to the real power demand, whereas the phase angle of the generator buses may be above the reference value depending on the amount of real power flowing into the bus. Thus, for the Gauss-Seidel method, an initial voltage estimate of $1.0 + j0.0$ for unknown voltages is satisfactory, and the converged solution correlates with the actual operating states.

For P-Q buses, the real and reactive powers P_i^{sch} and Q_i^{sch} are known. Starting with an initial estimate, (6.31) is solved for the real and imaginary components of voltage. For the voltage-controlled buses (P-V buses) where P_i^{sch} and $|V_i|$ are specified, first (6.33) is solved for $Q_i^{(k+1)}$, and then is used in (6.31) to solve for $V_i^{(k+1)}$. However, since $|V_i|$ is specified, only the imaginary part of $V_i^{(k+1)}$ is retained, and its real part is selected in order to satisfy

$$(e_i^{(k+1)})^2 + (f_i^{(k+1)})^2 = |V_i|^2 \tag{6.34}$$

or

$$e_i^{(k+1)} = \sqrt{|V_i|^2 - (f_i^{(k+1)})^2} \tag{6.35}$$

where $e_i^{(k+1)}$ and $f_i^{(k+1)}$ are the real and imaginary components of the voltage $V_i^{(k+1)}$ in the iterative sequence.

The rate of convergence is increased by applying an acceleration factor to the approximate solution obtained from each iteration.

$$V_i^{(k+1)} = V_i^{(k)} + \alpha(V_{i\,cal}^{(k)} - V_i^{(k)}) \tag{6.36}$$

where α is the acceleration factor. Its value depends upon the system. The range of 1.3 to 1.7 is found to be satisfactory for typical systems.

The updated voltages immediately replace the previous values in the solution of the subsequent equations. The process is continued until changes in the real and imaginary components of bus voltages between successive iterations are within a specified accuracy, i.e.,

$$\begin{aligned} |e_i^{(k+1)} - e_i^{(k)}| &\leq \epsilon \\ |f_i^{(k+1)} - f_i^{(k)}| &\leq \epsilon \end{aligned} \tag{6.37}$$

For the power mismatch to be reasonably small and acceptable, a very tight tolerance must be specified on both components of the voltage. A voltage accuracy in the range of 0.00001 to 0.00005 pu is satisfactory. In practice, the method for determining the completion of a solution is based on an accuracy index set up on the power mismatch. The iteration continues until the magnitude of the largest element in the ΔP and ΔQ columns is less than the specified value. A typical power mismatch accuracy is 0.001 pu

Once a solution is converged, the net real and reactive powers at the slack bus are computed from (6.32) and (6.33).

6.6 LINE FLOWS AND LOSSES

After the iterative solution of bus voltages, the next step is the computation of line flows and line losses. Consider the line connecting the two buses i and j in Figure 6.8. The line current I_{ij}, measured at bus i and defined positive in the direction

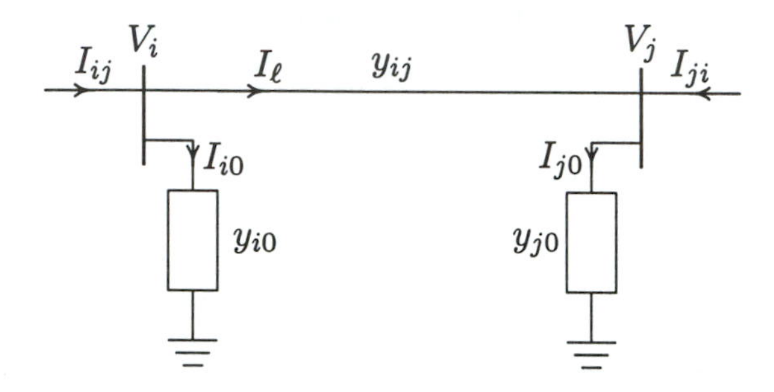

FIGURE 6.8
Transmission line model for calculating line flows.

$i \rightarrow j$ is given by

$$I_{ij} = I_\ell + I_{i0} = y_{ij}(V_i - V_j) + y_{i0}V_i \tag{6.38}$$

Similarly, the line current I_{ji} measured at bus j and defined positive in the direction $j \rightarrow i$ is given by

$$I_{ji} = -I_\ell + I_{j0} = y_{ij}(V_j - V_i) + y_{j0}V_j \tag{6.39}$$

The complex powers S_{ij} from bus i to j and S_{ji} from bus j to i are

$$S_{ij} = V_i I_{ij}^* \tag{6.40}$$
$$S_{ji} = V_j I_{ji}^* \tag{6.41}$$

The power loss in line $i - j$ is the algebraic sum of the power flows determined from (6.40) and (6.41), i.e.,

$$S_{L\,ij} = S_{ij} + S_{ji} \tag{6.42}$$

The power flow solution by the Gauss-Seidel method is demonstrated in the following two examples.

Example 6.7

Figure 6.9 shows the one-line diagram of a simple three-bus power system with generation at bus 1. The magnitude of voltage at bus 1 is adjusted to 1.05 per unit. The scheduled loads at buses 2 and 3 are as marked on the diagram. Line impedances are marked in per unit on a 100-MVA base and the line charging susceptances are neglected.

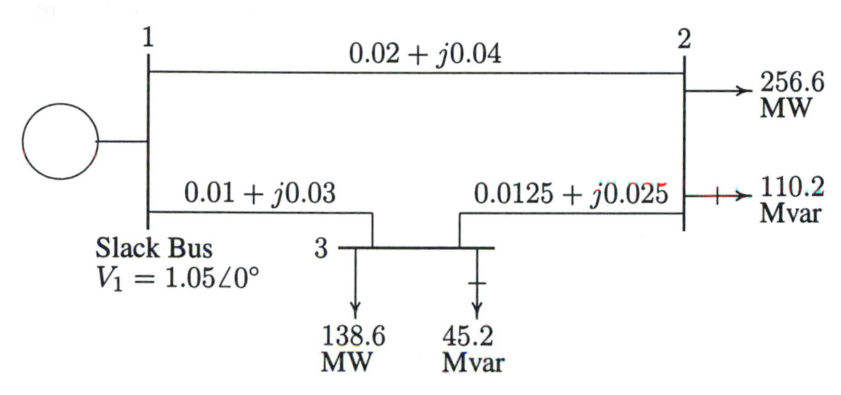

FIGURE 6.9
One-line diagram of Example 6.7 (impedances in pu on 100-MVA base).

(a) Using the Gauss-Seidel method, determine the phasor values of the voltage at the load buses 2 and 3 (P-Q buses) accurate to four decimal places.
(b) Find the slack bus real and reactive power.
(c) Determine the line flows and line losses. Construct a power flow diagram showing the direction of line flow.

(a) Line impedances are converted to admittances

$$y_{12} = \frac{1}{0.02 + j0.04} = 10 - j20$$

Similarly, $y_{13} = 10 - j30$ and $y_{23} = 16 - j32$. The admittances are marked on the network shown in Figure 6.10.
At the P-Q buses, the complex loads expressed in per units are

$$S_2^{sch} = -\frac{(256.6 + j110.2)}{100} = -2.566 - j1.102 \quad \text{pu}$$

$$S_3^{sch} = -\frac{(138.6 + j45.2)}{100} = -1.386 - j0.452 \quad \text{pu}$$

Since the actual admittances are readily available in Figure 6.10, for hand calculation, we use (6.28). Bus 1 is taken as reference bus (slack bus). Starting from an initial estimate of $V_2^{(0)} = 1.0 + j0.0$ and $V_3^{(0)} = 1.0 + j0.0$, V_2 and V_3 are computed from (6.28) as follows

$$V_2^{(1)} = \frac{\frac{P_2^{sch} - jQ_2^{sch}}{V_2^{*(0)}} + y_{12}V_1 + y_{23}V_3^{(0)}}{y_{12} + y_{23}}$$

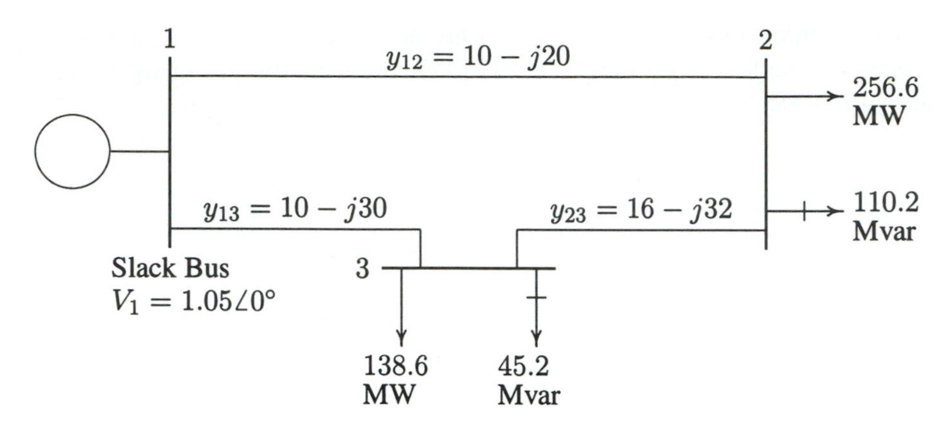

FIGURE 6.10
One-line diagram of Example 6.7 (admittances in pu on 100-MVA base).

$$= \frac{\frac{-2.566+j1.102}{1.0-j0} + (10 - j20)(1.05 + j0) + (16 - j32)(1.0 + j0)}{(26 - j52)}$$
$$= 0.9825 - j0.0310$$

and

$$V_3^{(1)} = \frac{\frac{P_3^{sch} - jQ_3^{sch}}{V_3^{*(0)}} + y_{13}V_1 + y_{23}V_2^{(1)}}{y_{13} + y_{23}}$$
$$= \frac{\frac{-1.386+j0.452}{1-j0} + (10 - j30)(1.05 + j0) + (16 - j32)(0.9825 - j0.0310)}{(26 - j62)}$$
$$= 1.0011 - j0.0353$$

For the second iteration we have

$$V_2^{(2)} = \frac{\frac{-2.566+j1.102}{0.9825+j0.0310} + (10 - j20)(1.05 + j0) + (16 - j32)(1.0011 - j0.0353)}{(26 - j52)}$$
$$= 0.9816 - j0.0520$$

and

$$V_3^{(2)} = \frac{\frac{-1.386+j0.452}{1.0011+j0.0353} + (10 - j30)(1.05 + j0) + (16 - j32)(0.9816 - j0.052)}{(26 - j62)}$$
$$= 1.0008 - j0.0459$$

The process is continued and a solution is converged with an accuracy of 5×10^{-5} per unit in seven iterations as given below.

$$V_2^{(3)} = 0.9808 - j0.0578 \qquad V_3^{(3)} = 1.0004 - j0.0488$$

$$V_2^{(4)} = 0.9803 - j0.0594 \qquad V_3^{(4)} = 1.0002 - j0.0497$$
$$V_2^{(5)} = 0.9801 - j0.0598 \qquad V_3^{(5)} = 1.0001 - j0.0499$$
$$V_2^{(6)} = 0.9801 - j0.0599 \qquad V_3^{(6)} = 1.0000 - j0.0500$$
$$V_2^{(7)} = 0.9800 - j0.0600 \qquad V_3^{(7)} = 1.0000 - j0.0500$$

The final solution is

$$V_2 = 0.9800 - j0.0600 = 0.98183\angle{-3.5035^\circ} \quad \text{pu}$$
$$V_3 = 1.0000 - j0.0500 = 1.00125\angle{-2.8624^\circ} \quad \text{pu}$$

(b) With the knowledge of all bus voltages, the slack bus power is obtained from (6.27)

$$\begin{aligned} P_1 - jQ_1 &= V_1^*[V_1(y_{12} + y_{13}) - (y_{12}V_2 + y_{13}V_3)] \\ &= 1.05[1.05(20 - j50) - (10 - j20)(0.98 - j.06) - \\ &\quad (10 - j30)(1.0 - j0.05)] \\ &= 4.095 - j1.890 \end{aligned}$$

or the slack bus real and reactive powers are $P_1 = 4.095$ pu $= 409.5$ MW and $Q_1 = 1.890$ pu $= 189$ Mvar.

(c) To find the line flows, first the line currents are computed. With line charging capacitors neglected, the line currents are

$$I_{12} = y_{12}(V_1 - V_2) = (10 - j20)[(1.05 + j0) - (0.98 - j0.06)] = 1.9 - j0.8$$
$$I_{21} = -I_{12} = -1.9 + j0.8$$
$$I_{13} = y_{13}(V_1 - V_3) = (10 - j30)[(1.05 + j0) - (1.0 - j0.05)] = 2.0 - j1.0$$
$$I_{31} = -I_{13} = -2.0 + j1.0$$
$$I_{23} = y_{23}(V_2 - V_3) = (16 - j32)[(0.98 - j0.06) - (1 - j0.05)] = -.64 + j.48$$
$$I_{32} = -I_{23} = 0.64 - j0.48$$

The line flows are

$$\begin{aligned} S_{12} &= V_1 I_{12}^* = (1.05 + j0.0)(1.9 + j0.8) = 1.995 + j0.84 \ \text{pu} \\ &= 199.5 \ \text{MW} + j84.0 \ \text{Mvar} \\ S_{21} &= V_2 I_{21}^* = (0.98 - j0.06)(-1.9 - j0.8) = -1.91 - j0.67 \ \text{pu} \\ &= -191.0 \ \text{MW} - j67.0 \ \text{Mvar} \\ S_{13} &= V_1 I_{13}^* = (1.05 + j0.0)(2.0 + j1.0) = 2.1 + j1.05 \ \text{pu} \\ &= 210.0 \ \text{MW} + j105.0 \ \text{Mvar} \end{aligned}$$

$$\begin{aligned}
S_{31} = V_3 I_{31}^* &= (1.0 - j0.05)(-2.0 - j1.0) = -2.05 - j0.90 \text{ pu} \\
&= -205.0 \text{ MW} - j90.0 \text{ Mvar} \\
S_{23} = V_2 I_{23}^* &= (0.98 - j0.06)(-0.656 + j0.48) = -0.656 - j0.432 \text{ pu} \\
&= -65.6 \text{ MW} - j43.2 \text{ Mvar} \\
S_{32} = V_3 I_{32}^* &= (1.0 - j0.05)(0.64 + j0.48) = 0.664 + j0.448 \text{ pu} \\
&= 66.4 \text{ MW} + j44.8 \text{ Mvar}
\end{aligned}$$

and the line losses are

$$\begin{aligned}
S_{L\,12} &= S_{12} + S_{21} = 8.5 \text{ MW} + j17.0 \text{ Mvar} \\
S_{L\,13} &= S_{13} + S_{31} = 5.0 \text{ MW} + j15.0 \text{ Mvar} \\
S_{L\,23} &= S_{23} + S_{32} = 0.8 \text{ MW} + j1.60 \text{ Mvar}
\end{aligned}$$

The power flow diagram is shown in Figure 6.11, where real power direction is indicated by $\rightarrow$ and the reactive power direction is indicated by $\mapsto$. The values within parentheses are the real and reactive losses in the line.

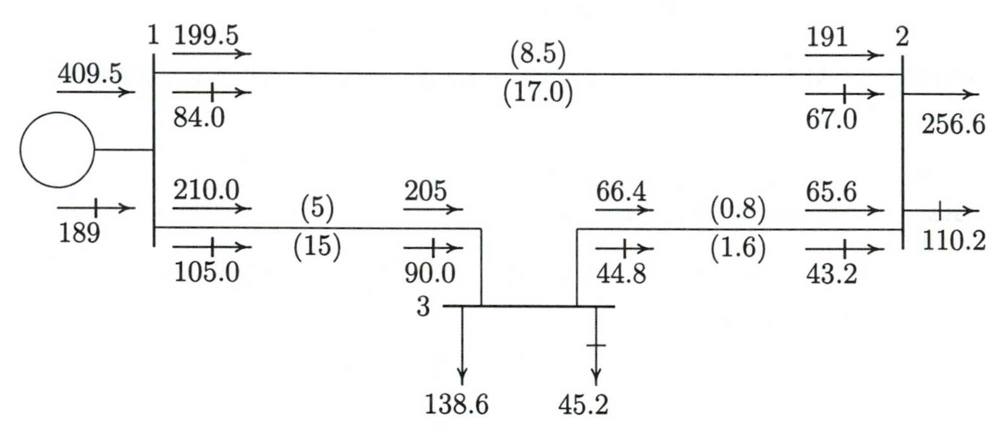

FIGURE 6.11
Power flow diagram of Example 6.7 (powers in MW and Mvar).

Example 6.8

Figure 6.12 shows the one-line diagram of a simple three-bus power system with generators at buses 1 and 3. The magnitude of voltage at bus 1 is adjusted to 1.05 pu. Voltage magnitude at bus 3 is fixed at 1.04 pu with a real power generation of 200 MW. A load consisting of 400 MW and 250 Mvar is taken from bus 2. Line impedances are marked in per unit on a 100 MVA base, and the line charging susceptances are neglected. Obtain the power flow solution by the Gauss-Seidel method including line flows and line losses.

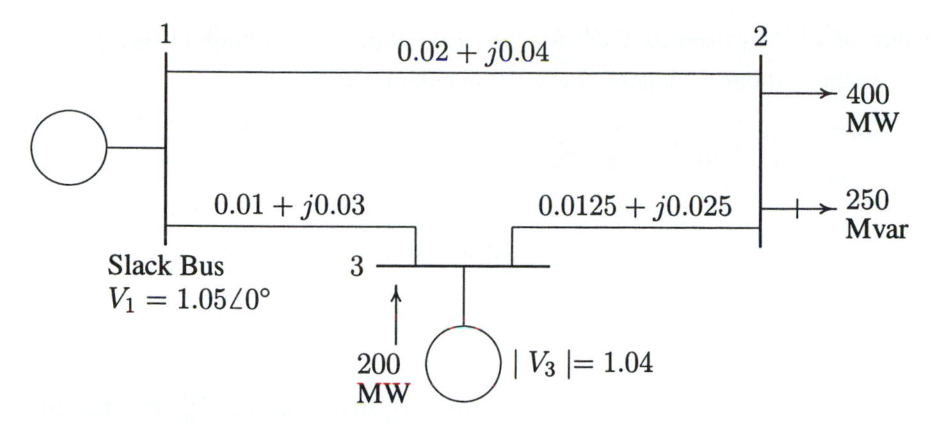

FIGURE 6.12
One-line diagram of Example 6.8 (impedances in pu on 100-MVA base).

Line impedances converted to admittances are $y_{12} = 10 - j20$, $y_{13} = 10 - j30$ and $y_{23} = 16 - j32$. The load and generation expressed in per units are

$$S_2^{sch} = -\frac{(400 + j250)}{100} = -4.0 - j2.5 \quad \text{pu}$$

$$P_3^{sch} = \frac{200}{100} = 2.0 \quad \text{pu}$$

Bus 1 is taken as the reference bus (slack bus). Starting from an initial estimate of $V_2^{(0)} = 1.0 + j0.0$ and $V_3^{(0)} = 1.04 + j0.0$, V_2 and V_3 are computed from (6.28).

$$\begin{aligned} V_2^{(1)} &= \frac{\frac{P_2^{sch} - jQ_2^{sch}}{V_2^{*(0)}} + y_{12}V_1 + y_{23}V_3^{(0)}}{y_{12} + y_{23}} \\ &= \frac{\frac{-4.0 + j2.5}{1.0 - j0} + (10 - j20)(1.05 + j0) + (16 - j32)(1.04 + j0)}{(26 - j52)} \\ &= 0.97462 - j0.042307 \end{aligned}$$

Bus 3 is a regulated bus where voltage magnitude and real power are specified. For the voltage-controlled bus, first the reactive power is computed from (6.30)

$$\begin{aligned} Q_3^{(1)} &= -\Im\{V_3^{*(0)}[V_3^{(0)}(y_{13} + y_{23}) - y_{13}V_1 - y_{23}V_2^{(1)}]\} \\ &= -\Im\{(1.04 - j0)[(1.04 + j0)(26 - j62) - (10 - j30)(1.05 + j0) - \\ &\quad (16 - j32)(0.97462 - j0.042307)]\} \\ &= 1.16 \end{aligned}$$

The value of $Q_3^{(1)}$ is used as Q_3^{sch} for the computation of voltage at bus 3. The complex voltage at bus 3, denoted by $V_{c3}^{(1)}$, is calculated

$$
\begin{aligned}
V_{c3}^{(1)} &= \frac{\frac{P_3^{sch} - jQ_3^{sch}}{V_3^{*(0)}} + y_{13}V_1 + y_{23}V_2^{(1)}}{y_{13} + y_{23}} \\
&= \frac{\frac{2.0 - j1.16}{1.04 - j0} + (10 - j30)(1.05 + j0) + (16 - j32)(0.97462 - j0.042307)}{(26 - j62)} \\
&= 1.03783 - j0.005170
\end{aligned}
$$

Since $|V_3|$ is held constant at 1.04 pu, only the imaginary part of $V_{c3}^{(1)}$ is retained, i.e, $f_3^{(1)} = -0.005170$, and its real part is obtained from

$$
e_3^{(1)} = \sqrt{(1.04)^2 - (0.005170)^2} = 1.039987
$$

Thus

$$
V_3^{(1)} = 1.039987 - j0.005170
$$

For the second iteration, we have

$$
\begin{aligned}
V_2^{(2)} &= \frac{\frac{P_2^{sch} - jQ_2^{sch}}{V_2^{*(1)}} + y_{12}V_1 + y_{23}V_3^{(1)}}{y_{12} + y_{23}} \\
&= \frac{\frac{-4.0 + j2.5}{.97462 + j.042307} + (10 - j20)(1.05) + (16 - j32)(1.039987 + j0.005170)}{(26 - j52)} \\
&= 0.971057 - j0.043432
\end{aligned}
$$

$$
\begin{aligned}
Q_3^{(2)} &= -\Im\{V_3^{*(1)}[V_3^{(1)}(y_{13} + y_{23}) - y_{13}V_1 - y_{23}V_2^{(2)}]\} \\
&= -\Im\{(1.039987 + j0.005170)[(1.039987 - j0.005170)(26 - j62) - \\
&\quad (10 - j30)(1.05 + j0) - (16 - j32)(0.971057 - j0.043432)]\} \\
&= 1.38796
\end{aligned}
$$

$$
\begin{aligned}
V_{c3}^{(2)} &= \frac{\frac{P_3^{sch} - jQ_3^{sch}}{V_3^{*(1)}} + y_{13}V_1 + y_{23}V_2^{(2)}}{y_{13} + y_{23}} \\
&= \frac{\frac{2.0 - j1.38796}{1.039987 + j0.00517} + (10 - j30)(1.05) + (16 - j32)(.971057 - j.043432)}{(26 - j62)} \\
&= 1.03908 - j0.00730
\end{aligned}
$$

Since $|V_3|$ is held constant at 1.04 pu, only the imaginary part of $V_{c3}^{(2)}$ is retained, i.e, $f_3^{(2)} = -0.00730$, and its real part is obtained from

$$e_3^{(2)} = \sqrt{(1.04)^2 - (0.00730)^2} = 1.039974$$

or

$$V_3^{(2)} = 1.039974 - j0.00730$$

The process is continued and a solution is converged with an accuracy of 5×10^{-5} pu in seven iterations as given below.

$$
\begin{array}{lll}
V_2^{(3)} = 0.97073 - j0.04479 & Q_3^{(3)} = 1.42904 & V_3^{(3)} = 1.03996 - j0.00833 \\
V_2^{(4)} = 0.97065 - j0.04533 & Q_3^{(4)} = 1.44833 & V_3^{(4)} = 1.03996 - j0.00873 \\
V_2^{(5)} = 0.97062 - j0.04555 & Q_3^{(5)} = 1.45621 & V_3^{(5)} = 1.03996 - j0.00893 \\
V_2^{(6)} = 0.97061 - j0.04565 & Q_3^{(6)} = 1.45947 & V_3^{(6)} = 1.03996 - j0.00900 \\
V_2^{(7)} = 0.97061 - j0.04569 & Q_3^{(7)} = 1.46082 & V_3^{(7)} = 1.03996 - j0.00903
\end{array}
$$

The final solution is

$$V_2 = 0.97168\angle{-2.6948^\circ} \quad \text{pu}$$

$$
\begin{aligned}
S_3 &= 2.0 + j1.4617 \quad \text{pu} \\
V_3 &= 1.04\angle{-.498^\circ} \quad \text{pu} \\
S_1 &= 2.1842 + j1.4085 \quad \text{pu}
\end{aligned}
$$

Line flows and line losses are computed as in Example 6.7, and the results expressed in MW and Mvar are

$$
\begin{array}{lll}
S_{12} = 179.36 + j118.734 & S_{21} = -170.97 - j101.947 & S_{L\,12} = 8.39 + j16.79 \\
S_{13} = 39.06 + j22.118 & S_{31} = -38.88 - j\,21.569 & S_{L\,13} = 0.18 + j0.548 \\
S_{23} = -229.03 - j148.05 & S_{32} = 238.88 + j167.746 & S_{L\,23} = 9.85 + j19.69
\end{array}
$$

The power flow diagram is shown in Figure 6.13, where real power direction is indicated by $\rightarrow$ and the reactive power direction is indicated by $\mapsto$. The values within parentheses are the real and reactive losses in the line.

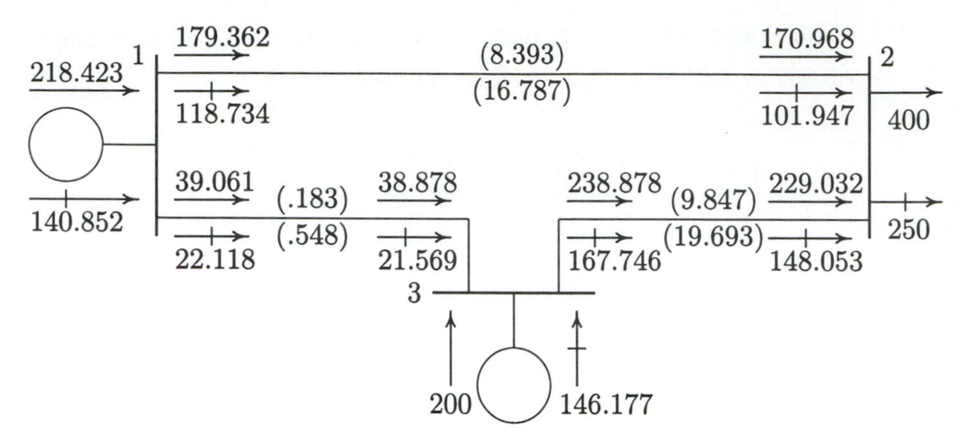

FIGURE 6.13
Power flow diagram of Example 6.8 (powers in MW and Mvar).

6.7 TAP CHANGING TRANSFORMERS

In Section 2.6 it was shown that the flow of real power along a transmission line is determined by the angle difference of the terminal voltages, and the flow of reactive power is determined mainly by the magnitude difference of terminal voltages. Real and reactive powers can be controlled by use of tap changing transformers and regulating transformers.

In a tap changing transformer, when the ratio is at the nominal value, the transformer is represented by a series admittance y_t in per unit. With off-nominal ratio, the per unit admittance is different from both sides of the transformer, and the admittance must be modified to include the effect of the off-nominal ratio. Consider a transformer with admittance y_t in series with an ideal transformer representing the off-nominal tap ratio 1:a as shown in Figure 6.14. y_t is the admittance in per unit based on the nominal turn ratio and a is the per unit off-nominal tap position allowing for small adjustment in voltage of usually ± 10 percent. In the case of phase shifting transformers, a is a complex number. Consider a fictitious bus x between the turn ratio and admittance of the transformer. Since the complex power on either side of the ideal transformer is the same, it follows that if the voltage goes through a positive phase angle shift, the current will go through a negative phase angle shift. Thus, for the assumed direction of currents, we have

$$V_x = \frac{1}{a} V_j \tag{6.43}$$

$$I_i = -a^* I_j \tag{6.44}$$

The current I_i is given by

$$I_i = y_t(V_i - V_x)$$

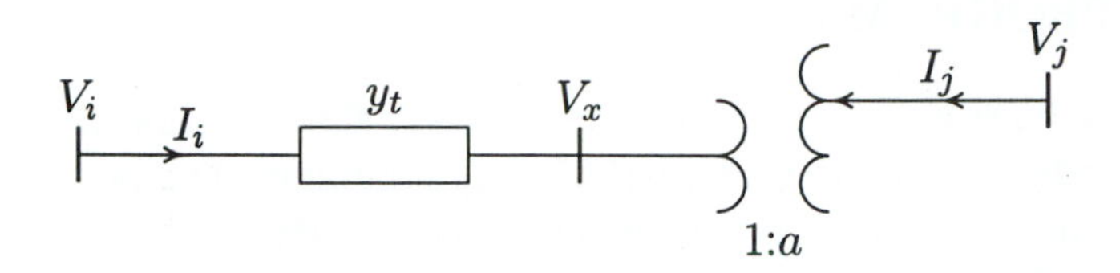

FIGURE 6.14
Transformer with tap setting ratio a:1

Substituting for V_x, we have

$$I_i = y_t V_i - \frac{y_t}{a} V_j \tag{6.45}$$

Also, from (6.44) we have

$$I_j = -\frac{1}{a^*} I_i$$

substituting for I_i from (6.45) we have

$$I_j = -\frac{y_t}{a^*} V_i + \frac{y_t}{|a|^2} V_j \tag{6.46}$$

writing (6.45) and (6.46) in matrix form results in

$$\begin{bmatrix} I_i \\ I_j \end{bmatrix} = \begin{bmatrix} y_t & -\frac{y_t}{a} \\ -\frac{y_t}{a^*} & \frac{y_t}{|a|^2} \end{bmatrix} \begin{bmatrix} V_i \\ V_j \end{bmatrix} \tag{6.47}$$

For the case when a is real, the π model shown in Figure 6.15 represents the admittance matrix in (6.47). In the π model, the left side corresponds to the non-tap side and the right side corresponds to the tap side of the transformer.

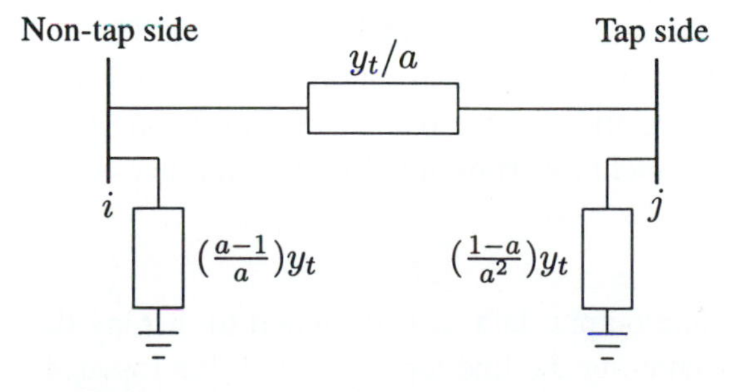

FIGURE 6.15
Equivalent circuit for a tap changing transformer.

6.8 POWER FLOW PROGRAMS

Several computer programs have been developed for the power flow solution of practical systems. Each method of solution consists of four programs. The program for the Gauss-Seidel method is **lfgauss**, which is preceded by **lfybus**, and is followed by **busout** and **lineflow**. Programs **lfybus**, **busout**, and **lineflow** are designed to be used with two more power flow programs. These are **lfnewton** for the Newton-Raphson method and **decouple** for the fast decoupled method. The following is a brief description of the programs used in the Gauss-Seidel method.

lfybus This program requires the line and transformer parameters and transformer tap settings specified in the input file named **linedata**. It converts impedances to admittances and obtains the bus admittance matrix. The program is designed to handle parallel lines.

lfgauss This program obtains the power flow solution by the Gauss-Seidel method and requires the files named **busdata** and **linedata**. It is designed for the direct use of load and generation in MW and Mvar, bus voltages in per unit, and angle in degrees. Loads and generation are converted to per unit quantities on the base MVA selected. A provision is made to maintain the generator reactive power of the voltage-controlled buses within their specified limits. The violation of reactive power limit may occur if the specified voltage is either too high or too low. After a few iterations (10th iteration in the Gauss method), the var calculated at the generator buses are examined. If a limit is reached, the voltage magnitude is adjusted in steps of 0.5 percent up to ± 5 percent to bring the var demand within the specified limits.

busout This program produces the bus output result in a tabulated form. The bus output result includes the voltage magnitude and angle, real and reactive power of generators and loads, and the shunt capacitor/reactor Mvar. Total generation and total load are also included as outlined in the sample case.

lineflow This program prepares the line output data. It is designed to display the active and reactive power flow entering the line terminals and line losses as well as the net power at each bus. Also included are the total real and reactive losses in the system. The output of this portion is also shown in the sample case.

6.9 DATA PREPARATION

In order to perform a power flow analysis by the Gauss-Seidel method in the *MATLAB* environment, the following variables must be defined: power system base MVA, power mismatch accuracy, acceleration factor, and maximum number of iterations. The name (in lowercase letters) reserved for these variables are **basemva**, **accuracy**, **accel**, and **maxiter**, respectively. Typical values are as follows:

```
basemva = 100;    accuracy = 0.001;
accel   = 1.6;    maxiter = 80;
```

The initial step in the preparation of input file is the numbering of each bus. Buses are numbered sequentially. Although the numbers are sequentially assigned, the buses need not be entered in sequence. In addition, the following data files are required.

BUS DATA FILE – busdata The format for the bus entry is chosen to facilitate the required data for each bus in a single row. The information required must be included in a matrix called **busdata**. Column 1 is the bus number. Column 2 contains the bus code. Columns 3 and 4 are voltage magnitude in per unit and phase angle in degrees. Columns 5 and 6 are load MW and Mvar. Column 7 through 10 are MW, Mvar, minimum Mvar and maximum Mvar of generation, in that order. The last column is the injected Mvar of shunt capacitors. The bus code entered in column 2 is used for identifying load, voltage-controlled, and slack buses as outlined below:

1 This code is used for the slack bus. The only necessary information for this bus is the voltage magnitude and its phase angle.

0 This code is used for load buses. The loads are entered positive in megawatts and megavars. For this bus, initial voltage estimate must be specified. This is usually 1 and 0 for voltage magnitude and phase angle, respectively. If voltage magnitude and phase angle for this type of bus are specified, they will be taken as the initial starting voltage for that bus instead of a flat start of 1 and 0.

2 This code is used for the voltage-controlled buses. For this bus, voltage magnitude, real power generation in megawatts, and the minimum and maximum limits of the megavar demand must be specified.

LINE DATA FILE – linedata Lines are identified by the node-pair method. The information required must be included in a matrix called **linedata**. Columns 1 and 2 are the line bus numbers. Columns 3 through 5 contain the line resistance, reactance, and one-half of the total line charging susceptance in per unit on the specified

MVA base. The last column is for the transformer tap setting; for lines, 1 must be entered in this column. The lines may be entered in any sequence or order with the only restriction being that if the entry is a transformer, the left bus number is assumed to be the tap side of the transformer.

The IEEE 30 bus system is used to demonstrate the data preparation and the use of the power flow programs by the Gauss-Seidel method.

Example 6.9

Figure 6.16 is part of the American Electric Power Service Corporation network which is being made available to the electric utility industry as a standard test case for evaluating various analytical methods and computer programs for the solution of power system problems. Use the **lfgauss** program to obtain the power solution by the Gauss-Seidel method. Bus 1 is taken as the slack bus with its voltage adjusted to $1.06\angle 0^\circ$ pu. The data for the voltage-controlled buses is

Regulated Bus Data			
Bus No.	Voltage Magnitude	Min. Mvar Capacity	Max. Mvar Capacity
2	1.043	-40	50
5	1.010	-40	40
8	1.010	-10	40
11	1.082	-6	24
13	1.071	-6	24

Transformer tap setting are given in the table below. The left bus number is assumed to be the tap side of the transformer.

Transformer Data	
Transformer Designation	Tap Setting pu
4 – 12	0.932
6 – 9	0.978
6 – 10	0.969
28 – 27	0.968

The data for the injected Q due to shunt capacitors is

Injected Q due to Capacitors	
Bus No.	Mvar
10	19
24	4.3

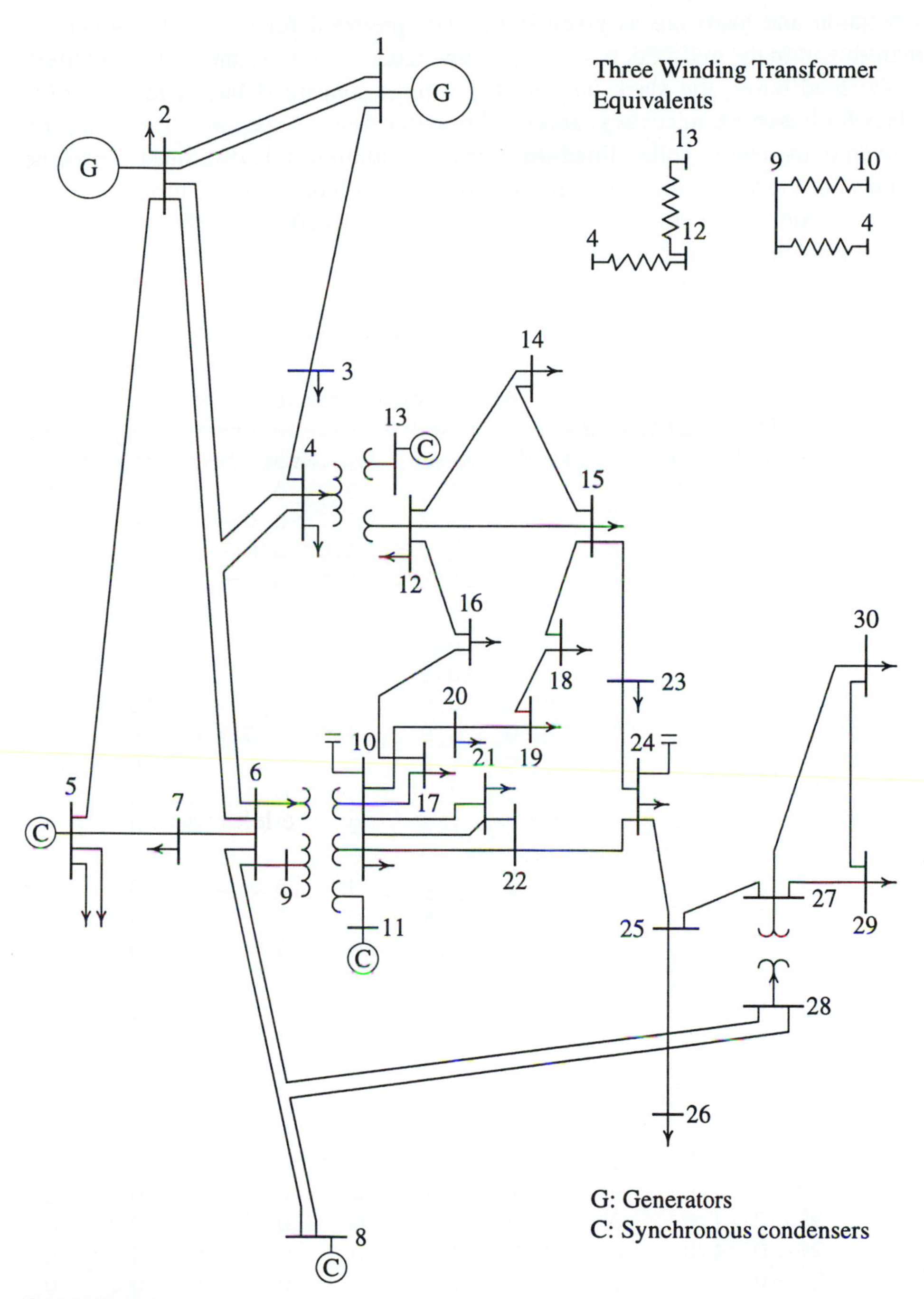

FIGURE 6.16
30-bus IEEE sample system.

Generation and loads are as given in the data prepared for use in the *MATLAB* environment in the matrix defined as **busdata**. Code 0, code 1, and code 2 are used for the load buses, the slack bus and the voltage-controlled buses, respectively. Values for **basemva**, **accuracy**, **accel** and **maxiter** must be specified. Line data are as given in the matrix called **linedata**. The last column of this data must contain 1 for lines, or the tap setting values for transformers with off-nominal turn ratio. The control commands required are **lfybus**, **lfgauss** and **lineflow**. A **diary** command may be used to save the output to the specified file name. The power flow data and the commands required are as follows.

```
clear          % clears all variables from workspace.
basemva = 100;  accuracy = 0.001; accel = 1.8; maxiter = 100;
%   IEEE 30-BUS TEST SYSTEM (American Electric Power)
%       Bus Bus Voltage Angle --Load--  ---Generator---Injected
%        No code Mag. Degree MW  Mvar   MW  Mvar Qmin Qmax Mvar
busdata=[1  1  1.06   0     0.0  0.0    0.0  0.0   0   0     0
         2  2  1.043  0    21.70 12.7  40.0  0.0 -40  50     0
         3  0  1.0    0     2.4   1.2   0.0  0.0   0   0     0
         4  0  1.06   0     7.6   1.6   0.0  0.0   0   0     0
         5  2  1.01   0    94.2  19.0   0.0  0.0 -40  40     0
         6  0  1.0    0     0.0   0.0   0.0  0.0   0   0     0
         7  0  1.0    0    22.8  10.9   0.0  0.0   0   0     0
         8  2  1.01   0    30.0  30.0   0.0  0.0 -10  40     0
         9  0  1.0    0     0.0   0.0   0.0  0.0   0   0     0
        10  0  1.0    0     5.8   2.0   0.0  0.0   0   0    19
        11  2  1.082  0     0.0   0.0   0.0  0.0  -6  24     0
        12  0  1.0    0    11.2   7.5   0    0     0   0     0
        13  2  1.071  0     0.0   0.0   0    0    -6  24     0
        14  0  1.0    0     6.2   1.6   0    0     0   0     0
        15  0  1.0    0     8.2   2.5   0    0     0   0     0
        16  0  1.0    0     3.5   1.8   0    0     0   0     0
        17  0  1.0    0     9.0   5.8   0    0     0   0     0
        18  0  1.0    0     3.2   0.9   0    0     0   0     0
        19  0  1.0    0     9.5   3.4   0    0     0   0     0
        20  0  1.0    0     2.2   0.7   0    0     0   0     0
        21  0  1.0    0    17.5  11.2   0    0     0   0     0
        22  0  1.0    0     0.0   0.0   0    0     0   0     0
        23  0  1.0    0     3.2   1.6   0    0     0   0     0
        24  0  1.0    0     8.7   6.7   0    0     0   0   4.3
        25  0  1.0    0     0.0   0.0   0    0     0   0     0
        26  0  1.0    0     3.5   2.3   0    0     0   0     0
        27  0  1.0    0     0.0   0.0   0    0     0   0     0
        28  0  1.0    0     0.0   0.0   0    0     0   0     0
        29  0  1.0    0     2.4   0.9   0    0     0   0     0
        30  0  1.0    0    10.6   1.9   0    0     0   0    0];
```

```
% Line Data
%
%          Bus  bus   R         X        1/2 B 1 for Line code or
%          nl   nr    pu        pu       pu    tap setting value
linedata=[1     2     0.0192    0.0575   0.02640      1
          1     3     0.0452    0.1852   0.02040      1
          2     4     0.0570    0.1737   0.01840      1
          3     4     0.0132    0.0379   0.00420      1
          2     5     0.0472    0.1983   0.02090      1
          2     6     0.0581    0.1763   0.01870      1
          4     6     0.0119    0.0414   0.00450      1
          5     7     0.0460    0.1160   0.01020      1
          6     7     0.0267    0.0820   0.00850      1
          6     8     0.0120    0.0420   0.00450      1
          6     9     0.0       0.2080   0.0      0.978
          6    10     0.0       0.5560   0.0      0.969
          9    11     0.0       0.2080   0.0          1
          9    10     0.0       0.1100   0.0          1
          4    12     0.0       0.2560   0.0      0.932
         12    13     0.0       0.1400   0.0          1
         12    14     0.1231    0.2559   0.0          1
         12    15     0.0662    0.1304   0.0          1
         12    16     0.0945    0.1987   0.0          1
         14    15     0.2210    0.1997   0.0          1
         16    17     0.0824    0.1923   0.0          1
         15    18     0.1073    0.2185   0.0          1
         18    19     0.0639    0.1292   0.0          1
         19    20     0.0340    0.0680   0.0          1
         10    20     0.0936    0.2090   0.0          1
         10    17     0.0324    0.0845   0.0          1
         10    21     0.0348    0.0749   0.0          1
         10    22     0.0727    0.1499   0.0          1
         21    22     0.0116    0.0236   0.0          1
         15    23     0.1000    0.2020   0.0          1
         22    24     0.1150    0.1790   0.0          1
         23    24     0.1320    0.2700   0.0          1
         24    25     0.1885    0.3292   0.0          1
         25    26     0.2544    0.3800   0.0          1
         25    27     0.1093    0.2087   0.0          1
         28    27     0.0000    0.3960   0.0      0.968
         27    29     0.2198    0.4153   0.0          1
         27    30     0.3202    0.6027   0.0          1
         29    30     0.2399    0.4533   0.0          1
          8    28     0.0636    0.2000   0.0214       1
          6    28     0.0169    0.0599   0.065        1];
```

```
%
    lfybus                  % Forms the bus admittance matrix
    lfgauss        % Power flow solution by Gauss-Seidel method
    busout       % Prints the power flow solution on the screen
    lineflow  % Computes and displays the line flow and losses
```

The **lfgauss**, **busout** and the **lineflow** produce the following tabulated results.

```
          Power Flow Solution by Gauss-Seidel Method
             Maximum Power mismatch = 0.000951884
                   No. of iterations = 34
```

Bus No.	Voltage Mag.	Angle Degree	-----Load----- MW	Mvar	--Generation-- MW	Mvar	Injected Mvar
1	1.060	0.000	0.000	0.000	260.950	-17.010	0.00
2	1.043	-5.496	21.700	12.700	40.000	48.826	0.00
3	1.022	-8.002	2.400	1.200	0.000	0.000	0.00
4	1.013	-9.659	7.600	1.600	0.000	0.000	0.00
5	1.010	-14.380	94.200	19.000	0.000	35.995	0.00
6	1.012	-11.396	0.000	0.000	0.000	0.000	0.00
7	1.003	-13.149	22.800	10.900	0.000	0.000	0.00
8	1.010	-12.114	30.000	30.000	0.000	30.759	0.00
9	1.051	-14.432	0.000	0.000	0.000	0.000	0.00
10	1.044	-16.024	5.800	2.000	0.000	0.000	19.00
11	1.082	-14.432	0.000	0.000	0.000	16.113	0.00
12	1.057	-15.301	11.200	7.500	0.000	0.000	0.00
13	1.071	-15.300	0.000	0.000	0.000	10.406	0.00
14	1.043	-16.190	6.200	1.600	0.000	0.000	0.00
15	1.038	-16.276	8.200	2.500	0.000	0.000	0.00
16	1.045	-15.879	3.500	1.800	0.000	0.000	0.00
17	1.039	-16.187	9.000	5.800	0.000	0.000	0.00
18	1.028	-16.881	3.200	0.900	0.000	0.000	0.00
19	1.025	-17.049	9.500	3.400	0.000	0.000	0.00
20	1.029	-16.851	2.200	0.700	0.000	0.000	0.00
21	1.032	-16.468	17.500	11.200	0.000	0.000	0.00
22	1.033	-16.455	0.000	0.000	0.000	0.000	0.00
23	1.027	-16.660	3.200	1.600	0.000	0.000	0.00
24	1.022	-16.829	8.700	6.700	0.000	0.000	4.30
25	1.019	-16.423	0.000	0.000	0.000	0.000	0.00
26	1.001	-16.835	3.500	2.300	0.000	0.000	0.00
27	1.026	-15.913	0.000	0.000	0.000	0.000	0.00
28	1.011	-12.056	0.000	0.000	0.000	0.000	0.00

```
29  1.006  -17.133    2.400    0.900    0.000    0.000    0.00
30  0.994  -18.016   10.600    1.900    0.000    0.000    0.00

Total                283.400  126.200  300.950  125.089   23.30

                    Line Flow and Losses

--Line--  Power at bus & line flow    --Line loss-- Transformer
from  to    MW       Mvar      MVA      MW      Mvar      tap

 1        260.950  -17.010  261.504
       2  177.743  -22.140  179.117   5.461   10.517
       3   83.197    5.125   83.354   2.807    7.079

 2         18.300   36.126   40.497
       1 -172.282   32.657  175.350   5.461   10.517
       4   45.702    2.720   45.783   1.106   -0.519
       5   82.990    1.704   83.008   2.995    8.178
       6   61.905   -0.966   61.913   2.047    2.263

 3         -2.400   -1.200    2.683
       1  -80.390    1.954   80.414   2.807    7.079
       4   78.034   -3.087   78.095   0.771    1.345

 4         -7.600   -1.600    7.767
       2  -44.596   -3.239   44.713   1.106   -0.519
       3  -77.263    4.432   77.390   0.771    1.345
       6   70.132  -17.624   72.313   0.605    1.181
      12   44.131   14.627   46.492   0.000    4.686     0.932

 5        -94.200   16.995   95.721
       2  -79.995    6.474   80.256   2.995    8.178
       7  -14.210   10.467   17.649   0.151   -1.687

 6          0.000    0.000    0.000
       2  -59.858    3.229   59.945   2.047    2.263
       4  -69.527   18.805   72.026   0.605    1.181
       7   37.537   -1.915   37.586   0.368   -0.598
       8   29.534   -3.712   29.766   0.103   -0.558
       9   27.687   -7.318   28.638   0.000    1.593     0.978
      10   15.828    0.656   15.842   0.000    1.279     0.969
      28   18.840   -9.575   21.134   0.060  -13.085

 7        -22.800  -10.900   25.272
       5   14.361  -12.154   18.814   0.151   -1.687
```

	6	-37.170	1.317	37.193	0.368	-0.598
8		-30.000	0.759	30.010		
	6	-29.431	3.154	29.599	0.103	-0.558
	28	-0.570	-2.366	2.433	0.000	-4.368
9		0.000	0.000	0.000		
	6	-27.687	8.911	29.086	0.000	1.593
	11	0.003	-15.653	15.653	-0.000	0.461
	10	27.731	6.747	28.540	0.000	0.811
10		-5.800	17.000	17.962		
	6	-15.828	0.623	15.840	0.000	1.279
	9	-27.731	-5.936	28.359	0.000	0.811
	20	9.018	3.569	9.698	0.081	0.180
	17	5.347	4.393	6.920	0.014	0.037
	21	15.723	9.846	18.551	0.110	0.236
	22	7.582	4.487	8.811	0.052	0.107
11		0.000	16.113	16.113		
	9	-0.003	16.114	16.114	-0.000	0.461
12		-11.200	-7.500	13.479		
	4	-44.131	-9.941	45.237	0.000	4.686
	13	-0.021	-10.274	10.274	0.000	0.132
	14	7.852	2.428	8.219	0.074	0.155
	15	17.852	6.968	19.164	0.217	0.428
	16	7.206	3.370	7.955	0.053	0.112
13		0.000	10.406	10.406		
	12	0.021	10.406	10.406	0.000	0.132
14		-6.200	-1.600	6.403		
	12	-7.778	-2.273	8.103	0.074	0.155
	15	1.592	0.708	1.742	0.006	0.006
15		-8.200	-2.500	8.573		
	12	-17.634	-6.540	18.808	0.217	0.428
	14	-1.586	-0.702	1.734	0.006	0.006
	18	6.009	1.741	6.256	0.039	0.079
	23	5.004	2.963	5.815	0.031	0.063
16		-3.500	-1.800	3.936		
	12	-7.152	-3.257	7.859	0.053	0.112
	17	3.658	1.440	3.931	0.012	0.027

17		-9.000	-5.800	10.707		
	16	-3.646	-1.413	3.910	0.012	0.027
	10	-5.332	-4.355	6.885	0.014	0.037
18		-3.200	-0.900	3.324		
	15	-5.970	-1.661	6.197	0.039	0.079
	19	2.779	0.787	2.888	0.005	0.010
19		-9.500	-3.400	10.090		
	18	-2.774	-0.777	2.881	0.005	0.010
	20	-6.703	-2.675	7.217	0.017	0.034
20		-2.200	-0.700	2.309		
	19	6.720	2.709	7.245	0.017	0.034
	10	-8.937	-3.389	9.558	0.081	0.180
21		-17.500	-11.200	20.777		
	10	-15.613	-9.609	18.333	0.110	0.236
	22	-1.849	-1.627	2.463	0.001	0.001
22		0.000	0.000	0.000		
	10	-7.531	-4.380	8.712	0.052	0.107
	21	1.850	1.628	2.464	0.001	0.001
	24	5.643	2.795	6.297	0.043	0.067
23		-3.200	-1.600	3.578		
	15	-4.972	-2.900	5.756	0.031	0.063
	24	1.771	1.282	2.186	0.006	0.012
24		-8.700	-2.400	9.025		
	22	-5.601	-2.728	6.230	0.043	0.067
	23	-1.765	-1.270	2.174	0.006	0.012
	25	-1.322	1.604	2.079	0.008	0.014
25		0.000	0.000	0.000		
	24	1.330	-1.590	2.073	0.008	0.014
	26	3.520	2.372	4.244	0.044	0.066
	27	-4.866	-0.786	4.929	0.026	0.049
26		-3.500	-2.300	4.188		
	25	-3.476	-2.306	4.171	0.044	0.066
27		0.000	0.000	0.000		
	25	4.892	0.835	4.963	0.026	0.049

	28	-18.192	-4.152	18.660	-0.000	1.310	
	29	6.178	1.675	6.401	0.086	0.162	
	30	7.093	1.663	7.286	0.162	0.304	
28		0.000	0.000	0.000			
	27	18.192	5.463	18.994	-0.000	1.310	0.968
	8	0.570	-2.003	2.082	0.000	-4.368	
	6	-18.780	-3.510	19.106	0.060	-13.085	
29		-2.400	-0.900	2.563			
	27	-6.093	-1.513	6.278	0.086	0.162	
	30	3.716	0.601	3.764	0.034	0.063	
30		-10.600	-1.900	10.769			
	27	-6.932	-1.359	7.064	0.162	0.304	
	29	-3.683	-0.537	3.722	0.034	0.063	
Total loss					17.594	22.233	

6.10 NEWTON-RAPHSON POWER FLOW SOLUTION

Because of its quadratic convergence, Newton's method is mathematically superior to the Gauss-Seidel method and is less prone to divergence with ill-conditioned problems. For large power systems, the Newton-Raphson method is found to be more efficient and practical. The number of iterations required to obtain a solution is independent of the system size, but more functional evaluations are required at each iteration. Since in the power flow problem real power and voltage magnitude are specified for the voltage-controlled buses, the power flow equation is formulated in polar form. For the typical bus of the power system shown in Figure 6.7, the current entering bus i is given by (6.24). This equation can be rewritten in terms of the bus admittance matrix as

$$I_i = \sum_{j=1}^{n} Y_{ij} V_j \tag{6.48}$$

In the above equation, j includes bus i. Expressing this equation in polar form, we have

$$I_i = \sum_{j=1}^{n} |Y_{ij}||V_j| \angle \theta_{ij} + \delta_j \tag{6.49}$$

The complex power at bus i is

$$P_i - jQ_i = V_i^* I_i \tag{6.50}$$

Substituting from (6.49) for I_i in (6.50),

$$P_i - jQ_i = |V_i|\angle -\delta_i \sum_{j=1}^{n} |Y_{ij}||V_j|\angle\theta_{ij} + \delta_j \tag{6.51}$$

Separating the real and imaginary parts,

$$P_i = \sum_{j=1}^{n} |V_i||V_j||Y_{ij}| \cos(\theta_{ij} - \delta_i + \delta_j) \tag{6.52}$$

$$Q_i = -\sum_{j=1}^{n} |V_i||V_j||Y_{ij}| \sin(\theta_{ij} - \delta_i + \delta_j) \tag{6.53}$$

Equations (6.52) and (6.53) constitute a set of nonlinear algebraic equations in terms of the independent variables, voltage magnitude in per unit, and phase angle in radians. We have two equations for each load bus, given by (6.52) and (6.53), and one equation for each voltage-controlled bus, given by (6.52). Expanding (6.52) and (6.53) in Taylor's series about the initial estimate and neglecting all higher order terms results in the following set of linear equations.

$$\left[\begin{array}{c} \Delta P_2^{(k)} \\ \vdots \\ \Delta P_n^{(k)} \\ \hline \Delta Q_2^{(k)} \\ \vdots \\ \Delta Q_n^{(k)} \end{array}\right] = \left[\begin{array}{ccc|ccc} \frac{\partial P_2}{\partial \delta_2}^{(k)} & \cdots & \frac{\partial P_2}{\partial \delta_n}^{(k)} & \frac{\partial P_2}{\partial |V_2|}^{(k)} & \cdots & \frac{\partial P_2}{\partial |V_n|}^{(k)} \\ \vdots & \ddots & \vdots & \vdots & \ddots & \vdots \\ \frac{\partial P_n}{\partial \delta_2}^{(k)} & \cdots & \frac{\partial P_n}{\partial \delta_n}^{(k)} & \frac{\partial P_n}{\partial |V_2|}^{(k)} & \cdots & \frac{\partial P_n}{\partial |V_n|}^{(k)} \\ \hline \frac{\partial Q_2}{\partial \delta_2}^{(k)} & \cdots & \frac{\partial Q_2}{\partial \delta_n}^{(k)} & \frac{\partial Q_2}{\partial |V_2|}^{(k)} & \cdots & \frac{\partial Q_2}{\partial |V_n|}^{(k)} \\ \vdots & \ddots & \vdots & \vdots & \ddots & \vdots \\ \frac{\partial Q_n}{\partial \delta_2}^{(k)} & \cdots & \frac{\partial Q_n}{\partial \delta_n}^{(k)} & \frac{\partial Q_n}{\partial |V_2|}^{(k)} & \cdots & \frac{\partial Q_n}{\partial |V_n|}^{(k)} \end{array}\right] \left[\begin{array}{c} \Delta \delta_2^{(k)} \\ \vdots \\ \Delta \delta_n^{(k)} \\ \hline \Delta |V_2^{(k)}| \\ \vdots \\ \Delta |V_n^{(k)}| \end{array}\right]$$

In the above equation, bus 1 is assumed to be the slack bus. The Jacobian matrix gives the linearized relationship between small changes in voltage angle $\Delta\delta_i^{(k)}$ and voltage magnitude $\Delta|V_i^{(k)}|$ with the small changes in real and reactive power $\Delta P_i^{(k)}$ and $\Delta Q_i^{(k)}$. Elements of the Jacobian matrix are the partial derivatives of (6.52) and (6.53), evaluated at $\Delta\delta_i^{(k)}$ and $\Delta|V_i^{(k)}|$. In short form, it can be written as

$$\begin{bmatrix} \Delta P \\ \Delta Q \end{bmatrix} = \begin{bmatrix} J_1 & J_2 \\ J_3 & J_4 \end{bmatrix} \begin{bmatrix} \Delta\delta \\ \Delta|V| \end{bmatrix} \tag{6.54}$$

For voltage-controlled buses, the voltage magnitudes are known. Therefore, if m buses of the system are voltage-controlled, m equations involving ΔQ and ΔV

and the corresponding columns of the Jacobian matrix are eliminated. Accordingly, there are $n-1$ real power constraints and $n-1-m$ reactive power constraints, and the Jacobian matrix is of order $(2n-2-m) \times (2n-2-m)$. $\boldsymbol{J_1}$ is of the order $(n-1) \times (n-1)$, $\boldsymbol{J_2}$ is of the order $(n-1) \times (n-1-m)$, $\boldsymbol{J_3}$ is of the order $(n-1-m) \times (n-1)$, and $\boldsymbol{J_4}$ is of the order $(n-1-m) \times (n-1-m)$.

The diagonal and the off-diagonal elements of $\boldsymbol{J_1}$ are

$$\frac{\partial P_i}{\partial \delta_i} = \sum_{j \neq i} |V_i||V_j||Y_{ij}| \sin(\theta_{ij} - \delta_i + \delta_j) \tag{6.55}$$

$$\frac{\partial P_i}{\partial \delta_j} = -|V_i||V_j||Y_{ij}| \sin(\theta_{ij} - \delta_i + \delta_j) \qquad j \neq i \tag{6.56}$$

The diagonal and the off-diagonal elements of $\boldsymbol{J_2}$ are

$$\frac{\partial P_i}{\partial |V_i|} = 2|V_i||Y_{ii}| \cos\theta_{ii} + \sum_{j \neq i} |V_j||Y_{ij}| \cos(\theta_{ij} - \delta_i + \delta_j) \tag{6.57}$$

$$\frac{\partial P_i}{\partial |V_j|} = |V_i||Y_{ij}| \cos(\theta_{ij} - \delta_i + \delta_j) \qquad j \neq i \tag{6.58}$$

The diagonal and the off-diagonal elements of $\boldsymbol{J_3}$ are

$$\frac{\partial Q_i}{\partial \delta_i} = \sum_{j \neq i} |V_i||V_j||Y_{ij}| \cos(\theta_{ij} - \delta_i + \delta_j) \tag{6.59}$$

$$\frac{\partial Q_i}{\partial \delta_j} = -|V_i||V_j||Y_{ij}| \cos(\theta_{ij} - \delta_i + \delta_j) \qquad j \neq i \tag{6.60}$$

The diagonal and the off-diagonal elements of $\boldsymbol{J_4}$ are

$$\frac{\partial Q_i}{\partial |V_i|} = -2|V_i||Y_{ii}| \sin\theta_{ii} - \sum_{j \neq i} |V_j||Y_{ij}| \sin(\theta_{ij} - \delta_i + \delta_j) \tag{6.61}$$

$$\frac{\partial Q_i}{\partial |V_j|} = -|V_i||Y_{ij}| \sin(\theta_{ij} - \delta_i + \delta_j) \qquad j \neq i \tag{6.62}$$

The terms $\Delta P_i^{(k)}$ and $\Delta Q_i^{(k)}$ are the difference between the scheduled and calculated values, known as the *power residuals*, given by

$$\Delta P_i^{(k)} = P_i^{sch} - P_i^{(k)} \tag{6.63}$$

$$\Delta Q_i^{(k)} = Q_i^{sch} - Q_i^{(k)} \tag{6.64}$$

The new estimates for bus voltages are

$$\delta_i^{(k+1)} = \delta_i^{(k)} + \Delta\delta_i^{(k)} \tag{6.65}$$

$$|V_i^{(k+1)}| = |V_i^{(k)}| + \Delta|V_i^{(k)}| \tag{6.66}$$

The procedure for power flow solution by the Newton-Raphson method is as follows:

1. For load buses, where P_i^{sch} and Q_i^{sch} are specified, voltage magnitudes and phase angles are set equal to the slack bus values, or 1.0 and 0.0, i.e., $|V_i^{(0)}| = 1.0$ and $\delta_i^{(0)} = 0.0$. For voltage-regulated buses, where $|V_i|$ and P_i^{sch} are specified, phase angles are set equal to the slack bus angle, or 0, i.e., $\delta_i^{(0)} = 0$.

2. For load buses, $P_i^{(k)}$ and $Q_i^{(k)}$ are calculated from (6.52) and (6.53) and $\Delta P_i^{(k)}$ and $\Delta Q_i^{(k)}$ are calculated from (6.63) and (6.64).

3. For voltage-controlled buses, $P_i^{(k)}$ and $\Delta P_i^{(k)}$ are calculated from (6.52) and (6.63), respectively.

4. The elements of the Jacobian matrix (J_1, J_2, J_3, and J_4) are calculated from (6.55) – (6.62).

5. The linear simultaneous equation (6.54) is solved directly by optimally ordered triangular factorization and Gaussian elimination.

6. The new voltage magnitudes and phase angles are computed from (6.65) and (6.66).

7. The process is continued until the residuals $\Delta P_i^{(k)}$ and $\Delta Q_i^{(k)}$ are less than the specified accuracy, i.e.,

$$\begin{aligned} |\Delta P_i^{(k)}| &\le \epsilon \\ |\Delta Q_i^{(k)}| &\le \epsilon \end{aligned} \tag{6.67}$$

The power flow solution by the Newton-Raphson method is demonstrated in the following example.

Example 6.10

Obtain the power flow solution by the Newton-Raphson method for the system of Example 6.8.

Line impedances converted to admittances are $y_{12} = 10 - j20$, $y_{13} = 10 - j30$, and $y_{23} = 16 - j32$. This results in the bus admittance matrix

$$Y_{bus} = \begin{bmatrix} 20 - j50 & -10 + j20 & -10 + j30 \\ -10 + j20 & 26 - j52 & -16 + j32 \\ -10 + j30 & -16 + j32 & 26 - j62 \end{bmatrix}$$

Converting the bus admittance matrix to polar form with angles in radian yields

$$Y_{bus} = \begin{bmatrix} 53.85165\angle{-1.9029} & 22.36068\angle 2.0344 & 31.62278\angle 1.8925 \\ 22.36068\angle 2.0344 & 58.13777\angle{-1.1071} & 35.77709\angle 2.0344 \\ 31.62278\angle 1.8925 & 35.77709\angle 2.0344 & 67.23095\angle{-1.1737} \end{bmatrix}$$

From (6.52) and (6.53), the expressions for real power at bus 2 and 3 and the reactive power at bus 2 are

$$\begin{aligned} P_2 &= |V_2||V_1||Y_{21}|\cos(\theta_{21} - \delta_2 + \delta_1) + |V_2^2||Y_{22}|\cos\theta_{22} + \\ &\quad |V_2||V_3||Y_{23}|\cos(\theta_{23} - \delta_2 + \delta_3) \\ P_3 &= |V_3||V_1||Y_{31}|\cos(\theta_{31} - \delta_3 + \delta_1) + |V_3||V_2||Y_{32}|\cos(\theta_{32} - \\ &\quad \delta_3 + \delta_2) + |V_3^2||Y_{33}|\cos\theta_{33} \\ Q_2 &= -|V_2||V_1||Y_{21}|\sin(\theta_{21} - \delta_2 + \delta_1) - |V_2^2||Y_{22}|\sin\theta_{22} - \\ &\quad |V_2||V_3||Y_{23}|\sin(\theta_{23} - \delta_2 + \delta_3) \end{aligned}$$

Elements of the Jacobian matrix are obtained by taking partial derivatives of the above equations with respect to δ_2, δ_3 and $|V_2|$.

$$\begin{aligned} \frac{\partial P_2}{\partial \delta_2} &= |V_2||V_1||Y_{21}|\sin(\theta_{21} - \delta_2 + \delta_1) + |V_2||V_3||Y_{23}| \\ &\quad \sin(\theta_{23} - \delta_2 + \delta_3) \\ \frac{\partial P_2}{\partial \delta_3} &= -|V_2||V_3||Y_{23}|\sin(\theta_{23} - \delta_2 + \delta_3) \\ \frac{\partial P_2}{\partial |V_2|} &= |V_1||Y_{21}|\cos(\theta_{21} - \delta_2 + \delta_1) + 2|V_2||Y_{22}|\cos\theta_{22} + \\ &\quad |V_3||Y_{23}|\cos(\theta_{23} - \delta_2 + \delta_3) \\ \frac{\partial P_3}{\partial \delta_2} &= -|V_3||V_2||Y_{32}|\sin(\theta_{32} - \delta_3 + \delta_2) \\ \frac{\partial P_3}{\partial \delta_3} &= |V_3||V_1||Y_{31}|\sin(\theta_{31} - \delta_3 + \delta_1) + |V_3||V_2||Y_{32}| \\ &\quad \sin(\theta_{32} - \delta_3 + \delta_2) \\ \frac{\partial P_3}{\partial |V_2|} &= |V_3||Y_{32}|\cos(\theta_{32} - \delta_3 + \delta_2) \end{aligned}$$

$$\frac{\partial Q_2}{\partial \delta_2} = |V_2||V_1||Y_{21}|\cos(\theta_{21} - \delta_2 + \delta_1) + |V_2||V_3||Y_{23}| \cos(\theta_{23} - \delta_2 + \delta_3)$$

$$\frac{\partial Q_2}{\partial \delta_3} = -|V_2||V_3||Y_{23}|\cos(\theta_{23} - \delta_2 + \delta_3)$$

$$\frac{\partial Q_2}{\partial |V_2|} = -|V_1||Y_{21}|\sin(\theta_{21} - \delta_2 + \delta_1) - 2|V_2||Y_{22}|\sin\theta_{22} - |V_3||Y_{23}|\sin(\theta_{23} - \delta_2 + \delta_3)$$

The load and generation expressed in per units are

$$S_2^{sch} = -\frac{(400 + j250)}{100} = -4.0 - j2.5 \quad \text{pu}$$

$$P_3^{sch} = \frac{200}{100} = 2.0 \quad \text{pu}$$

The slack bus voltage is $V_1 = 1.05\angle 0$ pu, and the bus 3 voltage magnitude is $|V_3| = 1.04$ pu. Starting with an initial estimate of $|V_2^{(0)}| = 1.0$, $\delta_2^{(0)} = 0.0$, and $\delta_3^{(0)} = 0.0$, the power residuals are computed from (6.63) and (6.64)

$$\Delta P_2^{(0)} = P_2^{sch} - P_2^{(0)} = -4.0 - (-1.14) = -2.8600$$

$$\Delta P_3^{(0)} = P_3^{sch} - P_3^{(0)} = 2.0 - (0.5616) = 1.4384$$

$$\Delta Q_2^{(0)} = Q_2^{sch} - Q_2^{(0)} = -2.5 - (-2.28) = -0.2200$$

Evaluating the elements of the Jacobian matrix with the initial estimate, the set of linear equations in the first iteration becomes

$$\begin{bmatrix} -2.8600 \\ 1.4384 \\ -0.2200 \end{bmatrix} = \begin{bmatrix} 54.28000 & -33.28000 & 24.86000 \\ -33.28000 & 66.04000 & -16.64000 \\ -27.14000 & 16.64000 & 49.72000 \end{bmatrix} \begin{bmatrix} \Delta\delta_2^{(0)} \\ \Delta\delta_3^{(0)} \\ \Delta|V_2^{(0)}| \end{bmatrix}$$

Obtaining the solution of the above matrix equation, the new bus voltages in the first iteration are

$$\Delta\delta_2^{(0)} = -0.045263 \qquad \delta_2^{(1)} = 0 + (-0.045263) = -0.045263$$

$$\Delta\delta_3^{(0)} = -0.007718 \qquad \delta_3^{(1)} = 0 + (-0.007718) = -0.007718$$

$$\Delta|V_2^{(0)}| = -0.026548 \qquad |V_2^{(1)}| = 1 + (-0.026548) = 0.97345$$

Voltage phase angles are in radians. For the second iteration, we have

$$\begin{bmatrix} -0.099218 \\ 0.021715 \\ -0.050914 \end{bmatrix} = \begin{bmatrix} 51.724675 & -31.765618 & 21.302567 \\ -32.981642 & 65.656383 & -15.379086 \\ -28.538577 & 17.402838 & 48.103589 \end{bmatrix} \begin{bmatrix} \Delta\delta_2^{(1)} \\ \Delta\delta_3^{(1)} \\ \Delta|V_2^{(1)}| \end{bmatrix}$$

and

$$\begin{array}{ll} \Delta\delta_2^{(1)} = -0.001795 & \delta_2^{(2)} = -0.045263 + (-0.001795) = -0.04706 \\ \Delta\delta_3^{(1)} = -0.000985 & \delta_3^{(2)} = -0.007718 + (-0.000985) = -0.00870 \\ \Delta|V_2^{(1)}| = -0.001767 & |V_2^{(2)}| = 0.973451 + (-0.001767) = 0.971684 \end{array}$$

For the third iteration, we have

$$\begin{bmatrix} -0.000216 \\ 0.000038 \\ -0.000143 \end{bmatrix} = \begin{bmatrix} 51.596701 & -31.693866 & 21.147447 \\ -32.933865 & 65.597585 & -15.351628 \\ -28.548205 & 17.396932 & 47.954870 \end{bmatrix} \begin{bmatrix} \Delta\delta_2^{(2)} \\ \Delta\delta_3^{(2)} \\ \Delta|V_2^{(2)}| \end{bmatrix}$$

and

$$\begin{array}{ll} \Delta\delta_2^{(2)} = -0.000038 & \delta_2^{(3)} = -0.047058 + (-0.0000038) = -0.04706 \\ \Delta\delta_3^{(2)} = -0.0000024 & \delta_3^{(3)} = -0.008703 + (-0.0000024) = 0.008705 \\ \Delta|V_2^{(2)}| = -0.0000044 & |V_2^{(3)}| = 0.971684 + (-0.0000044) = 0.97168 \end{array}$$

The solution converges in 3 iterations with a maximum power mismatch of 2.5×10^{-4} with $V_2 = 0.97168\angle{-2.696°}$ and $V_3 = 1.04\angle{-0.4988°}$. From (6.52) and (6.53), the expressions for reactive power at bus 3 and the slack bus real and reactive powers are

$$\begin{aligned} Q_3 &= -|V_3||V_1||Y_{31}|\sin(\theta_{31} - \delta_3 + \delta_1) - |V_3||V_2||Y_{32}| \\ &\quad \sin(\theta_{32} - \delta_3 + \delta_2) - |V_3|^2|Y_{33}|\sin\theta_{33} \\ P_1 &= |V_1|^2|Y_{11}|\cos\theta_{11} + |V_1||V_2||Y_{12}|\cos(\theta_{12} - \delta_1 + \delta_2) + |V_1||V_3| \\ &\quad |Y_{13}|\cos(\theta_{13} - \delta_1 + \delta_3) \\ Q_1 &= -|V_1|^2|Y_{11}|\sin\theta_{11} - |V_1||V_2||Y_{12}|\sin(\theta_{12} - \delta_1 + \delta_2) - |V_1||V_3| \\ &\quad |Y_{13}|\sin(\theta_{13} - \delta_1 + \delta_3) \end{aligned}$$

Upon substitution, we have

$$\begin{aligned} Q_3 &= 1.4617 \text{ pu} \\ P_1 &= 2.1842 \text{ pu} \\ Q_1 &= 1.4085 \text{ pu} \end{aligned}$$

Finally, the line flows are calculated in the same manner as the line flow calculations in the Gauss-Seidel method described in Example 6.7, and the power flow diagram is as shown in Figure 6.13.

A program named **lfnewton** is developed for power flow solution by the Newton-Raphson method for practical power systems. This program must be preceded by the **lfybus** program. **busout** and **lineflow** programs can be used to print the load flow solution and the line flow results. The format is the same as the Gauss-Seidel. The following is a brief description of the **lfnewton** program.

lfnewton This program obtains the power flow solution by the Newton-Raphson method and requires the **busdata** and the **linedata** files described in Section 6.9. It is designed for the direct use of load and generation in MW and Mvar, bus voltages in per unit, and angle in degrees. Loads and generation are converted to per unit quantities on the base MVA selected. A provision is made to maintain the generator reactive power of the voltage-controlled buses within their specified limits. The violation of reactive power limit may occur if the specified voltage is either too high or too low. In the second iteration, the var calculated at the generator buses are examined. If a limit is reached, the voltage magnitude is adjusted in steps of 0.5 percent up to ±5 percent to bring the var demand within the specified limits.

Example 6.11

Obtain the power flow solution for the IEEE-30 bus test system by the Newton-Raphson method.

The data required is the same as in Example 6.9 with the following commands

```
clear        % clears all variables from the workspace.
basemva = 100;  accuracy = 0.001;  maxiter = 12;

busdata =  [ same as in Example 6.9 ];
linedata = [ same as in Example 6.9 ];

     lfybus                    % Forms the bus admittance matrix
     lfnewton    % Power flow solution by Newton-Raphson method
     busout      % Prints the power flow solution on the screen
     lineflow % Computes and displays the line flow and losses
```

The output of **lfnewton** is

```
          Power Flow Solution by Newton-Raphson Method
             Maximum Power mismatch = 7.54898e-07
                    No. of iterations = 4

Bus Voltage  Angle     -----Load-----    --Generation-- Injected
No. Mag.     Degree    MW       Mvar      MW      Mvar     Mvar

1   1.060    0.000    0.000    0.000  260.998  -17.021     0.00
2   1.043   -5.497   21.700   12.700   40.000   48.822     0.00
3   1.022   -8.004    2.400    1.200    0.000    0.000     0.00
```

```
4   1.013   -9.661    7.600    1.600    0.000    0.000    0.00
5   1.010  -14.381   94.200   19.000    0.000   35.975    0.00
6   1.012  -11.398    0.000    0.000    0.000    0.000    0.00
7   1.003  -13.150   22.800   10.900    0.000    0.000    0.00
8   1.010  -12.115   30.000   30.000    0.000   30.826    0.00
9   1.051  -14.434    0.000    0.000    0.000    0.000    0.00
10  1.044  -16.024    5.800    2.000    0.000    0.000   19.00
11  1.082  -14.434    0.000    0.000    0.000   16.119    0.00
12  1.057  -15.302   11.200    7.500    0.000    0.000    0.00
13  1.071  -15.302    0.000    0.000    0.000   10.423    0.00
14  1.042  -16.191    6.200    1.600    0.000    0.000    0.00
15  1.038  -16.278    8.200    2.500    0.000    0.000    0.00
16  1.045  -15.880    3.500    1.800    0.000    0.000    0.00
17  1.039  -16.188    9.000    5.800    0.000    0.000    0.00
18  1.028  -16.884    3.200    0.900    0.000    0.000    0.00
19  1.025  -17.052    9.500    3.400    0.000    0.000    0.00
20  1.029  -16.852    2.200    0.700    0.000    0.000    0.00
21  1.032  -16.468   17.500   11.200    0.000    0.000    0.00
22  1.033  -16.455    0.000    0.000    0.000    0.000    0.00
23  1.027  -16.662    3.200    1.600    0.000    0.000    0.00
24  1.022  -16.830    8.700    6.700    0.000    0.000    4.30
25  1.019  -16.424    0.000    0.000    0.000    0.000    0.00
26  1.001  -16.842    3.500    2.300    0.000    0.000    0.00
27  1.026  -15.912    0.000    0.000    0.000    0.000    0.00
28  1.011  -12.057    0.000    0.000    0.000    0.000    0.00
29  1.006  -17.136    2.400    0.900    0.000    0.000    0.00
30  0.995  -18.015   10.600    1.900    0.000    0.000    0.00

Total                283.400  126.200  300.998  125.144   23.30
```

The output of the **lineflow** is the same as the line flow output of Example 6.9 with the power mismatch as dictated by the Newton-Raphson method.

6.11 FAST DECOUPLED POWER FLOW SOLUTION

Power system transmission lines have a very high X/R ratio. For such a system, real power changes ΔP are less sensitive to changes in the voltage magnitude and are most sensitive to changes in phase angle $\Delta\delta$. Similarly, reactive power is less sensitive to changes in angle and are mainly dependent on changes in voltage magnitude. Therefore, it is reasonable to set elements $\boldsymbol{J_2}$ and $\boldsymbol{J_3}$ of the Jacobian matrix to zero. Thus, (6.54) becomes

$$\begin{bmatrix} \Delta P \\ \Delta Q \end{bmatrix} = \begin{bmatrix} J_1 & 0 \\ 0 & J_4 \end{bmatrix} \begin{bmatrix} \Delta\delta \\ \Delta|V| \end{bmatrix} \tag{6.68}$$

or

$$\Delta P = J_1 \Delta\delta = [\frac{\partial P}{\partial \delta}]\ \Delta\delta \qquad (6.69)$$

$$\Delta Q = J_4 \Delta|V| = [\frac{\partial Q}{\partial |V|}]\ \Delta|V| \qquad (6.70)$$

(6.69) and (6.70) show that the matrix equation is separated into two decoupled equations requiring considerably less time to solve compared to the time required for the solution of (6.54). Furthermore, considerable simplification can be made to eliminate the need for recomputing J_1 and J_4 during each iteration. This procedure results in the decoupled power flow equations developed by Stott and Alsac[75–76]. The diagonal elements of J_1 described by (6.55) may be written as

$$\frac{\partial P_i}{\partial \delta_i} = \sum_{j=1}^{n} |V_i||V_j||Y_{ij}| \sin(\theta_{ij} - \delta_i + \delta_j) - |V_i|^2 |Y_{ii}| \sin\theta_{ii}$$

Replacing the first term of the above equation with $-Q_i$, as given by (6.53), results in

$$\frac{\partial P_i}{\partial \delta_i} = -Q_i - |V_i|^2 |Y_{ii}| \sin\theta_{ii}$$
$$= -Q_i - |V_i|^2 B_{ii}$$

Where $B_{ii} = |Y_{ii}| \sin\theta_{ii}$ is the imaginary part of the diagonal elements of the bus admittance matrix. B_{ii} is the sum of susceptances of all the elements incident to bus i. In a typical power system, the self-susceptance $B_{ii} \gg Q_i$, and we may neglect Q_i. Further simplification is obtained by assuming $|V_i|^2 \approx |V_i|$, which yields

$$\frac{\partial P_i}{\partial \delta_i} = -|V_i| B_{ii} \qquad (6.71)$$

Under normal operating conditions, $\delta_j - \delta_i$ is quite small. Thus, in (6.56) assuming $\theta_{ii} - \delta_i + \delta_j \approx \theta_{ii}$, the off-diagonal elements of J_1 becomes

$$\frac{\partial P_i}{\partial \delta_j} = -|V_i||V_j| B_{ij}$$

Further simplification is obtained by assuming $|V_j| \approx 1$

$$\frac{\partial P_i}{\partial \delta_j} = -|V_i| B_{ij} \qquad (6.72)$$

Similarly, the diagonal elements of J_4 described by (6.61) may be written as

$$\frac{\partial Q_i}{\partial |V_i|} = -|V_i||Y_{ii}|\sin\theta_{ii} - \sum_{j=1}^{n} |V_i||V_j||Y_{ij}|\sin(\theta_{ij} - \delta_i + \delta_j)$$

replacing the second term of the above equation with $-Q_i$, as given by (6.53), results in

$$\frac{\partial Q_i}{\partial |V_i|} = -|V_i||Y_{ii}|\sin\theta_{ii} + Q_i$$

Again, since $B_{ii} = Y_{ii}\sin\theta_{ii} \gg Q_i$, Q_i may be neglected and (6.61) reduces to

$$\frac{\partial Q_i}{\partial |V_i|} = -|V_i|B_{ii} \tag{6.73}$$

Likewise in (6.62), assuming $\theta_{ij} - \delta_i + \delta_j \approx \theta_{ij}$ yields

$$\frac{\partial Q_i}{\partial |V_j|} = -|V_i|B_{ij} \tag{6.74}$$

With these assumptions, equations (6.69) and (6.70) take the following form

$$\frac{\boldsymbol{\Delta P}}{|\boldsymbol{V_i}|} = -\boldsymbol{B'}\ \boldsymbol{\Delta\delta} \tag{6.75}$$

$$\frac{\boldsymbol{\Delta Q}}{|\boldsymbol{V_i}|} = -\boldsymbol{B''}\ \boldsymbol{\Delta|V|} \tag{6.76}$$

Here, $\boldsymbol{B'}$ and $\boldsymbol{B''}$ are the imaginary part of the bus admittance matrix $\boldsymbol{Y_{bus}}$. Since the elements of this matrix are constant, they need to be triangularized and inverted only once at the beginning of the iteration. $\boldsymbol{B'}$ is of order of $(n-1)$. For voltage-controlled buses where $|V_i|$ and P_i are specified and Q_i is not specified, the corresponding row and column of $\boldsymbol{Y_{bus}}$ are eliminated. Thus, $\boldsymbol{B''}$ is of order of $(n-1-m)$, where m is the number of voltage-regulated buses. Therefore, in the fast decoupled power flow algorithm, the successive voltage magnitude and phase angle changes are

$$\boldsymbol{\Delta\delta} = -[\boldsymbol{B'}]^{-1}\frac{\boldsymbol{\Delta P}}{|\boldsymbol{V}|} \tag{6.77}$$

$$\boldsymbol{\Delta|V|} = -[\boldsymbol{B''}]^{-1}\frac{\boldsymbol{\Delta Q}}{|\boldsymbol{V}|} \tag{6.78}$$

The fast decoupled power flow solution requires more iterations than the Newton-Raphson method, but requires considerably less time per iteration, and a power flow solution is obtained very rapidly. This technique is very useful in contingency analysis where numerous outages are to be simulated or a power flow solution is required for on-line control.

Example 6.12

Obtain the power flow solution by the fast decoupled method for the system of Example 6.8.

The bus admittance matrix of the system as obtained in Example 6.10 is

$$Y_{bus} = \begin{bmatrix} 20 - j50 & -10 + j20 & -10 + j30 \\ -10 + j20 & 26 - j52 & -16 + j32 \\ -10 + j30 & -16 + j32 & 26 - j62 \end{bmatrix}$$

In this system, bus 1 is the slack bus and the corresponding bus susceptance matrix for evaluation of phase angles $\Delta\delta_2$ and $\Delta\delta_3$ is

$$B' = \begin{bmatrix} -52 & 32 \\ 32 & -62 \end{bmatrix}$$

The inverse of the above matrix is

$$[B']^{-1} = \begin{bmatrix} -0.028182 & -0.014545 \\ -0.014545 & -0.023636 \end{bmatrix}$$

From (6.52) and (6.53), the expressions for real power at bus 2 and 3 and the reactive power at bus 2 are

$$\begin{aligned} P_2 &= |V_2||V_1||Y_{21}|\cos(\theta_{21} - \delta_2 + \delta_1) + |V_2^2||Y_{22}|\cos\theta_{22} \\ &\quad + |V_2||V_3||Y_{23}|\cos(\theta_{23} - \delta_2 + \delta_3) \\ P_3 &= |V_3||V_1||Y_{31}|\cos(\theta_{31} - \delta_3 + \delta_1) + |V_3||V_2||Y_{32}|\cos(\theta_{32} \\ &\quad - \delta_3 + \delta_2) + |V_3^2||Y_{33}|\cos\theta_{33} \\ Q_2 &= -|V_2||V_1||Y_{21}|\sin(\theta_{21} - \delta_2 + \delta_1) - |V_2^2||Y_{22}|\sin\theta_{22} \\ &\quad - |V_2||V_3||Y_{23}|\sin(\theta_{23} - \delta_2 + \delta_3) \end{aligned}$$

The load and generation expressed in per units are

$$S_2^{sch} = -\frac{(400 + j250)}{100} = -4.0 - j2.5 \quad \text{pu}$$

$$P_3^{sch} = \frac{200}{100} = 2.0 \quad \text{pu}$$

The slack bus voltage is $V_1 = 1.05\angle 0$ pu, and the bus 3 voltage magnitude is $|V_3| = 1.04$ pu. Starting with an initial estimate of $|V_2^{(0)}| = 1.0$, $\delta_2^{(0)} = 0.0$, and $\delta_3^{(0)} = 0.0$, the power residuals are computed from (6.63) and (6.64)

$$\Delta P_2^{(0)} = P_2^{sch} - P_2^{(0)} = -4.0 - (-1.14) = -2.86$$

$$\Delta P_3^{(0)} = P_3^{sch} - P_3^{(0)} = 2.0 - (0.5616) = 1.4384$$

$$\Delta Q_2^{(0)} = Q_2^{sch} - Q_2^{(0)} = -2.5 - (-2.28) = -0.22$$

The fast decoupled power flow algorithm given by (6.77) becomes

$$\begin{bmatrix} \Delta\delta_2^{(0)} \\ \Delta\delta_3^{(0)} \end{bmatrix} = -\begin{bmatrix} -0.028182 & -0.014545 \\ -0.014545 & -0.023636 \end{bmatrix} \begin{bmatrix} \frac{-2.8600}{1.0} \\ \frac{1.4384}{1.04} \end{bmatrix} = \begin{bmatrix} -0.060483 \\ -0.008909 \end{bmatrix}$$

Since bus 3 is a regulated bus, the corresponding row and column of $\boldsymbol{B'}$ are eliminated and we get

$$B'' = [-52]$$

From (6.78), we have

$$\Delta|V_2| = -\left[\frac{-1}{52}\right]\left[\frac{-.22}{1.0}\right] = -0.0042308$$

The new bus voltages in the first iteration are

$$\begin{aligned} \Delta\delta_2^{(0)} &= -0.060483 & \delta_2^{(1)} &= 0 + (-0.060483) = -0.060483 \\ \Delta\delta_3^{(0)} &= -0.008989 & \delta_3^{(1)} &= 0 + (-0.008989) = -0.008989 \\ \Delta|V_2^{(0)}| &= -0.0042308 & |V_2^{(1)}| &= 1 + (-0.0042308) = 0.995769 \end{aligned}$$

The voltage phase angles are in radians. The process is continued until power residuals are within a specified accuracy. The result is tabulated in the table below.

Iter	δ_2	δ_3	$\lvert V_2\rvert$	ΔP_2	ΔP_3	ΔQ_2
1	-0.060482	-0.008909	0.995769	-2.860000	1.438400	-0.220000
2	-0.056496	-0.007952	0.965274	0.175895	-0.070951	-1.579042
3	-0.044194	-0.008690	0.965711	0.640309	-0.457039	0.021948
4	-0.044802	-0.008986	0.972985	-0.021395	0.001195	0.365249
5	-0.047665	-0.008713	0.973116	-0.153368	0.112899	0.006657
6	-0.047614	-0.008645	0.971414	0.000520	0.002610	-0.086136
7	-0.046936	-0.008702	0.971333	0.035980	-0.026190	-0.004067
8	-0.046928	-0.008720	0.971732	0.000948	-0.001411	0.020119
9	-0.047087	-0.008707	0.971762	-0.008442	0.006133	0.001558
10	-0.047094	-0.008702	0.971669	-0.000470	0.000510	-0.004688
11	-0.047057	-0.008705	0.971660	0.001971	-0.001427	-0.000500
12	-0.047054	-0.008706	0.971681	0.000170	-0.000163	0.001087
13	-0.047063	-0.008706	0.971684	-0.000458	0.000330	0.000151
14	-0.047064	-0.008706	0.971680	-0.000053	0.000048	-0.000250

Converting phase angles to degrees the final solution is $V_2 = 0.97168\angle{-2.696°}$ and $V_3 = 1.04\angle{-0.4988°}$. Using (6.52) and (6.53) as in Example 6.10, the reactive power at bus 3 and the slack bus real and reactive powers are

$$Q_3 = 1.4617 \text{ pu}$$
$$P_1 = 2.1842 \text{ pu}$$
$$Q_1 = 1.4085 \text{ pu}$$

The fast decoupled power flow for this example has taken 14 iterations with the maximum power mismatch of 2.5×10^{-4} pu compared to the Newton-Raphson method which took only three iterations. The highest X/R ratio of the transmission lines in this example is 3. For systems with a higher X/R ratio, the fast decoupled power flow method converges in relatively fewer iterations. However, the number of iterations is a function of system size.

Finally, the line flows are calculated in the same manner as the line flow calculations in the Gauss-Seidel method described in Example 6.7, and the power flow diagram is as shown in Figure 6.13.

A program named **decouple** is developed for power flow solution by the fast decoupled method for practical power systems. This program must be preceded by the **lfybus** program. **busout** and **lineflow** programs can be used to print the load flow solution and the line flow results. The format is the same as the Gauss-Seidel method. The following is a brief description of the **decouple** program:

decouple This program finds the power flow solution by the fast decouple method and requires the **busdata** and the **linedata** files described in Section 6.9. It is designed for the direct use of load and generation in MW and Mvar, bus voltages in per unit, and angle in degrees. Loads and generation are converted to per unit quantities on the base MVA selected. A provision is made to maintain the generator reactive power of the voltage-controlled buses within their specified limits. The violation of reactive power limit may occur if the specified voltage is either too high or too low. In the 10th iteration, the vars calculated at the generator buses are examined. If a limit is reached, the voltage magnitude is adjusted in steps of 0.5 percent up to ± 5 percent to bring the var demand within the specified limits.

Example 6.13

Obtain the power flow solution for the IEEE-30 bus test system by the fast decoupled method.

Data required is the same as in Example 6.9 with the following commands

```
clear       % clears all variables from the workspace.
basemva = 100;  accuracy = 0.001;  maxiter = 20;
```

busdata = [*same as in Example 6.9*];
linedata = [*same as in Example 6.9*];

```
lfybus                   % Forms the bus admittance matrix
decouple   % Power flow solution by fast decoupled method
busout     % Prints the power flow solution on the screen
lineflow % Computes and displays the line flow and losses
```

The output of **decouple** is

```
          Power Flow Solution by Fast Decoupled Method
             Maximum Power mismatch = 0.000919582
                    No. of iterations = 15
```

Bus No.	Voltage Mag.	Angle Degree	-----Load----- MW	Mvar	--Generation-- MW	Mvar	Injected Mvar
1	1.060	0.000	0.000	0.000	260.998	-17.021	0.00
2	1.043	-5.497	21.700	12.700	40.000	48.822	0.00
3	1.022	-8.004	2.400	1.200	0.000	0.000	0.00
4	1.013	-9.662	7.600	1.600	0.000	0.000	0.00
5	1.010	-14.381	94.200	19.000	0.000	35.975	0.00
6	1.012	-11.398	0.000	0.000	0.000	0.000	0.00
7	1.003	-13.149	22.800	10.900	0.000	0.000	0.00
8	1.010	-12.115	30.000	30.000	0.000	30.828	0.00
9	1.051	-14.434	0.000	0.000	0.000	0.000	0.00
10	1.044	-16.024	5.800	2.000	0.000	0.000	19.00
11	1.082	-14.434	0.000	0.000	0.000	16.120	0.00
12	1.057	-15.303	11.200	7.500	0.000	0.000	0.00
13	1.071	-15.303	0.000	0.000	0.000	10.421	0.00
14	1.042	-16.198	6.200	1.600	0.000	0.000	0.00
15	1.038	-16.276	8.200	2.500	0.000	0.000	0.00
16	1.045	-15.881	3.500	1.800	0.000	0.000	0.00
17	1.039	-16.188	9.000	5.800	0.000	0.000	0.00
18	1.028	-16.882	3.200	0.900	0.000	0.000	0.00
19	1.025	-17.051	9.500	3.400	0.000	0.000	0.00
20	1.029	-16.852	2.200	0.700	0.000	0.000	0.00
21	1.032	-16.468	17.500	11.200	0.000	0.000	0.00
22	1.033	-16.454	0.000	0.000	0.000	0.000	0.00
23	1.027	-16.661	3.200	1.600	0.000	0.000	0.00
24	1.022	-16.829	8.700	6.700	0.000	0.000	4.30
25	1.019	-16.423	0.000	0.000	0.000	0.000	0.00
26	1.001	-16.840	3.500	2.300	0.000	0.000	0.00

```
27  1.026  -15.912    0.000    0.000    0.000    0.000    0.00
28  1.011  -12.057    0.000    0.000    0.000    0.000    0.00
29  1.006  -17.136    2.400    0.900    0.000    0.000    0.00
30  0.995  -18.014   10.600    1.900    0.000    0.000    0.00

Total                283.400  126.200  300.998  125.145   23.30
```

The output of the **lineflow** is the same as the line flow output of Example 6.9 with the power mismatch as dictated by the fast decoupled method.

PROBLEMS

6.1. A power system network is shown in Figure 6.17. The generators at buses 1 and 2 are represented by their equivalent current sources with their reactances in per unit on a 100-MVA base. The lines are represented by π model where series reactances and shunt reactances are also expressed in per unit on a 100 MVA base. The loads at buses 3 and 4 are expressed in MW and Mvar.

(a) Assuming a voltage magnitude of 1.0 per unit at buses 3 and 4, convert the loads to per unit impedances. Convert network impedances to admittances and obtain the bus admittance matrix by inspection.

(b) Use the function **Y = ybus(zdata)** to obtain the bus admittance matrix. The function argument **zdata** is a matrix containing the line bus numbers, resistance and reactance. (See Example 6.1.)

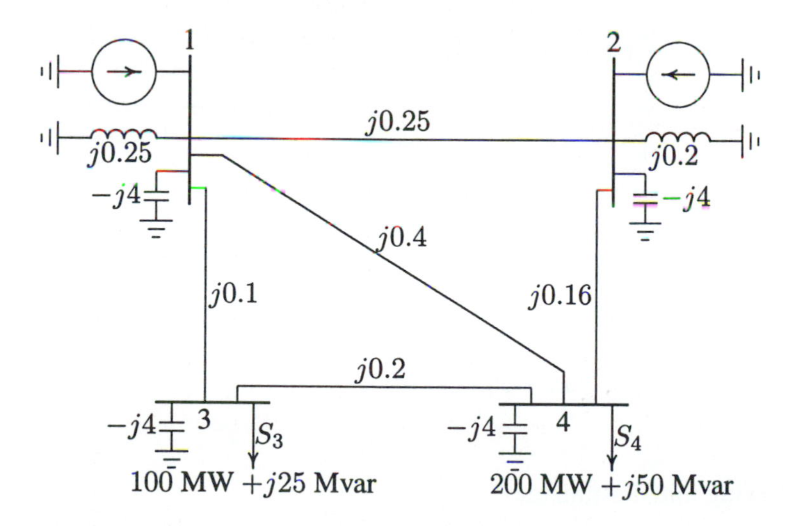

FIGURE 6.17
One-line diagram for Problem 6.1.

6.2. A power system network is shown in Figure 6.18. The values marked are impedances in per unit on a base of 100 MVA. The currents entering buses 1 and 2 are

$$I_1 = 1.38 - j2.72 \text{ pu}$$
$$I_2 = 0.69 - j1.36 \text{ pu}$$

(a) Determine the bus admittance matrix by inspection.
(b) Use the function **Y = ybus(zdata)** to obtain the bus admittance matrix. The function argument **zdata** is a matrix containing the line bus numbers, resistance and reactance. (See Example 6.1.) Write the necessary *MATLAB* commands to obtain the bus voltages.

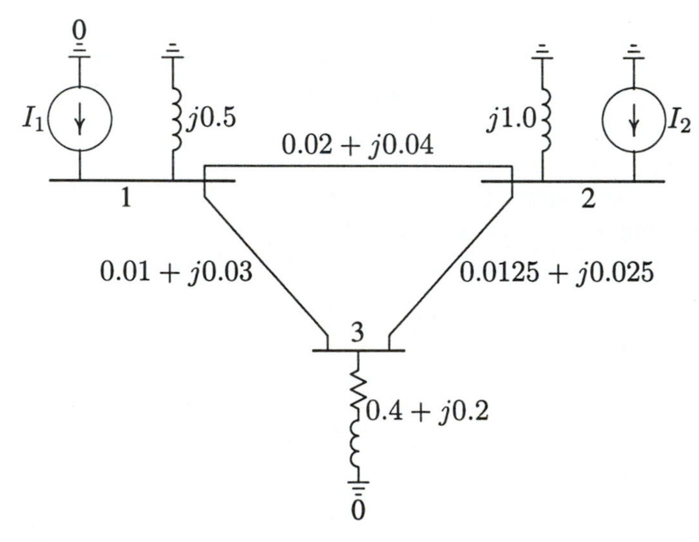

FIGURE 6.18
One-line diagram for Problem 6.2.

6.3. Use Gauss-Seidel method to find the solution of the following equations

$$x_1 + x_1x_2 = 10$$
$$x_1 + x_2 = 6$$

with the following initial estimates
(a) $x_1^{(0)} = 1$ and $x_2^{(0)} = 1$
(b) $x_1^{(0)} = 1$ and $x_2^{(0)} = 2$
Continue the iterations until $|\Delta x_1^{(k)}|$ and $|\Delta x_2^{(k)}|$ are less than 0.001.

6.4. A fourth-order polynomial equation is given by

$$x^4 - 21x^3 + 147x^2 - 379x + 252 = 0$$

(a) Use Newton-Raphson method and hand calculations to find one of the roots of the polynomial equation. Start with the initial estimate of $x^{(0)} = 0$ and continue until $|\Delta x^{(k)}| < 0.001$.
(b) Write a *MATLAB* program to find the roots of the above polynomial by Newton-Raphson method. The program should prompt the user to input the initial estimate. Run using the initial estimates of 0, 3, 6, 10.
(c) Check your answers using the *MATLAB* function **r = roots(A)**, where **A** is a row vector containing the polynomial coefficients in descending powers.

6.5. Use Newton-Raphson method and hand calculation to find the solution of the following equations:

$$x_1^2 - 2x_1 - x_2 = 3$$
$$x_1^2 + x_2^2 = 41$$

(a) Start with the initial estimates of $x_1^{(0)} = 2$, $x_2^{(0)} = 3$. Perform three iterations.
(b) Write a *MATLAB* program to find one of the solutions of the above equations by Newton-Raphson method. The program should prompt the user to input the initial estimates. Run the program with the above initial estimates.

6.6. In the power system network shown in Figure 6.19, bus 1 is a slack bus with $V_1 = 1.0\angle 0°$ per unit and bus 2 is a load bus with $S_2 = 280$ MW + $j60$ Mvar. The line impedance on a base of 100 MVA is $Z = 0.02 + j0.04$ per unit.
(a) Using Gauss-Seidel method, determine V_2 . Use an initial estimate of $V_2^{(0)} = 1.0 + j0.0$ and perform four iterations.
(b) If after several iterations voltage at bus 2 converges to $V_2 = 0.90 - j0.10$, determine S_1 and the real and reactive power loss in the line.

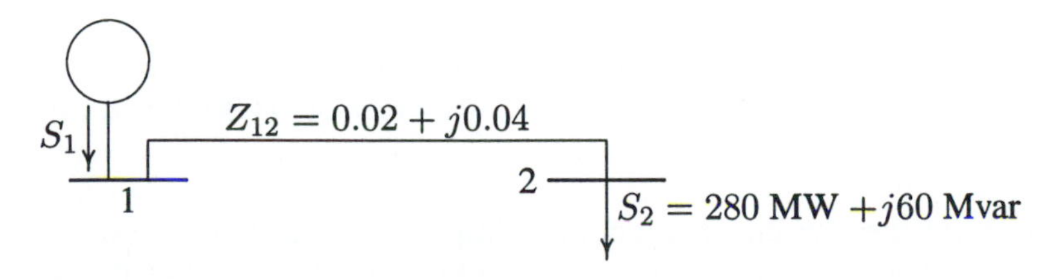

FIGURE 6.19
One-line diagram for Problem 6.6.

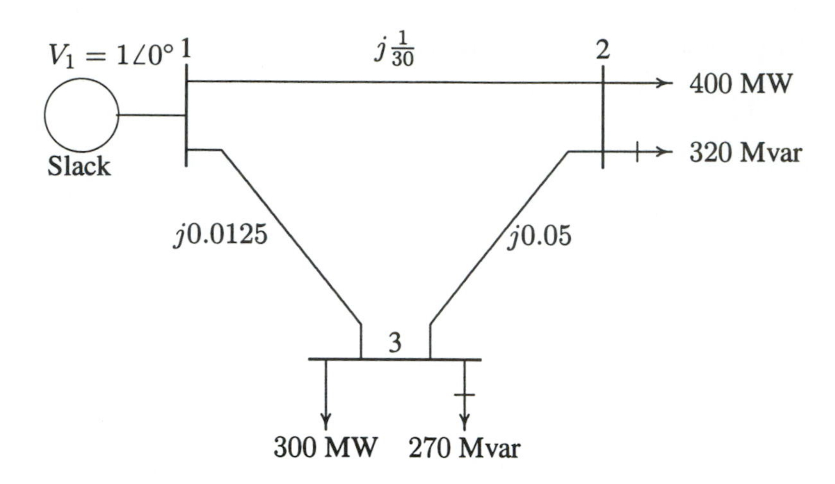

FIGURE 6.20
One-line diagram for Problem 6.7.

6.7. Figure 6.20 shows the one-line diagram of a simple three-bus power system with generation at bus 1. The voltage at bus 1 is $V_1 = 1.0\angle 0°$ per unit. The scheduled loads on buses 2 and 3 are marked on the diagram. Line impedances are marked in per unit on a 100-MVA base. For the purpose of hand calculations, line resistances and line charging susceptances are neglected.

(a) Using Gauss-Seidel method and initial estimates of $V_2^{(0)} = 1.0 + j0$ and $V_3^{(0)} = 1.0 + j0$, determine V_2 and V_3. Perform two iterations.

(b) If after several iterations the bus voltages converge to

$$V_2 = 0.90 - j0.10 \text{ pu}$$
$$V_3 = 0.95 - j0.05 \text{ pu}$$

determine the line flows and line losses and the slack bus real and reactive power. Construct a power flow diagram and show the direction of the line flows.

(c) Check the power flow solution using the **lfgauss** and other required programs. (Refer to Example 6.9.) Use a power accuracy of 0.00001 and an acceleration factor of 1.0.

6.8. Figure 6.21 shows the one-line diagram of a simple three-bus power system with generation at buses 1 and 3. The voltage at bus 1 is $V_1 = 1.025\angle 0°$ per unit. Voltage magnitude at bus 3 is fixed at 1.03 pu with a real power generation of 300 MW. A load consisting of 400 MW and 200 Mvar is taken from bus 2. Line impedances are marked in per unit on a 100-MVA base. For the

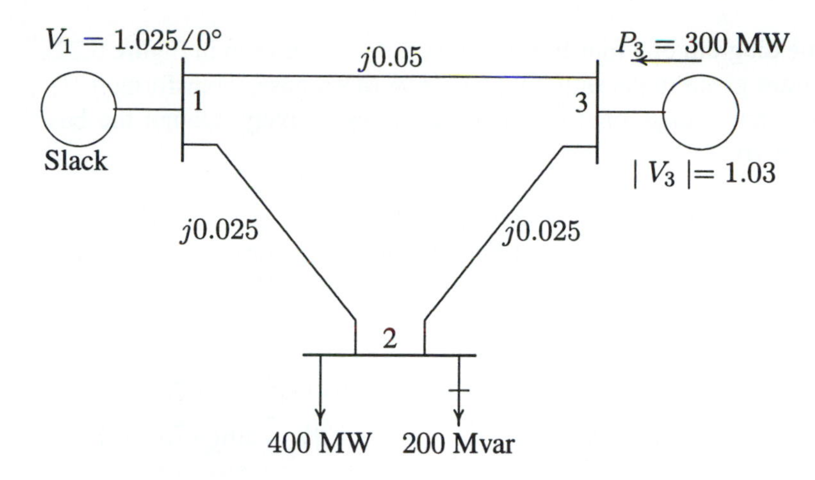

FIGURE 6.21
One-line diagram for Problem 6.8.

purpose of hand calculations, line resistances and line charging susceptances are neglected.

(a) Using Gauss-Seidel method and initial estimates of $V_2^{(0)} = 1.0 + j0$ and $V_3^{(0)} = 1.03 + j0$ and keeping $|V_3| = 1.03$ pu, determine the phasor values of V_2 and V_3 . Perform two iterations.

(b) If after several iterations the bus voltages converge to

$$V_2 = 1.001243\angle{-2.1°} = 1.000571 - j0.0366898 \text{ pu}$$
$$V_3 = 1.03\angle 1.36851° = 1.029706 + j0.0246 \text{ pu}$$

determine the line flows and line losses and the slack bus real and reactive power. Construct a power flow diagram and show the direction of the line flows.

(c) Check the power flow solution using the **lfgauss** and other required programs. (Refer to Example 6.9.)

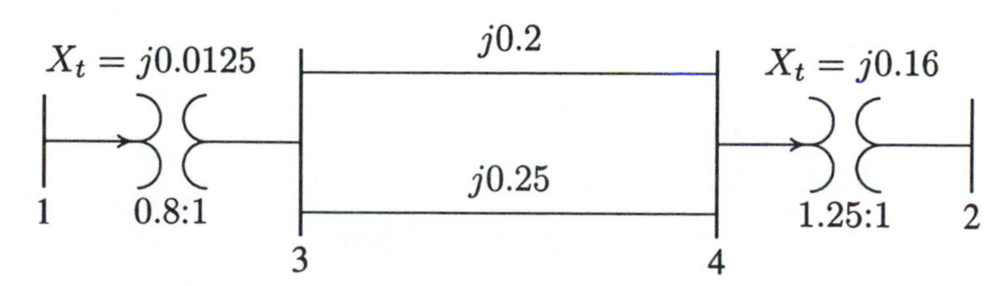

FIGURE 6.22
One-line diagram for Problem 6.9.

6.9. The one-line diagram of a four-bus power system is as shown in Figure 6.22. Reactances are given in per unit on a common MVA base. Transformers T_1 and T_2 have tap settings of 0.8:1, and 1.25:1 respectively. Obtain the bus admittance matrix.

6.10. In the two-bus system shown in Figure 6.23, bus 1 is a slack bus with $V_1 = 1.0\angle 0°$ pu. A load of 150 MW and 50 Mvar is taken from bus 2. The line admittance is $y_{12} = 10\angle -73.74°$ pu on a base of 100 MVA. The expression for real and reactive power at bus 2 is given by

$$\begin{aligned} P_2 &= \ \ 10|V_2||V_1|\cos(106.26° - \delta_2 + \delta_1) + 10|V_2|^2\cos(-73.74°) \\ Q_2 &= -10|V_2||V_1|\sin(106.26° - \delta_2 + \delta_1) - 10|V_2|^2\sin(-73.74°) \end{aligned}$$

Using Newton-Raphson method, obtain the voltage magnitude and phase angle of bus 2. Start with an initial estimate of $|V_2|^{(0)} = 1.0$ pu and $\delta_2{}^{(0)} = 0°$. Perform two iterations.

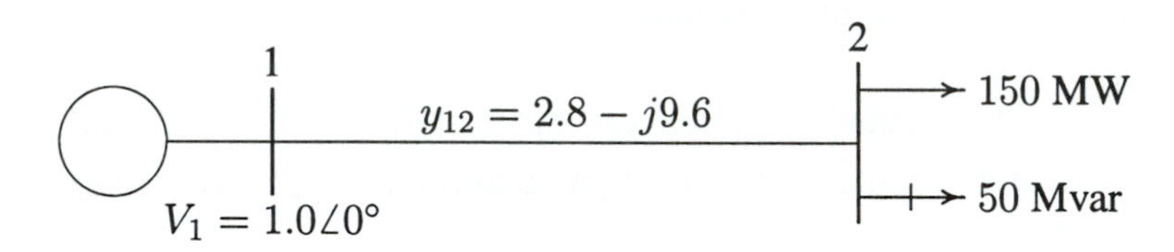

FIGURE 6.23
One-line diagram for Problem 6.10.

6.11. In the two-bus system shown in Figure 6.24, bus 1 is a slack bus with $V_1 = 1.0\angle 0°$ pu. A load of 100 MW and 50 Mvar is taken from bus 2. The line impedance is $z_{12} = 0.12 + j0.16$ pu on a base of 100 MVA. Using Newton-Raphson method, obtain the voltage magnitude and phase angle of bus 2. Start with an initial estimate of $|V_2|^{(0)} = 1.0$ pu and $\delta_2{}^{(0)} = 0°$. Perform two iterations.

1
2
$z_{12} = 0.12 + j0.16$
100 MW
50 Mvar
$V_1 = 1.0\angle 0°$

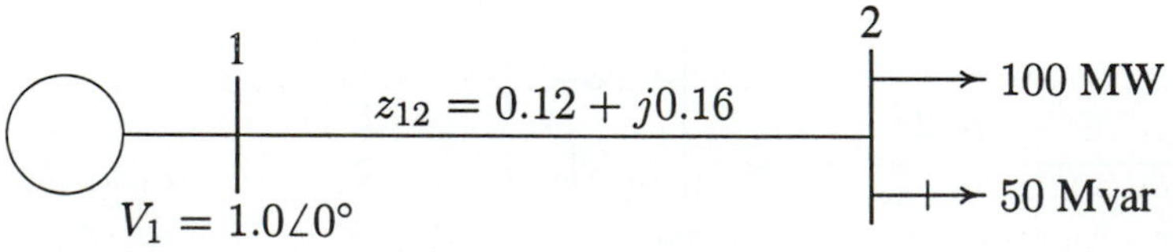

FIGURE 6.24
One-line diagram for Problem 6.11.

6.12. Figure 6.25 shows the one-line diagram of a simple three-bus power system with generation at buses 1 and 2. The voltage at bus 1 is $V = 1.0\angle 0°$ per unit. Voltage magnitude at bus 2 is fixed at 1.05 pu with a real power generation of 400 MW. A load consisting of 500 MW and 400 Mvar is taken from bus 3. Line admittances are marked in per unit on a 100 MVA base. For the purpose of hand calculations, line resistances and line charging susceptances are neglected.

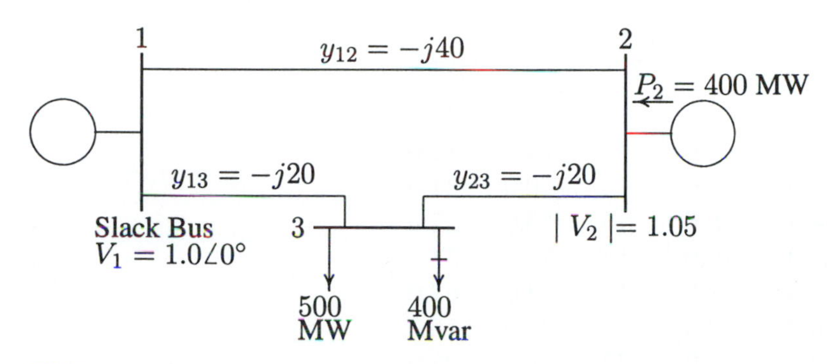

FIGURE 6.25
One-line diagram for Problem 6.12

(a) Show that the expression for the real power at bus 2 and real and reactive power at bus 3 are

$$P_2 = 40|V_2||V_1|\cos(90° - \delta_2 + \delta_1) + 20|V_2||V_3|\cos(90° - \delta_2 + \delta_3)$$
$$P_3 = 20|V_3||V_1|\cos(90° - \delta_3 + \delta_1) + 20|V_3||V_2|\cos(90° - \delta_3 + \delta_2)$$
$$Q_3 = -20|V_3||V_1|\sin(90° - \delta_3 + \delta_1) - 20|V_3||V_2|\sin(90° - \delta_3 + \delta_2) + 40|V_3|^2$$

(b) Using Newton-Raphson method, start with the initial estimates of ${V_2}^{(0)} = 1.0 + j0$ and ${V_3}^{(0)} = 1.0 + j0$, and keeping $|V_2| = 1.05$ pu, determine the phasor values of V_2 and V_3. Perform two iterations.
(c) Check the power flow solution for Problem 6.12 using **lfnewton** and other required programs. Assume the regulated bus (bus # 2) reactive power limits are between 0 and 600 Mvar.

6.13. For Problem 6.12:
(a) Obtain the power flow solution using the fast decoupled algorithm. Perform two iterations.
(b) Check the power flow solution for Problem 6.12 using **decouple** and other required programs. Assume the regulated bus (bus # 2) reactive power limits are between 0 and 600 Mvar.

6.14. The 26-bus power system network of an electric utility company is shown in Figure 6.26 (page 256). Obtain the power flow solution by the following

methods:
(a) Gauss-Seidel power flow (see Example 6.9).
(b) Newton-Raphson power flow (see Example 6.11).
(c) Fast decoupled power flow (see Example 6.13).

The load data is as follows.

LOAD DATA					
Bus	Load		Bus	Load	
No.	MW	Mvar	No.	MW	Mvar
1	51.0	41.0	14	24.0	12.0
2	22.0	15.0	15	70.0	31.0
3	64.0	50.0	16	55.0	27.0
4	25.0	10.0	17	78.0	38.0
5	50.0	30.0	18	153.0	67.0
6	76.0	29.0	19	75.0	15.0
7	0.0	0.0	20	48.0	27.0
8	0.0	0.0	21	46.0	23.0
9	89.0	50.0	22	45.0	22.0
10	0.0	0.0	23	25.0	12.0
11	25.0	15.0	24	54.0	27.0
12	89.0	48.0	25	28.0	13.0
13	31.0	15.0	26	40.0	20.0

Voltage magnitude, generation schedule, and the reactive power limits for the regulated buses are tabulated below. Bus 1, whose voltage is specified as $V_1 = 1.025\angle 0^\circ$, is taken as the slack bus.

GENERATION DATA				
Bus	Voltage	Generation	Mvar Limits	
No.	Mag.	MW	Min.	Max.
1	1.025			
2	1.020	79.0	40.0	250.0
3	1.025	20.0	40.0	150.0
4	1.050	100.0	40.0	80.0
5	1.045	300.0	40.0	160.0
26	1.015	60.0	15.0	50.0

The Mvar of the shunt capacitors installed at substations and the transformer tap settings are given below.

SHUNT CAPACITORS	
Bus No.	Mvar
1	4.0
4	2.0
5	5.0
6	2.0
11	1.5
12	2.0
15	0.5
19	5.0

TRANSFORMER TAP	
Designation	Tap Setting
2 – 3	0.960
2 – 13	0.960
3 – 13	1.017
4 – 8	1.050
4 – 12	1.050
6 – 19	0.950
7 – 9	0.950

The line and transformer data containing the series resistance and reactance in per unit and one-half the total capacitance in per unit susceptance on a 100-MVA base are tabulated below.

LINE AND TRANSFORMER DATA									
Bus No.	Bus No.	R, pu	X, pu	$\frac{1}{2}B$, pu	Bus No.	Bus No.	R, pu	X, pu	$\frac{1}{2}B$, pu
1	2	0.0005	0.0048	0.0300	10	22	0.0069	0.0298	0.005
1	18	0.0013	0.0110	0.0600	11	25	0.0960	0.2700	0.010
2	3	0.0014	0.0513	0.0500	11	26	0.0165	0.0970	0.004
2	7	0.0103	0.0586	0.0180	12	14	0.0327	0.0802	0.000
2	8	0.0074	0.0321	0.0390	12	15	0.0180	0.0598	0.000
2	13	0.0035	0.0967	0.0250	13	14	0.0046	0.0271	0.001
2	26	0.0323	0.1967	0.0000	13	15	0.0116	0.0610	0.000
3	13	0.0007	0.0054	0.0005	13	16	0.0179	0.0888	0.001
4	8	0.0008	0.0240	0.0001	14	15	0.0069	0.0382	0.000
4	12	0.0016	0.0207	0.0150	15	16	0.0209	0.0512	0.000
5	6	0.0069	0.0300	0.0990	16	17	0.0990	0.0600	0.000
6	7	0.0053	0.0306	0.0010	16	20	0.0239	0.0585	0.000
6	11	0.0097	0.0570	0.0001	17	18	0.0032	0.0600	0.038
6	18	0.0037	0.0222	0.0012	17	21	0.2290	0.4450	0.000
6	19	0.0035	0.0660	0.0450	19	23	0.0300	0.1310	0.000
6	21	0.0050	0.0900	0.0226	19	24	0.0300	0.1250	0.002
7	8	0.0012	0.0069	0.0001	19	25	0.1190	0.2249	0.004
7	9	0.0009	0.0429	0.0250	20	21	0.0657	0.1570	0.000
8	12	0.0020	0.0180	0.0200	20	22	0.0150	0.0366	0.000
9	10	0.0010	0.0493	0.0010	21	24	0.0476	0.1510	0.000
10	12	0.0024	0.0132	0.0100	22	23	0.0290	0.0990	0.000
10	19	0.0547	0.2360	0.0000	22	24	0.0310	0.0880	0.000
10	20	0.0066	0.0160	0.0010	23	25	0.0987	0.1168	0.000

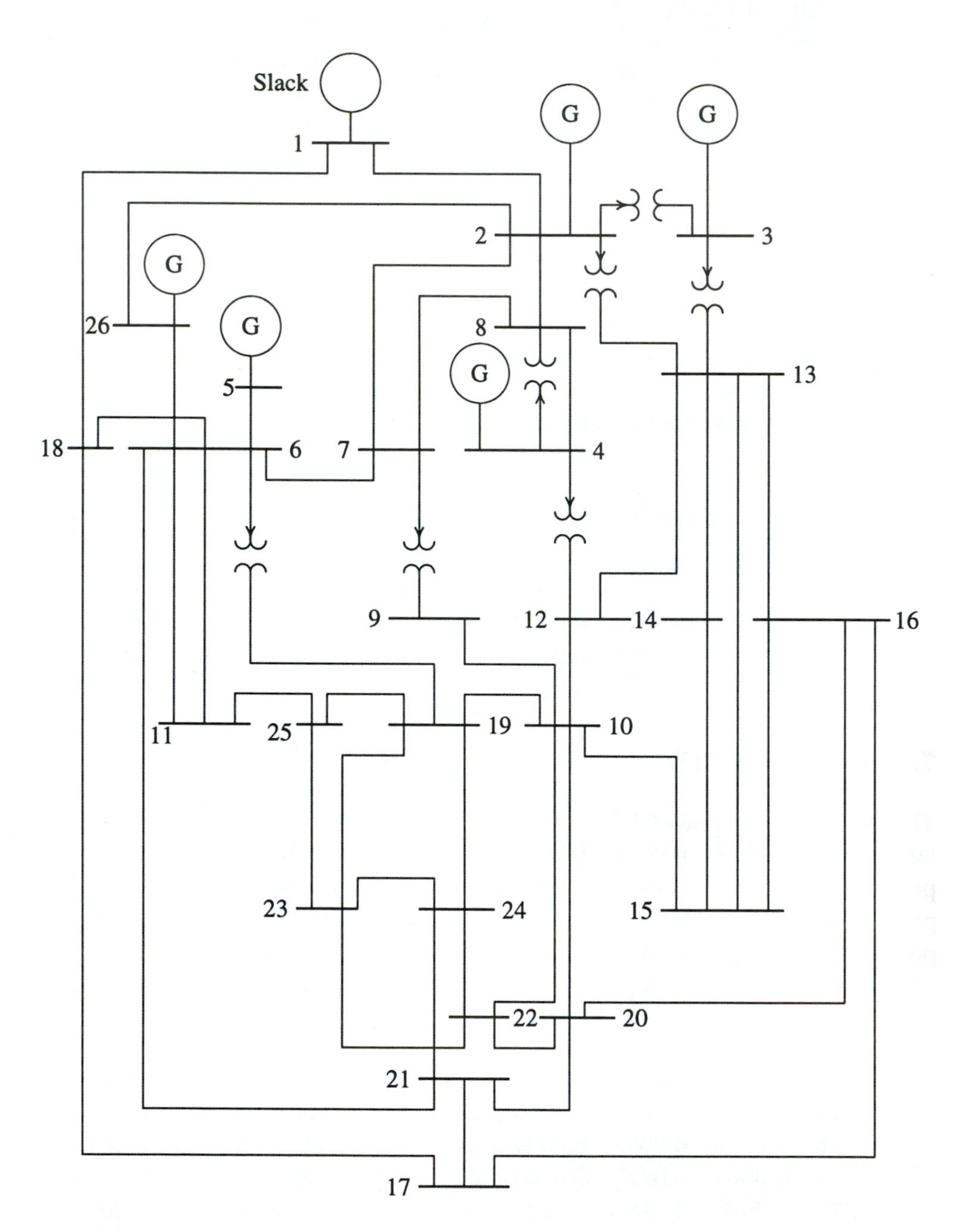

FIGURE 6.26
One-line diagram for Problem 6.14.

CHAPTER 7

OPTIMAL DISPATCH OF GENERATION

7.1 INTRODUCTION

The formulation of power flow problem and its solutions were discussed in Chapter 6. One type of bus in the power flow was the voltage-controlled bus, where real power generation and voltage magnitude were specified. The power flow solution provided the voltage phase angle and the reactive power generation. In a practical power system, the power plants are not located at the same distance from the center of loads and their fuel costs are different. Also, under normal operating conditions, the generation capacity is more than the total load demand and losses. Thus, there are many options for scheduling generation. In an interconnected power system, the objective is to find the real and reactive power scheduling of each power plant in such a way as to minimize the operating cost. This means that the generator's real and reactive power are allowed to vary within certain limits so as to meet a particular load demand with minimum fuel cost. This is called the *optimal power flow* (OPF) problem. The OPF is used to optimize the power flow solution of large scale power system. This is done by minimizing selected objective functions while maintaining an acceptable system performance in terms of generator capability limits and the output of the compensating devices. The objective functions, also known as *cost functions*, may present economic costs, system security, or other objectives. Efficient reactive power planning enhances economic operation as well as system security. The OPF has been studied by many researchers and many algorithms using different objective functions and methods have been presented [11, 12, 22, 23, 40, 42, 54, 78].

In this chapter, we will limit our analysis to the economic dispatch of real power generation. The classical optimization of continuous functions is introduced. The application of constraints to optimization problems is presented. Following this, the incremental production cost of generation is introduced. The economic dispatch of generation for minimization of the total operating cost with transmission losses neglected is obtained. Next, the transmission loss formula is derived and the economic dispatch of generation based on the loss formula is obtained. A program named **bloss** is developed for the evaluation of the transmission loss **B** coefficients which can be used following any one of the power flow programs **lfgauss**, **lfnewton**, or **decouple** discussed in Chapter 6. Also, a general program called **dispatch** is developed for the optimal scheduling of real power generation and can be used in conjunction with the **bloss** program.

7.2 NONLINEAR FUNCTION OPTIMIZATION

Unconstrained Parameter Optimization

Nonlinear function optimization is an important tool in computer-aided design and is part of a broader class of optimization called *nonlinear programming*. The underlying theory and the computational methods are discussed in many books. The basic goal is the minimization of some nonlinear objective cost function subject to nonlinear equality and inequality constraints.

The mathematical tools that are used to solve unconstrained parameter optimization problems come directly from multivariable calculus. The necessary condition to minimize the cost function

$$f(x_1, x_2, \ldots, x_n) \tag{7.1}$$

is obtained by setting derivative of f with respect to the variables equal to zero, i.e.,

$$\frac{\partial f}{\partial x_i} = 0 \qquad i = 1, \ldots, n \tag{7.2}$$

or

$$\nabla f = 0 \tag{7.3}$$

where

$$\nabla f = (\frac{\partial f}{\partial x_1}, \frac{\partial f}{\partial x_2}, \ldots, \frac{\partial f}{\partial x_n}) \tag{7.4}$$

which is known as the *gradient vector*. The terms associated with second derivatives is given by

$$H = \frac{\partial^2 f}{\partial x_i \, \partial x_j} \tag{7.5}$$

The above equation results in a symmetric matrix called the *Hessian matrix* of the function.

Once the derivative of f is vanished at local extrema $(\hat{x}_1, \hat{x}_2, \ldots, \hat{x}_n)$, for f to have a relative minimum, the Hessian matrix evaluated at $(\hat{x}_1, \hat{x}_2, \ldots, \hat{x}_n)$ must be a positive definite matrix. This condition requires that all the eigenvalues of the Hessian matrix evaluated at $(\hat{x}_1, \hat{x}_2, \ldots, \hat{x}_n)$ be positive.

In summary, the unconstrained minimum of a function is found by setting its partial derivatives (with respect to the parameters that may be varied) equal to zero and solving for the parameter values. Among the sets of parameter values obtained, those at which the matrix of second partial derivatives of the cost function is positive definite are local minima. If there is a single local minimum, it is also the global minimum; otherwise, the cost function must be evaluated at each of the local minima to determine which one is the global minimum.

Example 7.1

Find the minimum of

$$f(x_1, x_2, \ldots, x_n) = x_1^2 + 2x_2^2 + 3x_3^2 + x_1x_2 + x_2x_3 - 8x_1 - 16x_2 - 32x_3 + 110$$

Equating the first derivatives to zero, results in

$$\frac{\partial f}{\partial x_1} = 2x_1 + x_2 - 8 = 0$$
$$\frac{\partial f}{\partial x_2} = x_1 + 4x_2 + x_3 - 16 = 0$$
$$\frac{\partial f}{\partial x_3} = x_2 + 6x_3 - 32 = 0$$

or

$$\begin{bmatrix} 2 & 1 & 0 \\ 1 & 4 & 1 \\ 0 & 1 & 6 \end{bmatrix} \begin{bmatrix} x_1 \\ x_2 \\ x_3 \end{bmatrix} = \begin{bmatrix} 8 \\ 16 \\ 32 \end{bmatrix}$$

The solution of the above linear simultaneous equation is readily obtained (in *MATLAB* use $X = A\backslash B$) and is given by $(\hat{x}_1, \hat{x}_2, \hat{x}_3) = (3, 2, 5)$. The function evaluated at this point is $f(3, 2, 5) = 2$. To see if this point is a minimum, we evaluate the second derivatives and form the Hessian matrix

$$H(\hat{X}) = \begin{bmatrix} 2 & 1 & 0 \\ 1 & 4 & 1 \\ 0 & 1 & 6 \end{bmatrix}$$

Using the *MATLAB* function **eig(H)**, the eigenvalues are found to be 1.55, 4.0 and 6.45, which are all positive. Thus, the Hessian matrix is a positive definite matrix and (3, 2, 5) is a minimum point.

7.2.1 CONSTRAINED PARAMETER OPTIMIZATION: EQUALITY CONSTRAINTS

This type of problem arises when there are functional dependencies among the parameters to be chosen. The problem is to minimize the cost function

$$f(x_1, x_2, \ldots, x_n) \tag{7.6}$$

subject to the equality constraints

$$g_i(x_1, x_2, \ldots, x_n) = 0 \qquad i = 1, 2, \ldots, k \tag{7.7}$$

Such problems may be solved by the *Lagrange multiplier* method. This provides an augmented cost function by introducing k-vector λ of undetermined quantities. The unconstrained cost function becomes

$$\mathcal{L} = f + \sum_{i=1}^{k} \lambda_i g_i \tag{7.8}$$

The resulting necessary conditions for constrained local minima of $\mathcal{L}$ are the following:

$$\frac{\partial \mathcal{L}}{\partial x_i} = \frac{\partial f}{\partial x_i} + \sum_{i=1}^{k} \lambda_i \frac{\partial g_i}{\partial x_i} = 0 \tag{7.9}$$

$$\frac{\partial \mathcal{L}}{\partial \lambda_i} = g_i = 0 \tag{7.10}$$

Note that Equation (7.10) is simply the original constraints.

Example 7.2

Use the Lagrange multiplier method for solving constrained parameter optimizations to determine the minimum distance from origin of the xy plane to a circle described by

$$(x-8)^2 + (y-6)^2 = 25$$

The minimum distance is obtained by minimization of the distance square, given by

$$f(x,y) = x^2 + y^2$$

The *MATLAB* **plot** command is used to plot the circle as shown in Figure 7.1.

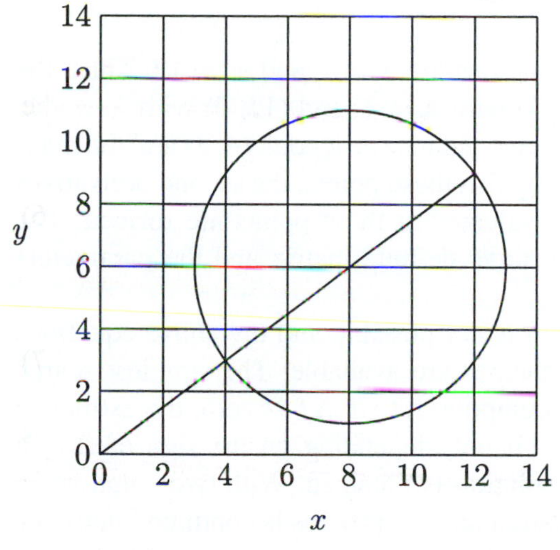

FIGURE 7.1
Constraint function of Example 7.2

From this graph, clearly the minimum distance is 5, located at point (4, 3).

Now let us use Lagrange multiplier to minimize $f(x,y)$ subject to the constraint described by the circle equation. Forming the Lagrange function, we obtain

$$\mathcal{L} = x^2 + y^2 + \lambda[(x-8)^2 + (y-6)^2 - 25]$$

The necessary conditions for extrema are

$$\frac{\partial \mathcal{L}}{\partial x} = 2x + \lambda(2x - 16) = 0 \quad \text{or} \quad 2x(\lambda + 1) = 16\lambda$$

$$\frac{\partial \mathcal{L}}{\partial y} = 2y + \lambda(2y - 12) = 0 \quad \text{or} \quad 2y(\lambda + 1) = 12\lambda$$
$$\frac{\partial \mathcal{L}}{\partial \lambda} = (x - 8)^2 + (y - 6)^2 - 25 = 0$$

The solution of the above three equations will provide optimal points. In this problem, a direct solution can be obtained as follows:

Eliminating λ from the first two equations results in

$$y = \frac{3}{4}x$$

Substituting for y in the third equation yields

$$\frac{25}{16}x^2 - 25x + 75 = 0$$

The solutions of the above quadratic equations are $x = 4$ and $x = 12$. Thus, the corresponding extrema are at points (4, 3) with $\lambda = 1$, and (12, 9) with $\lambda = -3$. From Figure 7.1, it is clear that the minimum distance is at point (4, 3) and the maximum distance is at point (12, 9). To distinguish these points, the second derivatives are obtained and the Hessian matrices evaluated at these points are formed. The matrix with positive eigenvalues is a positive definite matrix and the parameters correspond to the minimum point.

In many problems, a direct solution is not possible and the above equations are solved iteratively. Many iterative schemes are available. The simplest search method is to assume a value for λ and compute Δf. If Δf is zero, the estimated λ corresponds to the optimum solution. If not, depending on the sign of Δf, λ is increased or decreased, and another solution is obtained. With two solutions, a better value of λ is obtained by extrapolation and the process is continued until Δf is within a specified accuracy. A significantly superior method applicable to continuous functions is the Newton-Raphson method. One way to apply the Newton-Raphson method to the problem at hand is as follows: From the first two equations, x and y are found. These are

$$x = \frac{8\lambda}{\lambda + 1}$$
$$y = \frac{6\lambda}{\lambda + 1}$$

Substituting into the third equation results in

$$f(\lambda) = \frac{100\lambda^2}{(\lambda + 1)^2} - \frac{200\lambda}{\lambda + 1} + 75 = 0$$

This is a nonlinear equation in terms of λ and can be solved by the Newton-Raphson method. The Newton-Raphson method is a successive approximation procedure based on an initial estimate of the unknown and the use of Taylor's series expansion (see Chapter 6 for more details). For a one-dimensional case,

$$\Delta\lambda^{(k)} = \frac{-\Delta f(\lambda)^k}{(\frac{df}{d\lambda})^k} \tag{7.11}$$

and

$$\lambda^{(k+1)} = \lambda^{(k)} + \Delta\lambda^{(k)} \tag{7.12}$$

Starting with an estimated value of λ, a new value is found in the direction of steepest decent (negative gradient). The process is repeated in the direction of negative gradient until $\Delta f(\lambda)$ is less than a specified accuracy. This algorithm is known as the *gradient method*. For the above function, the gradient is

$$\frac{df(\lambda)}{d\lambda} = \frac{200\lambda}{(\lambda+1)^3} - \frac{200}{(\lambda+1)^2} = \frac{-200}{(\lambda+1)^3}$$

The following commands show the procedure for the solution of the given equation by the Newton-Raphson method.

```
iter = 0;                                    % Iteration counter
Df = 10;                     % Error in Df is set to a high value
Lambda = input('Enter estimated value of Lambda = ');
fprintf('\n ')
disp(['     Iter      Df         J      DLambda    Lambda' ...
'        x           y'])
while abs(Df)  >= 0.0001                  % Test for convergence
iter = iter + 1;                             % No. of iterations
x = 8*Lambda/(Lambda + 1);
y = 6*Lambda/(Lambda + 1);
Df = (x- 8)^2 + (y - 6)^2 - 25;                       % Residual
J = -200/(Lambda + 1)^3;
Delambda =-Df/J;                           % Change in variable
disp([iter, Df, J, Delambda, Lambda, x, y])
Lambda = Lambda + Delambda;               % Successive solution
end
```

When the program is run, the user is prompted to enter the initial estimate for λ. Using a value of $\lambda = 0.4$, the result is

```
Enter estimated value of Lambda = 0.4
```

Iter	Δf	J	$\Delta\lambda$	λ	x	y
1	26.0240	-72.8863	0.3570	0.4000	2.2857	1.7134
2	7.3934	-36.8735	0.2005	0.7570	3.4468	2.5851
3	1.0972	-26.6637	0.0411	0.9575	3.9132	2.9349
4	0.0337	-25.0505	0.0013	0.9987	3.9973	2.9980
5	0.0000	-25.0001	0.0000	1.0000	4.0000	3.0000

After five iterations, the solution converges to $\lambda = 1.0$, $x = 4$, and $y = 3$, corresponding to the minimum length. If the program is run with an initial estimate of -2, the solution converges to $\lambda = -3$, $x = 12$, $y = 9$, which corresponds to the maximum length.

7.2.2 CONSTRAINT PARAMETER OPTIMIZATION: INEQUALITY CONSTRAINTS

Practical optimization problems contain inequality constraints as well as equality constraints. The problem is to minimize the cost function

$$f(x_1, x_2, \ldots, x_n) \tag{7.13}$$

subject to the equality constraints

$$g_i(x_1, x_2, \ldots, x_n) = 0 \qquad i = 1, 2, \ldots, k \tag{7.14}$$

and the inequality constraints

$$u_j(x_1, x_2, \ldots, x_n) \leq 0 \qquad i = 1, 2, \ldots, m \tag{7.15}$$

The Lagrange multiplier is extended to include the inequality constraints by introducing m-vector μ of undetermined quantities. The unconstrained cost function becomes

$$\mathcal{L} = f + \sum_{i=1}^{k} \lambda_i g_i + \sum_{j=1}^{m} \mu_j u_j \tag{7.16}$$

The resulting necessary conditions for constrained local minima of $\mathcal{L}$ are the following:

$$\frac{\partial \mathcal{L}}{\partial x_i} = 0 \quad i = 1, \ldots, n \tag{7.17}$$

$$\frac{\partial \mathcal{L}}{\partial \lambda_i} = g_i = 0 \quad i = 1, \ldots, k \tag{7.18}$$

$$\frac{\partial \mathcal{L}}{\partial \mu_j} = u_j \leq 0 \quad j = 1, \ldots, m \tag{7.19}$$

$$\mu_j u_j = 0 \;\; \& \;\; \mu_j > 0 \quad j = 1, \ldots, m \tag{7.20}$$

Note that Equation (7.18) is simply the original equality constraints. Suppose $(\hat{x}_1, \hat{x}_2, \ldots, \hat{x}_n)$ is a relative minimum. The inequality constraints in (7.19) is said to be inactive if strict inequality holds at $(\hat{x}_1, \hat{x}_2, \ldots, \hat{x}_n)$ and $\mu_j = 0$. On the other hand, when strict equality holds, the constraint is active at this point, (i.e., if the constraint $\mu_j u_j(\hat{x}_1, \hat{x}_2, \ldots, \hat{x}_n) = 0$ and $\mu_j > 0$. This is known as the *Kuhn-Tucker* necessary condition.

Example 7.3

Solve Example 7.2 with an additional inequality constraint defined below. The problem is to find the minimum value of the function

$$f(x, y) = x^2 + y^2$$

subject to one equality constraint

$$g(x, y) = (x - 8)^2 + (y - 6)^2 - 25 = 0$$

and one inequality constraint,

$$u(x, y) = 2x + y \geq 12$$

The unconstrained cost function from (7.16) is

$$\mathcal{L} = x^2 + y^2 + \lambda[(x - 8)^2 + (y - 6)^2 - 25] + \mu(2x + y - 12)$$

The resulting necessary conditions for constrained local minima of $\mathcal{L}$ are

$$\begin{aligned}
\frac{\partial \mathcal{L}}{\partial x} &= 2x + 2\lambda(x - 8) + 2\mu = 0 \\
\frac{\partial \mathcal{L}}{\partial y} &= 2y + 2\lambda(y - 6) + \mu = 0 \\
\frac{\partial \mathcal{L}}{\partial \lambda} &= (x - 8)^2 + (y - 6)^2 - 25 = 0 \\
\frac{\partial \mathcal{L}}{\partial \mu} &= 2x + y - 12 = 0
\end{aligned}$$

Eliminating μ from the first two equations result in

$$(2x - 4y)(1 + \lambda) + 8\lambda = 0$$

From the fourth condition, we have

$$y = 12 - 2x$$

Substituting for y in the above equation, yields

$$x = \frac{4\lambda + 4.8}{1 + \lambda}$$

Now substituting for x in the previous equation, we get

$$y = \frac{4\lambda + 2.4}{1 + \lambda}$$

Substituting for x and y in the third condition (equality constraint) results in an equation in terms of λ

$$\left(\frac{4\lambda + 4.8}{1 + \lambda} - 8\right)^2 + \left(\frac{4\lambda + 2.4}{1 + \lambda} - 6\right)^2 - 25 = 0$$

from which we have the following equation

$$\lambda^2 + 2\lambda + 0.36 = 0$$

Roots of the above equation are $\lambda = -0.2$ and $\lambda = -1.8$. Substituting for these values of λ in the expression for x and y, the corresponding extrema are

$$(x, y) = (5, 2) \quad \text{for} \quad \lambda = -0.2, \quad \mu = -5.6$$
$$(x, y) = (3, 6) \quad \text{for} \quad \lambda = -1.8, \quad \mu = -12$$

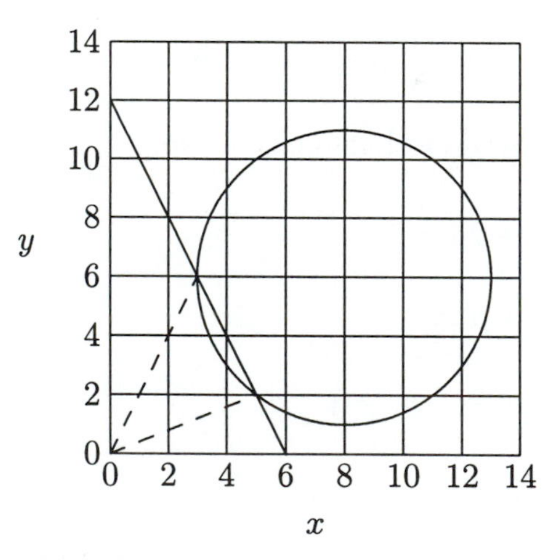

FIGURE 7.2
Constraint functions of Example 7.3.

The minimum distance from the cost function is 5.385, located at point $(5, 2)$, and the maximum distance is 6.71 located at point $(3, 6)$.

Adding the inequality constraint $2x + y \geq 12$ to the graphs in Figure 7.1, the solution is verified graphically as shown in Figure 7.2.

7.3 OPERATING COST OF A THERMAL PLANT

The factors influencing power generation at minimum cost are operating efficiencies of generators, fuel cost, and transmission losses. The most efficient generator in the system does not guarantee minimum cost as it may be located in an area where fuel cost is high. Also, if the plant is located far from the load center, transmission losses may be considerably higher and hence the plant may be overly uneconomical. Hence, the problem is to determine the generation of different plants such that the total operating cost is minimum. The operating cost plays an important role in the economic scheduling and are discussed here.

The input to the thermal plant is generally measured in Btu/h, and the output is measured in MW. A simplified input-output curve of a thermal unit known as *heat-rate* curve is given in Figure 7.3(a). Converting the ordinate of heat-rate

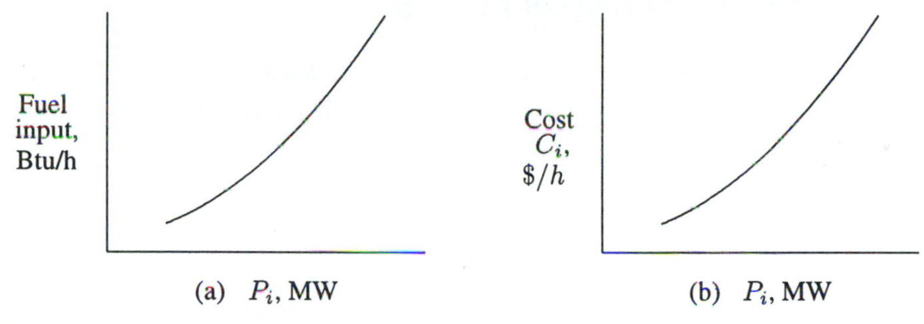

FIGURE 7.3
(a) Heat-rate curve. (b) Fuel-cost curve.

curve from Btu/h to $/h results in the *fuel-cost* curve shown in Figure 7.3(b). In all practical cases, the fuel cost of generator i can be represented as a quadratic function of real power generation

$$C_i = \alpha_i + \beta_i P_i + \gamma_i P_i^2 \tag{7.21}$$

An important characteristic is obtained by plotting the derivative of the fuel-cost curve versus the real power. This is known as the *incremental fuel-cost* curve shown in Figure 7.4.

$$\frac{dC_i}{dP_i} = 2\gamma_i P_i + \beta_i \tag{7.22}$$

The incremental fuel-cost curve is a measure of how costly it will be to produce the next increment of power. The total operating cost includes the fuel cost, and the cost of labor, supplies and maintenance. These costs are assumed to be a fixed percentage of the fuel cost and are generally included in the incremental fuel-cost curve.

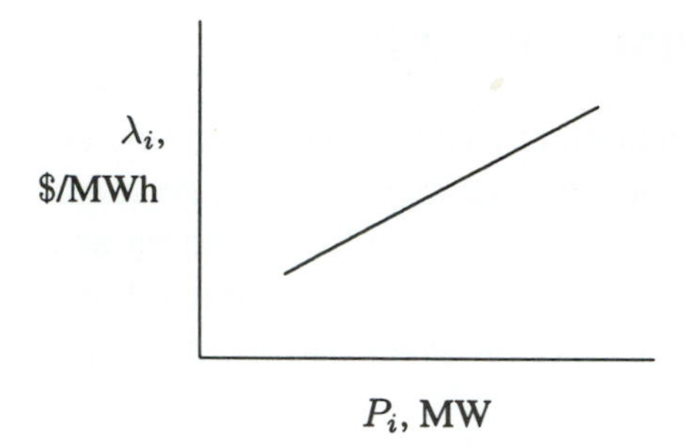

FIGURE 7.4
Typical incremental fuel-cost curve.

7.4 ECONOMIC DISPATCH NEGLECTING LOSSES AND NO GENERATOR LIMITS

The simplest economic dispatch problem is the case when transmission line losses are neglected. That is, the problem model does not consider system configuration and line impedances. In essence, the model assumes that the system is only one bus with all generation and loads connected to it as shown schematically in Figure 7.5.

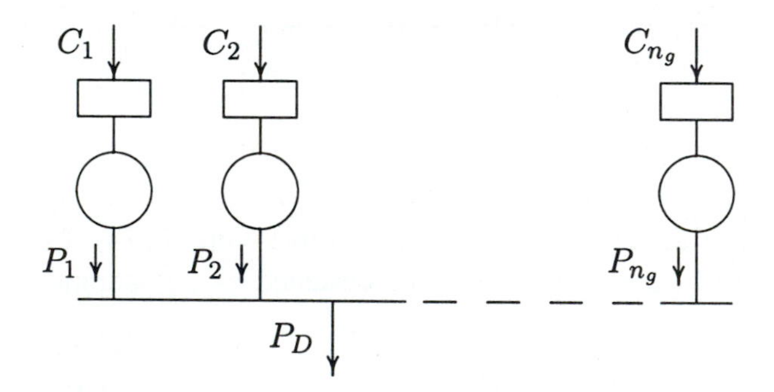

FIGURE 7.5
Plants connected to a common bus.

Since transmission losses are neglected, the total demand P_D is the sum of all generation. A cost function C_i is assumed to be known for each plant. The problem is to find the real power generation for each plant such that the objective function (i.e., total production cost) as defined by the equation

$$\begin{aligned} C_t &= \sum_{i=1}^{n_g} C_i \\ &= \sum_{i=1}^{n} \alpha_i + \beta_i P_i + \gamma_i P_i^2 \end{aligned} \tag{7.23}$$

is minimum, subject to the constraint

$$\sum_{i=1}^{n_g} P_i = P_D \tag{7.24}$$

where C_t is the total production cost, C_i is the production cost of ith plant, P_i is the generation of ith plant, P_D is the total load demand, and n_g is the total number of dispatchable generating plants.

A typical approach is to augment the constraints into objective function by using the Lagrange multipliers

$$\mathcal{L} = C_t + \lambda \left(P_D - \sum_{i=1}^{n_g} P_i \right) \tag{7.25}$$

The minimum of this unconstrained function is found at the point where the partials of the function to its variables are zero.

$$\frac{\partial \mathcal{L}}{\partial P_i} = 0 \tag{7.26}$$

$$\frac{\partial \mathcal{L}}{\partial \lambda} = 0 \tag{7.27}$$

First condition, given by (7.26), results in

$$\frac{\partial C_t}{\partial P_i} + \lambda(0 - 1) = 0$$

Since

$$C_t = C_1 + C_2 + \cdots + C_{n_g}$$

then

$$\frac{\partial C_t}{\partial P_i} = \frac{dC_i}{dP_i} = \lambda$$

and therefore the condition for optimum dispatch is

$$\frac{dC_i}{dP_i} = \lambda \qquad i = 1, \ldots, n_g \tag{7.28}$$

or

$$\beta_i + 2\gamma_i P_i = \lambda \tag{7.29}$$

Second condition, given by (7.27), results in

$$\sum_{i=1}^{n_g} P_i = P_D \tag{7.30}$$

Equation (7.30) is precisely the equality constraint that was to be imposed. In summary, when losses are neglected with no generator limits, for most economic operation, all plants must operate at equal incremental production cost while satisfying the equality constraint given by (7.30). In order to find the solution, (7.29) is solved for P_i

$$P_i = \frac{\lambda - \beta_i}{2\gamma_i} \tag{7.31}$$

The relations given by (7.31) are known as the *coordination equations*. They are functions of λ. An analytical solution can be obtained for λ by substituting for P_i in (7.30), i.e.,

$$\sum_{i=1}^{n_g} \frac{\lambda - \beta_i}{2\gamma_i} = P_D \tag{7.32}$$

or

$$\lambda = \frac{P_D + \sum_{i=1}^{n_g} \frac{\beta_i}{2\gamma_i}}{\sum_{i=1}^{n_g} \frac{1}{2\gamma_i}} \tag{7.33}$$

The value of λ found from (7.33) is substituted in (7.31) to obtain the optimal scheduling of generation.

The solution for economic dispatch neglecting losses was found analytically. However when losses are considered the resulting equations as seen in Section 7.6 are nonlinear and must be solved iteratively. Thus, an iterative procedure is introduced here and (7.31) is solved iteratively. In an iterative search technique, starting with two values of λ, a better value of λ is obtained by extrapolation, and the process is continued until ΔP_i is within a specified accuracy. However, as mentioned earlier, a rapid solution is obtained by the use of the gradient method. To do this, (7.32) is written as

$$f(\lambda) = P_D \tag{7.34}$$

Expanding the left-hand side of the above equation in Taylor's series about an operating point $\lambda^{(k)}$, and neglecting the higher-order terms results in

$$f(\lambda)^{(k)} + \left(\frac{df(\lambda)}{d\lambda}\right)^{(k)} \Delta\lambda^{(k)} = P_D \tag{7.35}$$

or

$$\Delta\lambda^{(k)} = \frac{\Delta P^{(k)}}{\left(\frac{df(\lambda)}{d\lambda}\right)^{(k)}} = \frac{\Delta P^{(k)}}{\sum\left(\frac{dP_i}{d\lambda}\right)^{(k)}} \tag{7.36}$$

or

$$\Delta\lambda^{(k)} = \frac{\Delta P^{(k)}}{\sum \frac{1}{2\gamma_i}} \tag{7.37}$$

and therefore,

$$\lambda^{(k+1)} = \lambda^{(k)} + \Delta\lambda^{(k)} \tag{7.38}$$

where

$$\Delta P^{(k)} = P_D - \sum_{i=1}^{n_g} P_i^{(k)} \tag{7.39}$$

The process is continued until $\Delta P^{(k)}$ is less than a specified accuracy.

Example 7.4

The fuel-cost functions for three thermal plants in $/h are given by

$$C_1 = 500 + 5.3P_1 + 0.004P_1^2$$
$$C_2 = 400 + 5.5P_2 + 0.006P_2^2$$
$$C_3 = 200 + 5.8P_3 + 0.009P_3^2$$

where P_1, P_2, and P_3 are in MW. The total load, P_D, is 800 MW. Neglecting line losses and generator limits, find the optimal dispatch and the total cost in $/h
(a) by analytical method using (7.33)
(b) by graphical demonstration.
(c) by iterative technique using the gradient method.

(a) From (7.33), λ is found to be

$$\lambda = \frac{800 + \frac{5.3}{0.008} + \frac{5.5}{0.012} + \frac{5.8}{0.018}}{\frac{1}{0.008} + \frac{1}{0.012} + \frac{1}{0.018}} = \frac{800 + 1443.0555}{263.8889} = 8.5 \quad \$/\text{MWh}$$

Substituting for λ in the coordination equation, given by (7.31), the optimal dispatch is

$$P_1 = \frac{8.5 - 5.3}{2(0.004)} = 400.0000$$
$$P_2 = \frac{8.5 - 5.5}{2(0.006)} = 250.0000$$
$$P_3 = \frac{8.5 - 5.8}{2(0.009)} = 150.0000$$

(b) From (7.28), the necessary conditions for optimal dispatch are

$$\frac{dC_1}{dP_1} = 5.3 + 0.008P_1 = \lambda$$
$$\frac{dC_2}{dP_2} = 5.5 + 0.012P_2 = \lambda$$
$$\frac{dC_3}{dP_3} = 5.8 + 0.018P_3 = \lambda$$

subject to

$$P_1 + P_2 + P_3 = P_D$$

To demonstrate the concept of equal incremental cost for optimal dispatch, we can use *MATLAB* **plot** command to plot the incremental cost of each plant on the same graph as shown in Figure 7.6. To obtain a solution, various values of λ could be tried until one is found which produces $\sum P_i = P_D$. For each λ, if $\sum P_i < P_D$, we increase λ otherwise, if $\sum P_i > P_D$, we reduce λ. Therefore, the horizontal dashed-line shown in the graph is moved up or down until at the optimum point $\hat{\lambda}$, $\sum P_i = P_D$. For this example, with $P_D = 800$ MW, the optimal dispatch is $P_1 = 400$, $P_2 = 250$, and $P_3 = 150$ at $\hat{\lambda} = 8.5$ \$/MWh.

(c) For the numerical solution using the gradient method, assume the initial value of $\lambda^{(1)} = 6.0$. From coordination equations, given by (7.31), P_1, P_2, and P_3 are

$$P_1^{(1)} = \frac{6.0 - 5.3}{2(0.004)} = 87.5000$$
$$P_2^{(1)} = \frac{6.0 - 5.5}{2(0.006)} = 41.6667$$
$$P_3^{(1)} = \frac{6.0 - 5.8}{2(0.009)} = 11.1111$$

Since $P_D = 800$ MW, the error ΔP from (7.39) is

$$\Delta P^{(1)} = 800 - (87.5 + 41.6667 + 11.1111) = 659.7222$$

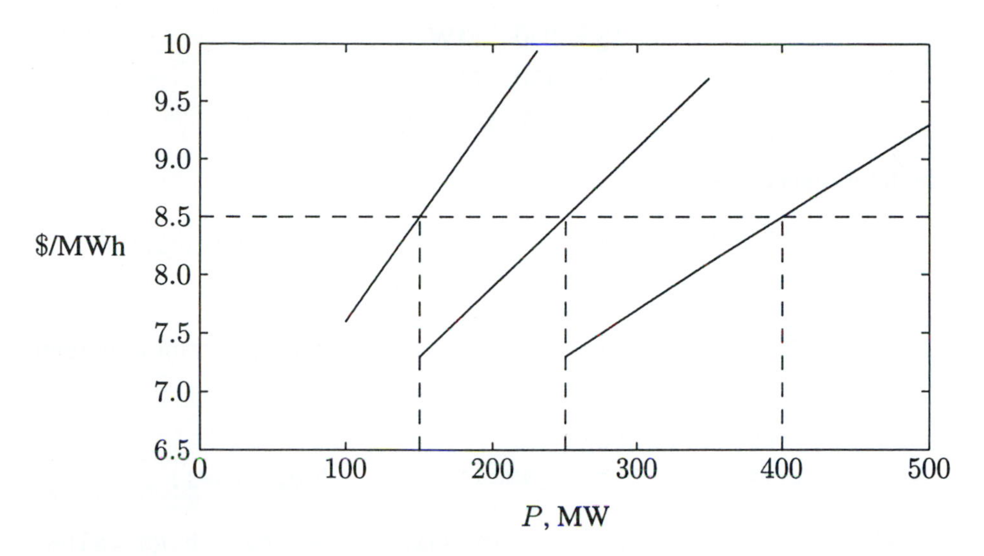

FIGURE 7.6
Illustrating the concept of equal incremental cost production cost.

From (7.37)

$$\Delta\lambda^{(1)} = \frac{659.7222}{\frac{1}{2(0.004)} + \frac{1}{2(0.006)} + \frac{1}{2(0.009)}} = \frac{659.7222}{263.8888} = 2.5$$

Therefore, the new value of λ is

$$\lambda^{(2)} = 6.0 + 2.5 = 8.5$$

Continuing the process, for the second iteration, we have

$$P_1^{(2)} = \frac{8.5 - 5.3}{2(0.004)} = 400.0000$$

$$P_2^{(2)} = \frac{8.5 - 5.5}{2(0.006)} = 250.0000$$

$$P_3^{(2)} = \frac{8.5 - 5.8}{2(0.009)} = 150.0000$$

and

$$\Delta P^{(2)} = 800 - (400 + 250 + 150) = 0.0$$

Since $\Delta P^{(2)} = 0$, the equality constraint is met in two iterations. Therefore, the optimal dispatch are

$$P_1 = 400 \quad \text{MW}$$

$$P_2 = 250 \quad \text{MW}$$
$$P_3 = 150 \quad \text{MW}$$
$$\hat{\lambda} = 8.5 \quad \$/\text{MWh}$$

and the total fuel cost is

$$C_t = 500 + 5.3(400) + 0.004(400)^2 + 400 + 5.5(250) + 0.006(250)^2$$
$$+200 + 5.8(150) + 0.009(150)^2 = 6,682.5 \quad \$/\text{h}$$

To demonstrate the above method, the following simple program is written for Example 7.4.

```
alpha =[500; 400; 200];
beta = [5.3; 5.5; 5.8]; gamma=[.004; .006; .009];
PD=800;
DelP = 10;            % Error in DelP is set to a high value
lambda = input('Enter estimated value of Lambda = ');
fprintf(' ')
disp(['   Lambda       P1        P2        P3        DP'...
 '     grad      Delambda'])
iter = 0;                                  % Iteration counter
while abs(DelP)  >= 0.001               % Test for convergence
iter = iter + 1;                           % No. of iterations
P = (lambda - beta)./(2*gamma);    % Coordination equation
DelP =PD - sum(P);                                % Residual
J = sum(ones(length(gamma), 1)./(2*gamma));% Gradient sum
Delambda = DelP/J;                        % Change in variable
disp([lambda, P(1), P(2), P(3), DelP, J, Delambda])
lambda = lambda + Delambda;            % Successive solution
end
totalcost = sum(alpha + beta.*P + gamma.*P.^2)
```

When the program is run, the result is

```
Enter estimated value of Lambda = 6

Lambda     P1       P2       P3        DP      grad    Delambda
6.0000  87.500  41.6667  11.1111  659.7222  263.8889  2.500
8.5000 400.000 250.0000 150.0000    0.0000  263.8889  0.000

   totalcost =

                6682.5
```

A general program called **dispatch** is developed for the optimal dispatch problem. The program returns the system λ, the optimal dispatch generation vector P, and the total cost. The following reserved variables are required by the **dispatch** program:

Pdt This reserved name must be used to specify the total load in MW. If **Pdt** is not specified the user is prompted to input the total load. If **dispatch** is used following any of the power flow programs, the total load is automatically passed by the power flow program.

cost This reserved name must be used to specify the cost function coefficients. The coefficients are arranged in the *MATLAB* matrix format. Each row contains the coefficients of the cost function in ascending powers of P.

mwlimits This name is reserved for the generator's real power limits and are discussed in Section 7.5. This entry is specified in matrix form with the first column representing the minimum value and the second column representing the maximum value. If **mwlimits** is not specified, the program obtains the optimal dispatch of generation with no limits.

B B0 B00 These names are reserved for the loss formula coefficient matrices and are discussed in Section 7.6. If these variables are not specified, optimal dispatch of generation is obtained neglecting losses.

The total generation cost of a thermal power system can be obtained with the aid of the **gencost** command. This program can be used following any of the power flow programs or the **dispatch** program, provided cost function matrix is defined.

Example 7.5

Neglecting generator limits and system losses, use **dispatch** program to obtain the optimal dispatch of generation for thermal plants specified in Example 7.4.

We use the following command:

```
cost = [500   5.3   0.004
        400   5.5   0.006
        200   5.8   0.009];
Pdt = 800;
dispatch
gencost
```

The result is

```
Incremental cost of delivered power(system lambda) = 8.5$/MWh
Optimal Dispatch of Generation:

     400.0000
     250.0000
     150.0000
   Total generation cost = 6682.50 $/h
```

7.5 ECONOMIC DISPATCH NEGLECTING LOSSES AND INCLUDING GENERATOR LIMITS

The power output of any generator should not exceed its rating nor should it be below that necessary for stable boiler operation. Thus, the generations are restricted to lie within given minimum and maximum limits. The problem is to find the real power generation for each plant such that the objective function (i.e., total production cost) as defined by (7.23) is minimum, subject to the constraint given by (7.24) and the inequality constraints given by

$$P_{i(min)} \leq P_i \leq P_{i(max)} \qquad i = 1, \ldots, n_g \tag{7.40}$$

Where $P_{i(min)}$ and $P_{i(max)}$ are the minimum and maximum generating limits respectively for plant i.

The Kuhn-Tucker conditions complement the Lagrangian conditions to include the inequality constraints as additional terms. The necessary conditions for the optimal dispatch with losses neglected becomes

$$\begin{aligned} \frac{dC_i}{dP_i} &= \lambda \qquad \text{for} \quad P_{i(min)} < P_i < P_{i(max)} \\ \frac{dC_i}{dP_i} &\leq \lambda \qquad \text{for} \quad P_i = P_{i(max)} \\ \frac{dC_i}{dP_i} &\geq \lambda \qquad \text{for} \quad P_i = P_{i(min)} \end{aligned} \tag{7.41}$$

The numerical solution is the same as before. That is, for an estimated λ, P_i are found from the coordination Equation (7.31) and iteration is continued until $\sum P_i = P_D$. As soon as any plant reaches a maximum or minimum, the plant becomes pegged at the limit. In effect, the plant output becomes a constant, and only the unviolated plants must operate at equal incremental cost.

Example 7.6

Find the optimal dispatch and the total cost in $/h for the thermal plants of Example 7.4 when the total load is 975 MW with the following generator limits (in MW):

$$\begin{aligned} 200 \le P_1 \le 450 \\ 150 \le P_2 \le 350 \\ 100 \le P_3 \le 225 \end{aligned}$$

Assume the initial value of $\lambda^{(1)} = 6.0$. From coordination equations given by (7.31), P_1, P_2, and P_3 are

$$\begin{aligned} P_1^{(1)} &= \frac{6.0 - 5.3}{2(0.004)} = 87.5000 \\ P_2^{(1)} &= \frac{6.0 - 5.5}{2(0.006)} = 41.6667 \\ P_3^{(1)} &= \frac{6.0 - 5.8}{2(0.009)} = 11.1111 \end{aligned}$$

Since $P_D = 975$ MW, the error ΔP from (7.39) is

$$\Delta P^{(1)} = 975 - (87.5 + 41.6667 + 11.1111) = 834.7222$$

From (7.37)

$$\Delta\lambda^{(1)} = \frac{834.7222}{\frac{1}{2(0.004)} + \frac{1}{2(0.006)} + \frac{1}{2(0.009)}} = \frac{834.7222}{263.8888} = 3.1632$$

Therefore, the new value of λ is

$$\lambda^{(2)} = 6.0 + 3.1632 = 9.1632$$

Continuing the process, for the second iteration, we have

$$\begin{aligned} P_1^{(2)} &= \frac{9.1632 - 5.3}{2(0.004)} = 482.8947 \\ P_2^{(2)} &= \frac{9.1632 - 5.5}{2(0.006)} = 305.2632 \\ P_3^{(2)} &= \frac{9.1632 - 5.8}{2(0.009)} = 186.8421 \end{aligned}$$

and

$$\Delta P^{(2)} = 975 - (482.8947 + 305.2632 + 186.8421) = 0.0$$

Since $\Delta P^{(2)} = 0$, the equality constraint is met in two iterations. However, P_1 exceeds its upper limit. Thus, this plant is pegged at its upper limit. Hence $P_1 = 450$ and is kept constant at this value. Thus, the new imbalance in power is

$$\Delta P^{(2)} = 975 - (450 + 305.2632 + 186.8421) = 32.8947$$

From (7.37)

$$\Delta\lambda^{(2)} = \frac{32.8947}{\frac{1}{2(0.006)} + \frac{1}{2(0.009)}} = \frac{32.8947}{138.8889} = 0.2368$$

Therefore, the new value of λ is

$$\lambda^{(3)} = 9.1632 + 0.2368 = 9.4$$

For the third iteration, we have

$$\begin{aligned}
P_1^{(3)} &= 450 \\
P_2^{(3)} &= \frac{9.4 - 5.5}{2(0.006)} = 325 \\
P_3^{(3)} &= \frac{9.4 - 5.8}{2(0.009)} = 200
\end{aligned}$$

and

$$\Delta P^{(3)} = 975 - (450 + 325 + 200) = 0.0$$

$\Delta P^{(3)} = 0$, and the equality constraint is met and P_2 and P_3 are within their limits. Thus, the optimal dispatch is

$$\begin{aligned}
P_1 &= 450 \quad \text{MW} \\
P_2 &= 325 \quad \text{MW} \\
P_3 &= 200 \quad \text{MW} \\
\hat{\lambda} &= 9.4 \quad \$/\text{MWh}
\end{aligned}$$

and the total fuel cost is

$$\begin{aligned}
C_t = 500 &+ 5.3(450) + 0.004(450)^2 + 400 + 5.5(325) + 0.006(325)^2 \\
&+ 200 + 5.8(200) + 0.009(200)^2 = 8,236.25 \quad \$/\text{h}
\end{aligned}$$

The following commands can be used to obtain the optimal dispatch of generation including generator limits.

```
cost = [500   5.3   0.004
        400   5.5   0.006
        200   5.8   0.009];
mwlimits=[200  450
          150  350
          100  225];
Pdt = 975;
dispatch
gencost
```

The result is

```
Incremental cost of delivered power(system lambda) = 9.4$/MWh
Optimal Dispatch of Generation:

      450
      325
      200

Total generation cost = 8236.25 $/h
```

7.6 ECONOMIC DISPATCH INCLUDING LOSSES

When transmission distances are very small and load density is very high, transmission losses may be neglected and the optimal dispatch of generation is achieved with all plants operating at equal incremental production cost. However, in a large interconnected network where power is transmitted over long distances with low load density areas, transmission losses are a major factor and affect the optimum dispatch of generation. One common practice for including the effect of transmission losses is to express the total transmission loss as a quadratic function of the generator power outputs. The simplest quadratic form is

$$P_L = \sum_{i=1}^{n_g} \sum_{j=1}^{n_g} P_i B_{ij} P_j \tag{7.42}$$

A more general formula containing a linear term and a constant term, referred to as *Kron's loss formula*, is

$$P_L = \sum_{i=1}^{n_g} \sum_{j=1}^{n_g} P_i B_{ij} P_j + \sum_{i=1}^{n_g} B_{0i} P_i + B_{00} \tag{7.43}$$

The coefficients B_{ij} are called *loss coefficients* or *B-coefficients*. B-coefficients are assumed constant, and reasonable accuracy can be expected provided the actual operating conditions are close to the base case where the B-constants were computed. There are various ways of arriving at a loss equation. A method for obtaining these B-coefficients is presented in Section 7.7.

The economic dispatching problem is to minimize the overall generating cost C_i, which is the function of plant output

$$C_t = \sum_{i=1}^{n_g} C_i$$
$$= \sum_{i=1}^{n} \alpha_i + \beta_i P_i + \gamma_i P_i^2 \qquad (7.44)$$

subject to the constraint that generation should equal total demands plus losses, i.e.,

$$\sum_{i=1}^{n_g} P_i = P_D + P_L \qquad (7.45)$$

satisfying the inequality constraints, expressed as follows:

$$P_{i(min)} \le P_i \le P_{i(max)} \qquad i = 1, \ldots, n_g \qquad (7.46)$$

where $P_{i(min)}$ and $P_{i(max)}$ are the minimum and maximum generating limits, respectively, for plant i.

Using the Lagrange multiplier and adding additional terms to include the inequality constraints, we obtain

$$\mathcal{L} = C_t + \lambda(P_D + P_L - \sum_{i=1}^{n_g} P_i) + \sum_{i=1}^{n_g} \mu_{i(max)}(P_i - P_{i(max)}) +$$
$$\sum_{i=1}^{n_g} \mu_{i(min)}(P_i - P_{i(min)}) \qquad (7.47)$$

The constraints should be understood to mean the $\mu_{i(max)} = 0$ when $P_i < P_{i(max)}$ and that $\mu_{i(min)} = 0$ when $P_i > P_{i(min)}$. In other words, if the constraint is not violated, its associated μ variable is zero and the corresponding term in (7.47) does not exist. The constraint only becomes active when violated. The minimum of this unconstrained function is found at the point where the partials of the function to its variables are zero.

$$\frac{\partial \mathcal{L}}{\partial P_i} = 0 \qquad (7.48)$$

$$\frac{\partial \mathcal{L}}{\partial \lambda} = 0 \tag{7.49}$$

$$\frac{\partial \mathcal{L}}{\partial \mu_{i(max)}} = P_i - P_{i(max)} = 0 \tag{7.50}$$

$$\frac{\partial \mathcal{L}}{\partial \mu_{i(min)}} = P_i - P_{i(min)} = 0 \tag{7.51}$$

Equations (7.50) and (7.51) imply that P_i should not be allowed to go beyond its limit, and when P_i is within its limits $\mu_{i(min)} = \mu_{i(max)} = 0$ and the Kuhn-Tucker function becomes the same as the Lagrangian one. First condition, given by (7.48), results in

$$\frac{\partial C_t}{\partial P_i} + \lambda(0 + \frac{\partial P_L}{\partial P_i} - 1) = 0$$

Since

$$C_t = C_1 + C_2 + \cdots + C_{n_g}$$

then

$$\frac{\partial C_t}{\partial P_i} = \frac{dC_i}{dP_i}$$

and therefore the condition for optimum dispatch is

$$\frac{dC_i}{dP_i} + \lambda \frac{\partial P_L}{\partial P_i} = \lambda \qquad i = 1, \ldots, n_g \tag{7.52}$$

The term $\frac{\partial P_L}{\partial P_i}$ is known as the incremental transmission loss. Second condition, given by (7.49), results in

$$\sum_{i=1}^{n_g} P_i = P_D + P_L \tag{7.53}$$

Equation (7.53) is precisely the equality constraint that was to be imposed.

Classically, Equation (7.52) is rearranged as

$$\left(\frac{1}{1 - \frac{\partial P_L}{\partial P_i}} \right) \frac{dC_i}{dP_i} = \lambda \qquad i = 1, \ldots, n_g \tag{7.54}$$

or

$$L_i \frac{dC_i}{dP_i} = \lambda \qquad i = 1, \ldots, n_g \tag{7.55}$$

where L_i is known as the *penalty factor* of plant i and is given by

$$L_i = \frac{1}{1 - \frac{\partial P_L}{\partial P_i}} \tag{7.56}$$

Hence, the effect of transmission loss is to introduce a penalty factor with a value that depends on the location of the plant. Equation (7.55) shows that the minimum cost is obtained when the incremental cost of each plant multiplied by its penalty factor is the same for all plants.

The incremental production cost is given by (7.22), and the incremental transmission loss is obtained from the loss formula (7.43) which yields

$$\frac{\partial P_L}{\partial P_i} = 2\sum_{j=1}^{n_g} B_{ij}P_j + B_{0i} \tag{7.57}$$

Substituting the expression for the incremental production cost and the incremental transmission loss in (7.52) results in

$$\beta_i + 2\gamma_i P_i + 2\lambda \sum_{j=1}^{n_g} B_{ij}P_j + B_{0i}\lambda = \lambda$$

or

$$\left(\frac{\gamma_i}{\lambda} + B_{ii}\right) P_i + \sum_{\substack{j=1 \\ j\neq i}}^{n_g} B_{ij}P_j = \frac{1}{2}\left(1 - B_{0i} - \frac{\beta_i}{\lambda}\right) \tag{7.58}$$

Extending (7.58) to all plants results in the following linear equations in matrix form

$$\begin{bmatrix} \frac{\gamma_1}{\lambda}+B_{11} & B_{12} & \cdots & B_{1n_g} \\ B_{21} & \frac{\gamma_2}{\lambda}+B_{22} & \cdots & B_{2n_g} \\ \vdots & \vdots & \ddots & \vdots \\ B_{n_g1} & B_{n_g2} & \cdots & \frac{\gamma_{n_g}}{\lambda}+B_{n_gn_g} \end{bmatrix} \begin{bmatrix} P_1 \\ P_2 \\ \vdots \\ P_{n_g} \end{bmatrix} = \frac{1}{2} \begin{bmatrix} 1-B_{01}-\frac{\beta_1}{\lambda} \\ 1-B_{02}-\frac{\beta_2}{\lambda} \\ \vdots \\ 1-B_{0n_g}-\frac{\beta_{n_g}}{\lambda} \end{bmatrix} \tag{7.59}$$

or in short form

$$\boldsymbol{EP} = \boldsymbol{D} \tag{7.60}$$

To find the optimal dispatch for an estimated value of $\lambda^{(1)}$, the simultaneous linear equation given by (7.60) is solved. In *MATLAB* use the command $\mathbf{P} = \mathbf{E}\backslash\mathbf{D}$.

Then the iterative process is continued using the gradient method. To do this, from (7.58), P_i at the kth iteration is expressed as

$$P_i^{(k)} = \frac{\lambda^{(k)}(1 - B_{0i}) - \beta_i - 2\lambda^{(k)} \sum_{j \neq i} B_{ij} P_j^{(k)}}{2(\gamma_i + \lambda^{(k)} B_{ii})} \tag{7.61}$$

Substituting for P_i from (7.61) in (7.53) results in

$$\sum_{i=1}^{n_g} \frac{\lambda^{(k)}(1 - B_{0i}) - \beta_i - 2\lambda^{(k)} \sum_{j \neq i} B_{ij} P_j^{(k)}}{2(\gamma_i + \lambda^{(k)} B_{ii})} = P_D + P_L^{(k)} \tag{7.62}$$

or

$$f(\lambda)^{(k)} = P_D + P_L^{(k)} \tag{7.63}$$

Expanding the left-hand side of the above equation in Taylor's series about an operating point $\lambda^{(k)}$, and neglecting the higher-order terms results in

$$f(\lambda)^{(k)} + \left(\frac{df(\lambda)}{d\lambda}\right)^{(k)} \Delta\lambda^{(k)} = P_D + P_L^{(k)} \tag{7.64}$$

or

$$\begin{aligned} \Delta\lambda^{(k)} &= \frac{\Delta P^{(k)}}{\left(\frac{df(\lambda)}{d\lambda}\right)^{(k)}} \\ &= \frac{\Delta P^{(k)}}{\sum \left(\frac{dP_i}{d\lambda}\right)^{(k)}} \end{aligned} \tag{7.65}$$

where

$$\sum_{i=1}^{n_g} \left(\frac{\partial P_i}{\partial \lambda}\right)^{(k)} = \sum_{i=1}^{n_g} \frac{\gamma_i(1 - B_{0i}) + B_{ii}\beta_i - 2\gamma_i \sum_{j \neq i} B_{ij} P_j^{(k)}}{2(\gamma_i + \lambda^{(k)} B_{ii})^2} \tag{7.66}$$

and therefore,

$$\lambda^{(k+1)} = \lambda^{(k)} + \Delta\lambda^{(k)} \tag{7.67}$$

where

$$\Delta P^{(k)} = P_D + P_L^{(k)} - \sum_{i=1}^{n_g} P_i^{(k)} \tag{7.68}$$

The process is continued until $\Delta P^{(k)}$ is less than a specified accuracy.

If an approximate loss formula expressed by

$$P_L = \sum_{i=1}^{n_g} B_{ii} P_i^2 \tag{7.69}$$

is used, $B_{ij} = 0$, $B_{00} = 0$, and solution of the simultaneous equation given by (7.61) reduces to the following simple expression

$$P_i^{(k)} = \frac{\lambda^{(k)} - \beta_i}{2(\gamma_i + \lambda^{(k)} B_{ii})} \tag{7.70}$$

and (7.66) reduces to

$$\sum_{i=1}^{n_g} \left(\frac{\partial P_i}{\partial \lambda} \right)^{(k)} = \sum_{i=1}^{n_g} \frac{\gamma_i + B_{ii}\beta_i}{2(\gamma_i + \lambda^{(k)} B_{ii})^2} \tag{7.71}$$

Example 7.7

The fuel cost in \$/h of three thermal plants of a power system are

$$C_1 = 200 + 7.0P_1 + 0.008P_1^2 \quad \$/\text{h}$$
$$C_2 = 180 + 6.3P_2 + 0.009P_2^2 \quad \$/\text{h}$$
$$C_3 = 140 + 6.8P_3 + 0.007P_3^2 \quad \$/\text{h}$$

where P_1, P_2, and P_3 are in MW. Plant outputs are subject to the following limits

$$10 \text{ MW} \leq 85 \text{ MW}$$
$$10 \text{ MW} \leq 80 \text{ MW}$$
$$10 \text{ MW} \leq 70 \text{ MW}$$

For this problem, assume the real power loss is given by the simplified expression

$$P_{L(pu)} = 0.0218P_{1(pu)}^2 + 0.0228P_{2(pu)}^2 + 0.0179P_{3(pu)}^2$$

where the loss coefficients are specified in per unit on a 100-MVA base. Determine the optimal dispatch of generation when the total system load is 150 MW.

In the cost function P_i is expressed in MW. Therefore, the real power loss in terms of MW generation is

$$P_L = \left[0.0218 \left(\frac{P_1}{100} \right)^2 + 0.0228 \left(\frac{P_2}{100} \right)^2 + 0.0179 \left(\frac{P_3}{100} \right)^2 \right] \times 100 \quad \text{MW}$$
$$= 0.000218P_1^2 + 0.000228P_2^2 + 0.000179P_3^2 \quad \text{MW}$$

For the numerical solution using the gradient method, assume the initial value of $\lambda^{(1)} = 8.0$. From coordination equations, given by (7.70), $P_1^{(1)}$, $P_2^{(1)}$, and $P_3^{(1)}$ are

$$P_1^{(1)} = \frac{8.0 - 7.0}{2(0.008 + 8.0 \times 0.000218)} = 51.3136 \quad \text{MW}$$
$$P_2^{(1)} = \frac{8.0 - 6.3}{2(0.009 + 8.0 \times 0.000228)} = 78.5292 \quad \text{MW}$$
$$P_3^{(1)} = \frac{8.0 - 6.8}{2(0.007 + 8.0 \times 0.000179)} = 71.1575 \quad \text{MW}$$

The real power loss is

$$P_L^{(1)} = 0.000218(51.3136)^2 + 0.000228(78.5292)^2 + 0.000179(71.1575)^2 = 2.886$$

Since $P_D = 150$ MW, the error $\Delta P^{(1)}$ from (7.68) is

$$\Delta P^{(1)} = 150 + 2.8864 - (51.3136 + 78.5292 + 71.1575) = -48.1139$$

From (7.71)

$$\sum_{i=1}^{3} \left(\frac{\partial P_i}{\partial \lambda}\right)^{(1)} = \frac{0.008 + 0.000218 \times 7.0}{2(0.008 + 8.0 \times 0.000218)^2} + \frac{0.009 + 0.000228 \times 6.3}{2(0.009 + 8.0 \times 0.000228)^2}$$
$$+ \frac{0.007 + 0.000179 \times 6.8}{2(0.007 + 8.0 \times 0.000179)^2} = 152.4924$$

From (7.65)

$$\Delta \lambda^{(1)} = \frac{-48.1139}{152.4924} = -0.31552$$

Therefore, the new value of λ is

$$\lambda^{(2)} = 8.0 - 0.31552 = 7.6845$$

Continuing the process, for the second iteration, we have

$$P_1^{(2)} = \frac{7.6845 - 7.0}{2(0.008 + 7.6845 \times 0.000218)} = 35.3728 \quad \text{MW}$$
$$P_2^{(2)} = \frac{7.6845 - 6.3}{2(0.009 + 7.6845 \times 0.000228)} = 64.3821 \quad \text{MW}$$
$$P_3^{(2)} = \frac{7.6845 - 6.8}{2(0.007 + 7.6845 \times 0.000179)} = 52.8015 \quad \text{MW}$$

The real power loss is

$$P_L^{(2)} = 0.000218(35.3728)^2 + 0.000228(64.3821)^2 + 0.000179(52.8015)^2 = 1.717$$

Since $P_D = 150$ MW, the error $\Delta P^{(2)}$ from (7.68) is

$$\Delta P^{(2)} = 150 + 1.7169 - (35.3728 + 64.3821 + 52.8015) = -0.8395$$

From (7.71)

$$\sum_{i=1}^{3}\left(\frac{\partial P_i}{\partial \lambda}\right)^{(2)} = \frac{0.008 + 0.000218 \times 7.0}{2(0.008 + 7.684 \times 0.000218)^2} + \frac{0.009 + 0.000228 \times 6.3}{2(0.009 + 7.684 \times 0.000228)^2}$$
$$+ \frac{0.007 + 0.000179 \times 6.8}{2(0.007 + 7.6845 \times 0.000179)^2} = 154.588$$

From (7.65)

$$\Delta\lambda^{(2)} = \frac{-0.8395}{154.588} = -0.005431$$

Therefore, the new value of λ is

$$\lambda^{(3)} = 7.6845 - 0.005431 = 7.679$$

For the third iteration, we have

$$P_1^{(3)} = \frac{7.679 - 7.0}{2(0.008 + 7.679 \times 0.000218)} = 35.0965 \quad \text{MW}$$
$$P_2^{(3)} = \frac{7.679 - 6.3}{2(0.009 + 7.679 \times 0.000228)} = 64.1369 \quad \text{MW}$$
$$P_3^{(3)} = \frac{7.679 - 6.8}{2(0.007 + 7.679 \times 0.000179)} = 52.4834 \quad \text{MW}$$

The real power loss is

$$P_L^{(3)} = 0.000218(35.0965)^2 + 0.000228(64.1369)^2 + 0.000179(52.4834)^2 = 1.699$$

Since $P_D = 150$ MW, the error $\Delta P^{(3)}$ from (7.68) is

$$\Delta P^{(3)} = 150 + 1.6995 - (35.0965 + 64.1369 + 52.4834) = -0.01742$$

From (7.71)

$$\sum_{i=1}^{3}\left(\frac{\partial P_i}{\partial \lambda}\right)^{(3)} = \frac{0.008 + 0.000218 \times 7.0}{2(0.008 + 7.679 \times 0.000218)^2} + \frac{0.009 + 0.000228 \times 6.3}{2(0.009 + 7.679 \times 0.000228)^2}$$
$$+ \frac{0.007 + 0.000179 \times 6.8}{2(0.007 + 7.679 \times 0.000179)^2} = 154.624$$

From (7.65)

$$\Delta\lambda^{(3)} = \frac{-0.01742}{154.624} = -0.0001127$$

Therefore, the new value of λ is

$$\lambda^{(4)} = 7.679 - 0.0001127 = 7.6789$$

Since $\Delta\lambda^{(3)}$, is small the equality constraint is met in four iterations, and the optimal dispatch for $\lambda = 7.6789$ are

$$P_1^{(4)} = \frac{7.6789 - 7.0}{2(0.008 + 7.679 \times 0.000218)} = 35.0907 \quad \text{MW}$$
$$P_2^{(4)} = \frac{7.6789 - 6.3}{2(0.009 + 7.679 \times 0.000228)} = 64.1317 \quad \text{MW}$$
$$P_3^{(4)} = \frac{7.6789 - 6.8}{2(0.007 + 7.679 \times 0.000179)} = 52.4767 \quad \text{MW}$$

The real power loss is

$$P_L^{(4)} = 0.000218(35.0907)^2 + 0.000228(64.1317)^2 + 0.000179(52.4767)^2 = 1.699$$

and the total fuel cost is

$$C_t = 200 + 7.0(35.0907) + 0.008(35.0907)^2 + 180 + 6.3(64.1317) + 0.009(64.1317)^2 + 140 + 6.8(52.4767) + 0.007(52.4767)^2 = 1592.65 \text{ \$/h}$$

The **dispatch** program can be used to find the optimal dispatch of generation. The program is designed for the loss coefficients to be expressed in per unit. The loss coefficients are arranged in a matrix form with the variable name B. The base MVA must be specified by the variable name **basemva**. If base mva is not specified, it is set to 100 MVA.

We use the following commands

```
cost = [200  7.0    0.008
        180  6.3    0.009
        140  6.8    0.007];
mwlimits =[10  85
           10  80
           10  70];
Pdt = 150;
B = [0.0218        0        0
```

```
          0     0.0228        0
          0          0   0.0179];
basemva = 100;
dispatch
gencost
```

The result is

```
Incremental cost of delivered power(system lambda) =
7.678935$/MWh
Optimal Dispatch of Generation:

   35.0907
   64.1317
   52.4767

Total system loss = 1.6991 MW
Total generation cost = 1592.65 $/h
```

Example 7.8

Figure 7.7 (page 295) shows the one-line diagram of a power system described in Example 7.9. The B matrices of the loss formula for this system are found in Example 7.9. They are given in per unit on a 100 MVA base as follows

$$\begin{aligned} B &= \begin{bmatrix} 0.0218 & 0.0093 & 0.0028 \\ 0.0093 & 0.0228 & 0.0017 \\ 0.0028 & 0.0017 & 0.0179 \end{bmatrix} \\ B_0 &= \begin{bmatrix} 0.0003 & 0.0031 & 0.0015 \end{bmatrix} \\ B_{00} &= 0.00030523 \end{aligned}$$

Cost functions, generator limits, and total loads are given in Example 7.7. Use **dispatch** program to obtain the optimal dispatch of generation.

We use the following commands.

```
cost = [200  7.0    0.008
        180  6.3    0.009
        140  6.8    0.007];
mwlimits =[10  85
           10  80
           10  70];
Pdt = 150;
```

```
B =  [0.0218   0.0093   0.0028
       0.0093   0.0228   0.0017
       0.0028   0.0017   0.0179];
B0 = [0.0003   0.0031   0.0015];
B00 = 0.00030523;
basemva = 100;
dispatch
gencost
```

The result is

```
Incremental cost of delivered power (system lambda) =
7.767785 $/MWh
Optimal Dispatch of Generation:

   33.4701
   64.0974
   55.1011

Total generation cost =    1599.98 $/h
```

7.7 DERIVATION OF LOSS FORMULA

One of the major steps in the optimal dispatch of generation is to express the system losses in terms of the generator's real power outputs. There are several methods of obtaining the loss formula. One method developed by Kron and adopted by Kirchmayer is the *loss coefficient* or *B-coefficient* method.

The total injected complex power at bus i, denoted by S_i, is given by

$$S_i = P_i + jQ_i = V_i I_i^* \tag{7.72}$$

The summation of powers over all buses gives the total system losses

$$P_L + jQ_L = \sum_{i=1}^{n} V_i I_i^* = V_{bus}^T I_{bus}^* \tag{7.73}$$

where P_L and Q_L are the real and reactive power loss of the system. V_{bus} is the column vector of the nodal bus voltages and I_{bus} is the column vector of the injected bus currents. The expression for the bus currents in terms of bus voltage was derived in Chapter 6 and is given by (6.2) as

$$I_{bus} = Y_{bus} V_{bus} \tag{7.74}$$

where Y_{bus} is the bus admittance matrix with ground as reference. Solving for V_{bus}, we have

$$\begin{aligned} V_{bus} &= Y_{bus}^{-1} I_{bus} \\ &= Z_{bus} I_{bus} \end{aligned} \tag{7.75}$$

The inverse of the bus admittance matrix is known as the *bus impedance matrix.* The bus admittance matrix is nonsingular if there are shunt elements (such as shunt capacitive susceptance) connected to the ground (bus number 0). As discussed in Chapter 6, the bus admittance matrix is sparse and its inverse can be expressed as a product of sparse matrix factors. Actually Z_{bus}, which is also required for short-circuit analysis, can be obtained directly by the method of *building algorithm* without the need for matrix inversion. This technique is discussed in Chapter 9.

Substituting for V_{bus} from (7.75) into (7.73), results in

$$\begin{aligned} P_L + jQ_L &= [Z_{bus} I_{bus}]^T I_{bus}^* \\ &= I_{bus}^T Z_{bus}^T I_{bus}^* \end{aligned} \tag{7.76}$$

Z_{bus} is a symmetrical matrix; therefore, $Z_{bus}^T = Z_{bus}$, and the total system loss becomes

$$P_L + jQ_L = I_{bus}^T Z_{bus} I_{bus}^* \tag{7.77}$$

The expression in (7.77) can also be expressed with the use of index notation as

$$P_L + jQ_L = \sum_{i=1}^{n} \sum_{j=1}^{n} I_i Z_{ij} I_j^* \tag{7.78}$$

Since the bus impedance matrix is symmetrical, i.e., $Z_{ij} = Z_{ji}$, the above equation may be rewritten as

$$P_L + jQ_L = \frac{1}{2} \sum_{i=1}^{n} \sum_{j=1}^{n} Z_{ij} (I_i I_j^* + I_j I_i^*) \tag{7.79}$$

The quantity inside the parentheses in (7.79) is real; thus the power loss can be broken into its real and imaginary components as

$$P_L = \frac{1}{2} \sum_{i=1}^{n} \sum_{j=1}^{n} R_{ij} (I_i I_j^* + I_j I_i^*) \tag{7.80}$$

$$Q_L = \frac{1}{2} \sum_{i=1}^{n} \sum_{j=1}^{n} X_{ij} (I_i I_j^* + I_j I_i^*) \tag{7.81}$$

where R_{ij} and X_{ij} are the real and imaginary elements of the bus impedance matrix, respectively. Again, since $R_{ij} = R_{ji}$, the real power loss equation can be converted back into

$$P_L = \sum_{i=1}^{n}\sum_{j=1}^{n} I_i R_{ij} I_j^* \tag{7.82}$$

Or in matrix form, the equation for the system real power loss becomes

$$P_L = I_{bus}^T R_{bus} I_{bus}^* \tag{7.83}$$

where R_{bus} is the real part of the bus impedance matrix. In order to obtain the general formula for the system power loss in terms of generator powers, we define the total load current as the sum of all individual load currents, i.e.,

$$I_{L1} + I_{L2} + \cdots + I_{Ln_d} = I_D \tag{7.84}$$

where n_d is the number of load buses and I_D is the total load currents. Now the individual bus currents are assumed to vary as a constant complex fraction of the total load current, i.e.,

$$I_{Lk} = \ell_k I_D \qquad k = 1, 2, \ldots, n_d \tag{7.85}$$

or

$$\ell_k = \frac{I_{Lk}}{I_D} \tag{7.86}$$

Assuming bus 1 to be the reference bus (slack bus), expanding the first row in (7.75) results in

$$V_1 = Z_{11}I_1 + Z_{12}I_2 + \cdots + Z_{1n}In \tag{7.87}$$

If n_g is the number of generator buses and n_d is the number of load buses, the above equation can be written in terms of the load currents and generator currents as

$$V_1 = \sum_{i=1}^{n_g} Z_{1i}I_{gi} + \sum_{k=1}^{n_d} Z_{1k}I_{Lk} \tag{7.88}$$

Substituting for I_{Lk} from (7.85) into (7.88), we have

$$\begin{aligned} V_1 &= \sum_{i=1}^{n_g} Z_{1i}I_{gi} + I_D \sum_{k=1}^{n_d} \ell_k Z_{1k} \\ &= \sum_{i=1}^{n_g} Z_{1i}I_{gi} + I_D T \end{aligned} \tag{7.89}$$

where

$$T = \sum_{k=1}^{n_d} \ell_k Z_{1k} \tag{7.90}$$

If I_0 is defined as the current flowing away from bus 1, with all other load currents set to zero, we have

$$V_1 = -Z_{11} I_0 \tag{7.91}$$

Substituting for V_1 in (7.89) and solving for I_D, we have

$$I_D = -\frac{1}{T} \sum_{i=1}^{n_g} Z_{1i} I_{gi} - \frac{1}{T} Z_{11} I_0 \tag{7.92}$$

Substituting for I_D from (7.92) into (7.85), the load currents become

$$I_{Lk} = -\frac{\ell_k}{T} \sum_{i=1}^{n_g} Z_{1i} I_{gi} - \frac{\ell_k}{T} Z_{11} I_0 \tag{7.93}$$

Let

$$\rho = -\frac{\ell_k}{T} \tag{7.94}$$

Then

$$I_{Lk} = \rho_k \sum_{i=1}^{n_g} Z_{1i} I_{gi} + \rho_k Z_{11} I_0 \tag{7.95}$$

Augmenting the generator currents with the above relation in matrix form, we have

$$\begin{bmatrix} I_{g1} \\ I_{g2} \\ \vdots \\ I_{gn_g} \\ I_{L1} \\ I_{L2} \\ \vdots \\ I_{Ln_d} \end{bmatrix} = \begin{bmatrix} 1 & 0 & \cdots & 0 & 0 \\ 0 & 1 & \cdots & 0 & 0 \\ \vdots & \vdots & \ddots & \vdots & \vdots \\ 0 & 0 & \cdots & 1 & 0 \\ \rho_1 Z_{11} & \rho_1 Z_{12} & \cdots & \rho_1 Z_{1n_g} & \rho_1 Z_{11} \\ \rho_2 Z_{11} & \rho_2 Z_{12} & \cdots & \rho_2 Z_{1n_g} & \rho_2 Z_{11} \\ \vdots & \vdots & \ddots & \vdots & \vdots \\ \rho_k Z_{11} & \rho_k Z_{12} & \cdots & \rho_k Z_{1n_g} & \rho_k Z_{11} \end{bmatrix} \begin{bmatrix} I_{g1} \\ I_{g2} \\ \vdots \\ I_{gn_g} \\ \\ \\ \\ I_0 \end{bmatrix} \tag{7.96}$$

Showing the above matrix by C, (7.96) becomes

$$I_{bus} = C I_{new} \tag{7.97}$$

Substituting for I_{bus} in (7.83), we have

$$\begin{aligned} P_L &= [CI_{new}]^T R_{bus} C^* I_{new}^* \\ &= I_{new}^T C^T R_{bus} C^* I_{new}^* \end{aligned} \tag{7.98}$$

If S_{gi} is the complex power at bus i, the generator current is

$$\begin{aligned} I_{gi} = \frac{S_{gi}^*}{V_i^*} &= \frac{P_{gi} - jQ_{gi}}{V_i^*} \\ &= \frac{1 - j\frac{Q_{gi}}{P_{gi}}}{V_i^*} P_{gi} \end{aligned} \tag{7.99}$$

or

$$I_{gi} = \psi_i P_{gi} \tag{7.100}$$

where

$$\psi_i = \frac{1 - j\frac{Q_{gi}}{P_{gi}}}{V_i^*} \tag{7.101}$$

Adding the current I_0 to the column vector current I_{gi} in (7.100) results in

$$\begin{bmatrix} I_{g1} \\ I_{g2} \\ \vdots \\ I_{gn_g} \\ I_0 \end{bmatrix} = \begin{bmatrix} \psi_1 & 0 & \cdots & 0 & 0 \\ 0 & \psi_2 & \cdots & 0 & 0 \\ \vdots & \vdots & \ddots & \vdots & \vdots \\ 0 & 0 & \cdots & \psi_{n_g} & 0 \\ 0 & 0 & \cdots & 0 & I_0 \end{bmatrix} \begin{bmatrix} P_{g1} \\ P_{g2} \\ \vdots \\ P_{gn_g} \\ 1 \end{bmatrix} \tag{7.102}$$

or in short form

$$I_{new} = \Psi P_{G1} \tag{7.103}$$

where

$$P_{G1} = \begin{bmatrix} P_{g1} \\ P_{g2} \\ \vdots \\ P_{gn_g} \\ 1 \end{bmatrix} \tag{7.104}$$

Substituting from (7.103) for I_{new} in (7.98), the loss equation becomes

$$\begin{aligned} P_L &= [\Psi P_{G1}]^T C^T R_{bus} C^* \Psi^* P_{G1}^* \\ &= P_{G1}^T \Psi^T C^T R_{bus} C^* \Psi^* P_{G1}^* \end{aligned} \tag{7.105}$$

The resultant matrix in the above equation is complex and the real power loss is found from its real part, thus

$$P_L = P_{G1}^T \Re[H] P_{G1}^* \tag{7.106}$$

where

$$H = \Psi^T C^T R_{bus} C^* \Psi^* \tag{7.107}$$

Since elements of the matrix H are complex, its real part must be used for computing the real power loss. It is found that H is a *Hermitian matrix*. This means that H is symmetrical and $H = H^*$. Thus, real part of H is found from

$$\Re[H] = \frac{H + H^*}{2} \tag{7.108}$$

The above matrix is partitioned as follows

$$\Re[H] = \begin{bmatrix} B_{11} & B_{12} & \cdots & B_{1n_g} & B_{01}/2 \\ B_{21} & B_{22} & \cdots & B_{2n_g} & B_{02}/2 \\ \vdots & \vdots & \ddots & \vdots & \vdots \\ B_{n_g1} & B_{n_g2} & \cdots & B_{n_gn_g} & B_{0n_g}/2 \\ B_{01}/2 & B_{02}/2 & \cdots & B_{0n_g}/2 & B_{00} \end{bmatrix} \tag{7.109}$$

Substituting for $\Re[H]$ into (7.106), yields

$$P_L = [P_{g1}\, P_{g2} \cdots P_{gn_g}\; 1] \begin{bmatrix} B_{11} & B_{12} & \cdots & B_{1n_g} & B_{01}/2 \\ B_{21} & B_{22} & \cdots & B_{2n_g} & B_{02}/2 \\ \vdots & \vdots & \ddots & \vdots & \vdots \\ B_{n_g1} & B_{n_g2} & \cdots & B_{n_gn_g} & B_{0n_g}/2 \\ B_{01}/2 & B_{02}/2 & \cdots & B_{0n_g}/2 & B_{00} \end{bmatrix} \begin{bmatrix} P_{g1} \\ P_{g2} \\ \cdots \\ P_{gn_g} \\ 1 \end{bmatrix} \tag{7.110}$$

or

$$P_L = \begin{bmatrix} P_{g1} & P_{g2} & \cdots & P_{gn_g} \end{bmatrix} \begin{bmatrix} B_{11} & B_{12} & \cdots & B_{1n_g} \\ B_{21} & B_{22} & \cdots & B_{2n_g} \\ \vdots & \vdots & \ddots & \vdots \\ B_{n_g1} & B_{n_g2} & \cdots & B_{n_gn_g} \end{bmatrix} \begin{bmatrix} P_{g1} \\ P_{g2} \\ \cdots \\ P_{gn_g} \end{bmatrix}$$

$$+ \begin{bmatrix} P_{g1} & P_{g2} & \cdots & P_{gn_g} \end{bmatrix} \begin{bmatrix} B_{01}/2 \\ B_{02}/2 \\ \cdots \\ B_{0n_g}/2 \end{bmatrix} + B_{00} \tag{7.111}$$

To find the loss coefficients, first a power flow solution is obtained for the initial operating state. This provides the voltage magnitude and phase angles at all buses. From these results, load currents I_{Lk}, the total load current I_D, and ℓ_k are obtained. Next the bus matrix Z_{bus} is found. This can be obtained by converting the bus admittance matrix found from **lfybus** or directly from the *building algorithm* described in Chapter 9. Next the transformation matrices C and Ψ and H are obtained. Finally the B-coefficients are evaluated from (7.109). It should be noted that the B-coefficients are functions of the system operating state. If a new scheduling of generation is not drastically different from the initial operating condition, the loss coefficients may be assumed constant. A program named **bloss** is developed for the computation of the B-coefficients. This program requires the power flow solution and can be used following any of the power flow programs such as **lfgauss**, **lfnewton**, or **decouple**. The B-coefficients obtained are based on the generation in per unit. When generation are expressed in MW, the loss coefficients are

$$B_{ij} = B_{ij\,pu}/S_B \qquad B_{0i} = B_{0i\,pu} \;\text{ and }\; B_{00} = B_{00\,pu} \times S_B$$

where S_B is the base MVA.

Example 7.9

Figure 7.7 shows the one-line diagram of a simple 5-bus power system with generator at buses 1, 2, and 3. Bus 1, with its voltage set at $1.06\angle 0°$ pu, is taken as the slack bus. Voltage magnitude and real power generation at buses 2 and 3 are 1.045 pu, 40 MW, and 1.030 pu, 30 MW, respectively.

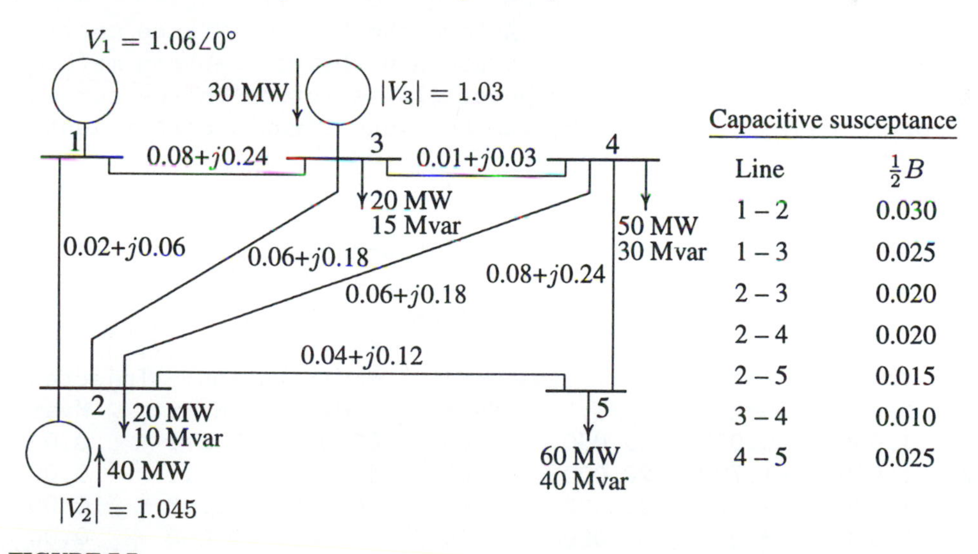

Capacitive susceptance	
Line	$\frac{1}{2}B$
1 – 2	0.030
1 – 3	0.025
2 – 3	0.020
2 – 4	0.020
2 – 5	0.015
3 – 4	0.010
4 – 5	0.025

FIGURE 7.7
One-line diagram of Example 7.9 (impedances in pu on 100-MVA base).

The load MW and Mvar values are shown on the diagram. Line impedances and one-half of the line capacitive susceptance are given in per unit on a 100-MVA base. Obtain the power flow solution and use the **bloss** program to obtain the loss coefficients in per unit.

We use the following commands

```
clear
basemva = 100;  accuracy = 0.0001;  maxiter = 10;

%       Bus Bus Voltage Angle -Load- ----Generator----Injected
%       No code Mag. Degree  MW   Mvar  MW Mvar  Qmin Qmax Mvar
busdata=[1   1  1.06   0.0    0    0     0   0    10   50   0
         2   2  1.045  0.0   20   10    40  30    10   50   0
         3   2  1.03   0.0   20   15    30  10    10   40   0
         4   0  1.00   0.0   50   30     0   0     0    0   0
         5   0  1.00   0.0   60   40     0   0     0    0  0];
%         Bus bus    R       X       1/2 B  1 for lines code or
%         nl  nr     pu      pu       pu    tap setting value
linedata=[1  2     0.02    0.06    0.030    1
          1  3     0.08    0.24    0.025    1
          2  3     0.06    0.18    0.020    1
          2  4     0.06    0.18    0.020    1
          2  5     0.04    0.12    0.015    1
          3  4     0.01    0.03    0.010    1
          4  5     0.08    0.24    0.025    1];
lfybus                          % form the bus admittance matrix
lfnewton         % Power flow solution by Newton-Raphson method
busout           % Prints the power flow solution on the screen
bloss                   % Obtains the loss formula coefficients
```

The result is

```
          Power Flow Solution by Newton-Raphson Method
             Maximum Power mismatch = 1.43025e-05
                    No. of iterations = 3

Bus Voltage  Angle    -----Load-----   --Generation-- Injected
No. Mag.     Degree    MW       Mvar     MW      Mvar     Mvar
1   1.060    0.000    0.000     0.00   83.051   7.271     0.00
2   1.045   -1.782   20.000    10.00   40.000  41.811     0.00
3   1.030   -2.664   20.000    15.00   30.000  24.148     0.00
4   1.019   -3.243   50.000    30.00    0.000   0.000     0.00
5   0.990   -4.405   60.000    40.00    0.000   0.000     0.00
Total               150.000   95.000  153.051  73.230     0.00
```

```
B =
      0.0218     0.0093     0.0028
      0.0093     0.0228     0.0017
      0.0028     0.0017     0.0179

B0 =
      0.0003     0.0031     0.0015
B00 =
      3.0523e-04
Total system loss = 3.05248 MW
```

As we have seen, any of the power flow programs, together with the **bloss** and **dispatch** programs can be used to obtain the optimal dispatch of generation. The **dispatch** program produces a variable named **dpslack**. This is the difference (absolute value) between the scheduled slack generation determined from the coordination equation, and the slack generation, obtained from the power flow solution. A power flow solution obtained with the new scheduling of generation results in a new loss coefficients, which can be used to solve the coordination equation again. This process can be continued until **dpslack** is within a specified tolerance. This procedure is demonstrated in the following example.

Example 7.10

The generation cost and the real power limits of the generators of the power system in Example 7.9 is given in Example 7.4 and Example 7.6. Obtain the optimal dispatch of generation. Continue the optimization process until the difference (absolute value) between the scheduled slack generation, determined from the coordination equation, and the slack generation, obtained from the power flow solution, is within 0.001 MW.

We use the following commands

```
clear
basemva = 100;  accuracy = 0.0001;  maxiter = 10;

%        Bus Bus  Voltage Angle --Load-- --Generator-- Injected
%        No  code Mag.   Degree MW   Mvar MW   Mvar Qmin Qmax Mvar

busdata=[1   1    1.06   0.0     0     0   0    0    10   50    0
         2   2    1.045  0.0    20    10  40   30    10   50    0
         3   2    1.03   0.0    20    15  30   10    10   40    0
         4   0    1.00   0.0    50    30   0    0     0    0    0
         5   0    1.00   0.0    60    40   0    0     0    0   0];
```

```
%         Bus bus    R       X       1/2 B  1 for lines code or
%         nl  nr     pu      pu       pu    tap setting value

linedata=[1   2     0.02    0.06    0.030      1
          1   3     0.08    0.24    0.025      1
          2   3     0.06    0.18    0.020      1
          2   4     0.06    0.18    0.020      1
          2   5     0.04    0.12    0.015      1
          3   4     0.01    0.03    0.010      1
          4   5     0.08    0.24    0.025      1];

     cost = [200  7.0    0.008
             180  6.3    0.009
             140  6.8    0.007];

     mwlimits =[10  85
                10  80
                10  70];

     lfybus                     % forms the bus admittance matrix
     lfnewton    % Power flow solution by Newton-Raphson method
     busout     % Prints the power flow solution on the screen
     bloss                % Obtains the loss formula coefficients
     gencost             % Computes the total generation cost $/h
     dispatch            % Obtains optimum dispatch of generation
          % dpslack is the difference (absolute value) between
          % the scheduled slack generation determined from the
             % coordination equation, and the slack generation
                     % obtained from the power flow solution.

     while dpslack > 0.001                  % Test for convergence
     lfnewton                              % New power flow solution
     bloss                             % Loss coefficients are updated
     dispatch %Optimum dispatch of gen.with new B-coefficients
     end
     busout                  % Prints the final power flow solution
     gencost % Generation cost with optimum scheduling of gen.
```

The result is

```
          Power Flow Solution by Newton-Raphson Method
             Maximum Power mismatch = 1.43025e-05
                    No. of iterations = 3
```

```
Bus Voltage  Angle    -----Load-----    --Generation-- Injected
No. Mag.     Degree    MW       Mvar      MW      Mvar      Mvar

1   1.060    0.000    0.000     0.00    83.051   7.271     0.00
2   1.045   -1.782   20.000    10.00    40.000  41.811     0.00
3   1.030   -2.664   20.000    15.00    30.000  24.148     0.00
4   1.019   -3.243   50.000    30.00     0.000   0.000     0.00
5   0.990   -4.405   60.000    40.00     0.000   0.000     0.00
Total                150.000   95.000  153.051  73.230     0.00

    B =
         0.0218    0.0093    0.0028
         0.0093    0.0228    0.0017
         0.0028    0.0017    0.0179

    B0 =
         0.0003    0.0031    0.0015
    B00 =
         3.0523e-04
    Total system loss = 3.05248 MW

    Total generation cost =    1633.24 $/h
    Incremental cost of delivered power (system lambda) =
    7.767608 $/MWh
    Optimal Dispatch of Generation:

       33.4558
       64.1101
       55.1005

    Absolute value of the slack bus real power mismatch,
    dpslack = 0.4960 pu
```

In this example the final optimal dispatch of generation was obtained in six iterations. The results for final loss coefficients and final optimal dispatch of generation is presented below

```
    B =
         0.0472    0.0130    0.0036
         0.0130    0.0130    0.0010
         0.0036    0.0010    0.0115

    B0 =
         0.0047    0.0012    0.0004
```

```
B00 =
     3.0516e-04

Total system loss = 2.15691 MW
Incremental cost of delivered power (system lambda) =
7.759051 $/MWh
Optimal Dispatch of Generation:

   23.5581
   69.5593
   59.0368

Absolute value of the slack bus real power mismatch,
dpslack = 0.0009 pu

     Power Flow Solution by Newton-Raphson Method
        Maximum Power mismatch = 1.90285e-08
               No. of iterations = 4
```

Bus No.	Voltage Mag.	Angle Degree	-----Load----- MW	Mvar	--Generation-- MW	Mvar	Injected Mvar
1	1.060	0.000	0.000	0.000	23.649	25.727	0.00
2	1.045	-0.282	20.000	10.000	69.518	30.767	0.00
3	1.030	-0.495	20.000	15.000	58.990	14.052	0.00
4	1.019	-1.208	50.000	30.000	0.000	0.000	0.00
5	0.990	-2.729	60.000	40.000	0.000	0.000	0.00
Total			150.000	95.000	152.154	70.545	0.00

```
Total generation cost =     1596.96 $/h
```

The total generation cost for the initial operating condition is 1, 633.24 \$/h and the total generation cost with optimal dispatch of generation is 1, 596.96 \$/h. This results in a savings of 36.27 \$/h.

Example 7.11

Figure 7.8 is the 26-bus power system network of Problem 6.14. Bus 1 is taken as the slack bus with its voltage adjusted to $1.025\angle 0^\circ$ pu. The data for the voltage-controlled buses is

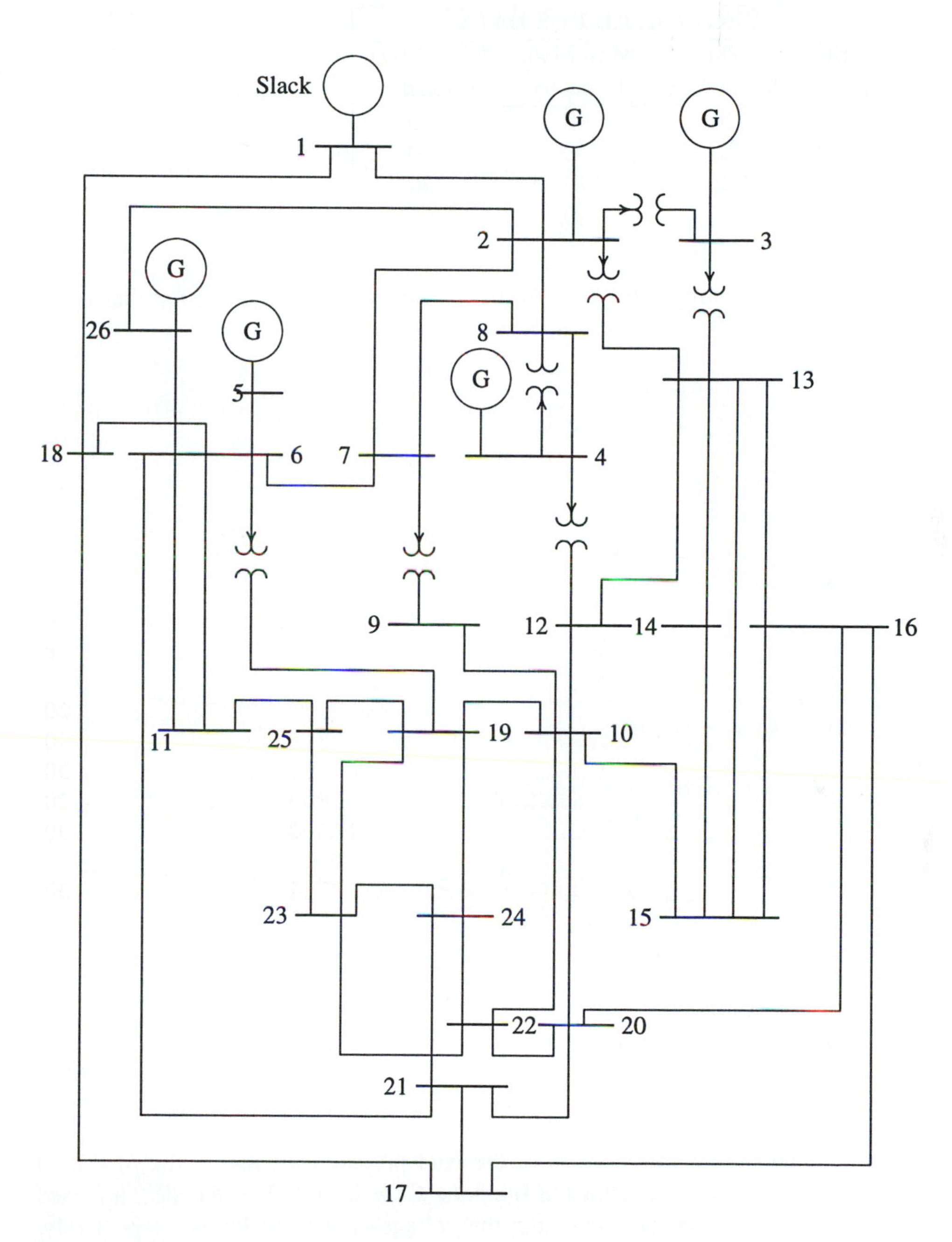

FIGURE 7.8
One-line diagram of Example 7.11.

REGULATED BUS DATA			
Bus No.	Voltage Magnitude	Min. Mvar Capacity	Max. Mvar Capacity
2	1.020	40	250
3	1.025	40	150
4	1.050	40	80
5	1.045	40	160
26	1.015	15	50

Transformer tap settings are given in the table below. The left bus number is assumed to be the tap side of the transformer.

TRANSFORMER DATA	
Transformer Designation	Tap Setting Per Unit
2 – 3	0.960
2 – 13	0.960
3 – 13	1.017
4 – 8	1.050
4 – 12	1.050
6 – 19	0.950
7 – 9	0.950

The shunt capacitive data is

SHUNT CAPACITOR DATA	
Bus No.	Mvar
1	4.0
4	2.0
5	5.0
6	2.0
9	3.0
11	1.5
12	2.0
15	0.5
19	5.0

Generation and loads are as given in the data prepared for use in the *MATLAB* environment in the matrix defined as **busdata**. Code 0, code 1, and code 2 are used for the load buses, the slack bus, and the voltage-controlled buses, respectively. Values for **basemva, accuracy**, and **maxiter** must be specified. Line data are as given in the matrix called **linedata**. The last column of this data must contain 1 for lines, or the tap setting values for transformers with off-nominal turn ratio. The generator's operating costs in \$/h, with P_i in MW are as follow:

$$
\begin{aligned}
C_1 &= 240 + 7.0P_1 + 0.0070P_1^2 \\
C_2 &= 200 + 10.0P_2 + 0.0095P_2^2 \\
C_3 &= 220 + 8.5P_3 + 0.0090P_3^2 \\
C_4 &= 200 + 11.0P_4 + 0.0090P_4^2 \\
C_5 &= 220 + 10.5P_5 + 0.0080P_5^2 \\
C_{26} &= 190 + 12.0P_{26} + 0.0075P_{26}^2
\end{aligned}
$$

The generator's real power limits are

GENERATOR REAL POWER LIMITS		
Gen.	Min. MW	Max. MW
1	100	500
2	50	200
3	80	300
4	50	150
5	50	200
5	50	120

Write the necessary commands to obtain the optimal dispatch of generation using **dispatch**. Continue the optimization process until the difference (absolute value) between the scheduled slack generation, determined from the coordination equation, and the slack generation, obtained from the power flow solution, is within 0.001 MW.

We use the following commands:

```
clear
basemva = 100;  accuracy = 0.0001;  maxiter = 10;

%       Bus Bus  Voltage Angle --Load-- --Generator---Injected
%       No  code Mag.   Degree MW Mvar  MW  Mvar Qmin Qmax Mvar
busdata=[1   1   1.025   0.0   51  41    0   0    0    0    4
         2   2   1.020   0.0   22  15   79   0   40  250    0
         3   2   1.025   0.0   64  50   20   0   40  150    0
         4   2   1.050   0.0   25  10  100   0   25   80    2
         5   2   1.045   0.0   50  30  300   0   40  160    5
         6   0   1.00    0.0   76  29    0   0    0    0    2
         7   0   1.00    0.0    0   0    0   0    0    0    0
         8   0   1.00    0.0    0   0    0   0    0    0    0
         9   0   1.00    0.0   89  50    0   0    0    0    3
        10   0   1.00    0.0    0   0    0   0    0    0    0
        11   0   1.00    0.0   25  15    0   0    0    0  1.5
```

```
        12   0   1.00    0.0   89  48    0   0     0   0   2
        13   0   1.00    0.0   31  15    0   0     0   0   0
        14   0   1.00    0.0   24  12    0   0     0   0   0
        15   0   1.00    0.0   70  31    0   0     0   0 0.5
        16   0   1.00    0.0   55  27    0   0     0   0   0
        17   0   1.00    0.0   78  38    0   0     0   0   0
        18   0   1.00    0.0  153  67    0   0     0   0   0
        19   0   1.00    0.0   75  15    0   0     0   0   5
        20   0   1.00    0.0   48  27    0   0     0   0   0
        21   0   1.00    0.0   46  23    0   0     0   0   0
        22   0   1.00    0.0   45  22    0   0     0   0   0
        23   0   1.00    0.0   25  12    0   0     0   0   0
        24   0   1.00    0.0   54  27    0   0     0   0   0
        25   0   1.00    0.0   28  13    0   0     0   0   0
        26   2   1.015   0.0   40  20   60   0    15  50  0];

%         Bus bus    R       X        1/2 B  1 for lines code or
%         nl  nr    pu      pu        pu    tap setting value
linedata=[1   2   0.00055  0.00480  0.03000   1
          1  18   0.00130  0.01150  0.06000   1
          2   3   0.00146  0.05130  0.05000   0.96
          2   7   0.01030  0.05860  0.01800   1
          2   8   0.00740  0.03210  0.03900   1
          2  13   0.00357  0.09670  0.02500   0.96
          2  26   0.03230  0.19670  0.00000   1
          3  13   0.00070  0.00548  0.00050   1.017
          4   8   0.00080  0.02400  0.00010   1.050
          4  12   0.00160  0.02070  0.01500   1.050
          5   6   0.00690  0.03000  0.09900   1
          6   7   0.00535  0.03060  0.00105   1
          6  11   0.00970  0.05700  0.00010   1
          6  18   0.00374  0.02220  0.00120   1
          6  19   0.00350  0.06600  0.04500   0.95
          6  21   0.00500  0.09000  0.02260   1
          7   8   0.00120  0.00693  0.00010   1
          7   9   0.00095  0.04290  0.02500   0.95
          8  12   0.00200  0.01800  0.02000   1
          9  10   0.00104  0.04930  0.00100   1
         10  12   0.00247  0.01320  0.01000   1
         10  19   0.05470  0.23600  0.00000   1
         10  20   0.00660  0.01600  0.00100   1
         10  22   0.00690  0.02980  0.00500   1
         11  25   0.09600  0.27000  0.01000   1
         11  26   0.01650  0.09700  0.00400   1
         12  14   0.03270  0.08020  0.00000   1
```

```
        12  15   0.01800  0.05980  0.00000   1
        13  14   0.00460  0.02710  0.00100   1
        13  15   0.01160  0.06100  0.00000   1
        13  16   0.01793  0.08880  0.00100   1
        14  15   0.00690  0.03820  0.00000   1
        15  16   0.02090  0.05120  0.00000   1
        16  17   0.09900  0.06000  0.00000   1
        16  20   0.02390  0.05850  0.00000   1
        17  18   0.00320  0.06000  0.03800   1
        17  21   0.22900  0.44500  0.00000   1
        19  23   0.03000  0.13100  0.00000   1
        19  24   0.03000  0.12500  0.00200   1
        19  25   0.11900  0.22490  0.00400   1
        20  21   0.06570  0.15700  0.00000   1
        20  22   0.01500  0.03660  0.00000   1
        21  24   0.04760  0.15100  0.00000   1
        22  23   0.02900  0.09900  0.00000   1
        22  24   0.03100  0.08800  0.00000   1
        23  25   0.09870  0.11680  0.00000   1];

cost = [240   7.0   0.0070
        200  10.0   0.0095
        220   8.5   0.0090
        200  11.0   0.0090
        220  10.5   0.0080
        190  12.0   0.0075];

mwlimits =[100  500
            50   200
            80   300
            50   150
            50   200
            50   120];

lfybus                      % Forms the bus admittance matrix
lfnewton    % Power flow solution by Newton-Raphson method
busout      % Prints the power flow solution on the screen
bloss             % Obtains the loss formula coefficients
gencost          % Computes the total generation cost $/h
dispatch         % Obtains optimum dispatch of generation
    % dpslack is the difference (absolute value) between
    % the scheduled slack generation determined from the
      % coordination equation, and the slack generation
              % obtained from the power flow solution.
```

```
while dpslack>.001%Repeat till dpslack is within tolerance
lfnewton                          % New power flow solution
bloss                       % Loss coefficients are updated
dispatch %Optimum dispatch of gen. with new B-coefficients
end
busout                 % Prints the final power flow solution
gencost  % Generation cost with optimum scheduling of gen.
```

The result is

```
              Power Flow Solution by Newton-Raphson Method
                 Maximum Power mismatch = 3.18289e-10
                        No. of iterations = 6
```

Bus No.	Voltage Mag.	Angle Degree	-----Load----- MW	Mvar	--Generation-- MW	Mvar	Injected Mvar
1	1.025	0.000	51.000	41.000	719.534	224.011	4.00
2	1.020	-0.931	22.000	15.000	79.000	125.354	0.00
3	1.035	-4.213	64.000	50.000	20.000	63.030	0.00
4	1.050	-3.582	25.000	10.000	100.000	49.223	2.00
5	1.045	1.129	50.000	30.000	300.000	124.466	5.00
6	0.999	-2.573	76.000	29.000	0.000	0.000	2.00
7	0.994	-3.204	0.000	0.000	0.000	0.000	0.00
8	0.997	-3.299	0.000	0.000	0.000	0.000	0.00
9	1.009	-5.393	89.000	50.000	0.000	0.000	3.00
10	0.989	-5.561	0.000	0.000	0.000	0.000	0.00
11	0.997	-3.218	25.000	15.000	0.000	0.000	1.50
12	0.993	-4.692	89.000	48.000	0.000	0.000	2.00
13	1.014	-4.430	31.000	15.000	0.000	0.000	0.00
14	1.000	-5.040	24.000	12.000	0.000	0.000	0.00
15	0.991	-5.538	70.000	31.000	0.000	0.000	0.50
16	0.983	-5.882	55.000	27.000	0.000	0.000	0.00
17	0.987	-4.985	78.000	38.000	0.000	0.000	0.00
18	1.007	-1.866	153.000	67.000	0.000	0.000	0.00
19	1.004	-6.397	75.000	15.000	0.000	0.000	5.00
20	0.980	-6.025	48.000	27.000	0.000	0.000	0.00
21	0.977	-5.778	46.000	23.000	0.000	0.000	0.00
22	0.978	-6.437	45.000	22.000	0.000	0.000	0.00
23	0.976	-7.087	25.000	12.000	0.000	0.000	0.00
24	0.968	-7.347	54.000	27.000	0.000	0.000	0.00
25	0.974	-6.775	28.000	13.000	0.000	0.000	0.00
26	1.015	-1.803	40.000	20.000	60.000	32.706	0.00
Total			1263.000	637.000	1278.534	618.791	25.00

```
 B =
      0.0014    0.0015    0.0009   -0.0001   -0.0004   -0.0002
      0.0015    0.0043    0.0050    0.0001   -0.0008   -0.0003
      0.0009    0.0050    0.0315   -0.0000   -0.0020   -0.0016
     -0.0001    0.0001   -0.0000    0.0029   -0.0006   -0.0009
     -0.0004   -0.0008   -0.0020   -0.0006    0.0085   -0.0001
     -0.0002   -0.0003   -0.0016   -0.0009   -0.0001    0.0176

 B0 =
     -0.0002   -0.0008    0.0067    0.0001    0.0000   -0.0012

B00 =
      0.0056

     Total system loss = 15.53 MW

     Total generation cost =    16760.73 $/h
     Incremental cost of delivered power (system lambda) =
     13.911780 $/MWh
     Optimal Dispatch of Generation:

       474.1196
       173.7886
       190.9515
       150.0000
       196.7196
       103.5772

     Absolute value of the slack bus real power mismatch,
     dpslack = 2.4541 pu
```

In this example the final optimal dispatch of generation was obtained in three iterations. The results for final loss coefficients and final optimal dispatch of generation is presented below

```
 B =
      0.0017    0.0012    0.0007   -0.0001   -0.0005   -0.0002
      0.0012    0.0014    0.0009    0.0001   -0.0006   -0.0001
      0.0007    0.0009    0.0031    0.0000   -0.0010   -0.0006
     -0.0001    0.0001    0.0000    0.0024   -0.0006   -0.0008
     -0.0005   -0.0006   -0.0010   -0.0006    0.0129   -0.0002
     -0.0002   -0.0001   -0.0006   -0.0008   -0.0002    0.0150
```

```
 B0 =
      1.0e-03 *
     -0.3908   -0.1297    0.7047    0.0591    0.2161   -0.6635

B00 =
      0.0056

     Total system loss = 12.807 MW
     Incremental cost of delivered power (system lambda) =
     13.538113 $/MWh
     Optimal Dispatch of Generation:
       447.6919
       173.1938
       263.4859
       138.8142
       165.5884
        87.0260

     Absolute value of the slack bus real power mismatch,
     dpslack =  0.0008 pu

          Power Flow Solution by Newton-Raphson Method
             Maximum Power mismatch = 2.33783e-05
                    No. of iterations = 3
```

Bus No.	Voltage Mag.	Angle Degree	-----Load----- MW	Mvar	--Generation-- MW	Mvar	Injected Mvar
1	1.025	0.000	51.000	41.000	447.611	250.582	4.00
2	1.020	-0.200	22.000	15.000	173.087	57.303	0.00
3	1.045	-0.639	64.000	50.000	263.363	78.280	0.00
4	1.050	-2.101	25.000	10.000	138.716	33.449	2.00
5	1.045	-1.453	50.000	30.000	166.099	142.890	5.00
6	1.001	-2.874	76.000	29.000	0.000	0.000	2.00
7	0.995	-2.406	0.000	0.000	0.000	0.000	0.00
8	0.998	-2.278	0.000	0.000	0.000	0.000	0.00
9	1.010	-4.387	89.000	50.000	0.000	0.000	3.00
10	0.991	-4.311	0.000	0.000	0.000	0.000	0.00
11	0.998	-2.824	25.000	15.000	0.000	0.000	1.50
12	0.994	-3.282	89.000	48.000	0.000	0.000	2.00
13	1.022	-1.261	31.000	15.000	0.000	0.000	0.00
14	1.008	-2.445	24.000	12.000	0.000	0.000	0.00
15	0.999	-3.229	70.000	31.000	0.000	0.000	0.50
16	0.990	-3.990	55.000	27.000	0.000	0.000	0.00
17	0.983	-4.366	78.000	38.000	0.000	0.000	0.00

```
18  1.007   -1.884  153.000   67.000     0.000     0.000     0.00
19  1.005   -6.074   75.000   15.000     0.000     0.000     5.00
20  0.983   -4.759   48.000   27.000     0.000     0.000     0.00
21  0.977   -5.411   46.000   23.000     0.000     0.000     0.00
22  0.980   -5.325   45.000   22.000     0.000     0.000     0.00
23  0.978   -6.388   25.000   12.000     0.000     0.000     0.00
24  0.969   -6.672   54.000   27.000     0.000     0.000     0.00
25  0.975   -6.256   28.000   13.000     0.000     0.000     0.00
26  1.015   -0.284   40.000   20.000    86.939    27.892     0.00

Total                1263.000  637.000 1275.800   590.396    25.00

Total generation cost =    15447.72 $/h
```

The total generation cost for the initial operating condition is $16,760.73$ \$/h and the total generation cost with optimal dispatch of generation is $15,447.72$ \$/h. This results in a savings of $1,313.01$ \$/h. That is, with this loading, the total annual savings is over \$11 million.

PROBLEMS

7.1. Find a rectangle of maximum perimeter that can be inscribed in a circle of unit radius given by

$$g(x,y) = x^2 + y^2 - 1 = 0$$

Check the eigenvalues for sufficient conditions.

7.2. Find the minimum of the function

$$f(x,y) = x^2 + 2y^2$$

subject to the equality constraint

$$g(x,y) = x + 2y + 4 = 0$$

Check for the sufficient conditions.

7.3. Use the Lagrangian multiplier method for solving constrained parameter optimization problems to determine an isosceles triangle of maximum area that may be inscribed in a circle of radius 1.

7.4. For a second-order bandpass filter with transfer function

$$H(s) = \frac{\omega_n^2}{s^2 + 2\zeta\omega_n s + \omega_n^2}$$

determine the values of the damping ratio and natural frequency, ζ and ω_n, corresponding to a Bode plot whose peak occurs at 7071.07 radians/sec and whose half-power bandwidth is 12,720.2 radians/sec.

7.5. Find the minimum value of the function

$$f(x, y) = x^2 + y^2$$

subject to the equality constraint

$$g(x, y) = x^2 - 6x - y^2 + 17 = 0$$

7.6. Find the minimum value of the function

$$f(x, y) = x^2 + y^2$$

subject to one equality constraint

$$g(x, y) = x^2 - 5x - y^2 + 20 = 0$$

and one inequality constraint

$$u(x, y) = 2x + y \geq 6$$

7.7. The fuel-cost functions in \$/h for two 800 MW thermal plants are given by

$$\begin{aligned} C_1 &= 400 + 6.0P_1 + 0.004P_1^2 \\ C_2 &= 500 + \ \beta P_2 + \quad \gamma P_2^2 \end{aligned}$$

where P_1 and P_2 are in MW.

(a) The incremental cost of power λ is \$8/MWh when the total power demand is 550 MW. Neglecting losses, determine the optimal generation of each plant.
(b) The incremental cost of power λ is \$10/MWh when the total power demand is 1300 MW. Neglecting losses, determine the optimal generation of each plant.
(c) From the results of (a) and (b) find the fuel-cost coefficients β and γ of the second plant.

7.8. The fuel-cost functions in \$/h for three thermal plants are given by

$$\begin{aligned} C_1 &= 350 + 7.20P_1 + 0.0040P_1^2 \\ C_2 &= 500 + 7.30P_2 + 0.0025P_2^2 \\ C_3 &= 600 + 6.74P_3 + 0.0030P_3^2 \end{aligned}$$

where P_1, P_2, and P_3 are in MW. The governors are set such that generators share the load equally. Neglecting line losses and generator limits, find the total cost in $/h when the total load is

(i) $P_D = 450$ MW
(ii) $P_D = 745$ MW
(iii) $P_D = 1335$ MW

7.9. Neglecting line losses and generator limits, determine the optimal scheduling of generation for each loading condition in Problem 7.8
(a) by analytical technique, using (7.33) and (7.31).
(b) using Iterative method. Start with an initial estimate of $\lambda = 7.5$ $/MWh.
(c) find the savings in $/h for each case compared to the costs in Problem 7.8 when the generators shared load equally.
Use the **dispatch** program to check your results.

7.10. Repeat Problem 7.9 (a) and (b), but this time consider the following generator limits (in MW)

$$122 \le P_1 \le 400$$
$$260 \le P_2 \le 600$$
$$50 \le P_3 \le 445$$

Use the **dispatch** program to check your results.

7.11. The fuel-cost function in $/h of two thermal plants are

$$C_1 = 320 + 6.2P_1 + 0.004P_1^2$$
$$C_2 = 200 + 6.0P_2 + 0.003P_2^2$$

where P_1 and P_2 are in MW. Plant outputs are subject to the following limits (in MW)

$$50 \le P_1 \le 250$$
$$50 \le P_2 \le 350$$

The per-unit system real power loss with generation expressed in per unit on a 100-MVA base is given by

$$P_{L(pu)} = 0.0125P_{1(pu)}^2 + 0.00625P_{2(pu)}^2$$

The total load is 412.35 MW. Determine the optimal dispatch of generation. Start with an initial estimate of $\lambda = 7$ $/MWh. Use the **dispatch** program to check your results.

7.12. The 9-bus power system network of an Electric Utility Company is shown in Figure 7.9. The load data is tabulated below. Voltage magnitude, generation schedule and the reactive power limits for the regulated buses are also tabulated below. Bus 1, whose voltage is specified as $V_1 = 1.03\angle 0°$, is taken as the slack bus.

LOAD DATA		
Bus	Load	
No.	MW	Mvar
1	0	0
2	20	10
3	25	15
4	10	5
5	40	20
6	60	40
7	10	5
8	80	60
9	100	80

GENERATION DATA				
Bus	Voltage	Generation	Mvar Limits	
No.	Mag.	MW	Min.	Max.
1	1.03			
2	1.04	80	0	250
7	1.01	120	0	100

The Mvar of the shunt capacitors installed at substations are given below

SHUNT CAPACITORS	
Bus No.	Mvar
3	1.0
4	3.0

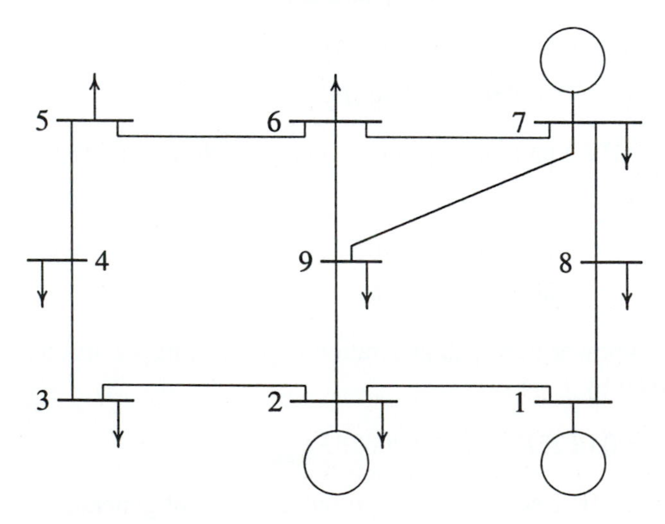

FIGURE 7.9
One-line diagram for Problem 7.12.

The line data containing the series resistance and reactance in per unit, and one-half of the total capacitance in per unit susceptance on a 100 MVA base is tabulated below.

LINE DATA				
Bus No.	Bus No.	R, PU	X, PU	$\frac{1}{2}$ B, PU
1	2	0.018	0.054	0.0045
1	8	0.014	0.036	0.0030
2	9	0.006	0.030	0.0028
2	3	0.013	0.036	0.0030
3	4	0.010	0.050	0.0000
4	5	0.018	0.056	0.0000
5	6	0.020	0.060	0.0000
6	7	0.015	0.045	0.0038
6	9	0.002	0.066	0.0000
7	8	0.032	0.076	0.0000
7	9	0.022	0.065	0.0000

The generator's operating costs in \$/h are as follows:

$$C_1 = 240 + 6.7P_1 + 0.009P_1^2$$
$$C_2 = 220 + 6.1P_2 + 0.005P_2^2$$
$$C_7 = 240 + 6.5P_7 + 0.008P_7^2$$

The generator's real power limits are

GENERATOR REAL POWER LIMITS		
Gen.	Min. MW	Max. MW
1	50	200
2	50	200
7	50	100

Write the necessary commands to obtain the optimal dispatch of generation using **dispatch**. Continue the optimization process until the difference (absolute value) between the scheduled slack generation, determined from the coordination equation, and the slack generation, obtained from the power flow solution, is within 0.001 MW.

CHAPTER 8

SYNCHRONOUS MACHINE TRANSIENT ANALYSIS

8.1 INTRODUCTION

The steady state performance of the synchronous machine was described in Chapter 3. Under balanced steady state operations, the rotor mmf and the resultant stator mmf are stationary with respect to each other. As a result, the flux linkages with the rotor circuit do not change with time, and no voltages are induced in the rotor circuits. The per phase equivalent circuit then becomes a constant generated emf in series with a simple impedance. In Chapter 3, for steady state operation the generator was represented with a constant emf behind the synchronous reactance X_s. For salient-pole rotor, because of the nonuniformity of the air gap, the generator was modeled with direct axis reactance X_d and the quadrature axis reactance X_q.

Under transient conditions, such as short circuits at the generator terminals, the flux linkages with the rotor circuits change with time. This result in transient currents in all the rotor circuits, which in turn reacts on the armature. For the transient analysis, the idealized synchronous machine is represented as a group of magnetically coupled circuits with inductances which depend on the angular position of the rotor. The resulting differential equations describing the machine have time-varying coefficients, and a closed form of solution in most cases is not feasible. A great simplification can be made by transformation of stator variables from phases a, b, and c into new variables the frame of reference of which moves with the rotor. The transformation is based on the so-called *two-axis theory*, which was pioneered by Blondel, Doherty, Nickle, and Park [20, 61]. The transformed equations are linear provided that the speed is assumed to be constant.

In this chapter, the voltage equation of a synchronous machine is first established. Reference frame theory is then used to establish the machine equations with the stator variables transformed to a reference frame fixed in the rotor (Park's equations). The Park's equations are solved numerically during balanced three-phase short circuit. If the speed deviation is taken into account, transformed equations become nonlinear and must be solved by numerical integration. In *MATLAB*, the nonlinear differential equations of the synchronous machine in matrix form can be simulated with ease. Also, there is the additional advantage that the original voltage equations can be used without the need for any transformations. In particular, the numerical solution is obtained for the line-to-line and the line-to-ground short circuits using direct-phase quantities.

Another objective of this chapter is to develop simple network models of the synchronous generator for the power system fault analysis and transient stability studies. For this purpose, the generator behavior is divided into three periods: the *subtransient period*, lasting only for the first few cycles; the *transient period* covering a relatively longer time; and, finally, the *steady state period*. Thus, the generator equivalent circuits during transient state are obtained.

8.2 TRANSIENT PHENOMENA

To better understand the synchronous machine transient phenomena, we first study the transient behavior of a simple RL circuit. Consider a sinusoidal voltage $v(t) = V_m \sin(\omega t + \alpha)$ applied to a simple RL circuit at time $t = 0$, as shown in Figure 8.1.

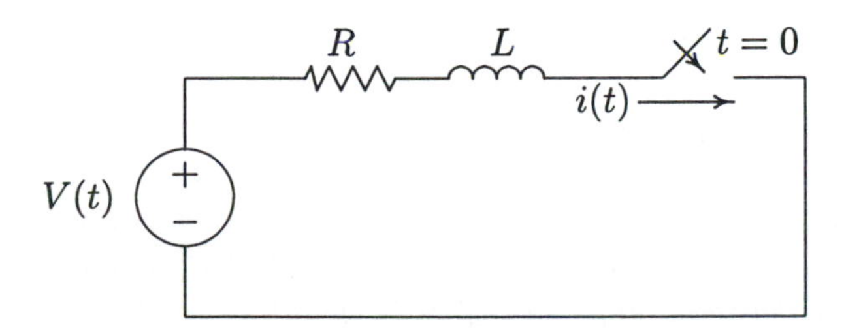

FIGURE 8.1
A simple series circuit with constant R and L.

The circuit consists of R in series with a constant L. The instantaneous voltage equation for the circuit is

$$Ri(t) + L\frac{di(t)}{dt} = V_m \sin(\omega t + \alpha) \tag{8.1}$$

The solution for the current may be shown to be

$$i(t) = I_m \sin(\omega t + \alpha - \gamma) - I_m e^{-t/\tau} \sin(\alpha - \gamma) \tag{8.2}$$

where $I_m = V_m/Z$, $\tau = L/R$, $\gamma = \tan^{-1} \omega L/R$, and $Z = \sqrt{R^2 + X^2}$. The first term is the steady state sinusoidal component. The second term is a dc transient component known as *dc offset* which decays exponentially. The dc and sinusoidal components are equal and opposite when $t = 0$, so that the condition for zero initial current is satisfied. The magnitude of the dc component depends on the instant of application of the voltage to the circuit, as defined by the angle α. The dc component is zero when $(\alpha = \gamma)$. This current waveform is shown in Figure 8.2(a). Similarly, the dc component will have a maximum initial value of V_m/Z which is the peak value of the alternating component, if the circuit is closed when $\alpha = \gamma - \pi/2$ radians. The current waveform with maximum dc offset is shown in Figure 8.2(b). If $\omega L \gg R$, then $\gamma \simeq \pi/2$, so that circuit closure at voltage maximum would give no dc component, and closure at voltage zero would cause the maximum dc transient current to flow.

Example 8.1

In the circuit of Figure 8.1, let $R = 0.125\ \Omega$, $L = 10$ mH, and the source voltage be given by $v(t) = 151 \sin(377t + \alpha)$. Determine the current response after closing the switch for the following cases.
(a) No dc offset.
(b) For maximum dc offset.

$$Z = 0.125 + j(377)(0.01) = 0.125 + j3.77 = 3.772\angle 88.1^\circ$$

$$I_m = \frac{151}{3.772} = 40 \text{ A}$$

and

$$\tau = \frac{L}{R} = 0.08 \text{ sec}$$

From (8.2) the response is

$$i(t) = 40 \sin(\omega t + \alpha - 88.1^\circ) - 40e^{-t/0.08} \sin(\alpha - 88.1^\circ)$$

The response has no dc offset if switch is closed when $\alpha = 88.1^\circ$, and it has the maximum dc offset when $\alpha = 88.1^\circ - 90^\circ = -1.9^\circ$. The following commands produce the responses shown in Figures 8.2(a) and 8.2(b).

```
alf1 = 88.1*pi/180;
alf2 = -1.9*pi/180;
gamma = 88.1*pi/180;
t = 0:.001:.3;
i1 = 40*sin(377*t+alf1-gamma)-40*exp(-t/.08).*sin(alf1-gamma);
i2 = 40*sin(377*t+alf2-gamma)-40*exp(-t/.08).*sin(alf2-gamma);
subplot(2,1,1), plot(t, i1)
xlabel('t, sec'), ylabel('i(t)')
subplot(2,1,2), plot(t, i2)
xlabel('t, sec'), ylabel('i(t)')
subplot(111)
```

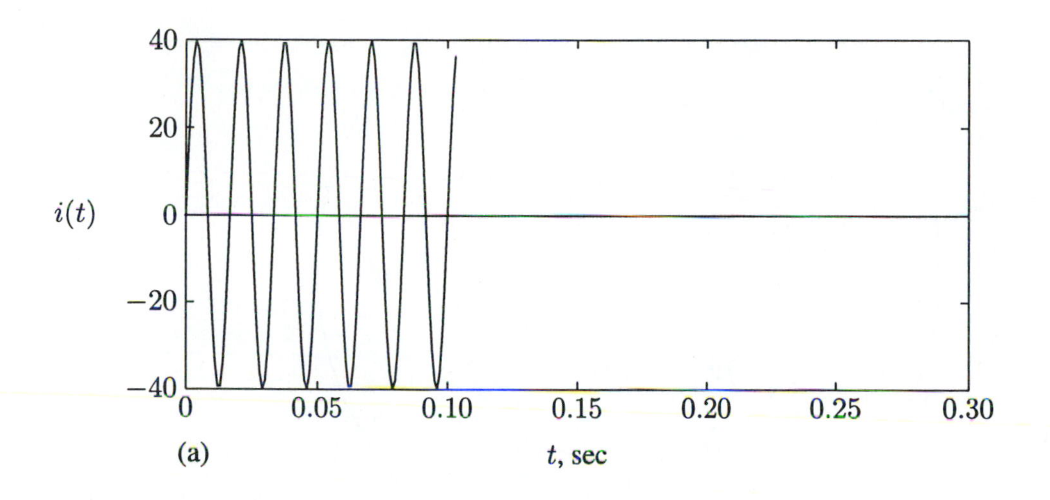

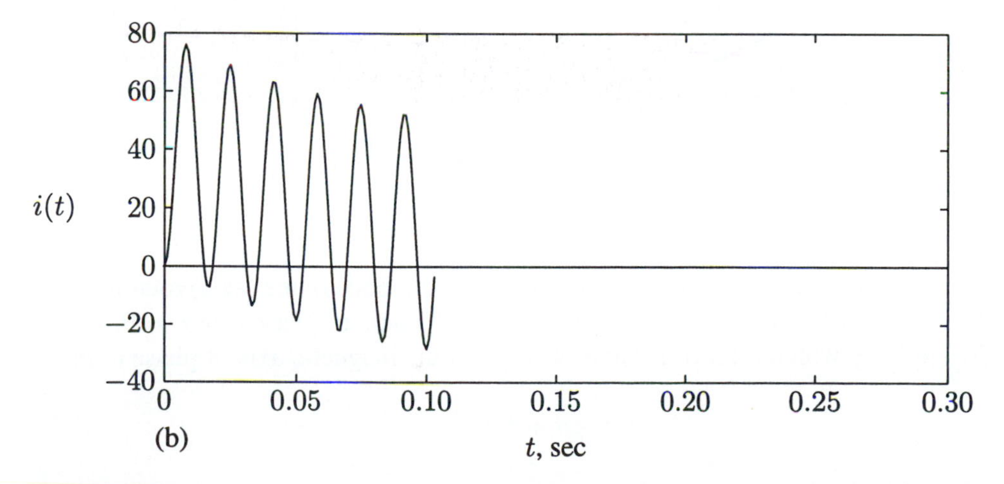

FIGURE 8.2
Current waveform, (a) with no dc offset, (b) with maximum dc offset.

8.3 SYNCHRONOUS MACHINE TRANSIENTS

The synchronous machine consists of three stator windings mounted on the stator, and one field winding mounted on the rotor. Two additional fictitious windings could be added to the rotor, one along the direct axis and one along the quadrature axis, which model the short-circuited paths of the damper windings. These windings are shown schematically in Figure 8.3.

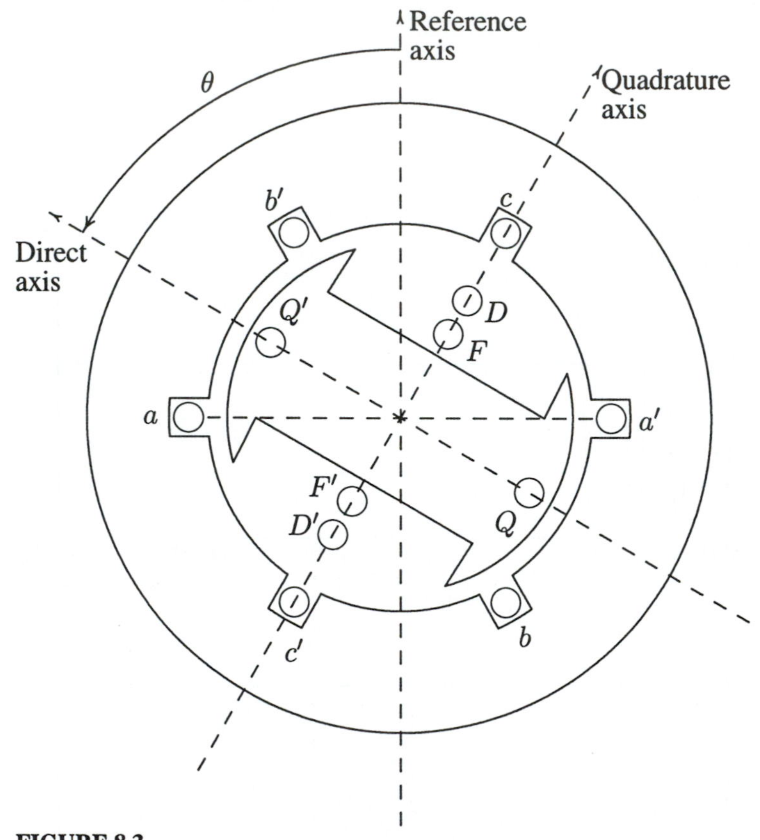

FIGURE 8.3
Schematic representation of a synchronous machine.

We shall assume a synchronously rotating reference frame (axis) rotating with the synchronous speed ω which will be along the axis of phase a at $t = 0$. If θ is the angle by which rotor direct axis is ahead of the magnetic axis of phase a, then

$$\theta = \omega t + \delta + \frac{\pi}{2} \tag{8.3}$$

where δ is the displacement of the quadrature axis from the synchronously rotating reference axis and $(\delta + \frac{\pi}{2})$ is the displacement of the direct axis.

In the classical method, the idealized synchronous machine is represented as a group of magnetically coupled circuits with inductances which depend on the angular position of the rotor. In addition, saturation is neglected and spatial distribution of armature mmf is assumed sinusoidal. The circuits are shown schematically in Figure 8.4.

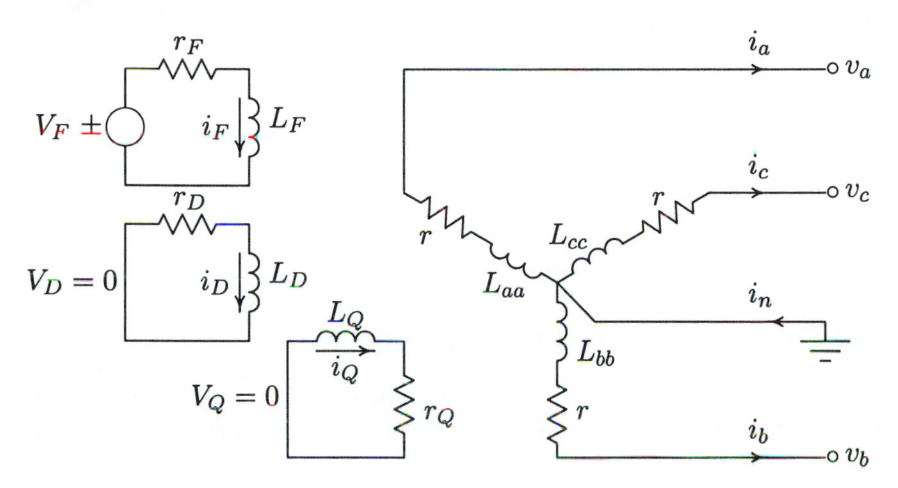

FIGURE 8.4
Schematic representation of mutually coupled circuits.

The stator currents are assumed to have a positive direction flowing out of the machine terminals. Since the machine is a generator, following the circuit passive sign convention, the voltage equation becomes

$$\begin{bmatrix} v_a \\ v_b \\ v_c \\ -v_F \\ 0 \\ 0 \end{bmatrix} = - \begin{bmatrix} r & 0 & 0 & 0 & 0 & 0 \\ 0 & r & 0 & 0 & 0 & 0 \\ 0 & 0 & r & 0 & 0 & 0 \\ 0 & 0 & 0 & r_F & 0 & 0 \\ 0 & 0 & 0 & 0 & r_D & 0 \\ 0 & 0 & 0 & 0 & 0 & r_Q \end{bmatrix} \begin{bmatrix} i_a \\ i_b \\ i_c \\ i_F \\ i_D \\ i_Q \end{bmatrix} - \frac{d}{dt} \begin{bmatrix} \lambda_a \\ \lambda_b \\ \lambda_c \\ \lambda_F \\ \lambda_D \\ \lambda_Q \end{bmatrix} \tag{8.4}$$

The above equation may be written in partitioned form as

$$\begin{bmatrix} \mathbf{v}_{abc} \\ \mathbf{v}_{FDQ} \end{bmatrix} = - \begin{bmatrix} \mathbf{R}_{abc} & 0 \\ 0 & \mathbf{R}_{FDQ} \end{bmatrix} \begin{bmatrix} \mathbf{i}_{abc} \\ \mathbf{i}_{FDQ} \end{bmatrix} - \frac{d}{dt} \begin{bmatrix} \lambda_{abc} \\ \lambda_{FDQ} \end{bmatrix} \tag{8.5}$$

where

$$\mathbf{v}_{FDQ} = \begin{bmatrix} -v_F \\ 0 \\ 0 \end{bmatrix} \quad \mathbf{i}_{FDQ} = \begin{bmatrix} i_F \\ i_D \\ i_Q \end{bmatrix} \quad \lambda_{FDQ} = \begin{bmatrix} \lambda_F \\ \lambda_D \\ \lambda_Q \end{bmatrix} \quad \text{etc.} \tag{8.6}$$

The flux linkages are functions of self- and mutual inductances given by

$$\begin{bmatrix} \lambda_a \\ \lambda_b \\ \lambda_c \\ \lambda_F \\ \lambda_D \\ \lambda_Q \end{bmatrix} = \begin{bmatrix} L_{aa} & L_{ab} & L_{ac} & L_{aF} & L_{aD} & L_{aQ} \\ L_{ba} & L_{bb} & L_{bc} & L_{bF} & L_{bD} & L_{bQ} \\ L_{ca} & L_{cb} & L_{cc} & L_{cF} & L_{cD} & L_{cQ} \\ L_{Fa} & L_{Fb} & L_{Fc} & L_{FF} & L_{FD} & L_{FQ} \\ L_{Da} & L_{Db} & L_{Dc} & L_{DF} & L_{DD} & L_{DQ} \\ L_{Qa} & L_{Qb} & L_{Qc} & L_{QF} & L_{QD} & L_{QQ} \end{bmatrix} \begin{bmatrix} i_a \\ i_b \\ i_c \\ i_F \\ i_D \\ i_Q \end{bmatrix} \tag{8.7}$$

or in compact form we have

$$\begin{bmatrix} \lambda_{abc} \\ \lambda_{FDQ} \end{bmatrix} = \begin{bmatrix} \mathbf{L}_{SS} & \mathbf{L}_{SR} \\ \mathbf{L}_{RS} & \mathbf{L}_{RR} \end{bmatrix} \begin{bmatrix} \mathbf{i}_{abc} \\ \mathbf{i}_{FDQ} \end{bmatrix} \tag{8.8}$$

8.3.1 INDUCTANCES OF SALIENT-POLE MACHINES

The self-inductance of any stator coil varies periodically from a maximum (when the direct axis coincides with the coil magnetic axis) to a minimum (when the quadrature axis is in line with the coil magnetic axis). The self-inductance L_{aa}, for example, will be a maximum for $\theta = 0$, a minimum for $\theta = 90°$ and maximum, again for $\theta = 180°$, and so on. That is, L_{aa} has a period of 180° and can be represented approximately by cosines of second harmonics. Because of the rotor symmetry, the diagonal elements of the submatrix $\mathbf{L}_{SS}$ are represented as

$$\begin{aligned} L_{aa} &= L_s + L_m \cos 2\theta \\ L_{bb} &= L_s + L_m \cos 2(\theta - 2\pi/3) \\ L_{cc} &= L_s + L_m \cos 2(\theta + 2\pi/3) \end{aligned} \tag{8.9}$$

where θ is the angle between the direct axis and the magnetic axis of phase a, as shown in Figure 8.3. The mutual inductances between any two stator phases are also periodic functions of rotor angular position because of the rotor saliency. We can conclude from the symmetry considerations that the mutual inductance between phase a and b should have a negative maximum when the pole axis is lined up 30° behind phase a or 30° ahead of phase b, and a negative minimum when it is midway between the two phases. Thus, the variations of stator mutual inductances, i.e., the off-diagonal elements of the submatrix $\mathbf{L}_{SS}$ can be represented as follows.

$$\begin{aligned} L_{ab} = L_{ba} &= -M_s - L_m \cos 2(\theta + \pi/6) \\ L_{bc} = L_{cb} &= -M_s - L_m \cos 2(\theta - \pi/2) \\ L_{ca} = L_{ac} &= -M_s - L_m \cos 2(\theta + 5\pi/6) \end{aligned} \tag{8.10}$$

All the rotor self-inductances are constant since the effects of stator slots and saturation are neglected. They are represented with single subscript notation.

$$L_{FF} = L_F \quad L_{DD} = L_D \quad L_{QQ} = L_Q \tag{8.11}$$

The mutual inductance between any two circuits both in direct axis (or both in quadrature axis) is constant. The mutual inductance between any rotor direct axis circuit and quadrature axis circuit vanishes. Thus, we have

$$L_{FD} = L_{DF} = M_R \quad L_{FQ} = L_{QF} = 0 \quad L_{DQ} = L_{QD} = 0 \tag{8.12}$$

Finally, let us consider the mutual inductances between stator and rotor circuits, which are periodic functions of rotor angular position. Because only the space-fundamental component of the produced flux links the sinusoidally distributed stator, all stator-rotor mutual inductances vary sinusoidally, reaching a maximum when the two coils in question align. Thus, their variations can be written as follows.

$$\begin{aligned}
L_{aF} &= L_{Fa} = M_F \cos\theta \\
L_{bF} &= L_{Fb} = M_F \cos(\theta - 2\pi/3) \\
L_{cF} &= L_{Fc} = M_F \cos(\theta + 2\pi/3) \\
L_{aD} &= L_{Da} = M_D \cos\theta \\
L_{bD} &= L_{Db} = M_D \cos(\theta - 2\pi/3) \\
L_{cD} &= L_{Dc} = M_D \cos(\theta + 2\pi/3) \\
L_{aQ} &= L_{Qa} = M_Q \sin\theta \\
L_{bQ} &= L_{Qb} = M_Q \sin(\theta - 2\pi/3) \\
L_{cQ} &= L_{Qc} = M_Q \sin(\theta + 2\pi/3)
\end{aligned} \tag{8.13}$$

The resulting differential equations (8.4) describing the behavior of the machine have time-varying coefficients given by (8.9)–(8.13), and we are not able to use Laplace transforms directly to obtain a closed form of solution.

8.4 THE PARK TRANSFORMATION

A great simplification can be made by transformation of stator variables from phases a, b, and c into new variables the frame of reference of which moves with the rotor. The transformation is based on the so called *two-axis theory*, which was pioneered by Blondel, Doherty, Nickle, and Park [20, 61].

The transformed quantities are obtained from the projection of the actual variables on three axes; one along the direct axis of the rotor field winding, called the direct axis; a second along the neutral axis of the field winding, called the quadrature axis; and the third on a stationary axis. For example, the three armature currents i_a, i_b, and i_c are replaced by three fictitious currents with the symbols i_d, i_q, and i_0. They are found such that, in a balanced condition, when $i_a + i_b + i_c = 0$, they produce the same flux, at any instant, as the actual phase currents in the armature. The third fictitious current i_0 is needed to make the transformation possible when the sum of the three-phase current is not zero.

The Park transformation for currents is as follows

$$\begin{bmatrix} i_0 \\ i_d \\ i_q \end{bmatrix} = \sqrt{2/3} \begin{bmatrix} 1/\sqrt{2} & 1/\sqrt{2} & 1/\sqrt{2} \\ \cos\theta & \cos(\theta - 2\pi/3) & \cos(\theta + 2\pi/3) \\ \sin\theta & \sin(\theta - 2\pi/3) & \sin(\theta + 2\pi/3) \end{bmatrix} \begin{bmatrix} i_a \\ i_b \\ i_c \end{bmatrix} \tag{8.14}$$

or, in matrix notation

$$\mathbf{i}_{0dq} = \mathbf{P}\mathbf{i}_{abc} \tag{8.15}$$

Similarly for voltages and flux linkages, we have

$$\mathbf{v}_{0dq} = \mathbf{P}\mathbf{v}_{abc} \tag{8.16}$$

$$\lambda_{0dq} = \mathbf{P}\lambda_{abc} \tag{8.17}$$

The Park transformation matrix is orthogonal, i.e., $\mathbf{P}^{-1} = \mathbf{P}^T$ and thus, it is a power invariant transformation matrix. For the inverse Park transformation matrix we get

$$\mathbf{P}^{-1} = \sqrt{2/3} \begin{bmatrix} 1/\sqrt{2} & \cos\theta & \sin\theta \\ 1/\sqrt{2} & \cos(\theta - 2\pi/3) & \sin(\theta - 2\pi/3) \\ 1/\sqrt{2} & \cos(\theta + 2\pi/3) & \sin(\theta + 2\pi/3) \end{bmatrix} \tag{8.18}$$

We now wish to transform the time-varying inductances to a rotor frame of reference with the original rotor quantities unaffected. Thus, in (8.17) we augment the $\mathbf{P}$ matrix with a 3×3 identity matrix $\mathbf{U}$ to get

$$\begin{bmatrix} \lambda_{0dq} \\ \lambda_{FDQ} \end{bmatrix} = \begin{bmatrix} \mathbf{P} & \mathbf{0} \\ \mathbf{0} & \mathbf{U} \end{bmatrix} \begin{bmatrix} \lambda_{abc} \\ \lambda_{FDQ} \end{bmatrix} \tag{8.19}$$

or

$$\begin{bmatrix} \lambda_{abc} \\ \lambda_{FDQ} \end{bmatrix} = \begin{bmatrix} \mathbf{P}^{-1} & \mathbf{0} \\ \mathbf{0} & \mathbf{U} \end{bmatrix} \begin{bmatrix} \lambda_{odq} \\ \lambda_{FDQ} \end{bmatrix} \tag{8.20}$$

Substituting in (8.8), we get

$$\begin{bmatrix} \mathbf{P}^{-1} & \mathbf{0} \\ \mathbf{0} & \mathbf{U} \end{bmatrix} \begin{bmatrix} \lambda_{odq} \\ \lambda_{FDQ} \end{bmatrix} = \begin{bmatrix} \mathbf{L}_{SS} & \mathbf{L}_{SR} \\ \mathbf{L}_{RS} & \mathbf{L}_{RR} \end{bmatrix} \begin{bmatrix} \mathbf{P}^{-1} & \mathbf{0} \\ \mathbf{0} & \mathbf{U} \end{bmatrix} \begin{bmatrix} \mathbf{i}_{odq} \\ \mathbf{i}_{FDQ} \end{bmatrix} \tag{8.21}$$

or

$$\begin{bmatrix} \lambda_{odq} \\ \lambda_{FDQ} \end{bmatrix} = \begin{bmatrix} \mathbf{P} & \mathbf{0} \\ \mathbf{0} & \mathbf{U} \end{bmatrix} \begin{bmatrix} \mathbf{L}_{SS} & \mathbf{L}_{SR} \\ \mathbf{L}_{RS} & \mathbf{L}_{RR} \end{bmatrix} \begin{bmatrix} \mathbf{P}^{-1} & \mathbf{0} \\ \mathbf{0} & \mathbf{U} \end{bmatrix} \begin{bmatrix} \mathbf{i}_{odq} \\ \mathbf{i}_{FDQ} \end{bmatrix} \tag{8.22}$$

Substituting for $\mathbf{P}$, $\mathbf{P}^{-1}$ and the inductances given by (8.9)–(8.13), the above equation reduces to

$$\begin{bmatrix} \lambda_0 \\ \lambda_d \\ \lambda_q \\ \lambda_F \\ \lambda_D \\ \lambda_Q \end{bmatrix} = \begin{bmatrix} L_0 & 0 & 0 & 0 & 0 & 0 \\ 0 & L_d & 0 & kM_F & KM_D & 0 \\ 0 & 0 & L_q & 0 & 0 & kM_Q \\ 0 & kM_F & 0 & L_F & M_R & 0 \\ 0 & kM_D & 0 & M_R & L_D & 0 \\ 0 & 0 & kM_Q & 0 & 0 & L_Q \end{bmatrix} \begin{bmatrix} i_0 \\ i_d \\ i_q \\ i_F \\ i_D \\ i_Q \end{bmatrix} \tag{8.23}$$

where we have introduced the following new parameters

$$L_0 = L_s - 2M_s \tag{8.24}$$

$$L_d = L_s + M_s + \frac{3}{2}L_m \tag{8.25}$$

$$L_q = L_s + M_s - \frac{3}{2}L_m \tag{8.26}$$

and $k = \sqrt{3/2}$.

Transforming the stator-based currents ($\mathbf{i}_{abc}$) into rotor-based currents ($\mathbf{i}_{0dq}$), with rotor currents unaffected, we obtain

$$\begin{bmatrix} \mathbf{i}_{0dq} \\ \mathbf{i}_{FDQ} \end{bmatrix} = \begin{bmatrix} \mathbf{P} & \mathbf{0} \\ \mathbf{0} & \mathbf{U} \end{bmatrix} \begin{bmatrix} \mathbf{i}_{abc} \\ \mathbf{i}_{FDQ} \end{bmatrix} \tag{8.27}$$

or

$$\begin{bmatrix} \mathbf{i}_{abc} \\ \mathbf{i}_{FDQ} \end{bmatrix} = \begin{bmatrix} \mathbf{P}^{-1} & \mathbf{0} \\ \mathbf{0} & \mathbf{U} \end{bmatrix} \begin{bmatrix} \mathbf{i}_{odq} \\ \mathbf{i}_{FDQ} \end{bmatrix} \tag{8.28}$$

and similarly for voltages, we get

$$\begin{bmatrix} \mathbf{v}_{abc} \\ \mathbf{v}_{FDQ} \end{bmatrix} = \begin{bmatrix} \mathbf{P}^{-1} & \mathbf{0} \\ \mathbf{0} & \mathbf{U} \end{bmatrix} \begin{bmatrix} \mathbf{v}_{odq} \\ \mathbf{v}_{FDQ} \end{bmatrix} \tag{8.29}$$

Substituting (8.20), (8.28), and (8.29) into (8.5), we get

$$\begin{aligned} \begin{bmatrix} \mathbf{P}^{-1} & \mathbf{0} \\ \mathbf{0} & \mathbf{U} \end{bmatrix} \begin{bmatrix} \mathbf{v}_{0dq} \\ \mathbf{v}_{FDQ} \end{bmatrix} &= - \begin{bmatrix} \mathbf{R}_{abc} & 0 \\ 0 & \mathbf{R}_{FDQ} \end{bmatrix} \begin{bmatrix} \mathbf{P}^{-1} & \mathbf{0} \\ \mathbf{0} & \mathbf{U} \end{bmatrix} \begin{bmatrix} \mathbf{i}_{0dq} \\ \mathbf{i}_{FDQ} \end{bmatrix} \\ &\quad - \frac{d}{dt} \begin{bmatrix} \mathbf{P}^{-1} & \mathbf{0} \\ \mathbf{0} & \mathbf{U} \end{bmatrix} \begin{bmatrix} \lambda_{0dq} \\ \lambda_{FDQ} \end{bmatrix} \end{aligned} \tag{8.30}$$

or

$$\begin{bmatrix} \mathbf{v}_{0dq} \\ \mathbf{v}_{FDQ} \end{bmatrix} = -\begin{bmatrix} \mathbf{P} & \mathbf{0} \\ \mathbf{0} & \mathbf{U} \end{bmatrix} \begin{bmatrix} \mathbf{R}_{abc} & 0 \\ 0 & \mathbf{R}_{FDQ} \end{bmatrix} \begin{bmatrix} \mathbf{P}^{-1} & \mathbf{0} \\ \mathbf{0} & \mathbf{U} \end{bmatrix} \begin{bmatrix} \mathbf{i}_{0dq} \\ \mathbf{i}_{FDQ} \end{bmatrix} - \begin{bmatrix} \mathbf{P} & \mathbf{0} \\ \mathbf{0} & \mathbf{U} \end{bmatrix} \frac{d}{dt} \begin{bmatrix} \mathbf{P}^{-1} & \mathbf{0} \\ \mathbf{0} & \mathbf{U} \end{bmatrix} \begin{bmatrix} \lambda_{0dq} \\ \lambda_{FDQ} \end{bmatrix} \tag{8.31}$$

Evaluating the first term, and obtaining the derivative of the second term in (8.31), yields

$$\begin{bmatrix} \mathbf{v}_{0dq} \\ \mathbf{v}_{FDQ} \end{bmatrix} = \begin{bmatrix} \mathbf{R}_{abc} & 0 \\ 0 & \mathbf{R}_{FDQ} \end{bmatrix} \begin{bmatrix} \mathbf{i}_{0dq} \\ \mathbf{i}_{FDQ} \end{bmatrix} - \begin{bmatrix} \mathbf{P}\frac{d}{dt}\mathbf{P}^{-1} & \mathbf{0} \\ \mathbf{0} & \mathbf{U} \end{bmatrix} \begin{bmatrix} \lambda_{0dq} \\ \lambda_{FDQ} \end{bmatrix} - \frac{d}{dt} \begin{bmatrix} \lambda_{0dq} \\ \lambda_{FDQ} \end{bmatrix} \tag{8.32}$$

Next, the expression for $\mathbf{P}\frac{d}{dt}\mathbf{P}^{-1}$ can be written as

$$\mathbf{P}\frac{d}{dt}\mathbf{P}^{-1} = \mathbf{P}\frac{d\theta}{dt}\frac{d}{d\theta}\mathbf{P}^{-1} = \omega\mathbf{P}\frac{d}{d\theta}\mathbf{P}^{-1} \tag{8.33}$$

Substituting for $\mathbf{P}$ from (8.14), and for the derivative of $\mathbf{P}^{-1}$ from (8.18), we get

$$\begin{aligned} \mathbf{P}\frac{d}{dt}\mathbf{P}^{-1} &= 2/3\omega \begin{bmatrix} 1/\sqrt{2} & 1/\sqrt{2} & 1/\sqrt{2} \\ \cos\theta & \cos(\theta - 2\pi/3) & \cos(\theta + 2\pi/3) \\ \sin\theta & \sin(\theta - 2\pi/3) & \sin(\theta + 2\pi/3) \end{bmatrix} \\ &\quad \begin{bmatrix} 0 & -\sin\theta & \cos\theta \\ 0 & -\sin(\theta - 2\pi/3) & \cos(\theta - 2\pi/3) \\ 0 & -\sin(\theta + 2\pi/3) & \cos(\theta + 2\pi/3) \end{bmatrix} \\ &= \omega \begin{bmatrix} 0 & 0 & 0 \\ 0 & 0 & 1 \\ 0 & -1 & 0 \end{bmatrix} \end{aligned} \tag{8.34}$$

Substituting (8.23) and (8.34) into (8.32), the machine equation in the rotor frame of reference becomes

$$\begin{bmatrix} v_0 \\ v_d \\ v_q \\ -v_F \\ 0 \\ 0 \end{bmatrix} = -\begin{bmatrix} r & 0 & 0 & 0 & 0 & 0 \\ 0 & r & \omega L_q & 0 & 0 & \omega k M_Q \\ 0 & -\omega L_d & r & -\omega k M_F & -\omega k M_D & 0 \\ 0 & 0 & 0 & r_F & 0 & 0 \\ 0 & 0 & 0 & 0 & r_D & 0 \\ 0 & 0 & 0 & 0 & 0 & r_Q \end{bmatrix} \begin{bmatrix} i_0 \\ i_d \\ i_q \\ i_F \\ i_D \\ i_Q \end{bmatrix}$$

$$-\begin{bmatrix} L_0 & 0 & 0 & 0 & 0 & 0 \\ 0 & L_d & 0 & kM_F & kM_D & 0 \\ 0 & 0 & L_q & 0 & 0 & kM_Q \\ 0 & kM_F & 0 & L_F & M_R & 0 \\ 0 & kM_D & 0 & M_R & L_D & 0 \\ 0 & 0 & kM_Q & 0 & 0 & L_Q \end{bmatrix} \frac{d}{dt} \begin{bmatrix} i_0 \\ i_d \\ i_q \\ i_F \\ i_D \\ i_Q \end{bmatrix} \tag{8.35}$$

We now make some observations regarding the nature of the above equations. The most important one is that they have constant coefficients provided that speed is assumed constant. Also, the first equation

$$v_0 = -ri_0 - L_0 \frac{di_0}{dt}$$

is not coupled to the other equations. Therefore, it can be treated separately. The variables v_0, L_0, and i_0 are known as the *zero-sequence variables*. The name originally comes from the theory of symmetrical components, as discussed in Chapter 10. Finally, we note that while the transformation technique is a mathematical process, it provides valuable insight into internal phenomena and gives the effects of transients. Furthermore, it provides physical meaning to the new quantities.

8.5 BALANCED THREE-PHASE SHORT CIRCUIT

Consider a three-phase synchronous generator operating at synchronous speed with constant excitation. We will explore the nature of the three armature currents and the field current following a three-phase short circuit at the armature terminals. The machine is assumed to be initially unloaded, i.e.,

$$i_a(0^+) = i_b(0^+) = i_c(0^+) = 0$$

With reference to (8.15), this condition results in

$$i_0(0^+) = i_d(0^+) = i_q(0^+) = 0$$

The initial value of the field current is

$$i_F(0^+) = \frac{V_F}{r_F}$$

For balanced three-phase short circuit at the terminals of the machine

$$v_a = v_b = v_c = 0$$

With reference to (8.16), this condition results in

$$v_0 = v_d = v_q = 0$$

Since $i_0 = 0$, the machine equation in the rotor reference frame following a three-phase short circuit becomes

$$\begin{bmatrix} v_d \\ -v_F \\ 0 \\ v_q \\ 0 \end{bmatrix} = - \begin{bmatrix} r & 0 & 0 & \omega L_q & \omega k M_Q \\ 0 & r_F & 0 & 0 & 0 \\ 0 & 0 & r_D & 0 & 0 \\ -\omega L_d & -\omega k M_F & -\omega k M_D & r & 0 \\ 0 & 0 & 0 & 0 & r_Q \end{bmatrix} \begin{bmatrix} i_d \\ i_F \\ i_D \\ i_q \\ i_Q \end{bmatrix}$$
$$- \begin{bmatrix} L_d & kM_F & kM_D & 0 & 0 \\ kM_F & L_F & M_R & 0 & 0 \\ kM_D & M_R & L_D & 0 & 0 \\ 0 & 0 & 0 & L_q & kM_Q \\ 0 & 0 & 0 & kM_Q & L_Q \end{bmatrix} \frac{d}{dt} \begin{bmatrix} i_d \\ i_F \\ i_D \\ i_q \\ i_Q \end{bmatrix} \tag{8.36}$$

This equation is in the state-space form and can be written in compact form as

$$\mathbf{v} = -\mathbf{R}\mathbf{i} - \mathbf{L}\frac{d}{dt}\mathbf{i} \tag{8.37}$$

or

$$\frac{d}{dt}\mathbf{i} = -\mathbf{L}^{-1}\mathbf{R}\mathbf{i} - \mathbf{L}^{-1}\mathbf{v} \tag{8.38}$$

If speed is assumed constant, the resulting state-space equation is linear and an analytical solution can be obtained by the Laplace transform technique. However, the availability of powerful simulation packages make it possible to simulate the nonlinear differential equations of the synchronous machine easily in matrix form. To consider the speed variation we need to include the dynamic equation of the machine. This is a second-order differential equation known as the *swing equation* which is described in Chapter 11. The swing equation can be expressed in the state-space form as two first-order differential equation and can easily be augmented with (8.36). Since the speed variation has very little effect in the momentary current immediately following the fault, speed variation may be neglected.

Once a solution is obtained for the direct axis and quadrature axis currents, the phase currents are obtained through the inverse Park transformation, i.e.,

$$\mathbf{i}_{abc} = \mathbf{P}^{-1}\mathbf{i}_{0dq} \tag{8.39}$$

Substituting for $\mathbf{P}^{-1}$ from (8.18), and noting $i_0 = 0$, the phase currents are

$$\begin{aligned} i_a &= i_d \cos\theta + i_q \sin\theta \\ i_b &= i_d \cos(\theta - 2\pi/3) + i_q \sin(\theta - 2\pi/3) \\ i_c &= i_d \cos(\theta + 2\pi/3) + i_q \sin(\theta + 2\pi/3) \end{aligned} \tag{8.40}$$

MATLAB provides two M-files named **ode23** and **ode45** for numerical solution of differential equations employing the Runge-Kutta-Fehlberg integration method. **ode23** uses a simple second and third order pair of formulas for medium accuracy and **ode45** uses a fourth and fifth order pair for higher accuracy. Synchronous machine simulation during balanced three-phase fault is demonstrated in the following example.

Example 8.2

A 500-MVA, 30-kV, 60-Hz synchronous generator is operating at no-load with a constant excitation voltage of 400 V. A three-phase short circuit occurs at the armature terminals. Use **ode45** to simulate (8.36), and obtain the transient waveforms for the current in each phase and the field current. Assume the short circuit is applied at the instant when the rotor direct axis is along the magnetic axis of phase a, i.e., $\delta = 0$. Also, assume that the rotor speed remains constant at the synchronous value. The machine parameters are

Generator Parameters for Example 8.2		
$L_d = 0.0072$ H	$L_q = 0.0070$ H	$L_F = 2.500$ H
$L_D = 0.0068$ H	$L_Q = 0.0016$ H	$M_F = 0.100$ H
$M_D = 0.0054$ H	$M_Q = 0.0026$ H	$M_R = 0.125$ H
$r = 0.0020\ \Omega$	$r_F = 0.4000\ \Omega$	$r_D = 0.015\ \Omega$
$r_Q = 0.0150\ \Omega$	$L_0 = 0.0010$ H	

The dc field voltage is $V_F = 400$ V. The derivatives of the state equation given by (8.38), together with the coefficient matrices in (8.36), are defined in a function file named **symshort.m**, which returns the state derivatives. The initial value of the field current is

$$i_F(0^+) = \frac{V_F}{r_F} = \frac{400}{0.4} = 1000 \ \ \text{A}$$

and since the machine is initially on no-load

$$i_0(0^+) = i_d(0^+) = i_q(0^0) = 0$$

The following file **chp8ex2.m** uses **ode45** to simulate the differential equations defined in **symshort** over the desired interval. The periodic nature of currents necessitates a very small step size for integration. The currents i_d and i_q are substituted in (8.40) and the phase currents are determined.

```
VF = 400; rF = 0.4; iF0 = VF/rF;
f = 60;   w=2.*pi*f;
```

```
d = 0;    d=d*pi/180;
t0 = 0 ; tfinal = 0.80;
tspan =[t0, tfinal];
i0 = [0; iF0; 0; 0; 0 ];          % Initial currents
[t,i] = ode45('symshort',tspan,i0);
theta = w*t + d + pi/2;
id = i(:,1), iq = i(:,4), iF = i(:,2);
ia = sqrt(2/3)*(id.*cos(theta) + iq.*sin(theta));
ib = sqrt(2/3)*(id.*cos(theta-2*pi/3)+ iq.*sin(theta-2*pi/3));
ic = sqrt(2/3)*(id.*cos(theta+2*pi/3)+ iq.*sin(theta+2*pi/3));
figure(1), plot(t,ia),  xlabel('Time - sec.'), ylabel('ia, A')
title(['Three-phase short circuit ia, ','delta =',num2str(d)])
figure(2), plot(t,ib),  xlabel('Time - sec.'), ylabel('ib, A')
title(['Three-phase short circuit ib, ','delta =',num2str(d)])
figure(3), plot(t,ic),  xlabel('Time - sec.'), ylabel('ic, A')
title(['Three-phase short circuit ic, ','delta =',num2str(d)])
figure(4), plot(t,iF),  xlabel('Time - sec.'), ylabel('iF, A')
title(['Three-phase short circuit iF,','delta = ',num2str(d)])
```

Results of the simulations are shown in Figure 8.5.

Armature currents in the various phases vary with time in a rather complicated way. Analysis of the waveforms show that they consist of

- A fundamental-frequency component.
- A dc component.
- A double-frequency component.

The fundamental-frequency component is symmetrical with respect to the time axis. Its superposition on the dc component will give an unsymmetrical waveform. The degree of asymmetry depends upon the point of the voltage cycle at which the short circuit takes place. The field current shown in Figure 8.5, like the stator current, consists of dc and ac components. The ac component is decaying and is comprised of a fundamental and a second harmonic. The second harmonic components in the field current as well as the armature currents are relatively small and are usually neglected. Furthermore, in Section 8.7 we see that during short circuit, the effective reactance of the machine may be assumed only along the direct axis and very simple models are obtained for power system fault studies and transient stability analysis. Before we obtain these simplified models, we consider the unbalanced short circuit of synchronous machine.

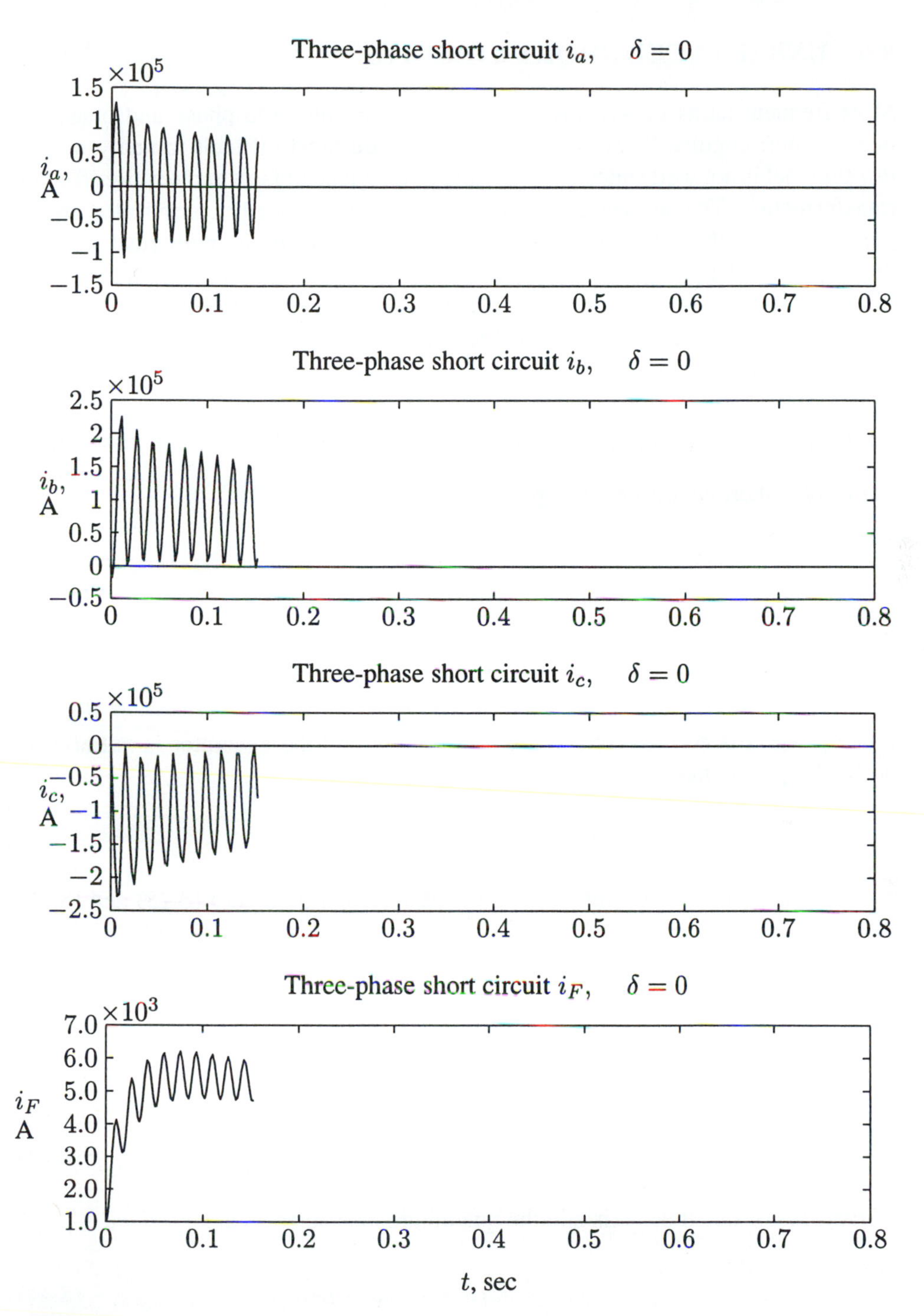

FIGURE 8.5
Balanced three-phase short-circuit current waveforms.

8.6 UNBALANCED SHORT CIRCUITS

Most frequent faults on synchronous machines are phase-to-phase and phase-to-neutral short circuits. These unbalanced faults are most difficult to analyze. The d-q-0 model is not well suited for the study of unbalanced fault and requires further transformation. The analytical solution becomes exceptionally complicated and at the end of it all the solutions are still only approximate. In the numerical solution the original voltage equations can be used without the need for any transformations. In the following section the machine equations are developed in direct-phase quantities for simulation of the synchronous machine for the line-to-line and the line-to-ground short circuits.

8.6.1 LINE-TO-LINE SHORT CIRCUIT

For a solid short circuit between phases a and b,

$$v_b = v_c = 0$$

and

$$i_b = -i_c$$

Since phase a is not involved in the short circuit and the generator is initially on no-load, $i_a = 0$, thus

$$i_0 = i_a + i_b + i_c = 0$$

and from (8.35), $v_0 = 0$. Substituting the above conditions into (8.15) and (8.16) yields

$$v_d \sin\theta - v_q \cos\theta = 0 \tag{8.41}$$

and

$$i_d = \sqrt{2}\, i_b \sin\theta \tag{8.42}$$

$$i_q = \sqrt{2}\, i_b \cos\theta \tag{8.43}$$

Derivatives of the direct axis and the quadrature axis currents are

$$\frac{di_d}{dt} = \sqrt{2}\,\frac{di_b}{dt}\sin\theta + \sqrt{2}\,\omega i_b \cos\theta \tag{8.44}$$

$$\frac{di_q}{dt} = \sqrt{2}\,\frac{di_b}{dt}\cos\theta - \sqrt{2}\,\omega i_b \sin\theta \tag{8.45}$$

Substituting (8.42)–(8.45) into (8.36) and applying (8.41) to the first and fourth equations in (8.36), the voltage equation for a line-to-line fault in direct-phase quantities is obtained.

$$\begin{bmatrix} -v_F \\ 0 \\ 0 \\ 0 \end{bmatrix} =$$

$$-\begin{bmatrix} \sqrt{2}k\omega M_F \cos\theta & r_F & 0 & 0 \\ \sqrt{2}k\omega M_D \cos\theta & r_D & 0 & 0 \\ \sqrt{2}k\omega M_Q \sin\theta & 0 & 0 & r_Q \\ \sqrt{2}[r + \omega(L_d - L_q)] \sin 2\theta & k\omega M_F \cos\theta & k\omega M_D \cos\theta & k\omega M_Q \sin\theta \end{bmatrix} \begin{bmatrix} i_b \\ i_F \\ i_D \\ i_Q \end{bmatrix}$$

$$-\begin{bmatrix} \sqrt{2}kM_F \sin\theta & L_F & M_R & 0 \\ \sqrt{2}kM_D \sin\theta & M_R & L_D & 0 \\ -\sqrt{2}kM_Q \cos\theta & 0 & 0 & L_Q \\ \sqrt{2}(L_d \sin^2\theta + L_q \cos^2\theta) & kM_F \sin\theta & kM_D \sin\theta & -kM_Q \cos\theta \end{bmatrix} \frac{d}{dt} \begin{bmatrix} i_b \\ i_F \\ i_D \\ i_Q \end{bmatrix} \tag{8.46}$$

This equation is in the state-space form and is written in compact form as (8.37). The state derivatives is given by (8.38), which is suitable for numerical integration.

Example 8.3

The synchronous generator in Problem 8.2 is operating at no-load with a constant excitation voltage of 400 V. A line-to-line short circuit occurs between phases b and c at the armature terminals. Use **ode45** to simulate (8.46), and obtain the waveforms for current in phase b and the field current. Assume the short circuit is applied at the instant when the rotor direct axis is along the magnetic axis of phase a, i.e., $\delta = 0$. Also, assume that the rotor speed remains constant at the synchronous value.

The dc field voltage is $V_F = 400$ V. The derivatives of the state equation given by (8.38), together with the coefficient matrices in (8.46) are defined in a function file named **llshort.m** which returns the state derivatives. The following file **chp8ex3.m** uses **ode45** to simulate the differential equation defined in **llshort** over the desired interval. The current in phase b and the field current are determined and their plots are shown in Figure 8.6.

```
VF = 400; rF = 0.4; iF0 = VF/rF;
f = 60;    w = 2.*pi*f;
d = 0;     d = d*pi/180;
t0 = 0 ; tfinal = 0.80;
tspan = [t0, tfinal];
```

```
i0 = [0; iF0; 0; 0;];                    % Initial currents
[t,i] = ode45('llshort', tspan, i0);
ib=i(:,1);  iF=i(:,2);
figure(1), plot(t,ib), xlabel('t, sec'), ylabel('ib, A')
title(['Line-line short circuit ib, ','delta = ', num2str(d)])
figure(2), plot(t,iF), xlabel('t, sec.'),ylabel('iF, A')
title(['Line-line short circuit iF,  ','delta = ',num2str(d)])
```

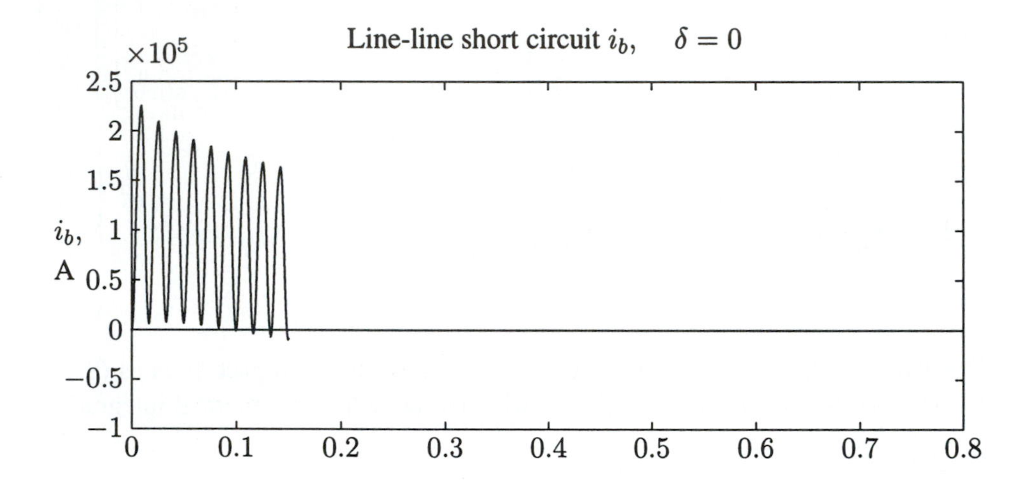

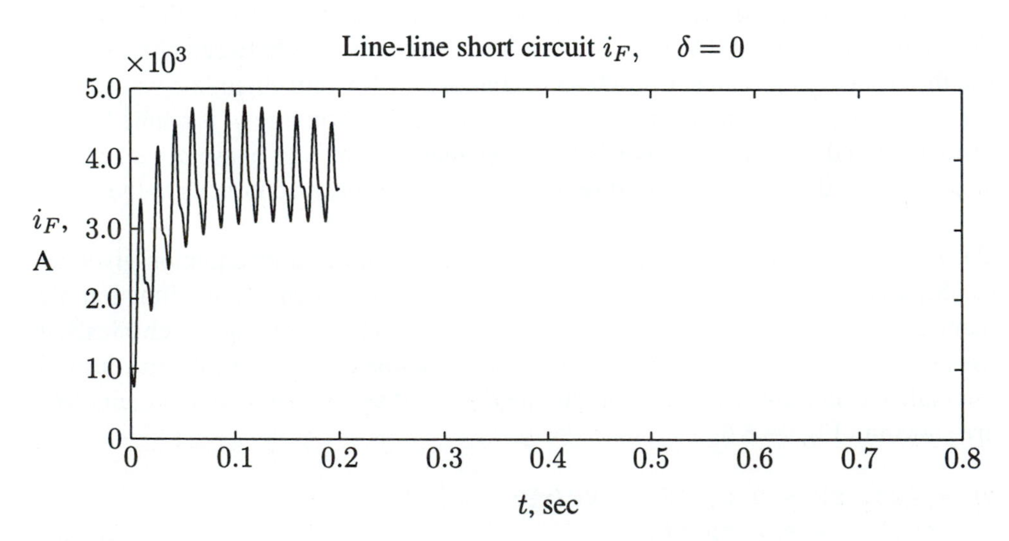

FIGURE 8.6
Line-to-line short-circuit current waveforms.

8.6.2 LINE-TO-GROUND SHORT CIRCUIT

For a solid short circuit between phases a and ground

$$v_a = 0$$

and with the machine initially on no-load

$$i_b = i_c = 0$$

A convenient way to obtain the voltage equation for line-to-ground short circuit is to start with (8.4), i.e., the machine voltage equation in direct phase quantities. Applying the short circuit condition to this equation and expressing the inductances in terms of the more commonly d-q-0 reactances, the following equation is obtained for the line-to-ground fault on phase a.

$$\begin{bmatrix} 0 \\ -v_F \\ 0 \\ 0 \end{bmatrix} = - \begin{bmatrix} r - 2\omega L_m \sin 2\theta & -\omega M_F \sin\theta & -\omega M_D \sin\theta & \omega M_Q \cos\theta \\ -\omega M_F \sin\theta & r_F & 0 & 0 \\ -\omega M_D \sin\theta & 0 & r_D & 0 \\ \omega M_Q \cos\theta & 0 & 0 & r_Q \end{bmatrix} \begin{bmatrix} i_a \\ i_F \\ i_D \\ i_Q \end{bmatrix} - \begin{bmatrix} L_s + L_m \cos 2\theta & M_F \cos\theta & M_D \cos\theta & M_Q \sin\theta \\ M_F \cos\theta & L_F & M_R & 0 \\ M_D \cos\theta & M_R & L_D & 0 \\ M_Q \sin\theta & 0 & 0 & L_Q \end{bmatrix} \frac{d}{dt} \begin{bmatrix} i_a \\ i_F \\ i_D \\ i_Q \end{bmatrix} \tag{8.47}$$

where

$$L_s = \frac{1}{3}(L_0 + L_d + L_q) \tag{8.48}$$

$$L_m = \frac{1}{3}(L_d - L_q) \tag{8.49}$$

Equation (8.47) is in the state-space form and is written in compact form as (8.37). The state derivatives is given by (8.38) which is suitable for numerical integration.

Example 8.4

The synchronous generator in Problem 8.2 is operating at no-load with a constant excitation voltage of 400 V. A line-to-ground short circuit occurs on phase a of the armature terminals. Use **ode45** to simulate (8.47), and obtain the waveforms for the current in phase a and the field current. Assume the short circuit is applied at the instant when the rotor direct axis is along the magnetic axis of phase a, i.e., $\delta = 0$. Also, assume that the rotor speed remains constant at the synchronous value.

The dc field voltage is $V_F = 400$ V. The derivative of the state equation given by (8.38), together with the coefficient matrices in (8.47), are defined in a function file named **lgshort.m** which returns the state derivatives. The following file **chp8ex4.m** uses **ode45** to simulate the differential equation defined in **lgshort** over the desired interval. The phase and the field currents are determined and their plots are shown in Figure 8.7.

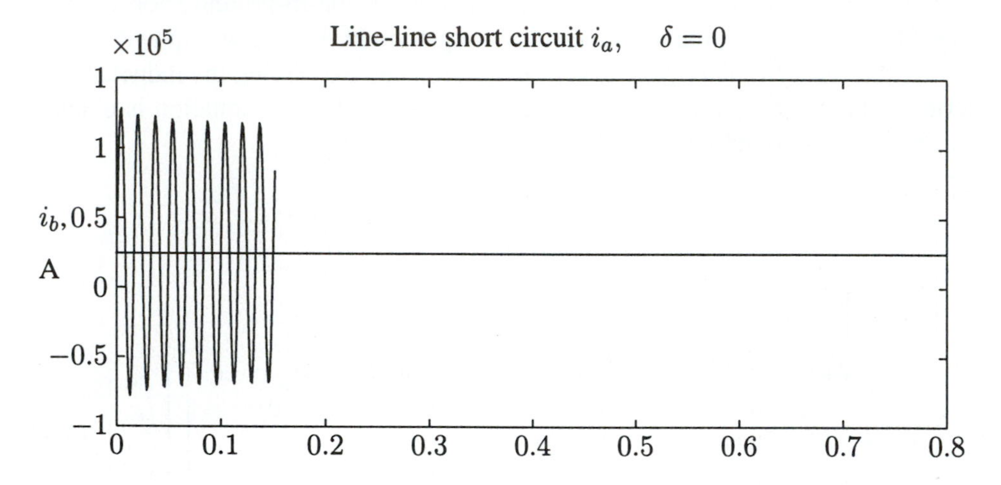

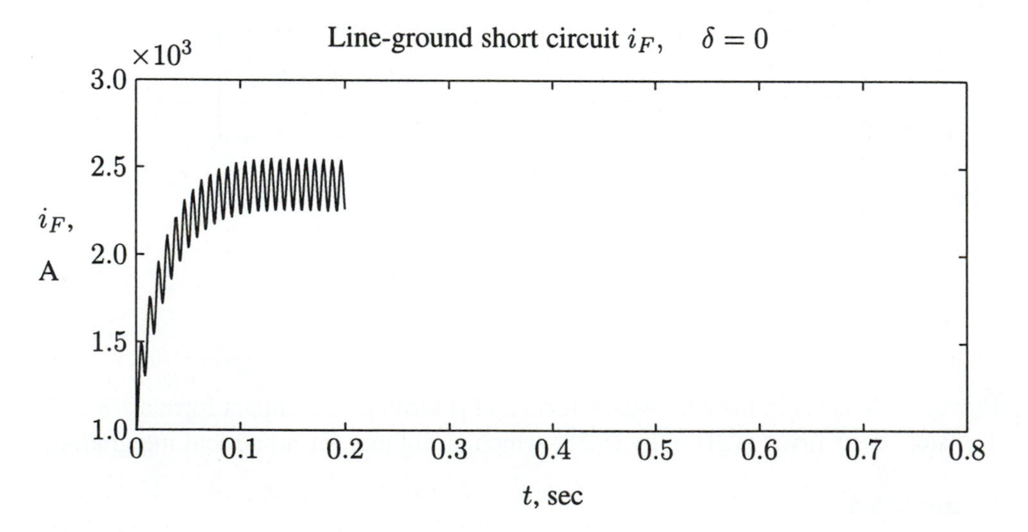

FIGURE 8.7
Line-to-ground short-circuit current waveforms.

```
VF = 400; rF = 0.4; iF0 = VF/rF;
f = 60;     w = 2.*pi*f;
d = 0;      d = d*pi/180;
t0 = 0 ; tfinal = 0.80; tspan = [t0, tfinal];
i0 = [0; iF0; 0; 0;];             % Initial currents
tol =  0.0001;                    % accuracy
[t,i] = ode45('lgshort', tspan, i0, tol);
ia=i(:,1);  iF=i(:,2);
figure(1), plot(t,ia), xlabel('t, sec'), ylabel('ia, A')
title(['Line-ground short circuit ia,','delta =', num2str(d)])
figure(2), plot(t,iF), xlabel('t, sec'), ylabel('iF, A')
title(['Line-ground short circuit iF,','delta = ',num2str(d)])
```

A three-phase model, which uses direct physical parameters, is well suited for simulation on a computer, and it is not necessary to go through complex transformations. The analysis can easily be extended to take the speed variation into account by including the dynamic equations of the machine.

8.7 SIMPLIFIED MODELS OF SYNCHRONOUS MACHINES FOR TRANSIENT ANALYSES

In Chapter 3, for steady-state operation, the generator was represented with a constant emf behind a synchronous reactance X_s. For salient-pole rotor, because of the nonuniformity of the air gap, the generator was modeled with direct axis reactance X_d and the quadrature axis reactance X_q. However, under short circuit conditions, the circuit reactance is much greater than the resistance. Thus, the stator current lags nearly $\pi/2$ radians behind the driving voltage, and the armature reaction mmf is centered almost on the direct axis. Therefore, during short circuit, the effective reactance of the machine may be assumed only along the direct axis.

The three-phase short circuit waveform shown in Figure 8.5 shows that the ac component of the armature current decays from a very high initial value to the steady state value. This is because the machine reactance changes due to the effect of the armature reaction. At the instant prior to short circuit, there is some flux on the direct axis linking both stator and rotor, due only to rotor mmf if the machine is on open circuit, or due to the resultant of rotor and stator mmf if some stator current is flowing. When there is a sudden increase of stator current on short circuit, the flux linking stator and rotor cannot change instantaneously due to eddy currents flowing in the rotor and damper circuits, which oppose this change. Since, the stator mmf is unable at first to establish any armature reaction, the reactance of armature reaction is negligible, and the initial reactance is very small and similar in value to the leakage reactance. As the eddy current in the damper circuit and eventually in the field circuit decays, the armature reaction is fully established. The armature reaction which is produced by a nearly zero power factor current provides mostly demagnetizing effect and the machine reactance increases to the direct axis synchronous reactance.

For purely qualitative purposes, a useful picture can be obtained by thinking of the field and damper windings as the secondaries of a transformer whose primary is the armature winding. During normal steady state conditions, there is no transformer action between stator and rotor windings of the synchronous machine as the resultant field produced by the stator and rotor both revolve with the same synchronous speed. This is similar to a transformer with open-circuited secondaries. For this condition, its primary may be described by the synchronous reactance X_d. During disturbance, the rotor speed is no longer the same as that of the revolving field produced by stator windings resulting in the transformer action. Thus, field and damper circuits resemble much more nearly as short-circuited secondaries. The equivalent circuit for this condition, referred to the stator side, is shown in Figure 8.8. Ignoring winding resistances, the equivalent reactance of Figure 8.8,

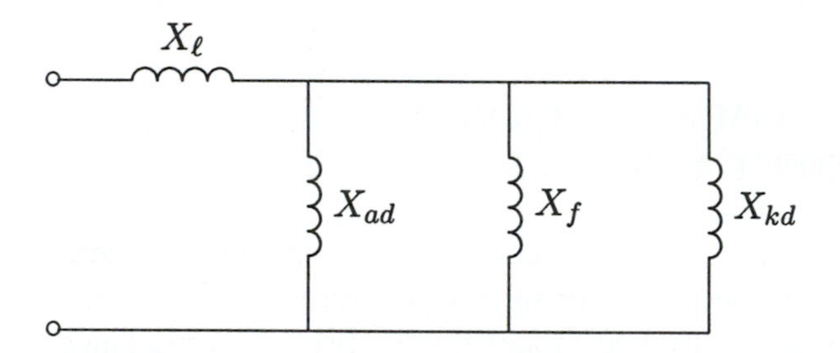

FIGURE 8.8
Equivalent circuit for the subtransient period.

known as the *direct axis subtransient reactance*, is

$$X''_d = X_\ell + \left(\frac{1}{X_{ad}} + \frac{1}{X_f} + \frac{1}{X_{kd}}\right)^{-1} \tag{8.50}$$

If the damper winding resistance R_k is inserted in Figure 8.8 and the Thévenin's inductance seen at the terminals of R_k is obtained, the circuit time constant, known as the *direct axis short circuit subtransient time constant*, becomes

$$\tau''_d = \frac{X_{kd} + \left(\frac{1}{X_\ell} + \frac{1}{X_f} + \frac{1}{X_{ad}}\right)^{-1}}{R_k} \tag{8.51}$$

In (8.51) reactances are assumed in per unit and, therefore, they have the same numerical values as inductances in per unit. For a 2-pole, turbo-alternators X''_d may be between 0.07 and 0.12 per unit, while for water-wheel alternators the range may be 0.1 to 0.35 per unit. The direct axis subtransient reactance X''_d is only used in calculations if the effect of the initial current is important, as for example, when determining the circuit breaker short-circuit rating.

Typically, the damper circuit has relatively high resistance and the direct axis short circuit subtransient time constant is very small, around 0.035 second. Thus, this component of current decays quickly. It is then permissible to ignore the branch of the equivalent circuit which takes account of the damper windings, and the equivalent circuit reduces to that of Figure 8.9.

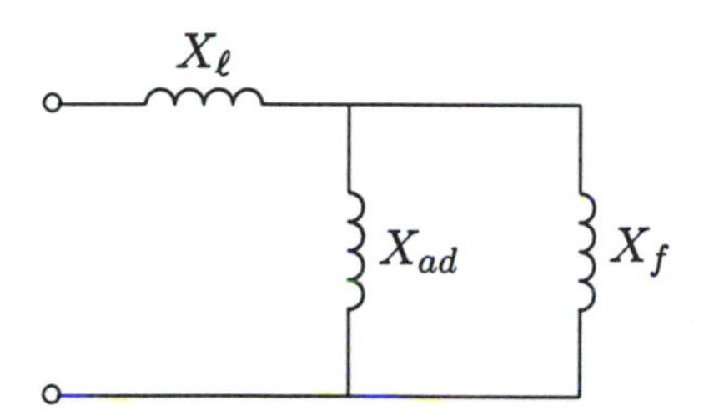

FIGURE 8.9
Equivalent circuit for the transient period.

Ignoring winding resistances, the equivalent reactance of Figure 8.9, known as the *direct axis short circuit transient reactance*, is

$$X'_d = X_\ell + \left(\frac{1}{X_{ad}} + \frac{1}{X_f}\right)^{-1} \tag{8.52}$$

If the field winding resistance R_f is inserted in Figure 8.9, and the Thévenin's inductance seen at the terminals of R_f is obtained, the circuit time constant, known as the *direct axis short circuit transient time constant*, becomes

$$\tau'_d = \frac{X_f + (\frac{1}{X_\ell} + \frac{1}{X_{ad}})^{-1}}{R_f} \tag{8.53}$$

The direct axis transient short circuit reactance X'_d may lie between 0.10 to 0.25 per unit. The short circuit transient time constant τ'_d is usually in order of 1 to 2 seconds.

The field time constant which characterizes the decay of transients with the armature open-circuited is called the *direct axis open circuit transient time constant*. This is given by

$$\tau'_{d0} = \frac{X_f}{R_f} \tag{8.54}$$

Typical values of the direct axis open circuit transient time constant are about 5 seconds. τ'_d is related to τ'_{d0} by

$$\tau'_d = \frac{X'_d}{X_d}\tau'_{d0} \tag{8.55}$$

Finally, when the disturbance is altogether over, there will be no hunting of the rotor, and, hence there will not be any transformer action between the stator and the rotor, and the circuit reduces to that of Figure 8.10.

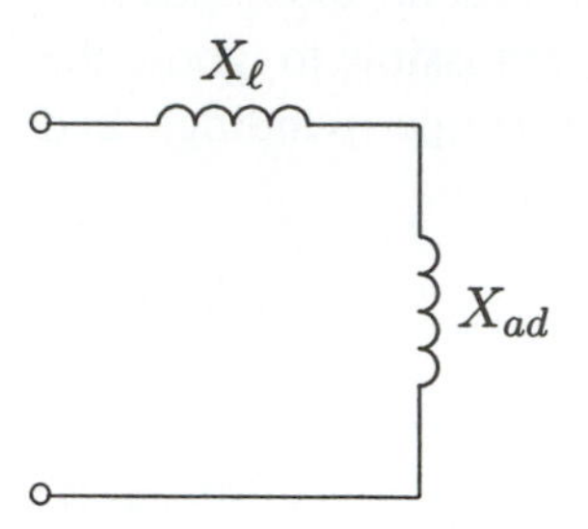

FIGURE 8.10
Equivalent circuit for the steady state.

The equivalent reactance becomes the direct axis synchronous reactance, given by

$$X_d = X_\ell + X_{ad} \tag{8.56}$$

Similar equivalent circuits are obtained for reactances along the quadrature axis. These reactances X_q'', X_q', and X_q may be considered for cases when the circuit resistance results in a power factor appreciably above zero and the armature reaction is not necessarily totally on the direct axis.

The fundamental-frequency component of armature current following the sudden application of a short circuit to the armature of an initially unloaded machine can be expressed as

$$i_{ac}(t) = \sqrt{2}E_0 \left[\left(\frac{1}{X_d''} - \frac{1}{X_d'} \right) e^{-t/\tau_d''} + \left(\frac{1}{X_d'} - \frac{1}{X_d} \right) e^{-t/\tau_d'} + \frac{1}{X_d} \right] \sin(\omega t + \delta) \tag{8.57}$$

A typical symmetric trace of the short circuit waveform obtained for the data of Example 8.5 is shown in Figure 8.11 (page 340).

It should be recalled that in the derivation of the above results, resistance was neglected except in consideration of the time constant. In addition, in the above treatment the dc and the second harmonic components corresponding to the decay of the trapped armature flux has been neglected. It should also be emphasized that the representation of the short-circuited paths of the damper windings and the solid iron rotor by a single equivalent damper circuit is an approximation to the actual situation. However, this approximation has been found to be quite valid in many cases. The synchronous machine reactances and time constants are provided by the manufacturers. These values can be obtained by a short circuit test described in the next section.

Example 8.5

A three-phase, 60-Hz synchronous machine is driven at constant synchronous speed by a prime mover. The armature windings are initially open-circuited and field voltage is adjusted so that the armature terminal voltage is at the rated value (i.e., 1.0 per unit). The machine has the following per unit reactances and time constants.

$$X''_d = 0.15 \text{ pu} \qquad \tau''_d = 0.035 \text{ sec}$$
$$X'_d = 0.40 \text{ pu} \qquad \tau'_d = 1.0 \text{ sec}$$
$$X_d = 1.20 \text{ pu}$$

a) Determine the steady state, transient and subtransient short circuit currents.
b) Obtain the fundamental-frequency waveform of the armature current for a three-phase short circuit at the terminals of the generator. Assume the short circuit is applied at the instant when the rotor direct axis is along the magnetic axis of phase a, i.e., $\delta = 0$.

$$I_d = \frac{E_0}{X_d} = \frac{1.0}{1.2} = 0.8333 \text{ pu}$$
$$I'_d = \frac{E_0}{X'_d} = \frac{1.0}{0.4} = 2.5 \text{ pu}$$
$$I''_d = \frac{E_0}{X''_d} = \frac{1.0}{0.15} = 6.666 \text{ pu}$$

To obtain the short circuit waveform, we write the following commands.

```
w0 = 2*pi*60;
E0 = 1.0;        delta = 0;
Xd2dash = 0.15;
Xddash = 0.4;
Xd = 1.2;
tau2dash = 0.035; taudash = 1.0;
t=0:1/(4*240):1.0;
iac = sqrt(2)*E0*((1/Xd2dash-1/Xddash)*exp(-t/tau2dash)+...
(1/Xddash-1/Xd)*exp(-t/taudash) + 1/Xd).*sin(w0*t + delta);
plot(t, iac), xlabel('t, sec'), ylabel('iac, A')
end
```

The result is shown in Figure 8.11.

The trace is obtained up to 1 second. If the simulation period is extended to about $5\tau'_d = 5.0$ seconds, the short circuit will reach to its steady state with a peak value of $I_{d_{(max)}} = \sqrt{2}\,E_0/X_d = \sqrt{2}\,(1.0)/1.2 = 1.1785$ per unit.

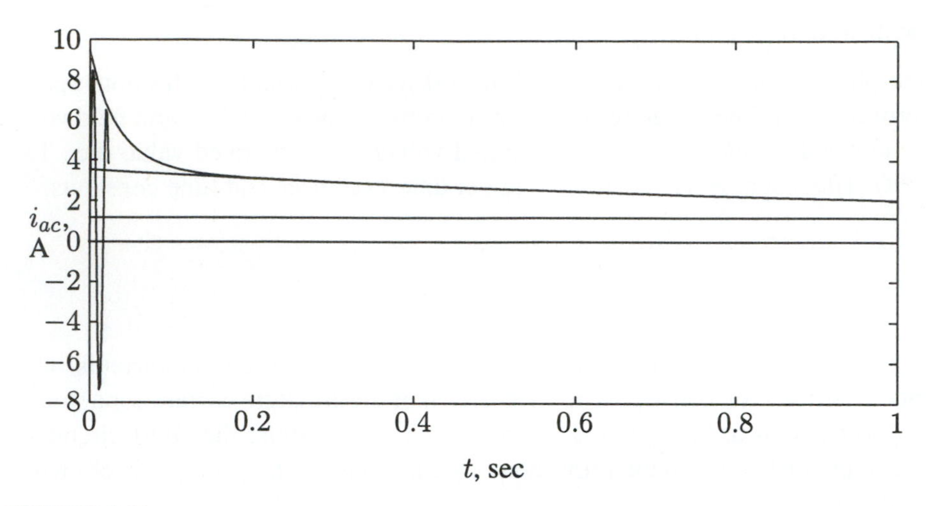

FIGURE 8.11
The 60-Hz component of the short-circuit current of a synchronous generator.

8.8 DC COMPONENTS OF STATOR CURRENTS

In the expression for the armature current as given by (8.57), the unidirectional transient component has not been taken into account. As seen from consideration of the simple RL circuit of Figure 8.1, there will in general be a dc offset depending on when the voltage is applied. Similarly, in the synchronous machine, the dc offset component depend on the instantaneous value of the stator voltage at the time of the short circuit. The rotor position is given by $\theta = \omega t + \delta + \pi/2$. The dc component depends on the rotor position δ when the short circuit occurs at time $t = 0$. The time constant associated with the decay of the dc component of the stator current is known as the *armature short circuit time constant*, τ_a. Most of the decay of the dc component occurs during the subtransient period. For this reason the average value of the direct axis and quadrature axis subtransient reactances is used for finding τ_a. It is approximately given by

$$\tau_a = \frac{X''_d + X''_q}{2R_a} \tag{8.58}$$

Typical values of the armature short circuit time constant is around 0.05 to 0.17 second.

Since the three-phase voltages are each separated by $2\pi/3$ radians, the amount of the dc component of the armature current is different in each phase and depends upon the point of the voltage cycle at which the short circuit occurs. The dc component for phase a is given by

$$I_{dc} = \sqrt{2}\frac{E_0}{X_d''}\sin\delta \; e^{-t/\tau_a} \tag{8.59}$$

The superposition of the dc component on the fundamental frequency component will give an asymmetrical waveform.

$$i_{asy}(t) = \sqrt{2}E_0\left[\left(\frac{1}{X_d''}-\frac{1}{X_d'}\right)e^{-t/\tau_d''}+\left(\frac{1}{X_d'}-\frac{1}{X_d}\right)e^{-t/\tau_d'}+\frac{1}{X_d}\right]\sin(\omega t+\delta)$$
$$+\sqrt{2}\frac{E_0}{X_d''}\sin\delta e^{-t/\tau_a} \tag{8.60}$$

The degree of asymmetry depends upon the point of the voltage cycle at which the fault takes place. The worst possible transient condition is $\delta = \pi/2$. The maximum possible initial magnitude of the dc component is

$$I_{dc_{(max)}} = \sqrt{2}\,\frac{E_0}{X_d''} \tag{8.61}$$

Therefore, the maximum rms current (ac plus dc) at the beginning of the short circuit is

$$I_{asy} = \sqrt{I_d''+I_{dc}^2} = \sqrt{\left(\frac{E_0}{X_d''}\right)^2+\left(\sqrt{2}\frac{E_0}{X_d''}\right)^2} \tag{8.62}$$

from which

$$I_{asy} = \sqrt{3}\,\frac{E_0}{X_d''} \tag{8.63}$$
$$= \sqrt{3}I_d''$$

In practice, the *momentary duty* of a circuit breaker is given in terms of the asymmetrical short circuit current.

Example 8.6

For the machine in Example 8.5, assume that a three-phase short circuit is applied at the instant when the rotor quadrature axis is along the magnetic axis of phase a, i.e., $\delta = \pi/2$ radians. Obtain the asymmetrical waveform of the armature current for phase a. The armature short circuit time constant is $\tau_a = 0.15$ sec.

In the *MATLAB* program of Example 8.5, we make $\delta = \pi/2$ and use (8.60) in place of the i_{ac} statement. Running this example results in the waveform shown in Figure 8.12.

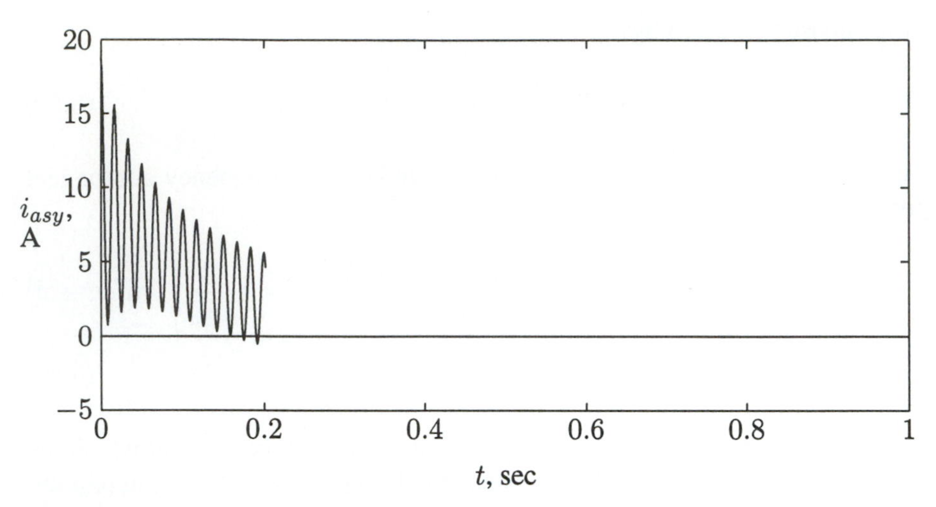

FIGURE 8.12
Synchronous generator asymmetrical short-circuit current $\delta = \pi/2$.

8.9 DETERMINATION OF TRANSIENT CONSTANTS

A sudden three-phase short circuit is applied to the terminals of an unloaded generator and the oscillogram of the current in one phase is obtained. The test is repeated until a symmetric waveform is obtained which does not contain the dc offset. This occurs when the voltage is near maximum at the instant of fault. The machine reactances X''_d, X'_d, and X_d and the time constants τ''_d and τ'_d are determined by analyzing the oscillogram waveform as follow.

The waveform is divided into three periods: the subtransient period, lasting only for the first two cycles, during which the current decrement is very rapid; the transient period, covering a relatively longer time, during which the current decrement is more moderate; and finally, the steady state period.

The no-load generated voltage E_0 is obtained by measuring the phase voltage and expressing it in per unit. The direct axis synchronous reactance X_d is determined from the part of the oscillogram where the envelope of the current has become constant. Denoting this amplitude with $I_{d_{(max)}}$, the rms value of the steady short circuit is $I_d = I_{d_{(max)}}/\sqrt{2}$. From this the direct axis synchronous reactance is found

$$X_d = \frac{E_0}{I_d} \tag{8.64}$$

The peak steady short circuit current is subtracted from two points after approximately the 10th cycle where the subtransient component has decayed. Dividing these values by $\sqrt{2}$ results in the following term

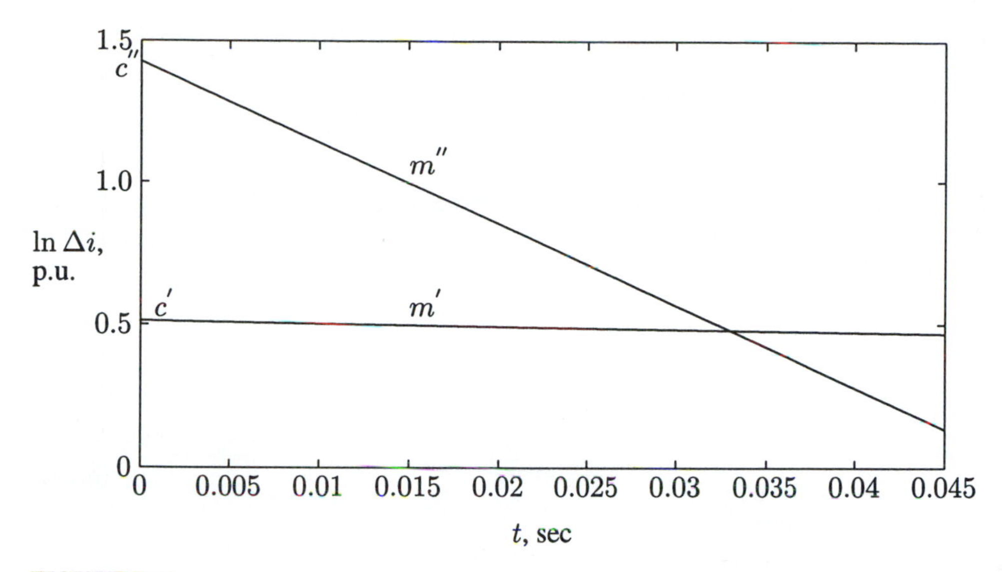

FIGURE 8.13
Current difference logarithm, $\ln \Delta i'$ and $\ln \Delta i''$.

$$\Delta i' = (I'_d - I_d)e^{-t/\tau'_d}$$

or

$$\begin{aligned}\ln \Delta i' &= \ln(I'_d - I_d) - t/\tau'_d \\ &= c' - m't\end{aligned}$$

If the points given by $\ln \Delta i'$ are plotted against a linear time scale, a straight line is obtained with y-intercept $c' = \ln(I'_d - I_d)$ and slope $-m'$, as shown in Figure 8.13. The rms transient component of current is obtained from

$$I'_d = e^{c'} + I_d \tag{8.65}$$

Transient reactance and time constant are then obtained by

$$X'_d = \frac{E_0}{I'_d} \tag{8.66}$$

and

$$\tau'_d = \frac{1}{m'} \tag{8.67}$$

To find the subtransient components, the peak current of the first two cycles are divided by $\sqrt{2}$. Subtracting the steady short circuit current and the rms transient currents found earlier from these points results in

$$\Delta i'' = (I''_d - I'_d)e^{-t/\tau''_d}$$

or

$$\ln \Delta i'' = \ln(I_d'' - I_d') - t/\tau_d''$$
$$= c'' - m''t$$

If the points given by $\ln \Delta i''$ are plotted against a linear time scale, a straight line is obtained with y-intercept $c'' = \ln(I_d'' - I_d')$ and slope $-m''$, shown in Figure 8.13. The rms subtransient component of current is given by

$$I_d'' = e^{c''} + I_d' \tag{8.68}$$

Subtransient reactance and time constant are

$$X_d'' = \frac{E_0}{I_d''} \tag{8.69}$$

and

$$\tau_d'' = \frac{1}{m''} \tag{8.70}$$

The above procedure is demonstrated in the following example.

Example 8.7

A three-phase, 60-Hz synchronous machine is driven at constant synchronous speed by a prime mover. The armature windings are initially open-circuited and field voltage is adjusted so that the armature terminal voltage is at the rated value (i.e., 1.0 per unit). The generator is suddenly subjected to a symmetrical three-phase short circuit at the instant when direct axis is along the magnetic axis of phase a, i.e., $\delta = 0$. An oscillogram of the short-circuited current is obtained. The peak values at the first two cycles, at the 20th and 21st cycles, and the steady value after a long time were recorded as tabulated in the following table.

I_{max}, pu	8.7569	6.7363	···	2.8893	2.8608	···	1.1785
Time, sec	0.0042	0.0208	···	0.3208	0.3375	···	5.0000

Determine the transient and the subtransient reactances and time constants.

The following statements are written with reference to the above procedure.

```
E0 = 1.0;
Im=[8.7569   6.7363  2.8893  2.8608   1.1785];
 t=[0.0042   0.0208  0.3208  0.3375   5.0000];
I = Im/sqrt(2);              % The rms value of the above envelope
id=I(5);                    % rms value of the steady short circuit
```

```
Dt2=[t(3) t(4)];                    % Time for 20th and 21st cycles
Di2=[I(3)-id I(4)-id];%Diff. between transient envelope and id
LDi2= log(Di2);                  %Natural log of the above two points
c2=polyfit(Dt2, LDi2, 1);
                        %Finds coefficients of a 1st-order polynomial
                    % i.e. the slope and intercept of a straight line
iddash=(exp(c2(2))+id)     % rms value of the transient current
Xddash=E0/iddash                    % Direct axis transient reactance
taudash=abs(1/c2(1))   %Direct axis sc transient time constant
Di=(iddash-id)*[exp(-t(1)/taudash)  exp(-t(2)/taudash)];
Di1=[I(1)-Di(1)-id   I(2)-Di(2)-id];    % Subtransient envelope
LDi1=log(Di1);
Dt1 =[t(1)  t(2)];           % Natural log of the first two points
c1=polyfit(Dt1, LDi1, 1);
                        %Finds coefficients of a 1st-order polynomial
                    % i.e. the slope and intercept of a straight line

id2dash=exp(c1(2))+iddash  %rms value of  subtransient current
Xd2dash= E0/id2dash          % Direct axis subtransient reactance
tau2dash=abs(1/c1(1))% direct axis sc subtransient time const.

t=0:.005:.045;
fit2 = polyval(c2, t); % line C2 evaluated for all values of t
fit1 = polyval(c1, t); % line C1 evaluated for all values of t
plot(t, fit1, t, fit2),grid % Logarithmic plot of id'' and id'
ylabel('ln(I)  pu' )        % intercepts are ln(Id'') and ln(Id')
xlabel('t, sec')         %slopes are reciprocal of time constants
```

The result is

$$I'_d = 2.5038 \text{ pu} \qquad X'_d = 0.3994; \text{ pu} \qquad \tau'_d = 0.9941 \text{ sec}$$
$$I''_d = 6.6728 \text{ pu} \qquad X''_d = 0.1499; \text{ pu} \qquad \tau''_d = 0.0348 \text{ sec}$$

Example 8.8

A 100-MVA, 13.8-kV, 60-Hz, Y-connected, three-phase synchronous generator is connected to a 13.8-kV/220-kV, 100-MVA, Δ-Y connected transformer. The reactances in per unit to the machine's own base are

$$X_d = 1.0 \text{ pu} \quad X'_d = 0.25 \text{ pu} \quad X''_d = 0.12 \text{ pu}$$

and its time constants are

$$\tau_a = 0.25 \text{ sec} \quad \tau_d'' = 0.4 \text{ sec} \quad \tau_d' = 1.1 \text{ pu}$$

The transformer reactance is 0.20 per unit on the same base. The generator is operating at the rated voltage and no-load when a three-phase fault occurs at the secondary terminals of the transformer as shown in Figure 8.14.

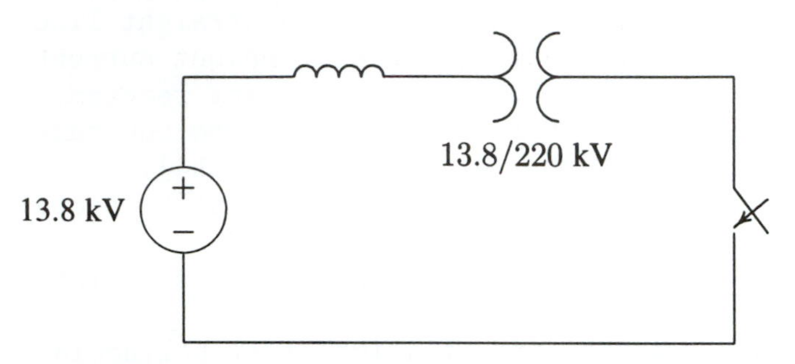

FIGURE 8.14
One-line diagram for Example 8.8.

(a) Find the subtransient, transient, and the steady state short circuit currents in per unit and actual amperes on both sides of the transformer.
(b) What is the maximum rms current (ac plus dc) at the beginning of the fault?
(c) Obtain the instantaneous expression for the short circuit current including the dc component. Assume $\delta = \pi/2$ radians.

(a) The base current on the generator side is

$$I_{B1} = \frac{S_B}{\sqrt{3}\,V_{B1}} = \frac{100 \times 10^3}{\sqrt{3}\,13.8} = 4184 \text{ A}$$

The base current on the secondary side of the transformer is

$$I_{B2} = \frac{13.8}{220}(4184) = 262.4 \text{ A}$$

the subtransient, transient and the steady state short circuit currents are

$$I_d'' = \frac{1.0}{0.12 + 0.2} = 3.125 \text{ pu} = 13,075 \text{ A} \quad \text{on the generator side}$$
$$= 820 \text{ A} \quad \text{on the 220-kV side}$$
$$I_d' = \frac{1.0}{0.25 + 0.2} = 2.22 \text{ pu} = 9,288.5 \text{ A} \quad \text{on the generator side}$$
$$= 582.5 \text{ A} \quad \text{on the 220-kV side}$$
$$I_d = \frac{1.0}{1.0 + 0.2} = 0.833 \text{ pu} = 3,486.6 \text{ A} \quad \text{on the generator side}$$
$$= 218.6 \text{ A} \quad \text{on the 220-kV side}$$

(b) The maximum rms current (ac plus dc) at the beginning of the fault is

$$I_{asy} = \sqrt{3}\, I_d'' = \sqrt{3}\,(3.125) = 5.4 \text{ pu} = 22,646 \text{ A} \quad \text{on the generator side}$$

(c) The instantaneous short circuit current including the dc component from (8.60), for $\delta = \pi/2$ is

$$i(t) = \sqrt{2}[(I_d'' - I_d')e^{-t/0.4} + (I_d' - I_d)e^{-t/1.1} + I_d] \sin(377t + \pi/2) + \sqrt{2} I_d'' e^{-t/0.25}$$

or

$$i(t) = [1.28e^{-2.5t} + 1.96e^{-0.91t} + 1.18] \sin(377t + \pi/2) + 4.42e^{-4t} \quad \text{pu}$$

Use *MATLAB* to obtain a plot of $i(t)$.

8.10 EFFECT OF LOAD CURRENT

If the fault occurs when the generator is delivering a prefault load current, two methods might be used in the solution of three-phase symmetrical fault currents.

(a) Use of internal voltages behind reactances

When there is a prefault load current, three fictitious internal voltages E'', E', and E_0 may be considered to be effective during the subtransient, transient, and the steady state periods, respectively. These voltages are known as the *voltage behind subtransient reactance*, *voltage behind transient reactance*, and *voltage behind synchronous reactance*. Consider the one-line diagram of a loaded generator shown in Figure 8.15(a). The internal voltages shown by the phasor diagram in Figure 8.15(b) are given by

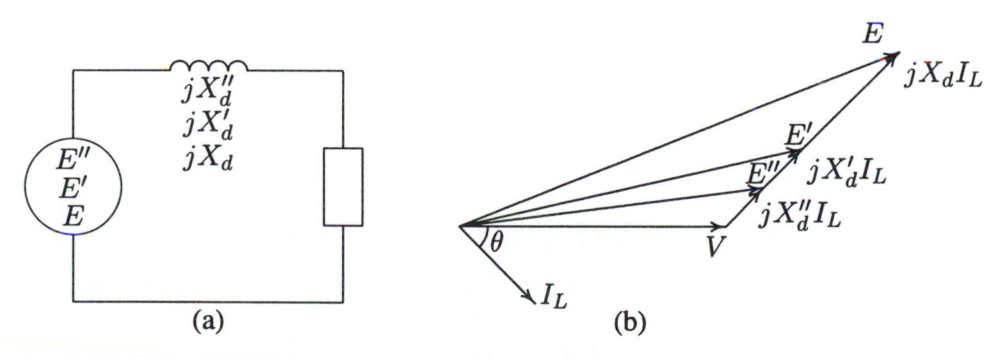

FIGURE 8.15
(a) One-line diagram of a loaded generator, (b) phasor diagram.

$$\begin{aligned} E'' &= V + jX''_d I_L \\ E' &= V + jX'_d I_L \\ E &= V + jX_d I_L \end{aligned} \tag{8.71}$$

Example 8.9

In Example 8.8, a three-phase load of 100 MVA, 0.8 power factor lagging is connected to the transformer secondary side as shown in Figure 8.16. The line-to-line voltage at the load terminals is 220 kV. A three-phase short circuit occurs at the load terminals. Find the generator transient current including the load current.

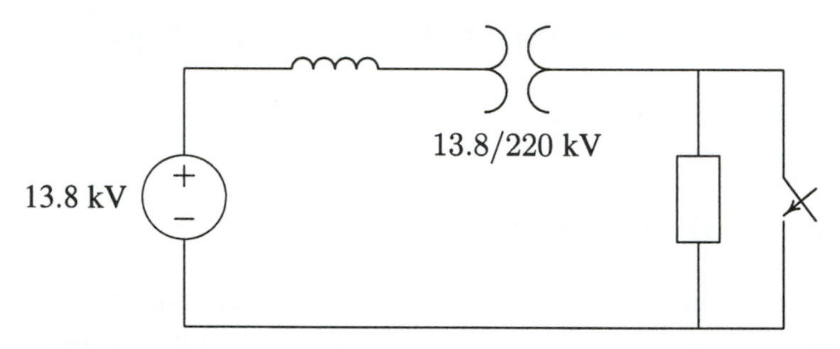

FIGURE 8.16
One-line diagram for Example 8.9.

The load may be represented by a per unit impedance as shown in Figure 8.16.

$$S_L = \frac{100\angle 36.87^\circ}{100} = 1\angle 36.87^\circ \ \text{pu}$$

$$V = \frac{220}{220} = 1\angle 0^\circ \ \text{pu}$$

$$Z_L = \frac{|V^2|}{S^*} = \frac{1}{1\angle -36.87^\circ} = 0.8 + j0.6 \ \text{pu}$$

Before fault, the load current is

$$I_L = \frac{V}{Z_L} = \frac{1.0\angle 0^\circ}{0.8 + j0.6} = 0.8 - j0.6 = 1\angle -36.87^\circ \ \text{pu}$$

The emf behind transient reactance is

$$\begin{aligned} E' &= V + j(X'_d + X_t)I_L \\ &= 1.0\angle 0^\circ + j(0.25 + 0.2)(0.8 - j0.6) = 1.27 + j0.36 = 1.32\angle 15.83^\circ \ \text{pu} \end{aligned}$$

When the fault is applied by closing switch S, the generator short circuit transient current is

$$I'_g = \frac{E'}{j(X'_d + X_t)} = \frac{1.32\angle 15.83^\circ}{j(0.25 + 0.2)} = 0.8 - j2.822 = 2.93\angle -74.17^\circ \ \text{pu}$$

(b) Using Thévenin's theorem and superposition with load current

The fault current is found in the absence of the load by obtaining the Thévenin's equivalent circuit to the point of fault. The total short circuit current is then given by superimposing the fault current with the load current.

Example 8.10

Find the generator transient current in Problem 8.9 using Thévenin's method.

The one-line diagram of Example 8.10 without the load is shown in Figure 8.17(a). The circuit for the Thévenin's equivalent impedance with respect to the point of fault is shown in Figure 8.17(b). The Thévenin's voltage is the prefault

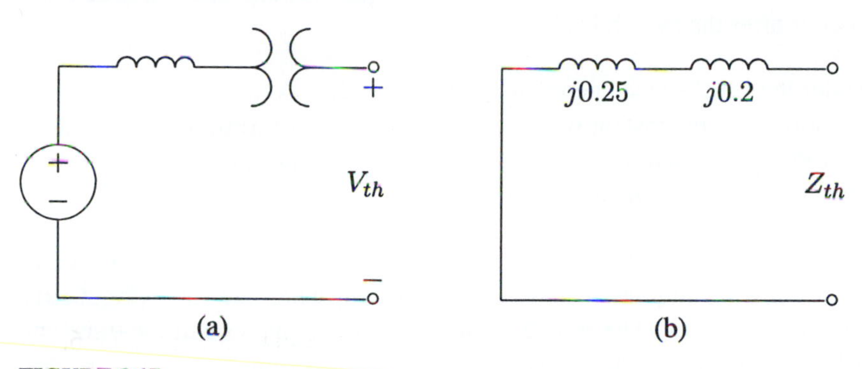

FIGURE 8.17
(a) One-line diagram for Example 8.10 without the load and (b) Thévenin's equivalent impedance to the point of fault.

terminal voltage, i.e.,

$$V_{th} = \frac{220}{220} = 1\angle 0^\circ \text{ pu}$$

and the Thévenin's impedance is

$$Z_{th} = j(0.25 + 0.2) = j0.45$$

The fault contribution is

$$I'_f = \frac{V_{th}}{Z_{th}} = \frac{1.0\angle 0^\circ}{j0.45} = -j2.222 \text{ pu}$$

Now superimposing the load current with the fault current results in

$$I'_g = I'_f + I_L = -j2.222 + (0.8 - j0.6) = 0.8 - j2.822 = 2.93\angle -74.17^\circ \text{ pu}$$

which checks with the result in Example 8.9.

PROBLEMS

8.1. A sinusoidal voltage given by $v(t) = 390\sin(315t + \alpha)$ is suddenly applied to a series RL circuit. $R = 32\ \Omega$ and $L = 0.4$ H.
(a) The switch is closed at such a time as to permit no transient current. What value of α corresponds to this instant of closing the switch? Obtain the instantaneous expression for $i(t)$. Use *MATLAB* to plot $i(t)$ up to 80 ms in steps of 0.01 ms.
(b) The switch is closed at such a time as to permit maximum transient current. What value of α corresponds to this instant of closing the switch? Obtain the instantaneous expression for $i(t)$. Use *MATLAB* to plot $i(t)$ up to 80 ms in steps of 0.01 ms.
(c) What is the maximum value of current in part (b) and at what time does this occur after the switch is closed?

8.2. Consider the synchronous generator in Example 8.2. A three-phase short circuit is applied at the instant when the rotor direct axis position is at $\delta = 30°$. Use **ode45** to simulate (8.36), and obtain and plot the transient waveforms for the current in each phase and the field current.

8.3. Consider the synchronous generator in Example 8.2. A line-to-line short circuit occurs between phases b and c at the instant when the rotor direct axis position is at $\delta = 30°$. Use **ode45** to simulate (8.46), and obtain the transient waveforms for the current in phase b and the field current.

8.4. Consider a line-to-ground short circuit between phase a and ground in a synchronous generator. Apply the short circuit conditions

$$v_a = 0$$
$$i_b = i_c = 0$$

to the voltage equation of the synchronous machine given by (8.4). Substitute for all the flux linkages in terms of the inductances given by (8.9)–(8.13) and verify Equation (8.47).

8.5. Consider the synchronous generator in Example 8.2. A line-to-ground short circuit occurs between phase a and ground at the instant when the rotor direct axis position is at $\delta = 30°$. Use **ode45** to simulate (8.47), and obtain the transient waveforms for the current in phase a and the field current.

8.6. A three-phase, 60-Hz synchronous machine is driven at constant synchronous speed by a prime mover. The armature windings are initially open-circuited and field voltage is adjusted so that the armature terminal voltage is at the rated value (i.e., 1.0 per unit). The machine has the following per unit reactances and time constants.

$X''_d = 0.25$ pu $\qquad \tau''_d = 0.04$ sec
$X'_d = 0.45$ pu $\qquad \tau'_d = 1.4$ sec
$X_d = 1.50$ pu

(a) Determine the steady state, transient, and subtransient short circuit currents.
(b) Obtain and plot the fundamental-frequency waveform of the armature current for a three-phase short circuit at the terminals of the generator. Assume the short circuit is applied at the instant when the rotor direct axis is along the magnetic axis of phase a, i.e., $\delta = 0$.

8.7. For the machine in Problem 8.6, assume that a three-phase short circuit is applied at the instant when the rotor quadrature axis is along the magnetic axis of phase a, i.e., $\delta = \pi/2$ radians. Obtain and plot the asymmetrical waveform of the armature current for phase a. The armature short circuit time constant is $\tau_a = 0.3$ sec.

8.8. A three-phase, 60-Hz synchronous machine is driven at constant synchronous speed by a prime mover. The armature windings are initially open-circuited and field voltage is adjusted so that the armature terminal voltage is at the rated value (i.e., 1.0 per unit). The generator is suddenly subjected to a symmetrical three-phase short circuit at the instant when direct axis is along the magnetic axis of phase a, i.e., $\delta = 0$. An oscillogram of the short-circuited current is obtained. The peak values at the first two cycles, at the 20th and 21st cycles, and the steady value after a long time were recorded as tabulated in the following table.

I_{max}, pu	5.4016	4.6037	···	2.6930	2.6720	···	0.9445
Time, sec	0.0042	0.0208	···	0.3208	0.3375	···	10.004

Determine the transient and the subtransient reactances and time constants.

8.9. A 100-MVA, three-phase, 60-Hz generator driven at constant speed has the following per unit reactances and time constants

$X''_d = 0.20$ pu $\qquad \tau''_d = 0.04$ sec
$X'_d = 0.30$ pu $\qquad \tau'_d = 1.0$ sec
$X_d = 1.20$ pu $\qquad \tau_a = 0.25$ sec

The armature windings are initially open-circuited and field voltage is adjusted so that the armature terminal voltage is at the rated value (i.e., 1.0

per unit). The generator is suddenly subjected to a symmetrical three-phase short circuit at the instant when $\delta = \pi/2$. Obtain and plot the asymmetrical waveform of the armature current for phase a. Determine

(a) The rms value of the ac component in phase a at $t = 0$.
(b) The dc component of the current in phase a at $t = 0$.
(c) The rms value of the asymmetrical current in phase a at $t = 0$.

8.10. A 100-MVA, 20-kV, 60-Hz three-phase synchronous generator is connected to a 100-MVA, 20/400 kV three-phase transformer. The machine has the following per unit reactances and time constants.

$$\begin{aligned} X''_d &= 0.15 \text{ pu} & \tau''_d &= 0.035 \text{ sec} \\ X'_d &= 0.25 \text{ pu} & \tau'_d &= 0.50 \text{ sec} \\ X_d &= 1.25 \text{ pu} & \tau_a &= 0.3 \text{ sec} \end{aligned}$$

The transformer reactance is 0.25 per unit. The generator is operating at the rated voltage and no-load when a three-phase short circuit occurs at the secondary terminals of the transformer.
(a) Find the subtransient, transient, and the steady state short circuit currents in per unit and actual amperes on both sides of the transformer.
(b) What is the maximum asymmetrical rms current (ac plus dc) at the beginning of the short circuit?
(c) Obtain and plot the instantaneous expression for the short circuit current including the dc component. Assume $\delta = \pi/2$ radians.

8.11. In Problem 8.10, a three-phase load of 80-MVA, 0.8 power factor lagging is connected to the transformer secondary side. The line-to-line voltage at the load terminals is 400 kV. A three-phase short circuit occurs at the load terminals. Find the generator transient current including the load current.

8.12. A 100-MVA, 20-kV synchronous generator is connected through a transmission line to a 100-MVA, 20-kV synchronous motor. The per unit transient reactances of the generator and motor are 0.25 and 0.20, respectively. The line reactance on the base of 100 MVA is 0.1 per unit. The motor is taking 50 MW at 0.8 power factor leading at a terminal voltage of 20 kV. A three-phase short circuit occurs at the generator terminals. Determine the transient currents in each of the two machines and in the short circuit.

CHAPTER

9

BALANCED FAULT

9.1 INTRODUCTION

Fault studies form an important part of power system analysis. The problem consists of determining bus voltages and line currents during various types of faults. Faults on power systems are divided into *three-phase balanced faults* and *unbalanced faults*. Different types of unbalanced faults are *single line-to-ground fault*, *line-to-line fault*, and *double line-to-ground fault*, which are dealt with in Chapter 10. The information gained from fault studies are used for proper relay setting and coordination. The three-phase balanced fault information is used to select and set phase relays, while the line-to-ground fault is used for ground relays. Fault studies are also used to obtain the rating of the protective switchgears.

The magnitude of the fault currents depends on the internal impedance of the generators plus the impedance of the intervening circuit. We have seen in Chapter 8 that the reactance of a generator under short circuit condition is not constant. For the purpose of fault studies, the generator behavior can be divided into three periods: the *subtransient period*, lasting only for the first few cycles; the *transient period*, covering a relatively longer time; and, finally, the *steady state period*. In this chapter, three-phase balanced faults are discussed. The bus impedance matrix by the *building algorithm* is formulated and is employed for the systematic computation of bus voltages and line currents during the fault. Two functions are developed for the formation of the bus impedance matrix. These function are **Zbus = zbuild(zdata)** and **Zbus = zbuildpi(linedata, gendata, yload)**. The latter one is compatible with power flow input/output files. A program named **symfault** is developed for systematic computation of three-phase balanced faults for a large interconnected power system.

9.2 BALANCED THREE-PHASE FAULT

This type of fault is defined as the simultaneous short circuit across all three phases. It occurs infrequently, but it is the most severe type of fault encountered. Because the network is balanced, it is solved on a per-phase basis. The other two phases carry identical currents except for the phase shift.

In Chapter 8 it was shown that the reactance of the synchronous generator under short-circuit conditions is a time-varying quantity, and for network analysis three reactances were defined. The subtransient reactance X''_d, for the first few cycles of the short circuit current, transient reactance X'_d, for the next (say) 30 cycles, and the synchronous reactance X_d, thereafter. Since the duration of the short circuit current depends on the time of operation of the protective system, it is not always easy to decide which reactance to use. Generally, the subtransient reactance is used for determining the interrupting capacity of the circuit breakers. In fault studies required for relay setting and coordination, transient reactance is used. Also, in typical transient stability studies, transient reactance is used.

A fault represents a structural network change equivalent with that caused by the addition of an impedance at the place of fault. If the fault impedance is zero, the fault is referred to as the *bolted fault* or the *solid fault*. The faulted network can be solved conveniently by the Thévenin's method. The procedure is demonstrated in the following example.

Example 9.1

The one-line diagram of a simple three-bus power system is shown in Figure 9.1. Each generator is represented by an emf behind the transient reactance. All impedances are expressed in per unit on a common 100 MVA base, and for simplicity, resistances are neglected. The following assumptions are made.

(*i*) Shunt capacitances are neglected and the system is considered on no-load.

(*ii*) All generators are running at their rated voltage and rated frequency with their emfs in phase.

Determine the fault current, the bus voltages, and the line currents during the fault when a balanced three-phase fault with a fault impedance $Z_f = 0.16$ per unit occurs on

(a) Bus 3.
(b) Bus 2.
(c) Bus 1.

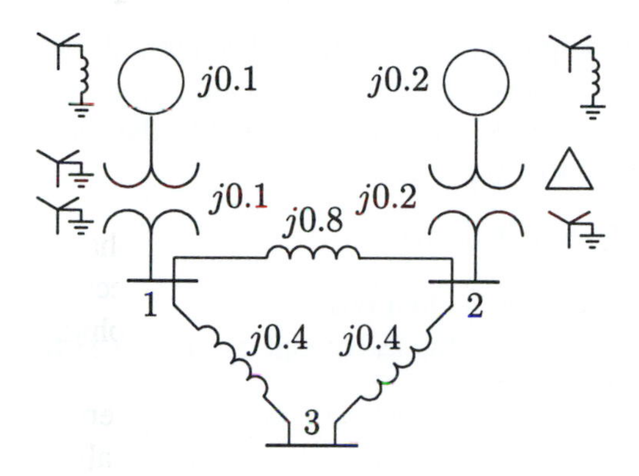

FIGURE 9.1
The impedance diagram of a simple power system.

The fault is simulated by switching on an impedance Z_f at bus 3 as shown in Figure 9.2(a). Thévenin's theorem states that the changes in the network voltage caused by the added branch (the fault impedance) shown in Figure 9.2(a) is equivalent to those caused by the added voltage $V_3(0)$ with all other sources short-circuited as shown in Figure 9.2(b).

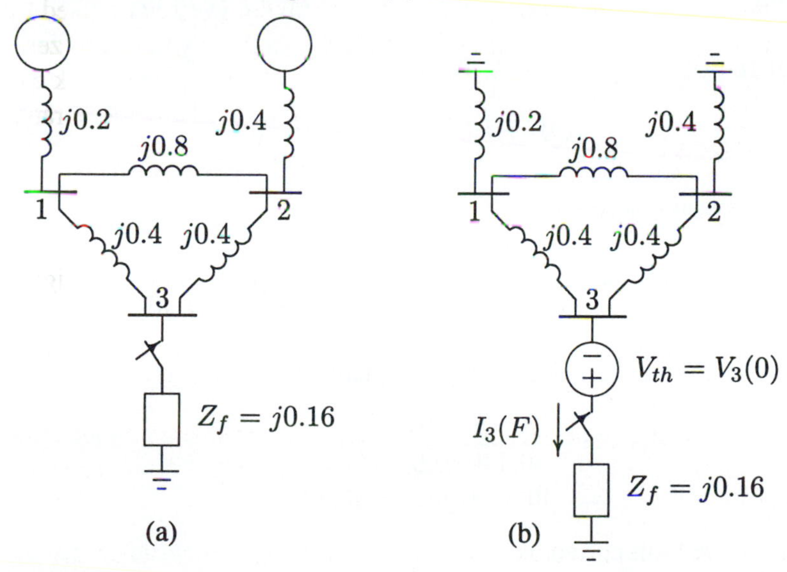

FIGURE 9.2
(a) The impedance network for fault at bus 3. (b) Thévenin's equivalent network.

(a) From 9.2(b), the fault current at bus 3 is

$$I_3(F) = \frac{V_3(0)}{Z_{33} + Z_f}$$

where $V_3(0)$ is the Thévenin's voltage or the prefault bus voltage. The prefault bus voltage can be obtained from the results of the power flow solution. In this example, since the loads are neglected and generator's emfs are assumed equal to the rated value, all the prefault bus voltages are equal to 1.0 per unit, i.e.,

$$V_1(0) = V_2(0) = V_3(0) = 1.0 \quad \text{pu}$$

Z_{33} is the Thévenin's impedance viewed from the faulted bus.

To find the Thévenin's impedance, we convert the Δ formed by buses 123 to an equivalent Y as shown in Figure 9.3(a).

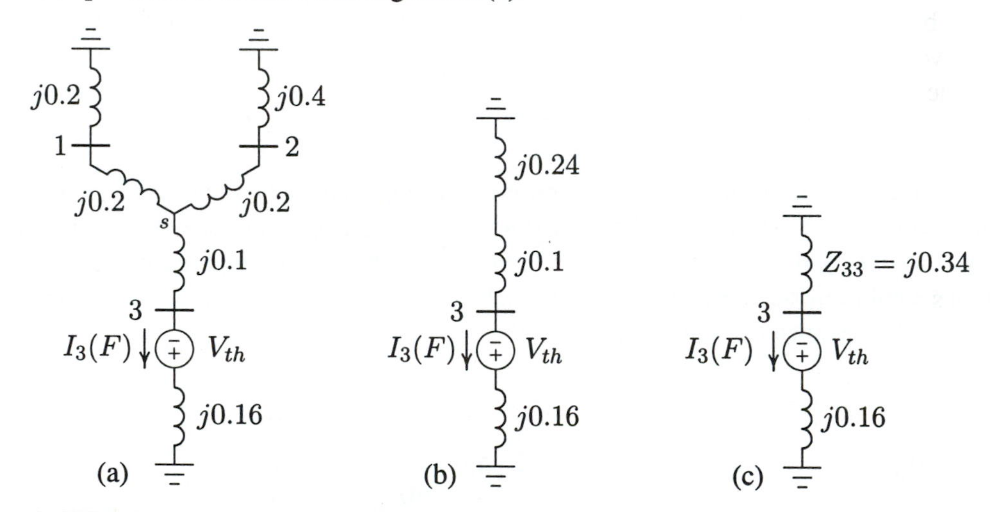

FIGURE 9.3
Reduction of Thévenin's equivalent network.

$$Z_{1s} = Z_{2s} = \frac{(j0.4)(j0.8)}{j1.6} = j0.2 \qquad Z_{3s} = \frac{(j0.4)(j0.4)}{j1.6} = j0.1$$

Combining the parallel branches, Thévenin's impedance is

$$\begin{aligned} Z_{33} &= \frac{(j0.4)(j0.6)}{j0.4 + j0.6} + j0.1 \\ &= j0.24 + j0.1 = j0.34 \end{aligned}$$

From Figure 9.3(c), the fault current is

$$I_3(F) = \frac{V_3(F)}{Z_{33} + Z_f} = \frac{1.0}{j0.34 + j0.16} = -j2.0 \quad \text{pu}$$

With reference to Figure 9.3(a), the current divisions between the two generators are

$$I_{G1} = \frac{j0.6}{j0.4 + j0.6} I_3(F) = -j1.2 \quad \text{pu}$$
$$I_{G2} = \frac{j0.4}{j0.4 + j0.6} I_3(F) = -j0.8 \quad \text{pu}$$

For the bus voltage changes from Figure 9.3(b), we get

$$\Delta V_1 = 0 - (j0.2)(-j1.2) \quad = -0.24 \quad \text{pu}$$
$$\Delta V_2 = 0 - (j0.4)(-j0.8) \quad = -0.32 \quad \text{pu}$$
$$\Delta V_3 = (j0.16)(-j2) - 1.0 = -0.68 \quad \text{pu}$$

The bus voltages during the fault are obtained by superposition of the prefault bus voltages and the changes in the bus voltages caused by the equivalent emf connected to the faulted bus, as shown in Figure 9.2(b), i.e.,

$$V_1(F) = V_1(0) + \Delta V_1 = 1.0 - 0.24 = 0.76 \quad \text{pu}$$
$$V_2(F) = V_2(0) + \Delta V_2 = 1.0 - 0.32 = 0.68 \quad \text{pu}$$
$$V_3(F) = V_3(0) + \Delta V_3 = 1.0 - 0.68 = 0.32 \quad \text{pu}$$

The short circuit-currents in the lines are

$$I_{12}(F) = \frac{V_1(F) - V_2(F)}{z_{12}} = \frac{0.76 - 0.68}{j0.8} = -j0.1 \quad \text{pu}$$
$$I_{13}(F) = \frac{V_1(F) - V_3(F)}{z_{13}} = \frac{0.76 - 0.32}{j0.4} = -j1.1 \quad \text{pu}$$
$$I_{23}(F) = \frac{V_2(F) - V_3(F)}{z_{23}} = \frac{0.68 - 0.32}{j0.4} = -j0.9 \quad \text{pu}$$

(b) The fault with impedance Z_f at bus 2 is depicted in Figure 9.4(a), and its Thévenin's equivalent circuit is shown in Figure 9.4(b). To find the Thévenin's impedance, we combine the parallel branches in Figure 9.4(b). Also, combining parallel branches from ground to bus 2 in Figure 9.5(a), results in

$$Z_{22} = \frac{(j0.6)(j0.4)}{j0.6 + j0.4} = j0.24$$

From Figure 9.5(b), the fault current is

$$I_2(F) = \frac{V_2(0)}{Z_{22} + Z_f} = \frac{1.0}{j0.24 + j0.16} = -j2.5 \quad \text{pu}$$

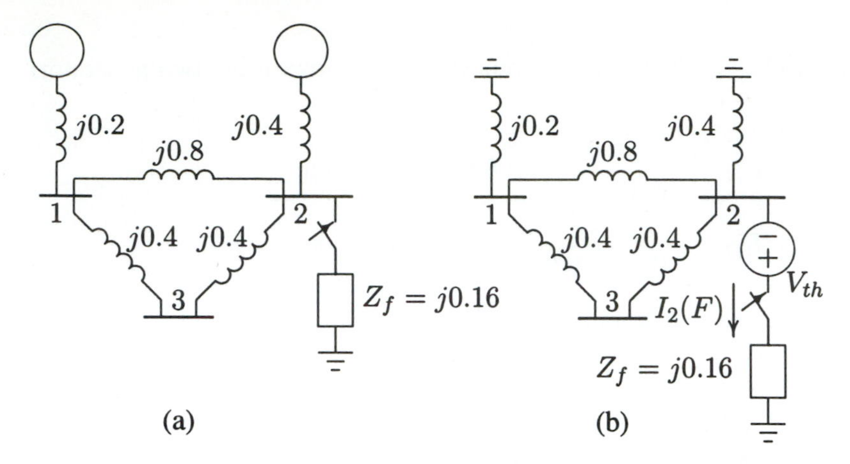

FIGURE 9.4
(a) The impedance network for fault at bus 2. (b) Thévenin's equivalent network.

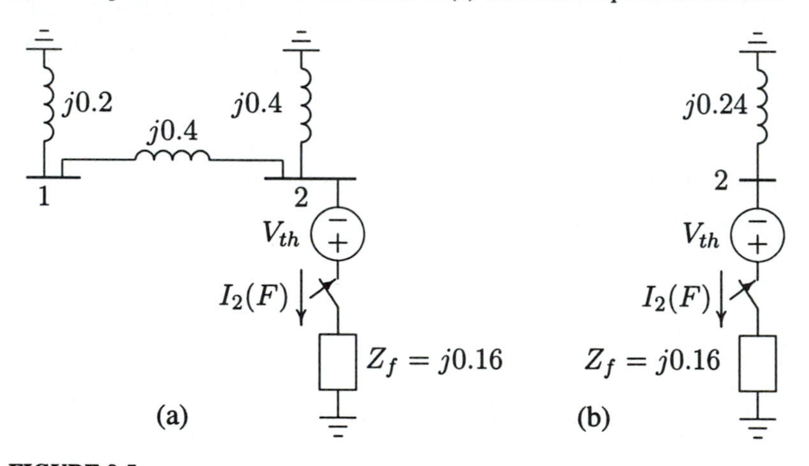

FIGURE 9.5
Reduction of Thévenin's equivalent network.

With reference to Figure 9.5(a), the current divisions between the generators are

$$I_{G1} = \frac{j0.4}{j0.4 + j0.6} I_2(F) = -j1.0 \quad \text{pu}$$

$$I_{G2} = \frac{j0.6}{j0.4 + j0.6} I_2(F) = -j1.5 \quad \text{pu}$$

For the bus voltage changes from Figure 9.4(a), we get

$$\Delta V_1 = 0 - (j0.2)(-j1.0) = -0.2 \quad \text{pu}$$

$$\Delta V_2 = 0 - (j0.4)(-j1.5) = -0.6 \quad \text{pu}$$

$$\Delta V_3 = -0.2 - (j0.4)(\frac{-j1.0}{2}) = -0.4 \quad \text{pu}$$

The bus voltages during the fault are obtained by superposition of the prefault bus voltages and the changes in the bus voltages caused by the equivalent emf connected to the faulted bus, as shown in Figure 9.4(b), i.e.,

$$V_1(F) = V_1(0) + \Delta V_1 = 1.0 - 0.2 = 0.8 \quad \text{pu}$$
$$V_2(F) = V_2(0) + \Delta V_2 = 1.0 - 0.6 = 0.4 \quad \text{pu}$$
$$V_3(F) = V_3(0) + \Delta V_3 = 1.0 - 0.4 = 0.6 \quad \text{pu}$$

The short circuit-currents in the lines are

$$I_{12}(F) = \frac{V_1(F) - V_2(F)}{z_{12}} = \frac{0.8 - 0.4}{j0.8} = -j0.5 \quad \text{pu}$$
$$I_{13}(F) = \frac{V_1(F) - V_3(F)}{z_{13}} = \frac{0.8 - 0.6}{j0.4} = -j0.5 \quad \text{pu}$$
$$I_{32}(F) = \frac{V_3(F) - V_3(F)}{z_{32}} = \frac{0.6 - 0.4}{j0.4} = -j0.5 \quad \text{pu}$$

(c) The fault with impedance Z_f at bus 1 is depicted in Figure 9.6(a), and its Thévenin's equivalent circuit is shown in Figure 9.6(b).

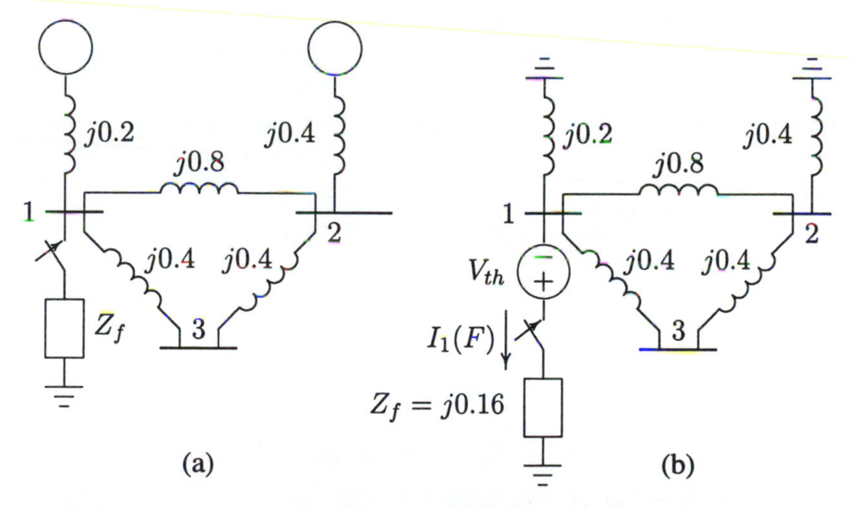

FIGURE 9.6
(a) The impedance network for fault at bus 1. (b) Thévenin's equivalent network.

To find the Thévenin's impedance, we combine the parallel branches in Figure 9.6(b). Also, combining parallel branches from ground to bus 1 in Figure 9.7(a),

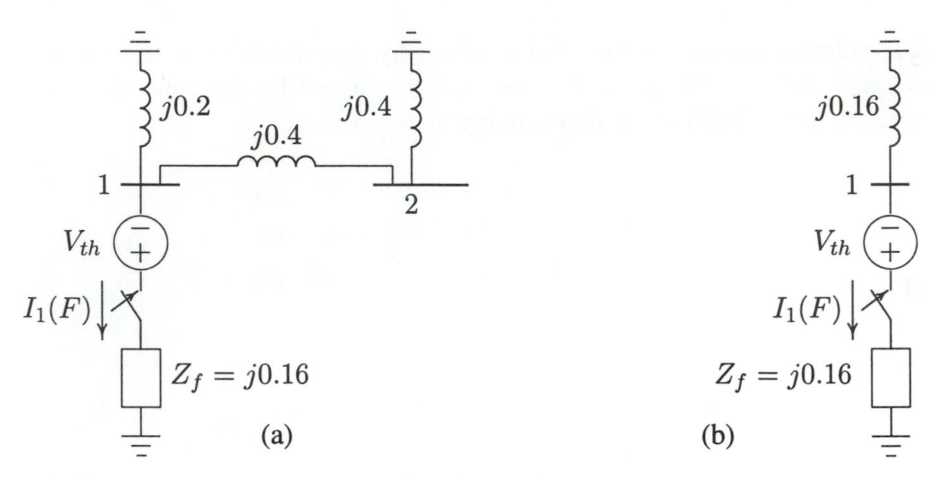

FIGURE 9.7
Reduction of Thévenin's equivalent network.

results in

$$Z_{11} = \frac{(j0.2)(j0.8)}{j0.2 + j0.8} = j0.16$$

From Figure 9.7(b), the fault current is

$$I_1(F) = \frac{V_1(0)}{Z_{11} + Z_f} = \frac{1.0}{j0.16 + j0.16} = -j3.125 \quad \text{pu}$$

With reference to Figure 9.7(a), the current divisions between the two generators are

$$I_{G1} = \frac{j0.8}{j0.2 + j0.8} I_2(F) = -j2.50 \quad \text{pu}$$
$$I_{G2} = \frac{j0.2}{j0.2 + j0.8} I_2(F) = -j0.625 \quad \text{pu}$$

For the bus voltage changes from Figure 9.6(b), we get

$$\Delta V_1 = 0 - (j0.2)(-j2.5) = -0.50 \quad \text{pu}$$
$$\Delta V_2 = 0 - (j0.4)(-j0.625) = -0.25 \quad \text{pu}$$
$$\Delta V_3 = -0.5 + (j0.4)(\frac{-j0.625}{2}) = -0.375 \quad \text{pu}$$

Bus voltages during the fault are obtained by superposition of the prefault bus voltages and the changes in the bus voltages caused by the equivalent emf connected to the faulted bus, as shown in Figure 9.6(b), i.e.,

$$\begin{aligned} V_1(F) &= V_1(0) + \Delta V_1 = 1.0 - 0.50 = 0.50 \quad \text{pu} \\ V_2(F) &= V_2(0) + \Delta V_2 = 1.0 - 0.25 = 0.75 \quad \text{pu} \\ V_3(F) &= V_3(0) + \Delta V_3 = 1.0 - 0.375 = 0.625 \quad \text{pu} \end{aligned}$$

The short-circuit currents in the lines are

$$\begin{aligned} I_{21}(F) &= \frac{V_2(F) - V_1(F)}{z_{21}} = \frac{0.75 - 0.5}{j0.8} = -j0.3125 \quad \text{pu} \\ I_{31}(F) &= \frac{V_3(F) - V_1(F)}{z_{31}} = \frac{0.625 - 0.5}{j0.4} = -j0.3125 \quad \text{pu} \\ I_{23}(F) &= \frac{V_2(F) - V_3(F)}{z_{23}} = \frac{0.75 - 0.625}{j0.4} = -j0.3125 \quad \text{pu} \end{aligned}$$

In the above example the load currents were neglected and all prefault bus voltages were assumed to be equal to 1.0 per unit. For more accurate calculation, the prefault bus voltages can be obtained from the power flow solution. As we have seen in Chapter 6, in a power system, loads are specified and the load currents are unknown. One way to include the effects of load currents in the fault analysis is to express the loads by a constant impedance evaluated at the prefault bus voltages. This is a very good approximation which results in linear nodal equations. The procedure is summarized in the following steps.

- The prefault bus voltages are obtained from the results of the power flow solution.

- In order to preserve the linearity feature of the network, loads are converted to constant admittances using the prefault bus voltages.

- The faulted network is reduced into a Thévenin's equivalent circuit as viewed from the faulted bus. Applying Thévenin's theorem, changes in the bus voltages are obtained.

- Bus voltages during the fault are obtained by superposition of the prefault bus voltages and the changes in the bus voltages computed in the previous step.

- The currents during the fault in all branches of the network are then obtained.

9.3 SHORT-CIRCUIT CAPACITY (SCC)

The short-circuit capacity at a bus is a common measure of the strength of a bus. The short-circuit capacity or the short-circuit MVA at bus k is defined as the product of the magnitudes of the rated bus voltage and the fault current. The short-circuit MVA is used for determining the dimension of a bus bar, and the *interrupting* capacity of a circuit breaker. The interrupting capacity is only one of many ratings of a circuit breaker and should not be confused with the *momentary duty* of the breaker described in (8.63).

Based on the above definition, the short-circuit capacity or the short-circuit MVA at bus k is given by

$$SCC = \sqrt{3}\, V_{Lk} I_k(F) \times 10^{-3} \quad \text{MVA} \tag{9.1}$$

where the line-to-line voltage V_{Lk} is expressed in kilovolts and $I_k(F)$ is expressed in amperes. The symmetrical three-phase fault current in per unit is given by

$$I_k(F)_{pu} = \frac{V_k(0)}{X_{kk}} \tag{9.2}$$

where $V_k(0)$ is the per unit prefault bus voltage, and X_{kk} is the per unit reactance to the point of fault. System resistance is neglected and only the inductive reactance of the system is allowed for. This gives minimum system impedance and maximum fault current and a pessimistic answer. The base current is given by

$$I_B = \frac{S_B \times 10^3}{\sqrt{3}\, V_B} \tag{9.3}$$

where S_B is the base MVA and V_B is the line-to-line base voltage in kilovolts. Thus, the fault current in amperes is

$$\begin{aligned} I_k(F) &= I_k(F)_{pu} I_B \\ &= \frac{V_k(0)}{X_{kk}} \frac{S_B \times 10^3}{\sqrt{3}\, V_B} \end{aligned} \tag{9.4}$$

Substituting for $I_k(F)$ from (9.4) into (9.1) results in

$$SCC = \frac{V_k(0) S_B}{X_{kk}} \frac{V_L}{V_B} \tag{9.5}$$

If the base voltage is equal to the rated voltage, i.e., $V_L = V_B$

$$SCC = \frac{V_k(0) S_B}{X_{kk}} \tag{9.6}$$

The prefault bus voltage is usually assumed to be 1.0 per unit, and we therefore obtain from (9.6) the following approximate formula for the short-circuit capacity or the short-circuit MVA.

$$SCC = \frac{S_B}{X_{kk}} \quad \text{MVA} \tag{9.7}$$

9.4 SYSTEMATIC FAULT ANALYSIS USING BUS IMPEDANCE MATRIX

The network reduction used in the preceding example is not efficient and is not applicable to large networks. In this section a more general fault circuit analysis using nodal method is obtained. We see that by utilizing the elements of the bus impedance matrix, the fault current as well as the bus voltages during fault are readily and easily calculated.

Consider a typical bus of an n-bus power system network as shown in Figure 9.8. The system is assumed to be operating under balanced condition and a per phase circuit model is used. Each machine is represented by a constant voltage source behind proper reactances which may be X_d'', X_d', or X_d. Transmission lines are represented by their equivalent π model and all impedances are expressed in per unit on a common MVA base. A balanced three-phase fault is to be applied at bus k through a fault impedance Z_f. The prefault bus voltages are obtained from the power flow solution and are represented by the column vector

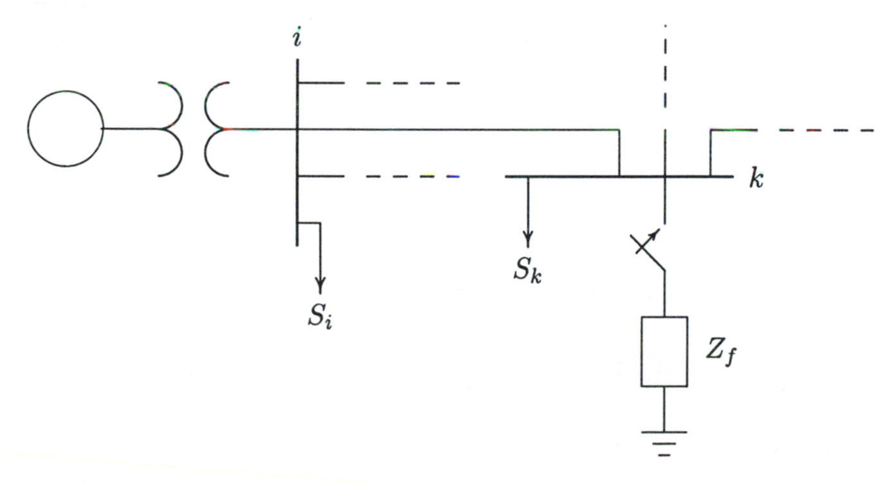

FIGURE 9.8
A typical bus of a power system.

$$\mathbf{V}_{bus}(0) = \begin{bmatrix} V_1(0) \\ \vdots \\ V_k(0) \\ \vdots \\ V_n(0) \end{bmatrix} \tag{9.8}$$

As already mentioned, short circuit currents are so much larger than the steady-state values that we may neglect the latter. However, a good approximation is to represent the bus load by a constant impedance evaluated at the prefault bus voltage, i.e.,

$$Z_{i\,L} = \frac{|V_i(0)|^2}{S_L^*} \tag{9.9}$$

The changes in the network voltage caused by the fault with impedance Z_f is equivalent to those caused by the added voltage $V_k(0)$ with all other sources short-circuited. Zeroing all voltage sources and representing all components and loads by their appropriate impedances, we obtain the Thévenin's circuit shown in Figure 9.9. The bus voltage changes caused by the fault in this circuit are represented by the column vector

$$\Delta\mathbf{V}_{bus} = \begin{bmatrix} \Delta V_1 \\ \vdots \\ \Delta V_k \\ \vdots \\ \Delta V_n \end{bmatrix} \tag{9.10}$$

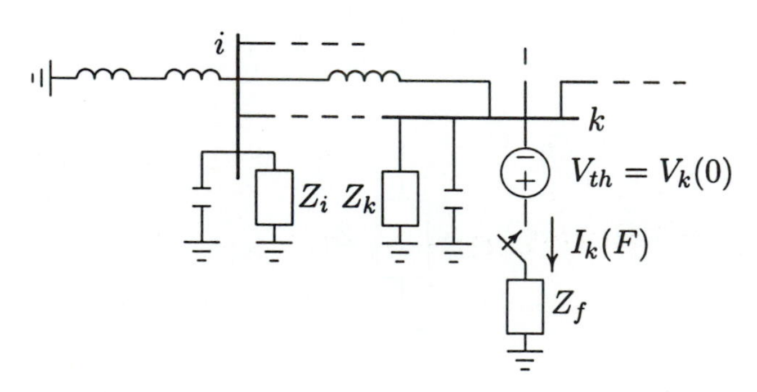

FIGURE 9.9
A typical bus of a power system.

From Thévenin's theorem bus voltages during the fault are obtained by superposition of the prefault bus voltages and the changes in the bus voltages given by

$$\mathbf{V}_{bus}(F) = \mathbf{V}_{bus}(0) + \mathbf{\Delta V}_{bus} \tag{9.11}$$

In Section 6.2, we obtained the node-voltage equation for an n-bus network. The injected bus currents are expressed in terms of the bus voltages (with bus 0 as reference), i.e.,

$$\mathbf{I}_{bus} = \mathbf{Y}_{bus}\mathbf{V}_{bus} \tag{9.12}$$

where $\mathbf{I}_{bus}$ is the bus current vector entering the bus and $\mathbf{Y}_{bus}$ is the bus admittance matrix. The diagonal element of each bus is the sum of admittances connected to it, i.e.,

$$Y_{ii} = \sum_{j=0}^{m} y_{ij} \qquad j \neq i \tag{9.13}$$

The off-diagonal element is equal to the negative of the admittance between the buses, i.e.,

$$Y_{ij} = Y_{ji} = -y_{ij} \tag{9.14}$$

where y_{ij} (lower case) is the actual admittance of the line i-j. For more details refer to Section 6.2.

In the Thévenin's circuit of Figure 9.9, current entering every bus is zero except at the faulted bus. Since the current at faulted bus is leaving the bus, it is taken as a negative current entering bus k. Thus the nodal equation applied to the Thévenin's circuit in Figure 9.9 becomes

$$\begin{bmatrix} 0 \\ \vdots \\ -I_k(F) \\ \vdots \\ 0 \end{bmatrix} = \begin{bmatrix} y_{11} & \cdots & y_{1k} & \cdots & y_{1n} \\ \vdots & \vdots & \vdots & \vdots & \vdots \\ y_{k1} & \cdots & y_{kk} & \cdots & y_{kn} \\ \vdots & \vdots & \vdots & \vdots & \vdots \\ y_{n1} & \cdots & y_{nk} & \cdots & y_{nn} \end{bmatrix} \begin{bmatrix} \Delta V_1 \\ \vdots \\ \Delta V_k \\ \vdots \\ \Delta V_n \end{bmatrix} \tag{9.15}$$

or

$$\mathbf{I}_{bus}(F) = \mathbf{Y}_{bus}\mathbf{\Delta V}_{bus} \tag{9.16}$$

Solving for $\mathbf{\Delta V}_{bus}$, we have

$$\mathbf{\Delta V}_{bus} = \mathbf{Z}_{bus}\mathbf{I}_{bus}(F) \tag{9.17}$$

where $\mathbf{Z}_{bus} = \mathbf{Y}_{bus}^{-1}$ is known as the *bus impedance matrix*. Substituting (9.17) into (9.11), the bus voltage vector during the fault becomes

$$\mathbf{V}_{bus\,(F)} = \mathbf{V}_{bus}(0) + \mathbf{Z}_{bus}\mathbf{I}_{bus}(F) \tag{9.18}$$

Writing the above matrix equation in terms of its elements, we have

$$\begin{bmatrix} V_1(F) \\ \vdots \\ V_k(F) \\ \vdots \\ V_n(F) \end{bmatrix} = \begin{bmatrix} V_1(0) \\ \vdots \\ V_k(0) \\ \vdots \\ V_n(0) \end{bmatrix} + \begin{bmatrix} Z_{11} & \cdots & Z_{1k} & \cdots & Z_{1n} \\ \vdots & \vdots & \vdots & \vdots & \vdots \\ Z_{k1} & \cdots & Z_{kk} & \cdots & Z_{kn} \\ \vdots & \vdots & \vdots & \vdots & \vdots \\ Z_{n1} & \cdots & Z_{nk} & \cdots & Z_{nn} \end{bmatrix} \begin{bmatrix} 0 \\ \vdots \\ -I_k(F) \\ \vdots \\ 0 \end{bmatrix} \tag{9.19}$$

Since we have only one single nonzero element in the current vector, the kth equation in (9.19) becomes

$$V_k(F) = V_k(0) - Z_{kk}I_k(F) \tag{9.20}$$

Also from the Thévenin's circuit shown in Figure 9.9, we have

$$V_k(F) = Z_f I_k(F) \tag{9.21}$$

For bolted fault, $Z_f = 0$ and $V_k(F) = 0$. Substituting for $V_k(F)$ from (9.21) into (9.20) and solving for the fault current, we get

$$I_k(F) = \frac{V_k(0)}{Z_{kk} + Z_f} \tag{9.22}$$

Thus for a fault at bus k we need only the Z_{kk} element of the bus impedance matrix. This element is indeed the Thévenin's impedance as viewed from the faulted bus. Also, writing the ith equation in (9.19) in terms of its element, we have

$$V_i(F) = V_i(0) - Z_{ik}I_k(F) \tag{9.23}$$

Substituting for $I_k(F)$, bus voltage during the fault at bus i becomes

$$V_i(F) = V_i(0) - \frac{Z_{ik}}{Z_{kk} + Z_f}V_k(0) \tag{9.24}$$

With the knowledge of bus voltages during the fault, we can calculate the fault current in all the lines. For the line connecting buses i and j with impedance z_{ij}, the short circuit current in this line (defined positive in the direction $i \rightarrow j$) is

$$I_{ij}(F) = \frac{V_i(F) - V_j(F)}{z_{ij}} \tag{9.25}$$

We note that with the knowledge of the bus impedance matrix, the fault current and bus voltages during the fault are readily obtained for any faulted bus in the network. This method is very simple and practical. Thus, all fault calculations are formulated in the bus frame of reference using bus impedance matrix Z_{bus}.

One way to find $\mathbf{Z}_{bus}$ is to formulate Y_{bus} matrix for the system and then find its inverse. The matrix inversion for a large power system with a large number of buses is not feasible. A computationally attractive and efficient method for finding Z_{bus} matrix is "building" or "assembling" the impedance matrix by adding one network element at a time. In effect, this is an indirect matrix inversion of the bus admittance matrix. The algorithm for building the bus impedance matrix is described in the next section.

Example 9.2

A three-phase fault with a fault impedance $Z_f = j0.16$ per unit occurs at bus 3 in the network of Example 9.1. Using the bus impedance matrix method, compute the fault current, the bus voltages, and the line currents during the fault.

In this example the bus impedance matrix is obtained by finding the inverse of the bus admittance matrix. In the next section, we describe an efficient method of finding the bus impedance matrix by the method of building algorithm.

To find the bus admittance matrix, the Thévenin's circuit in Figure 9.2(b) is redrawn with impedances converted to admittances as shown in Figure 9.10. The ith diagonal element of the bus admittance matrix is the sum of all admittances connected to bus i, and the ijth off-diagonal element is the negative of the admittance between buses i and j. Referring to Figure 9.10, the bus admittance matrix by inspection is

$$\mathbf{Y}_{bus} = \begin{bmatrix} -j8.75 & j1.25 & j2.5 \\ j1.25 & -j6.25 & j2.5 \\ j2.5 & j2.5 & -j5.0 \end{bmatrix}$$

Using *MATLAB* inverse function **inv**, the bus impedance matrix is obtained

$$\mathbf{Z}_{bus} = \begin{bmatrix} j0.16 & j0.08 & j0.12 \\ j0.08 & j0.24 & j0.16 \\ j0.12 & j0.16 & j0.34 \end{bmatrix}$$

From (9.22), for a fault at bus 3 with fault impedance $Z_f = j0.16$ per unit, the fault current is

$$I_3(F) = \frac{V_3(0)}{Z_{33} + Z_f} = \frac{1.0}{j0.34 + j0.16} = -j2.0 \quad \text{pu}$$

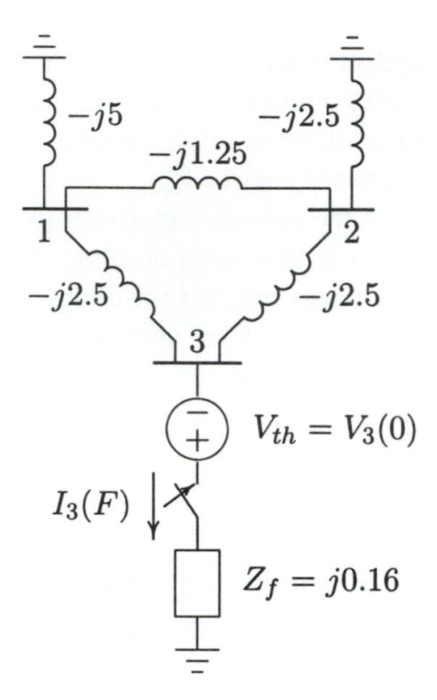

FIGURE 9.10
The admittance diagram for system of Figure 9.2 (b).

From (9.23), bus voltages during the fault are

$$\begin{aligned}
V_1(F) &= V_1(0) - Z_{13}I_3(F) = 1.0 - (j0.12)(-j2.0) = 0.76 \quad \text{pu} \\
V_2(F) &= V_2(0) - Z_{23}I_3(F) = 1.0 - (j0.16)(-j2.0) = 0.68 \quad \text{pu} \\
V_3(F) &= V_3(0) - Z_{33}I_3(F) = 1.0 - (j0.34)(-j2.0) = 0.32 \quad \text{pu}
\end{aligned}$$

From (9.25), the short circuit currents in the lines are

$$\begin{aligned}
I_{12}(F) &= \frac{V_1(F) - V_2(F)}{z_{12}} = \frac{0.76 - 0.68}{j0.8} = -j0.1 \quad \text{pu} \\
I_{13}(F) &= \frac{V_1(F) - V_3(F)}{z_{13}} = \frac{0.76 - 0.32}{j0.4} = -j1.1 \quad \text{pu} \\
I_{23}(F) &= \frac{V_2(F) - V_3(F)}{z_{23}} = \frac{0.68 - 0.32}{j0.4} = -j0.9 \quad \text{pu}
\end{aligned}$$

The results are exactly the same as the values found in Example 9.1(a). The reader is encouraged to repeat the above calculations for fault at buses 2 and 1, and compare the results with those obtained from parts (b) and (c) in Example 9.1.

Note that the values of the diagonal elements in the bus impedance matrix are the same as the Thévenin's impedances found in Example 9.1, thus eliminating the repeated need for network reduction for each fault location. Furthermore, the off-diagonal elements are utilized in (9.24) to obtain bus voltages during the fault. Therefore, the bus impedance matrix method is an indispensable tool for fault studies.

9.5 ALGORITHM FOR FORMATION OF THE BUS IMPEDANCE MATRIX

Before we present the building algorithm for the bus impedance matrix, a few definitions from the discipline of the graph theory are introduced. The *graph* of a network describes the geometrical structure of the network. The graph consists of redrawing the network, with a line representing each element of the network. The graph of the network for Figure 9.2(a) before the fault application is shown in Figure 9.11(a). The buses are represented by *nodes* or *vertices* and impedances by

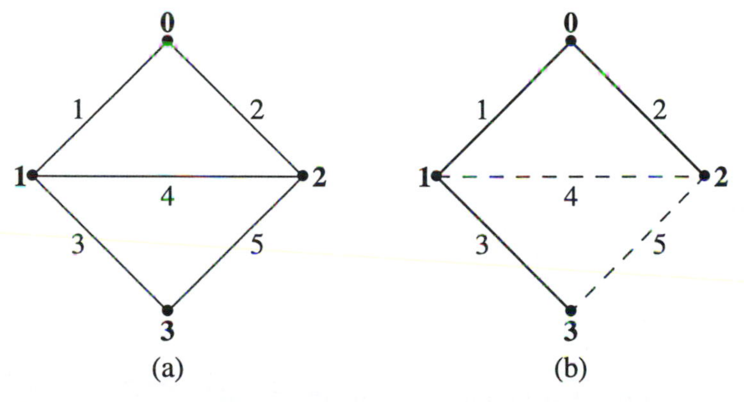

FIGURE 9.11
Graph, a selected tree, and a cotree for the network of Figure 9.2(b).

line segments called *elements* or *edges*. A *tree* of a connected graph is a connected subgraph connecting all the nodes without forming a loop. The elements of a tree are called *branches*. In general, a graph contains multiple trees. The number of branches in any selected tree denoted by b is always one less than the nodes, i.e.,

$$b = n - 1 \tag{9.26}$$

where n is the number of nodes including the reference node 0. Once a tree for a graph has been defined, the remaining elements are referred to as *links*. The collection of links is called a *cotree*. If e is the total number of elements in a graph, the number of links in a cotree is

$$l = e - b = e - n + 1 \tag{9.27}$$

A loop that contains one link is called a *basic loop*. The number of basic loops is unique; it equals the number of links and is the number of independent loop equations. A *cut set* is a minimal set of branches that, when cut, divides the graph into two connected subgraphs. A *fundamental cut set* is a cut set that contains only one branch. The number of fundamental cut sets is unique; it equals the number of branches and is the number of independent node equations. Figure 9.11(b) shows a tree of a graph with the tree branches highlighted by heavy lines and the cotree links by dashed lines.

The bus impedance matrix can be built up starting with a single element and the process is continued until all nodes and elements are included. Let us assume that $\mathbf{Z}_{bus}$ matrix exists for a partial network having m buses and a reference bus 0 as shown in Figure 9.12.

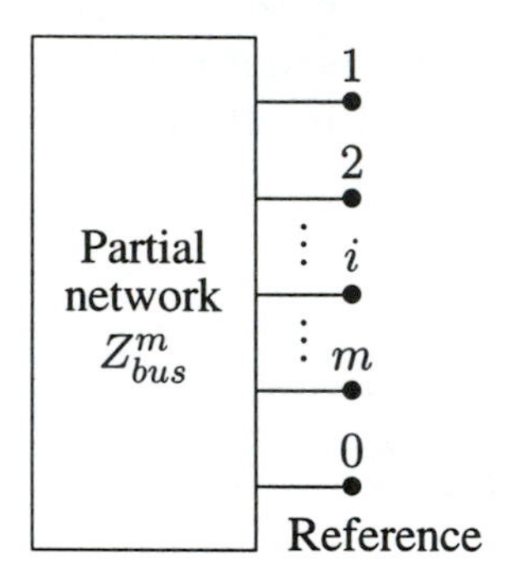

FIGURE 9.12
Partial network.

The corresponding network equation for this partial network is

$$\mathbf{V}_{bus} = \mathbf{Z}_{bus}\mathbf{I}_{bus} \tag{9.28}$$

For an n-bus system, m buses are included in the network and $\mathbf{Z}_{bus}$ is of order $m \times m$. We shall add one element at a time from the remaining portion of the network until all elements are included. The added element may be a branch or a link described as follows.

ADDITION OF A BRANCH

When the added element is a branch, a new bus is added to the partial network creating a new row and a column, and the new bus impedance matrix is of order $(m+1) \times (m+1)$. Let us add a branch with impedance z_{pq} from an existing bus p to a new bus q as shown in Figure 9.13(a). The network equation becomes

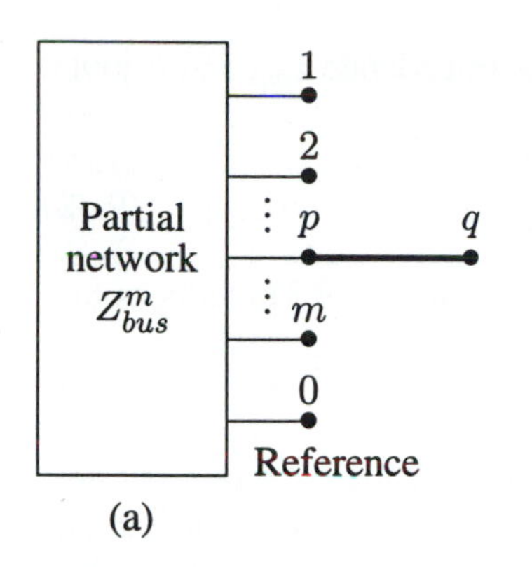

(a)

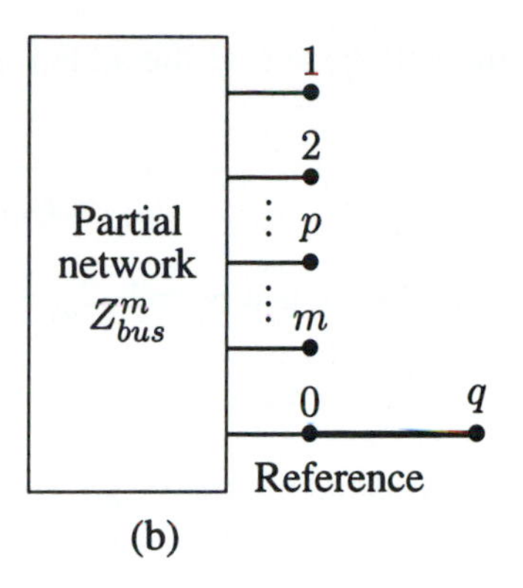

(b)

FIGURE 9.13
Addition of a branch p-q.

$$\begin{bmatrix} V_1 \\ V_2 \\ \vdots \\ V_p \\ \vdots \\ V_m \\ V_q \end{bmatrix} = \begin{bmatrix} Z_{11} & Z_{12} & \cdots & Z_{1p} & \cdots & Z_{1m} & Z_{1q} \\ Z_{21} & Z_{22} & \cdots & Z_{2p} & \cdots & Z_{2m} & Z_{2q} \\ \vdots & \vdots & \vdots & \vdots & \vdots & \vdots & \vdots \\ Z_{p1} & Z_{p2} & \cdots & Z_{pp} & \cdots & Z_{pm} & Z_{pq} \\ \vdots & \vdots & \vdots & \vdots & \vdots & \vdots & \vdots \\ Z_{m1} & Z_{m2} & \cdots & Z_{mp} & \cdots & Z_{mm} & Z_{mq} \\ Z_{q1} & Z_{q2} & \cdots & Z_{qp} & \cdots & Z_{qm} & Z_{qq} \end{bmatrix} \begin{bmatrix} I_1 \\ I_2 \\ \vdots \\ I_p \\ \vdots \\ I_m \\ I_q \end{bmatrix} \tag{9.29}$$

The addition of branch does not affect the original matrix, but requires the calculation of the elements in the q row and column. Since the elements of the power system network are linear and bilateral, $Z_{qi} = Z_{iq}$, for $q = 1, \ldots, m$.

First, let us compute the elements Z_{qi} for $i = 1, \ldots, m$ and $i \neq q$ (i.e., excluding diagonal element Z_{qq}). To calculate these elements we will apply a current source of 1 per unit at the ith bus, i.e., $I_i = 1$ pu, and keep remaining buses open-circuited, i.e., $I_k = 0$, $k = 1, \ldots, m$ and $k \neq i$. From (9.29), we get

$$\begin{aligned} V_1 &= Z_{1i} \\ V_2 &= Z_{2i} \\ &\vdots \\ V_p &= Z_{pi} \\ &\vdots \\ V_m &= Z_{mi} \\ V_q &= Z_{qi} \end{aligned} \tag{9.30}$$

From Figure 9.13(a)

$$V_q = V_p - v_{pq} \tag{9.31}$$

where v_{pq} is the voltage across the added branch with impedance z_{pq}, and is given by

$$v_{pq} = z_{pq} i_{pq} \tag{9.32}$$

Since added element p-q is a branch, $i_{pq} = 0$, thus $v_{pq} = 0$ and (9.31) reduces to

$$Z_{qi} = Z_{pi} \qquad i = 1, \ldots, m \qquad i \neq q \tag{9.33}$$

To calculate the diagonal element Z_{qq}, we will inject a current source of 1 per unit at the qth bus, i.e., $I_q = 1$ pu, and keep other buses open-circuited. From (9.29), we have

$$V_q = Z_{qq} \tag{9.34}$$

Since at the qth bus, the injected current flows from the bus q towards the bus p, $i_{pq} = -I_q = -1$. Hence, (9.32) reduces to

$$v_{pq} = -z_{pq} \tag{9.35}$$

Substituting for v_{pq} in (9.31), we get

$$V_q = V_p + z_{pq} \tag{9.36}$$

Now, since from (9.30) for $i = q$, $V_q = Z_{qq}$ and $V_p = Z_{pq}$, (9.36) becomes

$$Z_{qq} = Z_{pq} + z_{pq} \tag{9.37}$$

If node p is the reference node as shown in Figure 9.13(b), $V_p = 0$ and we obtain

$$Z_{qi} = Z_{pi} = V_p = 0 \qquad i = 1, \ldots, m \qquad i \neq q \tag{9.38}$$

From (9.37), the diagonal element becomes

$$Z_{qq} = z_{pq} \tag{9.39}$$

ADDITION OF A LINK

When the added element is a cotree link between the bus p and q, no new bus is created. The dimension of the $\mathbf{Z}_{bus}$ matrix remains the same but all the elements are required to be calculated. Let us add a link with impedance z_{pq} between two existing buses p and q as shown in Figure 9.14(a). If I_ℓ is the current through the added link in the direction shown in Figure 9.14(a), we have

$$z_{pq} I_\ell = V_p - V_q \tag{9.40}$$

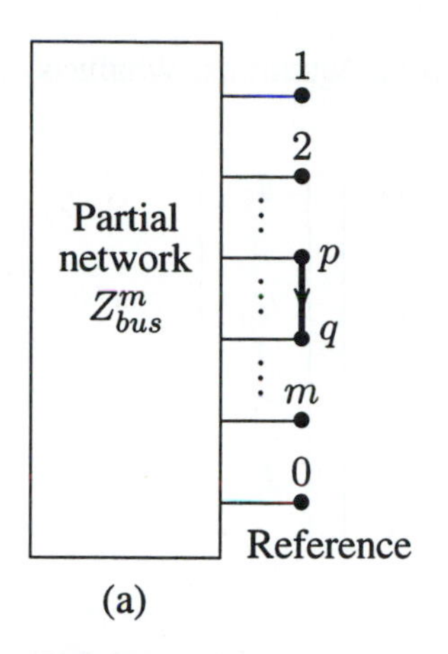

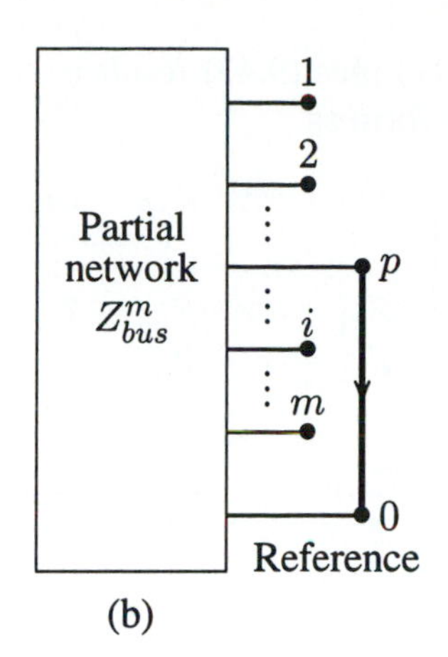

FIGURE 9.14
Addition of a link p-q.

or

$$V_q - V_p + z_{pq} I_\ell = 0 \tag{9.41}$$

The added link modifies the old current I_p to $(I_p - I_\ell)$ and the old current I_q to $(I_q + I_\ell)$ as shown in Figure 9.14(a), and the network equation becomes

$$\begin{aligned}
V_1 &= Z_{11}I_1 + \cdots + Z_{1p}(I_p - I_\ell) + Z_{1q}(I_q + I_\ell) + \cdots + Z_{1m}I_m \\
&\vdots \\
V_p &= Z_{p1}I_1 + \cdots + Z_{pp}(I_p - I_\ell) + Z_{pq}(I_q + I_\ell) + \cdots + Z_{pm}I_m \\
V_q &= Z_{q1}I_1 + \cdots + Z_{qp}(I_p - I_\ell) + Z_{qq}(I_q + I_\ell) + \cdots + Z_{qm}I_m \\
&\vdots \\
V_m &= Z_{m1}I_1 + \cdots + Z_{mp}(I_p - I_\ell) + Z_{mq}(I_q + I_\ell) + \cdots + Z_{mm}I_m
\end{aligned} \tag{9.42}$$

Substituting for V_p and V_q from (9.42) into (9.41) results in

$$\begin{aligned}
&(Z_{q1} - Z_{p1})I_1 + \cdots + (Z_{qp} - Z_{pp})I_p + \cdots + (Z_{qq} - Z_{pq})I_q + \cdots + \\
&(Z_{qm} - Z_{pm})I_m + (z_{pq} + Z_{pp} + Z_{qq} - 2Z_{pq})I_\ell = 0
\end{aligned} \tag{9.43}$$

Equations in (9.42) plus (9.43) result in $m+1$ simultaneous equations, which is written in matrix form as

$$\begin{bmatrix} V_1 \\ \vdots \\ V_p \\ V_q \\ \vdots \\ V_m \\ 0 \end{bmatrix} = \begin{bmatrix} Z_{11} & \cdots & Z_{1p} & Z_{1q} & \cdots & Z_{1m} & Z_{1\ell} \\ \vdots & \vdots & \vdots & \vdots & \vdots & \vdots & \vdots \\ Z_{p1} & \cdots & Z_{pp} & Z_{pq} & \cdots & Z_{pm} & Z_{p\ell} \\ Z_{q1} & \cdots & Z_{qp} & Z_{qq} & \cdots & Z_{qm} & Z_{q\ell} \\ \vdots & \vdots & \vdots & \vdots & \vdots & \vdots & \vdots \\ Z_{m1} & \cdots & Z_{mp} & Z_{mq} & \cdots & Z_{mm} & Z_{m\ell} \\ Z_{\ell 1} & \cdots & Z_{\ell p} & Z_{\ell q} & \cdots & Z_{\ell m} & Z_{\ell\ell} \end{bmatrix} \begin{bmatrix} I_1 \\ \vdots \\ I_p \\ I_q \\ \vdots \\ I_m \\ I_\ell \end{bmatrix} \tag{9.44}$$

where

$$Z_{\ell i} = Z_{i\ell} = Z_{iq} - Z_{ip} \tag{9.45}$$

and

$$Z_{\ell\ell} = z_{pq} + Z_{pp} + Z_{qq} - 2Z_{pq} \tag{9.46}$$

Now the link current I_ℓ can be eliminated. Equation (9.44) can be partitioned and rewritten in compact form as

$$\begin{bmatrix} \mathbf{V}_{bus} \\ 0 \end{bmatrix} = \begin{bmatrix} \mathbf{Z}_{bus}^{old} & \mathbf{\Delta Z} \\ \mathbf{\Delta Z}^T & Z_{\ell\ell} \end{bmatrix} \begin{bmatrix} \mathbf{I}_{bus} \\ I_\ell \end{bmatrix} \tag{9.47}$$

where

$$\mathbf{\Delta Z} = \begin{bmatrix} Z_{1\ell} & \cdots & Z_{p\ell} & Z_{q\ell} & \cdots & Z_{m\ell} \end{bmatrix}^T \tag{9.48}$$

Expanding (9.47), we get

$$\mathbf{V}_{bus} = \mathbf{Z}_{bus}^{old}\mathbf{I}_{bus} + \mathbf{\Delta Z} I_\ell \tag{9.49}$$

and

$$0 = \mathbf{\Delta Z}^T \mathbf{I}_{bus} + Z_{\ell\ell} I_\ell \tag{9.50}$$

or

$$I_\ell = -\frac{\mathbf{\Delta Z}^T}{Z_{\ell\ell}}\mathbf{I}_{bus} \tag{9.51}$$

Substituting from (9.51) for I_ℓ in (9.49), we have

$$\mathbf{V}_{bus} = \left[\mathbf{Z}_{bus}^{old} - \frac{\mathbf{\Delta Z}\,\mathbf{\Delta Z}^T}{Z_{\ell\ell}}\right]\mathbf{I}_{bus} \tag{9.52}$$

or

$$\mathbf{V}_{bus} = \mathbf{Z}_{bus}^{new}\mathbf{I}_{bus} \tag{9.53}$$

where

$$\mathbf{Z}_{bus}^{new} = \mathbf{Z}_{bus}^{old} - \frac{\Delta\mathbf{Z}\,\Delta\mathbf{Z}^T}{Z_{\ell\ell}} \tag{9.54}$$

Note that (9.54) reduces the matrix to its original size. The reason for this is that we have not added a new node but only linked two existing nodes.

The bus impedance matrix can be constructed with addition of branches and links in any sequence. However, it is best to select a tree that contains the elements connected to the reference node. If more than one element is connected between a given node and the reference node, only one element can be selected as a branch placing other elements in the cotree. The step-by-step procedure for building the bus impedance matrix which takes us from a given bus impedance matrix $\mathbf{Z}_{bus}^{old}$ to a new $\mathbf{Z}_{bus}^{new}$is summarized below.

Rule 1: Addition of a Tree Branch to the Reference

Start with the branches connected to the reference node. Addition of a branch z_{q0} between a new node q and the reference node 0 to the given $\mathbf{Z}_{bus}^{old}$ matrix of order $(m \times m)$, results in the $\mathbf{Z}_{bus}^{new}$ matrix of order $(m+1) \times (m+1)$. From the results of (9.38) and (9.39), we have

$$\mathbf{Z}_{bus}^{new} = \begin{bmatrix} Z_{11} & \cdots & Z_{1m} & 0 \\ \vdots & \ddots & 0 & 0 \\ 0 & \cdots & Z_{mm} & 0 \\ 0 & \cdots & 0 & z_{q0} \end{bmatrix} \tag{9.55}$$

This matrix is diagonal with the impedance values of the branches on the diagonal.

Rule 2: Addition of a Tree Branch from a New Bus to an Old Bus

Continue with the remaining branches of the tree connecting a new node to the existing node. Addition of a branch z_{pq} between a new node q and the existing node p to the given $\mathbf{Z}_{bus}^{old}$ matrix of order $(m \times m)$, results in the $\mathbf{Z}_{bus}^{new}$ matrix of order $(m+1) \times (m+1)$. From the results of (9.33) and (9.37), we have

$$\mathbf{Z}_{bus}^{new} = \begin{bmatrix} Z_{11} & \cdots & Z_{1p} & \cdots & Z_{1m} & Z_{1p} \\ \vdots & \vdots & \vdots & \vdots & \vdots & \vdots \\ Z_{p1} & \cdots & Z_{pp} & \cdots & Z_{pm} & Z_{pp} \\ \vdots & \vdots & \vdots & \vdots & \vdots & \vdots \\ Z_{m1} & \cdots & Z_{mp} & \cdots & Z_{mm} & Z_{mp} \\ Z_{p1} & \cdots & Z_{pp} & \cdots & Z_{pm} & Z_{pp} + z_{pq} \end{bmatrix} \tag{9.56}$$

Rule 3: Addition of a Cotree Link between two existing Buses

When a link with impedance z_{pq} is added between two existing nodes p and q, we augment the $\mathbf{Z}_{bus}^{old}$ matrix with a new row and a new column, and from (9.44) and (9.45) we have

$$\mathbf{Z}_{bus}^{new} = \begin{bmatrix} Z_{11} & Z_{1p} & Z_{1q} & \cdots & Z_{1m} & Z_{1q}-Z_{1p} \\ \vdots & \vdots & \vdots & \vdots & \vdots & \vdots \\ Z_{p1} & Z_{pp} & Z_{pq} & \cdots & Z_{pm} & Z_{pq}-Z_{pp} \\ Z_{q1} & Z_{qp} & Z_{qq} & \cdots & Z_{qm} & Z_{qq}-Z_{qp} \\ \vdots & \vdots & \vdots & \vdots & \vdots & \vdots \\ Z_{m1} & Z_{mp} & Z_{mq} & \cdots & Z_{mm} & Z_{mq}-Z_{mp} \\ Z_{q1}-Z_{p1} & Z_{qp}-Z_{pp} & Z_{qq}-Z_{pq} & \cdots & Z_{qm}-Z_{pm} & Z_{\ell\ell} \end{bmatrix} \tag{9.57}$$

where

$$Z_{\ell\ell} = z_{pq} + Z_{pp} + Z_{qq} - 2Z_{pq} \tag{9.58}$$

The new row and column is eliminated using the relation in (9.54), which is repeated here

$$\mathbf{Z}_{bus}^{new} = \mathbf{Z}_{bus}^{old} - \frac{\mathbf{\Delta Z}\,\mathbf{\Delta Z}^T}{Z_{\ell\ell}} \tag{9.59}$$

and $\mathbf{\Delta Z}$ is defined as

$$\mathbf{\Delta Z} = \begin{bmatrix} Z_{1q} - Z_{1p} \\ \vdots \\ Z_{pq} - Z_{pp} \\ Z_{qq} - Z_{qp} \\ \vdots \\ Z_{mq} - Z_{mp} \end{bmatrix} \tag{9.60}$$

When bus q is the reference bus, $Z_{qi} = Z_{iq} = 0$ (for $i = 1, m$), and (9.57) reduces to

$$\mathbf{Z}_{bus}^{new} = \begin{bmatrix} Z_{11} & \cdots & Z_{1p} & \cdots & Z_{1m} & -Z_{1p} \\ \vdots & \vdots & \vdots & \vdots & \vdots & \vdots \\ Z_{p1} & \cdots & Z_{pp} & \cdots & Z_{pm} & -Z_{pp} \\ \vdots & \vdots & \vdots & \vdots & \vdots & \vdots \\ Z_{m1} & \cdots & Z_{mp} & \cdots & Z_{mm} & -Z_{mp} \\ -Z_{p1} & \cdots & -Z_{pp} & \cdots & -Z_{pm} & Z_{\ell\ell} \end{bmatrix} \tag{9.61}$$

where $Z_{\ell\ell} = z_{pq} + Z_{pp}$, and

$$\Delta \mathbf{Z} = \begin{bmatrix} -Z_{1p} \\ \vdots \\ -Z_{pp} \\ \vdots \\ -Z_{mp} \end{bmatrix} \tag{9.62}$$

The algorithm to construct the $\mathbf{Z}_{bus}$ matrix by adding one element at a time can be used to remove lines or generators from the network. The procedure is identical to that of adding elements, except that the removed element is considered as negative impedance, in order to cancel the effect of the element.

Based on the above algorithm, two functions named **Zbus = zbuild(zdata)** and **Zbus = zbuild(linedata, gendata, yload)** are developed for the formation of the bus impedance matrix. These functions are described in Section 9.6. Before demonstrating this program, for the sake of better understanding the building algorithm, we shall demonstrate the hand calculation procedure for the simple three-bus network of Example 9.1.

Example 9.3

Construct the bus impedance matrix for the network in Example 9.1. The one-line impedance diagram is shown in Figure 9.15(a).

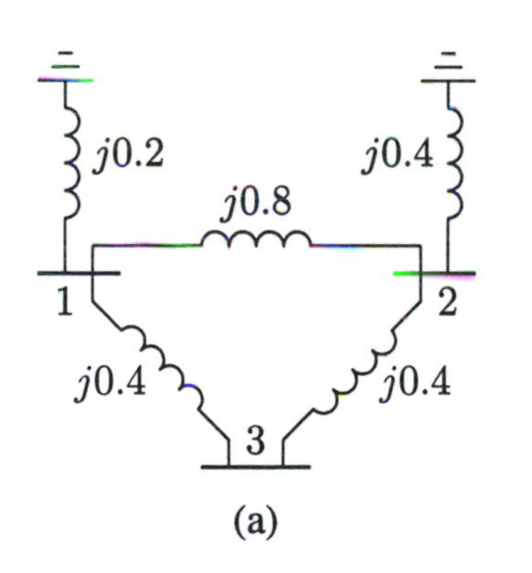

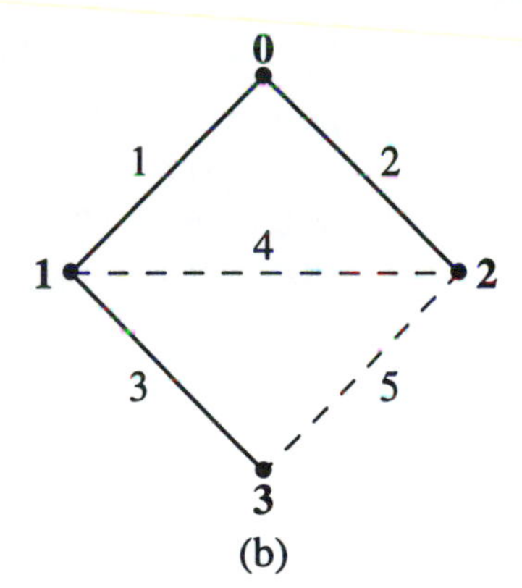

FIGURE 9.15
Impedance diagram of Example 9.1 and a proper tree.

The elements connected to the reference node are included in the proper tree as shown in Figure 9.15(b). We start with those branches of the tree connected to the reference node. Add branch 1, $z_{10} = j0.2$ between node $q = 1$ and reference node 0. According to rule 1, we have

$$\mathbf{Z}_{bus}^{(1)} = Z_{11} = z_{10} = j0.20$$

Next, add branch 2, $z_{20} = j0.4$ between node $q = 2$ and reference node 0

$$\mathbf{Z}_{bus}^{(2)} = \begin{bmatrix} Z_{11} & 0 \\ 0 & Z_{22} \end{bmatrix} = \begin{bmatrix} j0.2 & 0 \\ 0 & j0.4 \end{bmatrix}$$

Note that the off-diagonal elements of the bus impedance matrix are zero. This is because there is no connection between these buses other than to the reference. In this example, there are no more branches from a new bus to the reference. We continue with the remaining branches of the tree. Add branch 3, $z_{13} = j0.4$ between the new node $q = 3$ and the existing node $p = 1$. According to rule 2, we get

$$\mathbf{Z}_{bus}^{(3)} = \begin{bmatrix} Z_{11} & Z_{12} & Z_{11} \\ Z_{21} & Z_{22} & Z_{21} \\ Z_{11} & Z_{12} & Z_{11} + z_{13} \end{bmatrix} = \begin{bmatrix} j0.2 & 0 & j0.2 \\ 0 & j0.4 & 0 \\ j0.2 & 0 & j0.6 \end{bmatrix}$$

All tree branches are in place. We now proceed with the links. Add link 4, $z_{12} = j0.8$ between node $q = 2$ and node $p = 1$. From (9.57), we have

$$\mathbf{Z}_{bus}^{(4)} = \begin{bmatrix} Z_{11} & Z_{12} & Z_{13} & Z_{12} - Z_{11} \\ Z_{21} & Z_{22} & Z_{23} & Z_{22} - Z_{21} \\ Z_{31} & Z_{32} & Z_{33} & Z_{32} - Z_{31} \\ Z_{21} - Z_{11} & Z_{22} - Z_{21} & Z_{23} - Z_{13} & Z_{44} \end{bmatrix}$$

$$= \begin{bmatrix} j0.2 & 0 & j0.2 & -j0.2 \\ 0 & j0.4 & 0 & j0.4 \\ j0.2 & 0 & j0.6 & -j0.2 \\ -j0.2 & j0.4 & -j0.2 & Z_{44} \end{bmatrix}$$

From (9.58)

$$Z_{44} = z_{12} + Z_{11} + Z_{22} - 2Z_{12} = j0.8 + j0.2 + j0.4 - 2(j0) = j1.4$$

and

$$\frac{\Delta\mathbf{Z}\,\Delta\mathbf{Z}^T}{Z_{44}} = \frac{1}{j1.4}\begin{bmatrix} -j0.2 \\ j0.4 \\ -j0.2 \end{bmatrix} \begin{bmatrix} -j0.2 & j0.4 & -j0.2 \end{bmatrix}$$

$$= \begin{bmatrix} j0.02857 & -j0.05714 & j0.02857 \\ -j0.05714 & j0.11428 & -j0.05714 \\ j0.02857 & -j0.05714 & j0.02857 \end{bmatrix}$$

From (9.59), the new bus impedance matrix is

$$\mathbf{Z}_{bus}^{(4)} = \begin{bmatrix} j0.2 & 0 & j0.2 \\ 0 & j0.4 & 0 \\ j0.2 & 0 & j0.6 \end{bmatrix} - \begin{bmatrix} j0.02857 & -j0.05714 & j0.02857 \\ -j0.05714 & j0.11428 & -j0.05714 \\ j0.02857 & -j0.05714 & j0.02857 \end{bmatrix}$$

$$= \begin{bmatrix} j0.17143 & j0.05714 & j0.17143 \\ j0.05714 & j0.28571 & j0.05714 \\ j0.17143 & j0.05714 & j0.57143 \end{bmatrix}$$

Finally, we add link 5, $z_{23} = j0.4$ between node $q = 3$ and node $p = 2$. From (9.57), we have

$$\mathbf{Z}_{bus}^{(5)} = \begin{bmatrix} Z_{11} & Z_{12} & Z_{13} & Z_{13} - Z_{12} \\ Z_{21} & Z_{22} & Z_{23} & Z_{23} - Z_{22} \\ Z_{31} & Z_{32} & Z_{33} & Z_{33} - Z_{32} \\ Z_{31} - Z_{21} & Z_{32} - Z_{22} & Z_{33} - Z_{23} & Z_{44} \end{bmatrix}$$

$$= \begin{bmatrix} j0.17143 & j0.05714 & j0.17143 & j0.11429 \\ j0.05714 & j0.28571 & j0.05714 & -j0.22857 \\ j0.17143 & j0.05714 & j0.57143 & j0.51429 \\ j0.11429 & -j0.22857 & j0.51429 & Z_{44} \end{bmatrix}$$

From (9.58)

$$Z_{44} = z_{23} + Z_{22} + Z_{33} - 2Z_{23} = j0.4 + j0.28571 + j0.57143 - 2(j0.05714) = j1.14$$

and

$$\frac{\Delta\mathbf{Z}\,\Delta\mathbf{Z}^T}{Z_{44}} = \frac{1}{j1.4}\begin{bmatrix} j0.11429 \\ -j0.22857 \\ j0.51429 \end{bmatrix}\begin{bmatrix} j0.11429 & -j0.22857 & j0.51429 \end{bmatrix}$$

$$= \begin{bmatrix} j0.01143 & -j0.02286 & j0.05143 \\ -j0.02286 & j0.04571 & -j0.10286 \\ j0.05143 & -j0.10286 & j0.23143 \end{bmatrix}$$

From (9.59), the new bus impedance matrix is

$$\mathbf{Z}_{bus} = \begin{bmatrix} j0.17143 & j0.05714 & j0.17143 \\ j0.05714 & j0.28571 & j0.05714 \\ j0.17143 & j0.05714 & j0.57143 \end{bmatrix} - \begin{bmatrix} j0.01143 & -j0.02286 & j0.05143 \\ -j0.02286 & j0.04571 & -j0.10286 \\ j0.05143 & -j0.10286 & j0.23143 \end{bmatrix}$$

$$= \begin{bmatrix} j0.16 & j0.08 & j0.12 \\ j0.08 & j0.24 & j0.16 \\ j0.12 & j0.16 & j0.34 \end{bmatrix}$$

This is the desired bus impedance matrix $\mathbf{Z}_{bus}$, which is the same as the one obtained by inverting the $\mathbf{Y}_{bus}$ matrix in Example 9.2.

Example 9.4

The bus impedance matrix for the network shown in Figure 9.16 is found to be

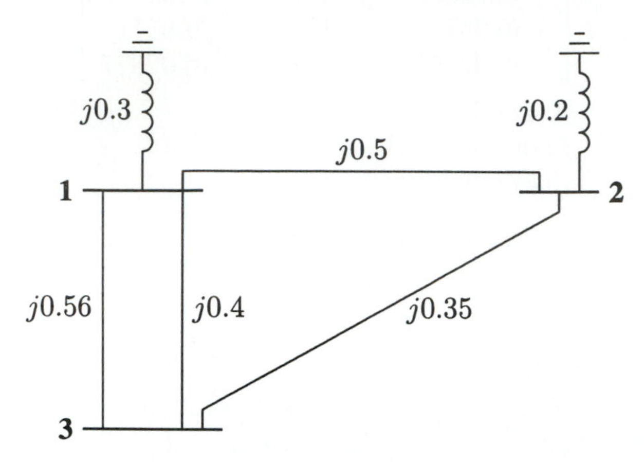

FIGURE 9.16
Impedance diagram for Example 9.4.

$$\mathbf{Z}_{bus} = \begin{bmatrix} j0.183 & j0.078 & j0.141 \\ j0.078 & j0.148 & j0.106 \\ j0.141 & j0.106 & j0.267 \end{bmatrix}$$

The line between buses 1 and 3 with impedance $Z_{13} = j0.56$ is removed by the simultaneous opening of breakers at both ends of the line. Determine the new bus impedance matrix.

The removal of an element is equivalent to connecting a link having an impedance equal to the negated value of the original impedance. Therefore, we add link $z_{13} = -j0.56$ between node $q = 3$ and node $p = 1$. From (9.57), we have

$$\mathbf{Z}_{bus} = \begin{bmatrix} Z_{11} & Z_{12} & Z_{13} & Z_{13} - Z_{11} \\ Z_{21} & Z_{22} & Z_{23} & Z_{23} - Z_{21} \\ Z_{31} & Z_{32} & Z_{33} & Z_{33} - Z_{31} \\ Z_{31} - Z_{11} & Z_{32} - Z_{12} & Z_{33} - Z_{13} & Z_{44} \end{bmatrix}$$

Thus, we get

$$Z_{bus} = \begin{bmatrix} j0.183 & j0.078 & j0.141 & -j0.042 \\ j0.078 & j0.148 & j0.106 & j0.028 \\ j0.141 & j0.106 & j0.267 & j0.126 \\ -j0.042 & j0.028 & j0.126 & Z_{44} \end{bmatrix}$$

From (9.58)

$$Z_{44} = z_{13} + Z_{11} + Z_{33} - 2Z_{13} = -j0.56 + j0.183 + j0.267 - 2(j0.141) = -j0.392$$

and

$$\frac{\Delta \mathbf{Z}\, \Delta \mathbf{Z}^T}{Z_{44}} = \frac{1}{-j0.392}\begin{bmatrix} -j0.042 \\ j0.028 \\ j0.126 \end{bmatrix}\begin{bmatrix} -j0.042 & j0.028 & j0.126 \end{bmatrix}$$

$$= \begin{bmatrix} -j0.0045 & j0.0030 & j0.0135 \\ j0.0030 & -j0.0020 & -j0.0090 \\ j0.0135 & -j0.0090 & -j0.0405 \end{bmatrix}$$

From (9.59), the new bus impedance matrix is

$$\mathbf{Z}_{bus} = \begin{bmatrix} j0.183 & j0.078 & j0.141 \\ j0.078 & j0.148 & j0.106 \\ j0.141 & j0.106 & j0.267 \end{bmatrix} - \begin{bmatrix} -j0.0045 & j0.0030 & j0.0135 \\ j0.0030 & -j0.0020 & -j0.0090 \\ j0.0135 & -j0.0090 & -j0.0405 \end{bmatrix}$$

$$= \begin{bmatrix} j0.1875 & j0.0750 & j0.1275 \\ j0.0750 & j0.1500 & j0.1150 \\ j0.1275 & j0.1150 & j0.3075 \end{bmatrix}$$

9.6 ZBUILD AND SYMFAULT PROGRAMS

Two functions are developed for the formation of the bus impedance matrix. One function is named **Zbus = zbuild(zdata)**, where the argument **zdata** is an $e \times 4$ matrix containing the impedance data of an e-element network. Columns 1 and 2 are the element bus numbers and columns 3 and 4 contain the element resistance and reactance, respectively, in per unit. Bus number 0 to generator buses contain generator impedances. These may be the subtransient, transient, or synchronous reactances. Also, any other shunt impedances such as capacitors and load impedance to ground (bus 0) may be included in this matrix.

The other function for the formation of the bus impedance matrix is **zbus = zbuildpi(linedata, gendata, yload)**, which is compatible with the power flow programs. The first argument **linedata** is consistent with the data required for the power flow solution. Columns 1 and 2 are the line bus numbers. Columns 3 through 5 contain line resistance, reactance, and one-half of the total line charging susceptance in per unit on the specified MVA base. The last column is for the transformer tap setting; for lines, 1 must be entered in this column. The lines may be entered in any sequence or order. The generator reactances are not included in the **linedata** of the power flow program and must be specified separately as required by the **gendata** in the second argument. **gendata** is an $n_g \times 4$ matrix, where each row contains bus 0, generator bus number, resistance and reactance. The last argument, **yload** is optional. This is a two-column matrix containing bus number and the complex load admittance. This data is provided by any of the power flow programs **lfgauss**, **lfnewton** or **decouple**. **yload** is automatically generated following the execution of any of the above power flow programs.

The **zbuild** and **zbuildpi** functions obtain the bus impedance matrix by the building algorithm method. These functions select a tree containing elements to the reference node. First, all branches connected to the reference node are processed. Then the remaining branches of the tree are connected, and finally the cotree links are added.

The program **symfault(zdata, Zbus, V)** is developed for the balanced three-phase fault studies. The function requires the **zdata** and the **Zbus** matrices. The third argument **V** is optional. If it is not included, the program sets all the prefault bus voltages to 1.0 per unit. If the variable **V** is included, the prefault bus voltages must be specified by the array **V** containing bus numbers and the complex bus voltage. The voltage vector **V** is automatically generated following the execution of any of the power flow programs. The use of the above functions are demonstrated in the following examples. When **symfault** is executed, it prompts the user to enter the faulted bus number and the fault impedance. The program computes the total fault current and tabulates the magnitude of the bus voltages and line currents during the fault.

Example 9.5

Use the function **zbus = zbuild(zdata)** to obtain the bus impedance matrix for the network in Example 9.3.

The network configuration containing resistances and reactances are specified and the **zbuild** function is used as follows.

```
zdata = [ 0   1   0    0.2
          0   2   0    0.4
          1   2   0    0.8
          1   3   0    0.4
          2   3   0    0.4];

Zbus = zbuild(zdata)
```

The result is

```
Zbus =
          0 + 0.16i      0 + 0.08i      0 + 0.12i
          0 + 0.08i      0 + 0.24i      0 + 0.16i
          0 + 0.12i      0 + 0.16i      0 + 0.34i
```

Example 9.6

A three-phase fault with a fault impedance $Z_f = j0.16$ per unit occurs at bus 3 in the network of Example 9.1. Use the **symfault** function to compute the fault current, the bus voltages and line currents during the fault.

In this example all shunt capacitances and loads are neglected and all the prefault bus voltages are assumed to be unity. The impedance diagram in Figure 9.2(b) is described by the variable **zdata** and the following commands are used.

```
zdata =     [ 0   1   0    0.2
              0   2   0    0.4
              1   2   0    0.8
              1   3   0    0.4
              2   3   0    0.4];

Zbus = zbuild(zdata)
symfault(zdata, Zbus)
```

The result is

```
Zbus =
          0 + 0.1600i      0 + 0.0800i      0 + 0.1200i
          0 + 0.0800i      0 + 0.2400i      0 + 0.1600i
          0 + 0.1200i      0 + 0.1600i      0 + 0.3400i

Enter Faulted Bus No. -> 3
Enter Fault Impedance Zf = R + j*X in
complex form (for bolted fault enter 0). Zf = j*0.16

Balanced three-phase fault at bus No. 3
Total fault current =   2.0000 Per unit

Bus Voltages during the fault in per unit
      Bus      Voltage       Angle
      No.      Magnitude     Degree
       1        0.7600       0.0000
       2        0.6800       0.0000
       3        0.3200       0.0000
```

```
Line currents for fault at bus No. 3
     From       To      Current         Angle
     Bus        Bus     Magnitude      Degree
      G          1       1.2000      -90.0000
      1          2       0.1000      -90.0000
      1          3       1.1000      -90.0000
      G          2       0.8000      -90.0000
      2          3       0.9000      -90.0000
      3          F       2.0000      -90.0000
```

Example 9.7

The 11-bus power system network of an electric utility company is shown in Figure 9.17.

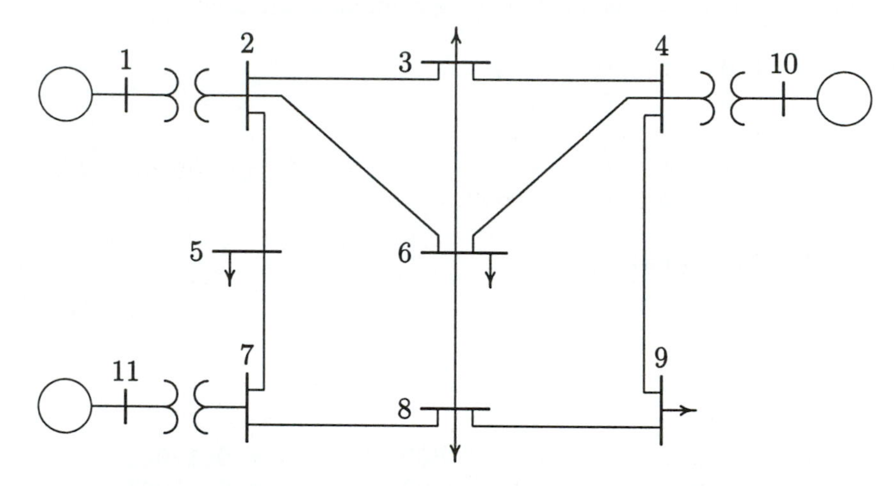

FIGURE 9.17
One-line diagram for Example 9.7

The transient impedance of the generators on a 100-MVA base are given below.

GEN. TRANSIENT IMPEDANCE PU		
Gen. No.	R_a	X'_d
1	0	0.20
10	0	0.15
11	0	0.25

The line and transformer data containing the series resistance and reactance in per unit, and one-half of the total capacitance in per unit susceptance on a 100-MVA base is tabulated below.

LINE AND TRANSFORMER DATA				
Bus No.	Bus No.	R, PU	X, PU	$\frac{1}{2}$ B, PU
1	2	0.00	0.06	0.0000
2	3	0.08	0.30	0.0004
2	5	0.04	0.15	0.0002
2	6	0.12	0.45	0.0005
3	4	0.10	0.40	0.0005
3	6	0.04	0.40	0.0005
4	6	0.15	0.60	0.0008
4	9	0.18	0.70	0.0009
4	10	0.00	0.08	0.0000
5	7	0.05	0.43	0.0003
6	8	0.06	0.48	0.0000
7	8	0.06	0.35	0.0004
7	11	0.00	0.10	0.0000
8	9	0.052	0.48	0.0000

Neglecting the shunt capacitors and the loads, use **zbuild(zdata)** function to obtain the bus impedance matrix. Assuming all the prefault bus voltages are equal to $1\angle 0°$, use **symfault** function to compute the fault current, bus voltages, and line currents for a bolted fault at bus 8. When using **zbuild** function, the generator reactances must be included in the impedance data with bus zero as the reference bus. The impedance data and the required commands are as follows.

```
%           Bus    Bus        R          X
%           No.    No.        pu         pu
zdata = [0      1      0.00       0.20
         0     10      0.00       0.15
         0     11      0.00       0.25
         1      2      0.00       0.06
         2      3      0.08       0.30
         2      5      0.04       0.15
         2      6      0.12       0.45
         3      4      0.10       0.40
         3      6      0.04       0.40
         4      6      0.15       0.60
         4      9      0.18       0.70
         4     10      0.00       0.08
         5      7      0.05       0.43
         6      8      0.06       0.48
         7      8      0.06       0.35
         7     11      0.00       0.10
         8      9      0.052      0.48
```

```
Zbus = zbuild(zdata)
symfault(zdata, Zbus)
```

The bus impedance matrix is displayed on the screen, and the three-phase short circuit result is

```
Enter Faulted Bus No. -> 8
Enter Fault Impedance Zf = R + j*X in
complex form (for bolted fault enter 0). Zf = 0
Balanced three-phase fault at bus No. 8
Total fault current =   3.3319 per unit

Bus Voltages during the fault in per unit
     Bus      Voltage       Angle
     No.      Magnitude     Degree
      1        0.8082      -1.8180
      2        0.7508      -2.5443
      3        0.6882      -1.5987
      4        0.7491      -2.4902
      5        0.7007      -2.3762
      6        0.5454      -1.0194
      7        0.5618      -3.8128
      8        0.0000       0.0000
      9        0.3008       2.4499
     10        0.8362      -1.4547
     11        0.6866      -2.2272
Line currents for fault at bus No. 8
     From       To      Current        Angle
     Bus        Bus     Magnitude      Degree
      G          1       0.9697      -82.4034
      1          2       0.9697      -82.4034
      2          3       0.2053      -87.8751
      2          5       0.3230      -79.9626
      2          6       0.4427      -81.6497
      3          6       0.3556      -88.0987
      4          3       0.1503      -88.4042
      4          6       0.3305      -82.3804
      4          9       0.6229      -81.3672
      5          7       0.3230      -79.9626
      6          8       1.1274      -83.8944
      7          8       1.5820      -84.0852
      8          F       3.3319      -83.5126
      9          8       0.6229      -81.3672
      G         10       1.1029      -82.6275
     10          4       1.1029      -82.6275
      G         11       1.2601      -85.1410
     11          7       1.2601      -85.1410
```

Example 9.8

In Example 9.7 consider the shunt capacitors and neglect the loads. Use **zbuildpi** function to obtain the bus impedance matrix. Assuming all the prefault bus voltages are equal to $1\angle 0°$, use **symfault** function to compute the fault current, bus voltages, and line currents for a bolted fault at bus 8.

The **zbuildpi(linedata, gendata, yload)** is designed to be compatible with the power flow programs. The first argument **linedata** is consistent with the data required for the power flow program. The generator reactances are not included in the **linedata** and must be specified separately by the **gendata**. The optional argument **yload** contains the load admittance which is generated from the power flow solution. The loads are neglected in this example, therefore, the argument **yload** is omitted. The impedance data and the required commands are as follows:

```
%          Bus   Bus     R        X         1/2B
%          No.   No.     pu       pu         pu
linedata=[1       2     0.00     0.06      0.0000
          2       3     0.08     0.30      0.0004
          2       5     0.04     0.15      0.0002
          2       6     0.12     0.45      0.0005
          3       4     0.10     0.40      0.0005
          3       6     0.04     0.40      0.0005
          4       6     0.15     0.60      0.0008
          4       9     0.18     0.70      0.0009
          4      10     0.00     0.08      0.0000
          5       7     0.05     0.43      0.0003
          6       8     0.06     0.48      0.0000
          7       8     0.06     0.35      0.0004
          7      11     0.00     0.10      0.0000
          8       9     0.052    0.48      0.0000];
%          Gen.  Ra      Xd'
gendata=[ 1      0      0.20
          10     0      0.15
          11     0      0.25];
Zbus=zbuildpi(linedata, gendata)
symfault(linedata, Zbus)
```

The bus impedance matrix is displayed on the screen, and the three-phase short circuit result is

```
Enter Faulted Bus No. -> 8
Enter Fault Impedance Zf = R + j*X in
complex form (for bolted fault enter 0). Zf = 0
Balanced three-phase fault at bus No. 8
```

Total fault current = 3.3301 per unit

Bus Voltages during the fault in per unit

Bus No.	Voltage Magnitude	Angle Degree
1	0.8080	-1.8188
2	0.7506	-2.5456
3	0.6879	-1.5986
4	0.7489	-2.4915
5	0.7006	-2.3774
6	0.5451	-1.0185
7	0.5617	-3.8137
8	0.0000	0.0000
9	0.3005	2.4564
10	0.8361	-1.4553
11	0.6866	-2.2276

Line currents for fault at bus No. 8

From Bus	To Bus	Current Magnitude	Angle Degree
1	2	0.9704	-82.4068
2	3	0.2056	-87.7898
2	5	0.3230	-79.9386
2	6	0.4429	-81.6055
3	6	0.3556	-88.0454
4	3	0.1505	-88.2647
4	6	0.3308	-82.2823
4	9	0.6232	-81.3096
5	7	0.3228	-79.9261
6	8	1.1269	-83.8935
7	8	1.5818	-84.0781
8	F	3.3301	-83.5110
9	8	0.6224	-81.3606
10	4	1.1038	-82.6316
11	7	1.2604	-85.1416

Example 9.9

Repeat the symmetrical three-phase short circuit analysis for Example 9.7 considering the prefault bus voltages and the effect of load currents. The load data is as follows:

LOAD DATA					
Bus	Load		Bus	Load	
No.	MW	Mvar	No.	MW	Mvar
1	0.0	0.0	7	0.0	0.0
2	0.0	0.0	8	110.0	90.0
3	150.0	120.0	9	80.0	50.0
4	0.0	0.0	10	0.0	0.0
5	120.0	60.0	11	0.0	0.0
6	140.0	90.0			

Voltage magnitude, generation schedule and the reactive power limits for the regulated buses are tabulated below. Bus 1, whose voltage is specified as $V_1 = 1.04\angle 0°$, is taken as the slack bus.

GENERATION DATA				
Bus	Voltage	Generation,	Mvar Limits	
No.	Mag.	MW	Min.	Max.
1	1.040			
10	1.035	200.0	0.0	180.0
11	1.030	160.0	0.0	120.0

Anyone of the power flow programs can be used to obtain the prefault bus voltages and the load admittance. The **lfnewton** program is used which returns the prefault bus voltage array **V** and the bus load admittance array **yload**. The required commands are as follows.

```
clear         % clears all variables from workspace.
basemva = 100;  accuracy = 0.0001;  maxiter = 10;
%       Bus Bus Voltage Angle --Load--  ---Generator---Injected
%        No code Mag. Degree MW   Mvar   MW  Mvar Qmin Qmax Mvar
busdata=[1  1  1.06   0      0.0  0.0    0.0  0.0   0    0   0
         2  0  1.0    0    0.0   0.0     0.0  0     0    0   0
         3  0  1.0    0  150.0 120.0     0.0  0     0    0   0
         4  0  1.0    0    0.0   0.0     0.0  0     0    0   0
         5  0  1.0    0  120.0  60.0     0.0  0     0    0   0
         6  0  1.0    0  140.0  90.0     0.0  0     0    0   0
         7  0  1.0    0    0.0   0.0     0.0  0     0    0   0
         8  0  1.0    0  110.0  90.0     0.0  0     0    0   0
         9  0  1.0    0   80.0  50.0     0.0  0     0    0   0
        10  2  1.035  0    0.0   0.0   200.0  0     0  180   0
        11  2  1.03   0    0.0   0.0   160.0  0     0  120   0];

%        Bus   Bus     R        X        1/2B
%        No.   No.     pu       pu        pu
```

```
linedata=[1      2      0.00      0.06      0.0000    1
          2      3      0.08      0.30      0.0004    1
          2      6      0.12      0.45      0.0005    1
          3      4      0.10      0.40      0.0005    1
          3      6      0.04      0.40      0.0005    1
          4      6      0.15      0.60      0.0008    1
          4      9      0.18      0.70      0.0009    1
          4     10      0.00      0.08      0.0000    1
          5      7      0.05      0.43      0.0003    1
          6      8      0.06      0.48      0.0000    1
          7      8      0.06      0.35      0.0004    1
          7     11      0.00      0.10      0.0000    1
          8      9      0.052     0.48      0.0000    1];
%         Gen.   Ra      Xd'
gendata=[ 1      0      0.20
          10     0      0.15
          11     0      0.25];
lfybus                       % Forms the bus admittance matrix
lfnewton          % Power flow solution by Newton-Raphson method
busout            % Prints the power flow solution on the screen
Zbus=zbuildpi(linedata, gendata, yload) %  Zbus including load
symfault(linedata,Zbus,V)%3-phase fault including load current
```

The result is

```
           Power Flow Solution by Newton-Raphson Method
              Maximum Power Mismatch = 0.0000533178
                     No. of Iterations = 3

Bus  Voltage  Angle     ----Load----    --Generation-- Injected
No.  Mag.     Degree     MW     Mvar       MW      Mvar    Mvar

1    1.040    0.000    0.000    0.000   248.622  149.163    0.0
2    1.031   -0.797    0.000    0.000     0.000    0.000    0.0
3    0.997   -2.619  150.000  120.000     0.000    0.000    0.0
4    1.024   -1.737    0.000    0.000     0.000    0.000    0.0
5    0.981   -7.414  120.000   60.000     0.000    0.000    0.0
6    0.992   -3.336  140.000   90.000     0.000    0.000    0.0
7    1.014   -4.614    0.000    0.000     0.000    0.000    0.0
8    0.981   -5.093  110.000   90.000     0.000    0.000    0.0
9    0.977   -4.842   80.000   50.000     0.000    0.000    0.0
10   1.035   -0.872    0.000    0.000   200.000  144.994    0.0
11   1.020   -3.737    0.000    0.000   160.000  161.121    0.0

Total                600.000  410.000   608.622  455.278    0.0
```

```
Enter Faulted Bus No. -> 8
Enter Fault Impedance Zf = R + j*X in
complex form (for bolted fault enter 0). Zf = 0
Balanced three-phase fault at bus No. 8
Total fault current =   3.3571 per unit

Bus Voltages during the in per unit
     Bus       Voltage         Angle
     No.      Magnitude       Degree
      1        0.8876        -0.9467
      2        0.8350        -2.0943
      3        0.7321        -2.5619
      4        0.7866        -3.1798
      5        0.5148        -8.3043
      6        0.5792        -2.4214
      7        0.5179        -8.2563
      8        0.0000         0.0000
      9        0.3156         0.9877
     10        0.8785        -1.7237
     11        0.6631        -5.7789

Line currents for fault at bus No. 8
     From        To      Current         Angle
     Bus         Bus     Magnitude      Degree
      1          2        0.9219       -73.3472
      2          3        0.3321       -73.7856
      2          6        0.5494       -76.3804
      3          6        0.3804       -87.3283
      4          3        0.1336       -87.2217
      4          6        0.3357       -81.1554
      4          9        0.6537       -81.4818
      6          8        1.1974       -85.2964
      7          5        0.0073       -82.5471
      7          8        1.4585       -88.5207
      8          F        3.3571       -85.4214
      9          8        0.6538       -82.8293
     10          4        1.1787       -79.4854
     11          7        1.4733       -87.0395
```

PROBLEMS

9.1. The system shown in Figure 9.18 is initially on no load with generators operating at their rated voltage with their emfs in phase. The rating of the generators and the transformers and their respective percent reactances are marked on the diagram. All resistances are neglected. The line impedance is $j160\ \Omega$. A three-phase balanced fault occurs at the receiving end of the transmission line. Determine the short-circuit current and the short-circuit MVA.

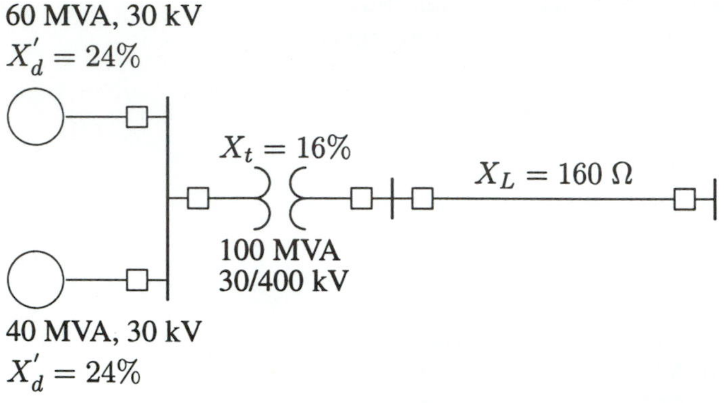

60 MVA, 30 kV
$X'_d = 24\%$
$X_t = 16\%$
$X_L = 160\ \Omega$
100 MVA
30/400 kV
40 MVA, 30 kV
$X'_d = 24\%$

FIGURE 9.18
One-line diagram for Problem 9.1.

9.2. The system shown in Figure 9.19 shows an existing plant consisting of a generator of 100 MVA, 30 kV, with 20 percent subtransient reactance and a generator of 50 MVA, 30 kV with 15 percent subtransient reactance, connected in parallel to a 30-kV bus bar. The 30-kV bus bar feeds a transmission line via the circuit breaker C which is rated at 1250 MVA. A grid supply is connected to the station bus bar through a 500-MVA, 400/30-kV transformer with 20 percent reactance. Determine the reactance of a current limiting reactor in ohm to be connected between the grid system and the existing bus bar such that the short-circuit MVA of the breaker C does not exceed.

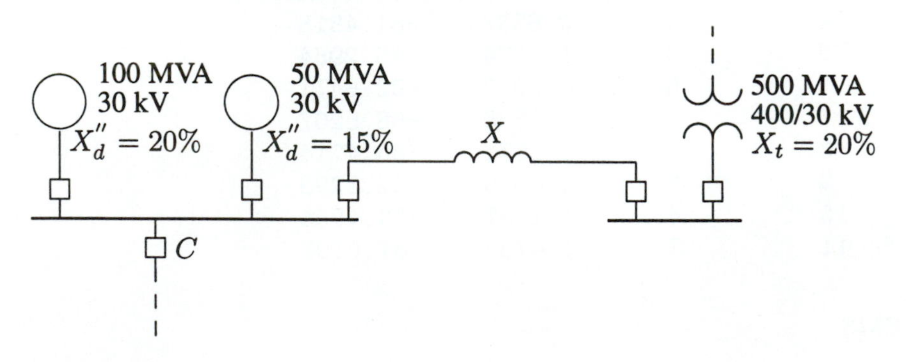

FIGURE 9.19
One-line diagram for Problem 9.2.

9.3. The one-line diagram of a simple power system is shown in Figure 9.20. Each generator is represented by an emf behind the transient reactance. All impedances are expressed in per unit on a common MVA base. All resistances and shunt capacitances are neglected. The generators are operating on no load at their rated voltage with their emfs in phase. A three-phase fault occurs at bus 1 through a fault impedance of $Z_f = j0.08$ per unit.
(a) Using Thévenin's theorem obtain the impedance to the point of fault and the fault current in per unit.
(b) Determine the bus voltages and line currents during fault.

3 $X_t = 0.1$ 1
$X_L = 0.2$
2
$X''_d = 0.1$
$X''_d = 0.1$

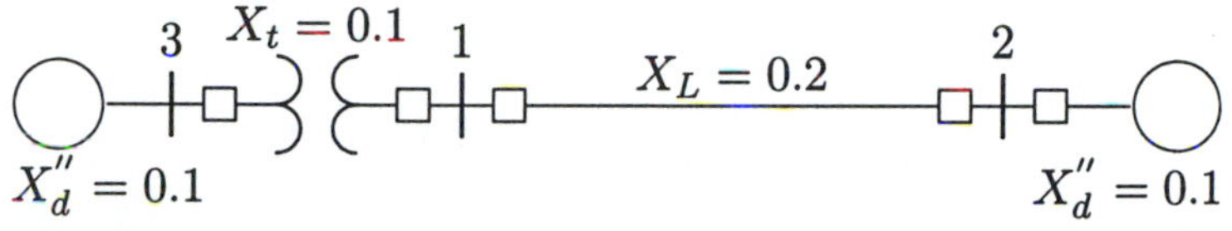

FIGURE 9.20
One-line diagram for Problem 9.3.

9.4. The one-line diagram of a simple three-bus power system is shown in Figure 9.21 Each generator is represented by an emf behind the subtransient reactance. All impedances are expressed in per unit on a common MVA base. All resistances and shunt capacitances are neglected. The generators are operating on no load at their rated voltage with their emfs in phase. A three-phase fault occurs at bus 3 through a fault impedance of $Z_f = j0.19$ per unit.
(a) Using Thévenin's theorem obtain the impedance to the point of fault and the fault current in per unit.
(b) Determine the bus voltages and line currents during fault.

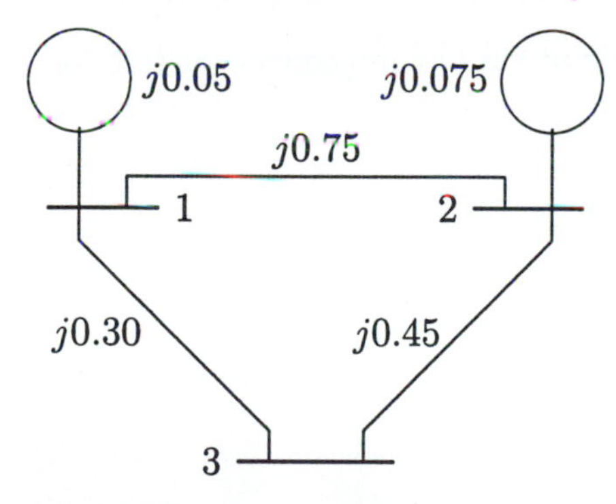

FIGURE 9.21
One-line diagram for Problem 9.4.

9.5. The one-line diagram of a simple four-bus power system is shown in Figure 9.22 Each generator is represented by an emf behind the transient reactance.

All impedances are expressed in per unit on a common MVA base. All resistances and shunt capacitances are neglected. The generators are operating on no load at their rated voltage with their emfs in phase. A bolted three-phase fault occurs at bus 4.

(a) Using Thévenin's theorem obtain the impedance to the point of fault and the fault current in per unit.

(b) Determine the bus voltages and line currents during fault.

(c) Repeat (a) and (b) for a fault at bus 2 with a fault impedance of $Z_f = j0.0225$.

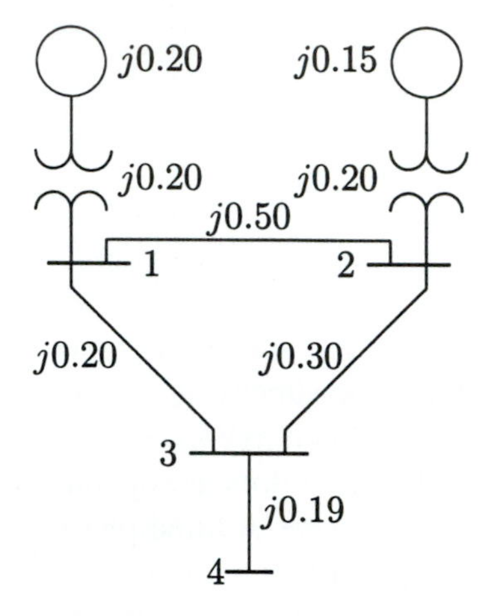

FIGURE 9.22
One-line diagram for Problem 9.5.

9.6. Using the method of building algorithm find the bus impedance matrix for the network shown in Figure 9.23.

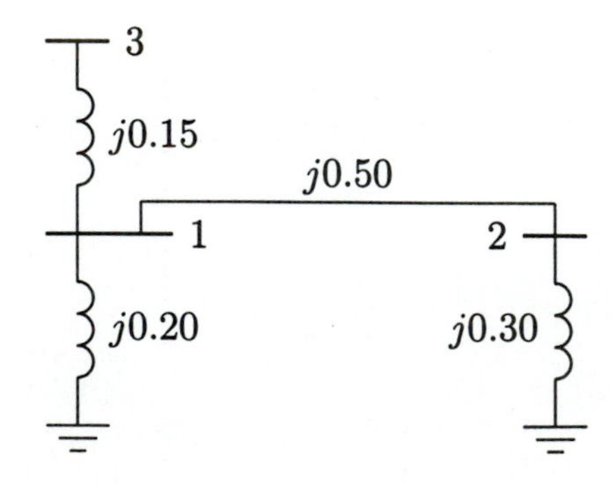

FIGURE 9.23
One-line diagram for Problem 9.6.

9.7. Obtain the bus impedance matrix for the network of Problem 9.3.

9.8. Obtain the bus impedance matrix for the network of Problem 9.4.

9.9. The bus impedance matrix for the network shown in Figure 9.24 is given by

$$Z_{bus} = j\begin{bmatrix} 0.300 & 0.200 & 0.275 \\ 0.200 & 0.400 & 0.250 \\ 0.275 & 0.250 & 0.41875 \end{bmatrix}$$

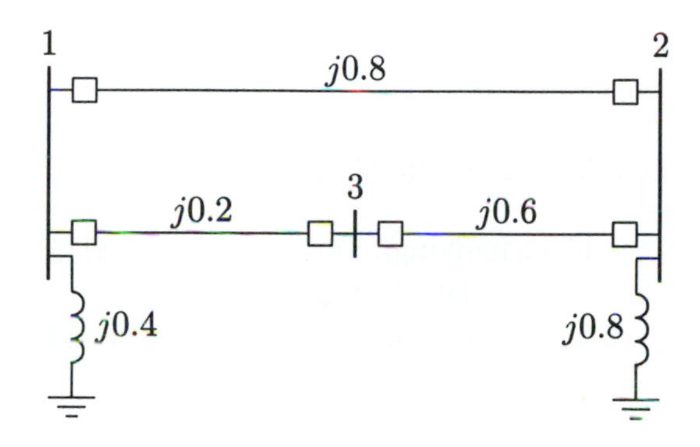

FIGURE 9.24
One-line diagram for Problem 9.9.

There is a line outage and the line from bus 1 to 2 is removed. Using the method of building algorithm determine the new bus impedance matrix.

9.10. The per unit bus impedance matrix for the power system of Problem 9.4 is given by

$$Z_{bus} = j\begin{bmatrix} 0.0450 & 0.0075 & 0.0300 \\ 0.0075 & 0.06375 & 0.0300 \\ 0.0300 & 0.0300 & 0.2100 \end{bmatrix}$$

A three-phase fault occurs at bus 3 through a fault impedance of $Z_f = j0.19$ per unit. Using the bus impedance matrix calculate the fault current, bus voltages, and line currents during fault. Check your result using the **Zbuild** and **symfault** programs.

9.11. The per unit bus impedance matrix for the power system of Problem 9.5 is given by

$$Z_{bus} = j\begin{bmatrix} 0.240 & 0.140 & 0.200 & 0.200 \\ 0.140 & 0.2275 & 0.175 & 0.175 \\ 0.200 & 0.175 & 0.310 & 0.310 \\ 0.200 & 0.1750 & 0.310 & 0.500 \end{bmatrix}$$

(a) A bolted three-phase fault occurs at bus 4. Using the bus impedance matrix calculate the fault current, bus voltages, and line currents during fault.
(b) Repeat (a) for a three-phase fault at bus 2 with a fault impedance of $Z_f = j0.0225$.
(c) Check your result using the **Zbuild** and **symfault** programs.

9.12. The per unit bus impedance matrix for the power system shown in Figure 9.25 is given by

$$Z_{bus} = j \begin{bmatrix} 0.150 & 0.075 & 0.140 & 0.135 \\ 0.075 & 0.1875 & 0.090 & 0.0975 \\ 0.140 & 0.090 & 0.2533 & 0.210 \\ 0.135 & 0.0975 & 0.210 & 0.2475 \end{bmatrix}$$

A three-phase fault occurs at bus4 through a fault impedance of $Z_f = j0.0025$ per unit. Using the bus impedance matrix calculate the fault current, bus voltages and line currents during fault. Check your result using the **Zbuild** and **symfault** programs.

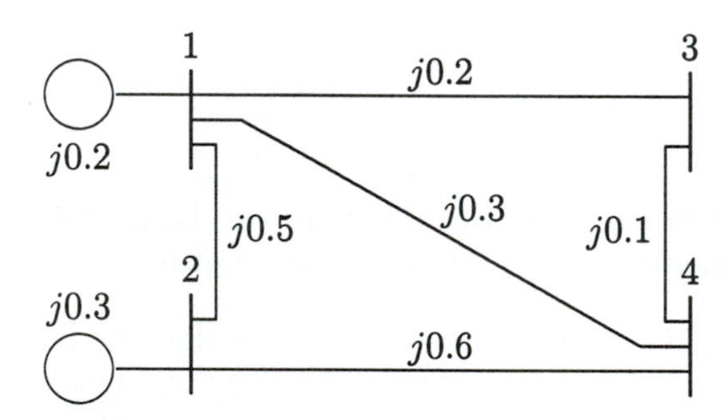

FIGURE 9.25
One-line diagram for Problem 9.12.

9.13. Repeat Example 9.7 for a bolted three-phase fault at bus 9.

9.14. Repeat Example 9.8 for a bolted three-phase fault at bus 9.

9.15. Repeat Example 9.9 for a bolted three-phase fault at bus 9.

9.16. The 6-bus power system network of an electric utility company is shown in Figure 9.26. The line and transformer data containing the series resistance and reactance in per unit, and one-half of the total capacitance in per unit susceptance on a 100-MVA base, is tabulated below.

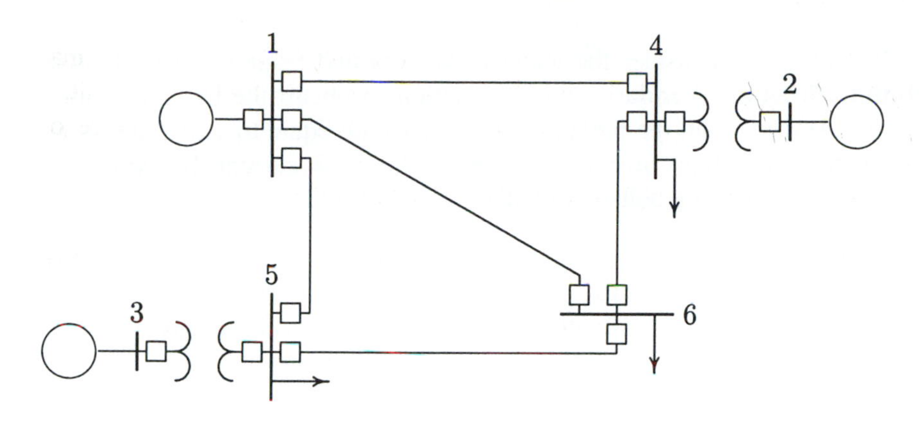

FIGURE 9.26
One-line diagram for Problem 9.16.

LINE AND TRANSFORMER DATA				
Bus No.	Bus No.	R, PU	X, PU	$\frac{1}{2}$ B, PU
1	4	0.035	0.225	0.0065
1	5	0.025	0.105	0.0045
1	6	0.040	0.215	0.0055
2	4	0.000	0.035	0.0000
3	5	0.000	0.042	0.0000
4	6	0.028	0.125	0.0035
5	6	0.026	0.175	0.0300

The transient impedance of the generators on a 100-MVA base are given below.

GEN. TRANSIENT IMPEDANCE, PU		
Gen. No.	R_a	X'_d
1	0	0.20
2	0	0.15
3	0	0.25

Neglecting the shunt capacitors and the loads, use **Zbus = zbuild(zdata)** function to obtain the bus impedance matrix. Assuming all the prefault bus voltages are equal to $1\angle 0°$, use **symfault(zdata, Zbus)** function to compute the fault current, bus voltages, and line currents for a bolted fault at bus 6. When using **Zbus = zbuild(zdata)** function, the generator reactances must be included in the **zdata** array with bus zero as the reference bus.

9.17. In Problem 9.16 consider the shunt capacitors and neglect the loads. use **zbuildpi(linedata, gendata, yload)** function to obtain the bus impedance matrix. Assuming all the prefault bus voltages are equal to $1\angle 0°$, use **symfault(linedata, Zbus)** function to compute the fault current, bus voltages, and line currents for a bolted fault at bus 6.

9.18. Repeat the symmetrical three-phase short circuit analysis for Problem 9.16 considering the prefault bus voltages and the effect of load currents. The load data is as follows.

LOAD DATA		
Bus	Load	
No.	MW	Mvar
1	0	0
2	0	0
3	0	0
4	100	70
5	90	30
6	160	110

Voltage magnitude, generation schedule, and the reactive power limits for the regulated buses are tabulated below. Bus 1, whose voltage is specified as $V_1 = 1.06\angle 0°$, is taken as the slack bus.

GENERATION DATA				
Bus	Voltage	Generation,	Mvar Limits	
No.	Mag.	MW	Min.	Max.
1	1.060			
2	1.040	150.0	0.0	140.0
3	1.030	100.0	0.0	90.0

Use anyone of the power flow programs to obtain the prefault bus voltages and the load admittance. The power flow program returns the prefault bus voltage array **V** and the bus load admittance array **yload**.

CHAPTER

10

SYMMETRICAL COMPONENTS AND UNBALANCED FAULT

10.1 INTRODUCTION

Different types of unbalanced faults are the *single line-to-ground fault*, *line-to-line fault*, and *double line-to-ground fault*.

The fault study that was presented in Chapter 9 has considered only three-phase balanced faults, which lends itself to a simple per phase approach. Various methods have been devised for the solution of unbalanced faults. However, since the one-line diagram simplifies the solution of the balanced three-phase problems, the method of symmetrical components that resolves the solution of unbalanced circuit into a solution of a number of balanced circuits is used. In this chapter, the symmetrical components method is discussed. It is then applied to the unbalanced faults, which allows once again the treatment of the problem on a simple per phase basis. Two functions are developed for the symmetrical components transformations. These are **abc2sc**, which provides transformation from phase quantities to symmetrical components, and **sc2abc** for the inverse transformation. In addition, these functions produce plots of unbalanced phasors and their symmetrical components. Finally, unbalanced faults are computed using the concept of symmetrical components. Three functions named **lgfault(zdata0, zbus0, zdata1, zbus1, zdata2, zbus2, V)**, **llfault(zdata1, zbus1, zdata2, zbus2, V)**, and **dlgfault(zdata0, zbus0, zdata1, zbus1, zdata2, zbus2, V)** are developed for the line-to-ground, line-to-line, and the double line-to-ground fault studies.

10.2 FUNDAMENTALS OF SYMMETRICAL COMPONENTS

Symmetrical components allow unbalanced phase quantities such as currents and voltages to be replaced by three separate balanced symmetrical components.

In three-phase system the phase sequence is defined as the order in which they pass through a positive maximum. Consider the phasor representation of a three-phase balanced current shown in Figure 10.1(a).

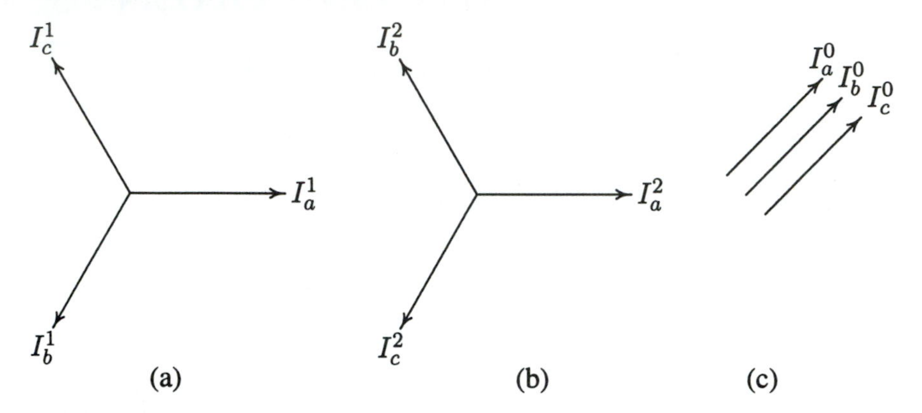

FIGURE 10.1
Representation of symmetrical components.

By convention, the direction of rotation of the phasors is taken to be counterclockwise. The three phasors are written as

$$\begin{aligned} I_a^1 &= I_a^1\angle 0^\circ &&= I_a^1 \\ I_b^1 &= I_a^1\angle 240^\circ &&= a^2 I_a^1 \\ I_c^1 &= I_a^1\angle 120^\circ &&= aI_a^1 \end{aligned} \tag{10.1}$$

where we have defined an operator a that causes a counterclockwise rotation of 120°, such that

$$\begin{aligned} a &= 1\angle 120^\circ = -0.5 + j0.866 \\ a^2 &= 1\angle 240^\circ = -0.5 - j0.866 \\ a^3 &= 1\angle 360^\circ = 1 + j0 \end{aligned} \tag{10.2}$$

From above, it is clear that

$$1 + a + a^2 = 0 \tag{10.3}$$

The order of the phasors is abc. This is designated the *positive phase sequence*. When the order is acb as in Figure 10.1(b), it is designated the *negative phase sequence*. The negative phase sequence quantities are represented as

$$\begin{aligned} I_a^2 &= I_a^2\angle 0^\circ &&= I_a^2 \\ I_b^2 &= I_a^2\angle 120^\circ &&= aI_a^2 \\ I_c^2 &= I_a^2\angle 240^\circ &&= a^2I_a^2 \end{aligned} \tag{10.4}$$

When analyzing certain types of unbalanced faults, it will be found that a third set of balanced phasors must be introduced. These phasors, known as the *zero phase sequence*, are found to be in phase with each other. Zero phase sequence currents, as in Figure 10.1(c), would be designated

$$I_a^0 = I_b^0 = I_c^0 \tag{10.5}$$

The superscripts 1, 2, and 0 are being used to represent positive, negative, and zero-sequence quantities, respectively. In some texts the notation 0, +, − is used instead of 0, 1, 2. The symmetrical components method was introduced by Dr. C. L. Fortescue in 1918. Based on his theory, three-phase unbalanced phasors of a three-phase system can be resolved into three balanced systems of phasors as follows.

1. Positive-sequence components consisting of a set of balanced three-phase components with a phase sequence *abc*.
2. Negative-sequence components consisting of a set of balanced three-phase components with a phase sequence *acb*.
3. Zero-sequence components consisting of three single-phase components, all equal in magnitude but with the same phase angles.

Consider the three-phase unbalanced currents I_a, I_b, and I_c shown in Figure 10.2 (page 405). We are seeking to find the three symmetrical components of the current such that

$$\begin{aligned} I_a &= I_a^0 + I_a^1 + I_a^2 \\ I_b &= I_b^0 + I_b^1 + I_b^2 \\ I_c &= I_c^0 + I_c^1 + I_c^2 \end{aligned} \tag{10.6}$$

According to the definition of the symmetrical components as given by (10.1), (10.4), and (10.5), we can rewrite (10.6) all in terms of phase a components.

$$\begin{aligned} I_a &= I_a^0 + I_a^1 + I_a^2 \\ I_b &= I_a^0 + a^2I_a^1 + aI_a^2 \\ I_c &= I_a^0 + aI_a^1 + a^2I_a^2 \end{aligned} \tag{10.7}$$

or

$$\begin{bmatrix} I_a \\ I_b \\ I_c \end{bmatrix} = \begin{bmatrix} 1 & 1 & 1 \\ 1 & a^2 & a \\ 1 & a & a^2 \end{bmatrix} \begin{bmatrix} I_a^0 \\ I_a^1 \\ I_a^2 \end{bmatrix} \tag{10.8}$$

We rewrite the above equation in matrix notation as

$$\mathbf{I}^{abc} = \mathbf{A}\, \mathbf{I}_a^{012} \tag{10.9}$$

where $\mathbf{A}$ is known as *symmetrical components transformation matrix* (SCTM) which transforms phasor currents $\mathbf{I}^{abc}$ into component currents $\mathbf{I}_a^{012}$, and is

$$\mathbf{A} = \begin{bmatrix} 1 & 1 & 1 \\ 1 & a^2 & a \\ 1 & a & a^2 \end{bmatrix} \tag{10.10}$$

Solving (10.9) for the symmetrical components of currents, we have

$$\mathbf{I}_a^{012} = \mathbf{A}^{-1}\, \mathbf{I}^{abc} \tag{10.11}$$

The inverse of $\mathbf{A}$ is given by

$$\mathbf{A}^{-1} = \frac{1}{3} \begin{bmatrix} 1 & 1 & 1 \\ 1 & a & a^2 \\ 1 & a^2 & a \end{bmatrix} \tag{10.12}$$

From (10.10) and (10.12), we conclude that

$$\mathbf{A}^{-1} = \frac{1}{3}\mathbf{A}^* \tag{10.13}$$

Substituting for $\mathbf{A}^{-1}$ in (10.11), we have

$$\begin{bmatrix} I_a^0 \\ I_a^1 \\ I_a^2 \end{bmatrix} = \frac{1}{3} \begin{bmatrix} 1 & 1 & 1 \\ 1 & a & a^2 \\ 1 & a^2 & a \end{bmatrix} \begin{bmatrix} I_a \\ I_b \\ I_c \end{bmatrix} \tag{10.14}$$

or in component form, the symmetrical components are

$$\begin{aligned} I_a^0 &= \tfrac{1}{3}(I_a + I_b + I_c) \\ I_a^1 &= \tfrac{1}{3}(I_a + aI_b + a^2 I_c) \\ I_a^2 &= \tfrac{1}{3}(I_a + a^2 I_b + aI_c) \end{aligned} \tag{10.15}$$

From (10.15), we note that the zero-sequence component of current is equal to one-third of the sum of the phase currents. Therefore, when the phase currents sum to zero, e.g., in a three-phase system with ungrounded neutral, the zero-sequence current cannot exist. If the neutral of the power system is grounded, zero-sequence current flows between the neutral and the ground.

Similar expressions exist for voltages. Thus the unbalanced phase voltages in terms of the symmetrical components voltages are

$$\begin{aligned} V_a &= V_a^0 + V_a^1 + V_a^2 \\ V_b &= V_a^0 + a^2V_a^1 + aV_a^2 \\ V_c &= V_a^0 + aV_a^1 + a^2V_a^2 \end{aligned} \tag{10.16}$$

or in matrix notation

$$\mathbf{V}^{abc} = \mathbf{A}\ \mathbf{V}_a^{012} \tag{10.17}$$

The symmetrical components in terms of the unbalanced voltages are

$$\begin{aligned} V_a^0 &= \tfrac{1}{3}(V_a + V_b + V_c) \\ V_a^1 &= \tfrac{1}{3}(V_a + aV_b + a^2V_c) \\ V_a^2 &= \tfrac{1}{3}(V_a + a^2V_b + aV_c) \end{aligned} \tag{10.18}$$

or in matrix notation

$$\mathbf{V}_a^{012} = \mathbf{A}^{-1}\ \mathbf{V}^{abc} \tag{10.19}$$

The apparent power may also be expressed in terms of the symmetrical components. The three-phase complex power is

$$S_{(3\phi)} = \mathbf{V}^{abc^T}\mathbf{I}^{abc*} \tag{10.20}$$

Substituting (10.9) and (10.17) in (10.20), we obtain

$$\begin{aligned} S_{(3\phi)} &= \left(\mathbf{A}\mathbf{V}_a^{012}\right)^T \left(\mathbf{A}\mathbf{I}_a^{012}\right)^* \\ &= \mathbf{V}_a^{012^T}\mathbf{A}^T\mathbf{A}^*\mathbf{I}_a^{012*} \end{aligned} \tag{10.21}$$

Since $\mathbf{A}^T = \mathbf{A}$, then from (10.13), $\mathbf{A}^T\mathbf{A}^* = 3$, and the complex power becomes

$$\begin{aligned} S_{(3\phi)} &= 3\left(\mathbf{V}^{012^T}\mathbf{I}^{012*}\right) \\ &= 3V_a^0I_a^{0*} + 3V_a^1I_a^{1*} + 3V_a^2I_a^{2*} \end{aligned} \tag{10.22}$$

Equation (10.22) shows that the total unbalanced power can be obtained from the sum of the symmetrical component powers. Often the subscript a of the symmetrical components are omitted, e.g., I^0, I^1, and I^2 are understood to refer to phase a.

Transformation from phase quantities to symmetrical components in *MATLAB* is very easy. Once the symmetrical components transformation matrix **A** is defined, its inverse is found using the *MATLAB* function **inv**. However, for quick calculations and graphical demonstration, the following functions are developed for symmetrical components analysis.

sctm The symmetrical components transformation matrix **A** is defined in this script file. Typing **sctm** defines **A**.

phasor(F) This function makes plots of phasors. The variable F may be expressed in an $n \times 1$ array in rectangular complex form or as an $n \times 2$ matrix. In the latter case, the first column is the phasor magnitude and the second column is its phase angle in degree.

F_{012} = **abc2sc**(F_{abc}) This function returns the symmetrical components of a set of unbalanced phasors in rectangular form. F_{abc} may be expressed in a 3×1 array in rectangular complex form or as a 3×2 matrix. In the latter case, the first column is the phasor magnitude and the second column is its phase angle in degree for a, b, and c phases. In addition, the function produces a plot of the unbalanced phasors and its symmetrical components.

F_{abc} = **sc2abc**(F_{012}) This function returns the unbalanced phasor in rectangular form when symmetrical components are specified. F_{012} may be expressed in a 3×1 array in rectangular complex form or as a 3×2 matrix. In the latter case, the first column is the phasor magnitude and the second column is its phase angle in degree for the zero-, positive-, and negative-sequence components, respectively. In addition, the function produces a plot of the unbalanced phasors and its symmetrical components.

Z_{012} = **zabc2sc**(Z_{abc}) This function transforms the phase impedance matrix to the sequence impedance matrix, given by (10.30).

F_p = **rec2pol**(F_r) This function converts the rectangular phasor F_r into polar form F_p.

F_r = **pol2rec**(F_p) This function converts the polar phasor F_p into rectangular form F_r.

Example 10.1

Obtain the symmetrical components of a set of unbalanced currents $I_a = 1.6\angle 25^\circ$, $I_b = 1.0\angle 180^\circ$, and $I_c = 0.9\angle 132^\circ$.

The commands

```
Iabc = [1.6     25
        1.0    180
        0.9    132];
I012 = abc2sc(Iabc);  % Symmetrical components of phase a
I012p= rec2pol(I012)          % Rectangular to polar form
```

result in

```
I012P =
        0.4512    96.4529
        0.9435    -0.0550
        0.6024    22.3157
```

and the plots of the phasors are shown in Figure 10.2.

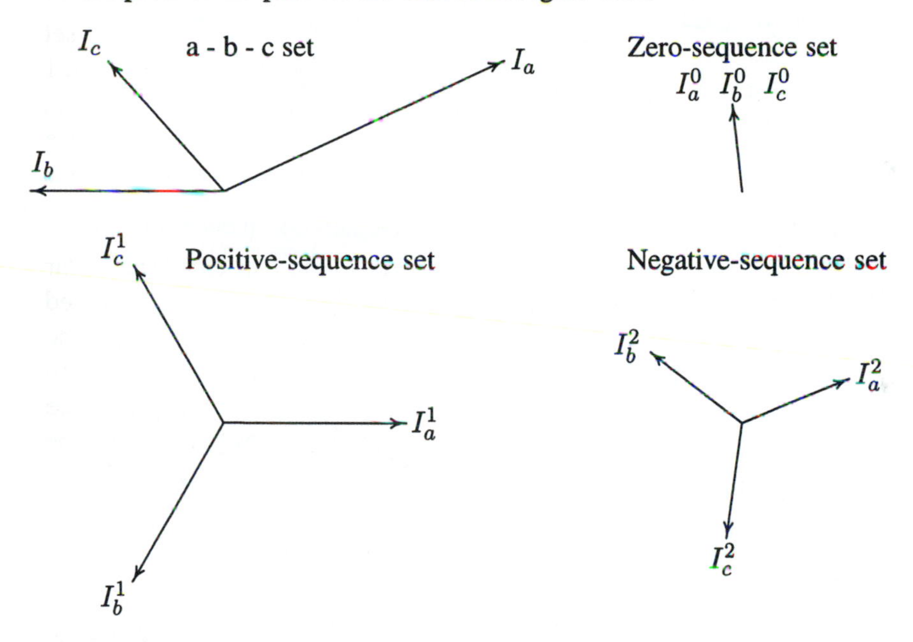

FIGURE 10.2
Resolution of unbalanced phasors into symmetrical components.

Example 10.2

The symmetrical components of a set of unbalanced three-phase voltages are $V_a^0 = 0.6\angle 90°$, $V_a^1 = 1.0\angle 30°$, and $V_a^2 = 0.8\angle -30°$. Obtain the original unbalanced phasors.

The commands

```
V012 = [0.6    90
        1.0    30
        0.8   -30];
Vabc = sc2abc(V012);%Unbalanced phasor to symmetrical comp.
Vabcp= rec2pol(Vabc)              % Rectangular to polar form
```

result in

```
Vabcp =
        1.7088    24.1825
        0.400     90.0000
        1.7088   155.8175
```

and the plots of the phasors are shown in Figure 10.3.

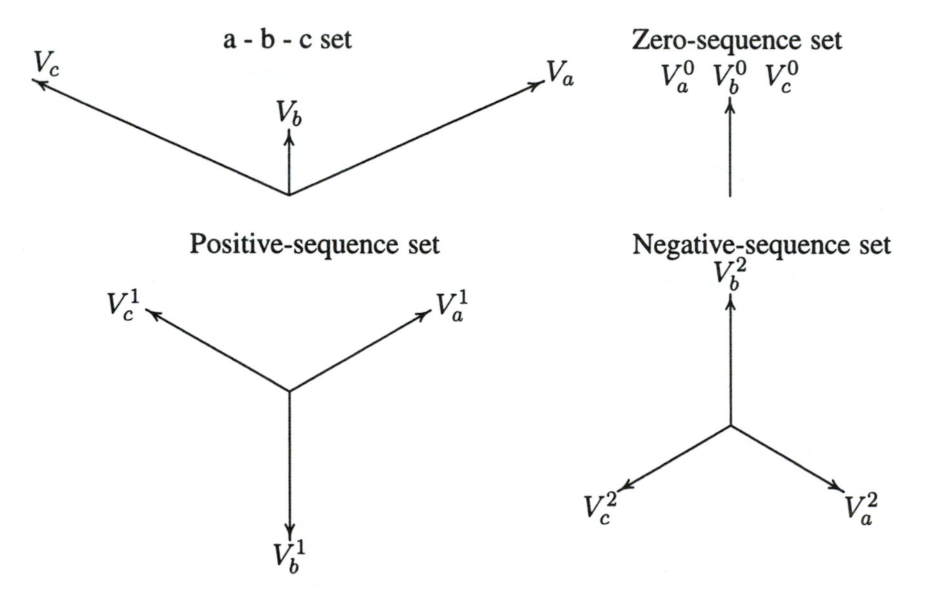

FIGURE 10.3
Transformation of the symmetrical components into phasor components.

10.3 SEQUENCE IMPEDANCES

This is the impedance of an equipment or component to the current of different sequences. The impedance offered to the flow of positive-sequence currents is known as the *positive-sequence impedance* and is denoted by Z^1. The impedance offered to the flow of negative-sequence currents is known as the *negative-sequence impedance*, shown by Z^2. When zero-sequence currents flow, the impedance is called the *zero-sequence impedance*, shown by Z^0. The sequence impedances of transmission lines, generators, and transformers are considered briefly here.

10.3.1 SEQUENCE IMPEDANCES OF Y-CONNECTED LOADS

A three-phase balanced load with self and mutual elements is shown in Figure 10.4. The load neutral is grounded through an impedance Z_n.

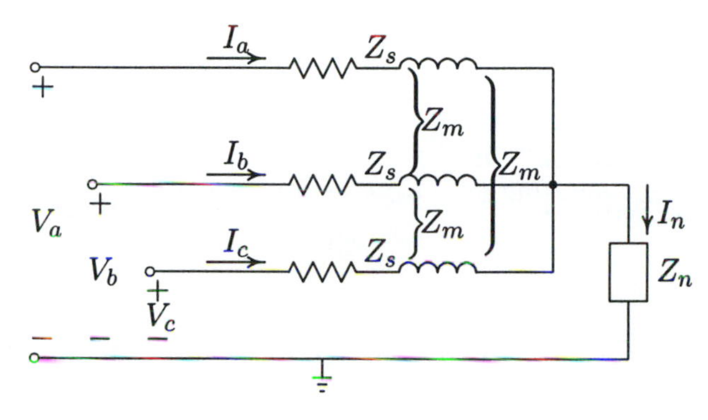

FIGURE 10.4
Balanced Y-connected load.

The line-to-ground voltages are

$$\begin{aligned} V_a &= Z_s I_a + Z_m I_b + Z_m I_c + Z_n I_n \\ V_b &= Z_m I_a + Z_s I_b + Z_m I_c + Z_n I_n \\ V_c &= Z_m I_a + Z_m I_b + Z_s I_c + Z_n I_n \end{aligned} \tag{10.23}$$

From Kirchhoff's current law, we have

$$I_n = I_a + I_b + I_c \tag{10.24}$$

Substituting for I_n from (10.24) into (10.23) and rewriting this equation in matrix form, yields

$$\begin{bmatrix} V_a \\ V_b \\ V_c \end{bmatrix} = \begin{bmatrix} Z_s + Z_n & Z_m + Z_n & Zm + Z_n \\ Z_m + Z_n & Z_s + Z_n & Zm + Z_n \\ Z_m + Z_n & Z_m + Z_n & Zs + Z_n \end{bmatrix} \begin{bmatrix} I_a \\ I_b \\ I_c \end{bmatrix} \tag{10.25}$$

or in compact form

$$\mathbf{V}^{abc} = \mathbf{Z}^{abc}\mathbf{I}^{abc} \tag{10.26}$$

where

$$\mathbf{Z}^{abc} = \begin{bmatrix} Z_s + Z_n & Z_m + Z_n & Z_m + Z_n \\ Z_m + Z_n & Z_s + Z_n & Z_m + Z_n \\ Z_m + Z_n & Z_m + Z_n & Z_s + Z_n \end{bmatrix} \tag{10.27}$$

Writing $\mathbf{V}^{abc}$ and $\mathbf{I}^{abc}$ in terms of their symmetrical components, we get

$$\mathbf{A}\mathbf{V}_a^{012} = \mathbf{Z}^{abc}\mathbf{A}\mathbf{I}_a^{012} \tag{10.28}$$

Multiplying (10.28) by $\mathbf{A}^{-1}$, we get

$$\begin{aligned} \mathbf{V}_a^{012} &= \mathbf{A}^{-1}\mathbf{Z}^{abc}\mathbf{A}\mathbf{I}_a^{012} \\ &= \mathbf{Z}^{012}\mathbf{I}_a^{012} \end{aligned} \tag{10.29}$$

where

$$\mathbf{Z}^{012} = \mathbf{A}^{-1}\mathbf{Z}^{abc}\mathbf{A} \tag{10.30}$$

Substituting for $\mathbf{Z}^{abc}$, $\mathbf{A}$, and $\mathbf{A}^{-1}$ from (10.27), (10.10), and (10.12), we have

$$\mathbf{Z}^{012} = \frac{1}{3}\begin{bmatrix} 1 & 1 & 1 \\ 1 & a & a^2 \\ 1 & a^2 & a \end{bmatrix} \begin{bmatrix} Z_s+Z_n & Z_m+Z_n & Z_m+Z_n \\ Z_m+Z_n & Z_s+Z_n & Z_m+Z_n \\ Z_m+Z_n & Z_m+Z_n & Z_s+Z_n \end{bmatrix} \begin{bmatrix} 1 & 1 & 1 \\ 1 & a^2 & a \\ 1 & a & a^2 \end{bmatrix} \tag{10.31}$$

Performing the above multiplications, we get

$$\mathbf{Z}^{012} = \begin{bmatrix} Z_s + 3Z_n + 2Z_m & 0 & 0 \\ 0 & Z_s - Z_m & 0 \\ 0 & 0 & Z_s - Z_m \end{bmatrix} \tag{10.32}$$

When there is no mutual coupling, we set $Z_m = 0$, and the impedance matrix becomes

$$\mathbf{Z}^{012} = \begin{bmatrix} Z_s + 3Z_n & 0 & 0 \\ 0 & Z_s & 0 \\ 0 & 0 & Z_s \end{bmatrix} \tag{10.33}$$

The impedance matrix has nonzero elements appearing only on the principal diagonal, and it is a diagonal matrix. Therefore, for a balanced load, the three sequences are independent. That is, currents of each phase sequence will produce voltage drops of the same phase sequence only. This is a very important property, as it permits the analysis of each sequence network on a per phase basis.

10.3.2 SEQUENCE IMPEDANCES OF TRANSMISSION LINES

Transmission line parameters were derived in Chapter 4. For static devices such as transmission lines, the phase sequence has no effect on the impedance, because the voltages and currents encounter the same geometry of the line, irrespective of the sequence. Thus, positive- and negative-sequence impedances are equal, i.e., $Z^1 = Z^2$.

In deriving the line parameters, the effect of ground and shielding conductors were neglected. Zero-sequence currents are in phase and flow through the a,b,c conductors to return through the grounded neutral. The ground or any shielding wire are effectively in the path of zero sequence. Thus, Z^0, which includes the effect of the return path through the ground, is generally different from Z^1 and Z^2. The determination of the zero sequence impedance with the presence of earth neutral wires is quite involved and the interested reader is referred to the Carson's formula [14]. To get an idea of the order of Z^0 we will consider the following simplified configuration. Consider 1-m length of a three-phase line with equilaterally spaced conductors as shown in Figure 10.5. The phase conductors carry zero-sequence (single-phase) currents with return paths through a grounded neutral. The ground surface is approximated to an equivalent fictitious conductor located at the average distance D_n from each of the three phases. Since conductor n carries the return current in opposite direction, we have

$$I_a^0 + I_b^0 + I_c^0 + I_n = 0 \tag{10.34}$$

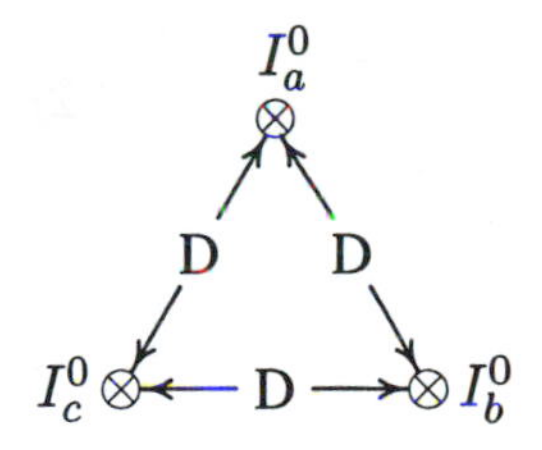

D_n

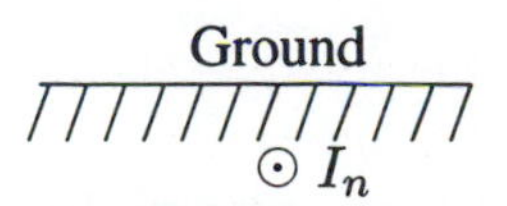

FIGURE 10.5
Zero-sequence current flow with earth return.

Since $I_a^0 = I_b^0 = I_c^0$, we have

$$I_n = -3I_a^0 \tag{10.35}$$

Utilizing the relation for the flux linkages of a conductor in a group expressed by (4.29), the total flux linkage of phase a conductor is

$$\lambda_{a0} = 2 \times 10^{-7} \left(I_a^0 \ln \frac{1}{r'} + I_b^0 \ln \frac{1}{D} + I_c^0 \ln \frac{1}{D} + I_n \ln \frac{1}{D_n} \right) \tag{10.36}$$

Substituting for I_b^0, I_c^0, and I_n in terms of I_a^0, we get

$$\begin{aligned}\lambda_{a0} &= 2 \times 10^{-7} I_a^0 \left(\ln \frac{1}{r'} + \ln \frac{1}{D} + \ln \frac{1}{D} - 3 \ln \frac{1}{D_n} \right) \\ &= 2 \times 10^{-7} I_a^0 \ln \frac{D_n^3}{r' D^2} \quad \text{Wb/m}\end{aligned} \tag{10.37}$$

Since $L_0 = \lambda_{a0}/I_a^0$, the zero sequence inductance per phase in mH per kilometer length is

$$\begin{aligned}L_0 &= 0.2 \ln \frac{D_n^3}{r' D^2} \\ &= 0.2 \ln \frac{D D_n^3}{r' D^3} \\ &= 0.2 \ln \frac{D}{r'} + 3 \left(0.2 \ln \frac{D_n}{D} \right) \quad \text{mH/Km}\end{aligned} \tag{10.38}$$

The first term above is the same as the positive-sequence inductance given by (4.33). Thus the zero sequence reactance can be expressed as

$$X^0 = X^1 + 3X_n \tag{10.39}$$

where

$$X_n = 2\pi f \left(0.2 \ln \frac{D_n}{D} \right) \quad \text{m}\Omega/\text{km} \tag{10.40}$$

The zero-sequence impedance of the transmission line is more than three times larger than the positive- or negative-sequence impedance.

10.3.3 SEQUENCE IMPEDANCES OF SYNCHRONOUS MACHINE

The inductances of a synchronous machine depend upon the phase order of the sequence current relative to the direction of rotation of the rotor. The positive-sequence generator impedance is the value found when positive-sequence current flows from the action of an imposed positive-sequence set of voltages. We have seen that the generator positive-sequence reactance varies, and in Section 9.2 one of the reactances X_d'', X_d', or X_d was used for the balanced three-phase fault studies.

When negative-sequence currents are impressed in the stator, the net flux in the air gap rotates at opposite direction to that of the rotor. That is, the net flux rotates at twice synchronous speed relative to the rotor. Since the field voltage is associated with the positive-sequence variables, the field winding has no influence. Consequently, only the damper winding produces an effect in the quadrature axis. Hence, there is no distinction between the transient and subtransient reactances in the quadrature axis as there is in the direct axis. The negative-sequence reactance is close to the positive-sequence subtransient reactance, i.e.,

$$X^2 \simeq X_d'' \tag{10.41}$$

Zero-sequence impedance is the impedance offered by the machine to the flow of the zero-sequence current. We recall that a set of zero sequence currents are all identical. Therefore, if the spatial distribution of mmf is assumed sinusoidal, the resultant air-gap flux would be zero, and there is no reactance due to armature reaction. The machine offers a very small reactance due to the leakage flux. Therefore, the zero-sequence reactance is approximated to the leakage reactance, i.e.,

$$X^0 \simeq X_\ell \tag{10.42}$$

10.3.4 SEQUENCE IMPEDANCES OF TRANSFORMER

In Chapter 3 we obtained the per phase equivalent circuit for a three-phase transformer. In power transformers, the core losses and the magnetization current are on the order of 1 percent of the rated value; therefore, the magnetizing branch is neglected. The transformer is modeled with the equivalent series leakage impedance. Since the transformer is a static device, the leakage impedance will not change if the phase sequence is changed. Therefore, the positive- and negative-sequence impedances are the same. Also, if the transformer permits zero-sequence current flow at all, the phase impedance to zero-sequence is equal to the leakage impedance, and we have

$$Z^0 = Z^1 = Z^2 = Z_\ell \tag{10.43}$$

From Section 3.9.1, we recall that in a Y-Δ, or a Δ-Y transformer, the positive-sequence line voltage on HV side leads the corresponding line voltage on the

LV side by $30°$. For the negative-sequence voltage the corresponding phase shift is $-30°$. The equivalent circuit for the zero-sequence impedance depends on the winding connections and also upon whether or not the neutrals are grounded. Figure 10.6 shows some of the more common transformer configurations and their zero-sequence equivalent circuits. We recall that in a transformer, when the core reluctance is neglected, there is an exact mmf balance between the primary and secondary. This means that current can flow in the primary only if there is a current in the secondary. Based on this observation we can check the validity of the zero-sequence circuits by applying a set of zero-sequence voltage to the primary and calculating the resulting currents.

(a) Y-Y connections with both neutrals grounded – We know that the zero sequence current equals the sum of phase currents. Since both neutrals are grounded, there is a path for the zero sequence current to flow in the primary and secondary, and the transformer exhibits the equivalent leakage impedance per phase as shown in Figure 10.6(a).

(b) Y-Y connection with the primary neutral grounded – The primary neutral is grounded, but since the secondary neutral is isolated, the secondary phase current must sum up to zero. This means that the zero-sequence current in the secondary is zero. Consequently, the zero sequence current in the primary is zero, reflecting infinite impedance or an open circuit as shown in Figure 10.6(b).

(c) Y-Δ with grounded neutral – In this configuration the primary currents can flow because there is zero-sequence circulating current in the Δ-connected secondary and a ground return path for the Y-connected primary. Note that no zero-sequence current can leave the Δ terminals, thus there is an isolation between the primary and secondary sides as shown in Figure 10.6(c).

(d) Y-Δ connection with isolated neutral – In this configuration, because the neutral is isolated, zero sequence current cannot flow and the equivalent circuit reflects an infinite impedance or an open as shown in Figure 10.6(d).

(e) Δ-Δ connection – In this configuration zero-sequence currents circulate in the Δ-connected windings, but no currents can leave the Δ terminals, and the equivalent circuit is as shown in Figure 10.6(e).

Notice that the neutral impedance plays an important part in the equivalent circuit. When the neutral is grounded through an impedance Z_n, because $I_n = 3I_0$, in the equivalent circuit the neutral impedance appears as $3Z_n$ in the path of I_0.

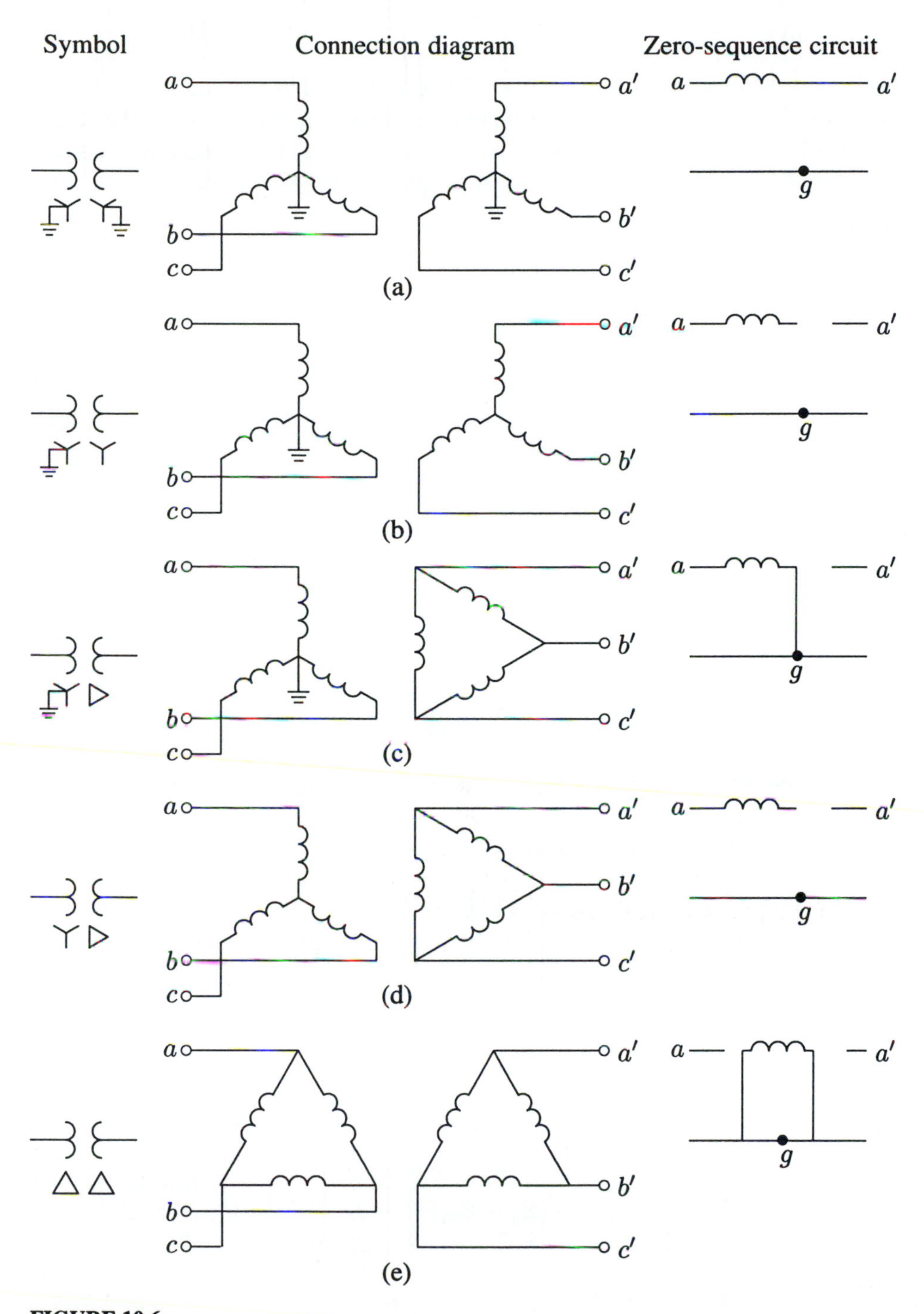

FIGURE 10.6
Transformer zero-sequence equivalent circuits.

Example 10.3

A balanced three-phase voltage of 100-V line-to-neutral is applied to a balanced Y-connected load with ungrounded neutral as shown in Figure 10.7. The three-phase load consists of three mutually-coupled reactances. Each phase has a series reactance of $Z_s = j12\ \Omega$, and the mutual coupling between phases is $Z_m = j4\ \Omega$.

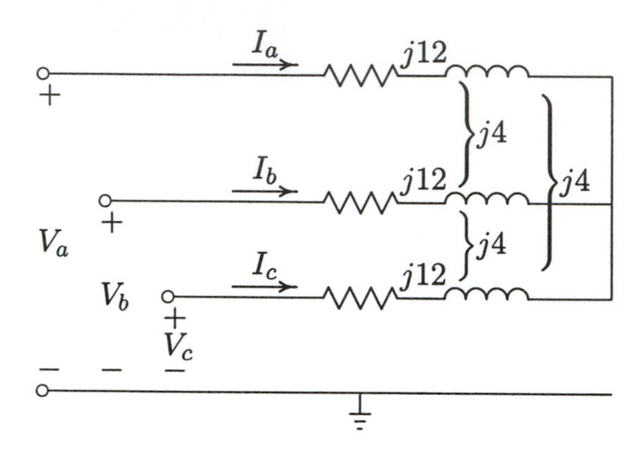

FIGURE 10.7
Circuit for Example 10.3.

(a) Determine the line currents by mesh analysis without using symmetrical components.
(b) Determine the line currents using symmetrical components.

(a) Applying KVL to the two independent mesh equations yields

$$Z_sI_a + Z_mI_b - Z_sI_b - Z_mI_a = V_a - V_b = |V_L|\angle\pi/6$$
$$Z_sI_b + Z_mI_c - Z_sI_c - Z_mI_b = V_b - V_c = |V_L|\angle -\pi/2$$

Also from KCL, we have

$$I_a + I_b + I_c = 0$$

Writing above equations in matrix form, results in

$$\begin{bmatrix} (Z_s - Z_m) & -(Z_s - Z_m) & 0 \\ 0 & (Z_s - Z_m) & -(Z_s - Z_m) \\ 1 & 1 & 1 \end{bmatrix} \begin{bmatrix} I_a \\ I_b \\ I_c \end{bmatrix} = \begin{bmatrix} |V_L|\angle\pi/6 \\ |V_L|\angle -\pi/2 \\ 0 \end{bmatrix}$$

or in compact form

$$\mathbf{Z}_{mesh}\mathbf{I}^{abc} = \mathbf{V}_{mesh}$$

Solving the above equations results in the line currents

$$\mathbf{I}^{abc} = \mathbf{Z}_{mesh}{}^{-1}\mathbf{V}_{mesh}$$

The following commands

```
% (a) Solution by mesh analysis
Zs=j*12; Zm=j*4; Va = 100; VL=Va*sqrt(3);
Z= [(Zs-Zm)  -(Zs-Zm)      0
       0      (Zs-Zm)  -(Zs-Zm)
       1         1         1   ];
V=[VL*cos(pi/6)+j*VL*sin(pi/6)
   VL*cos(-pi/2)+j*VL*sin(-pi/2)
             0                  ];
Y=inv(Z)
Iabc=Y*V;                      % Line currents (Rectangular form)
Iabcp=[abs(Iabc), angle(Iabc)*180/pi] %  Line currents (Polar)
```

result in

```
     Iabcp =
             12.5     -90.0
             12.5     150.0
             12.5      30.0
```

(b) Using the symmetrical components method, we have

$$\mathbf{V}^{012} = \mathbf{Z}^{012}\mathbf{I}^{012}$$

where

$$\mathbf{V}^{012} = \begin{bmatrix} 0 \\ V_a \\ 0 \end{bmatrix}$$

and from (10.32)

$$\mathbf{Z}^{012} = \begin{bmatrix} Z_s + 2Z_m & 0 & 0 \\ 0 & Z_s - Z_m & 0 \\ 0 & 0 & Z_s - Z_m \end{bmatrix}$$

for the sequence components of currents, we get

$$\mathbf{I}^{012} = [\mathbf{Z}^{012}]^{-1}\mathbf{V}^{012}$$

We write the following commands

```
% (b) Solution by symmetrical components method
Z012=[Zs+2*Zm   0      0          % Symmetrical components matrix
        0      Zs-Zm   0
        0       0     Zs-Zm];
V012=[0; Va ; 0];      %Symmetrical components of phase voltages
I012=inv(Z012)*V012; %Symmetrical components of line  currents
a=cos(2*pi/3)+j*sin(2*pi/3);
A=[ 1  1  1; 1 a^2  a; 1 a a^2];          % Transformation matrix
Iabc=A*I012;                        % Line currents (Rectangular form)
Iabcp=[abs(Iabc), angle(Iabc)*180/pi]  % Line currents (Polar)
```

which result in

```
    Iabcp =
            12.5     -90.0
            12.5     150.0
            12.5      30.0
```

This is the same result as in part (a).

Example 10.4

A three-phase unbalanced source with the following phase-to-neutral voltages

$$\mathbf{V}^{abc} = \begin{bmatrix} 200 & \angle 25^\circ \\ 100 & \angle -155^\circ \\ 80 & \angle 100^\circ \end{bmatrix}$$

is applied to the circuit in Figure 10.4 (page 407). The load series impedance per phase is $Z_s = 8 + j24$ and the mutual impedance between phases is $Z_m = j4$. The load and source neutrals are solidly grounded. Determine

(a) The load sequence impedance matrix $\mathbf{Z}^{012} = \mathbf{A}^{-1}\mathbf{Z}^{abc}\mathbf{A}$.
(b) The symmetrical components of voltage.
(c) The symmetrical components of current.
(d) The load phase currents.
(e) The complex power delivered to the load in terms of symmetrical components, $S_{3\phi} = 3(V_a^0 I_a^{0*} + V_a^1 I_a^{1*} + V_a^2 I_a^{2*})$.
(f) The complex power delivered to the load by summing up the power in each phase, $S_{3\phi} = V_a I_a^* + V_b I_b^* + V_c I_c^*$.

We write the following commands

```
Vabc = [200    25
        100  -155
         80   100];
Zabc = [8+j*24      j*4      j*4
           j*4  8+j*24      j*4
          j*4      j*4  8+j*24];
Z012 = zabc2sc(Zabc)     % Symmetrical components of impedance
V012 = abc2sc(Vabc);      % Symmetrical components of voltage
V012p= rec2pol(V012)                  % Rectangular to polar form
I012 = inv(Z012)*V012;    % Symmetrical components of current
I012p= rec2pol(I012)                  % Rectangular to polar form
Iabc = sc2abc(I012);                             % Phase currents
Iabcp= rec2pol(Iabc)                  % Rectangular to polar form
S3ph =3*(V012.')*conj(I012)%Power using symmetrical components
Vabcr = Vabc(:, 1).*(cos(pi/180*Vabc(:, 2)) +...
j*sin(pi/180*Vabc(:, 2)));
S3ph=(Vabcr.')*conj(Iabc)
                           % Power using phase currents and voltages
```

The result is

```
Z012 =
       8.00 + 32.00i   0.00 +  0.00i   0.00 +  0.00i
       0.00 +  0.00i   8.00 + 20.00i   0.00 +  0.00i
       0.00 -  0.00i   0.00 -  0.00i   8.00 + 20.00i

V012p =
        47.7739      57.6268
       112.7841      -0.0331
        61.6231      45.8825

I012p =
         1.4484     -18.3369
         5.2359     -68.2317
         2.8608     -22.3161

Iabcp =
         8.7507     -47.0439
         5.2292     143.2451
         3.0280      39.0675

S3ph =
       9.0471e+002+ 2.3373e+003i

S3ph =
       9.0471e+002+ 2.3373e+003i
```

10.4 SEQUENCE NETWORKS OF A LOADED GENERATOR

Figure 10.8 represents a three-phase synchronous generator with neutral grounded through an impedance Z_n. The generator is supplying a three-phase balanced load.

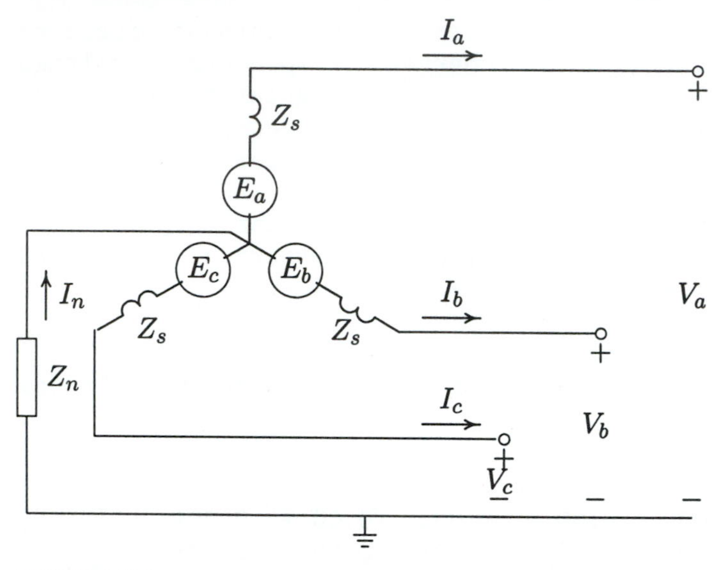

FIGURE 10.8
Three-phase balanced source and impedance.

The synchronous machine generates balanced three-phase internal voltages and is represented as a positive-sequence set of phasors

$$\mathbf{E}^{abc} = \begin{bmatrix} 1 \\ a^2 \\ a \end{bmatrix} E_a \tag{10.44}$$

The machine is supplying a three-phase balanced load. Applying Kirchhoff's voltage law to each phase we obtain

$$\begin{aligned} V_a &= E_a - Z_s I_a - Z_n I_n \\ V_b &= E_b - Z_s I_b - Z_n I_n \\ V_c &= E_c - Z_c I_c - Z_n I_n \end{aligned} \tag{10.45}$$

Substituting for $I_n = I_a + I_b + I_c$, and writing (10.45) in matrix form, we get

$$\begin{bmatrix} V_a \\ V_b \\ V_c \end{bmatrix} = \begin{bmatrix} E_a \\ E_b \\ E_c \end{bmatrix} - \begin{bmatrix} Z_s + Z_n & Z_n & Z_n \\ Z_n & Z_s + Z_n & Z_n \\ Z_n & Z_n & Z_s + Z_n \end{bmatrix} \begin{bmatrix} I_a \\ I_b \\ I_c \end{bmatrix} \tag{10.46}$$

or in compact form, we have

$$\mathbf{V}^{abc} = \mathbf{E}^{abc} - \mathbf{Z}^{abc}\mathbf{I}^{abc} \tag{10.47}$$

where $\mathbf{V}^{abc}$ is the phase terminal voltage vector and $\mathbf{I}^{abc}$ is the phase current vector. Transforming the terminal voltages and current phasors into their symmetrical components results in

$$\mathbf{A}\mathbf{V}_a^{012} = \mathbf{A}\mathbf{E}_a^{012} - \mathbf{Z}^{abc}\mathbf{A}\mathbf{I}_a^{012} \tag{10.48}$$

Multiplying (10.48) by $\mathbf{A}^{-1}$, we get

$$\begin{aligned}\mathbf{V}_a^{012} &= \mathbf{E}_a^{012} - \mathbf{A}^{-1}\mathbf{Z}^{abc}\mathbf{A}\mathbf{I}_a^{012}\\ &= \mathbf{E}_a^{012} - \mathbf{Z}^{012}\mathbf{I}_a^{012}\end{aligned} \tag{10.49}$$

where

$$\mathbf{Z}^{012} = \frac{1}{3}\begin{bmatrix} 1 & 1 & 1\\ 1 & a & a^2\\ 1 & a^2 & a\end{bmatrix}\begin{bmatrix} Z_s + Z_n & Z_n & Z_n\\ Z_n & Z_s + Z_n & Z_n\\ Z_n & Z_n & Z_s + Z_n\end{bmatrix}\begin{bmatrix} 1 & 1 & 1\\ 1 & a^2 & a\\ 1 & a & a^2\end{bmatrix} \tag{10.50}$$

Performing the above multiplications, we get

$$\mathbf{Z}^{012} = \begin{bmatrix} Z_s + 3Z_n & 0 & 0\\ 0 & Z_s & 0\\ 0 & 0 & Z_s\end{bmatrix} = \begin{bmatrix} Z^0 & 0 & 0\\ 0 & Z^1 & 0\\ 0 & 0 & Z^2\end{bmatrix} \tag{10.51}$$

Since the generated emf is balanced, there is only positive-sequence voltage, i.e.,

$$\mathbf{E}_a^{012} = \begin{bmatrix} 0\\ E_a\\ 0\end{bmatrix} \tag{10.52}$$

Substituting for $\mathbf{E}_a^{012}$ and $\mathbf{Z}^{012}$ in (10.49), we get

$$\begin{bmatrix} V_a^0\\ V_a^1\\ V_a^2\end{bmatrix} = \begin{bmatrix} 0\\ E_a\\ 0\end{bmatrix} - \begin{bmatrix} Z^0 & 0 & 0\\ 0 & Z^1 & 0\\ 0 & 0 & Z^2\end{bmatrix}\begin{bmatrix} I_a^0\\ I_a^1\\ I_a^2\end{bmatrix} \tag{10.53}$$

Since the above equation is very important, we write it in component form, and we get

$$\begin{aligned} V_a^0 &= 0 - Z^0 I_a^0\\ V_a^1 &= E_a - Z^1 I_a^1\\ V_a^2 &= 0 - Z^2 I_a^2\end{aligned} \tag{10.54}$$

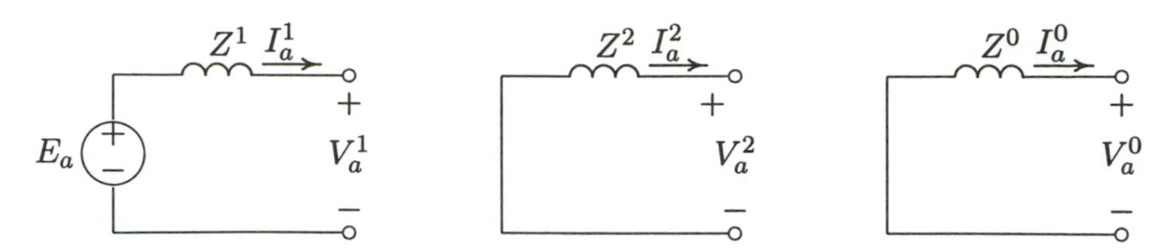

FIGURE 10.9
Sequence networks: (a) Positive-sequence; (b) negative-sequence; (c) zero-sequence.

The three equations given by (10.54) can be represented by the three equivalent sequence networks shown in Figure 10.9.

We make the following important observations.

- The three sequences are independent.
- The positive-sequence network is the same as the one-line diagram used in studying balanced three-phase currents and voltages.
- Only the positive-sequence network has a voltage source. Therefore, the positive-sequence current causes only positive-sequence voltage drops.
- There is no voltage source in the negative- or zero-sequence networks.
- Negative- and zero-sequence currents cause negative- and zero-sequence voltage drops only.
- The neutral of the system is the reference for positive-and negative-sequence networks, but ground is the reference for the zero-sequence networks. Therefore, the zero-sequence current can flow only if the circuit from the system neutrals to ground is complete.
- The grounding impedance is reflected in the zero sequence network as $3Z_n$.
- The three-sequence systems can be solved separately on a per phase basis. The phase currents and voltages can then be determined by superposing their symmetrical components of current and voltage respectively.

We are now ready with mathematical tools to analyze various types of unbalanced faults. First, the fault current is obtained using Thévenin's method and algebraic manipulation of sequence networks. The analysis will then be extended to find the bus voltages and fault current during fault, for different types of faults using the bus impedance matrix.

10.5 SINGLE LINE-TO-GROUND FAULT

Figure 10.10 illustrates a three-phase generator with neutral grounded through impedance Z_n.

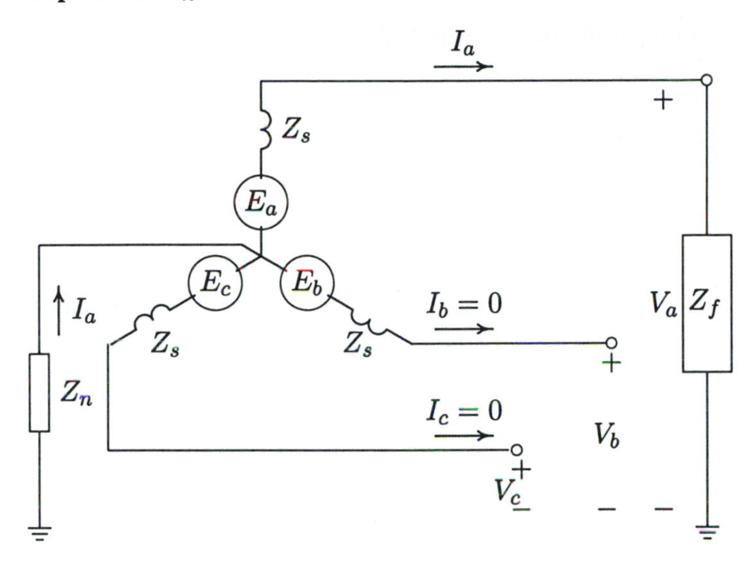

FIGURE 10.10
Line-to-ground fault on phase a.

Suppose a line-to-ground fault occurs on phase a through impedance Z_f. Assuming the generator is initially on no-load, the boundary conditions at the fault point are

$$V_a = Z_f I_a \tag{10.55}$$

$$I_b = I_c = 0 \tag{10.56}$$

Substituting for $I_b = I_c = 0$, the symmetrical components of currents from (10.14) are

$$\begin{bmatrix} I_a^0 \\ I_a^1 \\ I_a^2 \end{bmatrix} = \frac{1}{3} \begin{bmatrix} 1 & 1 & 1 \\ 1 & a & a^2 \\ 1 & a^2 & a \end{bmatrix} \begin{bmatrix} I_a \\ 0 \\ 0 \end{bmatrix} \tag{10.57}$$

From the above equation, we find that

$$I_a^0 = I_a^1 = I_a^2 = \frac{1}{3} I_a \tag{10.58}$$

Phase a voltage in terms of symmetrical components is

$$V_a = V_a^0 + V_a^1 + V_a^2 \tag{10.59}$$

Substituting for V_a^0, V_a^1, and V_a^2 from (10.54) and noting $I_a^0 = I_a^1 = I_a^2$, we get

$$V_a = E_a - (Z^1 + Z^2 + Z^0)I_a^0 \tag{10.60}$$

where $Z^0 = Z_s + 3Z_n$. Substituting for V_a from (10.55), and noting $I_a = 3I_a^0$, we get

$$3Z_f I_a^0 = E_a - (Z^1 + Z^2 + Z^0)I_a^0 \tag{10.61}$$

or

$$I_a^0 = \frac{E_a}{Z^1 + Z^2 + Z^0 + 3Z_f} \tag{10.62}$$

The fault current is

$$I_a = 3I_a^0 = \frac{3E_a}{Z^1 + Z^2 + Z^0 + 3Z_f} \tag{10.63}$$

Substituting for the symmetrical components of currents in (10.54), the symmetrical components of voltage and phase voltages at the point of fault are obtained.

Equations (10.58) and (10.62) can be represented by connecting the sequence networks in series as shown in the equivalent circuit of Figure 10.11. Thus, for line-to-ground faults, the Thévenin impedance to the point of fault is obtained for each sequence network, and the three sequence networks are placed in series. In many practical applications, the positive- and negative-sequence impedances are found to be equal. If the generator neutral is solidly grounded, $Z_n = 0$ and for bolted faults $Z_f = 0$.

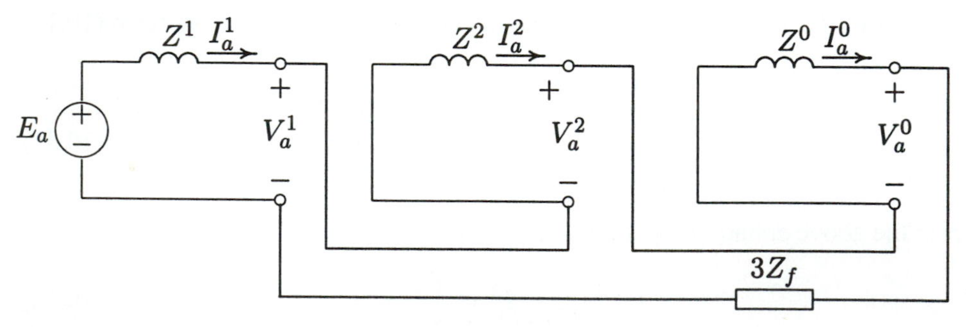

FIGURE 10.11
Sequence network connection for line-to-ground fault.

10.6 LINE-TO-LINE FAULT

Figure 10.12 shows a three-phase generator with a fault through an impedance Z_f between phases b and c. Assuming the generator is initially on no-load, the boundary conditions at the fault point are

$$V_b - V_c = Z_f I_b \tag{10.64}$$

$$I_b + I_c = 0 \tag{10.65}$$

$$I_a = 0 \tag{10.66}$$

Substituting for $I_a = 0$, and $I_c = -I_b$, the symmetrical components of currents from (10.14) are

$$\begin{bmatrix} I_a^0 \\ I_a^1 \\ I_a^2 \end{bmatrix} = \frac{1}{3}\begin{bmatrix} 1 & 1 & 1 \\ 1 & a & a^2 \\ 1 & a^2 & a \end{bmatrix}\begin{bmatrix} 0 \\ I_b \\ -I_b \end{bmatrix} \tag{10.67}$$

From the above equation, we find that

$$I_a^0 = 0 \tag{10.68}$$

$$I_a^1 = \frac{1}{3}(a - a^2)I_b \tag{10.69}$$

$$I_a^2 = \frac{1}{3}(a^2 - a)I_b \tag{10.70}$$

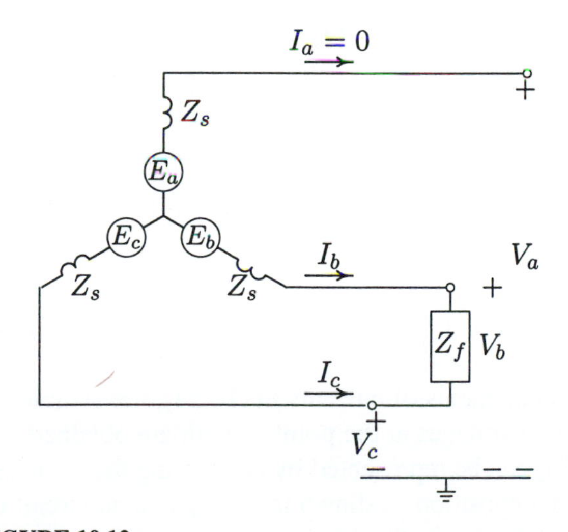

FIGURE 10.12
Line-to-line fault between phase b and c.

Also, from (10.69) and (10.70), we note that

$$I_a^1 = -I_a^2 \tag{10.71}$$

From (10.16), we have

$$\begin{aligned} V_b - V_c &= (a^2 - a)(V_a^1 - V_a^2) \\ &= Z_f I_b \end{aligned} \tag{10.72}$$

Substituting for V_a^1 and V_a^2 from (10.54) and noting $I_a^2 = -I_a^1$, we get

$$(a^2 - a)[E_a - (Z^1 + Z^2)I_a^1] = Z_f I_b \tag{10.73}$$

Substituting for I_b from (10.69), we get

$$E_a - (Z^1 + Z^2)I_a^1 = Z_f \frac{3I_a^1}{(a - a^2)(a^2 - a)} \tag{10.74}$$

Since $(a - a^2)(a^2 - a) = 3$, solving for I_a^1 results in

$$I_a^1 = \frac{E_a}{Z^1 + Z^2 + Z_f} \tag{10.75}$$

The phase currents are

$$\begin{bmatrix} I_a \\ I_b \\ I_c \end{bmatrix} = \begin{bmatrix} 1 & 1 & 1 \\ 1 & a^2 & a \\ 1 & a & a^2 \end{bmatrix} \begin{bmatrix} 0 \\ I_a^1 \\ -I_a^1 \end{bmatrix} \tag{10.76}$$

The fault current is

$$I_b = -I_c = (a^2 - a)I_a^1 \tag{10.77}$$

or

$$I_b = -j\sqrt{3}\, I_a^1 \tag{10.78}$$

Substituting for the symmetrical components of currents in (10.54), the symmetrical components of voltage and phase voltages at the point of fault are obtained.

Equations (10.71) and (10.75) can be represented by connecting the positive- and negative-sequence networks in opposition as shown in the equivalent circuit of Figure 10.13. In many practical applications, the positive- and negative-sequence impedances are found to be equal. For a bolted fault, $Z_f = 0$.

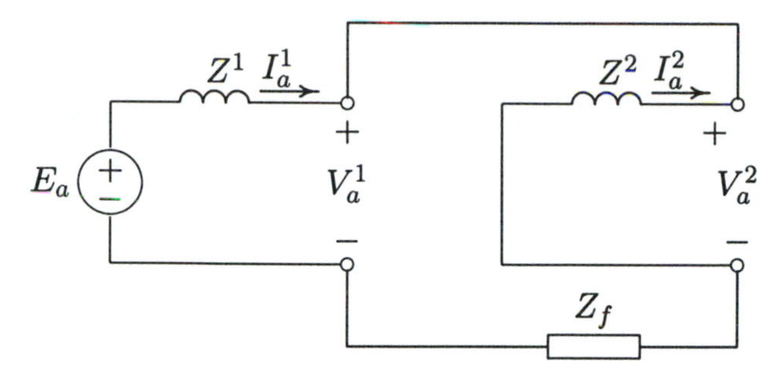

FIGURE 10.13
Sequence network connection for line-to-line fault.

10.7 DOUBLE LINE-TO-GROUND FAULT

Figure 10.14 shows a three-phase generator with a fault on phases b and c through an impedance Z_f to ground. Assuming the generator is initially on no-load, the boundary conditions at the fault point are

$$V_b = V_c = Z_f(I_b + I_c) \tag{10.79}$$

$$I_a = I_a^0 + I_a^1 + I_a^2 = 0 \tag{10.80}$$

From (10.16), the phase voltages V_b and V_c are

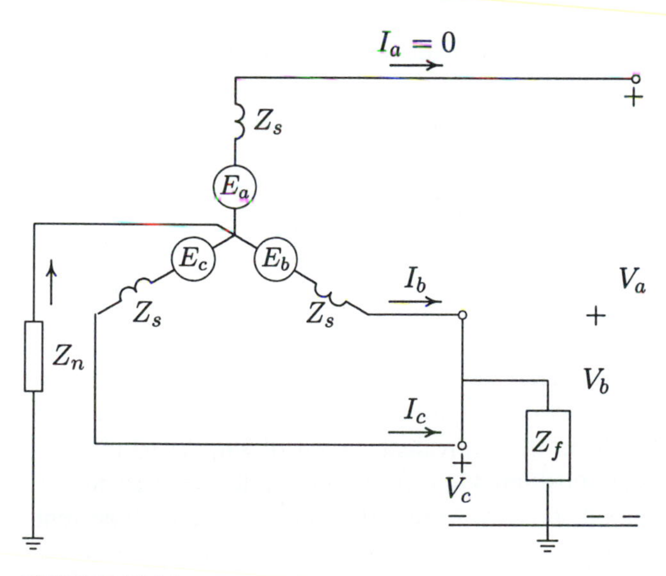

FIGURE 10.14
Double line-to-ground fault.

$$V_b = V_a^0 + a^2 V_a^1 + aV_a^2 \tag{10.81}$$
$$V_c = V_a^0 + aV_a^1 + a^2 V_a^2 \tag{10.82}$$

Since $V_b = V_c$, from above we note that

$$V_a^1 = V_a^2 \tag{10.83}$$

Substituting for the symmetrical components of currents in (10.79), we get

$$\begin{aligned} V_b &= Z_f(I_a^0 + a^2 I_a^1 + aI_a^2 + I_a^0 + aI_a^1 + a^2 I_a^2) \\ &= Z_f(2I_a^0 - I_a^1 - I_a^2) \\ &= 3Z_f I_a^0 \end{aligned} \tag{10.84}$$

Substituting for V_b from (10.84) and for V_a^2 from (10.83) into (10.81), we have

$$\begin{aligned} 3Z_f I_a^0 &= V_a^0 + (a^2 + a)V_a^1 \\ &= V_a^0 - V_a^1 \end{aligned} \tag{10.85}$$

Substituting for the symmetrical components of voltage from (10.54) into (10.85) and solving for I_a^0, we get

$$I_a^0 = -\frac{E_a - Z^1 I_a^1}{Z^0 + 3Z_f} \tag{10.86}$$

Also, substituting for the symmetrical components of voltage in (10.83), we obtain

$$I_a^2 = -\frac{E_a - Z^1 I_a^1}{Z^2} \tag{10.87}$$

Substituting for I_a^0 and I_a^2 into (10.80) and solving for I_a^1, we get

$$I_a^1 = \frac{E_a}{Z^1 + \frac{Z^2(Z^0+3Z_f)}{Z^2+Z^0+3Z_f}} \tag{10.88}$$

Equations (10.86)–(10.88) can be represented by connecting the positive-sequence impedance in series with the parallel combination of the negative-sequence and zero-sequence networks as shown in the equivalent circuit of Figure 10.15. The value of I_a^1 found from (10.88) is substituted in (10.86) and (10.87), and I_a^0 and I_a^2 are found. The phase currents are then found from (10.8). Finally, the fault current is obtained from

$$I_f = I_b + I_c = 3I_a^0 \tag{10.89}$$

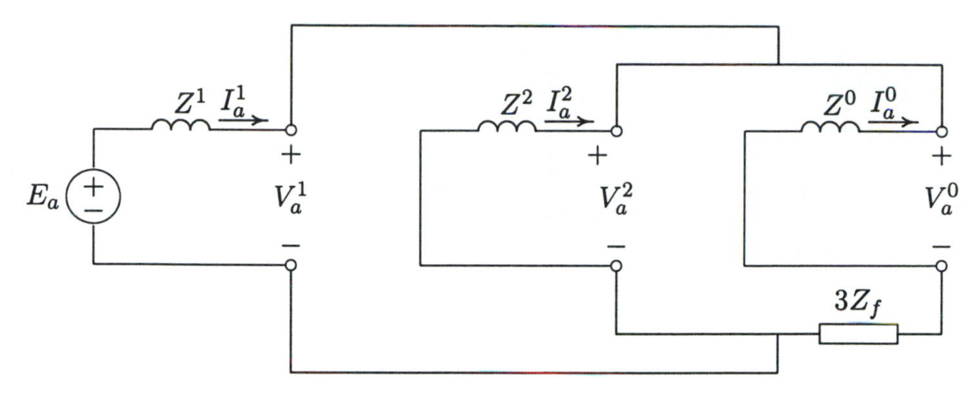

FIGURE 10.15
Sequence network connection for double line-to-ground fault.

Example 10.5

The one-line diagram of a simple power system is shown in Figure 10.16. The neutral of each generator is grounded through a current-limiting reactor of 0.25/3 per unit on a 100-MVA base. The system data expressed in per unit on a common 100-MVA base is tabulated below. The generators are running on no-load at their rated voltage and rated frequency with their emfs in phase.

Determine the fault current for the following faults.

(a) A balanced three-phase fault at bus 3 through a fault impedance $Z_f = j0.1$ per unit.

(b) A single line-to-ground fault at bus 3 through a fault impedance $Z_f = j0.10$ per unit.

(c) A line-to-line fault at bus 3 through a fault impedance $Z_f = j0.1$ per unit.

(d) A double line-to-ground fault at bus 3 through a fault impedance $Z_f = j0.1$ per unit.

Item	Base MVA	Voltage Rating	X^1	X^2	X^0
G_1	100	20 kV	0.15	0.15	0.05
G_2	100	20 kV	0.15	0.15	0.05
T_1	100	20/220 kV	0.10	0.10	0.10
T_2	100	20/220 kV	0.10	0.10	0.10
L_{12}	100	220 kV	0.125	0.125	0.30
L_{13}	100	220 kV	0.15	0.15	0.35
L_{23}	100	220 kV	0.25	0.25	0.7125

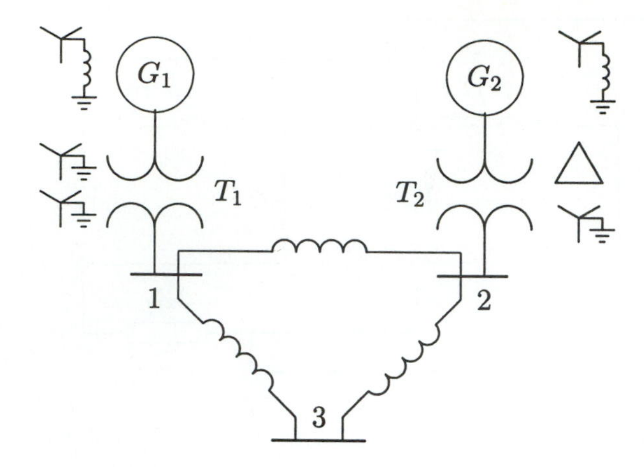

FIGURE 10.16
The one-line diagram for Example 10.5.

The positive-sequence impedance network is shown in Figure 10.17.

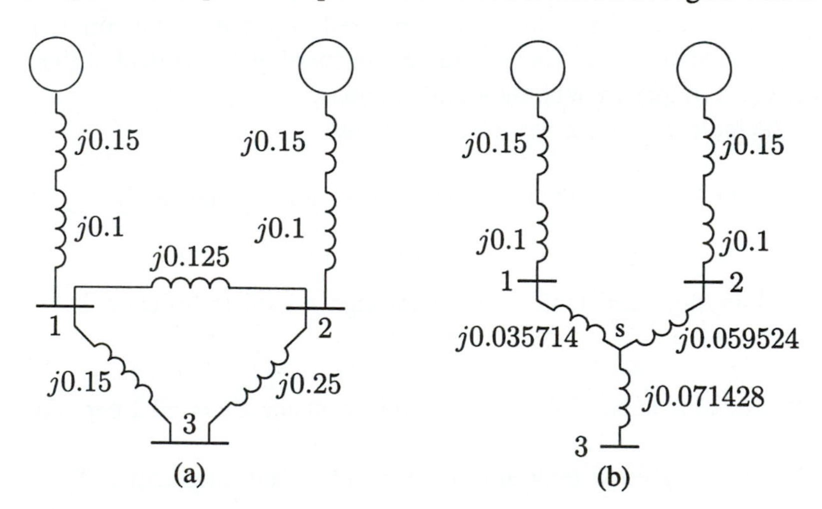

FIGURE 10.17
Positive-sequence impedance diagram for Example 10.5.

To find Thévenin impedance viewed from the faulted bus (bus 3), we convert the delta formed by buses 123 to an equivalent Y as shown in Figure 10.17(b).

$$Z_{1s} = \frac{(j0.125)(j0.15)}{j0.525} = j0.0357143$$

$$Z_{2s} = \frac{(j0.125)(j0.25)}{j0.525} = j0.0595238$$

$$Z_{3s} = \frac{(j0.15)(j0.25)}{j0.525} = j0.0714286$$

Combining the parallel branches, the positive-sequence Thévenin impedance is

$$\begin{aligned} Z_{33}^{1} &= \frac{(j0.2857143)(j0.3095238)}{j0.5952381} + j0.0714286 \\ &= j0.1485714 + j0.0714286 = j0.22 \end{aligned}$$

This is shown in Figure 10.18(a).

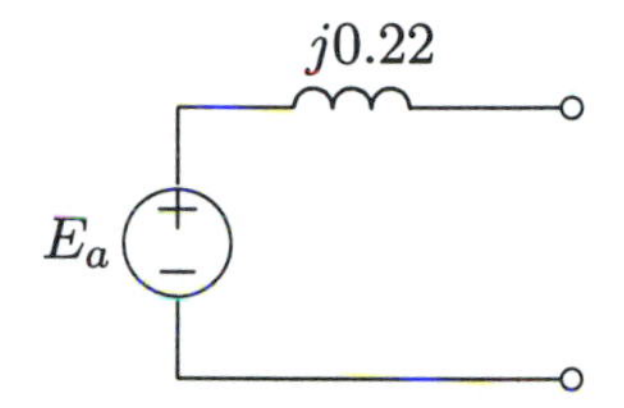

(a) Positive-sequence network

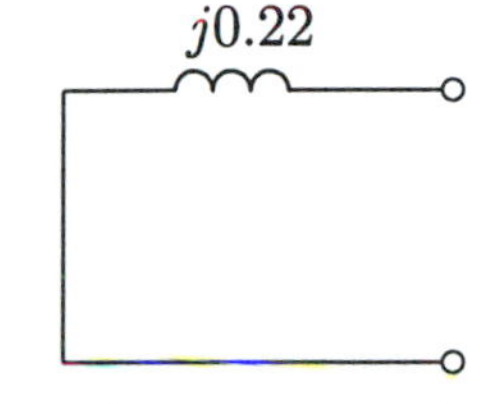

(b) Negative-sequence network

FIGURE 10.18
Reduction of the positive-sequence Thévenin equivalent network.

Since the negative-sequence impedance of each element is the same as the positive-sequence impedance, we have

$$Z_{33}^{2} = Z_{33}^{1} = j0.22$$

and the negative-sequence network is as shown in Figure 10.18(b). The equivalent circuit for the zero-sequence network is constructed according to the transformer winding connections of Figure 10.6 and is shown in Figure 10.19.

To find Thévenin impedance viewed from the faulted bus (bus 3), we convert the delta formed by buses 123 to an equivalent Y as shown in Figure 10.19(b).

$$Z_{1s} = \frac{(j0.30)(j0.35)}{j1.3625} = j0.0770642$$

$$Z_{2s} = \frac{(j0.30)(j0.7125)}{j1.3625} = j0.1568807$$

$$Z_{3s} = \frac{(j0.35)(j0.7125)}{j1.3625} = j0.1830257$$

Combining the parallel branches, the zero-sequence Thévenin impedance is

$$\begin{aligned} Z_{33}^{0} &= \frac{(j0.4770642)(j0.2568807)}{j0.7339449} + j0.1830275 \\ &= j0.1669725 + j0.1830275 = j0.35 \end{aligned}$$

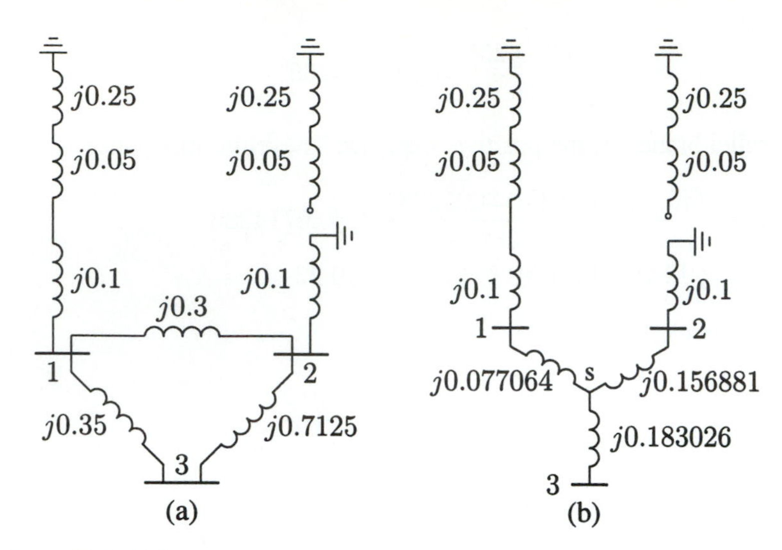

FIGURE 10.19
Zero-sequence impedance diagram for Example 10.5.

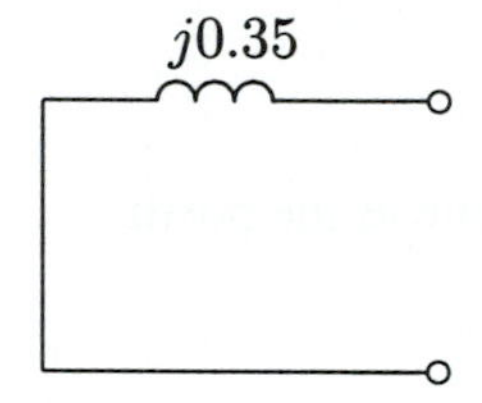

FIGURE 10.20
Zero-sequence network for Example 10.5.

The zero-sequence impedance diagram is shown in Figure 10.20.

(a) Balanced three-phase fault at bus 3.

Assuming the no-load generated emfs are equal to 1.0 per unit, the fault current is

$$I_3^a(F) = \frac{V_{3\,(0)}^a}{Z_{33}^1 + Z_f} = \frac{1.0}{j0.22 + j0.1} = -j3.125 \quad \text{pu}$$
$$= 820.1\angle{-90^\circ} \quad \text{A}$$

(b) Single line-to-ground fault at bus 3.

From (10.62), the sequence components of the fault current are

$$\begin{aligned} I_3^0 = I_3^1 = I_3^2 &= \frac{V_3^a(0)}{Z_{33}^1 + Z_{33}^2 + Z_{33}^0 + 3Z_f} \\ &= \frac{1.0}{j0.22 + j0.22 + j0.35 + 3(j0.1)} \\ &= -j0.9174 \ \text{pu} \end{aligned}$$

The fault current is

$$\begin{bmatrix} I_3^a \\ I_3^b \\ I_3^c \end{bmatrix} = \begin{bmatrix} 1 & 1 & 1 \\ 1 & a^2 & a \\ 1 & a & a^2 \end{bmatrix} \begin{bmatrix} I_3^0 \\ I_3^0 \\ I_3^0 \end{bmatrix} = \begin{bmatrix} 3I_3^0 \\ 0 \\ 0 \end{bmatrix} = \begin{bmatrix} -j2.7523 \\ 0 \\ 0 \end{bmatrix} \ \text{pu}$$

(c) Line-to line fault at bus 3.

The zero-sequence component of current is zero, i.e.,

$$I_3^0 = 0$$

From (10.75), the positive- and negative-sequence components of the fault current are

$$I_3^1 = -I_3^2 = \frac{V_{3(0)}^a}{Z_{33}^1 + Z_{33}^2 + Z_f} = \frac{1}{j0.22 + j0.22 + j0.1} = -j1.8519 \ \text{pu}$$

The fault current is

$$\begin{bmatrix} I_3^a \\ I_3^b \\ I_3^c \end{bmatrix} = \begin{bmatrix} 1 & 1 & 1 \\ 1 & a^2 & a \\ 1 & a & a^2 \end{bmatrix} \begin{bmatrix} 0 \\ -j1.8519 \\ j1.8519 \end{bmatrix} = \begin{bmatrix} 0 \\ -3.2075 \\ 3.2075 \end{bmatrix}$$

(d) Double line-to line-fault at bus 3.

From (10.88), the positive-sequence component of the fault current is

$$I_3^1 = \frac{V_{3(0)}^a}{Z_{33}^1 + \frac{Z_{33}^2(Z_{33}^0 + 3Z_f)}{Z_{33}^2 + Z_{33}^0 + 3Z_f}} = \frac{1}{j0.22 + \frac{j0.22(j0.35 + j0.3)}{j0.22 + j0.35 + j.3}} = -j2.6017 \ \text{pu}$$

The negative-sequence component of current from (10.87) is

$$I_3^2 = -\frac{V_{3(0)}^a - Z_{33}^1 I_3^1}{Z_{33}^2} = -\frac{1 - (j0.22)(-j2.6017)}{j0.22} = j1.9438 \ \text{pu}$$

The zero-sequence component of current from (10.86) is

$$I_3^0 = -\frac{V_{3(0)}^a - Z_{33}^1 I_3^1}{Z_{33}^0 + 3Z_f} = -\frac{1 - (j0.22)(-j2.6017)}{j0.35 + j0.3} = j0.6579 \ \text{pu}$$

and the phase currents are

$$\begin{bmatrix} I_3^a \\ I_3^b \\ I_3^c \end{bmatrix} = \begin{bmatrix} 1 & 1 & 1 \\ 1 & a^2 & a \\ 1 & a & a^2 \end{bmatrix} \begin{bmatrix} j0.6579 \\ -j2.6017 \\ j1.9438 \end{bmatrix} = \begin{bmatrix} 0 \\ 4.058\angle 165.93^\circ \\ 4.058\angle 14.07^\circ \end{bmatrix}$$

The fault current is

$$I_3(F) = I_3^b + I_3^c = 1.9732\angle 90^\circ$$

10.8 UNBALANCED FAULT ANALYSIS USING BUS IMPEDANCE MATRIX

We have seen that when the network is balanced, the symmetrical components impedances are diagonal, so that it is possible to calculate $\mathbf{Z}_{bus}$ separately for zero-, positive-, and negative-sequence networks. Also, we have observed that for a fault at bus k, the diagonal element in the k axis of the bus impedance matrix $\mathbf{Z}_{bus}$ is the Thévenin impedance to the point of fault. In order to obtain a solution for the unbalanced faults, the bus impedance matrix for each sequence network is obtained separately, then the sequence impedances $Z^0{}_{kk}$, $Z^1{}_{kk}$, and $Z^2{}_{kk}$ are connected together as described in Figures 10.11, 10.13, and 10.15. The fault formulas for various unbalanced faults is summarized below. In writing the symmetrical components of voltage and currents, the subscript a is left out and the symmetrical components are understood to refer to phase a.

10.8.1 SINGLE LINE-TO-GROUND FAULT USING $\mathbf{Z}_{bus}$

Consider a fault between phase a and ground through an impedance Z_f at bus k as shown in Figure 10.21. The line-to-ground fault requires that positive-, negative-, and zero-sequence networks for phase a be placed in series in order to compute the zero-sequence fault current as given by (10.62). Thus, in general, for a fault at bus k, the symmetrical components of fault current is

$$I_k^0 = I_k^1 = I_k^2 = \frac{V_k(0)}{Z_{kk}^1 + Z_{kk}^2 + Z_{kk}^0 + 3Z_f} \tag{10.90}$$

where Z_{kk}^1, Z_{kk}^2, and Z_{kk}^0 are the diagonal elements in the k axis of the corresponding bus impedance matrix and $V_k(0)$ is the prefault voltage at bus k. The fault phase current is

$$I_k^{abc} = \mathbf{A}I_k^{012} \tag{10.91}$$

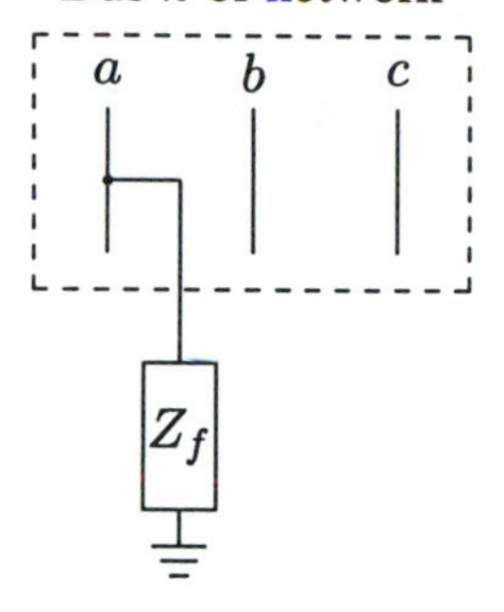

FIGURE 10.21
Line-to-ground fault at bus k.

10.8.2 LINE-TO-LINE FAULT USING $\mathbf{Z}_{bus}$

Consider a fault between phases b and c through an impedance Z_f at bus k as shown in Figure 10.22.

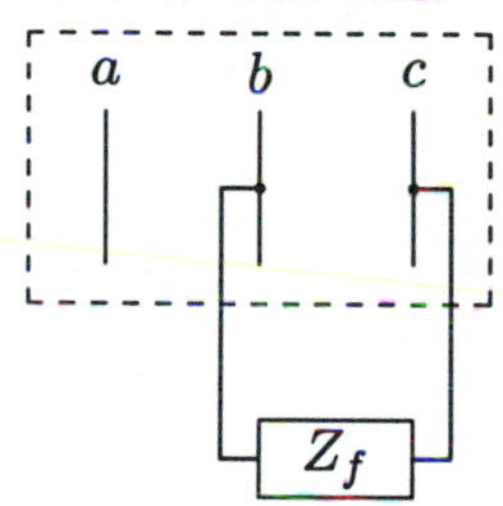

FIGURE 10.22
Line-to-line fault at bus k.

The phase a sequence network of Figure 10.13 is applicable here, where the positive- and negative-sequence networks are placed in opposition. The symmetrical components of the fault current as given from (10.68), (10.71), and (10.75) are

$$I_k^0 = 0 \tag{10.92}$$

$$I_k^1 = -I_k^2 = \frac{V_k(0)}{Z_{kk}^1 + Z_{kk}^2 + Z_f} \tag{10.93}$$

where Z_{kk}^1, and Z_{kk}^2 are the diagonal elements in the k axis of the corresponding bus impedance matrix. The fault phase current is then obtained from (10.91).

10.8.3 DOUBLE LINE-TO-GROUND FAULT USING $\mathbf{Z}_{bus}$

Consider a fault between phases b and c through an impedance Z_f to ground at bus k as shown in Figure 10.23.

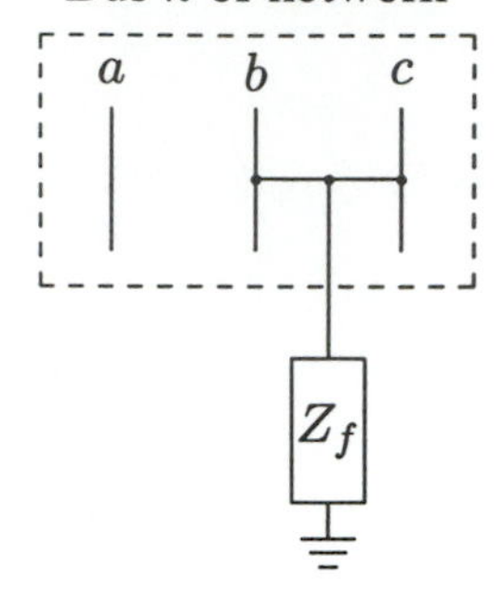

FIGURE 10.23
Double line-to-ground fault at bus k.

The phase a sequence network of Figure 10.15 is applicable here, where the positive-sequence impedance is placed in series with the parallel combination of the negative- and zero-sequence networks. The symmetrical components of the fault current as given from (10.86)–(10.88) are

$$I_k^1 = \frac{V_k(0)}{Z_{kk}^1 + \frac{Z_{kk}^2(Z_{kk}^0+3Z_f)}{Z_{kk}^2+Z_{kk}^0+3Z_f}} \tag{10.94}$$

$$I_k^2 = -\frac{V_k(0) - Z_{kk}^1 I_k^1}{Z_{kk}^2} \tag{10.95}$$

$$I_k^0 = -\frac{V_k(0) - Z_{kk}^1 I_k^1}{Z_{kk}^0 + 3Z_f} \tag{10.96}$$

where Z_{kk}^1, and Z_{kk}^2, and Z_{kk}^0 are the diagonal elements in the k axis of the corresponding bus impedance matrix. The phase currents are obtained from (10.91), and the fault current is

$$I_k(F) = I_k^b + I_k^c \tag{10.97}$$

10.8.4 BUS VOLTAGES AND LINE CURRENTS DURING FAULT

Using the sequence components of the fault current given by the formulas in (10.54), the symmetrical components of the ith bus voltages during fault are obtained

$$V_i^0(F) = 0 - Z_{ik}^0 I_k^0$$

$$V_i^1(F) = V_i^1(0) - Z_{ik}^1 I_k^1 \tag{10.98}$$
$$V_i^2(F) = 0 - Z_{ik}^2 I_k^2$$

where $V_i^1(0) = V_i(0)$ is the prefault phase voltage at bus i. The phase voltages during fault are

$$V_i^{abc} = \mathbf{A} V_i^{012} \tag{10.99}$$

The symmetrical components of fault current in line i to j is given by

$$I_{ij}^0 = \frac{V_i^0(F) - V_j^0(F)}{z_{ij}^0}$$
$$I_{ij}^1 = \frac{V_i^1(F) - V_j^1(F)}{z_{ij}^1} \tag{10.100}$$
$$I_{ij}^2 = \frac{V_i^2(F) - V_j^2(F)}{z_{ij}^2}$$

where z_{ij}^0, z_{ij}^1, and z_{ij}^2, are the zero-, positive-, and negative-sequence components of the actual line impedance between buses i and j. Having obtained the symmetrical components of line current, the phase fault current in line i to j is

$$I_{ij}^{abc} = \mathbf{A} I_{ij}^{012} \tag{10.101}$$

Example 10.6

Solve Example 10.5 using the bus impedance matrix. In addition, for each type of fault determine the bus voltages and line currents during fault.

Using the function **Zbus = zbuild(zdata)**, $\mathbf{Z}_{bus}^1$ and $\mathbf{Z}_{bus}^0$ are found for the positive-sequence network of Figure 10.17 and the zero-sequence network of Figure 10.19. The positive-sequence bus impedance matrix is

$$\mathbf{Z}_{bus}^1 = \begin{bmatrix} j0.1450 & j0.1050 & j0.1300 \\ j0.1050 & j0.1450 & j0.1200 \\ j0.1300 & j0.1200 & j0.2200 \end{bmatrix}$$

and the zero-sequence bus impedance matrix is

$$\mathbf{Z}_{bus}^0 = \begin{bmatrix} j0.1820 & j0.0545 & j0.1400 \\ j0.0545 & j0.0864 & j0.0650 \\ j0.1400 & j0.0650 & j0.3500 \end{bmatrix}$$

Since positive- and negative-sequence reactances for the system in Example 10.5 are identical, $\mathbf{Z}_{bus}^1 = \mathbf{Z}_{bus}^2$.

(a) Balanced three-phase fault at bus 3 through a fault impedance $Z_f = j0.1$.

The symmetrical components of fault current is given by

$$I_3^{012}(F) = \begin{bmatrix} 0 \\ \frac{1}{Z_{33}^1+Z_f} \\ 0 \end{bmatrix} = \begin{bmatrix} 0 \\ \frac{1}{j0.22+j0.1} \\ 0 \end{bmatrix} = \begin{bmatrix} 0 \\ -j3.125 \\ 0 \end{bmatrix}$$

The fault current is

$$I_3^{abc}(F) = \begin{bmatrix} 1 & 1 & 1 \\ 1 & a^2 & a \\ 1 & a & a^2 \end{bmatrix} \begin{bmatrix} 0 \\ -j3.125 \\ 0 \end{bmatrix} = \begin{bmatrix} 3.125\angle-90^\circ \\ 3.125\angle150^\circ \\ 3.125\angle30^\circ \end{bmatrix}$$

For balanced fault we only have the positive-sequence component of voltage. Thus, from (10.98), bus voltages during fault for phase a are

$$\begin{aligned} V_1(F) &= 1 - Z_{13}^1 I_3(F) = 1 - j0.13(-j3.125) = 0.59375 \\ V_2(F) &= 1 - Z_{23}^1 I_3(F) = 1 - j0.12(-j3.125) = 0.62500 \\ V_3(F) &= 1 - Z_{33}^1 I_3(F) = 1 - j0.22(-j3.125) = 0.31250 \end{aligned}$$

Fault currents in lines for phase a are

$$\begin{aligned} I_{21}(F) &= \frac{V_2(F) - V_1(F)}{z_{12}^1} = \frac{0.62500 - 0.59375}{j0.125} = 0.2500\angle-90^\circ \\ I_{13}(F) &= \frac{V_1(F) - V_3(F)}{z_{13}^1} = \frac{0.59375 - 0.31250}{j0.15} = 0.1875\angle-90^\circ \\ I_{23}(F) &= \frac{V_2(F) - V_3(F)}{z_{23}^1} = \frac{0.62500 - 0.31250}{j0.25} = 0.125\angle-90^\circ \end{aligned}$$

(b) Single line-to-ground fault at bus 3 through a fault impedance $Z_f = j0.1$.

From (10.90), the symmetrical components of fault current is given by

$$\begin{aligned} I_3^0(F) = I_3^1(F) = I_3^2(F) &= \frac{1.0}{Z_{33}^1 + Z_{33}^2 + Z_{33}^0 + 3Z_f} \\ &= \frac{1.0}{j0.22 + j0.22 + j0.35 + j3(0.1)} = -j0.9174 \end{aligned}$$

The fault current is

$$I_3^{abc}(F) = \begin{bmatrix} 1 & 1 & 1 \\ 1 & a^2 & a \\ 1 & a & a^2 \end{bmatrix} \begin{bmatrix} -j0.9174 \\ -j0.9174 \\ -j0.9174 \end{bmatrix} = \begin{bmatrix} 2.7523\angle-90^\circ \\ 0\angle0^\circ \\ 0\angle0^\circ \end{bmatrix}$$

From (10.98), the symmetrical components of bus voltages during fault are

$$V_1^{012}(F) = \begin{bmatrix} 0 - Z_{13}^0 I_3^0 \\ V_1^1(0) - Z_{13}^1 I_3^1 \\ 0 - Z_{13}^2 I_3^2 \end{bmatrix} = \begin{bmatrix} 0 - j0.140(-j0.9174) \\ 1 - j0.130(-j0.9174) \\ 0 - j0.130(-j0.9174) \end{bmatrix} = \begin{bmatrix} -0.1284 \\ 0.8807 \\ -0.1193 \end{bmatrix}$$

$$V_2^{012}(F) = \begin{bmatrix} 0 - Z_{23}^0 I_3^0 \\ V_2^1(0) - Z_{23}^1 I_3^1 \\ 0 - Z_{23}^2 I_3^2 \end{bmatrix} = \begin{bmatrix} 0 - j0.065(-j0.9174) \\ 1 - j0.120(-j0.9174) \\ 0 - j0.120(-j0.9174) \end{bmatrix} = \begin{bmatrix} -0.0596 \\ 0.8899 \\ -0.1101 \end{bmatrix}$$

$$V_3^{012}(F) = \begin{bmatrix} 0 - Z_{33}^0 I_3^0 \\ V_3^1(0) - Z_{33}^1 I_3^1 \\ 0 - Z_{33}^2 I_3^2 \end{bmatrix} = \begin{bmatrix} 0 - j0.350(-j0.9174) \\ 1 - j0.220(-j0.9174) \\ 0 - j0.220(-j0.9174) \end{bmatrix} = \begin{bmatrix} -0.3211 \\ 0.7982 \\ -0.2018 \end{bmatrix}$$

Bus voltages during fault are

$$V_1^{abc}(F) = \begin{bmatrix} 1 & 1 & 1 \\ 1 & a^2 & a \\ 1 & a & a^2 \end{bmatrix} \begin{bmatrix} -0.1284 \\ 0.8807 \\ -0.1193 \end{bmatrix} = \begin{bmatrix} 0.633\angle 0^\circ \\ 1.0046\angle -120.45^\circ \\ 1.0046\angle +120.45^\circ \end{bmatrix}$$

$$V_2^{abc}(F) = \begin{bmatrix} 1 & 1 & 1 \\ 1 & a^2 & a \\ 1 & a & a^2 \end{bmatrix} \begin{bmatrix} -0.0596 \\ 0.8899 \\ -0.1101 \end{bmatrix} = \begin{bmatrix} 0.7207\angle 0^\circ \\ 0.9757\angle -117.43^\circ \\ 0.9757\angle +117.43^\circ \end{bmatrix}$$

$$V_3^{abc}(F) = \begin{bmatrix} 1 & 1 & 1 \\ 1 & a^2 & a \\ 1 & a & a^2 \end{bmatrix} \begin{bmatrix} -0.3211 \\ 0.7982 \\ -0.2018 \end{bmatrix} = \begin{bmatrix} 0.2752\angle 0^\circ \\ 1.0647\angle -125.56^\circ \\ 1.0647\angle +125.56^\circ \end{bmatrix}$$

The symmetrical components of fault currents in lines for phase a are

$$I_{21}^{012} = \begin{bmatrix} \frac{V_2^0(F) - V_1^0(F)}{z_{12}^0} \\ \frac{V_2^1(F) - V_1^1(F)}{z_{12}^1} \\ \frac{V_2^2(F) - V_1^2(F)}{z_{12}^2} \end{bmatrix} = \begin{bmatrix} \frac{-0.0596-(-0.1284)}{j0.3} \\ \frac{0.8899-0.8807)}{j0.125} \\ \frac{-0.1101-(-0.1193)}{j0.125} \end{bmatrix} = \begin{bmatrix} 0.2294\angle -90^\circ \\ 0.0734\angle -90^\circ \\ 0.0734\angle -90^\circ \end{bmatrix}$$

$$I_{13}^{012} = \begin{bmatrix} \frac{V_1^0(F) - V_3^0(F)}{z_{13}^0} \\ \frac{V_1^1(F) - V_3^1(F)}{z_{13}^1} \\ \frac{V_1^2(F) - V_3^2(F)}{z_{13}^2} \end{bmatrix} = \begin{bmatrix} \frac{-0.1284-(-0.3211)}{j0.35} \\ \frac{0.8807-0.7982)}{j0.15} \\ \frac{-0.1193-(-0.2018)}{j0.15} \end{bmatrix} = \begin{bmatrix} 0.5505\angle -90^\circ \\ 0.5505\angle -90^\circ \\ 0.5505\angle -90^\circ \end{bmatrix}$$

$$I_{23}^{012} = \begin{bmatrix} \frac{V_2^0(F)-V_3^0(F)}{z_{23}^0} \\ \frac{V_2^1(F)-V_3^1(F)}{z_{23}^1} \\ \frac{V_2^2(F)-V_3^2(F)}{z_{23}^2} \end{bmatrix} = \begin{bmatrix} \frac{-0.0596-(-0.3211)}{j0.7125} \\ \frac{0.8899-0.7982)}{j0.25} \\ \frac{-0.1101-(-0.2018)}{j0.25} \end{bmatrix} = \begin{bmatrix} 0.3670\angle-90^\circ \\ 0.3670\angle-90^\circ \\ 0.3670\angle-90^\circ \end{bmatrix}$$

The line fault currents are

$$I_{21}^{abc}(F) = \begin{bmatrix} 1 & 1 & 1 \\ 1 & a^2 & a \\ 1 & a & a^2 \end{bmatrix} \begin{bmatrix} 0.2294\angle-90^\circ \\ 0.0734\angle-90^\circ \\ 0.0734\angle-90^\circ \end{bmatrix} = \begin{bmatrix} 0.3761\angle-90^\circ \\ 0.1560\angle-90^\circ \\ 0.1560\angle-90^\circ \end{bmatrix}$$

$$I_{13}^{abc}(F) = \begin{bmatrix} 1 & 1 & 1 \\ 1 & a^2 & a \\ 1 & a & a^2 \end{bmatrix} \begin{bmatrix} 0.5505\angle-90^\circ \\ 0.5505\angle-90^\circ \\ 0.5505\angle-90^\circ \end{bmatrix} = \begin{bmatrix} 1.6514\angle-90^\circ \\ 0 \\ 0 \end{bmatrix}$$

$$I_{23}^{abc}(F) = \begin{bmatrix} 1 & 1 & 1 \\ 1 & a^2 & a \\ 1 & a & a^2 \end{bmatrix} \begin{bmatrix} 0.3670\angle-90^\circ \\ 0.3670\angle-90^\circ \\ 0.3670\angle-90^\circ \end{bmatrix} = \begin{bmatrix} 1.1009\angle-90^\circ \\ 0 \\ 0 \end{bmatrix}$$

(c) Line-to-line fault at bus 3 through a fault impedance $Z_f = j0.1$.

From (10.92) and (10.93), the symmetrical components of fault current are

$$I_3^0 = 0$$

$$I_3^1 = -I_3^2 = \frac{V_3(0)}{Z_{33}^1 + Z_{33}^2 + Z_f} = \frac{1}{j0.22 + j0.22 + j0.1} = -j1.8519$$

The fault current is

$$I_3^{abc}(F) = \begin{bmatrix} 1 & 1 & 1 \\ 1 & a^2 & a \\ 1 & a & a^2 \end{bmatrix} \begin{bmatrix} 0 \\ -j1.8519 \\ j1.8519 \end{bmatrix} = \begin{bmatrix} 0 \\ -3.2075 \\ 3.2075 \end{bmatrix}$$

From (10.98), the symmetrical components of bus voltages during fault are

$$V_1^{012}(F) = \begin{bmatrix} 0 \\ V_1^1(0) - Z_{13}^1 I_3^1 \\ 0 - Z_{13}^2 I_3^2 \end{bmatrix} = \begin{bmatrix} 0 \\ 1 - j0.130(-j1.8519) \\ 0 - j0.130(j1.8519) \end{bmatrix} = \begin{bmatrix} 0 \\ 0.7593 \\ 0.2407 \end{bmatrix}$$

$$V_2^{012}(F) = \begin{bmatrix} 0 \\ V_2^1(0) - Z_{23}^1 I_3^1 \\ 0 - Z_{23}^2 I_3^2 \end{bmatrix} = \begin{bmatrix} 0 \\ 1 - j0.120(-j1.8519 \\ 0 - j0.120(j1.8519) \end{bmatrix} = \begin{bmatrix} 0 \\ 0.7778 \\ 0.2222 \end{bmatrix}$$

$$V_3^{012}(F) = \begin{bmatrix} 0 \\ V_3^1(0) - Z_{33}^1 I_3^1 \\ 0 - Z_{33}^2 I_3^2 \end{bmatrix} = \begin{bmatrix} 0 \\ 1 - j0.220(-j1.8519) \\ 0 - j0.220(j1.8519) \end{bmatrix} = \begin{bmatrix} 0 \\ 0.5926 \\ 0.4074 \end{bmatrix}$$

Bus voltages during fault are

$$V_1^{abc}(F) = \begin{bmatrix} 1 & 1 & 1 \\ 1 & a^2 & a \\ 1 & a & a^2 \end{bmatrix} \begin{bmatrix} 0 \\ 0.7593 \\ 0.2407 \end{bmatrix} = \begin{bmatrix} 1\angle 0^\circ \\ 0.672\angle -138.07^\circ \\ 0.672\angle +138.07^\circ \end{bmatrix}$$

$$V_2^{abc}(F) = \begin{bmatrix} 1 & 1 & 1 \\ 1 & a^2 & a \\ 1 & a & a^2 \end{bmatrix} \begin{bmatrix} 0 \\ 0.7778 \\ 0.2222 \end{bmatrix} = \begin{bmatrix} 1\angle 0^\circ \\ 0.6939\angle -136.10^\circ \\ 0.6939\angle +136.10^\circ \end{bmatrix}$$

$$V_3^{abc}(F) = \begin{bmatrix} 1 & 1 & 1 \\ 1 & a^2 & a \\ 1 & a & a^2 \end{bmatrix} \begin{bmatrix} 0 \\ 0.5926 \\ 0.4074 \end{bmatrix} = \begin{bmatrix} 1\angle 0^\circ \\ 0.5251\angle -162.21^\circ \\ 0.5251\angle +162.21^\circ \end{bmatrix}$$

The symmetrical components of fault currents in lines for phase a are

$$I_{21}^{012} = \begin{bmatrix} 0 \\ \frac{V_2^1(F) - V_1^1(F)}{z_{12}^1} \\ \frac{V_2^2(F) - V_1^2(F)}{z_{12}^2} \end{bmatrix} = \begin{bmatrix} 0 \\ \frac{0.7778-0.7593)}{j0.125} \\ \frac{0.2222-0.2407}{j0.125} \end{bmatrix} = \begin{bmatrix} 0 \\ 0.148\angle -90^\circ \\ 0.148\angle +90^\circ \end{bmatrix}$$

$$I_{13}^{012} = \begin{bmatrix} 0 \\ \frac{V_1^1(F) - V_3^1(F)}{z_{13}^1} \\ \frac{V_1^2(F) - V_3^2(F)}{z_{13}^2} \end{bmatrix} = \begin{bmatrix} 0 \\ \frac{0.7593-0.5926}{j0.15} \\ \frac{0.2407-0.4074}{j0.15} \end{bmatrix} = \begin{bmatrix} 0 \\ 1.1111\angle -90^\circ \\ 1.1111\angle +90^\circ \end{bmatrix}$$

$$I_{23}^{012} = \begin{bmatrix} 0 \\ \frac{V_2^1(F) - V_3^1(F)}{z_{23}^1} \\ \frac{V_2^2(F) - V_3^2(F)}{z_{23}^2} \end{bmatrix} = \begin{bmatrix} 0 \\ \frac{0.7778-0.5926}{j0.25} \\ \frac{0.2222-0.4074}{j0.25} \end{bmatrix} = \begin{bmatrix} 0 \\ 0.7407\angle -90^\circ \\ 0.7407\angle +90^\circ \end{bmatrix}$$

The line fault currents are

$$I_{21}^{abc}(F) = \begin{bmatrix} 1 & 1 & 1 \\ 1 & a^2 & a \\ 1 & a & a^2 \end{bmatrix} \begin{bmatrix} 0 \\ 0.148\angle -90^\circ \\ 0.148 \end{bmatrix} = \begin{bmatrix} 0 \\ -0.2566 \\ 0.2566 \end{bmatrix}$$

$$I_{13}^{abc}(F) = \begin{bmatrix} 1 & 1 & 1 \\ 1 & a^2 & a \\ 1 & a & a^2 \end{bmatrix} \begin{bmatrix} 0 \\ 1.1111\angle{-90°} \\ 1.1111\angle{+90°} \end{bmatrix} = \begin{bmatrix} 0 \\ -1.9245 \\ 1.9245 \end{bmatrix}$$

$$I_{23}^{abc}(F) = \begin{bmatrix} 1 & 1 & 1 \\ 1 & a^2 & a \\ 1 & a & a^2 \end{bmatrix} \begin{bmatrix} 0 \\ 0.7407\angle{-90°} \\ 0.7407\angle{+90°} \end{bmatrix} = \begin{bmatrix} 0 \\ -1.283 \\ 1.283 \end{bmatrix}$$

(d) Double line-to-ground fault at bus 3 through a fault impedance $Z_f = j0.1$.

From (10.94)–(10.96), the symmetrical components of fault current is given by

$$I_3^1 = \frac{V_3(0)}{Z_{33}^1 + \frac{Z_{33}^2(Z_{33}^0 + 3Z_f)}{Z_{33}^2 + Z_{33}^0 + 3Z_f}} = -\frac{1}{j0.22 + \frac{j0.22(j0.35 + j0.3)}{j0.22 + j0.35 + j0.3}} = -j2.6017$$

$$I_3^2 = -\frac{V_3(0) - Z_{33}^1 I_{33}^1}{Z_{33}^2} - \frac{1 - j0.22(-j2.6017)}{j0.22} = j1.9438$$

$$I_3^0 = -\frac{V_3(0) - Z_{33}^1 I_{33}^1}{Z_{33}^0 + 3Z_f} = -\frac{1 - j0.22(-j2.6017)}{j0.35 + j0.3} = j0.6579$$

The phase currents at the faulted bus are

$$I_3^{abc}(F) = \begin{bmatrix} 1 & 1 & 1 \\ 1 & a^2 & a \\ 1 & a & a^2 \end{bmatrix} \begin{bmatrix} j0.6579 \\ -j2.6017 \\ j1.9438 \end{bmatrix} = \begin{bmatrix} 0 \\ 4.0583\angle 165.93° \\ 4.0583\angle 14.07° \end{bmatrix}$$

and the total fault current is

$$I_3^b + I_3^c = 4.0583\angle 165.93° - 4.0583\angle 14.07° = 1.9732\angle 90°$$

From (10.98), the symmetrical components of bus voltages during fault are

$$V_1^{012}(F) = \begin{bmatrix} 0 - Z_{13}^0 I_3^0 \\ V_1^1(0) - Z_{13}^1 I_3^1 \\ 0 - Z_{13}^2 I_3^2 \end{bmatrix} = \begin{bmatrix} 0 - j0.140(j0.6579) \\ 1 - j0.130(-j2.6017) \\ 0 - j0.130(j1.9438) \end{bmatrix} = \begin{bmatrix} 0.0921 \\ 0.6618 \\ 0.2527 \end{bmatrix}$$

$$V_2^{012}(F) = \begin{bmatrix} 0 - Z_{23}^0 I_3^0 \\ V_2^1(0) - Z_{23}^1 I_3^1 \\ 0 - Z_{23}^2 I_3^2 \end{bmatrix} = \begin{bmatrix} 0 - j0.065(0.6579) \\ 1 - j0.120(-j2.6017 \\ 0 - j0.120(j1.9438) \end{bmatrix} = \begin{bmatrix} 0.0428 \\ 0.6878 \\ 0.2333 \end{bmatrix}$$

$$V_3^{012}(F) = \begin{bmatrix} 0 - Z_{33}^0 I_3^0 \\ V_3^1(0) - Z_{33}^1 I_3^1 \\ 0 - Z_{33}^2 I_3^2 \end{bmatrix} = \begin{bmatrix} 0 - j0.350(0.6579) \\ 1 - j0.220(-j2.6017) \\ 0 - j0.220(j1.9438) \end{bmatrix} = \begin{bmatrix} 0.2303 \\ 0.4276 \\ 0.4276 \end{bmatrix}$$

Bus voltages during fault are

$$V_1^{abc}(F) = \begin{bmatrix} 1 & 1 & 1 \\ 1 & a^2 & a \\ 1 & a & a^2 \end{bmatrix} \begin{bmatrix} 0.0921 \\ 0.6618 \\ 0.2527 \end{bmatrix} = \begin{bmatrix} 1.0066\angle 0^\circ \\ 0.5088\angle -135.86^\circ \\ 0.5088\angle +135.86^\circ \end{bmatrix}$$

$$V_2^{abc}(F) = \begin{bmatrix} 1 & 1 & 1 \\ 1 & a^2 & a \\ 1 & a & a^2 \end{bmatrix} \begin{bmatrix} 0.0428 \\ 0.6878 \\ 0.2333 \end{bmatrix} = \begin{bmatrix} 0.9638\angle 0^\circ \\ 0.5740\angle -136.70^\circ \\ 0.5740\angle +136.70^\circ \end{bmatrix}$$

$$V_3^{abc}(F) = \begin{bmatrix} 1 & 1 & 1 \\ 1 & a^2 & a \\ 1 & a & a^2 \end{bmatrix} \begin{bmatrix} 0.2303 \\ 0.4276 \\ 0.4276 \end{bmatrix} = \begin{bmatrix} 1.0855\angle 0^\circ \\ 0.1974\angle 180^\circ \\ 0.1974\angle +180^\circ \end{bmatrix}$$

The symmetrical components of fault currents in lines for phase a are

$$I_{12}^{012} = \begin{bmatrix} \frac{V_1^0(F)-V_2^0(F)}{z_{12}^0} \\ \frac{V_1^1(F)-V_2^1(F)}{z_{12}^1} \\ \frac{V_1^2(F)-V_2^2(F)}{z_{12}^2} \end{bmatrix} = \begin{bmatrix} \frac{0.0921-0.0428}{j0.3} \\ \frac{0.6618-0.687)}{j0.125} \\ \frac{0.2527-0.2333}{j0.125} \end{bmatrix} = \begin{bmatrix} 0.1645\angle -90^\circ \\ 0.2081\angle +90^\circ \\ 0.1555\angle -90^\circ \end{bmatrix}$$

$$I_{13}^{012} = \begin{bmatrix} \frac{V_1^0(F)-V_3^0(F)}{z_{13}^0} \\ \frac{V_1^1(F)-V_3^1(F)}{z_{13}^1} \\ \frac{V_1^2(F)-V_3^2(F)}{z_{13}^2} \end{bmatrix} = \begin{bmatrix} \frac{0.0921-0.2303}{j0.35} \\ \frac{0.6618-0.4276}{j0.15} \\ \frac{0.2527-0.4276}{j0.15} \end{bmatrix} = \begin{bmatrix} 0.3947\angle +90^\circ \\ 1.5610\angle -90^\circ \\ 1.1663\angle +90^\circ \end{bmatrix}$$

$$I_{23}^{012} = \begin{bmatrix} \frac{V_2^0(F)-V_3^0(F)}{z_{23}^0} \\ \frac{V_2^1(F)-V_3^1(F)}{z_{23}^1} \\ \frac{V_2^2(F)-V_3^2(F)}{z_{23}^2} \end{bmatrix} = \begin{bmatrix} \frac{0.0428-0.2303}{j0.7125} \\ \frac{0.6878-0.4276}{j0.25} \\ \frac{0.2333-0.4276}{j0.25} \end{bmatrix} = \begin{bmatrix} 0.2632\angle +90^\circ \\ 1.0407\angle -90^\circ \\ 0.7775\angle +90^\circ \end{bmatrix}$$

The line fault currents are

$$I_{12}^{abc}(F) = \begin{bmatrix} 1 & 1 & 1 \\ 1 & a^2 & a \\ 1 & a & a^2 \end{bmatrix} \begin{bmatrix} 0.1645\angle -90^\circ \\ 0.2081\angle +90^\circ \\ 0.1555\angle -90^\circ \end{bmatrix} = \begin{bmatrix} 0.1118\angle -90^\circ \\ 0.3682\angle -31.21 \\ 0.3682\angle -148.79^\circ \end{bmatrix}$$

$$I_{13}^{abc}(F) = \begin{bmatrix} 1 & 1 & 1 \\ 1 & a^2 & a \\ 1 & a & a^2 \end{bmatrix} \begin{bmatrix} 0.3947\angle +90^\circ \\ 1.5610\angle -90^\circ \\ 1.1663\angle +90^\circ \end{bmatrix} = \begin{bmatrix} 0 \\ 2.435\angle 165.93^\circ \\ 2.435\angle 14.07^\circ \end{bmatrix}$$

$$I_{23}^{abc}(F) = \begin{bmatrix} 1 & 1 & 1 \\ 1 & a^2 & a \\ 1 & a & a^2 \end{bmatrix} \begin{bmatrix} 0.2632\angle +90^\circ \\ 1.0407\angle -90^\circ \\ 0.7775\angle +90^\circ \end{bmatrix} = \begin{bmatrix} 0 \\ 1.6233\angle 165.93^\circ \\ 1.6233\angle 14.07^\circ \end{bmatrix}$$

10.9 UNBALANCED FAULT PROGRAMS

Three functions are developed for the unbalanced fault analysis. These functions are are **lgfault(zdata0, Zbus0, zdata1, Zbus1, zdata2, Zbus2, V)**, **llfault(zdata1, Zbus1,zdata2, Zbus2, V)**, and **dlgfault(zdata0, Zbus0, zdata1, Zbus1, zdata2, Zbus2, V)**. **lgfault** is designed for the single line-to-ground fault analysis, **llfault** for the line-to-line fault analysis, and **dlgfault** for the double line-to-ground fault analysis of a power system network. **lgfault** and **dlgfault** require the positive-, negative-, and zero-sequence bus impedance matrices **Zbus0**, **Zbus1**, and **Zbus2**, and **llfault** requires the positive- and negative-sequence bus impedance matrices **Zbus1**, and **Zbus2**. The last argument **V** is optional. If it is not included, the program sets all the prefault bus voltages to 1.0 per unit. If the variable **V** is included, the prefault bus voltages must be specified by the array **V** containing bus numbers and the complex bus voltage. The voltage vector **V** is automatically generated following the execution of any of the power flow programs.

The bus impedance matrices may be obtained from **Zbus0=zbuild(zdata0)**, and **Zbus1 = zbuild(zdata1)**. The argument **zdata1** contains the positive-sequence network impedances. **zdata0** contains the zero-sequence network impedances. Arguments **zdata0**, **zdata1** and **zdata2** are an $e\times 4$ matrices containing the impedance data of an e-element network. Columns 1 and 2 are the element bus numbers and columns 3 and 4 contain the element resistance and reactance, respectively, in per unit. Bus number 0 to generator buses contain generator impedances. These may be the subtransient, transient, or synchronous reactances. Also, any other shunt impedances such as capacitors and load impedances to ground (bus 0) may be included in this matrix.

The negative-sequence network has the same topology as the positive-sequence network. The line and transformer negative-sequence impedances are the same as the positive-sequence impedances, however, the generator negative-sequence reactances are different from the positive-sequence values. In the fault analysis of large power system usually the negative-sequence network impedances are assumed to be identical to the positive-sequence impedances. The zero-sequence network topology is different from the positive-sequence network. The zero-sequence network must be constructed according to the transformer winding connections of Figure 10.6. All transformer connections except Y-Y with both neutral grounded result in isolation between the primary and secondary in the zero-sequence network. For these connections the corresponding resistance and reactance columns in the zero-sequence data must be filled with **inf**. For grounded Y-Δ connections, additional entries must be included to represent the transformer impedance from bus 0 to the grounded Y-side. In case the neutral is grounded through an impedance Z_n, an impedance of $3Z_n$ must be added to the transformer reactance. The reader is reminded of the 30° phase shift in a Y-Δ or Δ-Y transformer. According to the ASA convention, the positive-sequence voltage is advanced by 30° when stepping up from the low-voltage side to the high-voltage side. Similarly, the negative-sequence voltage is retarded by 30° when stepping up from low-voltage to the high-voltage side. The phase shifts due to Δ-Y transformers have no effect on the bus voltages and line currents in that part of the system where the fault occurs. However, on the other side of the Δ-Y transformers, the sequence voltages, and currents must be shifted in phase before transforming to the phase quantities. The unbalanced fault programs presently ignores the 30° phase shift in the Δ-Y transformers.

The other function for the formation of the bus impedance matrix is **Zbus = zbuildpi(linedata, gendata, yload)**, which is compatible with the power flow programs. The first argument **linedata** is consistent with the data required for the power flow solution. Columns 1 and 2 are the line bus numbers. Columns 3 through 5 contain the line resistance, reactance, and one-half of the total line charging susceptance in per unit on the specified MVA base. The last column is for the transformer tap setting; for lines, 1 must be entered in this column. The generator reactances are not included in the **linedata** for the power flow program and must be specified separately as required by the **gendata** in the second argument. **gendata** is an $e_g \times 4$ matrix, where each row contains bus 0, generator bus number, resistance and reactance. The last argument **yload** is optional. This is a two-column matrix containing bus number and the complex load admittance. This data is provided by any of the power flow programs **lfgauss**, **lfnewton** or **decouple**. **yload** is automatically generated following the execution of the above power flow programs.

The program prompts the user to enter the faulted bus number and the fault impedance **Zf**. The program obtains the total fault current, bus voltages and line currents during the fault. The use of the above functions are demonstrated in the following examples.

Example 10.7

Use the **lgfault**, **llfault**, and **dlgfault** functions to compute the fault current, bus voltages and line currents in the circuit given in Example 10.5 for the following fault.

(a) A balanced three-phase fault at bus 3 through a fault impedance $Z_f = j0.1$ per unit.
(b) A single-line-to-ground fault at bus 3 through a fault impedance $Z_f = j0.1$ per unit.
(c) A line-to-line fault at bus 3 through a fault impedance $Z_f = j0.1$ per unit.
(d) A double line-to-ground fault at bus 3 through a fault impedance $Z_f = j0.1$ per unit.

In this example all shunt capacitances and loads are neglected and all the prefault bus voltages are assumed to be unity. The positive-sequence impedance diagram in Figure 10.17 is described by the variable **zdata1** and the zero-sequence impedance diagram in Figure 10.19 is described by the variable **zdata0**. The negative-sequence data is assumed to be the same as the positive-sequence data. We use the following commands.

```
zdata1 = [0   1   0    0.25
          0   2   0    0.25
          1   2   0    0.125
          1   3   0    0.15
          2   3   0    0.25];

zdata0 = [0   1   0    0.40
          0   2   0    0.10
          1   2   0    0.30
          1   3   0    0.35
          2   3   0    0.7125];

zdata2 = zdata1;
Zbus1 = zbuild(zdata1)
Zbus0 = zbuild(zdata0)
Zbus2 = Zbus1;
symfault(zdata1, Zbus1)
lgfault(zdata0, Zbus0, zdata1, Zbus1, zdata2, Zbus2)
llfault(zdata1, Zbus1, zdata2, Zbus2)
dlgfault(zdata0, Zbus0, zdata1, Zbus1, zdata2, Zbus2)
```

The result is

```
Three-phase balanced fault analysis
Enter Faulted Bus No. -> 3
Enter Fault Impedance Zf = R + j*X in
complex form (for bolted fault enter 0). Zf = j*0.1
Balanced three-phase fault at bus No. 3
Total fault current =   3.1250 per unit

Bus Voltages during fault in per unit
     Bus      Voltage       Angle
     No.      Magnitude     Degree
      1        0.5938       0.0000
      2        0.6250       0.0000
      3        0.3125       0.0000

Line currents for fault at bus No. 3
```

```
            From        To      Current       Angle
            Bus         Bus     Magnitude     Degree
             G           1        1.6250     -90.0000
             1           3        1.8750     -90.0000
             G           2        1.5000     -90.0000
             2           1        0.2500     -90.0000
             2           3        1.2500     -90.0000
             3           F        3.1250     -90.0000

Another fault location?
Enter 'y' or 'n' within single quote -> 'n'

Line-to-ground fault analysis
Enter Faulted Bus No. -> 3
Enter Fault Impedance Zf = R + j*X in
complex form (for bolted fault enter 0). Zf = j*0.1
Single line to-ground fault at bus No. 3
Total fault current =     2.7523 per unit

Bus Voltages during the fault in per unit
     Bus     -------Voltage Magnitude-------
     No.     Phase a      Phase b      Phase c
     1       0.6330       1.0046       1.0046
     2       0.7202       0.9757       0.9757
     3       0.2752       1.0647       1.0647

Line currents for fault at bus No. 3
     From       To        -----Line Current Magnitude----
     Bus        Bus       Phase a      Phase b      Phase c
      1          3        1.6514       0.0000       0.0000
      2          1        0.3761       0.1560       0.1560
      2          3        1.1009       0.0000       0.0000
      3          F        2.7523       0.0000       0.0000

Another fault location?
Enter 'y' or 'n' within single quote -> 'n'

Line-to-line fault analysis
Enter Faulted Bus No. -> 3
Enter Fault Impedance Zf = R + j*X in
complex form (for bolted fault enter 0). Zf = j*0.1
Line-to-line fault at bus No. 3
Total fault current =     3.2075 per unit

Bus Voltages during the fault in per unit
```

```
     Bus    -------Voltage Magnitude-------
     No.    Phase a     Phase b     Phase c
     1      1.0000      0.6720      0.6720
     2      1.0000      0.6939      0.6939
     3      1.0000      0.5251      0.5251

Line currents for fault at bus No. 3
     From      To      -----Line Current Magnitude----
     Bus       Bus     Phase a     Phase b     Phase c
      1         3      0.0000      1.9245      1.9245
      2         1      0.0000      0.2566      0.2566
      2         3      0.0000      1.2830      1.2830
      3         F      0.0000      3.2075      3.2075

Another fault location?
Enter 'y' or 'n' within single quote -> 'n'

Double line-to-ground fault analysis
Enter Faulted Bus No. -> 3
Enter Fault Impedance Zf = R + j*X in
complex form (for bolted fault enter 0). Zf = j*0.1
Double line-to-ground fault at bus No. 3
Total fault current =    1.9737 per unit

Bus Voltages during the fault in per unit
     Bus    -------Voltage Magnitude-------
     No.    Phase a     Phase b     Phase c
     1      1.0066      0.5088      0.5088
     2      0.9638      0.5740      0.5740
     3      1.0855      0.1974      0.1974

Line currents for fault at bus No. 3
     From      To      -----Line Current Magnitude----
     Bus       Bus     Phase a     Phase b     Phase c
      1         3      0.0000      2.4350      2.4350
      2         1      0.1118      0.3682      0.3682
      2         3      0.0000      1.6233      1.6233
      3         F      0.0000      4.0583      4.0583

Another fault location?
Enter 'y' or 'n' within single quote -> 'n'
```

Example 10.8

The 11-bus power system network of an electric utility company is shown in Figure 10.24. The positive- and zero-sequence reactances of the lines and transformers in per unit on a 100-MVA base is tabulated below. The transformer connections are shown in Figure 10.24. The Δ-Y transformer between buses 11 and 7 is grounded through a reactor of reactance 0.08 per unit. The generators positive-, and zero-sequence reactances including the reactance of grounding neutrals on a 100-MVA base is also tabulated below. Resistances, shunt reactances, and loads are neglected, and all negative-sequence reactances are assumed equal to the positive-sequence reactances. Use **zbuild** function to obtain the positive- and zero-sequence bus impedance matrices. Assuming all the prefault bus voltages are equal to $1\angle 0°$, use **lgfault**, **llfault**, and **dlgfault** to compute the fault current, bus voltages, and line currents for the following unbalanced faults.

(a) A bolted single line-to-ground fault at bus 8.
(b) A bolted line-to-line fault at bus 8.
(c) A bolted double line-to-ground fault at bus 8.

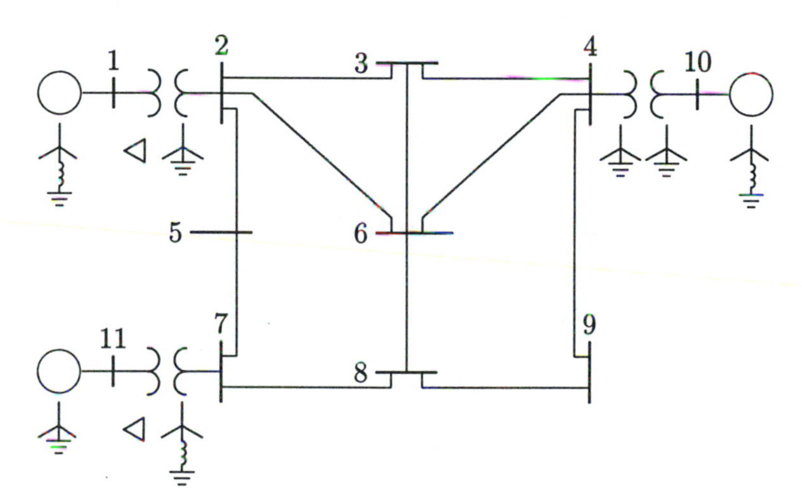

FIGURE 10.24
One-line diagram for Example 10.8.

GENERATOR TRANSIENT IMPEDANCE, PU			
Gen. No.	X^1	X^0	X_n
1	0.20	0.06	0.05
10	0.15	0.04	0.05
11	0.25	0.08	0.00

LINE AND TRANSFORMER DATA			
Bus No.	Bus No.	X^1, PU	X^0, PU
1	2	0.06	0.06
2	3	0.30	0.60
2	5	0.15	0.30
2	6	0.45	0.90
3	4	0.40	0.80
3	6	0.40	0.80
4	6	0.60	1.00
4	9	0.70	1.10
4	10	0.08	0.08
5	7	0.43	0.80
6	8	0.48	0.95
7	8	0.35	0.70
7	11	0.10	0.10
8	9	0.48	0.90

The equivalent circuit for the zero-sequence network is constructed according to the transformer winding connections of Figure 10.6 and is shown in Figure 10.25.

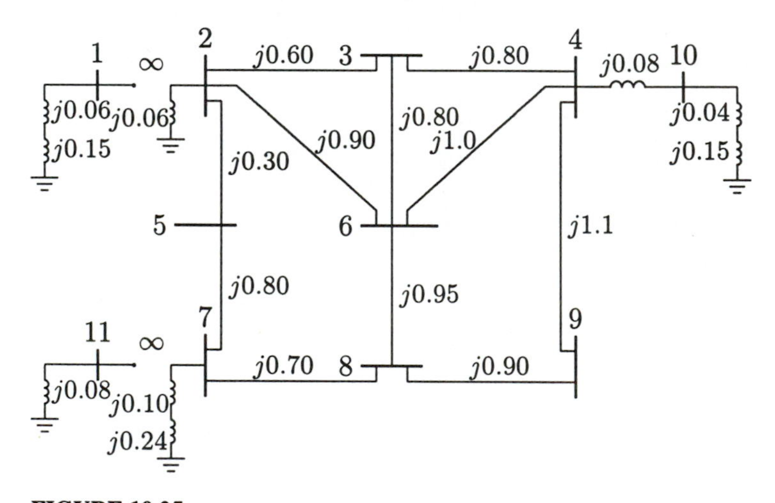

FIGURE 10.25
Zero-sequence network for Example 10.8.

When using **zbuild** function, the generator reactances must be included in the impedance data with bus zero as the reference bus.

The Δ-Y transformers result in isolation between the primary and secondary in the zero-sequence network. For these connections **inf** is entered in the corresponding resistance and reactance columns in the zero-sequence data. For grounded Y-Δconnections, additional entries are included to represent the transformer impedance from bus 0 to the grounded Y-side. The generators and transformers neutral reactor are included in the zero-sequence circuit each with a reactance of $3X_n$.

The positive- and zero-sequence impedance data and the required commands are as follows.

```
zdata1 = [ 0    1    0.00    0.20
           0   10    0.00    0.15
           0   11    0.00    0.25
           1    2    0.00    0.06
           2    3    0.00    0.30
           2    5    0.00    0.15
           2    6    0.00    0.45
           3    4    0.00    0.40
           3    6    0.00    0.40
           4    6    0.00    0.60
           4    9    0.00    0.70
           4   10    0.00    0.08
           5    7    0.00    0.43
           6    8    0.00    0.48
           7    8    0.00    0.35
           7   11    0.00    0.10
           8    9    0.00    0.48];

zdata0 = [ 0    1    0.00    0.06+3*0.05
           0   10    0.00    0.04+3*0.05
           0   11    0.00    0.08
           0    2    0.00    0.06
           0    7    0.00    0.10+3*.08
           1    2     inf     inf
           2    3    0.00    0.60
           2    5    0.00    0.30
           2    6    0.00    0.90
           3    4    0.00    0.80
           3    6    0.00    0.80
           4    6    0.00    1.00
           4    9    0.00    1.10
           4   10    0.00    0.08
           5    7    0.00    0.80
           6    8    0.00    0.95
           7    8    0.00    0.70
           7   11     inf     inf
           8    9    0.00   0.90];
```

```
zdata2=zdata1;
Zbus0 = zbuild(zdata0)
Zbus1 = zbuild(zdata1)
Zbus2 = Zbus1;
lgfault(zdata0, Zbus0, zdata1, Zbus1, zdata2, Zbus2)
llfault(zdata1, Zbus1,zdata2, Zbus2)
dlgfault(zdata0, Zbus0, zdata1, Zbus1, zdata2, Zbus2)
```

The result is

```
Line-to-ground fault analysis
Enter Faulted Bus No. -> 8
Enter Fault Impedance Zf = R + j*X in
complex form (for bolted fault enter 0). Zf = 0
Single line to-ground fault at bus No. 8
Total fault current =    2.8135 per unit

Bus Voltages during the fault in per unit
Bus     -------Voltage Magnitude-------
No.     Phase a      Phase b      Phase c
1       0.8907       0.9738       0.9738
2       0.8377       0.9756       0.9756
3       0.7451       0.9954       0.9954
4       0.7731       1.0063       1.0063
5       0.7824       0.9823       0.9823
6       0.5936       1.0123       1.0123
7       0.6295       0.9995       0.9995
8       0.0000       1.0898       1.0898
9       0.3299       1.0453       1.0453
10      0.8612       0.9995       0.9995
11      0.8231       0.9588       0.9588

Line currents for fault at bus No. 8
From      To      -----Line Current Magnitude----
Bus       Bus     Phase a      Phase b      Phase c
 1         2      0.5464       0.2732       0.2732
 2         3      0.2113       0.0407       0.0407
 2         6      0.3966       0.0207       0.0207
 3         6      0.2877       0.0073       0.0073
 4         3      0.0764       0.0479       0.0479
 4         6      0.2540       0.0255       0.0255
 4         9      0.5311       0.0023       0.0023
 5         2      0.2753       0.0023       0.0023
 6         8      0.9383       0.0121       0.0121
 7         5      0.2753       0.0023       0.0023
```

```
 7          8       1.3441       0.0098       0.0098
 8          F       2.8135       0.0000       0.0000
 9          8       0.5311       0.0023       0.0023
10          4       0.8615       0.0711       0.0711
11          7       0.7075       0.3538       0.3538

Another fault location?
Enter 'y' or 'n' within single quote -> 'n'

Line-to-line fault analysis
Enter Faulted Bus No. -> 8
Enter Fault Impedance Zf = R + j*X in
complex form (for bolted fault enter 0). Zf = 0
Line-to-line fault at bus No. 8
Total fault current =    2.9060 per unit

Bus Voltages during the fault in per unit
Bus     -------Voltage Magnitude-------
No.     Phase a      Phase b      Phase c
 1      1.0000       0.8576       0.8576
 2      1.0000       0.8168       0.8168
 3      1.0000       0.7757       0.7757
 4      1.0000       0.8157       0.8157
 5      1.0000       0.7838       0.7838
 6      1.0000       0.6871       0.6871
 7      1.0000       0.6947       0.6947
 8      1.0000       0.5000       0.5000
 9      1.0000       0.5646       0.5646
10      1.0000       0.8778       0.8778
11      1.0000       0.7749       0.7749

Line currents for fault at bus No. 8
From       To       -----Line Current Magnitude----
Bus        Bus      Phase a      Phase b      Phase c
 1          2       0.0000       0.8465       0.8465
 2          3       0.0000       0.1762       0.1762
 2          5       0.0000       0.2820       0.2820
 2          6       0.0000       0.3883       0.3883
 3          6       0.0000       0.3047       0.3047
 4          3       0.0000       0.1285       0.1285
 4          6       0.0000       0.2887       0.2887
 4          9       0.0000       0.5461       0.5461
 5          7       0.0000       0.2820       0.2820
 6          8       0.0000       0.9817       0.9817
 7          8       0.0000       1.3782       1.3782
```

```
 8          F        0.0000       2.9060       2.9060
 9          8        0.0000       0.5461       0.5461
10          4        0.0000       0.9633       0.9633
11          7        0.0000       1.0962       1.0962

Another fault location?
Enter 'y' or 'n' within single quote -> 'n'

Double line-to-ground fault analysis
Enter Faulted Bus No. -> 8
Enter Fault Impedance Zf = R + j*X in
complex form (for bolted fault enter 0). Zf = 0
Double line-to-ground fault at bus No. 8
Total fault current =    2.4222 per unit

Bus Voltages during the fault in per unit
Bus     -------Voltage Magnitude-------
No.     Phase a      Phase b      Phase c
1       0.9530       0.8441       0.8441
2       0.9562       0.7884       0.7884
3       0.9919       0.7122       0.7122
4       1.0107       0.7569       0.7569
5       0.9686       0.7365       0.7365
6       1.0208       0.5666       0.5666
7       0.9992       0.5907       0.5907
8       1.1391       0.0000       0.0000
9       1.0736       0.3151       0.3151
10      0.9991       0.8455       0.8455
11      0.9239       0.7509       0.7509

Line currents for fault at bus No. 8
From        To       -----Line Current Magnitude----
Bus         Bus      Phase a      Phase b      Phase c
 1          2        0.2352       0.8546       0.8546
 2          3        0.0350       0.2069       0.2069
 2          5        0.0020       0.3063       0.3063
 2          6        0.0178       0.4278       0.4278
 3          6        0.0063       0.3277       0.3277
 4          3        0.0413       0.1290       0.1290
 4          6        0.0220       0.3050       0.3050
 4          9        0.0020       0.5924       0.5924
 5          7        0.0020       0.3063       0.3063
 6          8        0.0104       1.0596       1.0596
 7          8        0.0084       1.4963       1.4963
 8          F        0.0000       3.1483       3.1483
```

```
    9          8       0.0020       0.5924       0.5924
   10          4       0.0612       1.0217       1.0217
   11          7       0.3046       1.1067       1.1067

    Another fault location?
Enter 'y' or 'n' within single quote -> 'n'
```

PROBLEMS

10.1. Obtain the symmetrical components for the set of unbalanced voltages $V_a = 300\angle-120°$, $V_b = 200\angle 90°$, and $V_c = 100\angle-30°$.

10.2. The symmetrical components of a set of unbalanced three-phase currents are $I_a^0 = 3\angle-30°$, $I_a^1 = 5\angle 90°$, and $I_a^2 = 4\angle 30°$. Obtain the original unbalanced phasors.

10.3. The operator a is defined as $a = 1\angle 120°$; show that

(a) $\frac{(1+a)}{(1+a^2)} = 1\angle 120°$

(b) $\frac{(1-a)^2}{(1+a)^2} = 3\angle-180°$

(c) $(a - a^2)(a^2 - a) = 3\angle 0°$

(d) $V_{an}^1 = \frac{1}{\sqrt{3}} V_{bc}^1 \angle 90°$

(e) $V_{an}^2 = \frac{1}{\sqrt{3}} V_{bc}^2 \angle-90°$

10.4. The line-to-line voltages in an unbalanced three-phase supply are $V_{ab} = 1000\angle 0°$, $V_{bc} = 866.0254\angle-150°$, and $V_{ca} = 500\angle 120°$. Determine the symmetrical components for line and phase voltages, then find the phase voltages V_{an}, V_{bn}, and V_{cn}.

10.5. In the three-phase system shown in Figure 10.26, phase a is on no load and phases b and c are short-circuited to ground.

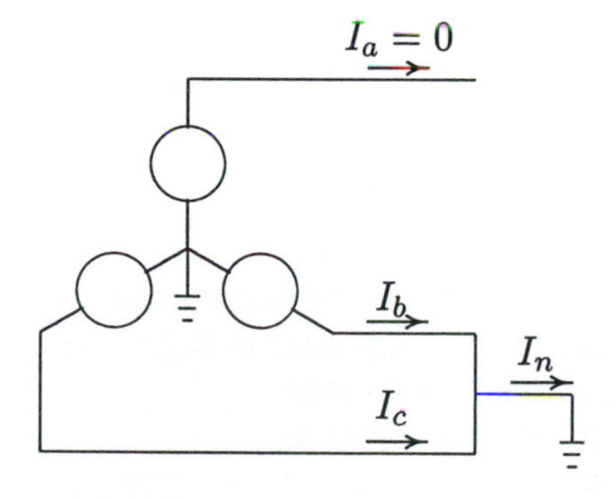

FIGURE 10.26
Circuit for Problem 10.5.

The following currents are given:
$I_b = 91.65\angle 160.9°$
$I_n = 60.00\angle 90°$
Find the symmetrical components of current I_a^0, I_a^1, and I_a^2.

10.6. A balanced three-phase voltage of 360-V line-to-neutral is applied to a balanced Y-connected load with ungrounded neutral, as shown in Figure 10.27. The three-phase load consists of three mutually-coupled reactances. Each phase has a series reactance of $Z_s = j24\ \Omega$, and the mutual coupling between phases is $Z_m = j6\ \Omega$.
(a) Determine the line currents by mesh analysis without using symmetrical components.
(b) Determine the line currents using symmetrical components.

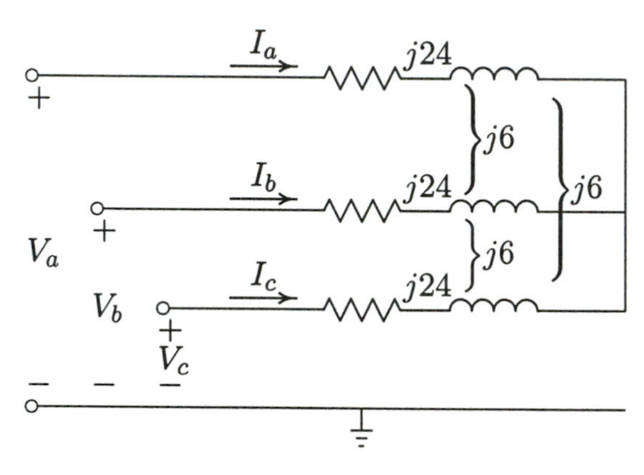

FIGURE 10.27
Circuit for Problem 10.6.

10.7. A three-phase unbalanced source with the following phase-to-neutral voltages

$$\mathbf{V}^{abc} = \begin{bmatrix} 300 & \angle -120° \\ 200 & \angle 90° \\ 100 & \angle -30° \end{bmatrix}$$

is applied to the circuit in Figure 10.28. The load series impedance per phase is $Z_s = 10 + j40$ and the mutual impedance between phases is $Z_m = j5$. The load and source neutrals are solidly grounded. Determine
(a) The load sequence impedance matrix, $\mathbf{Z}^{012} = \mathbf{A}^{-1}\mathbf{Z}^{abc}\mathbf{A}$.
(b) The symmetrical components of voltage.
(c) The symmetrical components of current.
(d) The load phase currents.

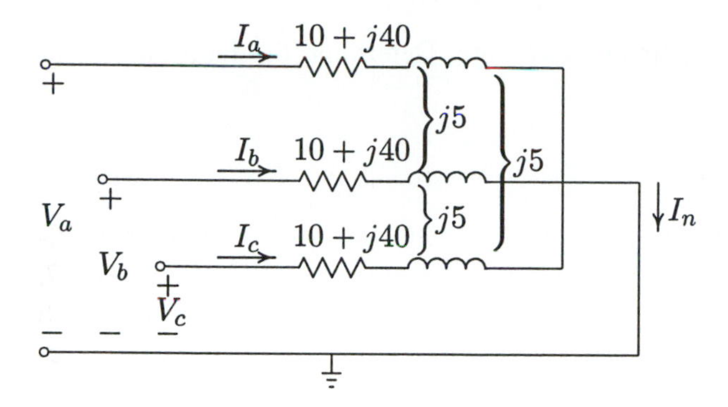

FIGURE 10.28
Circuit for Problem 10.7.

(e) The complex power delivered to the load in terms of symmetrical components, $S_{3\phi} = 3(V_a^0 I_a^{0*} + V_a^1 I_a^{1*} + V_a^2 I_a^{2*})$.
(f) The complex power delivered to the load by summing up the power in each phase, $S_{3\phi} = V_a I_a^* + V_b I_b^* + V_c I_c^*$.

10.8. The line-to-line voltages in an unbalanced three-phase supply are $V_{ab} = 600\angle 36.87°$, $V_{bc} = 800\angle 126.87°$, and $V_{ca} = 1000\angle -90°$. A Y-connected load with a resistance of 37 Ω per phase is connected to the supply. Determine
(a) The symmetrical components of voltage.
(b) The phase voltages.
(c) The line currents.

10.9. A generator having a solidly grounded neutral and rated 50-MVA, 30-kV has positive-, negative-, and zero-sequence reactances of 25, 15, and 5 percent, respectively. What reactance must be placed in the generator neutral to limit the fault current for a bolted line-to-ground fault to that for a bolted three-phase fault?

10.10. What reactance must be placed in the neutral of the generator of Problem 9 to limit the magnitude of the fault current for a bolted double line-to-ground fault to that for a bolted three-phase fault?

10.11. Three 15-MVA, 30-kV synchronous generators A, B, and C are connected via three reactors to a common bus bar, as shown in Figure 10.29. The neutrals of generators A and B are solidly grounded, and the neutral of generator C is grounded through a reactor of 2.0 Ω. The generator data and the reactance of the reactors are tabulated below. A line-to-ground fault occurs on phase a of the common bus bar. Neglect prefault currents and assume generators are operating at their rated voltage. Determine the fault current in phase a.

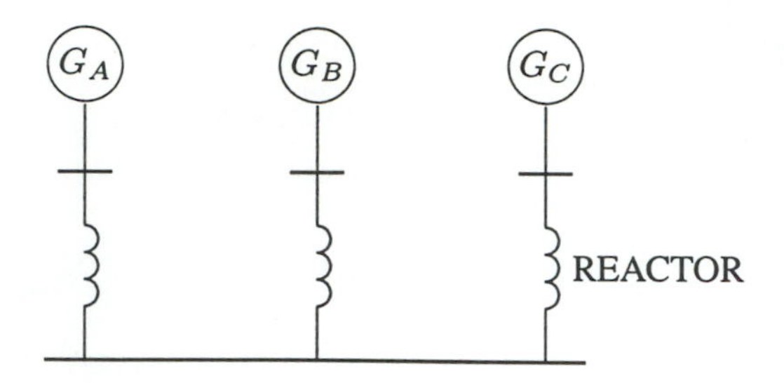

FIGURE 10.29
Circuit for Problem 10.11.

Item	X^1	X^2	X^0
G_A	0.25 pu	0.155 pu	0.056 pu
G_B	0.20 pu	0.155 pu	0.056 pu
G_C	0.20 pu	0.155 pu	0.060 pu
Reactor	6.0 Ω	6.0 Ω	6.0 Ω

10.12. Repeat Problem 10.11 for a bolted line-to-line fault between phases b and c.

10.13. Repeat Problem 10.11 for a bolted double line-to-ground fault on phases b and c.

10.14. The zero-, positive-, and negative-sequence bus impedance matrices for a three-bus power system are

$$\mathbf{Z}_{bus}^0 = j\begin{bmatrix} 0.20 & 0.05 & 0.12 \\ 0.05 & 0.10 & 0.08 \\ 0.12 & 0.08 & 0.30 \end{bmatrix} \quad \text{pu}$$

$$\mathbf{Z}_{bus}^1 = \mathbf{Z}_{bus}^2 = j\begin{bmatrix} 0.16 & 0.10 & 0.15 \\ 0.10 & 0.20 & 0.12 \\ 0.15 & 0.12 & 0.25 \end{bmatrix} \quad \text{pu}$$

Determine the per unit fault current and the bus voltages during fault for
(a) A bolted three-phase fault at bus 2.
(b) A bolted single line-to-ground fault at bus 2.
(c) A bolted line-to-line fault at bus 2.
(d) A bolted double line-to-ground fault at bus 2.

10.15. The reactance data for the power system shown in Figure 10.30 in per unit on a common base is as follows:

Item	X^1	X^2	X^0
G_1	0.10	0.10	0.05
G_2	0.10	0.10	0.05
T_1	0.25	0.25	0.25
T_2	0.25	0.25	0.25
Line 1–2	0.30	0.30	0.50

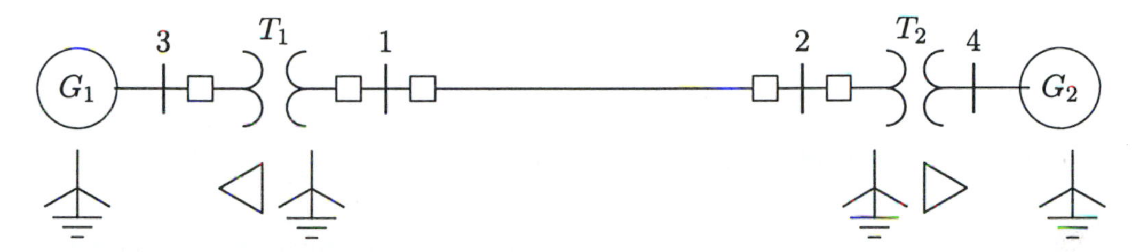

FIGURE 10.30
Circuit for Problem 10.15.

Obtain the Thévenin sequence impedances for the fault at bus 1 and compute the fault current in per unit for the following faults:

(a) A bolted three-phase fault at bus 1.
(b) A bolted single line-to-ground fault at bus 1.
(c) A bolted line-to-line fault at bus 1.
(d) A bolted double line-to-ground fault at bus 1.

10.16. For Problem 10.15, obtain the bus impedance matrices for the sequence networks. A bolted single line-to-ground fault occurs at bus 1. Find the fault current, the three-phase bus voltages during fault, and the line currents in each phase. Check your results using the **zbuild** and **lgfault** programs.

10.17. Repeat Problem 10.16 for a bolted line-to-line fault. Check your results using the **zbuild** and **llfault** programs.

10.18. Repeat Problem 10.16 for a bolted double line-to-ground fault. Check your results using the **zbuild** and **dlgfault** programs.

10.19. The positive-sequence reactances for the power system shown in Figure 10.31 are in per unit on a common MVA base. Resistances are neglected and the negative-sequence impedances are assumed to be the same as the positive-sequence impedances. A bolted line-to-line fault occurs between phases b and c at bus 2. Before the fault occurrence, all bus voltages are 1.0 per unit. Obtain the positive-sequence bus impedance matrix. Find the fault current, the three-phase bus voltages during fault, and the line currents in each phase. Check your results using the **zbuild** and **llfault** programs.

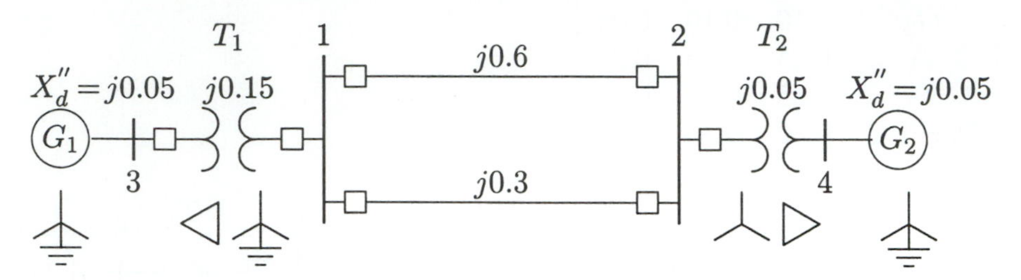

FIGURE 10.31
Circuit for Problem 10.19.

10.20. Use the **lgfault**, **llfault**, and **dlgfault** functions to compute the fault current, bus voltages, and line currents in the circuit given in Example 10.8 for the following unbalanced fault.

(a) A bolted single line-to-ground fault at bus 9.
(b) A bolted line-to-line fault at bus 9.
(c) A bolted double line-to-ground fault at bus 9.

All shunt capacitances and loads are neglected and the negative-sequence data is assumed to be the same as the positive-sequence data. All the prefault bus voltages are assumed to be unity.

10.21. The six-bus power system network of an electric utility company is shown in Figure 10.32. The positive- and zero-sequence reactances of the lines and transformers in per unit on a 100-MVA base is tabulated below.

LINE AND TRANSFORMER DATA			
Bus No.	Bus No.	X^1, PU	X^0, PU
1	4	0.225	0.400
1	5	0.105	0.200
1	6	0.215	0.390
2	4	0.035	0.035
3	5	0.042	0.042
4	6	0.125	0.250
5	6	0.175	0.350

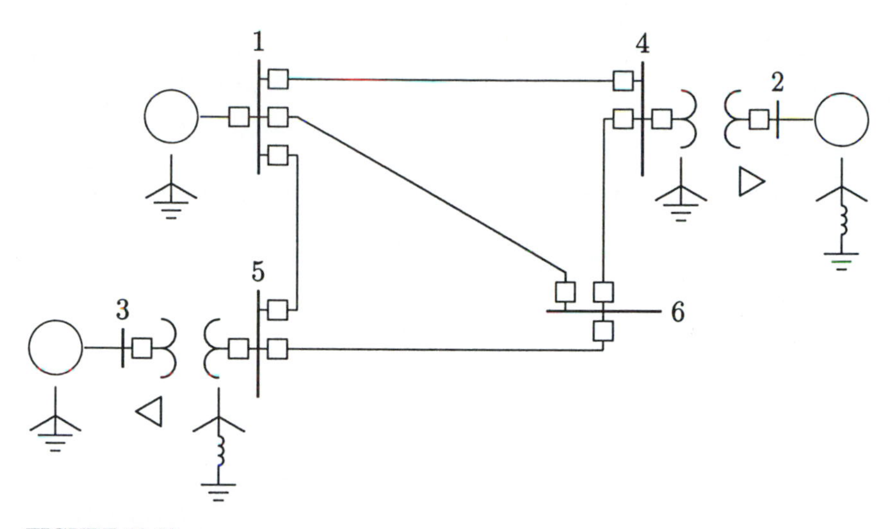

FIGURE 10.32
One-line diagram for Problem 10.32.

The transformer connections are shown in Figure 10.32. The Δ-Y transformer between buses 3 and 5 is grounded through a reactor of reactance 0.10 per unit. The generator's positive- and zero-sequence reactances including the reactance of grounding neutrals on a 100-MVA base is tabulated below.

GENERATOR TRANSIENT IMPEDANCE, PU			
Gen. No.	X^1	X^0	X_n
1	0.20	0.06	0.00
2	0.15	0.04	0.05
3	0.25	0.08	0.00

Resistances, shunt reactances, and loads are neglected, and all negative-sequence reactances are assumed equal to the positive-sequence reactances. Use **zbuild** function to obtain the positive- and zero-sequence bus impedance matrices. Assume all the prefault bus voltages are equal to $1\angle 0°$, use **lgfault**, **llfault**, and **dlgfault** to compute the fault current, bus voltages, and line currents for the following unbalanced faults.

(a) A bolted single line-to-ground fault at bus 6.
(b) A bolted line-to-line fault at bus 6.
(c) A bolted double line-to-ground fault at bus 6.

CHAPTER 11

STABILITY

11.1 INTRODUCTION

The tendency of a power system to develop restoring forces equal to or greater than the disturbing forces to maintain the state of equilibrium is known as *stability*. If the forces tending to hold machines in synchronism with one another are sufficient to overcome the disturbing forces, the system is said to remain stable (to stay in synchronism).

The stability problem is concerned with the behavior of the synchronous machines after a disturbance. For convenience of analysis, stability problems are generally divided into two major categories — *steady-state stability* and *transient stability*. Steady-state stability refers to the ability of the power system to regain synchronism after small and slow disturbances, such as gradual power changes. An extension of the steady-state stability is known as the *dynamic stability*. The dynamic stability is concerned with small disturbances lasting for a long time with the inclusion of automatic control devices. Transient stability studies deal with the effects of large, sudden disturbances such as the occurrence of a fault, the sudden outage of a line or the sudden application or removal of loads. Transient stability studies are needed to ensure that the system can withstand the transient condition following a major disturbance. Frequently, such studies are conducted when new generating and transmitting facilities are planned. The studies are helpful in determining such things as the nature of the relaying system needed, critical clearing time of circuit breakers, voltage level of, and transfer capability between systems.

11.2 SWING EQUATION

Under normal operating conditions, the relative position of the rotor axis and the resultant magnetic field axis is fixed. The angle between the two is known as the *power angle* or *torque angle*. During any disturbance, rotor will decelerate or accelerate with respect to the synchronously rotating air gap mmf, and a relative motion begins. The equation describing this relative motion is known as the *swing equation*. If, after this oscillatory period, the rotor locks back into synchronous speed, the generator will maintain its stability. If the disturbance does not involve any net change in power, the rotor returns to its original position. If the disturbance is created by a change in generation, load, or in network conditions, the rotor comes to a new operating power angle relative to the synchronously revolving field.

In order to understand the significance of the power angle we refer to the combined phasor/vector diagram of a two-pole cylindrical rotor generator illustrated in Figure 3.2. From this figure we see that the power angle δ_r is the angle between the rotor mmf F_r and the resultant air gap mmf F_{sr}, both rotating at synchronous speed. It is also the angle between the no-load generated emf E and the resultant stator voltage E_{sr}. If the generator armature resistance and leakage flux are neglected, the angle between E and the terminal voltage V, denoted by δ, is considered as the power angle.

Consider a synchronous generator developing an electromagnetic torque T_e and running at the synchronous speed ω_{sm}. If T_m is the driving mechanical torque, then under steady-state operation with losses neglected we have

$$T_m = T_e \tag{11.1}$$

A departure from steady state due to a disturbance results in an accelerating ($T_m > T_e$) or decelerating ($T_m < T_e$) torque T_a on the rotor.

$$T_a = T_m - T_e \tag{11.2}$$

If J is the combined moment of inertia of the prime mover and generator, neglecting frictional and damping torques, from law's of rotation we have

$$J\frac{d^2\theta_m}{dt^2} = T_a = T_m - T_e \tag{11.3}$$

where θ_m is the angular displacement of the rotor with respect to the stationary reference axis on the stator. Since we are interested in the rotor speed relative to synchronous speed, the angular reference is chosen relative to a synchronously rotating reference frame moving with constant angular velocity ω_{sm}, that is

$$\theta_m = \omega_{sm}t + \delta_m \tag{11.4}$$

where δ_m is the rotor position before disturbance at time $t = 0$, measured from the synchronously rotating reference frame. Derivative of (11.4) gives the rotor angular velocity

$$\omega_m = \frac{d\theta_m}{dt} = \omega_{ms} + \frac{d\delta_m}{dt} \tag{11.5}$$

and the rotor acceleration is

$$\frac{d^2\theta_m}{dt^2} = \frac{d^2\delta_m}{dt^2} \tag{11.6}$$

Substituting (11.6) into (11.3), we have

$$J\frac{d^2\delta_m}{dt^2} = T_m - T_e \tag{11.7}$$

Multiplying (11.7) by ω_m, results in

$$J\omega_m\frac{d^2\delta_m}{dt^2} = \omega_m T_m - \omega_m T_e \tag{11.8}$$

Since angular velocity times torque is equal to the power, we write the above equation in terms of power

$$J\omega_m\frac{d^2\delta_m}{dt^2} = P_m - P_e \tag{11.9}$$

The quantity $J\omega_m$ is called the inertia constant and is denoted by M. It is related to kinetic energy of the rotating masses, W_k.

$$W_k = \frac{1}{2}J\omega_m^2 = \frac{1}{2}M\omega_m \tag{11.10}$$

or

$$M = \frac{2W_k}{\omega_m} \tag{11.11}$$

Although M is called inertia constant, it is not really constant when the rotor speed deviates from the synchronous speed. However, since ω_m does not change by a large amount before stability is lost, M is evaluated at the synchronous speed and is considered to remain constant, i.e.,

$$M = \frac{2W_k}{\omega_{sm}} \tag{11.12}$$

The swing equation in terms of the inertia constant becomes

$$M\frac{d^2\delta_m}{dt^2} = P_m - P_e \tag{11.13}$$

It is more convenient to write the swing equation in terms of the electrical power angle δ. If p is the number of poles of a synchronous generator, the electrical power angle δ is related to the mechanical power angle δ_m by

$$\delta = \frac{p}{2}\delta_m \tag{11.14}$$

also,

$$\omega = \frac{p}{2}\omega_m \tag{11.15}$$

Swing equation in terms of electrical power angle is

$$\frac{2}{p}M\frac{d^2\delta}{dt^2} = P_m - P_e \tag{11.16}$$

Since power system analysis is done in per unit system, the swing equation is usually expressed in per unit. Dividing (11.16) by the base power S_B, and substituting for M from (11.12) results in

$$\frac{2}{p}\frac{2W_K}{\omega_{sm}S_B}\frac{d^2\delta}{dt^2} = \frac{P_m}{S_B} - \frac{P_e}{S_B} \tag{11.17}$$

We now define the important quantity known as the H *constant* or *per unit inertia constant*.

$$H = \frac{\text{kinetic energy in MJ at rated speed}}{\text{machine rating in MVA}} = \frac{W_K}{S_B} \tag{11.18}$$

The unit of H is seconds. The value of H ranges from 1 to 10 seconds, depending on the size and type of machine. Substituting in (11.17), we get

$$\frac{2}{p}\frac{2H}{\omega_{sm}}\frac{d^2\delta}{dt^2} = P_{m(pu)} - P_{e(pu)} \tag{11.19}$$

where $P_{m(pu)}$ and $P_{e(pu)}$ are the per unit mechanical power and electrical power, respectively. The electrical angular velocity is related to the mechanical angular velocity by $\omega_{sm} = (2/p)\omega_s$. (11.19) in terms of electrical angular velocity is

$$\frac{2H}{\omega_s}\frac{d^2\delta}{dt^2} = P_{m\,(pu)} - P_{e\,(pu)} \tag{11.20}$$

The above equation is often expressed in terms of frequency f_0, and to simplify the notation, the subscript *pu* is omitted and the powers are understood to be in per unit.

$$\frac{H}{\pi f_0}\frac{d^2\delta}{dt^2} = P_m - P_e \tag{11.21}$$

where δ is in electrical radian. If δ is expressed in electrical degrees, the swing equation becomes

$$\frac{H}{180 f_0}\frac{d^2\delta}{dt^2} = P_m - P_e \tag{11.22}$$

11.3 SYNCHRONOUS MACHINE MODELS FOR STABILITY STUDIES

The representation of a synchronous machine during transient conditions was discussed in Chapter 8. In Section 8.6 the cylindrical rotor machine was modeled with a constant voltage source behind proper reactances, which may be X_d'', X_d', or X_d. The simplest model for stability analysis is the classical model, where saliency is ignored, and the machine is represented by a constant voltage E' behind the direct axis transient reactance X_d'.

Consider a generator connected to a major substation of a very large system through a transmission line as shown in Figure 11.1.

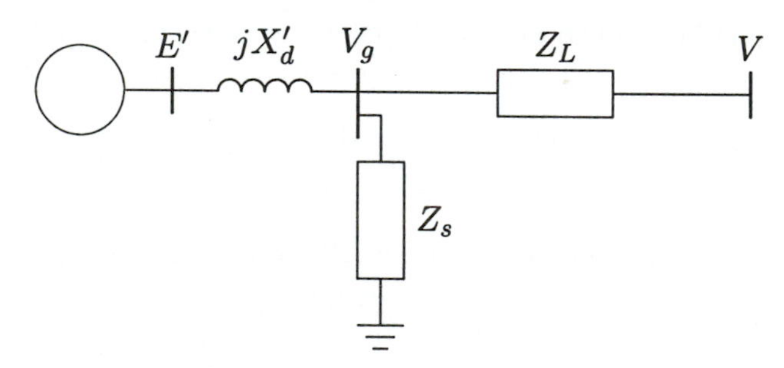

FIGURE 11.1
One machine connected to an infinite bus.

The substation bus voltage and frequency is assumed to remain constant. This is commonly referred to as an *infinite bus*, since its characteristics do not change regardless of the power supplied or consumed by any device connected to it. The generator is represented by a constant voltage behind the direct axis transient reactance X_d'. The node representing the generator terminal voltage V_g can be eliminated by converting the Y-connected impedances to an equivalent Δ with admittances given by

$$\begin{aligned}
y_{10} &= \frac{Z_L}{jX'_d Z_s + jX'_d Z_L + Z_L Z_s} \\
y_{20} &= \frac{jX'_d}{jX'_d Z_s + jX'_d Z_L + Z_L Z_s} \\
y_{12} &= \frac{Z_s}{jX'_d Z_s + jX'_d Z_L + Z_L Z_s}
\end{aligned} \tag{11.23}$$

The equivalent circuit with internal voltage represented by node 1 and the infinite bus by node 2 is shown in Figure 11.2. Writing the nodal equations, we have

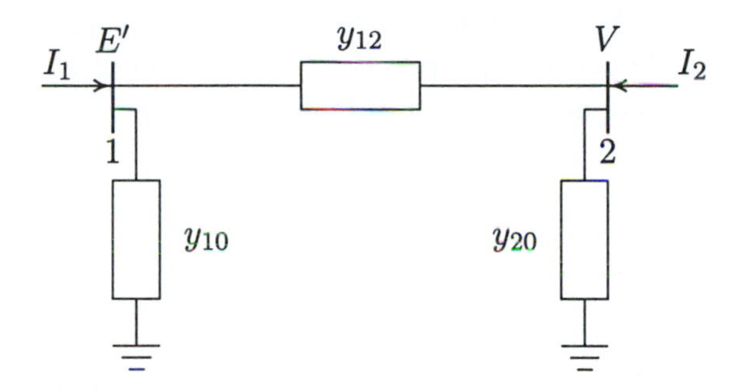

FIGURE 11.2
Equivalent circuit of one machine connected to an infinite bus.

$$\begin{aligned}
I_1 &= (y_{10} + y_{12})E' - y_{12}V \\
I_2 &= -y_{12}E' + (y_{20} + y_{12})V
\end{aligned} \tag{11.24}$$

The above equations can be written in terms of the bus admittance matrix as

$$\begin{bmatrix} I_1 \\ I_2 \end{bmatrix} = \begin{bmatrix} Y_{11} & Y_{12} \\ Y_{21} & Y_{22} \end{bmatrix} \begin{bmatrix} E' \\ V \end{bmatrix} \tag{11.25}$$

The diagonal elements of the bus admittance matrix are $Y_{11} = y_{10} + y_{12}$, and $Y_{22} = y_{20} + y_{12}$. The off-diagonal elements are $Y_{12} = Y_{21} = -y_{12}$. Expressing the voltages and admittances in polar form, the real power at node 1 is given by

$$\begin{aligned}
P_e &= \Re[E' I_1^*] \\
&= \Re[|E'|\angle\delta(|Y_{11}|\angle{-\theta_{11}}|E'|\angle{-\delta} + |Y_{12}|\angle{-\theta_{12}}|V|\angle 0)]
\end{aligned}$$

or

$$P_e = |E'|^2 |Y_{11}| \cos\theta_{11} + |E'||V||Y_{12}| \cos(\delta - \theta_{12}) \tag{11.26}$$

The power flow equation given by (6.25) when applied to the above two-bus power system results in the same expression as (11.26). In most systems, Z_L and Z_s are predominantly inductive. If all resistances are neglected, $\theta_{11} = \theta_{12} = 90°$, $Y_{12} = B_{12} = 1/X_{12}$, and we obtain a simplified expression for power

$$P_e = |E'||V||B_{12}|\cos(\delta - 90°)$$

or

$$P_e = \frac{|E'||V|}{X_{12}} \sin\delta \tag{11.27}$$

This is the simplest form of the power flow equation and is basic to an understanding of all stability problems. The relation shows that the power transmitted depends upon the transfer reactance and the angle between the two voltages. The curve P_e versus δ is known as the *power angle curve* and is shown in Figure 11.3.

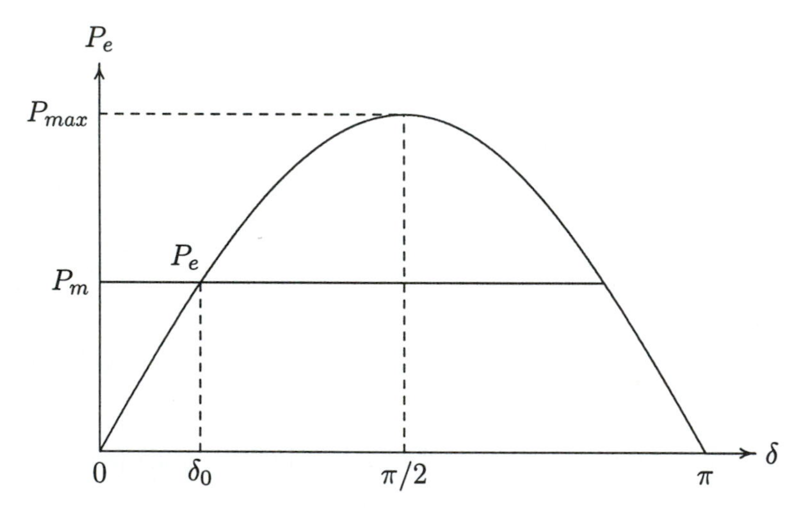

FIGURE 11.3
Power-angle curve.

The gradual increase of the generator power output is possible until the maximum electrical power is transferred. This maximum power is referred to as the *steady-state stability limit*, and occurs at an angular displacement of 90°.

$$P_{max} = \frac{|E'||V|}{X_{12}} \tag{11.28}$$

If an attempt were made to advance δ further by further increasing the shaft input, the electrical power output will decrease from the P_{max} point. The machine will accelerate, causing loss of synchronism with the infinite bus bar. The electric power equation in terms of P_{max} is

$$P_e = P_{max} \sin \delta \tag{11.29}$$

When a generator is suddenly short-circuited, the current during the transient period is limited by its transient reactance X_d'. Thus, for transient stability problems, with the saliency neglected, the machine is represented by the voltage E' behind the reactance X_d'. If V_g is the generator terminal voltage and I_a is the prefault steady state generator current, E' is computed from

$$E' = V_g + jX_d' I_a \tag{11.30}$$

Since the field winding has a small resistance, the field flux linkages will tend to remain constant during the initial disturbance, and thus the voltage E' is assumed constant. The transient power-angle curve has the same general form as the steady-state curve; however, it attains larger peak compared to the steady-state peak value.

11.3.1 SYNCHRONOUS MACHINE MODEL INCLUDING SALIENCY

In Section 3.4 we developed the two-axis model of a synchronous machine under steady state conditions taking into account the effect of saliency. The phasor diagram of the salient-pole machine under steady state conditions, with armature resistance neglected, was presented in Figure 3.8. This phasor diagram is represented in Figure 11.4.

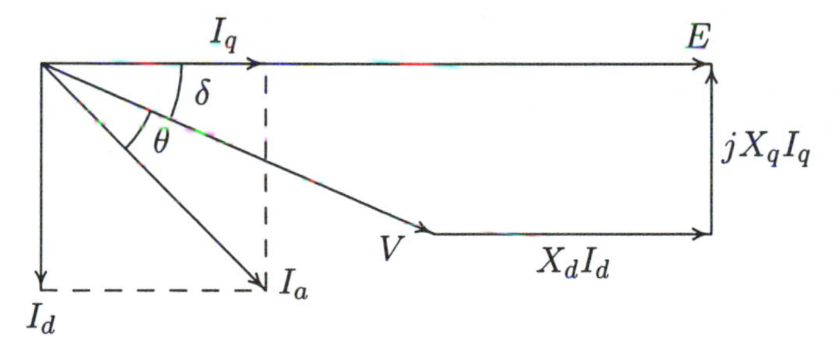

FIGURE 11.4
Phasor diagram during transient period.

The power-angle equation was given by (3.32). This equation presented in per unit is

$$P_e = \frac{|E||V|}{X_d} \sin \delta + |V|^2 \frac{X_d - X_q}{2X_d X_q} \sin 2\delta \tag{11.31}$$

and

$$|E| = |V|\cos\delta + X_d I_d$$

or

$$|E| = |V|\cos\delta + X_d|I_a|\sin(\delta+\theta) \tag{11.32}$$

where E is the no-load generated emf in per unit and V is the generator terminal voltage in per unit. X_d and X_q are direct and quadrature axis reactances of the synchronous machine. For a derivation of the above formula, refer to Section 3.4. For a given power delivered at a given terminal voltage, we must compute E. In order to do that, we must first compute δ as follows:

$$\begin{aligned} |V|\sin\delta &= X_q I_q \\ &= X_q|I_a|\cos(\delta+\theta) \\ &= X_q|I_a|(\cos\delta\cos\theta - \sin\delta\sin\theta) \end{aligned}$$

From the above relation, δ is found to be

$$\delta = \tan^{-1}\frac{X_q|I_a|\cos\theta}{|V| + X_q|I_a|\sin\theta} \tag{11.33}$$

Substituting for δ from (11.33) into (11.32) will result in the voltage E.

A logical extension of the model would be to include the effect of transient saliency. Since the machine circuits are largely inductive, the flux linkages tend to remain constant during the early part of the transient period. During the transient period, the direct axis transient reactance is X_d'. Since the field is on the direct axis X_q', the quadrature axis transient reactance remains the same as X_q. The phasor diagram under transient condition is shown in Figure 11.5. Following the procedure

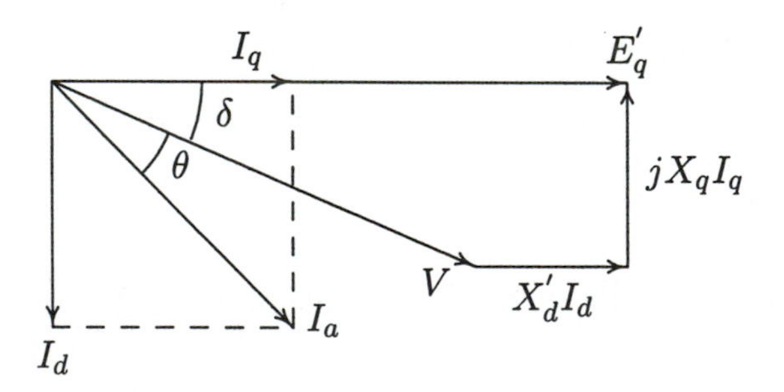

FIGURE 11.5
Phasor diagram during transient period.

in Section 3.4, the transient power-angle equation expressed in per unit becomes

$$P_e = \frac{|E'_q||V|}{X'_d}\sin\delta + |V|^2\frac{X'_d - X_q}{2X'_dX_q}\sin 2\delta \tag{11.34}$$

This equation represents the approximate behavior of the synchronous machine during the early part of transient period. We now have to determine E'_q. From the phasor diagram in Figure 11.5, we have

$$|E'_q| = |V|\cos\delta + X'_dI_d$$

or

$$|E'_q| = |V|\cos\delta + X'_d|I_a|\sin(\delta + \theta)$$

From (11.32), we find

$$|I_a|\sin(\delta + \theta) = \frac{|E| - |V|\cos\delta}{X_d}$$

and substitute it in the above equation to get

$$|E'_q| = \frac{X'_d|E| + (X_d - X'_d)|V|\cos\delta}{X_d} \tag{11.35}$$

The prefault excitation voltage and power angles are computed from (11.32) and (11.33).

In this section we presented two simple models for cylindrical rotor and salient rotor synchronous machines. The choice of model for a given situation must, in general, depend upon the type of study being conducted as well as the data available. Although these models are useful for many stability studies, it is not adequate for many situations. More accurate models must include the effects of the various rotor circuits.

Example 11.1

Consider a synchronous machine characterized by the following parameters:

$$X_d = 1.0 \quad X_q = 0.6 \quad X'_d = 0.3 \quad \text{per unit}$$

and negligible armature resistance. The machine is connected directly to an infinite bus of voltage 1.0 per unit. The generator is delivering a real power of 0.5 per unit at 0.8 power factor lagging. Determine the voltage behind transient reactance and the transient power-angle equation for the following cases.

(a) Neglecting the saliency effect
(b) Including the effect of saliency

$$\theta = \cos^{-1} 0.8 = 36.87^\circ$$
$$S = \frac{0.5}{0.8}\angle 36.87^\circ = 0.625\angle 36.87^\circ \quad \text{pu}$$

The prefault steady state current is

$$I_a = \frac{S^*}{V^*} = 0.625\angle{-36.87^\circ} \quad \text{pu}$$

(a) With saliency neglected, the voltage behind transient reactance is

$$E' = V + jX'_d I_a = 1.0 + (j0.3)(0.625\angle{-36.87^\circ}) = 1.1226\angle 7.679^\circ \quad \text{pu}$$

The transient power-angle curve is given by

$$P_e = \frac{|E'||V|}{X'_d}\sin\delta = \frac{(1.1226)(1)}{0.3}\sin\delta$$

or

$$P_e = 3.7419 \sin\delta$$

(b) When the saliency effect is considered, the initial steady state power angle given by (11.33) is

$$\delta = \tan^{-1}\frac{X_q|I_a|\cos\theta}{|V| + X_q|I_a|\sin\theta} = \tan^{-1}\frac{(0.6)(0.625)(0.8)}{1.0 + (0.6)(.625)(.6)} = 13.7608^\circ$$

The steady state excitation voltage E, given by (11.32), is

$$\begin{aligned}|E| &= |V|\cos\delta + X_d|I_a|\sin(\delta + \theta)\\ &= (1.0)\cos(13.7608^\circ) + (1.0)(0.625)\sin(13.7608^\circ + 36.87^\circ) = 1.4545 \quad \text{pu}\end{aligned}$$

The transient voltage E'_q given by (11.35) is

$$\begin{aligned}|E'_q| &= \frac{X'_d|E| + (X_d - X'_d)|V|\cos\delta}{X_d}\\ &= \frac{(0.3)(1.4545) + (1.0 - 0.3)(1.0)(\cos 13.7608^\circ)}{1.0} = 1.1162 \quad \text{pu}\end{aligned}$$

and from (11.34) the transient power-angle equation is

$$P_e = \frac{(1.1162)(1)}{0.3}\sin\delta + \frac{(1.0)^2(0.3 - 0.6)}{2(0.3)(0.6)}\sin 2\delta$$

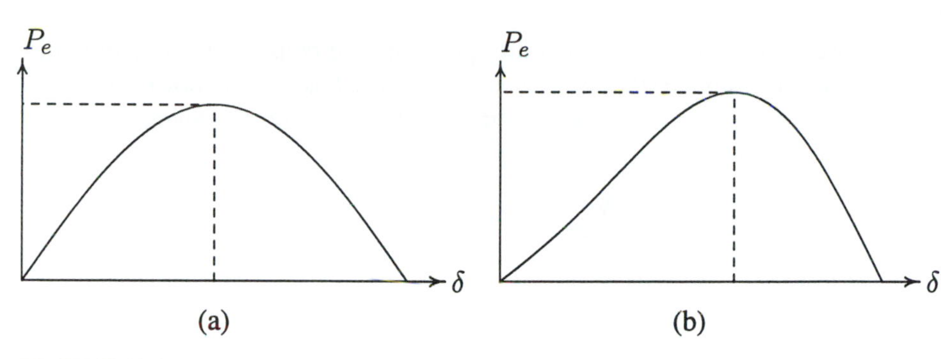

FIGURE 11.6
Transient power-angle curve for Example 11.1.

or

$$P_e = 3.7208 \sin \delta - 0.8333 \sin 2\delta$$

Using *MATLAB*, the power-angle equations obtained in (a) and (b) are plotted as shown in Figure 11.6. We use the function **[Pmax, k] = max(P)** to find the maximum power in case (b). The maximum power is found to be 4.032, occurring at angle $\delta(k) = 110.01°$.

The coefficient of $\sin 2\delta$ is relatively small, and since $X'_d < X_q$, it is negative. Thus, the $\sin 2\delta$ term has the property of subtracting from the $\sin \delta$ term in the region $0° < \delta < 90°$, but adding to it in the region $90° < \delta < 180°$. During sudden impact, when δ swings from its initial value to the maximum value for marginal stability, the overall effect of the $\sin 2\delta$ term has the tendency to average out to zero. For this reason, the $\sin 2\delta$ term is often ignored in the approximate power-angle equation.

11.4 STEADY-STATE STABILITY — SMALL DISTURBANCES

The steady-state stability refers to the ability of the power system to remain in synchronism when subjected to small disturbances. It is convenient to assume that the disturbances causing the changes disappear. The motion of the system is free, and stability is assured if the system returns to its original state. Such a behavior can be determined in a linear system by examining the characteristic equation of the system. It is assumed that the automatic controls, such as voltage regulator and governor, are not active. The actions of governor and excitation system and control devices are discussed in Chapter 12 when dealing with dynamic stability.

To illustrate the steady-state stability problem, we consider the dynamic behavior of a one-machine system connected an infinite bus bar as shown in Figure 11.1. Substituting for the electrical power from (11.29) into the swing equation given in (11.21) results in

$$\frac{H}{\pi f_0}\frac{d^2\delta}{dt^2} = P_m - P_{max}\sin\delta \tag{11.36}$$

The swing equation is a nonlinear function of the power angle. However, for small disturbances, the swing equation may be linearized with little loss of accuracy as follows. Consider a small deviation $\Delta\delta$ in power angle from the initial operating point δ_0, i.e.,

$$\delta = \delta_0 + \Delta\delta \tag{11.37}$$

Substituting in (11.36), we get

$$\frac{H}{\pi f_0}\frac{d^2(\delta_0 + \Delta\delta)}{dt^2} = P_m - P_{max}\sin(\delta_0 + \Delta\delta)$$

or

$$\frac{H}{\pi f_0}\frac{d^2\delta_0}{dt^2} + \frac{H}{\pi f_0}\frac{d^2\Delta\delta}{dt^2} = P_m - P_{max}(\sin\delta_0\cos\Delta\delta + \cos\delta_0\sin\Delta\delta)$$

Since $\Delta\delta$ is small, $\cos\Delta\delta \cong 1$ and $\sin\Delta\delta \cong \Delta\delta$, and we have

$$\frac{H}{\pi f_0}\frac{d^2\delta_0}{dt^2} + \frac{H}{\pi f_0}\frac{d^2\Delta\delta}{dt^2} = P_m - P_{max}\sin\delta_0 - P_{max}\cos\delta_0\,\Delta\delta$$

Since at the initial operating state

$$\frac{H}{\pi f_0}\frac{d^2\delta_0}{dt^2} = P_m - P_{max}\sin\delta_0$$

The above equation reduces to linearized equation in terms of incremental changes in power angle, i.e.,

$$\frac{H}{\pi f_0}\frac{d^2\Delta\delta}{dt^2} + P_{max}\cos\delta_0\,\Delta\delta = 0 \tag{11.38}$$

The quantity $P_{max}\cos\delta_0$ in (11.38) is the slope of the power-angle curve at δ_0. It is known as the *synchronizing coefficient*, denoted by P_s. This coefficient plays an important part in determining the system stability, and is given by

$$P_s = \left.\frac{dP}{d\delta}\right|_{\delta_0} = P_{max}\cos\delta_0 \tag{11.39}$$

Substituting in (11.38), we have

$$\frac{H}{\pi f_0}\frac{d^2\Delta\delta}{dt^2} + P_s\Delta\delta = 0 \tag{11.40}$$

The solution of the above second-order differential equation depends on the roots of the characteristic equation given by

$$s^2 = -\frac{\pi f_0}{H}P_s \tag{11.41}$$

When P_s is negative, we have one root in the right-half s-plane, and the response is exponentially increasing and stability is lost. When P_s is positive, we have two roots on the j-ω axis, and the motion is oscillatory and undamped. The system is marginally stable with a natural frequency of oscillation given by

$$\omega_n = \sqrt{\frac{\pi f_0}{H}P_s} \tag{11.42}$$

It can be seen from Figure 11.3 that the range where P_s (i.e., the slope $dP/d\delta$) is positive lies between 0 and 90° with a maximum value at no-load ($\delta_0 = 0$).

As long as there is a difference in angular velocity between the rotor and the resultant rotating air gap field, induction motor action will take place between them, and a torque will be set up on the rotor tending to minimize the difference between the two angular velocities. This is called the *damping torque*. The damping power is approximately proportional to the speed deviation.

$$P_d = D\frac{d\delta}{dt} \tag{11.43}$$

The damping coefficient D may be determined either from design data or by test. Additional damping torques are caused by the speed/torque characteristic of the prime mover and the load dynamic, which are not considered here. When the synchronizing power coefficient P_s is positive, because of the damping power, oscillations will damp out eventually, and the operation at the equilibrium angle will be restored. No loss of synchronism occurs and the system is stable.

If damping is accounted for, the linearized swing equation becomes

$$\frac{H}{\pi f_0}\frac{d^2\Delta\delta}{dt^2} + D\frac{d\Delta\delta}{dt} + P_s\,\Delta\delta = 0 \tag{11.44}$$

or

$$\frac{d^2\Delta\delta}{dt^2} + \frac{\pi f_0}{H}D\frac{d\Delta\delta}{dt} + \frac{\pi f_0}{H}P_s\,\Delta\delta = 0 \tag{11.45}$$

or in terms of the standard second-order differential equation, we have

$$\frac{d^2\Delta\delta}{dt^2} + 2\zeta\omega_n \frac{d\Delta\delta}{dt} + \omega_n^2\,\Delta\delta = 0 \tag{11.46}$$

where ω_n, the natural frequency of oscillation is given by (11.42), and ζ is defined as the dimensionless damping ratio, given by

$$\zeta = \frac{D}{2}\sqrt{\frac{\pi f_0}{HP_s}} \tag{11.47}$$

The characteristic equation is

$$s^2 + 2\zeta\omega_n s + \omega_n^2 = 0 \tag{11.48}$$

For normal operating conditions, $\zeta = D/2\sqrt{\frac{\pi f_0}{HP_s}} < 1$, and roots of the characteristic equation are complex

$$\begin{aligned} s_1, s_2 &= -\zeta\omega_n \pm j\omega_n\sqrt{1-\zeta^2} \\ &= -\zeta\omega_n + j\omega_d \end{aligned} \tag{11.49}$$

where ω_d is the damped frequency of oscillation given by

$$\omega_d = \omega_n\sqrt{1-\zeta^2} \tag{11.50}$$

It is clear that for positive damping, roots of the characteristic equation have negative real part if synchronizing power coefficient P_s is positive. The response is bounded and the system is stable.

We now write (11.46) in state variable form. This makes it possible to extend the analysis to multimachine systems. Let

$$\begin{aligned} &x_1 = \Delta\delta \quad \text{and} \quad x_2 = \Delta\omega = \dot{\Delta\delta} \quad \text{then} \\ &\dot{x}_1 = x_2 \quad \text{and} \quad \dot{x}_2 = -\omega_n^2 x_1 - 2\zeta\omega_n x_2 \end{aligned}$$

Writing the above equations in matrix, we have

$$\begin{bmatrix} \dot{x}_1 \\ \dot{x}_2 \end{bmatrix} = \begin{bmatrix} 0 & 1 \\ -\omega_n^2 & -2\zeta\omega_n \end{bmatrix} \begin{bmatrix} x_1 \\ x_2 \end{bmatrix} \tag{11.51}$$

or

$$\dot{\mathbf{x}}(t) = \mathbf{A}\mathbf{x}(t) \tag{11.52}$$

where

$$\mathbf{A} = \begin{bmatrix} 0 & 1 \\ -\omega_n^2 & -2\zeta\omega_n \end{bmatrix} \tag{11.53}$$

This is the unforced state variable equation or the *homogeneous* state equation. If the state variables x_1 and x_2 are the desired response, we define the output vector $\mathbf{y}(t)$ as

$$\mathbf{y}(t) = \begin{bmatrix} 1 & 0 \\ 0 & 1 \end{bmatrix} \begin{bmatrix} x_1 \\ x_2 \end{bmatrix} \tag{11.54}$$

or

$$\mathbf{y}(t) = \mathbf{C}\mathbf{x}(t) \tag{11.55}$$

Taking the Laplace transform, we have

$$s\mathbf{X}(s) - \mathbf{x}(0) = \mathbf{A}\mathbf{X}(s)$$

or

$$\mathbf{X}(s) = (s\mathbf{I} - \mathbf{A})^{-1}\mathbf{x}(0) \tag{11.56}$$

where

$$(s\mathbf{I} - \mathbf{A}) = \begin{bmatrix} s & -1 \\ \omega_n^2 & s + 2\zeta\omega_n \end{bmatrix} \tag{11.57}$$

Substituting for $(s\mathbf{I} - \mathbf{A})^{-1}$, we have

$$\mathbf{X}(s) = \frac{\begin{bmatrix} s + 2\zeta\omega_n & 1 \\ -\omega_n^2 & s \end{bmatrix} \mathbf{x}(0)}{s^2 + 2\zeta\omega_n s + \omega_n^2}$$

When the rotor is suddenly perturbed by a small angle $\Delta\delta_0$, $x_1(0) = \Delta\delta_0$ and $x_2(0) = \Delta\omega_0 = 0$, and we obtain

$$\Delta\delta(s) = \frac{(s + 2\zeta\omega_n)\Delta\delta_0}{s^2 + 2\zeta\omega_n s + \omega_n^2}$$

and

$$\Delta\omega(s) = -\frac{\omega_n^2 \Delta\delta_0}{s^2 + 2\zeta\omega_n s + \omega_n^2}$$

Taking inverse Laplace transforms results in the *zero-input* response

$$\Delta\delta = \frac{\Delta\delta_0}{\sqrt{1-\zeta^2}} e^{-\zeta\omega_n t}\sin(\omega_d t + \theta) \tag{11.58}$$

and

$$\Delta\omega = -\frac{\omega_n\,\Delta\delta_0}{\sqrt{1-\zeta^2}} e^{-\zeta\omega_n t}\sin\omega_d t \tag{11.59}$$

where ω_d is the damped frequency of oscillation, and θ is given by

$$\theta = \cos^{-1}\zeta \tag{11.60}$$

The motion of rotor relative to the synchronously revolving field is

$$\delta = \delta_0 + \frac{\Delta\delta_0}{\sqrt{1-\zeta^2}} e^{-\zeta\omega_n t}\sin(\omega_d t + \theta) \tag{11.61}$$

and the rotor angular frequency is

$$\omega = \omega_0 - \frac{\omega_n\,\Delta\delta_0}{\sqrt{1-\zeta^2}} e^{-\zeta\omega_n t}\sin\omega_d t \tag{11.62}$$

The response time constant is

$$\tau = \frac{1}{\zeta\omega_n} = \frac{2H}{\pi f_0 D} \tag{11.63}$$

and the response settles in approximately four time constants, and the settling time is

$$t_s \cong 4\tau \tag{11.64}$$

From (11.42) and (11.47), we note that as inertia constant H increases, the natural frequency and the damping ratio decreases, resulting in a longer settling time. An increase in the synchronizing power coefficient P_s results in an increase in the natural frequency and a decrease in the damping ratio.

Example 11.2

A 60-Hz synchronous generator having inertia constant H = 9.94 MJ/MVA and a transient reactance $X'_d = 0.3$ per unit is connected to an infinite bus through a purely reactive circuit as shown in Figure 11.7. Reactances are marked on the diagram on a common system base. The generator is delivering real power of 0.6 per unit, 0.8 power factor lagging to the infinite bus at a voltage of $V = 1$ per unit.

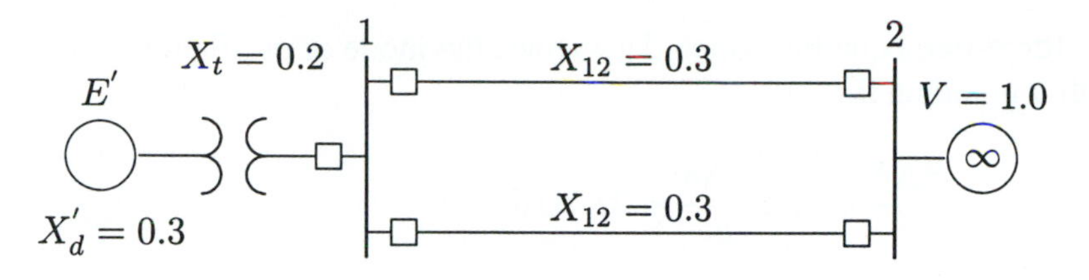

FIGURE 11.7
One-line diagram for Example 11.2.

Assume the per unit damping power coefficient is $D = 0.138$. Consider a small disturbance of $\Delta\delta = 10° = 0.1745$ radian. For example, the breakers open and then quickly close. Obtain equations describing the motion of the rotor angle and the generator frequency.

The transfer reactance between the generated voltage and the infinite bus is

$$X = 0.3 + 0.2 + \frac{0.3}{2} = 0.65$$

The per unit apparent power is

$$S = \frac{0.6}{0.8}\angle\cos^{-1} 0.8 = 0.75\angle 36.87°$$

The current is

$$I = \frac{S^*}{V^*} = \frac{0.75\angle -36.87°}{1.0\angle 0°} = 0.75\angle -36.87°$$

The excitation voltage is

$$E' = V + jXI = 1.0\angle 0° + (j0.65)(0.75\angle -36.87°) = 1.35\angle 16.79°$$

Thus, the initial operating power angle is $16.79° = 0.2931$ radian. The synchronizing power coefficient given by (11.39) is

$$P_s = P_{max}\cos\delta_0 \quad = \quad \frac{(1.35)(1)}{0.65}\cos 16.79° = 1.9884$$

The undamped angular frequency of oscillation and damping ratio are

$$\omega_n = \sqrt{\frac{\pi f_0}{H} P_s} = \sqrt{\frac{(\pi)(60)}{9.94} 1.9884} = 6.1405 \ \text{ rad/sec}$$

$$\zeta = \frac{D}{2}\sqrt{\frac{\pi f_0}{H P_s}} = \frac{0.138}{2}\sqrt{\frac{(\pi)(60)}{(9.94)(1.9884)}} = 0.2131$$

The linearized force-free equation which determines the mode of oscillation given by (11.46) with δ in radian is

$$\frac{d^2\Delta\delta}{dt^2} + 2.62\frac{d\Delta\delta}{dt} + 37.7\Delta\delta = 0$$

From (11.50), the damped angular frequency of oscillation is

$$\omega_d = \omega_n\sqrt{1-\zeta^2} = 6.1405\sqrt{1-(0.2131)^2} = 6.0 \text{ rad/sec}$$

corresponding to a damped oscillation frequency of

$$f_d = \frac{6.0}{2\pi} = 0.9549 \text{ Hz}$$

From (11.61) and (11.62), the motion of rotor relative to the synchronously revolving field in electrical degrees and the frequency excursion in Hz are given by equations

$$\delta = 16.79^\circ + 10.234e^{-1.3t}\sin(6.0t + 77.6966^\circ)$$
$$f = 60 - 0.1746e^{-1.3t}\sin 6.0t$$

The above equations are written in *MATLAB* commands as follows

```
E=1.35; V=1.0; H=9.94; X=0.65; Pm=0.6; D= 0.138; f0 = 60;
Pmax = E*V/X, d0 = asin(Pm/Pmax)                  % Max. power
Ps = Pmax*cos(d0)         % Synchronizing power coefficient
wn = sqrt(pi*60/H*Ps)% Undamped frequency of  oscillation
z = D/2*sqrt(pi*60/(H*Ps))                 % Damping ratio
wd=wn*sqrt(1-z^2), fd=wd/(2*pi) %Damped frequency oscill.
tau = 1/(z*wn)                             % Time constant
th = acos(z)                           % Phase angle theta
Dd0 = 10*pi/180;                   % Initial angle in radian
t = 0:.01:3;
Dd = Dd0/sqrt(1-z^2)*exp(-z*wn*t).*sin(wd*t + th);
d = (d0+Dd)*180/pi;                  % Power angle in degree
Dw = -wn*Dd0/sqrt(1-z^2)*exp(-z*wn*t).*sin(wd*t);
f = f0 + Dw/(2*pi);                        % Frequency in Hz
subplot(2,1,1), plot(t, d), grid
xlabel('t  sec'), ylabel('Delta    degree')
subplot(2,1,2), plot(t,f), grid
xlabel('t  sec'), ylabel('Frequency    Hz')
subplot(111)
```

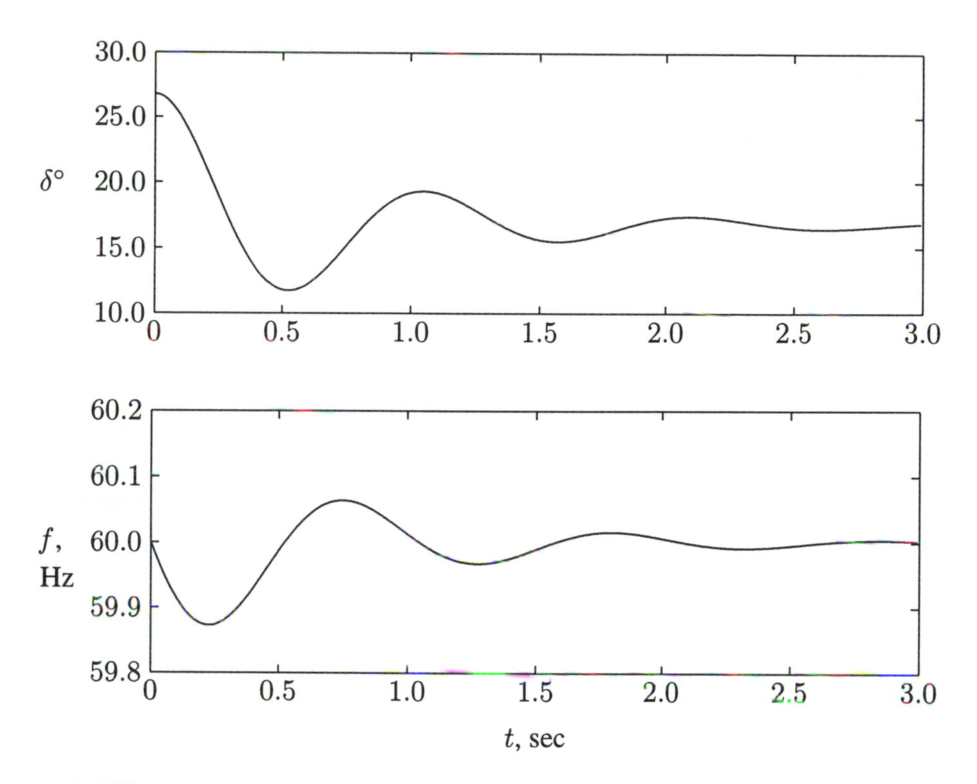

FIGURE 11.8
Natural responses of the rotor angle and frequency for machine of Example 11.2.

The result is shown in Figure 11.8.

The response shows that a small disturbance will be followed by a relatively slowly damped oscillation, or swing, of the rotor, before steady state operation at synchronous speed is resumed. In the case of a steam turbine generator, oscillations subside in a matter of two to three seconds. In the above example, the response settles in about $t_s \simeq 4\tau = 4(1/1.3) \simeq 3.1$ seconds. We also observe that the oscillations are fairly low in frequency, in the order of 0.955 Hz.

The formulation of the one-machine system with all control devices inactive resulted in a second-order differential equation or a two-dimensional state equation. Later on, when the analysis is extended to a multimachine system, an n-dimensional state variables equation is obtained. *MATLAB* Control Toolbox provides a function named **initial** for simulating continuous-time linear systems due to an initial condition on the states. Given the system

$$\begin{aligned} \dot{\mathbf{x}}(t) &= \mathbf{AX}(t) + \mathbf{Bu}(t) \\ \mathbf{y} &= \mathbf{Cx}(t) + \mathbf{Du}(t) \end{aligned} \tag{11.65}$$

$[\mathbf{y}, \mathbf{x}] = \mathbf{initial}(\mathbf{A}, \mathbf{B}, \mathbf{C}, \mathbf{D}, \mathbf{x_0}, \mathbf{t})$ returns the output and state responses of the system to the initial condition $\mathbf{x_0}$. The matrices $\mathbf{y}$ and $\mathbf{x}$ contain the output and state response of the system at the regularly spaced time vector t.

From (11.52)–(11.54), the zero-input state equation for Example 11.2 is

$$\begin{bmatrix} \dot{x}_1 \\ \dot{x}_2 \end{bmatrix} = \begin{bmatrix} 0 & 1 \\ -37.705 & -2.617 \end{bmatrix} \begin{bmatrix} x_1 \\ x_2 \end{bmatrix}$$

and

$$y = \begin{bmatrix} 1 & 0 \\ 0 & 1 \end{bmatrix} \begin{bmatrix} x_1 \\ x_2 \end{bmatrix}$$

The initial state variables are $\Delta\delta_0 = 10° = 0.1745$ radian, and $\Delta\dot{\delta}_0 = 0$. The following *MATLAB* commands are used to obtain the zero-input response for Example 11.2.

```
A = [0  1; -37.705  -2.617];
B = [0; 0];                            % Column B zero-input
C=[1 0; 0 1];%Unity matrix defining output y as x1 and x2
D = [0; 0];
Dx0 = [0.1745; 0];                     % Initial conditions
[y, x] = initial(A, B, C, D, Dx0, t);
Dd = x(:, 1); Dw = x(:, 2);   % State variables x1 and x2
d = (d0 + Dd)*180/pi;                 % Power angle in degree
f = f0 + Dw/(2*pi);                      % Frequency in Hz
subplot(2,1,1), plot(t, d), grid
xlabel('t  sec'), ylabel('Delta    Degree')
subplot(2,1,2), plot(t, f), grid
xlabel('t  sec'), ylabel('Frequency    Hz'),subplot(111)
```

The simulation results are exactly the same as the graphs shown in Figure 11.8.

Although it is convenient to assume that the disturbances causing the changes disappear, we will now investigate the system response to small power impacts. Assume the power input is increased by a small amount ΔP. Then the linearized swing equation becomes

$$\frac{H}{\pi f_0}\frac{d^2\Delta\delta}{dt^2} + D\frac{d\Delta\delta}{dt} + P_s\,\Delta\delta = \Delta P \tag{11.66}$$

or

$$\frac{d^2\Delta\delta}{dt^2} + \frac{\pi f_0}{H}D\frac{d\Delta\delta}{dt} + \frac{\pi f_0}{H}P_s\,\Delta\delta = \frac{\pi f_0}{H}\Delta P \tag{11.67}$$

or in terms of the standard second-order differential equation, we have

$$\frac{d^2\Delta\delta}{dt^2} + 2\zeta\omega_n\frac{d\Delta\delta}{dt} + \omega_n^2\Delta\delta = \Delta u \tag{11.68}$$

where

$$\Delta u = \frac{\pi f_0}{H}\Delta P \tag{11.69}$$

and ω_n and ζ are given by (11.42) and (11.47), respectively. Transforming to the state variable form, we have

$$\begin{aligned} x_1 &= \Delta\delta \quad \text{and} \quad x_2 = \Delta\omega = \dot{\Delta\delta} \quad \text{then} \\ \dot{x}_1 &= x_2 \quad \text{and} \quad \dot{x}_2 = -\omega_n^2 x_1 - 2\zeta\omega_n x_2 \end{aligned}$$

Writing the above equations in matrix, we have

$$\begin{bmatrix} \dot{x}_1 \\ \dot{x}_2 \end{bmatrix} = \begin{bmatrix} 0 & 1 \\ -\omega_n^2 & -2\zeta\omega_n \end{bmatrix}\begin{bmatrix} x_1 \\ x_2 \end{bmatrix} + \begin{bmatrix} 0 \\ 1 \end{bmatrix}\Delta u \tag{11.70}$$

or

$$\dot{\mathbf{x}}(\mathbf{t}) = \mathbf{A}\mathbf{x}(t) + \mathbf{B}\Delta u(t) \tag{11.71}$$

This is the forced state variable equation or the *zero-state* equation, and with x_1 and x_2 the desired response, the output vector $\mathbf{y}(t)$ is given by (11.55). Taking the Laplace transform of the state equation (11.71) with zero initial states results in

$$s\mathbf{X}(s) = \mathbf{A}\mathbf{X}(s) + \mathbf{B}\,\Delta U(s)$$

or

$$\mathbf{X}(s) = (s\mathbf{I} - \mathbf{A})^{-1}\mathbf{B}\,\Delta U(s) \tag{11.72}$$

where

$$\Delta U(s) = \frac{\Delta u}{s}$$

Substituting for $(s\mathbf{I} - \mathbf{A})^{-1}$, we have

$$\mathbf{X}(s) = \frac{\begin{bmatrix} s + 2\zeta\omega_n & 1 \\ -\omega_n^2 & s \end{bmatrix}\begin{bmatrix} 0 \\ 1 \end{bmatrix}\frac{\Delta u}{s}}{s^2 + 2\zeta\omega_n s + \omega_n^2}$$

or

$$\Delta\delta(s) = \frac{\Delta u}{s(s^2 + 2\zeta\omega_n s + \omega_n^2)}$$

and

$$\Delta\omega(s) = \frac{\Delta u}{s^2 + 2\zeta\omega_n + \omega_n^2}$$

Taking inverse Laplace transforms results in the step response

$$\Delta\delta = \frac{\Delta u}{\omega_n^2}[1 - \frac{1}{\sqrt{1-\zeta^2}} e^{-\zeta\omega_n t} \sin(\omega_d t + \theta)] \tag{11.73}$$

where $\theta = \cos^{-1}\zeta$ and

$$\Delta\omega = \frac{\Delta u}{\omega_n\sqrt{1-\zeta^2}} e^{-\zeta\omega_n t} \sin\omega_d t \tag{11.74}$$

Substituting for Δu from (11.69), the motion of rotor relative to the synchronously revolving field in electrical radian becomes

$$\delta = \delta_0 + \frac{\pi f_0 \Delta P}{H\omega_n^2}[1 - \frac{1}{\sqrt{1-\zeta^2}} e^{-\zeta\omega_n t} \sin(\omega_d t + \theta)] \tag{11.75}$$

and the rotor angular frequency in radian per second is

$$\omega = \omega_0 + \frac{\pi f_0 \Delta P}{H\omega_n\sqrt{1-\zeta^2}} e^{-\zeta\omega_n t} \sin\omega_d t \tag{11.76}$$

Example 11.3

The generator of Example 11.2 is operating in the steady state at $\delta_0 = 16.79°$ when the input power is increased by a small amount $\Delta P = 0.2$ per unit. The generator excitation and the infinite bus bar voltage are the same as before, i.e., $E' = 1.35$ per unit and $V = 1.0$ per unit.
(a) Using (11.75) and (11.76), obtain the step response for the rotor angle and the generator frequency.
(b) Obtain the response using the *MATLAB* **step** function.
(c) Obtain a *SIMULINK* block diagram representation of the state-space model and simulate to obtain the response.

(a) Substituting for H, δ_0, ζ, and ω_n evaluated in Example 11.2 in (11.75), and expressing the power angle in degree, we get

$$\delta = 16.79° + \frac{(180)(60)(0.2)}{(9.94)(6.1405)^2}[1 - \frac{1}{\sqrt{1-(0.2131)^2}} e^{-1.3t} \sin(6t + 77.6966°)]$$

or

$$\delta = 16.79° + 5.7631[1 - 1.0235e^{-1.3t}\sin(6t + 77.6966°)]$$

Also, substituting the values in (11.76) and expressing the frequency in Hz, we get

$$f = 60 + \frac{(60)(0.2)}{2(9.94)(6.1405)\sqrt{1-(0.2131)^2}}e^{-1.3t}\sin 6t$$

or

$$f = 60 + 0.10e^{-1.3t}\sin 6t$$

The above functions are plotted over a range of 0 to 3 seconds and the result is shown in Figure 11.9.

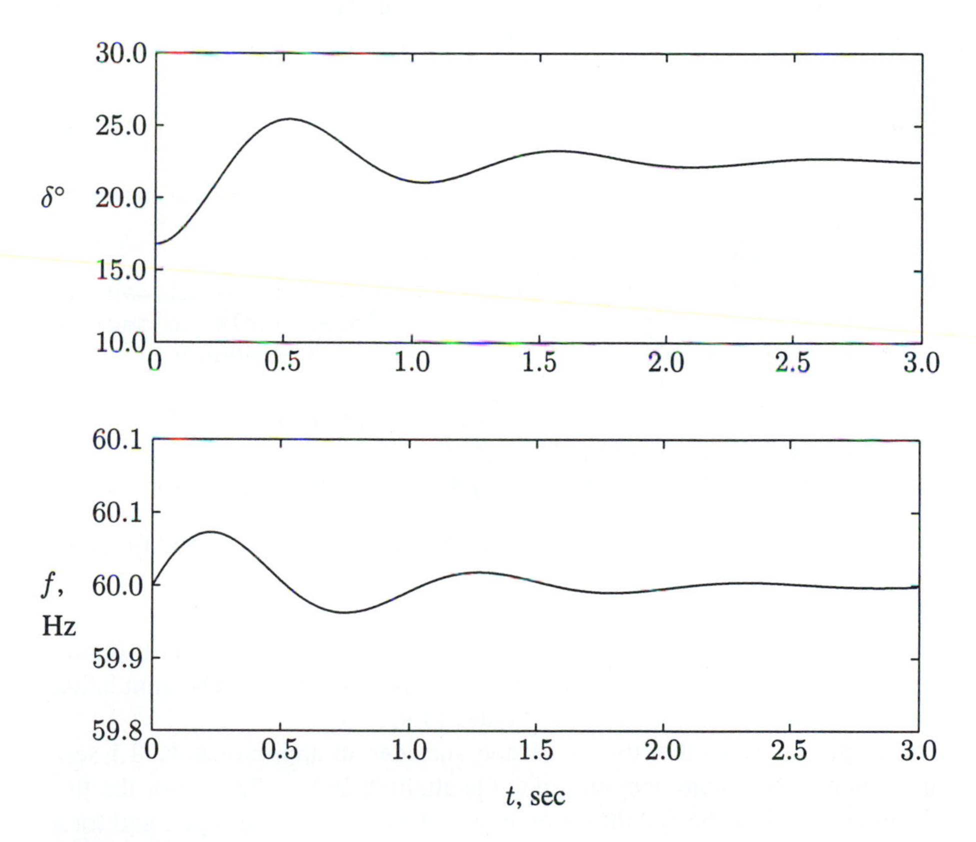

FIGURE 11.9
Step responses of the rotor angle and frequency for machine of Example 11.3.

(b) The step response of the state equation can be obtained conveniently, using the *MATLAB* Control Toolbox **[y, x] = lsim(A, B, C, D, u, t)** function or **[y,x] = step(A, B, C, D, iu, t)** function. These functions are particularly useful when dealing with multimachine systems. $[\mathbf{y}, \mathbf{x}] = \mathbf{step}(\mathbf{A}, \mathbf{B}, \mathbf{C}, \mathbf{D}, \mathbf{iu}, \mathbf{t})$ returns the output and step responses of the system. The index **iu** specifies which input to be used for the step response. With only one input **iu** = 1. The matrices **y** and **x** contain the output and state response of the system at the regularly spaced time vector t.

From (11.70), the state equation for Example 11.3 is

$$\begin{bmatrix} \dot{x_1} \\ \dot{x_2} \end{bmatrix} = \begin{bmatrix} 0 & 1 \\ -37.705 & -2.617 \end{bmatrix} \begin{bmatrix} x_1 \\ x_2 \end{bmatrix} + \begin{bmatrix} 0 \\ 1 \end{bmatrix} \Delta u$$

and

$$y = \begin{bmatrix} 1 & 0 \\ 0 & 1 \end{bmatrix} \begin{bmatrix} x_1 \\ x_2 \end{bmatrix}$$

From (11.69), $\Delta u = (60\pi/9.94)(0.2) = 3.79$. The following *MATLAB* commands are used to obtain the step response for Example 11.3.

```
A = [0  1; -37.705  -2.617];
Dp = 0.2; Du = 3.79;    % Small step change in power input
B = [0; 1]*Du;
C=[1 0; 0 1];%Unity matrix defining output y as x1 and x2
D = [0; 0];
[y, x] = step(A, B, C, D, 1, t);
Dd = x(:, 1); Dw = x(:, 2);    % State variables x1 and x2
d = (d0 + Dd)*180/pi;                % Power angle in degree
f = f0 + Dw/(2*pi);                        % Frequency in Hz
subplot(2,1,1), plot(t, d), grid
xlabel('t  sec'), ylabel('Delta    degree')
subplot(2,1,2), plot(t, f), grid
xlabel('t  sec'), ylabel('Frequency    Hz'),subplot(111)
```

The simulation results are exactly the same as the analytical solution and the plots are shown in Figure 11.9.

(c) A *SIMULINK* model named **sim11ex3.mdl** is constructed as shown in Figure 11.10. The file is opened and is run in the *SIMULINK WINDOW*. The simulation results in the same response as shown in Figure 11.9.

The response shows that the oscillation subsides in approximately 3.1 seconds and a new steady state operating point is attained at $\delta = 22.5°$. For the linearized swing equation, the stability is entirely independent of the input, and for a positive damping coefficient the system is always stable as long as the synchronizing power coefficient is positive. Theoretically, power can be increased gradually

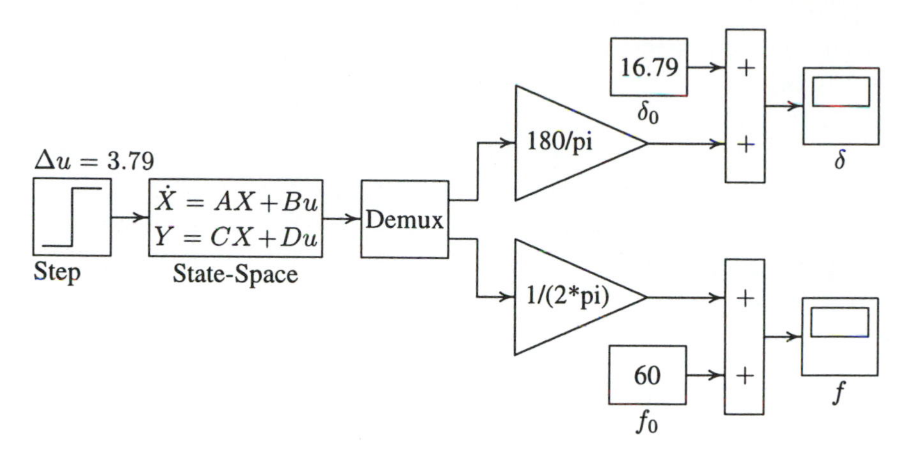

FIGURE 11.10
Simulation block diagram for Example 11.3.

up to the steady-state limit. It is important to note that the linearized equation is only valid for very small power impact and deviation from the operating state. Indeed, for a large sudden impact the nonlinear equation may result in unstable solution and stability is lost even if the impact is less than the steady-state power limit.

An important characteristic of the linear system is that the response is asymptotically stable if all roots of the characteristic equation have negative real part. The polynomial characteristic equation is obtained from the determinant of $(s\mathbf{I} - \mathbf{A})$ or eigenvalues of A. The eigenvalues provide very important results regarding the nature of response. The reciprocal of the real component of the eigenvalues gives the time constants, and the imaginary component gives the damped frequency of oscillations. Thus, the linear system expressed in the state variable form is asymptotically stable if and only if all of the eigenvalues of $\mathbf{A}$ lie in the left half of the complex plane. Therefore, to investigate the system stability of a multimachine system when subjected to small disturbances, all we need to do is to examine the eigenvalues of the $\mathbf{A}$ matrix. If the homogeneous state equation is written as

$$\dot{\mathbf{x}} = \mathbf{f}(\mathbf{x}) \tag{11.77}$$

we note that matrix $\mathbf{A}$ is the *Jacobian matrix* whose elements are partial derivatives of rows of $\mathbf{f}(\mathbf{x})$ with respect to state variables $x_1, x_2, \cdots, x_n$, evaluated at the equilibrium point. In *MATLAB* we can use the function $\mathbf{r} = \text{eig}(\mathbf{A})$, which returns the eigenvalues of the $\mathbf{A}$ matrix. In Example 11.2, the $\mathbf{A}$ matrix was found to be

$$\mathbf{A} = \begin{bmatrix} 0 & 1 \\ -\omega_n^2 & -2\zeta\omega_n \end{bmatrix} = \begin{bmatrix} 0 & 1 \\ -37.705 & -2.617 \end{bmatrix}$$

and the commands

```
A = [0  1; -37.705  -2.617];
r = eig(A)
```

result in

```
r =
     -1.3  +  6.00i
     -1.3  +  6.00i
```

The linearized model for small disturbances is very useful when the system is extended to include the governor action and the effect of automatic voltage regulators in a multimachine system. The linearized model allows the application of the linear control system analysis and compensation, which will be dealt with in Chapter 12.

11.5 TRANSIENT STABILITY — EQUAL-AREA CRITERION

The transient stability studies involve the determination of whether or not synchronism is maintained after the machine has been subjected to severe disturbance. This may be sudden application of load, loss of generation, loss of large load, or a fault on the system. In most disturbances, oscillations are of such magnitude that linearization is not permissible and the nonlinear swing equation must be solved.

A method known as the *equal-area criterion* can be used for a quick prediction of stability. This method is based on the graphical interpretation of the energy stored in the rotating mass as an aid to determine if the machine maintains its stability after a disturbance. The method is only applicable to a one-machine system connected to an infinite bus or a two-machine system. Because it provides physical insight to the dynamic behavior of the machine, application of the method to analysis of a single machine connected to a large system is considered here.

Consider a synchronous machine connected to an infinite bus. The swing equation with damping neglected as given by (11.21) is

$$\frac{H}{\pi f_0}\frac{d^2\delta}{dt^2} = P_m - P_e = P_a$$

where P_a is the accelerating power. From the above equation, we have

$$\frac{d^2\delta}{dt^2} = \frac{\pi f_0}{H}(P_m - P_e)$$

Multiplying both sides of the above equation by $2d\delta/dt$, we get

$$2\frac{d\delta}{dt}\frac{d^2\delta}{dt^2} = \frac{2\pi f_0}{H}(P_m - P_e)\frac{d\delta}{dt}$$

This may be written as

$$\frac{d}{dt}\left[\left(\frac{d\delta}{dt}\right)^2\right] = \frac{2\pi f_0}{H}(P_m - P_e)\frac{d\delta}{dt}$$

or

$$d\left[\left(\frac{d\delta}{dt}\right)^2\right] = \frac{2\pi f_0}{H}(P_m - P_e)d\delta$$

Integrating both sides,

$$\left(\frac{d\delta}{dt}\right)^2 = \frac{2\pi f_0}{H}\int_{\delta_0}^{\delta}(P_m - P_e)d\delta$$

or

$$\frac{d\delta}{dt} = \sqrt{\frac{2\pi f_0}{H}\int_{\delta_0}^{\delta}(P_m - P_e)d\delta} \tag{11.78}$$

Equation (11.78) gives the relative speed of the machine with respect to the synchronously revolving reference frame. For stability, this speed must become zero at some time after the disturbance. Therefore, from (11.78), we have for the stability criterion,

$$\int_{\delta_0}^{\delta}(P_m - P_e)d\delta = 0 \tag{11.79}$$

Consider the machine operating at the equilibrium point δ_0, corresponding to the mechanical power input $P_{m0} = P_{e0}$ as shown in Figure 11.11. Consider a sudden step increase in input power represented by the horizontal line P_{m1}. Since $P_{m1} > P_{e0}$, the accelerating power on the rotor is positive and the power angle δ increases. The excess energy stored in the rotor during the initial acceleration is

$$\int_{\delta_0}^{\delta_1}(P_{m1} - P_e)d\delta = \text{area } abc = \text{area } A_1 \tag{11.80}$$

With increase in δ, the electrical power increases, and when $\delta = \delta_1$, the electrical power matches the new input power P_{m1}. Even though the accelerating power is zero at this point, the rotor is running above synchronous speed; hence, δ and

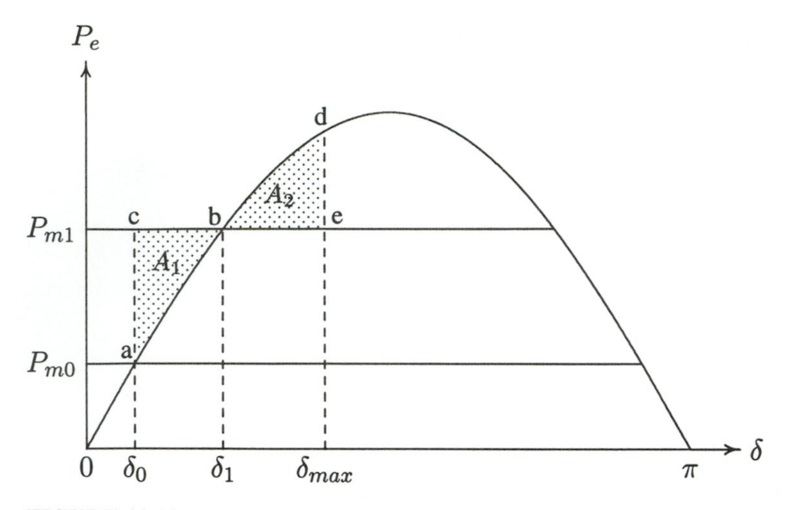

FIGURE 11.11
Equal-area criterion—sudden change of load.

electrical power P_e will continue to increase. Now $P_m < P_e$, causing the rotor to decelerate toward synchronous speed until $\delta = \delta_{max}$. According to (11.79), the rotor must swing past point b until an equal amount of energy is given up by the rotating masses. The energy given up by the rotor as it decelerates back to synchronous speed is

$$\int_{\delta_1}^{\delta_{max}} (P_{m1} - P_e)d\delta = \text{area } bde = \text{area } A_2 \tag{11.81}$$

The result is that the rotor swings to point b and the angle δ_{max}, at which point

$$|\text{area } A_1| = |\text{area } A_2| \tag{11.82}$$

This is known as the *equal-area criterion*. The rotor angle would then oscillate back and forth between δ_0 and δ_{max} at its natural frequency. The damping present in the machine will cause these oscillations to subside and the new steady state operation would be established at point b.

11.5.1 APPLICATION TO SUDDEN INCREASE IN POWER INPUT

The equal-area criterion is used to determine the maximum additional power P_m which can be applied for stability to be maintained. With a sudden change in the power input, the stability is maintained only if area A_2 at least equal to A_1 can be located above P_m. If area A_2 is less than area A_1, the accelerating momentum can never be overcome. The limit of stability occurs when δ_{max} is at the intersection of line P_m and the power-angle curve for $90° < \delta < 180°$, as shown in Figure 11.12.

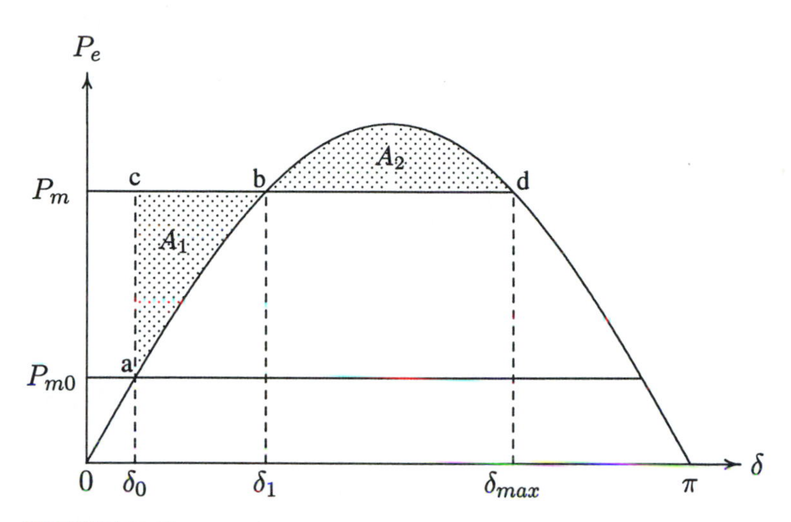

FIGURE 11.12
Equal-area criterion—maximum power limit.

Applying the equal-area criterion to Figure 11.12, we have

$$P_m(\delta_1 - \delta_0) - \int_{\delta_0}^{\delta_1} P_{max} \sin \delta \, d\delta = \int_{\delta_1}^{\delta_{max}} P_{max} \sin \delta \, d\delta \; - \; P_m(\delta_{max} - \delta_1)$$

Integrating the above expression yields

$$(\delta_{max} - \delta_0)P_m = P_{max}(\cos \delta_0 - \cos \delta_{max})$$

Substituting for P_m, from

$$P_m = P_{max} \sin \delta_{max}$$

into the above equation results in

$$(\delta_{max} - \delta_0) \sin \delta_{max} + \cos \delta_{max} = \cos \delta_0 \tag{11.83}$$

The above nonlinear algebraic equation can be solved by an iterative technique for δ_{max}. Once δ_{max} is obtained, the maximum permissible power or the transient stability limit is found from

$$P_m = P_{max} \sin \delta_1 \tag{11.84}$$

where

$$\delta_1 = \pi - \delta_{max} \tag{11.85}$$

Equation (11.83) is a nonlinear function of angle δ_{max}, written as

$$f(\delta_{max}) = c \tag{11.86}$$

An iterative solution is obtained, using the Newton-Raphson method, described in Section 6.3. Starting with an initial estimate of $\pi/2 < \delta_{max}^{(k)} < \pi$, the Newton-Raphson algorithm gives

$$\Delta\delta_{max}^{(k)} = \frac{c - f(\delta_{max}^{(k)})}{\left.\frac{df}{d\delta_{max}}\right|_{\delta_{max}^{(k)}}} \tag{11.87}$$

where $df/d\delta_{max}$ is the derivative of (11.83) and is given by

$$\left.\frac{df}{d\delta_{max}}\right|_{\delta_{max}^{(k)}} = (\delta_{max}^{(k)} - \delta_0)\cos\delta_{max}^{(k)} \tag{11.88}$$

and

$$\delta_{max}^{(k+1)} = \delta_{max}^{(k)} + \Delta\delta_{max}^{(k)} \tag{11.89}$$

A solution is obtained when the difference between the absolute value of the successive iteration is less than a specified accuracy, i.e.,

$$|\delta_{max}^{(k+1)} - \delta_{max}^{(k)}| \leq \epsilon \tag{11.90}$$

A function named **eacpower**(P_0, E, V, X) is developed for a one-machine system connected to an infinite bus. The function uses the above algorithm to find the sudden maximum permissible power that can be applied for critical stability. The function plots the power-angle curve and displays the shaded equal-areas. P_0, E, V, and X are the initial power, the transient internal voltage, the infinite bus bar voltage, and the transfer reactance, respectively, all in per unit. If **eacpower** is used without arguments, the user is prompted to enter the above quantities.

Example 11.4

The machine of Example 11.2 is delivering a real power of 0.6 per unit, at 0.8 power factor lagging to the infinite bus bar. The infinite bus bar voltage is 1.0 per unit. Determine

(a) The maximum power input that can be applied without loss of synchronism.

(b) Repeat (a) with zero initial power input. Assume the generator internal voltage remains constant at the value computed in (a).

In Example 11.2. the transfer reactance and the generator internal voltage were found to be $X = 0.65$ pu, and $E' = 1.35$ pu

(a) We use the following command:

```
P0 = 0.6; E = 1.35; V = 1.0; X = 0.65;
eacpower(P0, E, V, X)
```

which displays the graph shown in Figure 11.13 and results in

```
Initial power                        =   0.600 pu
Initial power angle                  =  16.791 degree
Sudden initial power                 =   1.084 pu
Total power for critical stability   =   1.684 pu
Maximum angle swing                  = 125.840 pu
New operating angle                  =  54.160 degree
```

(b) The initial power input is set to zero, i.e., $P_o = 0$, and using **eacpower(Po, E, V, X)** displays the graph shown in Figure 11.14 with the following results:

```
Initial power                        =   0.00 pu
Initial power angle                  =   0.00 degree
Sudden initial power                 =   1.505 pu
Total power for critical stability   =   1.505 pu
Maximum angle swing                  = 133.563 pu
New operating angle                  =  46.437 degree
```

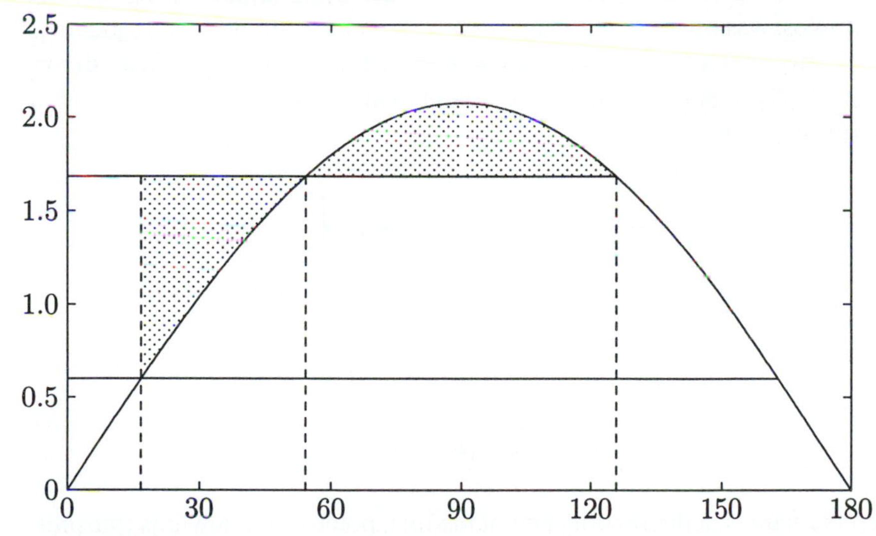

FIGURE 11.13
Maximum power limit by equal-area criterion for Example 11.4 (a).

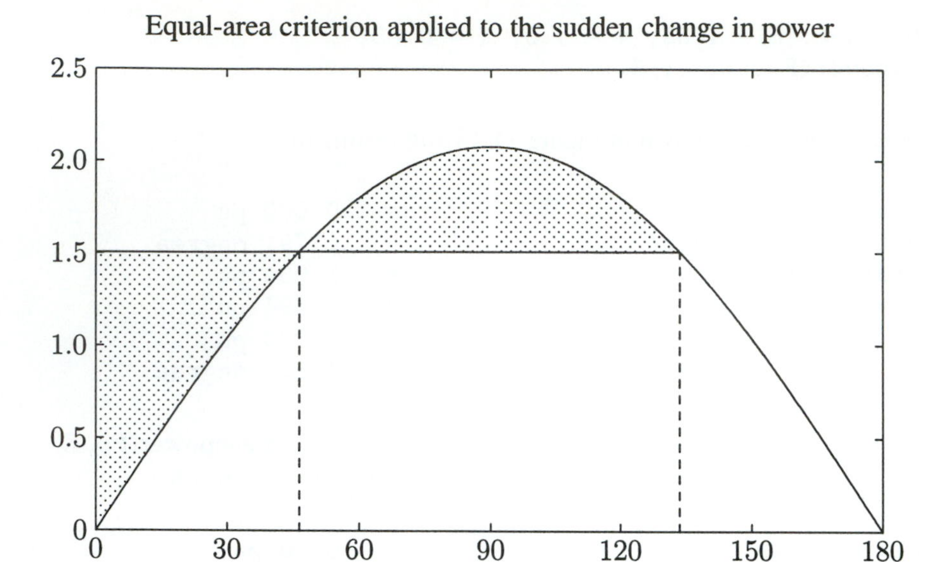

FIGURE 11.14
Maximum power limit by equal-area criterion for Example 11.4 (b).

11.6 APPLICATION TO THREE-PHASE FAULT

Consider Figure 11.15 where a generator is connected to an infinite bus bar through two parallel lines. Assume that the input power P_m is constant and the machine is operating steadily, delivering power to the system with a power angle δ_0 as shown in Figure 11.16. A temporary three-phase bolted fault occurs at the sending end of one of the line at bus 1.

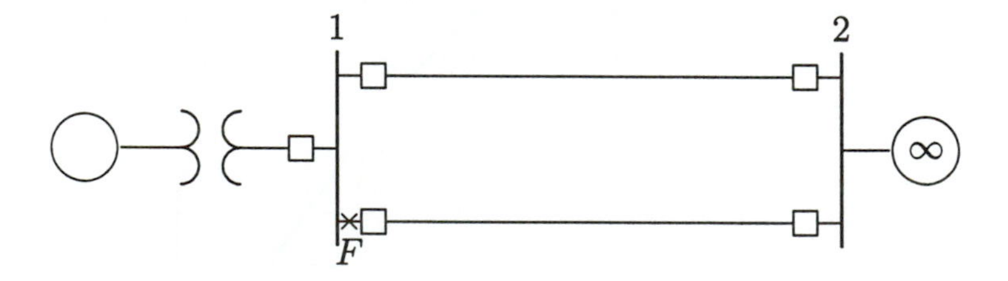

FIGURE 11.15
One-machine system connected to infinite bus, three-phase fault at F.

When the fault is at the sending end of the line, point F, no power is transmitted to the infinite bus. Since the resistances are neglected, the electrical power P_e is zero, and the power-angle curve corresponds to the horizontal axis. The machine accelerates with the total input power as the accelerating power, thereby increasing

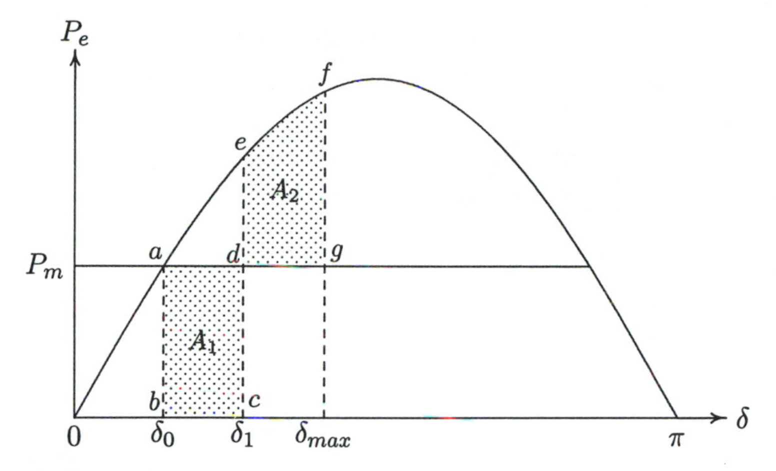

FIGURE 11.16
Equal-area criterion for a three-phase fault at the sending end.

its speed, storing added kinetic energy, and increasing the angle δ. When the fault is cleared, both lines are assumed to be intact. The fault is cleared at δ_1, which shifts the operation to the original power-angle curve at point e. The net power is now decelerating, and the previously stored kinetic energy will be reduced to zero at point f when the shaded area ($defg$), shown by A_2, equals the shaded area ($abcd$), shown by A_1. Since P_e is still greater than P_m, the rotor continues to decelerate and the path is retraced along the power-angle curve passing through points e and a. The rotor angle would then oscillate back and forth around δ_0 at its natural frequency. Because of the inherent damping, oscillation subsides and the operating point returns to the original power angle δ_0.

The critical clearing angle is reached when any further increase in δ_1 causes the area A_2, representing decelerating energy to become less than the area representing the accelerating energy. This occurs when δ_{max}, or point f, is at the intersection of line P_m and curve P_e, as shown in Figure 11.17. Applying equal-area criterion to Figure 11.17, we have

$$\int_{\delta_0}^{\delta_c} P_m \, d\delta = \int_{\delta_c}^{\delta_{max}} (P_{max} \sin\delta - P_m) d\delta$$

Integrating both sides, we have

$$P_m(\delta_c - \delta_0) = P_{max}(\cos\delta_c - \cos\delta_{max}) - P_m(\delta_{max} - \delta_c)$$

Solving for δ_c, we get

$$\cos\delta_c = \frac{P_m}{P_{max}}(\delta_{max} - \delta_0) + \cos\delta_{max} \tag{11.91}$$

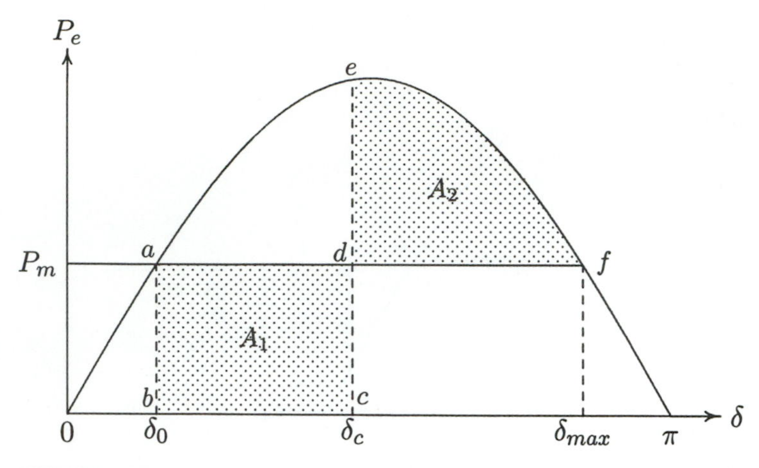

FIGURE 11.17
Equal-area criterion for critical clearing angle.

The application of equal-area criterion made it possible to find the critical clearing angle for the machine to remain stable. To find the critical clearing time, we still need to solve the nonlinear swing equation. For this particular case where the electrical power P_e during fault is zero, an analytical solution for critical clearing time can be obtained. The swing equation as given by (11.21), during fault with $P_e = 0$ becomes

$$\frac{H}{\pi f_0}\frac{d^2\delta}{dt^2} = P_m$$

or

$$\frac{d^2\delta}{dt^2} = \frac{\pi f_0}{H}P_m$$

Integrating both sides

$$\frac{d\delta}{dt} = \frac{\pi f_0}{H}P_m\int_0^t dt = \frac{\pi f_0}{H}P_m t$$

Integrating again, we get

$$\delta = \frac{\pi f_0}{2H}P_m t^2 + \delta_0$$

Thus, if δ_c is the critical clearing angle, the corresponding critical clearing time is

$$t_c = \sqrt{\frac{2H(\delta_c - \delta_0)}{\pi f_0 P_m}} \tag{11.92}$$

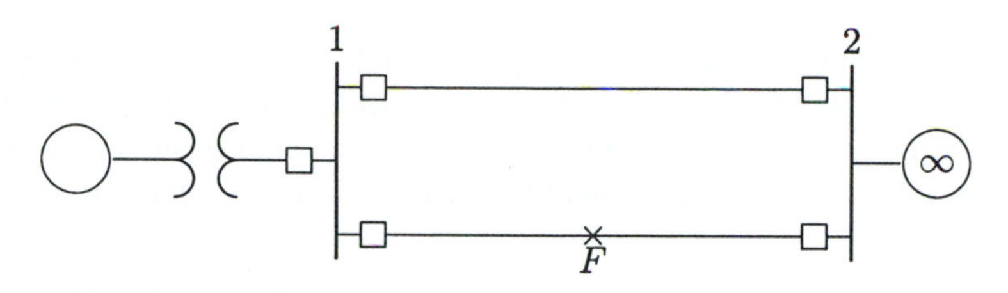

FIGURE 11.18
One-machine system connected to infinite bus, three-phase fault at F.

Now consider the fault location F at some distance away from the sending end as shown in Figure 11.18. Assume that the input power P_m is constant and the machine is operating steadily, delivering power to the system with a power angle δ_0 as shown in Figure 11.19. The power-angle curve corresponding to the prefault condition is given by curve A.

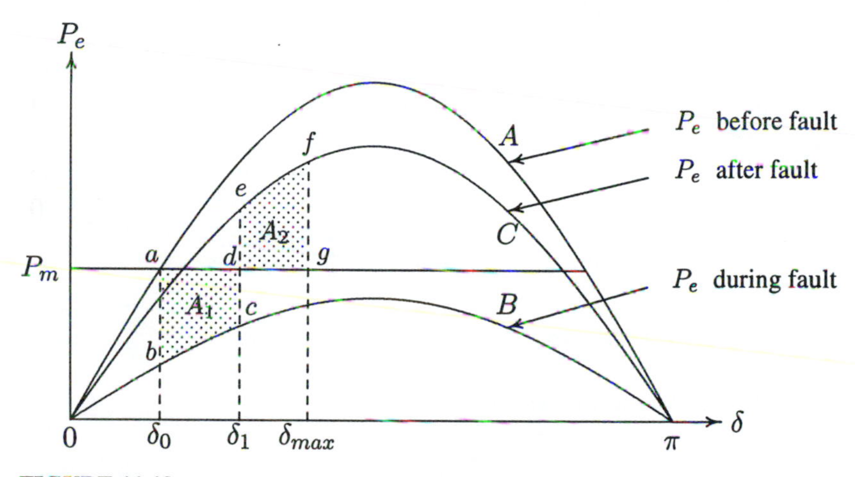

FIGURE 11.19
Equal-area criterion for a three-phase fault at the away from the sending end.

With fault location at F, away from the sending end, the equivalent transfer reactance between bus bars is increased, lowering the power transfer capability and the power-angle curve is represented by curve B. Finally, curve C represents the postfault power-angle curve, assuming the faulted line is removed. When the three-phase fault occurs, the operating point shifts immediately to point b on curve B. An excess of the mechanical input over its electrical output accelerates the rotor, thereby storing excess kinetic energy, and the angle δ increases. Assume the fault is cleared at δ_1 by isolating the faulted line. This suddenly shifts the operating point to e on curve C. The net power is now decelerating, and the previously stored kinetic energy will be reduced to zero at point f when the shaded area ($defg$) equals the shaded area ($abcd$). Since P_e is still greater than P_m, the rotor continues to decelerate, and the path is retraced along the power-angle curve passing through point e. The rotor angle will then oscillate back and forth around e at its natural frequency. The damping present in the machine will cause these oscillations to subside and a new steady state operation will be established at the intersection of P_m and curve C.

The critical clearing angle is reached when any further increase in δ_1 causes the area A_2, representing decelerating energy, to become less than the area representing the accelerating energy. This occurs when δ_{max}, or point f, is at the intersection of line P_m and curve C as shown in Figure 11.20.

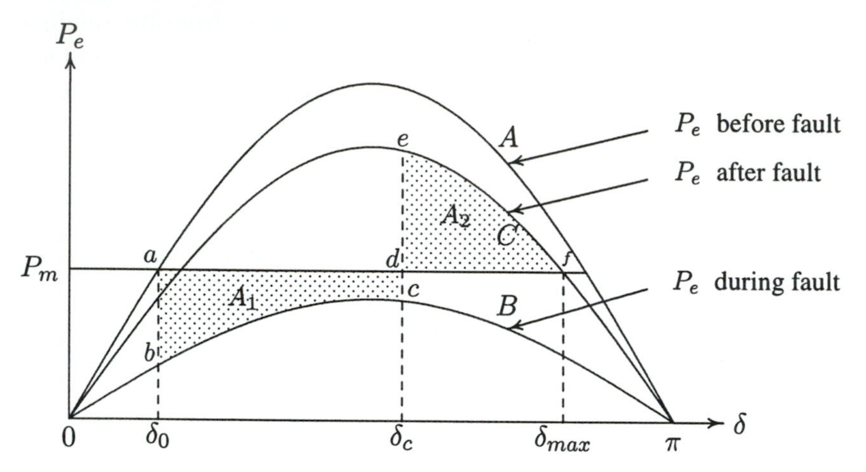

FIGURE 11.20
Equal-area criterion for critical clearing angle.

Applying equal-area criterion to Figure 11.20, we get

$$P_m(\delta_c - \delta_0) - \int_{\delta_0}^{\delta_c} P_{2\,max} \sin\delta \, d\delta = \int_{\delta_c}^{\delta_{max}} P_{3\,max} \sin\delta \, d\delta - P_m(\delta_{max} - \delta_c)$$

Integrating both sides, and solving for δ_c, we obtain

$$\cos\delta_c = \frac{P_m(\delta_{max} - \delta_0) + P_{3\,max}\cos\delta_{max} - P_{2\,max}\cos\delta_0}{P_{3\,max} - P_{2\,max}} \tag{11.93}$$

The application of equal-area criterion gives the critical clearing angle to maintain stability. However, because of the nonlinearity of the swing equation, an analytical solution for critical clearing time is not possible. In the next section we will discuss the numerical solution, which can readily be extended to large systems.

A function named **eacfault**(P_0, E, V, X_1, X_2, X_3) is developed for a one-machine system connected to an infinite bus. This function obtains the power-angle curve before fault, during fault, and after the fault clearance. The function uses equal-area criterion to find the critical clearing angle. For the case when power transfer during fault is zero, (11.92) is used to find the critical clearing time. The function plots the power-angle curve and displays the shaded equal-areas. P_0, E, and V are the initial power, the generator transient internal voltage, and the infinite bus bar voltage, all in per unit. X_1 is the transfer reactance before fault. X_2 is the transfer reactance during fault. If power transfer during fault is zero, **inf** must be used for X_2. Finally, X_3 is the postfault transfer reactance. If **eacfault** is used without arguments, the user is prompted to enter the above quantities.

Example 11.5

A 60-Hz synchronous generator having inertia constant $H = 5$ MJ/MVA and a direct axis transient reactance $X'_d = 0.3$ per unit is connected to an infinite bus through a purely reactive circuit as shown in Figure 11.21. Reactances are marked on the diagram on a common system base. The generator is delivering real power $P_e = 0.8$ per unit and $Q = 0.074$ per unit to the infinite bus at a voltage of $V = 1$ per unit.

(a) A temporary three-phase fault occurs at the sending end of the line at point F. When the fault is cleared, both lines are intact. Determine the critical clearing angle and the critical fault clearing time.

(b) A three-phase fault occurs at the middle of one of the lines, the fault is cleared, and the faulted line is isolated. Determine the critical clearing angle.

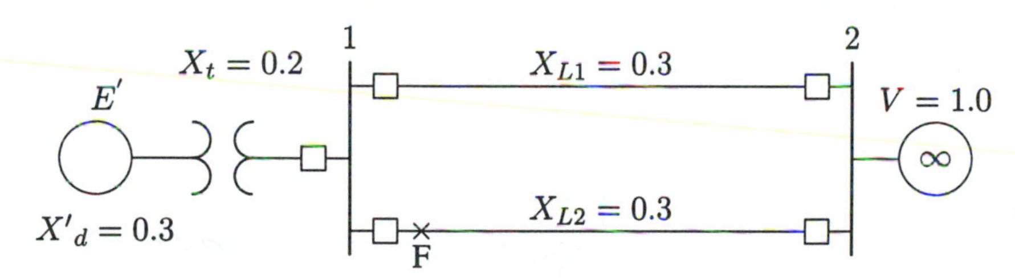

FIGURE 11.21
One-line diagram for Example 11.5.

The current flowing into the infinite bus is

$$I = \frac{S^*}{V^*} = \frac{0.8 - j0.074}{1.0\angle 0^\circ} = 0.8 - j0.074 \text{ pu}$$

The transfer reactance between internal voltage and the infinite bus before fault is

$$X_1 = 0.3 + 0.2 + \frac{0.3}{2} = 0.65$$

The transient internal voltage is

$$E' = V + jX_1 I = 1.0 + (j0.65)(0.8 - j0.074) = 1.17\angle 26.387^\circ \text{ pu}$$

(a) Since both lines are intact when the fault is cleared, the power-angle equation before and after the fault is

$$P_{max}\sin\delta = \frac{(1.17)(1.0)}{0.65}\sin\delta = 1.8\sin\delta$$

The initial operating angle is given by

$$1.8\sin\delta_0 = 0.8$$

or

$$\delta_0 = 26.388^\circ = 0.46055 \text{ rad}$$

and referring to Figure 11.17

$$\delta_{max} = 180^\circ - \delta_0 = 153.612^\circ = 2.681 \text{ rad}$$

Since the fault is at the beginning of the transmission line, the power transfer during fault is zero, and the critical clearing angle as given by (11.91) is

$$\cos\delta_c = \frac{0.8}{1.8}(2.681 - 0.46055) + \cos 153.61^\circ = 0.09106$$

Thus, the critical clearing angle is

$$\delta_c = \cos^{-1}(0.09106) = 84.775^\circ = 1.48 \text{ rad}$$

From (11.92), the critical clearing time is

$$t_c = \sqrt{\frac{2H(\delta_c - \delta_0)}{\pi f_0 P_m}} = \sqrt{\frac{(2)(5)(1.48 - 0.46055)}{(\pi)(60)(.8)}} = 0.26 \text{ second}$$

The use of function **eacfault**(P_m, E, V, X_1, X_2, X_3) to solve the above problem and to display power-angle plot with the shaded equal-areas is demonstrated below. We use the following commands

```
Pm = 0.8; E = 1.17; V = 1.0;
X1 = 0.65; X2 = inf; X3 = 0.65;
eacfault(Pm, E, V, X1, X2, X3)
```

The graph is displayed as shown in Figure 11.22 and the result is

```
Initial power angle      =  26.388
Maximum angle swing      = 153.612
Critical clearing angle  =  84.775
Critical clearing time   =   0.260 sec
```

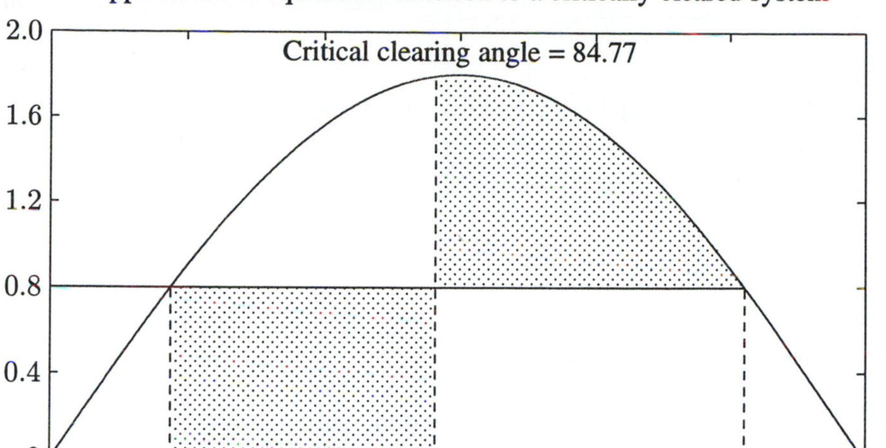

FIGURE 11.22
Equal-area criterion for Example 11.5 (a).

(b) The power-angle curve before the occurrence of the fault is the same as before, given by

$$P_{1\,max} = 1.8 \sin \delta$$

and the generator is operating at the initial power angle $\delta_0 = 26.4° = 0.4605$ rad. The fault occurs at point F at the middle of one line, resulting in the circuit shown in Figure 11.23. The transfer reactance during fault may be found most readily by converting the Y-circuit ABF to an equivalent delta, eliminating junction C. The resulting circuit is shown in Figure 11.24.

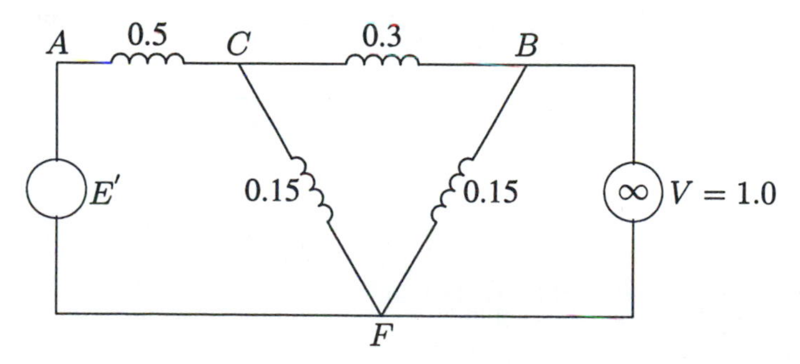

FIGURE 11.23
Equivalent circuit with three-phase fault at the middle of one line.

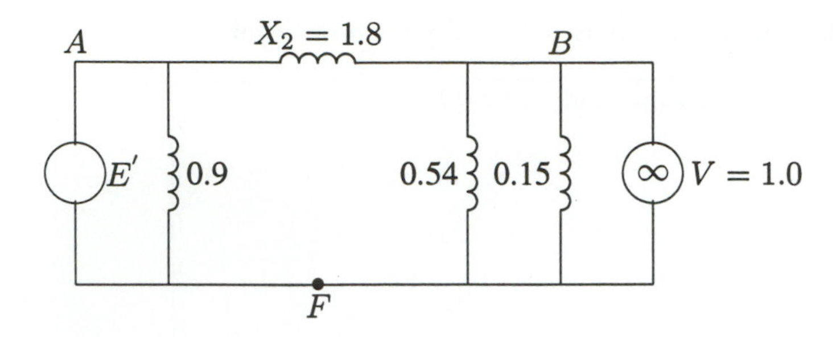

FIGURE 11.24
Equivalent circuit after Y-Δ transformation.

The equivalent reactance between generator and the infinite bus is

$$X_2 = \frac{(0.5)(0.3) + (0.5)(0.15) + (0.3)(0.15)}{0.15} = 1.8 \text{ pu}$$

Thus, the power-angle curve during fault is

$$P_{2\,max} \sin\delta = \frac{(1.17)(1.0)}{1.8} \sin\delta = 0.65 \sin\delta$$

When fault is cleared the faulted line is isolated. Therefore, the postfault transfer reactance is

$$X_3 = 0.3 + 0.2 + 0.3 = 0.8 \text{ pu}$$

and the power-angle curve is

$$P_{3\,max} \sin\delta = \frac{(1.17)(1.0)}{0.8} \sin\delta = 1.4625 \sin\delta$$

Referring to Figure 11.20

$$\delta_{max} = 180° - \sin^{-1}\left(\frac{0.8}{1.4625}\right) = 146.838° = 2.5628 \text{ rad}$$

Applying (11.93), the critical clearing angle is given by

$$\begin{aligned}\cos\delta_c &= \frac{0.8(2.5628 - 0.46055) + 1.4625 \cos 146.838° - 0.65 \cos 26.388°}{1.4625 - 0.65} \\ &= -0.15356\end{aligned}$$

Thus, the critical clearing angle is

$$\delta_c = \cos^{-1}(-0.15356) = 98.834°$$

Function **eacfault**(P_m, E, V, X_1, X_2, X_3) is used to solve part (b) and to display power-angle plot. We use the following commands

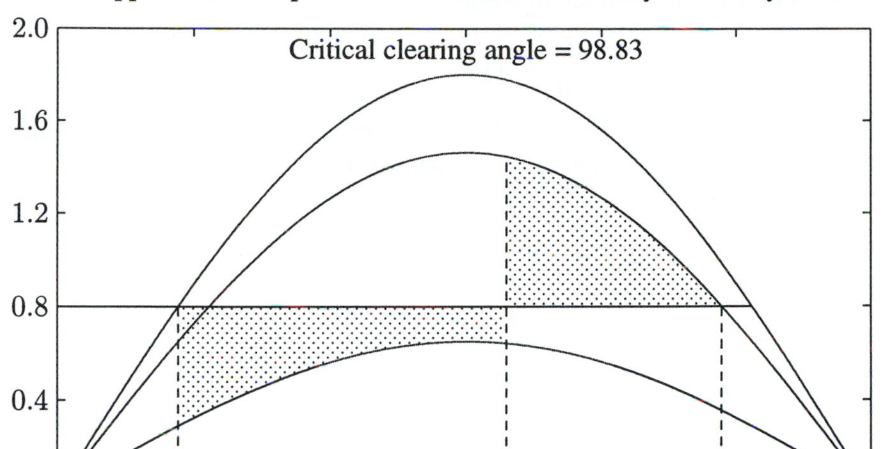

FIGURE 11.25
Equal-area criterion for Example 11.5 (b).

```
Pm = 0.8; E = 1.17; V = 1.0;
X1 = 0.65; X2 = 1.8; X3 = 0.8;
eacfault(Pm, E, V, X1, X2, X3)
```

The graph is displayed as shown in Figure 11.25 and the result is

```
Initial power angle      =  26.388
Maximum angle swing      = 146.838
Critical clearing angle  =  98.834
```

11.7 NUMERICAL SOLUTION OF NONLINEAR EQUATION

Numerical integration techniques can be applied to obtain approximate solutions of nonlinear differential equations. Many algorithms are available for numerical integration. Euler's method is the simplest and the least accurate of all numerical methods. It is presented here because of its simplicity. By studying this method, we will be able to grasp the basic ideas involved in numerical solutions of ODE and can more easily understand the more powerful methods such as the Runge-Kutta procedure.

Consider the first-order differential equation

$$\frac{dx}{dt} = f(x) \tag{11.94}$$

Euler's method is illustrated in Figure 11.26, where the curve shown represents the solution $x(t)$. If at t_0 the value of $x(t_0)$ denoted by x_0 is given, the curve can

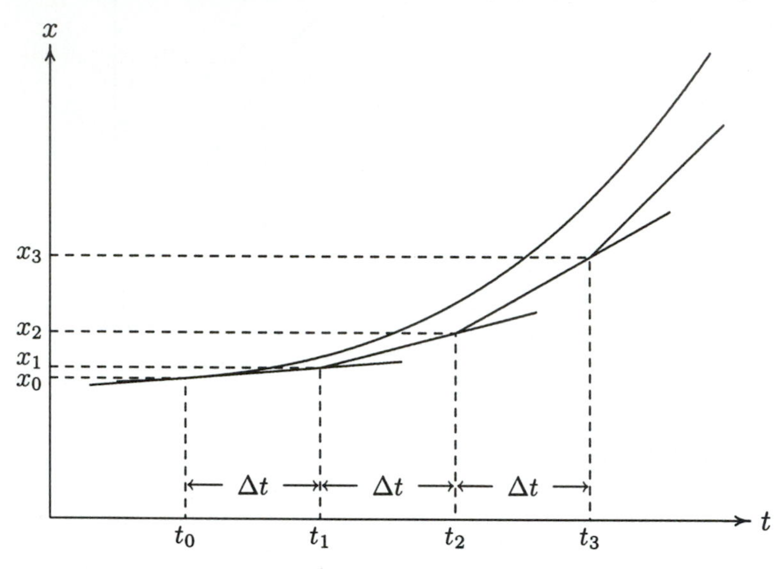

FIGURE 11.26
Graphical interpretation of Euler's method.

be approximated by the tangent evaluated at this point. For a small increment in t denoted by Δt, the increment in x is given by

$$\Delta x \approx \left.\frac{dx}{dt}\right|_{x_0} \Delta t$$

where $\left.\frac{dx}{dt}\right|_{x_0}$ is the slope of the curve at (t_0, x_0), which can be determined from (11.94). Thus, the value of x at $t_0 + \Delta t$ is

$$x_1 = x_0 + \Delta x = x_0 + \left.\frac{dx}{dt}\right|_{x_0} \Delta t$$

The above approximation is the Taylor series expansion of x around point (t_0, x_0), where higher-order terms have been discarded.

The subsequent values of x can be similarly determined. Hence, the computational algorithm is

$$x_{i+1} = x_i + \left.\frac{dx}{dt}\right|_{x_i} \Delta t \tag{11.95}$$

By applying the above algorithm successively, we can find approximate values of $x(t)$ at enough points from an initial state (t_0, x_0) to a final state (t_f, x_f). A graphical illustration is shown in Figure 11.26. Euler's method assumes that the slope is constant over the entire interval Δt causing the points to fall below the curve. An improvement can be obtained by calculating the slope at both the beginning and end of interval, and then averaging these slopes. This procedure is known as the modified Euler's method and is described as follows.

By using the derivative at the beginning of the step, the value at the end of the step ($t_1 = t_0 + \Delta t$) is predicted from

$$x_1^p = x_0 + \left.\frac{dx}{dt}\right|_{x_0} \Delta t$$

Using the predicted value of x_1^p, the derivative at the end of interval is determined by

$$\left.\frac{dx}{dt}\right|_{x_1^p} = f(t_1, x_1^p)$$

Then, the average value of the two derivatives is used to find the corrected value

$$x_1^c = x_0 + \left(\frac{\left.\frac{dx}{dt}\right|_{x_0} + \left.\frac{dx}{dt}\right|_{x_1^p}}{2} \right) \Delta t$$

Hence, the computational algorithm for the successive values is

$$x_{i+1}^c = x_i + \left(\frac{\left.\frac{dx}{dt}\right|_{x_i} + \left.\frac{dx}{dt}\right|_{x_{i+1}^p}}{2} \right) \Delta t \tag{11.96}$$

The problem with Euler's method is that there is numerical error introduced when discarding the higher-order terms in the series Taylor expansion. But by using a reasonably small value of Δt, we can decrease the error between successive points. If the step size is decreased too much, the number of steps increases and the computer round-off error increases directly with the number of operations. Thus, the step size must be selected small enough to obtain a reasonably accurate solution, but at the same time, large enough to avoid the numerical limitations of the computer.

The above technique can be applied to the solution of higher-order differential equations. An nth order differential equation can be expressed in terms of n first-order differential equations by the introduction of auxiliary variables. These variables are referred to as *state variables*, which may be physical quantities in a system. For example, given the second-order differential equation

$$a_2 \frac{d^2x}{dt^2} + a_1 \frac{dx}{dt} + a_0 x = c$$

and the initial conditions x_0 and $\left.\frac{dx}{dt}\right|_{x_0}$ at t_0, we introduce the following state variables.

$$x_1 = x$$
$$x_2 = \frac{dx}{dt}$$

Thus, the above second-order differential equation can be written as the two following simultaneous first-order differential equations.

$$\dot{x}_1 = x_2$$
$$\dot{x}_2 = \frac{c}{a_2} - \frac{a_0}{a_2} x_1 - \frac{a_1}{a_2} x_2$$

There are many other more powerful techniques for the numerical solution of nonlinear equations. A popular technique is the Runge-Kutta method, which is based on formulas derived by using an approximation to replace the truncated Taylor series expansion. The interested reader should refer to textbooks on numerical techniques. *MATLAB* provides two powerful functions for the numerical solution of differential equations employing the Runge-Kutta-Fehlberg methods. These are **ode23** and **ode45**, based on the Fehlberg second- and third-order pair of formulas for medium accuracy and forth- and fifth-order pair for higher accuracy. The nth-order differential equation must be transformed into n first order differential equations and must be placed in an M-file that returns the state derivative of the equations. The formats for these functions are

[t, x] = ode23('xprime', tspan, x0)
[t, x] = ode45('xprime', tspan, x0)

where **tspan =[t0, tfinal]** is the time interval for the integration and **x0** is a column vector of initial conditions at time **t0**. **xprime** is the state derivative of the equations, defined in a file named **xprime.m**

11.8 NUMERICAL SOLUTION OF THE SWING EQUATION

To demonstrate the solution of the swing equation, consider Figure 11.18 where a generator is connected to an infinite bus bar through two parallel lines. Assume that the input power P_m is constant. Under steady state operation $P_e = P_m$, and the initial power angle is given by

$$\delta_0 = \sin^{-1} \frac{P_m}{P_{1\,max}}$$

where

$$P_{1\,max} = \frac{|E'||V|}{X_1}$$

and X_1 is the transfer reactance before the fault. The rotor is running at synchronous speed, and the change in the angular velocity is zero, i.e.,

$$\Delta\omega_0 = 0$$

Now consider a three-phase fault at the middle of one line as shown in Figure 11.18. The equivalent transfer reactance between bus bars is increased, lowering the power transfer capability, and the amplitude of the power-angle equation becomes

$$P_{2\,max} = \frac{|E'||V|}{X_2}$$

where X_2 is the transfer reactance during fault. The swing equation given by (11.21) is

$$\frac{d^2\delta}{dt^2} = \frac{\pi f_0}{H}(P_m - P_{2\,max} \sin\delta) = \frac{\pi f_0}{H} P_a$$

The above swing equation is transformed into the state variable form as

$$\begin{aligned} \frac{d\delta}{dt} &= \Delta\omega \\ \frac{d\Delta\omega}{dt} &= \frac{\pi f_0}{H} P_a \end{aligned} \qquad (11.97)$$

We now apply the modified Euler's method to the above equations. By using the derivatives at the beginning of the step, the value at the end of the step ($t_1 = t_0 + \Delta t$) is predicted from

$$\begin{aligned} \delta_{i+1}^p &= \delta_i + \left.\frac{d\delta}{dt}\right|_{\Delta\omega_i} \Delta t \\ \Delta\omega_{i+1}^p &= \Delta\omega_i + \left.\frac{d\Delta\omega}{dt}\right|_{\delta_i} \Delta t \end{aligned}$$

Using the predicted value of δ_{i+1}^p, and $\Delta\omega_{i+1}^p$ the derivatives at the end of interval are determined by

$$\left.\frac{d\delta}{dt}\right|_{\Delta\omega_{i+1}^p} = \Delta\omega_{i+1}^p$$

$$\left.\frac{d\Delta\omega}{dt}\right|_{\delta_{i+1}^p} = \frac{\pi f_0}{H} P_a\Big|_{\delta_{i+1}^p}$$

Then, the average value of the two derivatives is used to find the corrected value

$$\delta_{i+1}^c = \delta_i + \left(\frac{\left.\frac{d\delta}{dt}\right|_{\Delta\omega_i} + \left.\frac{d\delta}{dt}\right|_{\Delta\omega_{i+1}^p}}{2} \right) \Delta t$$

$$\Delta\omega_{i+1}^c = \Delta\omega_i + \left(\frac{\left.\frac{d\Delta\omega}{dt}\right|_{\delta_i} + \left.\frac{d\Delta\omega}{dt}\right|_{\delta_{i+1}^p}}{2} \right) \Delta t \qquad (11.98)$$

Based on the above algorithm, a function named **swingmeu(**P_m, E, V, X_1, X_2, X_3, H, f, t_c, t_f, Dt**)** is written for the transient stability analysis of a one-machine system. The function arguments are

- **P_m** Per unit mechanical power input, assumed to remain constant
- **E** Constant voltage back of transient reactance in per unit
- **V** Infinite bus bar voltage in per unit
- **X_1** Per unit reactance between buses E and V before fault
- **X_2** Per unit reactance between buses E and V during fault
- **X_3** Per unit reactance between buses E and V after fault clearance
- **H** Generator inertia constant in second, (MJ/MVA)
- **f** System nominal frequency
- **t_c** Fault clearing time
- **t_f** Final time for integration
- **Dt** Integration time interval, required for modified Euler

If **swingmeu** is used without the arguments, the user is prompted to enter the required data. In addition, based on the *MATLAB* automatic step size Runge-Kutta **ode23** and **ode45** functions, two more functions are developed for the transient stability analysis of a one-machine system. These are **swingrk2(**P_m, E, V, X_1, X_2, X_3, H, f, t_c, t_f**)**, based on **ode23**, and **swingrk4(**P_m, E, V, X_1, X_2, X_3, H, f, t_c, t_f**)**, based on **ode45**. The function arguments are as defined above, except since these techniques use automatic step size, the argument **Dt** is not required. Again, if **swingrk2** and **swingrk4** are used without arguments, the user is prompted to enter the required data. All the functions above use a function named **cctime(**P_m, E, V, X_1, X_2, X_3, H, f**)**, which obtains the critical clearing time of fault for critical stability.

Example 11.6

In the system of Example 11.5 a three-phase fault at the middle of one line is cleared by isolating the faulted circuit simultaneously at both ends.

(a) The fault is cleared in 0.3 second. Obtain the numerical solution of the swing equation for 1.0 second using the modified Euler method (function **swingmeu**) with a step size of $\Delta t = 0.01$ second. From the swing curve, determine the system stability.
(b) The **swingmeu** function automatically calls upon the **cctime** function and determines the critical clearing time. Repeat the simulation and obtain the swing plots for the critical clearing time, and when fault is cleared in 0.5 second.
(c) Obtain a *SIMULINK* block diagram model for the swing equation, and simulate for a fault clearing time of 0.3 and 0.5 second. Repeat the simulation until a critical clearing time is obtained.

(a) For the purpose of understanding the procedure, the computations are performed for one step. From Example 11.5, the power-angle curve before the occurrence of the fault is given by

$$P_{1\,max} = 1.8 \sin \delta$$

and the generator is operating at the initial power angle

$$\delta_0 = 26.388^\circ = 0.46055 \text{ rad}$$
$$\Delta\omega_0 = 0$$

The fault occurs at point F at the middle of one line, resulting in the circuit shown in Figure 11.23 (page 499). From the results obtained in Example 11.5, the accelerating power equation is

$$P_a = 0.8 - 0.65 \sin \delta$$

Applying the modified Euler's method, the derivatives at the beginning of the step are

$$\left.\frac{d\delta}{dt}\right|_{\Delta\omega_0} = 0$$
$$\left.\frac{d\Delta\omega}{dt}\right|_{\delta_0} = \frac{\pi(60)}{5}(0.8 - 0.65 \sin 26.388^\circ) = 19.2684 \ rad/sec^2$$

At the end of the first step ($t_1 = 0.01$), the predicted values are

$$\delta_1^p = 0.46055 + (0)(0.01) = 0.46055 \text{ rad} = 26.388^\circ$$
$$\Delta\omega_1^p = 0 + (19.2684)(0.01) = 0.1927 \text{ rad/sec}$$

Using the predicted value of δ_1^p, and $\Delta\omega_1^p$, the derivatives at the end of interval are determined by

$$\left.\frac{d\delta}{dt}\right|_{\Delta\omega_1^p} = \Delta\omega_1^p = 0.1927 \ \text{rad/sec}$$

$$\left.\frac{d\Delta\omega}{dt}\right|_{\delta_1^p} = \frac{\pi(60)}{5}(0.8 - \sin 26.388^\circ) = 19.2684 \ \text{rad/sec}^2$$

Then, the average value of the two derivatives is used to find the corrected value

$$\delta_1^c = 0.46055 + \frac{0 + 0.1927}{2}(0.01) = 0.4615 \ \text{rad}$$

$$\Delta\omega_1^c = 0.0 + \frac{19.2684 + 19.2684}{2}(0.01) = 0.1927 \ \text{rad/sec}$$

The process is continued for the successive steps, until at $t = 0.3$ second when the fault is cleared. From Example 11.5, the postfault accelerating power equation is

$$P_a = 0.8 - 1.4625 \sin\delta$$

The process is continued with the new accelerating equation until the specified final time $t_f = 1.0$ second. The complete computations are obtained using the **swingmeu** function as follows

```
Pm = 0.80;  E = 1.17;  V = 1.0;
X1 = 0.65; X2 = 1.80; X3 = 0.8;
H = 5; f = 60; tc = 0.3; tf = 1.0; Dt = 0.01;
swingmeu(Pm, E, V, X1, X2, X3, H, f, tc, tf, Dt)
```

The time interval and the corresponding power angle δ in degrees and the speed deviation $\Delta\omega$ in rad/sec are displayed in a tabular form. The swing plot is displayed as shown in Figure 11.27.

The swing curve shows that the power angle returns after a maximum swing indicating that with inclusion of system damping, the oscillations will subside and a new operating angle is attained. Hence, the system is found to be stable for this fault clearing time. The critical clearing time is determined by the program to be

```
Critical clearing time  = 0.4 second
Critical clearing angle = 98.83 degrees
```

(b) The above program is run for a clearing time of $t_c = 0.4$ second and $t_c = 0.5$ second with the results shown in Figure 11.28. The swing curve for $t_c = 0.4$ second corresponds to the critical clearing time. The swing curve for $t_c = 0.5$

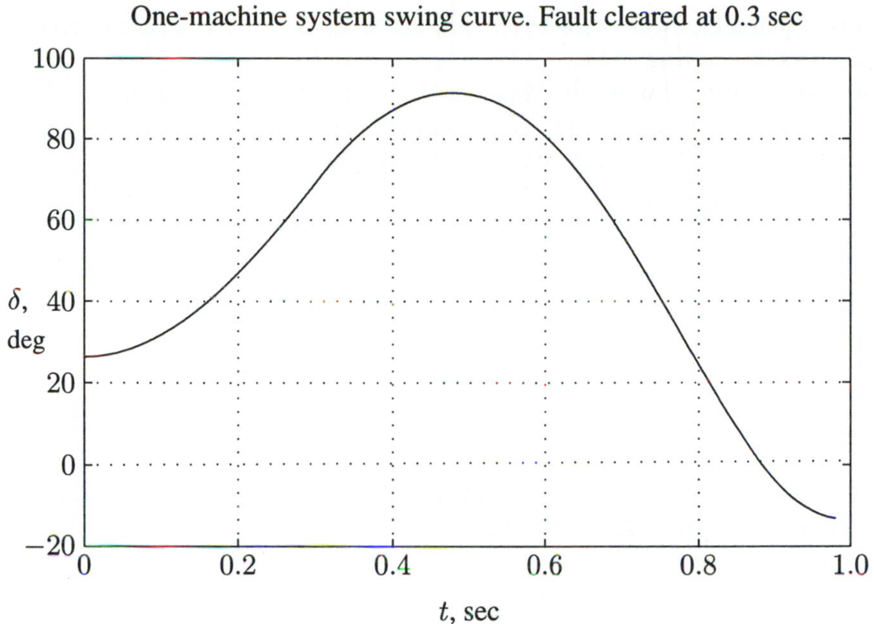

FIGURE 11.27
Swing curve for machine of Example 11.6. Fault cleared at 0.3 sec.

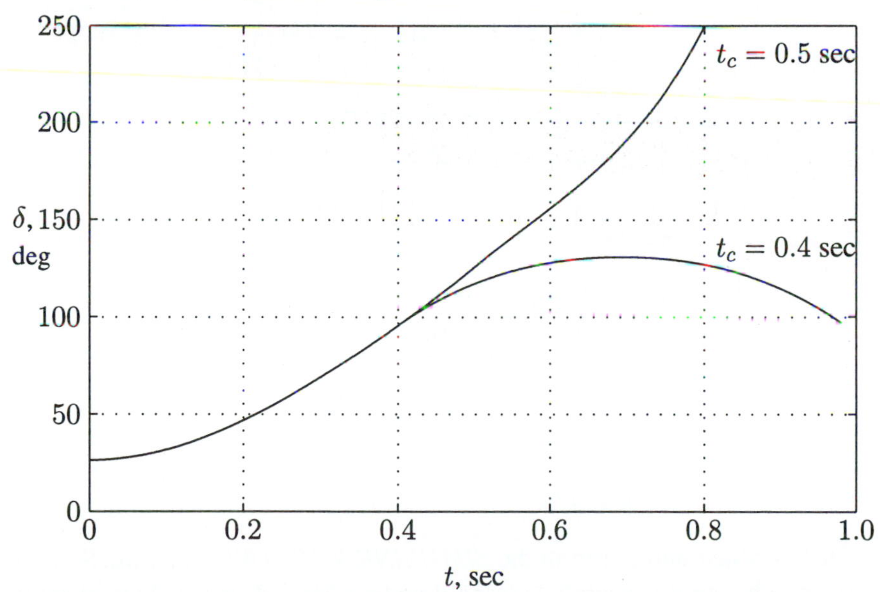

FIGURE 11.28
Swing curves for machine of Example 11.6, for fault clearance at 0.4 sec and 0.5 sec.

second shows that the power angle δ is increasing without limit. Hence, the system is unstable for this clearing time.

The swing curves for the machine in Problem 11.6 are obtained for fault clearing times of $t_c = 0.3$, $t_c = 0.4$, and $t_c = 0.5$, with **swingrk4** function, which uses the *MATLAB* **ode45** function. We use the following statements.

```
Pm = 0.80; E = 1.17; V = 1.0; H = 5.0; f = 60;
X1 = 0.65; X2 = 1.80; X3 = 0.8;
tc = 0.3; tf = 1;
swingrk4(Pm, E, V, X1, X2, X3, H, f, tc, tf)
tc = .5;
swingrk4(Pm, E, V, X1, X2, X3, H, f, tc, tf)
tc = .4;
swingrk4(Pm, E, V, X1, X2, X3, H, f, tc, tf)
```

The same numerical solutions are obtained and the swing curves are the same as the ones shown in Figures 11.27 and 11.28.

(c) Using the state-space representation of the swing equation, given in (11.97), a *SIMULINK* model named **sim11ex6.mdl** is constructed as shown in Figure 11.29.

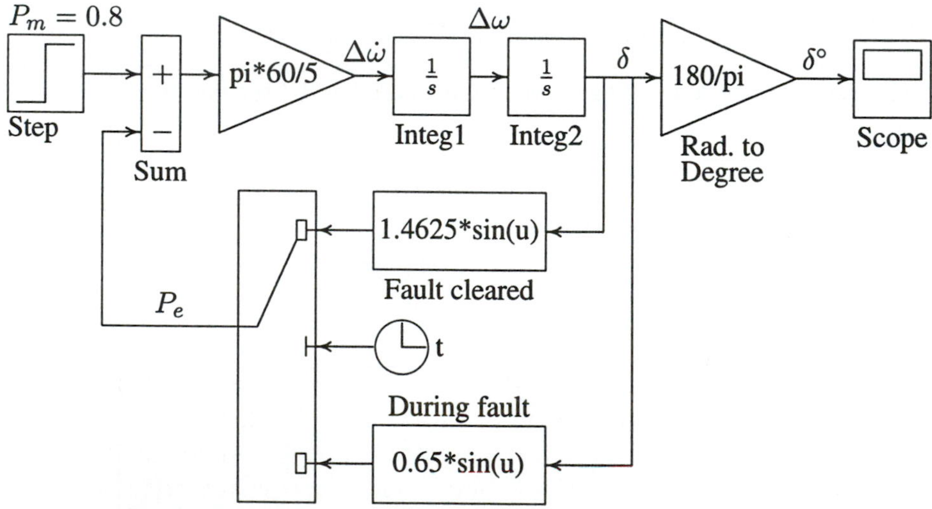

FIGURE 11.29
Simulation block diagram for Example 11.6.

The file is opened and is run in the *SIMULINK WINDOW*. Open the Switch Dialog Box and change the Switch Threshold setting for different values of fault clearing time. The simulation results in the same response as shown in Figure 11.27.

11.9 MULTIMACHINE SYSTEMS

Multimachine equations can be written similar to the one-machine system connected to the infinite bus. In order to reduce the complexity of the transient stability analysis, similar simplifying assumptions are made as follows.

1. Each synchronous machine is represented by a constant voltage source behind the direct axis transient reactance. This representation neglects the effect of saliency and assumes constant flux linkages.
2. The governor's actions are neglected and the input powers are assumed to remain constant during the entire period of simulation.
3. Using the prefault bus voltages, all loads are converted to equivalent admittances to ground and are assumed to remain constant.
4. Damping or asynchronous powers are ignored.
5. The mechanical rotor angle of each machine coincides with the angle of the voltage behind the machine reactance.
6. Machines belonging to the same station swing together and are said to be *coherent*. A group of coherent machines is represented by one equivalent machine.

The first step in the transient stability analysis is to solve the initial load flow and to determine the initial bus voltage magnitudes and phase angles. The machine currents prior to disturbance are calculated from

$$I_i = \frac{S_i^*}{V_i^*} = \frac{P_i - jQ_i}{V_i^*} \qquad i = 1, 2, \ldots, m \tag{11.99}$$

where m is the number of generators. V_i is the terminal voltage of the ith generator, P_i and Q_i are the generator real and reactive powers. All unknown values are determined from the initial power flow solution. The generator armature resistances are usually neglected and the voltages behind the transient reactances are then obtained.

$$E_i' = V_i + jX_d' I_i \tag{11.100}$$

Next, all loads are converted to equivalent admittances by using the relation

$$y_{i0} = \frac{S_i^*}{|V_i|^2} = \frac{P_i - jQ_i}{|V_i|^2} \tag{11.101}$$

To include voltages behind transient reactances, m buses are added to the n-bus power system network. The equivalent network with all loads converted to admittances is shown in Figure 11.30.

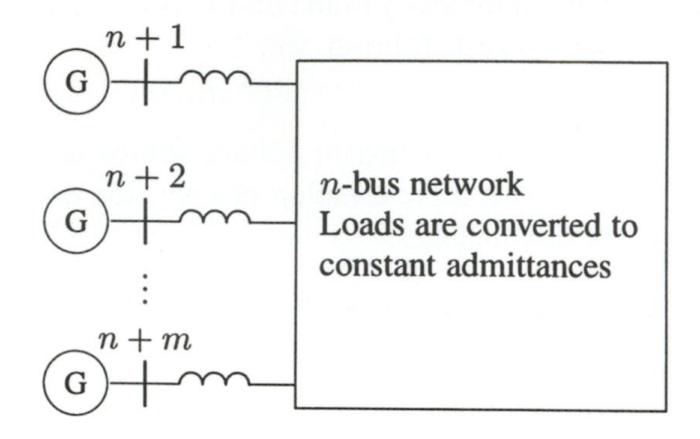

FIGURE 11.30
Power system representation for transient stability analysis.

Nodes $n+1, n+2, \ldots, n+m$ are the internal machine buses, i.e., the buses behind the transient reactances. The node voltage equation with node 0 as reference for this network, as given by (6.2), is

$$\left[\begin{array}{c} I_1 \\ I_2 \\ \vdots \\ I_n \\ \hline I_{n+1} \\ \vdots \\ I_{n+m} \end{array}\right] = \left[\begin{array}{ccc|ccc} Y_{11} & \cdots & Y_{1n} & Y_{1(n+1)} & \cdots & Y_{1(n+m)} \\ Y_{21} & \cdots & Y_{2n} & Y_{2(n+1)} & \cdots & Y_{2(n+m)} \\ \vdots & \ddots & \vdots & \vdots & \ddots & \vdots \\ Y_{n1} & \cdots & Y_{nn} & Y_{n(n+1)} & \cdots & Y_{n(n+m)} \\ \hline Y_{(n+1)1} & \cdots & Y_{(n+1)n} & Y_{(n+1)(n+1)} & \cdots & Y_{(n+1)(n+m)} \\ \vdots & \ddots & \vdots & \vdots & \ddots & \vdots \\ Y_{(n+m)1} & \cdots & Y_{(n+m)n} & Y_{(n+m)(n+1)} & \cdots & Y_{(n+m)(n+m)} \end{array}\right] \left[\begin{array}{c} V_1 \\ V_2 \\ \vdots \\ V_n \\ \hline E'_{n+1} \\ \vdots \\ E'_{n+m} \end{array}\right] \tag{11.102}$$

or

$$\mathbf{I}_{bus} = \mathbf{Y}_{bus}\,\mathbf{V}_{bus} \tag{11.103}$$

where $\mathbf{I}_{bus}$ is the vector of the injected bus currents and $\mathbf{V}_{bus}$ is the vector of bus voltages measured from the reference node. The diagonal elements of the bus admittance matrix are the sum of admittances connected to it, and the off-diagonal elements are equal to the negative of the admittance between the nodes. This is similar to the **lfybus** used in the power flow analysis. The difference is that additional nodes are added to include the machine voltages behind transient reactances. Also, diagonal elements are modified to include the load admittances.

To simplify the analysis, all nodes other than the generator internal nodes are eliminated using the Kron reduction formula. To eliminate the load buses, the bus admittance matrix in (11.102) is partitioned such that the n buses to be removed are represented in the upper n rows. Since no current enters or leaves the load buses, currents in the n rows are zero. The generator currents are denoted by the vector $\mathbf{I}_m$ and the generator and load voltages are represented by the vectors $\mathbf{E'}_m$ and $\mathbf{V}_n$, respectively. Then, Equation (11.102), in terms of submatrices, becomes

$$\begin{bmatrix} \mathbf{0} \\ \mathbf{I}_m \end{bmatrix} = \begin{bmatrix} \mathbf{Y}_{nn} & \mathbf{Y}_{nm} \\ \mathbf{Y}^t_{nm} & \mathbf{Y}_{mm} \end{bmatrix} \begin{bmatrix} \mathbf{V}_n \\ \mathbf{E'}_m \end{bmatrix} \tag{11.104}$$

The voltage vector $\mathbf{V}_n$ may be eliminated by substitution as follows.

$$\mathbf{0} = \mathbf{Y}_{nn}\mathbf{V}_n + \mathbf{Y}_{nm}\mathbf{E'}_m \tag{11.105}$$

$$\mathbf{I}_m = \mathbf{Y}^t_{nm}\mathbf{V}_n + \mathbf{Y}_{mm}\mathbf{E'}_m \tag{11.106}$$

from (11.105),

$$\mathbf{V}_n = -\mathbf{Y}_{nn}{}^{-1}\mathbf{Y}_{nm}\mathbf{E'}_m \tag{11.107}$$

Now substituting into (11.106), we have

$$\begin{aligned} \mathbf{I}_m &= [\mathbf{Y}_{mm} - \mathbf{Y}^t_{nm}\mathbf{Y}_{nn}{}^{-1}\mathbf{Y}_{nm}]\mathbf{E'}_m \\ &= \mathbf{Y}^{red}_{bus}\mathbf{E'}_m \end{aligned} \tag{11.108}$$

The reduced admittance matrix is

$$\mathbf{Y}^{red}_{bus} = \mathbf{Y}_{mm} - \mathbf{Y}^t_{nm}\mathbf{Y}_{nn}{}^{-1}\mathbf{Y}_{nm} \tag{11.109}$$

The reduced bus admittance matrix has the dimensions $(m \times m)$, where m is the number of generators.

The electrical power output of each machine can now be expressed in terms of the machine's internal voltages.

$$S^*_{ei} = E'^*_i I_i$$

or

$$P_{ei} = \Re[E'^*_i I_i] \tag{11.110}$$

where

$$I_i = \sum_{j=1}^{m} E'_j Y_{ij} \tag{11.111}$$

Expressing voltages and admittances in polar form, i.e., $E'_i = |E'_i|\angle\delta_i$ and $Y_{ij} = |Y_{ij}|\angle\theta_{ij}$, and substituting for I_i in (11.110), results in

$$P_{ei} = \sum_{j=1}^{m} |E'_i||E'_j||Y_{ij}|\cos(\theta_{ij} - \delta_i + \delta_j) \tag{11.112}$$

The above equation is the same as the power flow equation given by (6.52). Prior to disturbance, there is equilibrium between the mechanical power input and the electrical power output, and we have

$$P_{mi} = \sum_{j=1}^{m} |E'_i||E'_j||Y_{ij}|\cos(\theta_{ij} - \delta_i + \delta_j) \tag{11.113}$$

11.10 MULTIMACHINE TRANSIENT STABILITY

The classical transient stability study is based on the application of a three-phase fault. A solid three-phase fault at bus k in the network results in $V_k = 0$. This is simulated by removing the kth row and column from the prefault bus admittance matrix. The new bus admittance matrix is reduced by eliminating all nodes except the internal generator nodes. The generator excitation voltages during the fault and postfault modes are assumed to remain constant. The electrical power of the ith generator in terms of the new reduced bus admittance matrices are obtained from (11.112). The swing equation with damping neglected, as given by (11.21), for machine i becomes

$$\frac{H_i}{\pi f_0}\frac{d^2\delta_i}{dt^2} = P_{mi} - \sum_{j=1}^{m} |E'_i||E'_j||Y_{ij}|\cos(\theta_{ij} - \delta_i + \delta_j) \tag{11.114}$$

where Y_{ij} are the elements of the faulted reduced bus admittance matrix, and H_i is the inertia constant of machine i expressed on the common MVA base S_B. If H_{Gi} is the inertia constant of machine i expressed on the machine rated MVA S_{Gi}, then H_i is given by

$$H_i = \frac{S_{Gi}}{S_B}H_{Gi} \tag{11.115}$$

Showing the electrical power of the ith generator by P_e^f and transforming (11.114) into state variable model yields

$$\begin{aligned} \frac{d\delta_i}{dt} &= \Delta\omega_i \qquad i = 1, \ldots, m \\ \frac{d\Delta\omega_i}{dt} &= \frac{\pi f_0}{H_i}(P_m - P_e^f) \end{aligned} \tag{11.116}$$

We have two state equations for each generator, with initial power angles δ_{0_i} and $\Delta\omega_{0_i} = 0$. The *MATLAB* function **ode23** is employed to solve the above $2m$ first-order differential equations. When the fault is cleared, which may involve the removal of the faulty line, the bus admittance matrix is recomputed to reflect the change in the network. Next, the postfault reduced bus admittance matrix is evaluated and the postfault electrical power of the ith generator shown by P_i^{pf} is readily determined from (11.112). Using the postfault power P_i^{pf}, the simulation is continued to determine the system stability, until the plots reveal a definite trend as to stability or instability. Usually the slack generator is selected as the reference, and the phase angle difference of all other generators with respect to the reference machine are plotted. Usually, the solution is carried out for two swings to show that the second swing is not greater than the first one. If the angle differences do not increase, the system is stable. If any of the angle differences increase indefinitely, the system is unstable.

Based on the above procedure, a program named **trstab** is developed for transient stability analysis of a multimachine network subjected to a balanced three-phase fault. The program **trstab** must be preceded by the power flow program. Any of the power flow programs **lfgauss**, **lfnewton**, or **decouple** can be used. In addition to the power flow data, generator data must be specified in a matrix named **gendata**. The first column contains the generator bus number terminal. Columns 2 and 3 contain resistance and transient reactance in per unit on the specified common MVA base, and the last column contain the machine inertia constant in seconds, expressed on the common MVA base. The program **trstab** automatically adds additional buses to include the generator impedances in the power flow line data. Also, the bus admittance matrix is modified to include the load admittances **yload**, returned by the power flow program. The program prompts the user to enter the faulted bus number, fault clearing time, and the line numbers of the removed faulty line. The program displays the prefault, faulted, and postfault reduced bus admittance matrices. The machine phase angles are tabulated and a plot of the swing curves is obtained. The program inquires for other fault clearing times and fault locations. The use of **trstab** program is demonstrated in the following example.

Example 11.7

The power system network of an electric utility company is shown in Figure 11.31. The load data and voltage magnitude, generation schedule, and the reactive power limits for the regulated buses are tabulated on the next page. Bus 1, whose voltage is specified as $V_1 = 1.06\angle 0°$, is taken as the slack bus. The line data containing the series resistance and reactance in per unit, and one-half of the total capacitance in per unit susceptance on a 100-MVA base is also tabulated as shown.

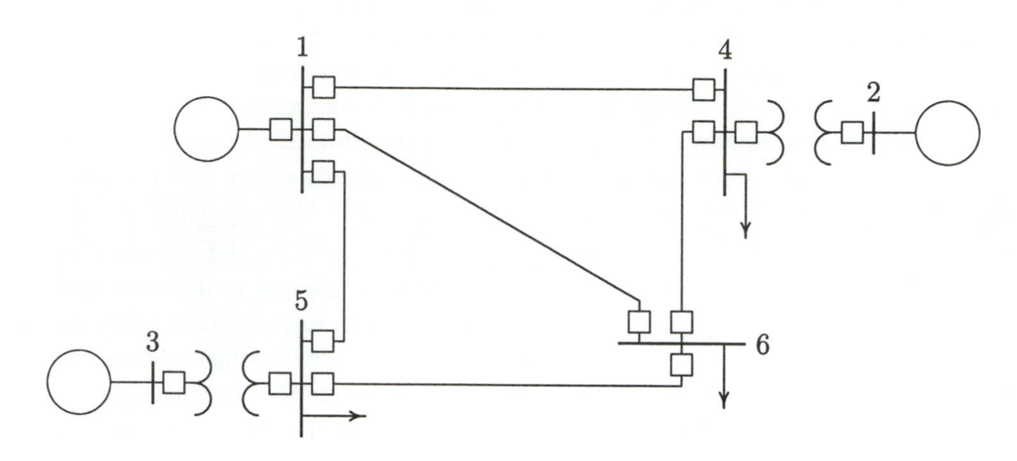

FIGURE 11.31
One-line diagram for Problem 11.7.

LOAD DATA		
Bus	Load	
No.	MW	Mvar
1	0	0
2	0	0
3	0	0
4	100	70
5	90	30
6	160	110

GENERATION SCHEDULE				
Bus No.	Voltage Mag.	Generation, MW	Mvar Limits Min.	Max.
1	1.06			
2	1.04	150	0	140
3	1.03	100	0	90

LINE DATA				
Bus No.	Bus No.	R, PU	X, PU	$\frac{1}{2}B$, PU
1	4	0.035	0.225	0.0065
1	5	0.025	0.105	0.0045
1	6	0.040	0.215	0.0055
2	4	0.000	0.035	0.0000
3	5	0.000	0.042	0.0000
4	6	0.028	0.125	0.0035
5	6	0.026	0.175	0.0300

The generator's armature resistances and transient reactances in per unit, and the inertia constants in seconds expressed on a 100-MVA base are given below:

MACHINE DATA			
Gen.	R_a	X'_d	H
1	0	0.20	20
2	0	0.15	4
3	0	0.25	5

A three-phase fault occurs on line 5–6 near bus 6, and is cleared by the simultaneous opening of breakers at both ends of the line. Using the **trstab** program, perform a transient stability analysis. Determine the system stability for
(a) When the fault is cleared in 0.4 second
(b) When the fault is cleared in 0.5 second
(c) Repeat the simulation to determine the critical clearing time.

The required data and commands are as follows:

```
basemva = 100;  accuracy = 0.0001;  maxiter = 10;
%       Bus Bus  Voltage Angle --Load-- --Generator-- Injected
%        No code Mag. degree MW   Mvar  MW   Mvar Qmin Qmax Mvar
busdata=[1  1    1.06   0       0     0    0    0    0     0   0
         2  2    1.04   0       0     0  150    0    0   140   0
         3  2    1.03   0       0     0  100    0    0    90   0
         4  0    1.0    0     100    70    0    0    0     0   0
         5  0    1.0    0      90    30    0    0    0     0   0
         6  0    1.0    0     160   110    0    0    0     0   0];
% Line data
%          Bus   bus    R       X       1/B     1 for line code or
%          nl    nr     pu      pu      pu       tap setting value
linedata=[1     4    0.035   0.225   0.0065     1.0
          1     5    0.025   0.105   0.0045     1.0
          1     6    0.040   0.215   0.0055     1.0
          2     4    0.000   0.035   0.0000     1.0
          3     5    0.000   0.042   0.0000     1.0
          4     6    0.028   0.125   0.0035     1.0
          5     6    0.026   0.175   0.0300     1.0];
lfybus            % form the bus admittance matrix for power flow
lfnewton           % Power flow solution by Newton-Raphson method
busout             % Prints the power flow solution on the screen
% Generator data
%          Gen.  Ra     Xd'     H
gendata=[ 1     0     0.20    20
          2     0     0.15     4
          3     0     0.25     5];
trstab                              % Performs the stability analysis.
       % User is prompted to enter the clearing time of fault.
```

The power flow result is

```
              Power Flow Solution by Newton-Raphson Method
                 Maximum Power Mismatch = 1.80187e-007
                        No. of Iterations = 4

Bus  Voltage  Angle   -----Load-----  ---Generation--- Injected
No.  Mag.     degree   MW       Mvar    MW       Mvar     Mvar

1    1.060    0.000    0.000    0.00  105.287  107.335   0.00
2    1.040    1.470    0.000    0.00  150.000   99.771   0.00
3    1.030    0.800    0.000    0.00  100.000   35.670   0.00
4    1.008   -1.401  100.000   70.00    0.000    0.000   0.00
5    1.016   -1.499   90.000   30.00    0.000    0.000   0.00
6    0.941   -5.607  160.000  110.00    0.000    0.000   0.00

Total                350.000  210.00  355.287  242.776   0.00
```

The **trstab** result is

```
Reduced prefault bus admittance matrix

  Ybf =
        0.3517 - 2.8875i   0.2542 + 1.1491i   0.1925 + 0.9856i
        0.2542 + 1.1491i   0.5435 - 2.8639i   0.1847 + 0.6904i
        0.1925 + 0.9856i   0.1847 + 0.6904i   0.2617 - 2.2835i

      G(i)    E'(i)     d0(i)     Pm(i)
       1     1.2781    8.9421    1.0529
       2     1.2035   11.8260    1.5000
       3     1.1427   13.0644    1.0000

Enter faulted bus No. -> 6

Reduced faulted bus admittance matrix

  Ypf =
        0.1913 - 3.5849i   0.0605 + 0.3644i   0.0523 + 0.4821i
        0.0605 + 0.3644i   0.3105 - 3.7467i   0.0173 + 0.1243i
        0.0523 + 0.4821i   0.0173 + 0.1243i   0.1427 - 2.6463i

Fault is cleared by opening a line.  The bus to bus numbers of
line to be removed must be entered within brackets, e.g. [5,7]
Enter the bus to bus Nos. of line to be removed -> [5, 6]
```

```
Reduced postfault bus admittance matrix

  Yaf =
        0.3392 - 2.8879i   0.2622 + 1.1127i   0.1637 + 1.0251i
        0.2622 + 1.1127i   0.6020 - 2.7813i   0.1267 + 0.5401i
        0.1637 + 1.0251i   0.1267 + 0.5401i   0.2859 - 2.0544i

Enter clearing time of fault in sec. tc = 0.4
Enter final simulation time in sec.  tf = 1.5
```

The phase angle differences of each machine with respect to the slack bus are printed in a tabular form on the screen, which is not presented here. The program also obtains a plot of the swing curves which is presented in Figure 11.32.

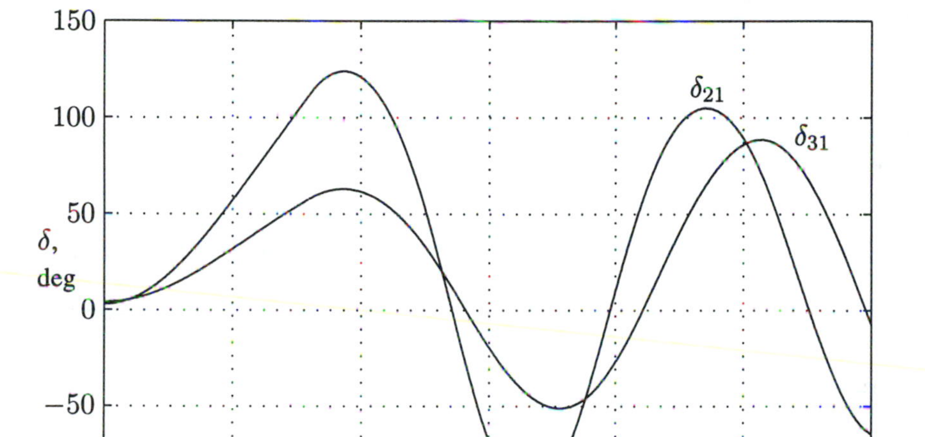

FIGURE 11.32
Plots of angle differences for machines 2 and 3 of Example 11.7.

Again the tabulated result is printed on the screen, and plots of the swing curves are obtained as shown in Figure 11.33. Figure 11.32 shows that the phase angle differences, after reaching a maximum of $\delta_{21} = 123.9^\circ$ and $\delta_{31} = 62.95^\circ$ will decrease, and the machines swing together. Hence, the system is found to be stable when fault is cleared in 0.4 second.

The program inquires for another fault clearing time, and the results continue as follows:

```
Another clearing time of fault?
Enter 'y' or 'n' within quotes -> 'y'
Enter clearing time of fault in sec. tc = 0.5
Enter final simulation time in sec.  tf = 1.5
```

The swing curves shown in Figure 11.33 show that machine 2 phase angle increases without limit. Thus, the system is unstable when fault is cleared in 0.5 second. The simulation is repeated for a clearing time of 0.45 second, which is found to be critically stable.

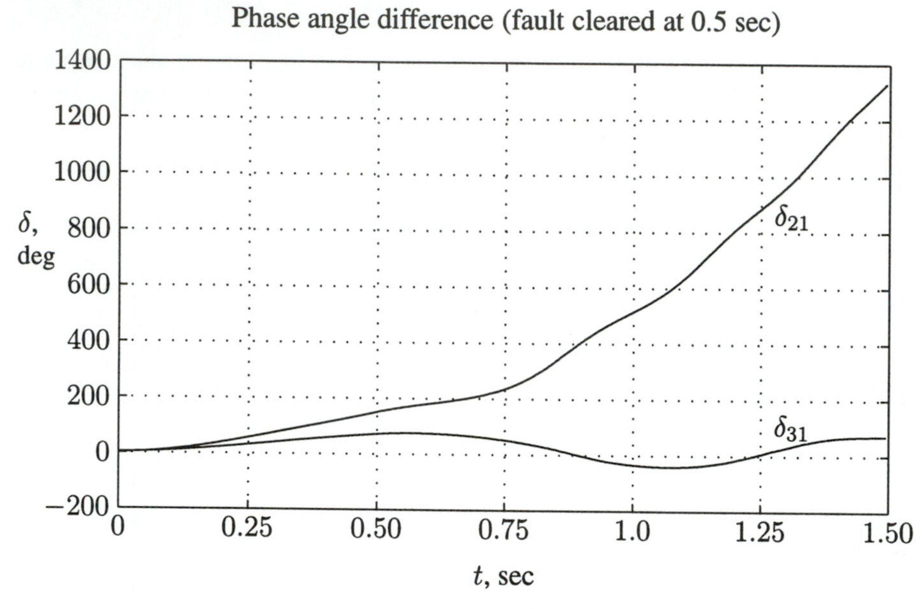

FIGURE 11.33
Plots of angle differences for machines 2 and 3 of Example 11.7.

PROBLEMS

11.1. A four-pole, 60-Hz synchronous generator has a rating of 200 MVA, 0.8 power factor lagging. The moment of inertia of the rotor is 45,100 kg·m^2. Determine M and H.

11.2. A two-pole, 60-Hz synchronous generator has a rating of 250 MVA, 0.8 power factor lagging. The kinetic energy of the machine at synchronous speed is 1080 MJ. The machine is running steadily at synchronous speed and delivering 60 MW to a load at a power angle of 8 electrical degrees. The load is suddenly removed. Determine the acceleration of the rotor. If the acceleration computed for the generator is constant for a period of 12 cycles, determine the value of the power angle and the rpm at the end of this time.

11.3. Determine the kinetic energy stored by a 250-MVA, 60-Hz, two-pole synchronous generator with an inertia constant H of 5.4 MJ/MVA. Assume the machine is running steadily at synchronous speed with a shaft input of 331,100 hp The electrical power developed suddenly changes from its normal value to a value of 200 MW. Determine the acceleration or deceleration of the rotor. If the acceleration computed for the generator is constant for a period of 9 cycles, determine the change in the power angle in that period and the rpm at the end of 9 cycles.

11.4. The swing equations of two interconnected synchronous machines are written as

$$\frac{H_1}{\pi f_0}\frac{d^2\delta_1}{dt^2} = P_{m1} - P_{e1}$$

$$\frac{H_2}{\pi f_0}\frac{d^2\delta_2}{dt^2} = P_{m2} - P_{e2}$$

Denote the relative power angle between the two machines by $\delta = \delta_1 - \delta_2$. Obtain a swing equation equivalent to that of a single machine in terms of δ, and show that

$$\frac{H}{\pi f_0}\frac{d^2\delta}{dt^2} = P_m - P_e$$

where

$$H = \frac{H_1 H_2}{H_1 + H_2}$$

$$P_m = \frac{H_2 P_{m1} - H_1 P_{m2}}{H_1 + H_2}$$

$$P_e = \frac{H_2 P_{e1} - H_1 P_{e2}}{H_1 + H_2}$$

11.5. Two synchronous generators represented by a constant voltage behind transient reactance are connected by a pure reactance $X = 0.3$ per unit, as shown in Figure 11.34. The generator inertia constants are $H_1 = 4.0$ MJ/MVA and $H_2 = 6$ MJ/MVA, and the transient reactances are $X_1' = 0.16$ and $X_2' = 0.20$ per unit. The system is operating in the steady state with $E_1' = 1.2$, $P_{m1} = 1.5$ and $E_2' = 1.1$, $P_{m2} = 1.0$ per unit. Denote the relative power

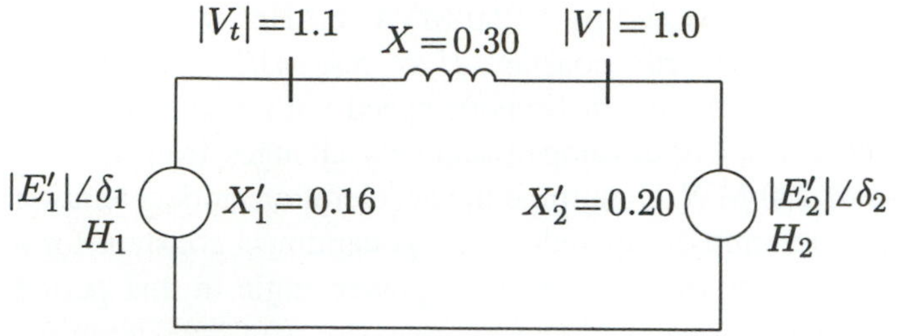

FIGURE 11.34
System of Problem 11.5.

angle between the two machines by $\delta = \delta_1 - \delta_2$. Referring to Problem 11.4, reduce the two-machine system to an equivalent one-machine against an infinite bus. Find the inertia constant of the equivalent machine, the mechanical input power, and the amplitude of its power angle curve, and obtain the equivalent swing equation in terms of δ.

11.6. A 60-Hz synchronous generator has a transient reactance of 0.2 per unit and an inertia constant of 5.66 MJ/MVA. The generator is connected to an infinite bus through a transformer and a double circuit transmission line, as shown in Figure 11.35. Resistances are neglected and reactances are expressed on a common MVA base and are marked on the diagram. The generator is delivering a real power of 0.77 per unit to bus bar 1. Voltage magnitude at bus 1 is 1.1. The infinite bus voltage $V = 1.0\angle 0°$ per unit. Determine the generator excitation voltage and obtain the swing equation as given by (11.36).

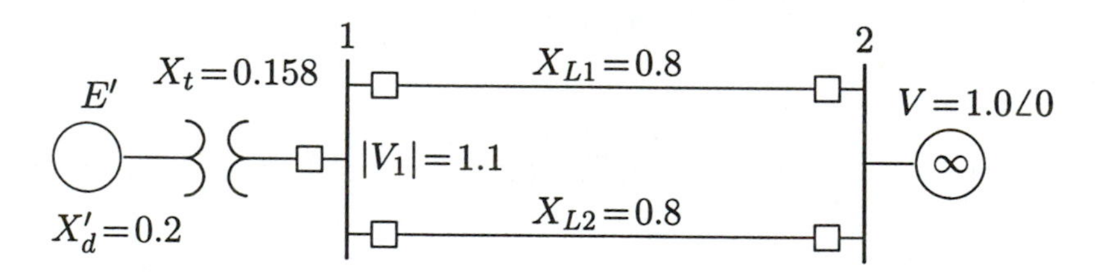

FIGURE 11.35
System of Problem 11.6.

11.7. A three-phase fault occurs on the system of Problem 11.6 at the sending end of the transmission lines. The fault occurs through an impedance of 0.082 per unit. Assume the generator excitation voltage remains constant at $E' = 1.25$ per unit. Obtain the swing equation during the fault.

11.8. The power-angle equation for a salient-pole generator is given by

$$P_e = P_{max} \sin \delta + P_K \sin 2\delta$$

Consider a small deviation in power angle from the initial operating point δ_0, i.e., $\delta = \delta + \delta_0$. Obtain an expression for the synchronizing power coefficient, similar to (11.39). Also, find the linearized swing equation in terms of $\Delta\delta$.

11.9. Consider the displacement x for a unit mass supported by a nonlinear spring as shown in Figure 11.36. The equation of motion is described by

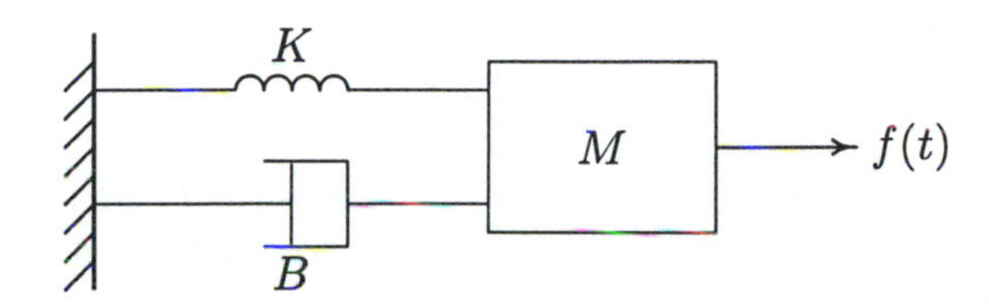

FIGURE 11.36
System of Problem 11.9.

$$M\frac{d^2x(t)}{dt^2} + B\frac{dx(t)}{dt} + K \sin x(t) = f(t)$$

where M is the mass, B is the frictional coefficient and K is the spring constant. The system is at the steady state $x(0) = 0$ for $f(0) = 0$. A small perturbation $f(t) = f(0) + \Delta f(t)$ results in the displacement $x(t) = x(0) + \Delta x(t)$.

(a) Obtain a linearized expression for the motion of the system in terms of the system parameters, $\Delta x(t)$ and $\Delta f(t)$.
(b) For $M = 1.6$, $B = 9.6$, and $K = 40$, find the damping ratio ζ and the damped frequency of oscillation ω_d.

11.10. The machine in the power system of Problem 11.6 has a per unit damping coefficient of $D = 0.15$. The generator excitation voltage is $E' = 1.25$ per unit and the generator is delivering a real power of 0.77 per unit to the infinite bus at a voltage of $V = 1.0$ per unit. Write the linearized swing equation for this power system. Use (11.61) and (11.62) to find the equations describing the motion of the rotor angle and the generator frequency for a small disturbance of $\Delta\delta = 15°$. Use *MATLAB* to obtain the plots of rotor angle and frequency.

11.11. Write the linearized swing equation of Problem 11.10 in state variable form. Use $[y, x]$ = **initial**(A, B, C, D, x_0, t) and **plot** commands to obtain the zero-input response for the initial conditions $\Delta\delta = 15°$, and $\Delta\omega_n = 0$.

11.12. The generator of Problem 11.10 is operating in the steady state at $\delta_0 = 27.835°$ when the input power is increased by a small amount $\Delta P = 0.15$ per unit. The generator excitation and the infinite bus voltage are the same as before. Use (11.75) and (11.76) to find the equations describing the motion of the rotor angle and the generator frequency for a small disturbance of $\Delta P = 0.15$ per unit. Use *MATLAB* to obtain the plots of rotor angle and frequency.

11.13. Write the linearized swing equation of Problem 11.10 in state variable form. Use $[y, x]$ = **step**$(A, B, C, D, 1, t)$ and **plot** commands to obtain the zero-state response when the input power is increased by a small amount $\Delta P = 0.15$ per unit.

11.14. The machine of Problem 11.6 is delivering a real power input of 0.77 per unit to the infinite bus at a voltage of 1.0 per unit. The generator excitation voltage is $E' = 1.25$ per unit. Use **eacpower**(P_m, E, V, X) to find
(a) The maximum power input that can be added without loss of synchronism.
(b) Repeat (a) with zero initial power input. Assume the generator internal voltage remains constant at the value computed in (a).

11.15. The machine of Problem 11.6 is delivering a real power input of 0.77 per unit to the infinite bus at a voltage of 1.0 per unit. The generator excitation voltage is $E' = 1.25$ per unit.
(a) A temporary three-phase fault occurs at the sending end of one of the transmission lines. When the fault is cleared, both lines are intact. Using equal area criterion, determine the critical clearing angle and the critical fault clearing time. Use **eacfault**$(P_m, E, V, X_1, X_2, X_3)$ to check the result and to display the power-angle plot.
(b) A three-phase fault occurs at the middle of one of the lines, the fault is cleared, and the faulted line is isolated. Determine the critical clearing angle. Use **eacfault**$(P_m, E, V, X_1, X_2, X_3)$ to check the results and to display the power-angle plot.

11.16. The machine of Problem 11.6 is delivering a real power input of 0.77 per unit to the infinite bus at a voltage of 1.0 per unit. The generator excitation voltage is $E' = 1.25$ per unit. A three-phase fault at the middle of one line is cleared by isolating the faulted circuit simultaneously at both ends.
(a) The fault is cleared in 0.2 second. Obtain the numerical solution of the swing equation for 1.5 seconds. Select one of the functions **swingmeu, swingrk2,** or **swingrk4.**
(b) Repeat the simulation and obtain the swing plots when fault is cleared in 0.4 second, and for the critical clearing time.

11.17. Consider the power system network of Example 11.7 with the described operating condition. A three-phase fault occurs on line 1–5 near bus 5 and is cleared by the simultaneous opening of breakers at both ends of the line. Using the **trstab** program, perform a transient stability analysis. Determine the system stability for

(a) When the fault is cleared in 0.2 second

(b) When the fault is cleared in 0.4 second

(c) Repeat the simulation to determine the critical clearing time.

11.18. The power system network of an electric company is shown in Figure 11.37. The load data is as follows.

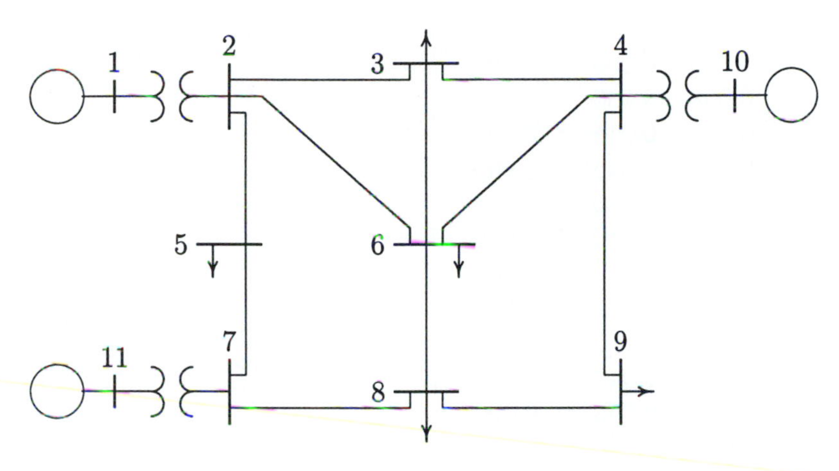

FIGURE 11.37
System of Problem 11.18.

LOAD DATA					
Bus	Load		Bus	Load	
No.	MW	Mvar	No.	MW	Mvar
1	0.0	0.0	7	0.0	0.0
2	0.0	0.0	8	110.0	90.0
3	150.0	120.0	9	80.0	50.0
4	0.0	0.0	10	0.0	0.0
5	120.0	60.0	11	0.0	0.0
6	140.0	90.0			

Voltage magnitude, generation schedule, and the reactive power limits for the regulated buses are tabulated below. Bus 1, whose voltage is specified as $V_1 = 1.04\angle 0°$, is taken as the slack bus.

GENERATION SCHEDULE				
Bus No.	Voltage Mag.	Generation, MW	Mvar Limits Min.	Max.
1	1.040			
10	1.035	200.0	0.0	180.0
11	1.030	160.0	0.0	120.0

The line data containing the per unit series impedance, and one-half of the shunt capacitive susceptance on a 100-MVA base is tabulated below.

LINE AND TRANSFORMER DATA				
Bus No.	Bus No.	R, PU	X, PU	$\frac{1}{2}B$, PU
1	2	0.000	0.006	0.000
2	3	0.008	0.030	0.004
2	5	0.004	0.015	0.002
2	6	0.012	0.045	0.005
3	4	0.010	0.040	0.005
3	6	0.004	0.040	0.005
4	6	0.015	0.060	0.008
4	9	0.018	0.070	0.009
4	10	0.000	0.008	0.000
5	7	0.005	0.043	0.003
6	8	0.006	0.048	0.000
7	8	0.006	0.035	0.004
7	11	0.000	0.010	0.000
8	9	0.005	0.048	0.000

The generator's armature resistance and transient reactances in per unit, and the inertia constants expressed on a 100-MVA base are given below.

MACHINE DATA			
Gen.	R_a	X'_d	H
1	0	0.20	12
10	0	0.15	10
11	0	0.25	9

A three-phase fault occurs on line 4–9, near bus 4, and is cleared by the simultaneous opening of breakers at both ends of the line. Using the **trstab** program, perform a transient stability analysis. Determine the stability for

(a) When the fault is cleared in 0.4 second

(b) When the fault is cleared in 0.8 second

(c) Repeat the simulation to determine the critical clearing time.

CHAPTER

12

POWER SYSTEM CONTROL

12.1 INTRODUCTION

So far, this text has concentrated on the problems of establishing a normal operating state and optimum scheduling of generation for a power system. This chapter deals with the control of active and reactive power in order to keep the system in the steady-state. In addition, simple models of the essential components used in the control systems are presented. The objective of the control strategy is to generate and deliver power in an interconnected system as economically and reliably as possible while maintaining the voltage and frequency within permissible limits.

Changes in real power affect mainly the system frequency, while reactive power is less sensitive to changes in frequency and is mainly dependent on changes in voltage magnitude. Thus, real and reactive powers are controlled separately. The *load frequency control* (LFC) loop controls the real power and frequency and the *automatic voltage regulator* (AVR) loop regulates the reactive power and voltage magnitude. Load frequency control (LFC) has gained in importance with the growth of interconnected systems and has made the operation of interconnected systems possible. Today, it is still the basis of many advanced concepts for the control of large systems.

The methods developed for control of individual generators, and eventually control of large interconnections, play a vital role in modern energy control centers. Modern energy control centers (ECC) are equipped with on-line computers performing all signal processing through the remote acquisition systems known as *supervisory control and data acquisition* (SCADA) systems. Only an introduction to power system control is presented here. This chapter utilizes some of the concepts of feedback control systems. Some students may not be fully versed in feedback theory. Therefore, a brief review of the fundamentals of linear control systems analysis and design is included in Appendix B. The use of *MATLAB CONTROL TOOLBOX* functions and some useful custom-made functions are also described in this appendix.

The role of *automatic generation control* (AGC) in power system operation, with reference to tie-line power control under normal operating conditions, is first analyzed. Typical responses to real power demand are illustrated using the latest simulation technique available by the *MATLAB SIMULINK* package. Finally, the requirement of reactive power and voltage regulation and the influence on stability of both speed and excitation controls, with use of suitable feedback signals, are examined.

12.2 BASIC GENERATOR CONTROL LOOPS

In an interconnected power system, load frequency control (LFC) and automatic voltage regulator (AVR) equipment are installed for each generator. Figure 12.1 represents the schematic diagram of the load frequency control (LFC) loop and the automatic voltage regulator (AVR) loop. The controllers are set for a particular operating condition and take care of small changes in load demand to maintain the frequency and voltage magnitude within the specified limits. Small changes in real power are mainly dependent on changes in rotor angle δ and, thus, the frequency. The reactive power is mainly dependent on the voltage magnitude (i.e., on the generator excitation). The excitation system time constant is much smaller than the prime mover time constant and its transient decay much faster and does not affect the LFC dynamic. Thus, the cross-coupling between the LFC loop and the AVR loop is negligible, and the load frequency and excitation voltage control are analyzed independently.

12.3 LOAD FREQUENCY CONTROL

The operation objectives of the LFC are to maintain reasonably uniform frequency, to divide the load between generators, and to control the tie-line interchange schedules. The change in frequency and tie-line real power are sensed, which is a measure of the change in rotor angle δ, i.e., the error $\Delta\delta$ to be corrected. The error signal, i.e., Δf and ΔP_{tie}, are amplified, mixed, and transformed into a real power command signal ΔP_V, which is sent to the prime mover to call for an increment in the torque.

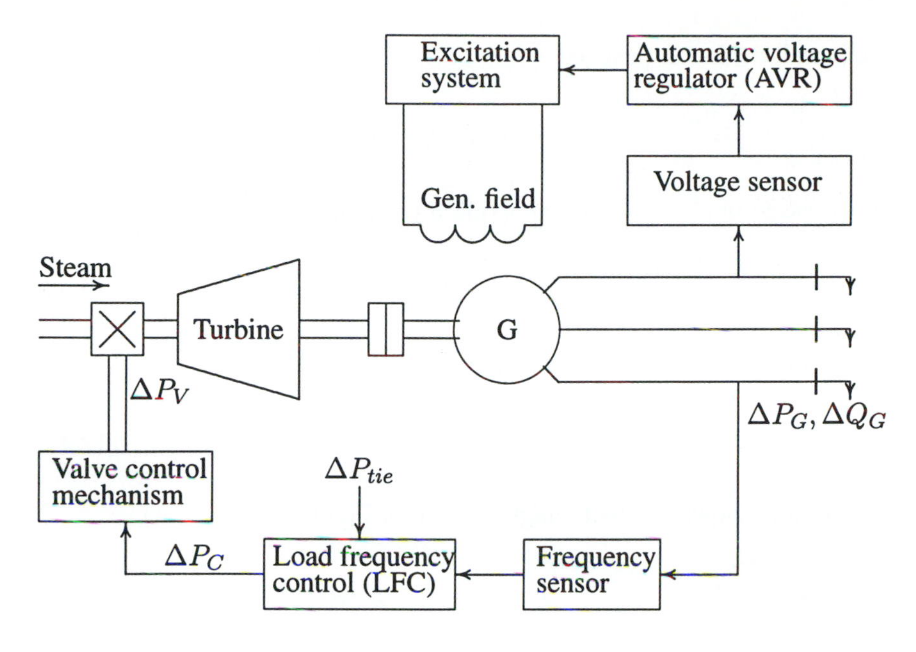

FIGURE 12.1
Schematic diagram of LFC and AVR of a synchronous generator.

The prime mover, therefore, brings change in the generator output by an amount ΔP_g which will change the values of Δf and ΔP_{tie} within the specified tolerance.

The first step in the analysis and design of a control system is mathematical modeling of the system. The two most common methods are the transfer function method and the state variable approach. The state variable approach can be applied to portray linear as well as nonlinear systems. In order to use the transfer function and linear state equations, the system must first be linearized. Proper assumptions and approximations are made to linearize the mathematical equations describing the system, and a transfer function model is obtained for the following components.

12.3.1 GENERATOR MODEL

Applying the swing equation of a synchronous machine given by (11.21) to small perturbation, we have

$$\frac{2H}{\omega_s}\frac{d^2\Delta\delta}{dt^2} = \Delta P_m - \Delta P_e \tag{12.1}$$

or in terms of small deviation in speed

$$\frac{d\Delta \frac{\omega}{\omega_s}}{dt} = \frac{1}{2H}(\Delta P_m - \Delta P_e) \tag{12.2}$$

With speed expressed in per unit, without explicit per unit notation, we have

$$\frac{d\Delta\omega}{dt} = \frac{1}{2H}(\Delta P_m - \Delta P_e) \tag{12.3}$$

Taking Laplace transform of (12.3), we obtain

$$\Delta\Omega(s) = \frac{1}{2Hs}[\Delta P_m(s) - \Delta P_e(s)] \tag{12.4}$$

The above relation is shown in block diagram form in Figure 12.2.

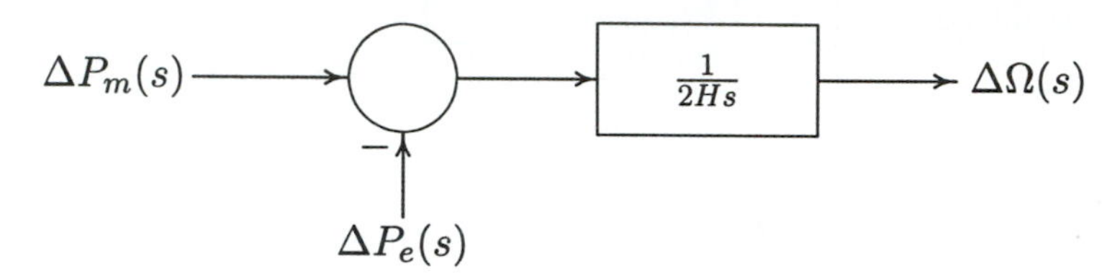

FIGURE 12.2
Generator block diagram.

12.3.2 LOAD MODEL

The load on a power system consists of a variety of electrical devices. For resistive loads, such as lighting and heating loads, the electrical power is independent of frequency. Motor loads are sensitive to changes in frequency. How sensitive it is to frequency depends on the composite of the speed-load characteristics of all the driven devices. The speed-load characteristic of a composite load is approximated by

$$\Delta P_e = \Delta P_L + D\Delta\omega \tag{12.5}$$

where ΔP_L is the nonfrequency-sensitive load change, and $D\Delta\omega$ is the frequency-sensitive load change. D is expressed as percent change in load divided by percent change in frequency. For example, if load is changed by 1.6 percent for a 1 percent change in frequency, then $D = 1.6$. Including the load model in the generator block diagram, results in the block diagram of Figure 12.3. Eliminating the simple feedback loop in Figure 12.3, results in the block diagram shown in Figure 12.4.

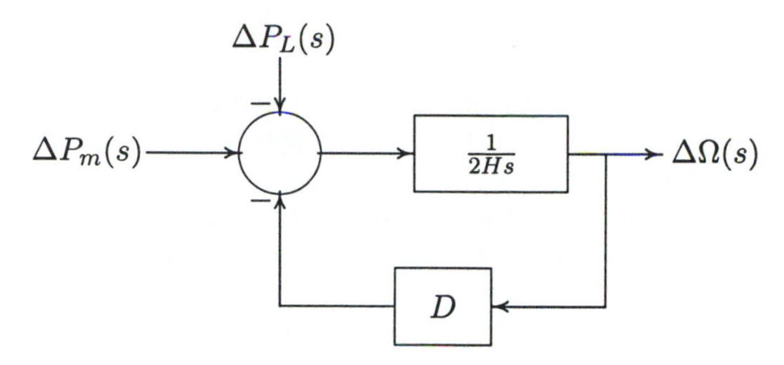

FIGURE 12.3
Generator and load block diagram.

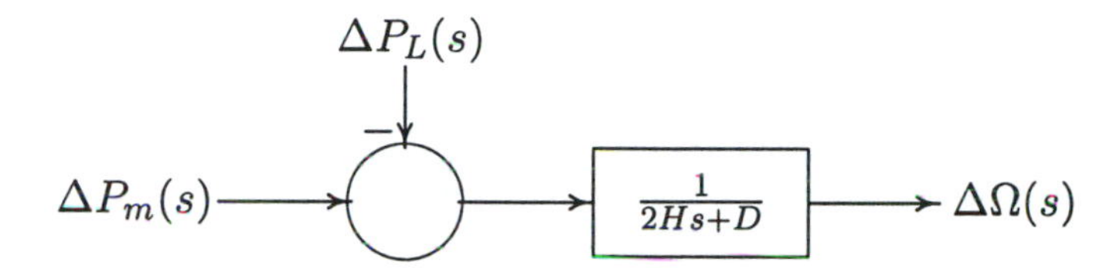

FIGURE 12.4
Generator and load block diagram.

12.3.3 PRIME MOVER MODEL

The source of mechanical power, commonly known as the *prime mover*, may be hydraulic turbines at waterfalls, steam turbines whose energy comes from the burning of coal, gas, nuclear fuel, and gas turbines. The model for the turbine relates changes in mechanical power output ΔP_m to changes in steam valve position ΔP_V. Different types of turbines vary widely in characteristics. The simplest prime mover model for the nonreheat steam turbine can be approximated with a single time constant τ_T, resulting in the following transfer function

$$G_T(s) = \frac{\Delta P_m(s)}{\Delta P_V(s)} = \frac{1}{1 + \tau_T s} \tag{12.6}$$

The block diagram for a simple turbine is shown in Figure 12.5.

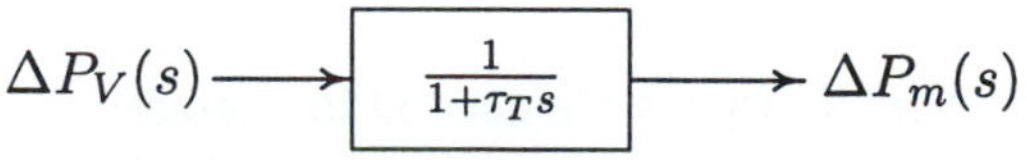

$\Delta P_V(s)$ → $\frac{1}{1+\tau_T s}$ → $\Delta P_m(s)$

FIGURE 12.5
Block diagram for a simple nonreheat steam turbine.

The time constant τ_T is in the range of 0.2 to 2.0 seconds.

12.3.4 GOVERNOR MODEL

When the generator electrical load is suddenly increased, the electrical power exceeds the mechanical power input. This power deficiency is supplied by the kinetic energy stored in the rotating system. The reduction in kinetic energy causes the turbine speed and, consequently, the generator frequency to fall. The change in speed is sensed by the turbine governor which acts to adjust the turbine input valve to change the mechanical power output to bring the speed to a new steady-state. The earliest governors were the Watt governors which sense the speed by means of rotating *flyballs* and provides mechanical motion in response to speed changes. However, most modern governors use electronic means to sense speed changes. Figure 12.6 shows schematically the essential elements of a conventional Watt governor, which consists of the following major parts.

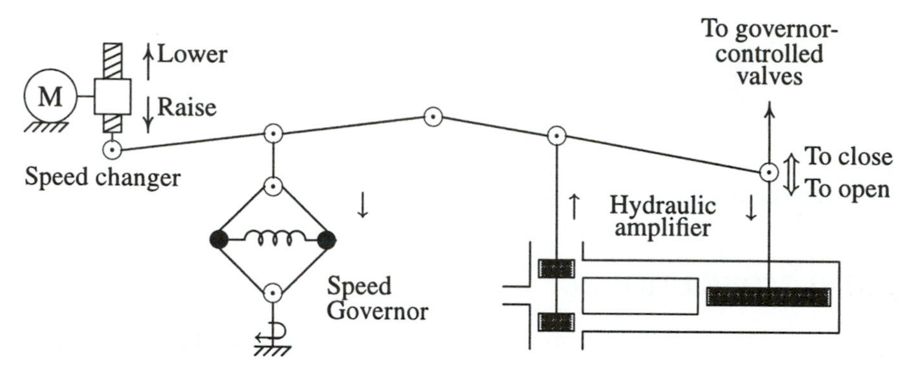

FIGURE 12.6
Speed governing system.

1. Speed Governor: The essential part are centrifugal flyballs driven directly or through gearing by the turbine shaft. The mechanism provides upward and downward vertical movements proportional to the change in speed.

2. Linkage Mechanism: These are links for transforming the flyballs movement to the turbine valve through a hydraulic amplifier and providing a feedback from the turbine valve movement.

3. Hydraulic Amplifier: Very large mechanical forces are needed to operate the steam valve. Therefore, the governor movements are transformed into high power forces via several stages of hydraulic amplifiers.

4. Speed Changer: The speed changer consists of a servomotor which can be operated manually or automatically for scheduling load at nominal frequency.

By adjusting this set point, a desired load dispatch can be scheduled at nominal frequency.

For stable operation, the governors are designed to permit the speed to drop as the load is increased. The steady-state characteristics of such a governor is shown in Figure 12.7.

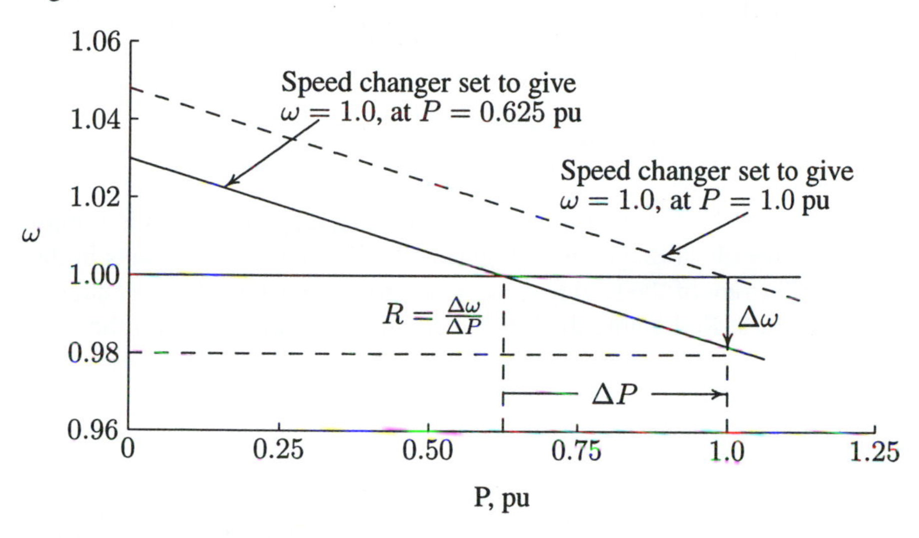

FIGURE 12.7
Governor steady-state speed characteristics.

The slope of the curve represents the speed regulation R. Governors typically have a speed regulation of 5–6 percent from zero to full load. The speed governor mechanism acts as a comparator whose output ΔP_g is the difference between the reference set power ΔP_{ref} and the power $\frac{1}{R}\Delta\omega$ as given from the governor speed characteristics, i.e.,

$$\Delta P_g = \Delta P_{ref} - \frac{1}{R}\Delta\omega \tag{12.7}$$

or in s-domain

$$\Delta P_g(s) = \Delta P_{ref}(s) - \frac{1}{R}\Delta\Omega(s) \tag{12.8}$$

The command ΔP_g is transformed through the hydraulic amplifier to the steam valve position command ΔP_V. Assuming a linear relationship and considering a simple time constant τ_g, we have the following s-domain relation

$$\Delta P_V(s) = \frac{1}{1+\tau_g}\Delta P_g(s) \tag{12.9}$$

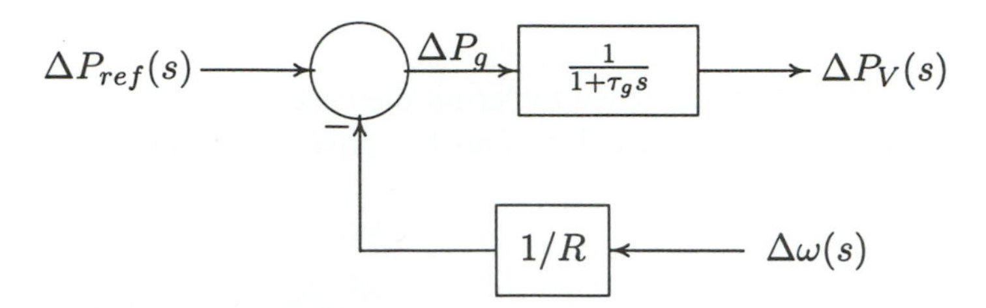

FIGURE 12.8
Block diagram representation of speed governing system for steam turbine.

Equations (12.8) and (12.9) are represented by the block diagram shown in Figure 12.8. Combining the block diagrams of Figures 12.4, 12.5, and 12.8 results in the complete block diagram of the load frequency control of an isolated power station shown in Figure 12.9. Redrawing the block diagram of Figure 12.9 with the load

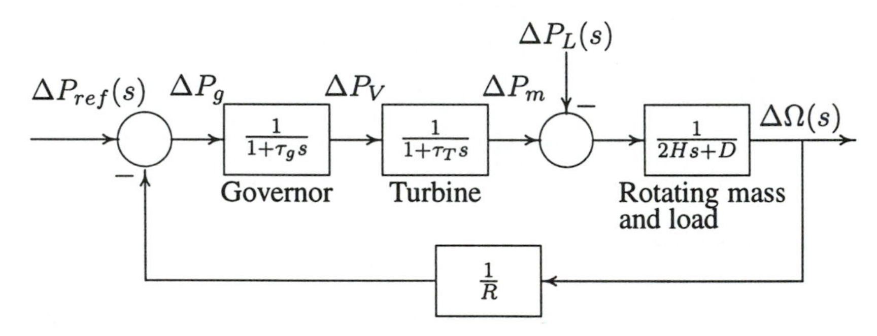

FIGURE 12.9
Load frequency control block diagram of an isolated power system.

change $-\Delta P_L(s)$ as the input and the frequency deviation $\Delta\Omega(s)$ as the output results in the block diagram shown in Figure 12.10. The open-loop transfer function of the block diagram in Figure 12.10 is

$$KG(s)H(s) = \frac{1}{R}\frac{1}{(2Hs + D)(1 + \tau_g s)(1 + \tau_T s)} \tag{12.10}$$

and the closed-loop transfer function relating the load change ΔP_L to the frequency deviation $\Delta\Omega$ is

$$\frac{\Delta\Omega(s)}{-\Delta P_L(s)} = \frac{(1 + \tau_g s)(1 + \tau_T s)}{(2Hs + D)(1 + \tau_g s)(1 + \tau_T s) + 1/R} \tag{12.11}$$

or

$$\Delta\Omega(s) = -\Delta P_L(s)T(s) \tag{12.12}$$

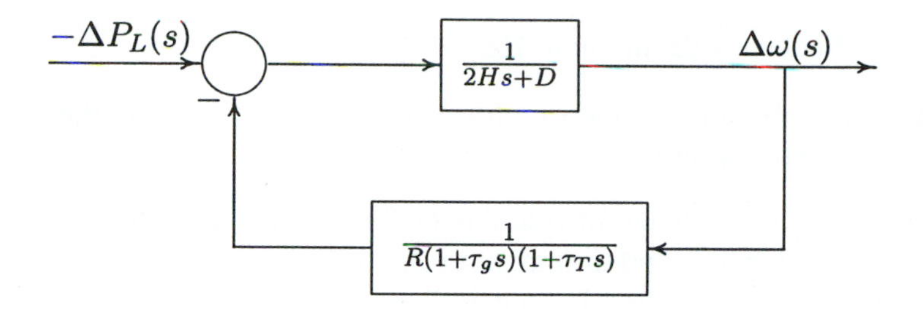

FIGURE 12.10
LFC block diagram with input $\Delta P_L(s)$ and output $\Delta\Omega(s)$.

The load change is a step input, i.e., $\Delta P_L(s) = \Delta P_L/s$. Utilizing the final value theorem, the steady-state value of $\Delta\omega$ is

$$\Delta\omega_{ss} = \lim_{s\to 0} s\Delta\Omega(s) = (-\Delta P_L)\frac{1}{D + 1/R} \tag{12.13}$$

It is clear that for the case with no frequency-sensitive load (i.e., with $D = 0$), the steady-state deviation in frequency is determined by the governor speed regulation, and is

$$\Delta\omega_{ss} = (-\Delta P_L)R \tag{12.14}$$

When several generators with governor speed regulations $R_1, R_2, \ldots, R_n$ are connected to the system, the steady-state deviation in frequency is given by

$$\Delta\omega_{ss} = (-\Delta P_L)\frac{1}{D + 1/R_1 + 1/R2 + \cdots 1/R_n} \tag{12.15}$$

Example 12.1

An isolated power station has the following parameters

Turbine time constant $\tau_T = 0.5$ sec
Governor time constant $\tau_g = 0.2$ sec
Generator inertia constant $H = 5$ sec
Governor speed regulation = R per unit

The load varies by 0.8 percent for a 1 percent change in frequency, i.e., $D = 0.8$
(a) Use the Routh-Hurwitz array (Appendix B.2.1) to find the range of R for control system stability.
(b) Use *MATLAB* **rlocus** function to obtain the root locus plot.
(c) The governor speed regulation of Example 12.1 is set to $R = 0.05$ per unit. The turbine rated output is 250 MW at nominal frequency of 60 Hz. A sudden load change of 50 MW ($\Delta P_L = 0.2$ per unit) occurs.

(i) Find the steady-state frequency deviation in Hz.

(ii) Use *MATLAB* to obtain the time-domain performance specifications and the frequency deviation step response.

(d) Construct the *SIMULINK* block diagram (see Appendix A.17) and obtain the frequency deviation response for the condition in part (c).

Substituting the system parameters in the LFC block diagram of Figure 12.10 results in the block diagram shown in Figure 12.11. The open-loop transfer function

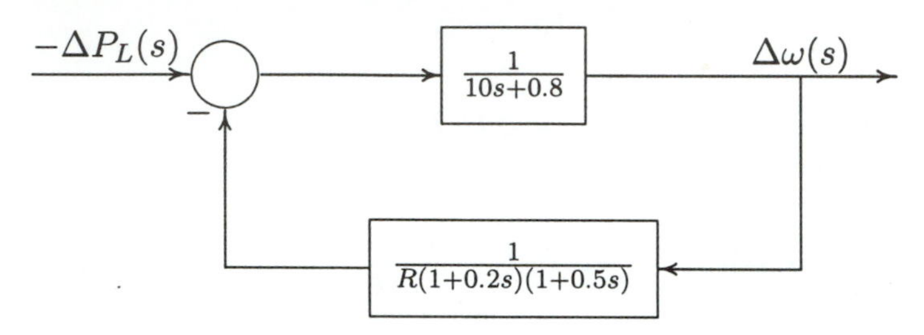

FIGURE 12.11
LFC block diagram for Example 12.1.

is

$$KG(s)H(s) = \frac{K}{(10s+0.8)(1+0.2s)(1+0.5s)}$$
$$= \frac{K}{s^3 + 7.08s^2 + 10.56s + 0.8}$$

where $K = \frac{1}{R}$

(a) The characteristic equation is given by

$$1 + KG(s)H(s) = 1 + \frac{K}{s^3 + 7.08s^2 + 10.56s + 0.8} = 0$$

which results in the characteristic polynomial equation

$$s^3 + 7.08s^2 + 10.56s + 0.8 + K = 0$$

The Routh-Hurwitz array for this polynomial is then (see Appendix B.2.1)

$$\begin{array}{c|ll}
s^3 & 1 & 10.56 \\
s^2 & 7.08 & 0.8 + K \\
s^1 & \frac{73.965-K}{7.08} & 0 \\
s^0 & 0.8 + K & 0
\end{array}$$

From the s^1 row, we see that for control system stability, K must be less than 73.965. Also from the s^0 row, K must be greater than -0.8. Thus, with positive values of K, for control system stability

$$K < 73.965$$

Since $R = \frac{1}{K}$, for control system stability, the governor speed regulation must be

$$R > \frac{1}{73.965} \qquad \text{or} \qquad R > 0.0135$$

For $K = 73.965$, the auxiliary equation from the s^2 row is

$$7.08s^2 + 74.765 = 0$$

or $s = \pm j3.25$. That is, for $R = 0.0135$, we have a pair of conjugate poles on the $j\omega$ axis, and the control system is marginally stable.

(b) To obtain the root-locus, we use the following commands.

```
num=1;
den = [1  7.08  10.56 .8];
figure(1), rlocus(num, den)
```

The result is shown in Figure 12.12.

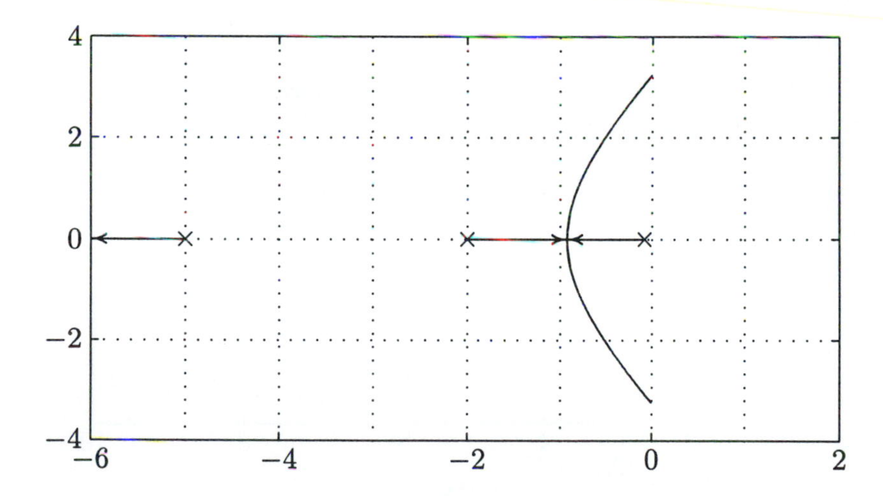

FIGURE 12.12
Root-locus plot for Example 12.1.

The loci intersect the $j\omega$ axis at $s = \pm j3.25$ for $K = 73.965$. Thus, the system is marginally stable for $R = \frac{1}{73.965} = 0.0135$.

(c) The closed-loop transfer function of the system shown in Figure 12.11 is

$$\frac{\Delta\Omega(s)}{-\Delta P_L(s)} = T(s) = \frac{(1+0.2s)(1+0.5s)}{(10s+0.8)(1+0.2s)(1+0.5s)+1/.05}$$
$$= \frac{0.1s^2+0.7s+1}{s^3+7.08s^2+10.56s+20.8}$$

(i) The steady-state frequency deviation due to a step input is

$$\Delta\omega_{ss} = \lim_{s\to 0} s\Delta\Omega(s) = \frac{1}{20.8}(-0.2) = -0.0096 \text{ pu}$$

Thus, the steady-state frequency deviation in hertz due to the sudden application of a 50-MW load is $\Delta f = (-0.0096)(60) = 0.576$ Hz.

(ii) To obtain the step response and the time-domain performance specifications, we use the following commands

```
PL = 0.2; numc = [0.1  0.7  1];
denc = [1   7.08  10.56  20.8];
t = 0:.02:10; c = -PL*step(num, den, t);
figure(2), plot(t, c), xlabel('t,  sec'), ylabel('pu')
title('Frequency deviation step response'), grid
  timespec(num, den)
```

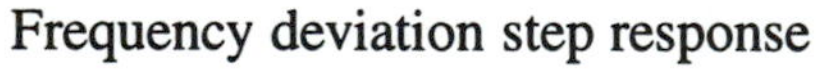

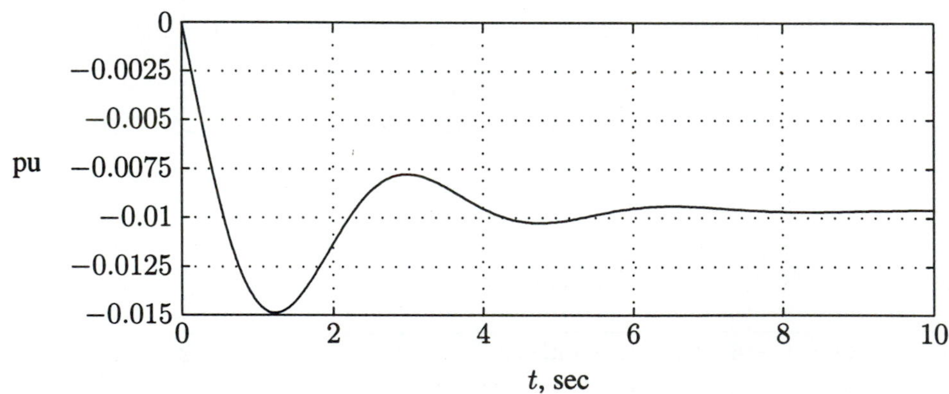

FIGURE 12.13
Frequency deviation step response for Example 12.1.

The frequency deviation step response is shown in Figure 12.13, and the time-domain performance specifications are

```
Peak time = 1.223     Percent overshoot = 54.80
Rise time = 0.418
Settling time = 6.8
```

(d) A *SIMULINK* model named **sim12ex1.mdl** is constructed as shown in Figure 12.14. The file is opened and is run in the *SIMULINK* window. The simulation results in the same response as shown in Figure 12.13.

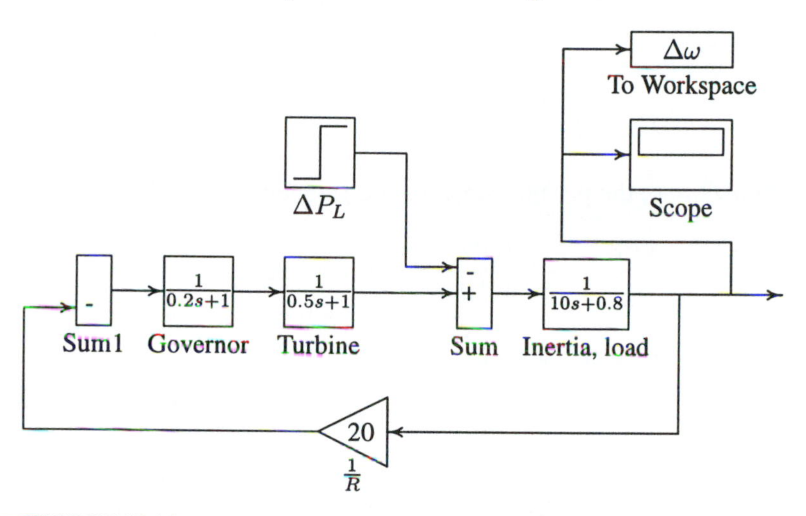

FIGURE 12.14
Simulation block diagram for Example 12.1.

Example 12.2

A single area consists of two generating units with the following characteristics.

Unit	Rating	Speed regulation R (pu on unit MVA base)
1	600 MVA	6%
2	500 MVA	4%

The units are operating in parallel, sharing 900 MW at the nominal frequency. Unit 1 supplies 500 MW and unit 2 supplies 400 MW at 60 Hz. The load is increased by 90 MW.

(a) Assume there is no frequency-dependent load, i.e., $D = 0$. Find the steady-state frequency deviation and the new generation on each unit.

(b) The load varies 1.5 percent for every 1 percent change in frequency, i.e., $D = 1.5$. Find the steady-state frequency deviation and the new generation on each unit.

First we express the governor speed regulation of each unit to a common MVA base. Select 1000 MVA for the apparent power base, then

$$R_1 = \frac{1000}{600}(0.06) = 0.1 \text{ pu}$$

$$R_2 = \frac{1000}{500}(0.05) = 0.08 \text{ pu}$$

The per unit load change is

$$\Delta P_L = \frac{90}{1000} = 0.09 \text{ pu}$$

(a) From (12.15) with $D = 0$, the per unit steady-state frequency deviation is

$$\Delta\omega_{ss} = \frac{-\Delta P_L}{\frac{1}{R_1} + \frac{1}{R_2}} = \frac{-0.09}{10 + 12.5} = -0.004 \text{ pu}$$

Thus, the steady-state frequency deviation in Hz is

$$\Delta f = (-0.004)(60) = -0.24 \text{ Hz}$$

and the new frequency is

$$f = f_0 + \Delta f = 60 - 0.24 = 59.76 \text{ Hz}$$

The change in generation for each unit is

$$\begin{aligned} \Delta P_1 = -\frac{\Delta\omega}{R_1} = -\frac{-0.004}{0.1} &= 0.04 \text{ pu} \\ &= 40 \text{ MW} \end{aligned}$$

$$\begin{aligned} \Delta P_2 = -\frac{\Delta\omega}{R_2} = -\frac{-0.004}{0.08} &= 0.05 \text{ pu} \\ &= 50 \text{ MW} \end{aligned}$$

Thus, unit 1 supplies 540 MW and unit 2 supplies 450 MW at the new operating frequency of 59.76 Hz.

MATLAB is used to plot the per unit speed characteristics of each governor as shown in Figure 12.15. As we can see from this figure, the initial generations are 0.5 and 0.40 per unit at the nominal frequency of 1.0 per unit. With the addition of 0.09 per unit power speed drops to 0.996 per unit. The new generations are 0.54

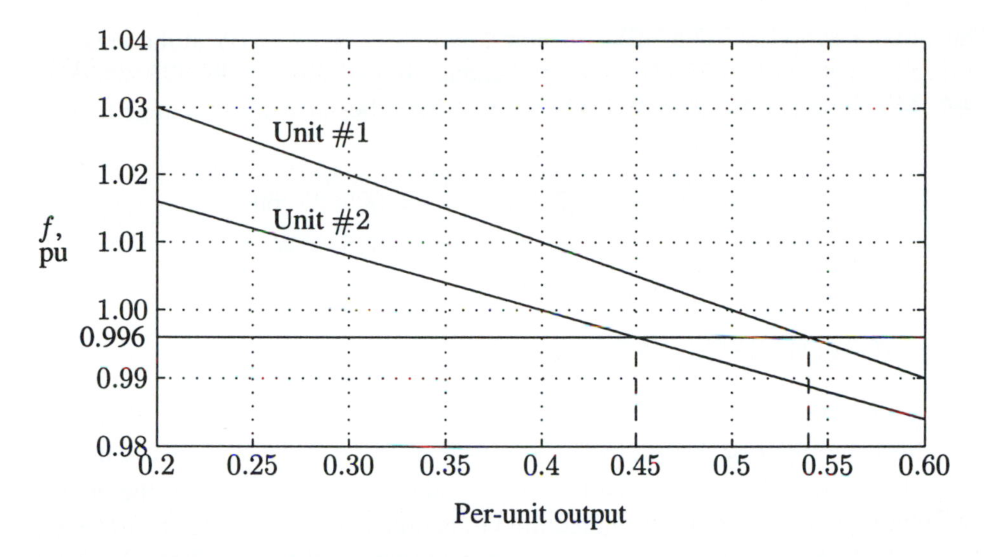

FIGURE 12.15
Load division between the two units of Example 12.2

and 0.45 per unit.

(b) For $D = 1.5$, the per unit steady-state frequency deviation is

$$\Delta\omega_{ss} = \frac{-\Delta P_L}{\frac{1}{R_1} + \frac{1}{R_2} + D} = \frac{-0.09}{10 + 12.5 + 1.5} = -0.00375 \ \text{pu}$$

Thus, the steady-state frequency deviation in Hz is

$$\Delta f = (-0.00375)(60) = -0.225 \ \text{Hz}$$

and the new frequency is

$$f = f_0 + \Delta f = 60 - 0.225 = 59.775 \ \text{Hz}$$

The change in generation for each unit is

$$\Delta P_1 = -\frac{\Delta\omega}{R_1} = -\frac{-0.00375}{0.1} = 0.0375 \ \text{pu}$$
$$= 37.500 \ \text{MW}$$

$$\Delta P_2 = -\frac{\Delta\omega}{R_2} = -\frac{-0.00375}{0.08} = 0.046875 \ \text{pu}$$
$$= 46.875 \ \text{MW}$$

Thus, unit 1 supplies 537.5 MW and unit 2 supplies 446.875 MW at the new operating frequency of 59.775 Hz. The total change in generation is 84.375, which is 5.625 MW less than the 90 MW load change. This is because of the change in load due to frequency drop which is given by

$$\begin{aligned} \Delta\omega D = (-0.00375)(1.5) &= -0.005625 \ \ \text{pu} \\ &= -5.625 \ \ \text{MW} \end{aligned}$$

12.4 AUTOMATIC GENERATION CONTROL

If the load on the system is increased, the turbine speed drops before the governor can adjust the input of the steam to the new load. As the change in the value of speed diminishes, the error signal becomes smaller and the position of the governor flyballs gets closer to the point required to maintain a constant speed. However, the constant speed will not be the set point, and there will be an offset. One way to restore the speed or frequency to its nominal value is to add an integrator. The integral unit monitors the average error over a period of time and will overcome the offset. Because of its ability to return a system to its set point, integral action is also known as the *rest action*. Thus, as the system load changes continuously, the generation is adjusted automatically to restore the frequency to the nominal value. This scheme is known as the *automatic generation control* (AGC). In an interconnected system consisting of several pools, the role of the AGC is to divide the loads among system, stations, and generators so as to achieve maximum economy and correctly control the scheduled interchanges of tie-line power while maintaining a reasonably uniform frequency. Of course, we are implicitly assuming that the system is stable, so the steady-state is achievable. During large transient disturbances and emergencies, AGC is bypassed and other emergency controls are applied. In the following section, we consider the AGC in a single area system and in an interconnected power system.

12.4.1 AGC IN A SINGLE AREA SYSTEM

With the primary LFC loop, a change in the system load will result in a steady-state frequency deviation, depending on the governor speed regulation. In order to reduce the frequency deviation to zero, we must provide a reset action. The rest action can be achieved by introducing an integral controller to act on the load reference setting to change the speed set point. The integral controller increases the system type by 1 which forces the final frequency deviation to zero. The LFC system, with the addition of the secondary loop, is shown in Figure 12.16. The integral controller gain K_I must be adjusted for a satisfactory transient response. Combining the parallel branches results in the equivalent block diagram shown in

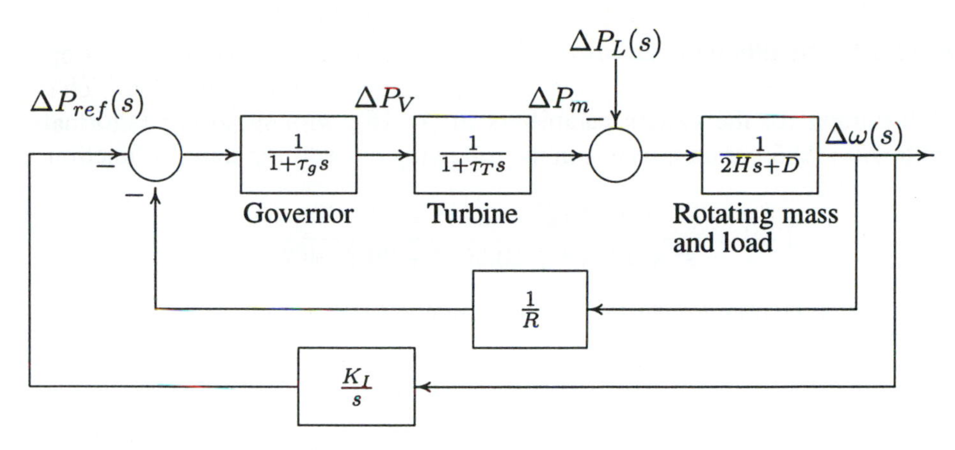

FIGURE 12.16
AGC for an isolated power system.

Figure 12.17.

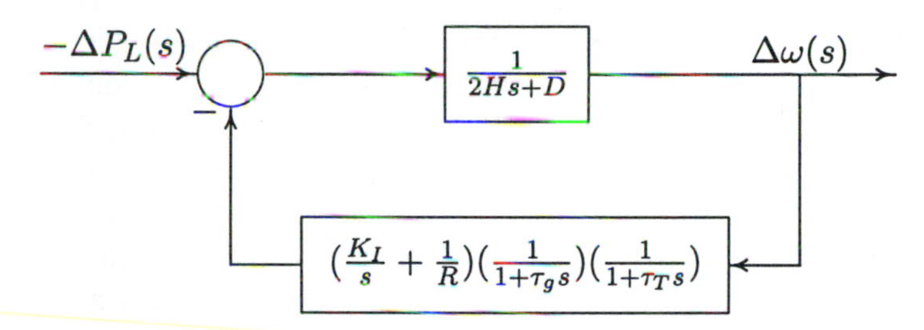

FIGURE 12.17
The equivalent block diagram of AGC for an isolated power system.

The closed-loop transfer function of the control system shown in Figure 12.17 with only $-\Delta P_L$ as input becomes

$$\frac{\Delta\Omega(s)}{-\Delta P_L(s)} = \frac{s(1+\tau_g s)(1+\tau_T s)}{s(2Hs+D)(1+\tau_g s)(1+\tau_T s) + K_I + s/R} \tag{12.16}$$

Example 12.3

The LFC system in Example 12.1 is equipped with the secondary integral control loop for automatic generation control.
(a) Use the *MATLAB* **step** function to obtain the frequency deviation step response for a sudden load change of $\Delta P_L = 0.2$ per unit. Set the integral controller gain to $K_I = 7$.
(b) Construct the *SIMULINK* block diagram and obtain the frequency deviation re-

sponse for the condition in part (a).

(a) Substituting for the system parameters in (12.16), with speed regulation adjusted to $R = 0.05$ per unit, results in the following closed-loop transfer function

$$T(s) = \frac{0.1s^3 + 0.7s^2 + s}{s^4 + 7.08s^3 + 10.56s^2 + 20.8s + 7}$$

To find the step response, we use the following commands

```
PL = 0.2;
KI = 7;
num = [0.1  0.7  1 0];
den = [1   7.08  10.56  20.8 KI];
t = 0:.02:12;
c = -PL*step(num, den, t);
plot(t, c), grid
xlabel('t, sec'), ylabel('pu')
title('Frequency deviation step response')
```

The step response is shown in Figure 12.18.

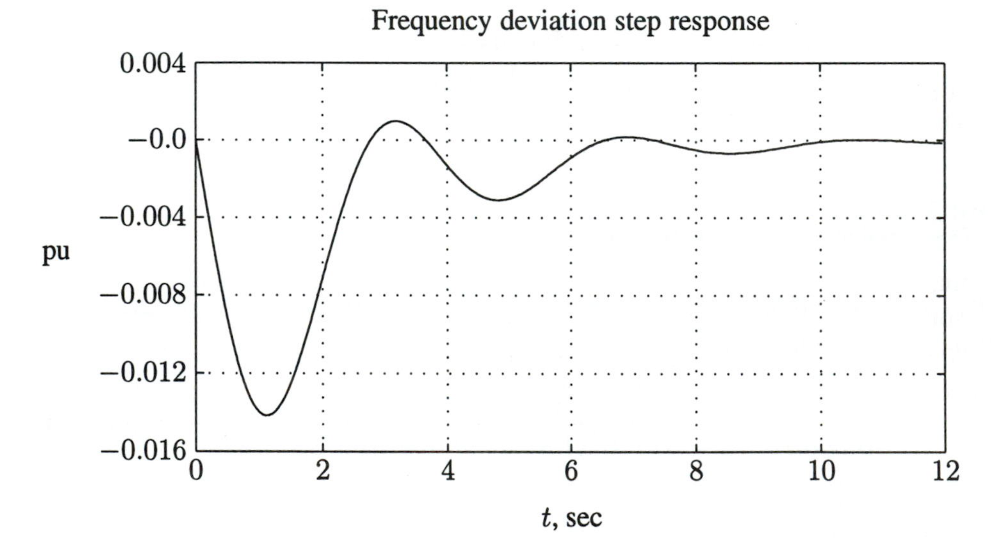

FIGURE 12.18
Frequency deviation step response for Example 12.3.

From the step response, we observe that the steady-state frequency deviation $\Delta\omega_{ss}$ is zero, and the frequency returns to its nominal value in approximately 10 seconds.

(b) A *SIMULINK* model named **sim12ex3.mdl** is constructed as shown in Figure 12.19. The file is opened and is run in the *SIMULINK* window. The simulation results in the same response as shown in Figure 12.18.

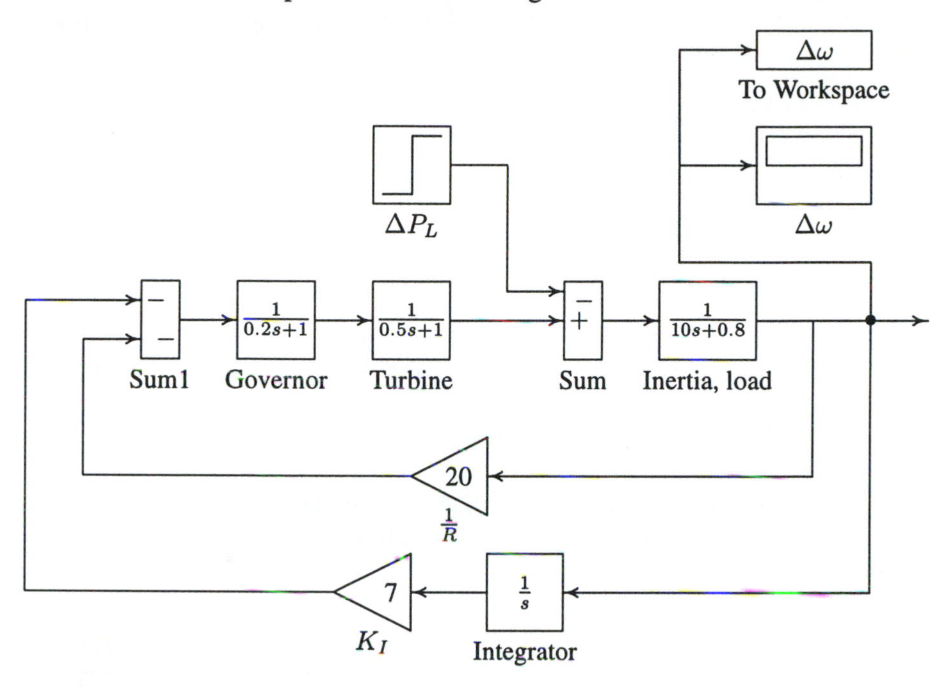

FIGURE 12.19
Simulation block diagram for Example 12.3.

12.4.2 AGC IN THE MULTIAREA SYSTEM

In many cases, a group of generators are closely coupled internally and swing in unison. Furthermore, the generator turbines tend to have the same response characteristics. Such a group of generators are said be *coherent*. Then it is possible to let the LFC loop represent the whole system, which is referred to as a *control area*. The AGC of a multiarea system can be realized by studying first the AGC for a two-area system. Consider two areas represented by an equivalent generating unit interconnected by a lossless tie line with reactance X_{tie}. Each area is represented by a voltage source behind an equivalent reactance as shown in Figure 12.20.

During normal operation, the real power transferred over the tie line is given by

$$P_{12} = \frac{|E_1||E_2|}{X_{12}} sin\delta_{12} \tag{12.17}$$

where $X_{12} = X_1 + X_{tie} + X_2$, and $\delta_{12} = \delta_1 - \delta_2$. Equation (12.17) can be linearized

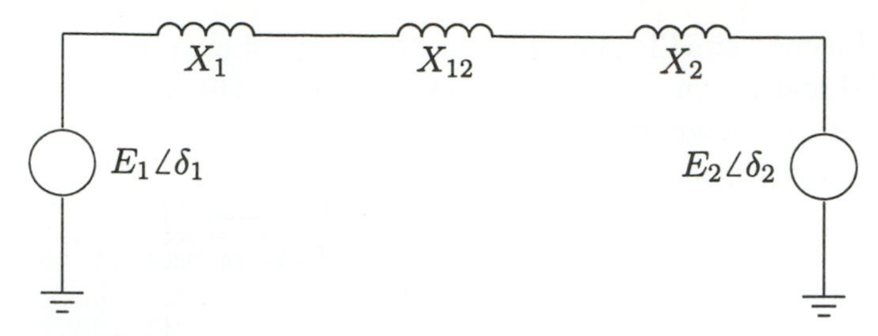

FIGURE 12.20
Equivalent network for a two-area power system.

for a small deviation in the tie-line flow ΔP_{12} from the nominal value, i.e.,

$$\begin{aligned}\Delta P_{12} &= \left.\frac{dP_{12}}{d\delta_{12}}\right|_{\delta_{12_0}} \Delta\delta_{12} \\ &= P_s \Delta\delta_{12}\end{aligned} \tag{12.18}$$

The quantity P_s is the slope of the power angle curve at the initial operating angle $\delta_{12_0} = \delta_{1_0} - \delta_{2_0}$. This was defined as the synchronizing power coefficient by (11.39) in Section 11.4. Thus we have

$$P_s = \left.\frac{dP_{12}}{d\delta_{12}}\right|_{\delta_{12_0}} = \frac{|E_1||E_2|}{X_{12}} \cos \Delta\delta_{12_0} \tag{12.19}$$

The tie-line power deviation then takes on the form

$$\Delta P_{12} = P_s(\Delta\delta_1 - \Delta\delta_2) \tag{12.20}$$

The tie-line power flow appears as a load increase in one area and a load decrease in the other area, depending on the direction of the flow. The direction of flow is dictated by the phase angle difference; if $\Delta\delta_1 > \Delta\delta_2$, the power flows from area 1 to area 2. A block diagram representation for the two-area system with LFC containing only the primary loop is shown in Figure 12.21.

Let us consider a load change ΔP_{L1} in area 1. In the steady-state, both areas will have the same steady-state frequency deviation, i.e.,

$$\Delta\omega = \Delta\omega_1 = \Delta\omega_2 \tag{12.21}$$

and

$$\begin{aligned}\Delta P_{m1} - \Delta P_{12} - \Delta P_{L1} &= \Delta\omega D_1. \\ \Delta P_{m2} + \Delta P_{12} &= \Delta\omega D_2\end{aligned} \tag{12.22}$$

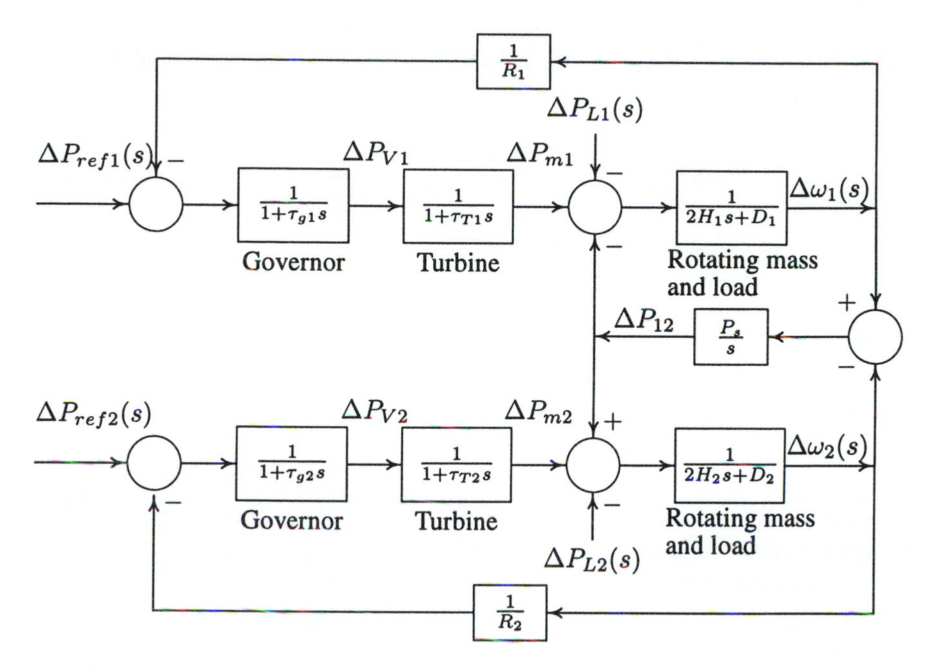

FIGURE 12.21
Two-area system with only primary LFC loop.

The change in mechanical power is determined by the governor speed characteristics, given by

$$\Delta P_{m1} = \frac{-\Delta\omega}{R_1} \tag{12.23}$$
$$\Delta P_{m2} = \frac{-\Delta\omega}{R_2}$$

Substituting from (12.23) into (12.22), and solving for $\Delta\omega$, we have

$$\Delta\omega = \frac{-\Delta P_{L1}}{(\frac{1}{R_1} + D_1) + (\frac{1}{R_2} + D_2)} \tag{12.24}$$
$$= \frac{-\Delta P_{L1}}{B_1 + B2}$$

where

$$B_1 = \frac{1}{R_1} + D_1 \tag{12.25}$$
$$B_2 = \frac{1}{R_2} + D_2$$

B_1 and B_2 are known as the *frequency bias factors*. The change in the tie-line power is

$$\Delta P_{12} = -\frac{(\frac{1}{R_2} + D_2)\Delta P_{L1}}{(\frac{1}{R_1} + D_1)(\frac{1}{R_2} + D_2)} \tag{12.26}$$

$$= \frac{B_2}{B_1 + B_2}(-\Delta P_{L1})$$

Example 12.4

A two-area system connected by a tie line has the following parameters on a 1000-MVA common base

Area	1	2
Speed regulation	$R_1 = 0.05$	$R_2 = 0.0625$
Frequency-sens. load coeff.	$D_1 = 0.6$	$D_2 = 0.9$
Inertia constant	$H_1 = 5$	$H_2 = 4$
Base power	1000 MVA	1000 MVA
Governor time constant	$\tau_{g1} = 0.2$ sec	$\tau_{g2} = 0.3$ sec
Turbine time constant	$\tau_{T1} = 0.5$ sec	$\tau_{T2} = 0.6$ sec

The units are operating in parallel at the nominal frequency of 60 Hz. The synchronizing power coefficient is computed from the initial operating condition and is given to be $P_s = 2.0$ per unit. A load change of 187.5 MW occurs in area 1.
(a) Determine the new steady-state frequency and the change in the tie-line flow.
(b) Construct the *SIMULINK* block diagram and obtain the frequency deviation response for the condition in part (a).

(a) The per unit load change in area 1 is

$$\Delta P_{L1} = \frac{187.5}{1000} = 0.1875 \text{ pu}$$

The per unit steady-state frequency deviation is

$$\Delta\omega_{ss} = \frac{-\Delta P_{L1}}{(\frac{1}{R_1} + D_1) + (\frac{1}{R_2} + D_2)} = \frac{-0.1875}{(20 + 0.6) + (16 + 0.9)} = -0.005 \text{ pu}$$

Thus, the steady-state frequency deviation in Hz is

$$\Delta f = (-0.005)(60) = -0.3 \text{ Hz}$$

and the new frequency is

$$f = f_0 + \Delta f = 60 - 0.3 = 59.7 \text{ Hz}$$

The change in mechanical power in each area is

$$\Delta P_{m1} = -\frac{\Delta\omega}{R_1} = -\frac{-0.005}{0.05} = 0.10 \ \text{pu}$$
$$= 100 \ \text{MW}$$

$$\Delta P_{m2} = -\frac{\Delta\omega}{R_2} = -\frac{-0.005}{0.0625} = 0.080 \ \text{pu}$$
$$= 80 \ \text{MW}$$

Thus, area 1 increases the generation by 100 MW and area 2 by 80 MW at the new operating frequency of 59.7 Hz. The total change in generation is 180 MW, which is 7.5 MW less than the 187.5 MW load change because of the change in the area loads due to frequency drop.

The change in the area 1 load is $\Delta\omega D_1 = (-0.005)(0.6) = -0.003$ per unit (-3.0 MW), and the change in the area 2 load is $\Delta\omega D_1 = (-0.005)(0.9) = -0.0045$ per unit (-4.5 MW). Thus, the change in the total area load is -7.5 MW. The tie-line power flow is

$$\Delta P_{12} = \Delta\omega\left(\frac{1}{R_2} + D2\right) = -0.005(16.9) = 0.0845 \ \text{pu}$$
$$= -84.5 \ \text{MW}$$

That is, 84.5 MW flows from area 2 to area 1. 80 MW comes from the increased generation in area 2, and 4.5 MW comes from the reduction in area 2 load due to frequency drop.

(b) A *SIMULINK* model named **sim12ex4.mdl** is constructed as shown in Figure 12.22. The file is opened and is run in the *SIMULINK* window. The simulation result is shown in Figure 12.23. The simulation diagram returns the vector DP, containing t, P_{m1}, P_{m2}, and P_{12}. A plot of the per unit power response is obtained in *MATLAB* as shown in Figure 12.24.

12.4.3 TIE-LINE BIAS CONTROL

In Example 12.4, where LFCs were equipped with only the primary control loop, a change of power in area 1 was met by the increase in generation in both areas associated with a change in the tie-line power, and a reduction in frequency. In the normal operating state, the power system is operated so that the demands of areas are satisfied at the nominal frequency. A simple control strategy for the normal mode is

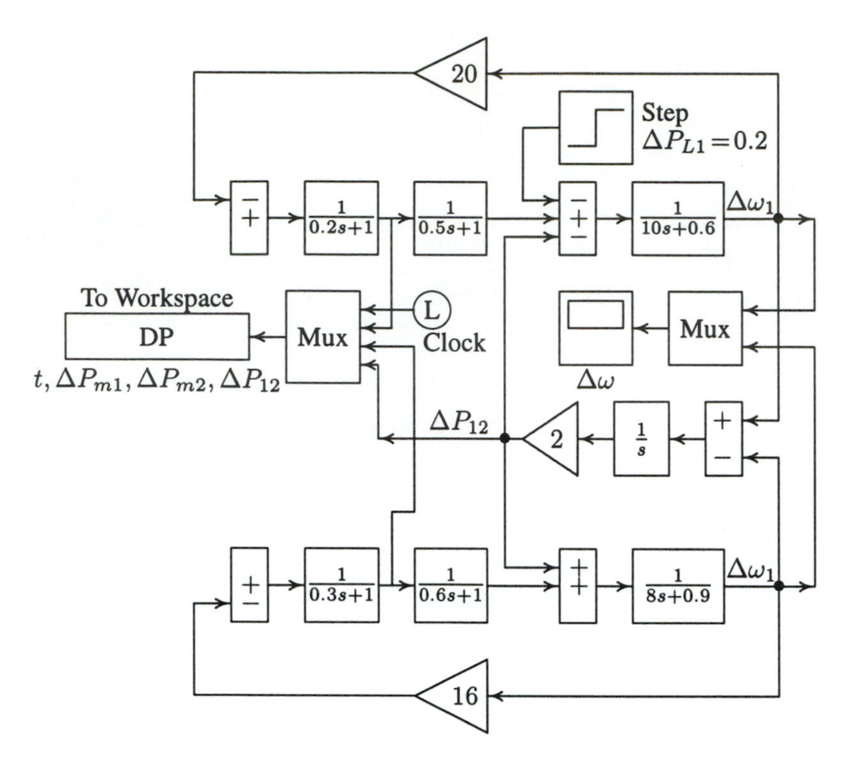

FIGURE 12.22
Simulation block diagram for Example 12.4.

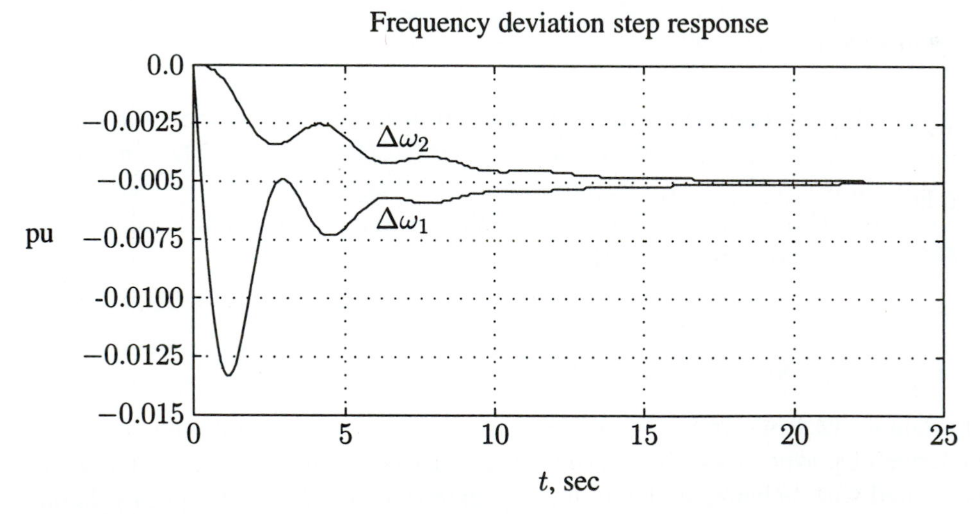

FIGURE 12.23
Frequency deviation step response for Example 12.4.

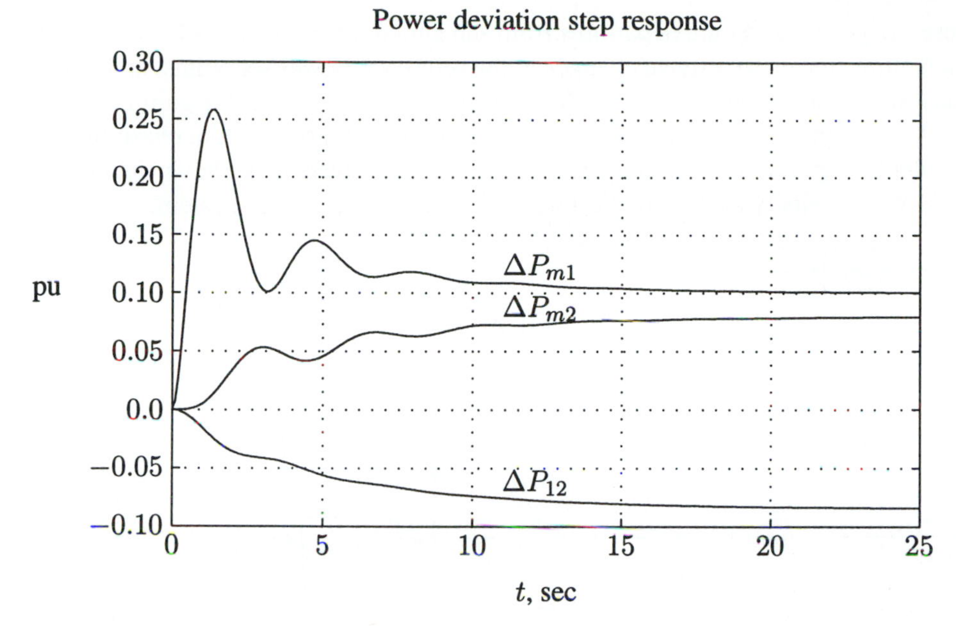

FIGURE 12.24
Power deviation step response for Example 12.4.

- Keep frequency approximately at the nominal value (60 Hz).
- Maintain the tie-line flow at about schedule.
- Each area should absorb its own load changes.

Conventional LFC is based upon tie-line bias control, where each area tends to reduce the area control error (ACE) to zero. The control error for each area consists of a linear combination of frequency and tie-line error.

$$\text{ACE}_i = \sum_{j=1}^{n} \Delta P_{ij} + K_i \, \Delta\omega \tag{12.27}$$

The area bias K_i determines the amount of interaction during a disturbance in the neighboring areas. An overall satisfactory performance is achieved when K_i is selected equal to the frequency bias factor of that area, i.e., $B_i = \frac{1}{R_i} + D_i$. Thus, the ACEs for a two-area system are

$$\begin{aligned} \text{ACE}_1 &= \Delta P_{12} + B_1 \, \Delta\omega_1 \\ \text{ACE}_2 &= \Delta P_{21} + B_2 \, \Delta\omega_2 \end{aligned} \tag{12.28}$$

where ΔP_{12} and ΔP_{21} are departures from scheduled interchanges. ACEs are used as actuating signals to activate changes in the reference power set points, and when steady-state is reached, ΔP_{12} and $\Delta\omega$ will be zero. The integrator gain constant must be chosen small enough so as not to cause the area to go into a chase mode. The block diagram of a simple AGC for a two-area system is shown in Figure 12.25. We can easily extend the tie-line bias control to an n-area system.

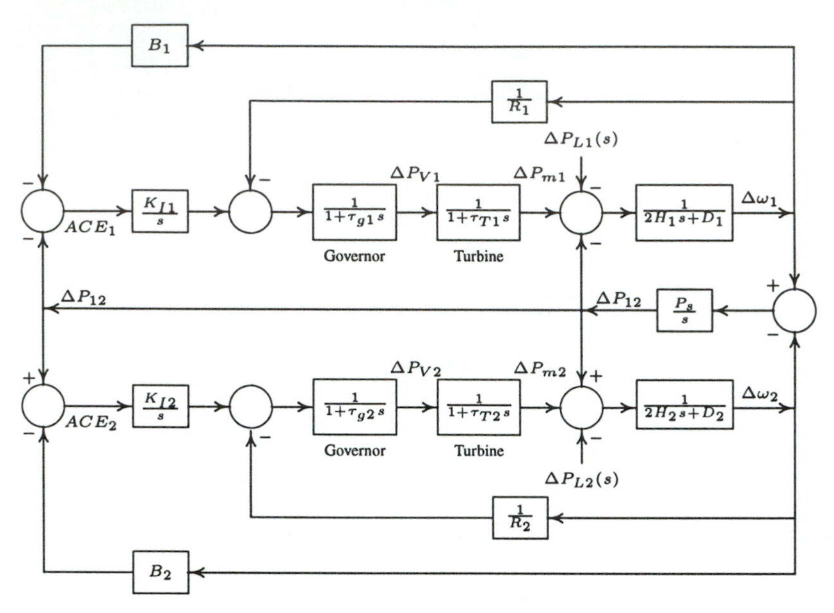

FIGURE 12.25
AGC block diagram for a two-area system.

Example 12.5

Construct the *SIMULINK* model for the two-area system of Example 12.4 with the inclusion of the ACEs, and obtain the frequency and power response for each area.

A *SIMULINK* model named **sim12ex5.mdl** is constructed as shown in Figure 12.26.

The file is opened and is run in the *SIMULINK* window. The integrator gain constants are adjusted for a satisfactory response. The simulation result for $K_{I1} = K_{I2} = 0.3$ is shown in Figure 12.27. The simulation diagram returns the vector ΔP, containing t, ΔP_{m1}, ΔP_{m2}, and ΔP_{12}. A plot of the per unit power response is obtained in *MATLAB* as shown in Figure 12.28. As we can see from Figure 12.27, the frequency deviation returns to zero with a settling time of approximately 20 seconds. Also, the tie-line power change reduces to zero, and the increase in area 1 load is met by the increase in generation ΔP_{m1}.

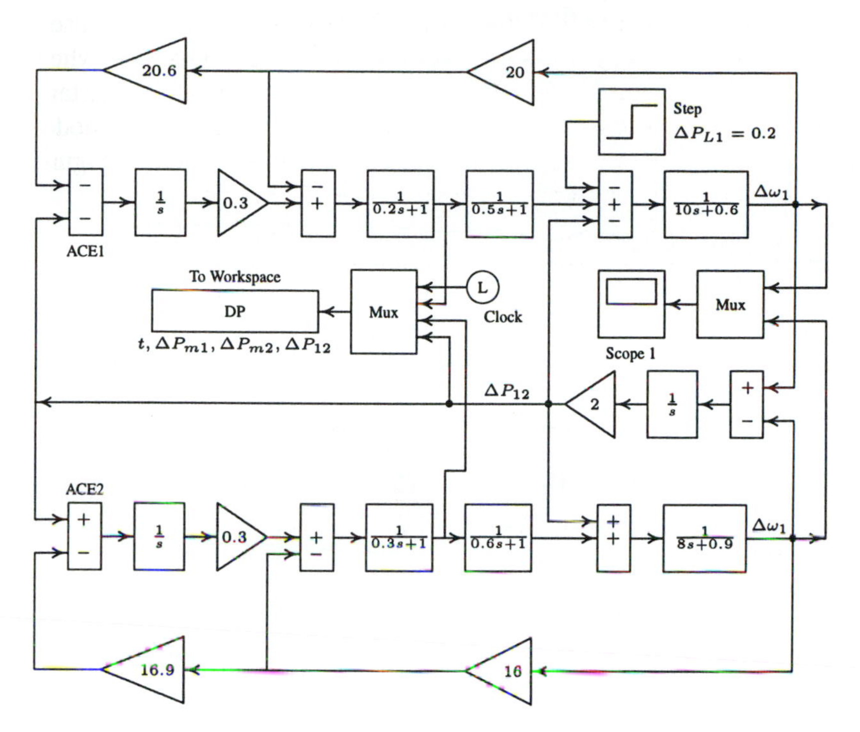

FIGURE 12.26
Simulation block diagram for Example 12.5.

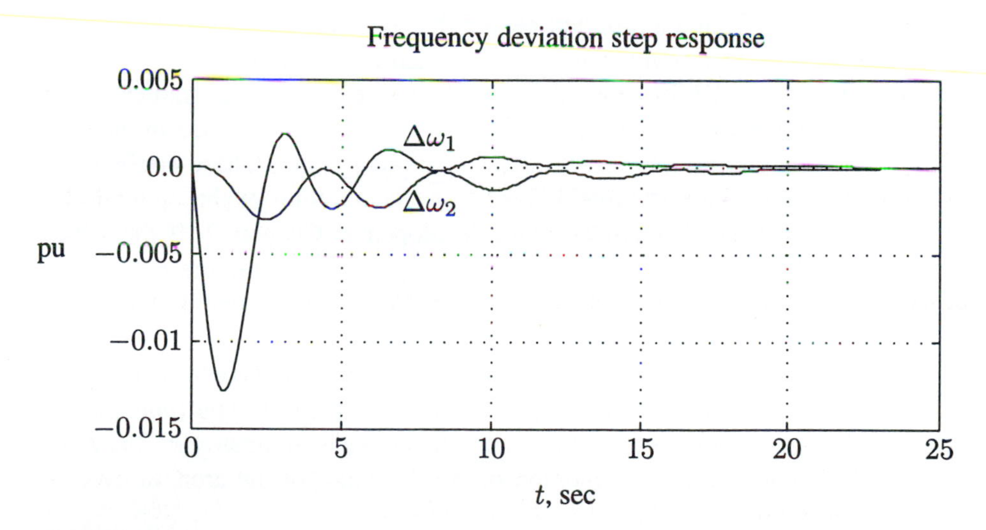

FIGURE 12.27
Frequency deviation step response for Example 12.5.

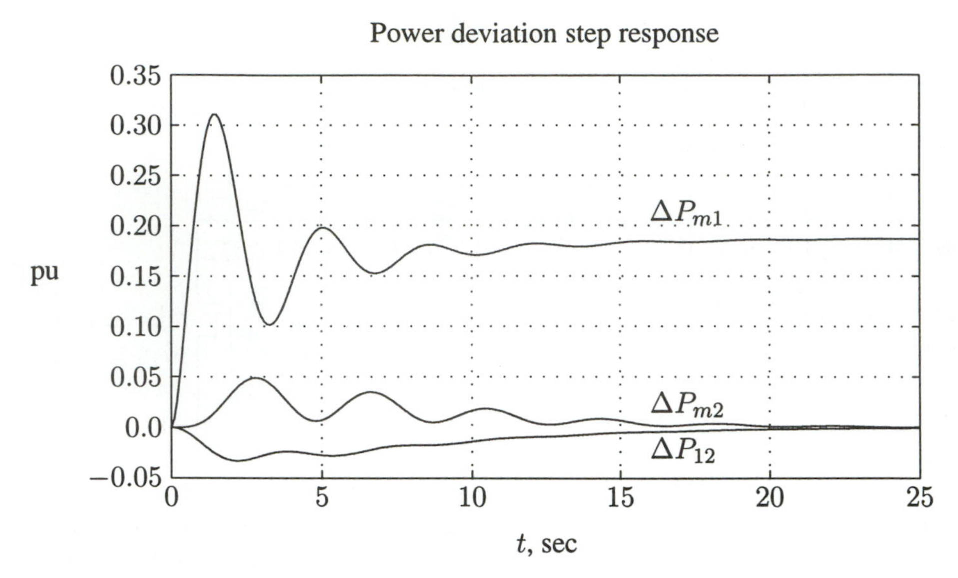

FIGURE 12.28
Power deviation step response for Example 12.5.

12.5 AGC WITH OPTIMAL DISPATCH OF GENERATION

The factors influencing power generation at minimum cost are operating efficiencies, fuel cost, and transmission losses. The optimal dispatch of generation was discussed in Chapter 7, and a program named **dispatch** was developed to find the optimal dispatch of generation for an interconnected power system.

The optimal dispatch of generation may be treated within the framework of LFC. In direct digital control systems, the digital computer is included in the control loop which scans the unit generation and tie-line flows. These settings are compared with the optimal settings derived from the solution of the optimal dispatch program, such as **dispatch** program developed in Chapter 7. If the actual settings are off from the optimal values, the computer generates the raise/lower pulses which are sent to the individual units. The allocation program will also take into account the tie-line power contracts between the areas.

With the development of modern control theory, several concepts are included in the AGC which go beyond the simple tie-line bias control. The fundamental approach is the use of more extended mathematical models. In retrospect, the AGC can be used to include the representation of the dynamics of the area, or even of the complete system.

Other concepts of the modern control theory are being employed, such as state estimation and optimal control with linear regulator utilizing constant feedback gains. In addition to the structures which aim at the control of deterministic signals and disturbances, there are schemes which employ stochastic control concepts, e.g., minimization of some expected value of an integral quadratic error criterion. Usually, this results in the design of the Kalman filter, which is of value for the control of small random disturbances.

12.6 REACTIVE POWER AND VOLTAGE CONTROL

The generator excitation system maintains generator voltage and controls the reactive power flow. The generator excitation of older systems may be provided through slip rings and brushes by means of dc generators mounted on the same shaft as the rotor of the synchronous machine. However, modern excitation systems usually use ac generators with rotating rectifiers, and are known as *brushless excitation*.

As we have seen, a change in the real power demand affects essentially the frequency, whereas a change in the reactive power affects mainly the voltage magnitude. The interaction between voltage and frequency controls is generally weak enough to justify their analysis separately.

The sources of reactive power are generators, capacitors, and reactors. The generator reactive powers are controlled by field excitation. Other supplementary methods of improving the voltage profile on electric transmission systems are transformer load-tap changers, switched capacitors, step-voltage regulators, and static var control equipment. The primary means of generator reactive power control is the generator excitation control using *automatic voltage regulator* (AVR), which is discussed in this chapter. The role of an (AVR) is to hold the terminal voltage magnitude of a synchronous generator at a specified level. The schematic diagram of a simplified AVR is shown in Figure 12.29.

An increase in the reactive power load of the generator is accompanied by a drop in the terminal voltage magnitude. The voltage magnitude is sensed through a potential transformer on one phase. This voltage is rectified and compared to a dc set point signal. The amplified error signal controls the exciter field and increases the exciter terminal voltage. Thus, the generator field current is increased, which results in an increase in the generated emf. The reactive power generation is increased to a new equilibrium, raising the terminal voltage to the desired value. We will look briefly at the simplified models of the component involved in the AVR system.

12.6.1 AMPLIFIER MODEL

The excitation system amplifier may be a magnetic amplifier, rotating amplifier, or modern electronic amplifier. The amplifier is represented by a gain K_A and a time

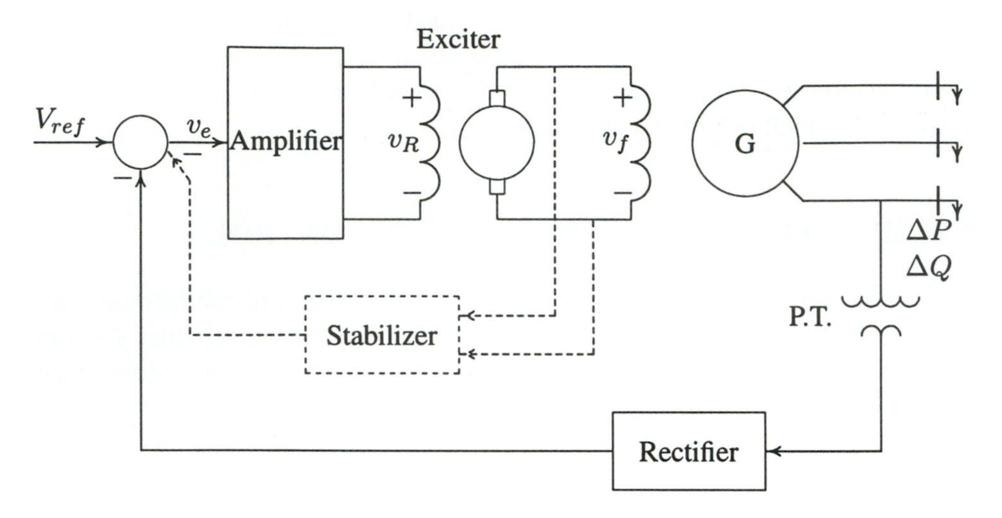

FIGURE 12.29
A typical arrangement of a simple AVR.

constant τ_A, and the transfer function is

$$\frac{V_R(s)}{V_e(s)} = \frac{K_A}{1 + \tau_A s} \tag{12.29}$$

Typical values of K_A are in the range of 10 to 400. The amplifier time constant is very small, in the range of 0.02 to 0.1 second, and often is neglected.

12.6.2 EXCITER MODEL

There is a variety of different excitation types. However, modern excitation systems uses ac power source through solid-state rectifiers such as SCR. The output voltage of the exciter is a nonlinear function of the field voltage because of the saturation effects in the magnetic circuit. Thus, there is no simple relationship between the terminal voltage and the field voltage of the exciter. Many models with various degrees of sophistication have been developed and are available in the IEEE recommendation publications. A reasonable model of a modern exciter is a linearized model, which takes into account the major time constant and ignores the saturation or other nonlinearities. In the simplest form, the transfer function of a modern exciter may be represented by a single time constant τ_E and a gain K_E, i.e.,

$$\frac{V_F(s)}{V_R(s)} = \frac{K_E}{1 + \tau_E s} \tag{12.30}$$

The time constant of modern exciters are very small.

12.6.3 GENERATOR MODEL

The synchronous machine generated emf is a function of the machine magnetization curve, and its terminal voltage is dependent on the generator load. In the linearized model, the transfer function relating the generator terminal voltage to its field voltage can be represented by a gain K_G and a time constant τ_G, and the transfer function is

$$\frac{V_t(s)}{V_F(s)} = \frac{K_G}{1 + \tau_G s} \tag{12.31}$$

These constants are load dependent, K_G may vary between 0.7 to 1, and τ_G between 1.0 and 2.0 seconds from full-load to no-load.

12.6.4 SENSOR MODEL

The voltage is sensed through a potential transformer and, in one form, it is rectified through a bridge rectifier. The sensor is modeled by a simple first order transfer function, given by

$$\frac{V_S(s)}{V_t(s)} = \frac{K_R}{1 + \tau_R s} \tag{12.32}$$

τ_R is very small, and we may assume a range of 0.01 to 0.06 second. Utilizing the above models results in the AVR block diagram shown in Figure 12.30.

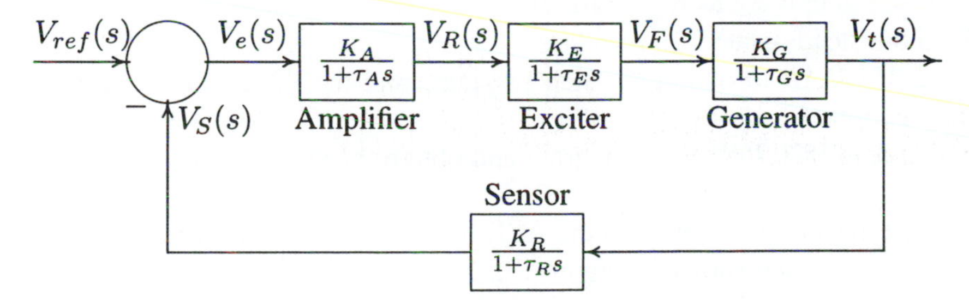

FIGURE 12.30
A simplified automatic voltage regulator block diagram.

The open-loop transfer function of the block diagram in Figure 12.30 is

$$KG(s)H(s) = \frac{K_A K_E K_G K_R}{(1 + \tau_A s)(1 + \tau_E s)(1 + \tau_G s)(1 + \tau_R s)} \tag{12.33}$$

and the closed-loop transfer function relating the generator terminal voltage $V_t(s)$ to the reference voltage $V_{ref}(s)$ is

$$\frac{V_t(s)}{V_{ref}(s)} = \frac{K_A K_E K_G K_R (1 + \tau_R s)}{(1 + \tau_A s)(1 + \tau_E s)(1 + \tau_G s)(1 + \tau_R s) + K_A K_E K_G K_R} \tag{12.34}$$

or

$$V_t(s) = T(s)V_{ref}(s) \tag{12.35}$$

For a step input $V_{ref}(s) = \frac{1}{s}$, using the final value theorem, the steady-state response is

$$V_{t_{ss}} = \lim_{s \to 0} sV_t(s) = \frac{K_A}{1 + K_A} \tag{12.36}$$

Example 12.6

The AVR system of a generator has the following parameters

	Gain	Time constant
Amplifier	K_A	$\tau_A = 0.1$
Exciter	$K_E = 1$	$\tau_E = 0.4$
Generator	$K_G = 1$	$\tau_G = 1.0$
Sensor	$K_R = 1$	$\tau_R = 0.05$

(a) Use the Routh-Hurwitz array (Appendix B.2.1) to find the range of K_A for control system stability.
(b) Use *MATLAB* **rlocus** function to obtain the root locus plot.
(c) The amplifier gain is set to $K_A = 10$
  (i) Find the steady-state step response.
  (ii) Use *MATLAB* to obtain the step response and the time-domain performance specifications.
(d) Construct the *SIMULINK* block diagram and obtain the step response.

Substituting the system parameters in the AVR block diagram of Figure 12.30 results in the block diagram shown in Figure 12.31.

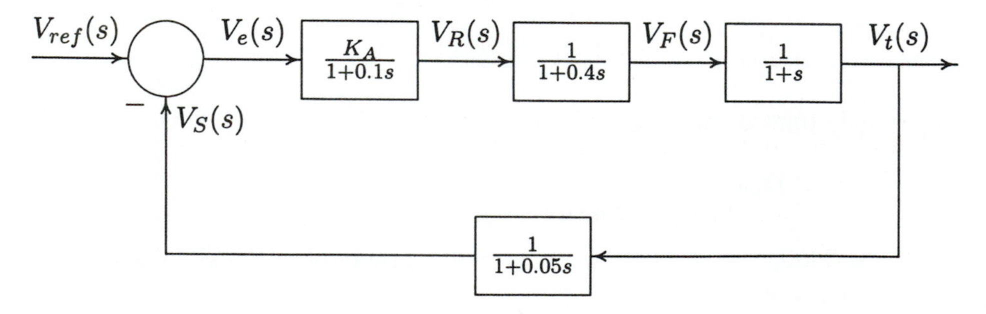

FIGURE 12.31
AVR block diagram for Example 12.6.

The open-loop transfer function of the AVR system shown in Figure 12.31 is

$$\begin{aligned}KG(s)H(s) &= \frac{K_A}{(1+0.1s)(1+0.4s)(1+s)(1+0.05s)}\\ &= \frac{500K_A}{(s+10)(s+2.5)(s+1)(s+20)}\\ &= \frac{500K_A}{s^4+33.5s^3+307.5s^2+775s+500}\end{aligned}$$

(a) The characteristic equation is given by

$$1+KG(s)H(s) = 1+\frac{500K_A}{s^4+33.5s^3+307.5s^2+775ss+500} = 0$$

which results in the characteristic polynomial equation

$$s^4+33.5s^3+307.5s^2+775s+500+500K_A = 0$$

The Routh-Hurwitz array for this polynomial is then (see Appendix B.2.1)

$$\begin{array}{c|lll} s^4 & 1 & 307.5 & 500+500K_A \\ s^3 & 33.5 & 775 & 0 \\ s^2 & 284.365 & 500+500K_A & 0 \\ s^1 & 58.9K_A-716.1 & 0 & 0 \\ s^0 & 500+500K_A & & \end{array}$$

From the s^1 row we see that, for control system stability, K_A must be less than 12.16, also from the s^0 row, K_A must be greater than -1. Thus, with positive values of K_A, for control system stability, the amplifier gain must be

$$K_A < 12.16$$

For $K = 12.16$, the auxiliary equation from the s^2 row is

$$284.365s^2 + 6580 = 0$$

or $s = \pm j4.81$. That is, for $K = 12.16$, we have a pair of conjugate poles on the $j\omega$ axis, and the control system is marginally stable.

(b) To obtain the root-locus plot for the range of K from 0 to 12.16, we use the following commands.

```
num=500;
den=[1  33.5  307.5  775  500];
figure(1), rlocus(num, den);
```

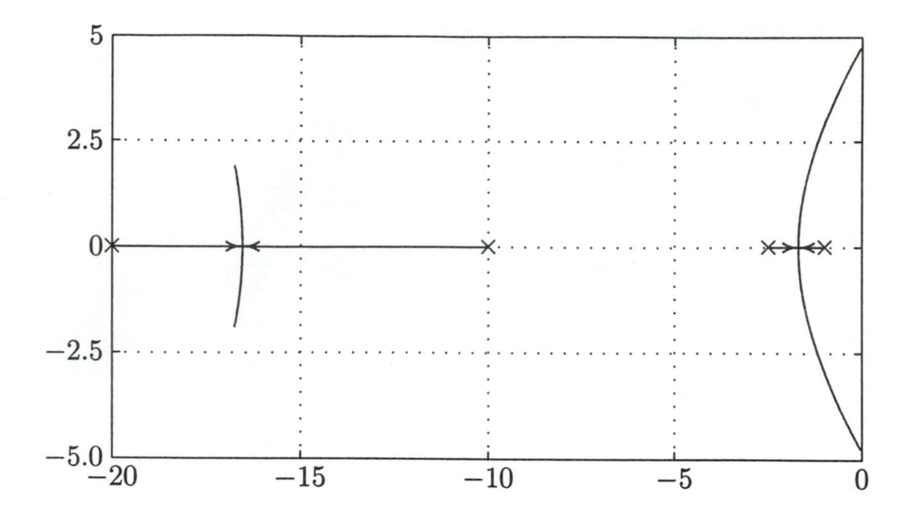

FIGURE 12.32
Root-locus plot for Example 12.6.

The result is shown in Figure 12.32. The loci intersect the $j\omega$ axis at $s = \pm j4.81$ for $K_A = 12.16$. Thus, the system is marginally stable for $K_A = 12.16$.
(c) The closed-loop transfer function of the system shown in Figure 12.31 is

$$\frac{V_t(s)}{V_{ref}(s)} = \frac{25K_A(s+20)}{s^4 + 33.5s^3 + 307.5s^2 + 775s + 500 + 500K_A}$$

(i) The steady-state response is

$$V_{t_{ss}} = \lim_{s \to 0} sV_t(s) = \frac{K_A}{1 + K_A}$$

For the amplifier gain of $K_A = 10$, the steady-state response is

$$V_{t_{ss}} = \frac{10}{1+10} = 0.909$$

and the steady-state error is

$$V_{e_{ss}} = 1.0 - 0.909 = 0.091$$

In order to reduce the steady-state error, the amplifier gain must be increased, which results in an unstable control system.
(ii) To obtain the step response and the time-domain performance specifications, we use the following commands

```
KA= 10;
numc=KA*[25  500];
denc=[1  33.5  307.5  775  500+ 500*KA];
t=0:.05:20;
c=step(numc, denc, t);
figure(2), plot(t, c), grid
timespec(numc, denc)
```

The time-domain performance specifications are

```
Peak time = 0.791        Percent overshoot = 82.46
Rise time = 0.247
Settling time = 19.04
```

The terminal voltage step response is shown in Figure 12.33.

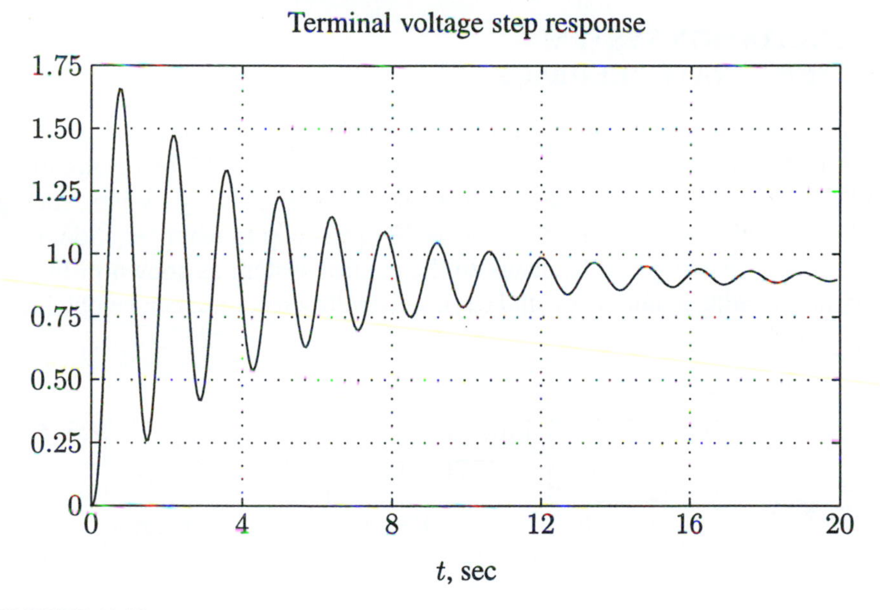

FIGURE 12.33
Terminal voltage step response for Example 12.6.

(d) A *SIMULINK* model named **sim12ex6.mdl** is constructed as shown in Figure 12.34. The file is opened and is run in the *SIMULINK* window. The simulation results in the same response as shown in Figure 12.33. From the results, we see that for an amplifier gain $K_A = 10$, the response is highly oscillatory, with a very large overshoot and a long settling time. Furthermore, the steady-state error is over 9 percent. We cannot have a small steady-state error and a satisfactory transient response at the same time.

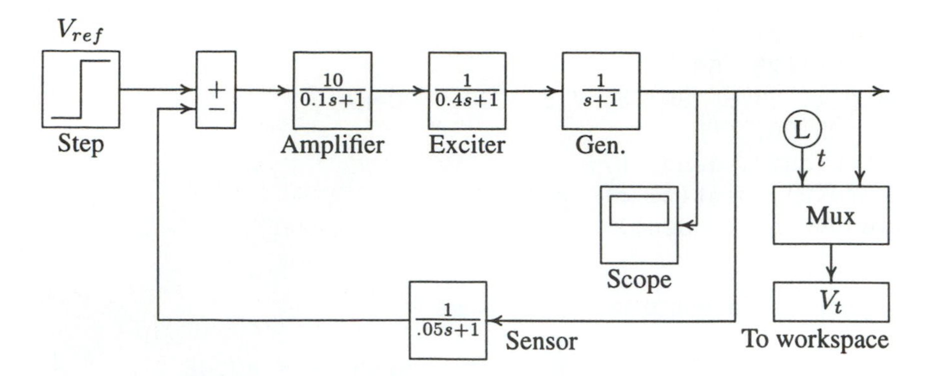

FIGURE 12.34
Simulation block diagram for Example 12.6.

12.6.5 EXCITATION SYSTEM STABILIZER — RATE FEEDBACK

As we have seen in Example 12.6, even for an small amplifier gain of $K_A = 10$, AVR step response is not satisfactory, and a value exceeding 12.16 results in an unbounded response. Thus, we must increase the relative stability by introducing a controller, which would add a zero to the AVR open-loop transfer function. One way to do this is to add a rate feedback to the control system as shown in Figure 12.35. By proper adjustment of K_F and τ_F, a satisfactory response can be obtained.

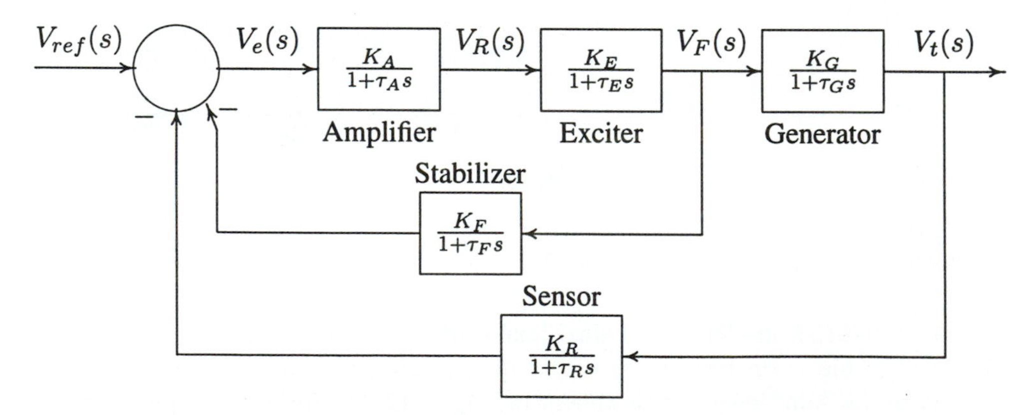

FIGURE 12.35
Block diagram of the compensated AVR system.

Example 12.7

A rate feedback stabilizer is added to the AVR system of Example 12.6. The stabilizer time constant is $\tau_F = 0.04$ second, and the derivative gain is adjusted to $K_F = 2$.
(a) Obtain the step response and the time-domain performance specifications.
(b) Construct the *SIMULINK* model and obtain the step response.

(a) Substituting for the parameters in the block diagram of Figure 12.35 and applying the Mason's gain formula, we obtain the closed-loop transfer function

$$\frac{V_t(s)}{V_{ref}(s)} = \frac{250(s^2 + 45s + 500)}{s^5 + 58.5s^4 + 13,645s^3 + 270,962.5s^2 + 274,875s + 137,500}$$

(i) The steady-state response is

$$V_{t_{ss}} = \lim_{s\to 0} sV_t(s) = \frac{(250)(500)}{137,500} = 0.909$$

To find the step response, we use the following commands

```
numc=250*[1 45 500];
denc=[1  58.5  13645  270962.5  274875 137500];
t=0:.05:10;
c=step(numc, denc, t); plot(t, c), grid
timespec(numc, denc)
```

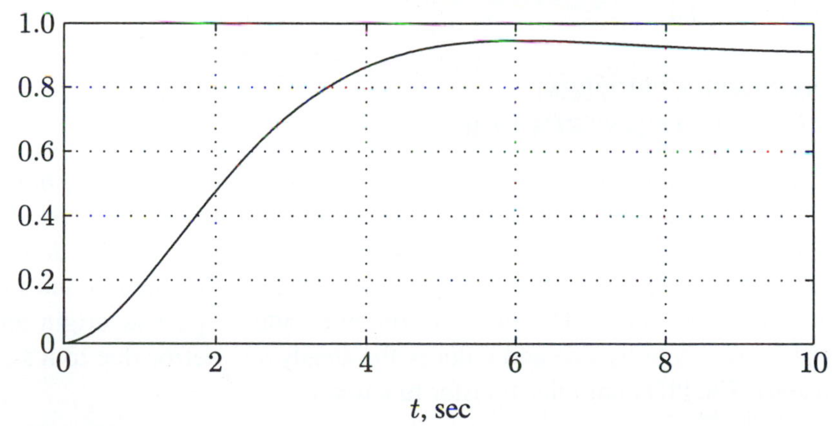

FIGURE 12.36
Terminal voltage step response for Example 12.7.

The step response is shown in Figure 12.36. The time-domain performance specifications are

```
Peak time = 6.08          Percent overshoot = 4.13
Rise time = 2.95
Settling time = 8.08
```

(b) A *SIMULINK* model named **sim12ex7.mdl** is constructed as shown in Figure 12.37.

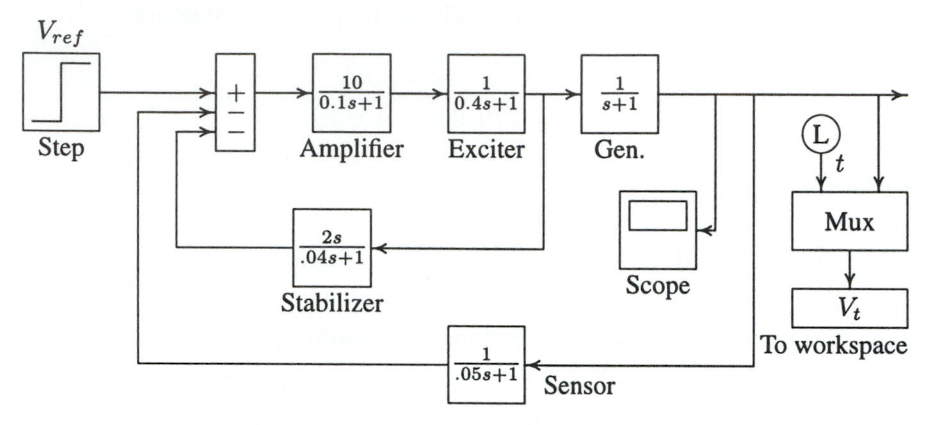

FIGURE 12.37
Simulation block diagram for Example 12.7.

The file is opened and is run in the *SIMULINK* window. The simulation results in the same response as shown in Figure 12.36. The results show a very satisfactory transient response with an overshoot of 4.13 percent and a settling time of approximately 8 seconds.

12.6.6 EXCITATION SYSTEM STABILIZER — PID CONTROLLER

One of the most common controllers available commercially is the *proportional integral derivative* (PID) controller. The PID controller is used to improve the dynamic response as well as to reduce or eliminate the steady-state error. The derivative controller adds a finite zero to the open-loop plant transfer function and improves the transient response. The integral controller adds a pole at origin and increases the system type by one and reduces the steady-state error due to a step function to zero. The PID controller transfer function is

$$G_C(s) = K_P + \frac{K_I}{s} + K_D s \tag{12.37}$$

The block diagram of an AVR compensated with a PID controller is shown in Figure 12.38.

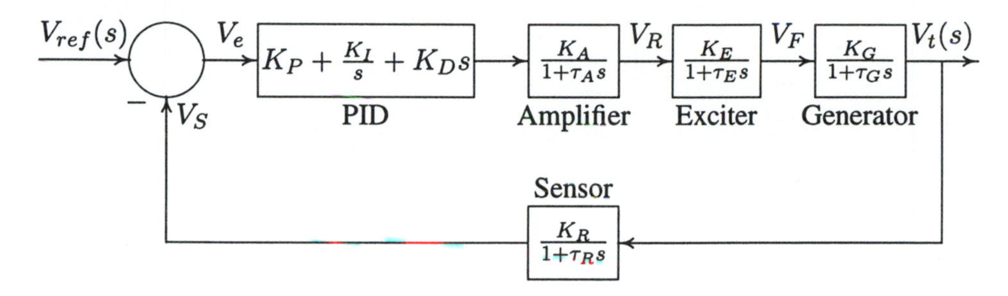

FIGURE 12.38
AVR system with PID controller.

Example 12.8

A PID controller is added in the forward path of the AVR system of Example 12.6 as shown in Figure 12.38. Construct the *SIMULINK* model. Set the proportional gain K_P to 1.0 and adjust K_I and K_D until a step response with a minimum overshoot and a very small settling time is obtained.

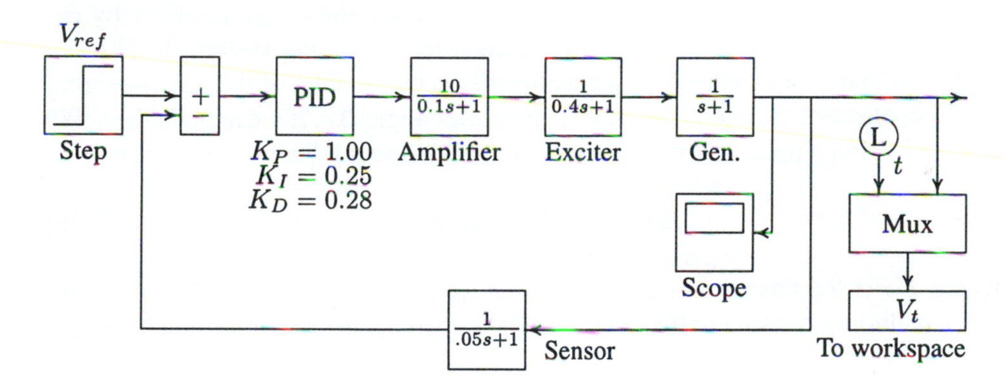

FIGURE 12.39
Simulation block diagram for Example 12.8.

A *SIMULINK* model named **sim12ex8.mdl** is constructed as shown in Figure 12.39. The file is opened and is run in the *SIMULINK* window. An integral gain of $K_I = 0.25$ and a derivative gain of $K_D = 0.28$ is found to be satisfactory. The response settles in about 1.4 seconds with a negligibly small overshoot. Note that the PID controller reduces the steady-state error to zero. The simulation result for the above settings is shown in Figure 12.40.

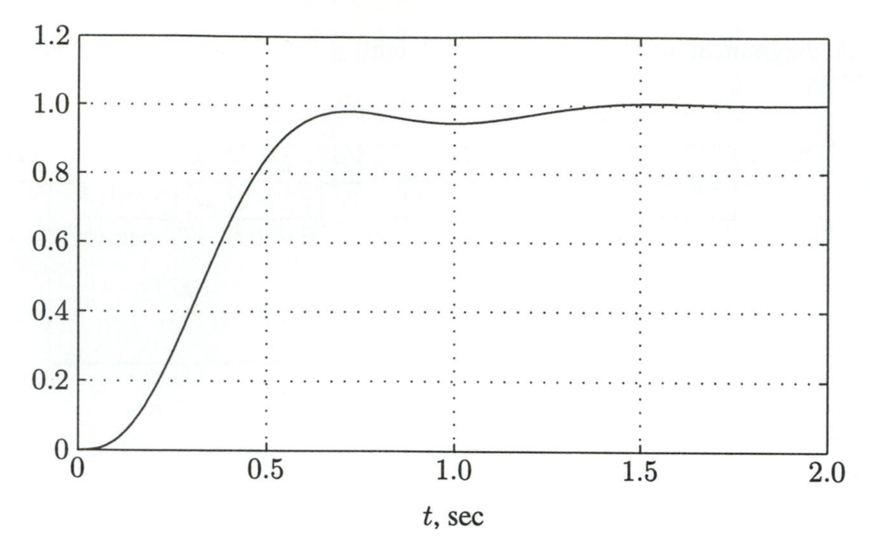

FIGURE 12.40
Terminal voltage step response for Example 12.8.

12.7 AGC INCLUDING EXCITATION SYSTEM

Since there is a weak coupling between the LFC and AVR systems, the frequency and voltage were controlled separately. We can study the coupling effect by extending the linearized AGC system to include the excitation system. In (12.17), we found that a small change in the real power is the product of the synchronizing power coefficient P_S and the change in the power angle $\Delta\delta$. If we include the small effect of voltage upon real power, we obtain the following linearized equation

$$\Delta P_e = P_s\,\Delta\delta + K_2 E' \tag{12.38}$$

where K_2 is the change in electrical power for a small change in the stator emf. Also, including the small effect of rotor angle upon the generator terminal voltage, we may write

$$\Delta V_t = K_5\,\Delta\delta + K_6 E' \tag{12.39}$$

where K_5 is the change in the terminal voltage for a small change in rotor angle at constant stator emf, and K_6 is the change in terminal voltage for a small change in the stator emf at constant rotor angle. Finally, modifying the generator field transfer function to include the effect of rotor angle, we may express the stator emf as

$$E' = \frac{K_G}{1 + \tau_G}(V_f - K_4\,\Delta\delta) \tag{12.40}$$

The above constants depend upon the network parameters and the operating conditions. For the detailed derivation, see references 2 and 52. For a stable system, P_S is positive. Also, K_2, K_4, and K_6 are positive, but K_5 may be negative. Including (12.38)–(12.40) in the AGC system of Figure 12.16 and the AVR system of Figure 12.38, a linearized model for the combined LFC and AVR systems is obtained. A combined simulation block diagram is constructed in Example 12.9.

Example 12.9

An isolated power station has the following parameters

	Gain	Time constant
Turbine	$K_T = 1$	$\tau_T = 0.5$
Governor	$K_g = 1$	$\tau_g = 0.2$
Amplifier	$K_A = 10$	$\tau_A = 0.1$
Exciter	$K_E = 1$	$\tau_E = 0.4$
Generator	$K_G = 0.8$	$\tau_G = 1.4$
Sensor	$K_R = 1$	$\tau_R = 0.05$
Inertia	$H = 5$	
Regulation	$R = 0.05$	

The load varies by 0.8 percent for a 1 percent change in frequency, i.e., $D = 0.8$. Assume the synchronizing coefficient P_S is 1.5, and the voltage coefficient K_6 is 0.5. Also, the coupling constants are $K_2 = 0.2$, $K_4 = 1.4$, and $K_5 = -0.1$. Construct the combined *SIMULINK* block diagram and obtain the frequency deviation and terminal voltage responses for a load change of $\Delta P_{L1} = 0.2$ per unit.

A *SIMULINK* model named **sim12ex9.mdl** is constructed as shown in Figure 12.41. The file is opened and is run in the *SIMULINK* window. The integrator gain in the secondary LFC loop is set to a value of 6.0. The excitation PID controller is tuned for $K_P = 1$, $K_I = 0.25$, and $K_D = 0.3$. The speed deviation step response and the terminal voltage step response are shown in Figures 12.42 and 12.43. It is observed that when the coupling coefficients are set to zero, there is little change in the transient response. Thus, separate treatment of frequency and voltage control loops is justified.

12.8 INTRODUCTORY MODERN CONTROL APPLICATION

The classical design techniques used so far are based on the root-locus that utilize only the plant output for feedback with a dynamic controller. In this section, we employ modern control designs that require the use of all state variables to form a linear static controller.

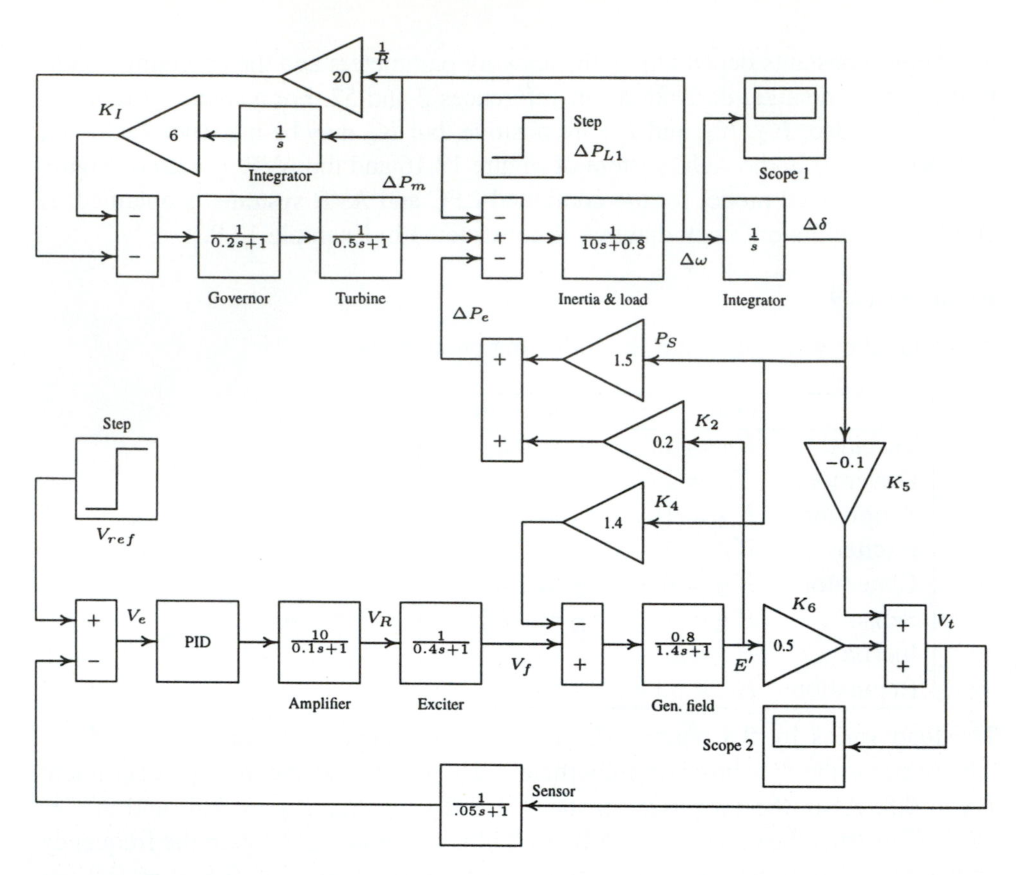

FIGURE 12.41
Simulation block diagram for Example 12.9.

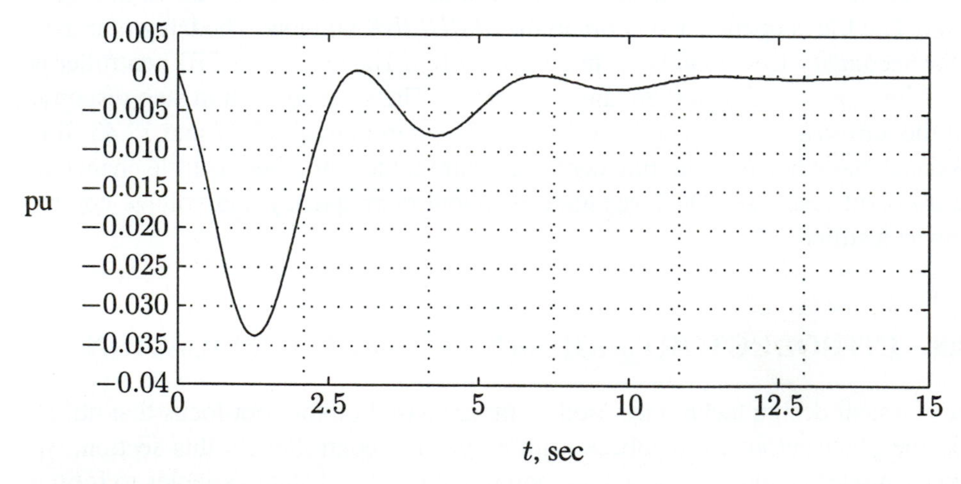

FIGURE 12.42
Frequency deviation step response for Example 12.8.

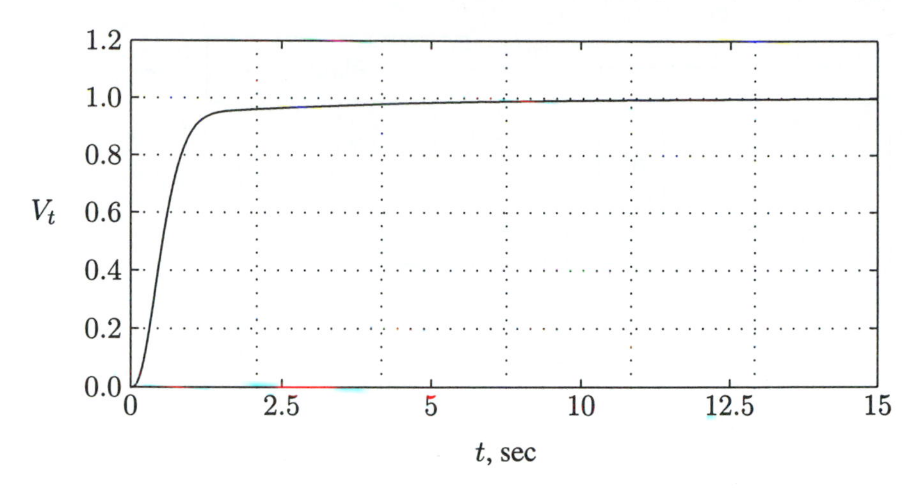

FIGURE 12.43
Terminal voltage step response for Example 12.8.

Modern control design is especially useful in multivariable systems. One approach in modern control systems accomplished by the use of state feedback is known as *pole-placement design.* The pole-placement design allows all roots of the system characteristic equation to be placed in desired locations. This results in a regulator with constant gain vector **K**.

The state-variable feedback concept requires that all states be accessible in a physical system. For systems in which all states are not available for feedback, a state estimator (observer) may be designed to implement the pole-placement design. The other approach to the design of regulator systems is the optimal control problem, where a specified mathematical performance criterion is minimized.

12.8.1 POLE-PLACEMENT DESIGN

The control is achieved by feeding back the state variables through a regulator with constant gains. Consider the control system presented in the state-variable form

$$\begin{aligned}\dot{\mathbf{x}}(t) &= \mathbf{A}\mathbf{x}(t) + \mathbf{B}u(t) \\ y(t) &= \mathbf{C}\mathbf{x}(\mathbf{t})\end{aligned} \tag{12.41}$$

Now consider the block diagram of the system shown in Figure 12.44 with the following state feedback control

$$u(t) = -\mathbf{K}\mathbf{x}(t) \tag{12.42}$$

where **K** is a $1 \times n$ vector of constant feedback gains. The control system input $r(t)$ is assumed to be zero. The purpose of this system is to return all state variables to values of zero when the states have been perturbed.

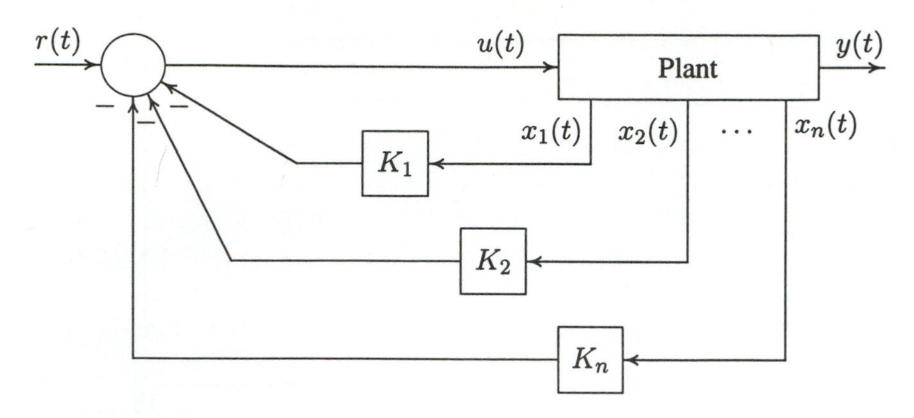

FIGURE 12.44
Control system design via pole placement.

Substituting (12.42) into (12.41), the compensated system state-variable representation becomes

$$\dot{\mathbf{x}}(t) = (\mathbf{A} - \mathbf{BK})\mathbf{x}(t) = \mathbf{A_f}\mathbf{x}(t) \tag{12.43}$$

The compensated system characteristic equation is

$$|s\mathbf{I} - \mathbf{A} + \mathbf{BK}| = 0 \tag{12.44}$$

Assume the system is represented in the phase variable canonical form as follows.

$$\begin{bmatrix} \dot{x}_1 \\ \dot{x}_2 \\ \vdots \\ \dot{x}_{n-1} \\ \dot{x}_n \end{bmatrix} = \begin{bmatrix} 0 & 1 & 0 & \dots & 0 \\ 0 & 0 & 1 & \dots & 0 \\ \vdots & & & & \\ 0 & 0 & 0 & \dots & 1 \\ -a_0 & -a_1 & -a_2 & \dots & -a_{n-1} \end{bmatrix} \begin{bmatrix} x_1 \\ x_2 \\ \vdots \\ x_{n-1} \\ x_n \end{bmatrix} + \begin{bmatrix} 0 \\ 0 \\ \vdots \\ 0 \\ 1 \end{bmatrix} u(t) \tag{12.45}$$

Substituting for $\mathbf{A}$ and $\mathbf{B}$ into (12.44), the compensated characteristic equation for the control system is found.

$$|s\mathbf{I}-\mathbf{A}+\mathbf{BK}| = s^n + (a_{n-1}+k_n)s^{n-1} + \cdots + (a_1+k_2)s + (a_0+k_1) = 0 \tag{12.46}$$

For the specified closed-loop pole locations $-\lambda_1, \ldots, -\lambda_n$, the desired characteristic equation is

$$\alpha_c(s) = (s+\lambda_1)\cdots(s+\lambda_n) = s^n + \alpha_{n-1}s^{n-1} + \cdots + \alpha_1 s + \alpha_0 = 0 \tag{12.47}$$

The design objective is to find the gain matrix $\mathbf{K}$ such that the characteristic equation for the controlled system is identical to the desired characteristic equation.

Thus, the gain vector $\mathbf{K}$ is obtained by equating coefficients of equations (12.46) and (12.47), and for the ith coefficient we get

$$k_i = \alpha_{i-1} - a_{i-1} \tag{12.48}$$

If the state model is not in the phase-variable canonical form, we can use the transformation technique to transform the given state model to the phase-variable canonical form. The gain factor is obtained for this model and then transformed back to conform with the original model. This procedure results in the following formula, known as *Ackermann's formula.*

$$\mathbf{K} = \begin{bmatrix} 0 & 0 & \cdots & 0 & 1 \end{bmatrix} \mathbf{S}^{-1} \alpha_c(\mathbf{A}) \tag{12.49}$$

where the matrix $\mathbf{S}$ is given by

$$\mathbf{S} = \begin{bmatrix} B & AB & A^2B & \ldots & A^{n-1}B \end{bmatrix} \tag{12.50}$$

and the notation $\alpha_c(\mathbf{A})$ is given by

$$\alpha_c(\mathbf{A}) = \mathbf{A}^n + \alpha_{n-1}\mathbf{A}^{n-1} + \cdots + \alpha_1\mathbf{A} + \alpha_0\mathbf{I} \tag{12.51}$$

The function $[\mathbf{K}, \mathbf{A_f}] = \mathbf{placepol}(\mathbf{A}, \mathbf{B}, \mathbf{C}, \mathbf{p})$ is developed for the pole-placement design. $\mathbf{A}, \mathbf{B}, \mathbf{C}$ are system matrices and $\mathbf{p}$ is a row vector containing the desired closed-loop poles. This function returns the gain vector $\mathbf{K}$ and the closed-loop system matrix $\mathbf{A_f}$. Also, the *MATLAB Control System Toolbox* contains two functions for pole-placement design. Function $\mathbf{K} = \mathbf{acker}(\mathbf{A}, \mathbf{B}, \mathbf{p})$ is for single input systems, and function $\mathbf{K} = \mathbf{place}(\mathbf{A}, \mathbf{B}, \mathbf{p})$, which uses a more reliable algorithm, is for multiinput systems. The condition that must exist to place the closed-loop poles at the desired location is to be able to transform the given state model into phase-variable canonical form.

We demonstrate the use of pole-placement design by applying it to the LFC of an isolated power system considered before, which is represented again in Figure 12.45. The s-domain equations describing the block diagram shown in Figure 12.45 are

$$\begin{aligned} (1 + \tau_g s)\Delta P_V(s) &= \Delta P_{ref} - \frac{1}{R}\Delta\Omega(s) \\ (1 + \tau_T s)\Delta P_m(s) &= \Delta P_V \\ (2Hs + D)\Delta\Omega(s) &= \Delta P_m - \Delta P_L \end{aligned} \tag{12.52}$$

Solving for the first derivative term, we have

$$\begin{aligned} s\Delta P_V(s) &= -\frac{1}{\tau_g}\Delta P_V - \frac{1}{R\tau_g}\Delta\Omega(s) + \frac{1}{\tau_g}\Delta P_{ref}(s) \\ s\Delta P_m(s) &= \frac{1}{\tau_T}\Delta P_V - \frac{1}{\tau_T}\Delta P_m \\ s\Delta\Omega(s) &= \frac{1}{2H}\Delta P_m - \frac{D}{2H}\Delta\Omega(s) - \frac{1}{2H}\Delta P_L \end{aligned} \tag{12.53}$$

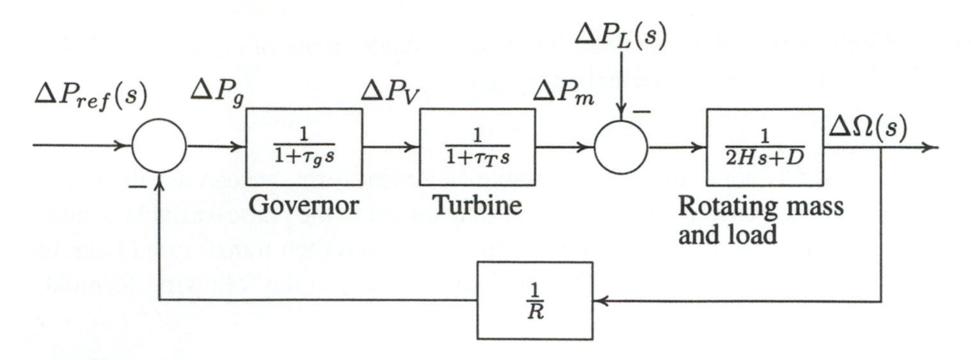

FIGURE 12.45
Load frequency control block diagram of an isolated power system.

Transforming into time-domain and expressing in matrix form the state equation becomes

$$\begin{bmatrix} \Delta\dot{P}_V \\ \Delta\dot{P}_m \\ \Delta\dot{\omega} \end{bmatrix} = \begin{bmatrix} \frac{-1}{\tau_g} & 0 & \frac{-1}{R\tau_g} \\ \frac{1}{\tau_T} & \frac{-1}{\tau_T} & 0 \\ 0 & \frac{1}{2H} & \frac{-D}{2H} \end{bmatrix} \begin{bmatrix} \Delta P_V \\ \Delta P_m \\ \Delta\omega \end{bmatrix} + \begin{bmatrix} 0 \\ 0 \\ \frac{-1}{2H} \end{bmatrix} \Delta P_L + \begin{bmatrix} \frac{1}{\tau_g} \\ 0 \\ 0 \end{bmatrix} \Delta P_{ref} \quad (12.54)$$

Example 12.10

Obtain the state variable representation of the LFC system of Example 12.1 with one input ΔP_L and perform the following analysis.
(a) Use the *MATLAB* **step** function to obtain the frequency deviation step response for a sudden load change of $\Delta P_L = 0.2$ per unit.
(b) Construct the *SIMULINK* block diagram and obtain the frequency deviation response for the condition in part (a).
(c) Use **placepole(A, B, C, p)** function to place the compensated closed-loop pole at $-2\pm j6$ and -3. Obtain the frequency deviation step response of the compensated system.
(d) Construct the *SIMULINK* block diagram and obtain the frequency deviation response for the condition in part (c).

Substituting the parameters of the system in Example 12.1 in the state equation (12.51) with $\Delta P_{ref} = 0$, we have

$$\dot{\mathbf{x}} = \begin{bmatrix} -5 & 0 & -100 \\ 2 & -2 & 0 \\ 0 & 0.1 & -0.08 \end{bmatrix} \mathbf{x} + \begin{bmatrix} 0 \\ 0 \\ -0.1 \end{bmatrix} u$$

and the output equation is

$$y = \begin{bmatrix} 0 & 0 & 1 \end{bmatrix} \mathbf{x}$$

where $y = \Delta\omega$ and

$$\mathbf{x} = \begin{bmatrix} \Delta P_V \\ \Delta P_m \\ \Delta\omega \end{bmatrix}$$

(a) We use the following commands

```
PL = 0.2;
A = [-5  0  -100; 2  -2  0; 0  0.1  -0.08];
B = [0;  0;  -0.1]; BPL = B*PL;
C = [0  0  1];  D = 0;
t=0:0.02:10;
[y, x] = step(A, BPL, C, D, 1, t);
figure(1), plot(t, y), grid
xlabel('t, sec'), ylabel('pu')
r =eig(A)
```

The frequency deviation step response result is shown in Figure 12.46, which is the same as the response obtained in Figure 12.13 using the transfer function method.

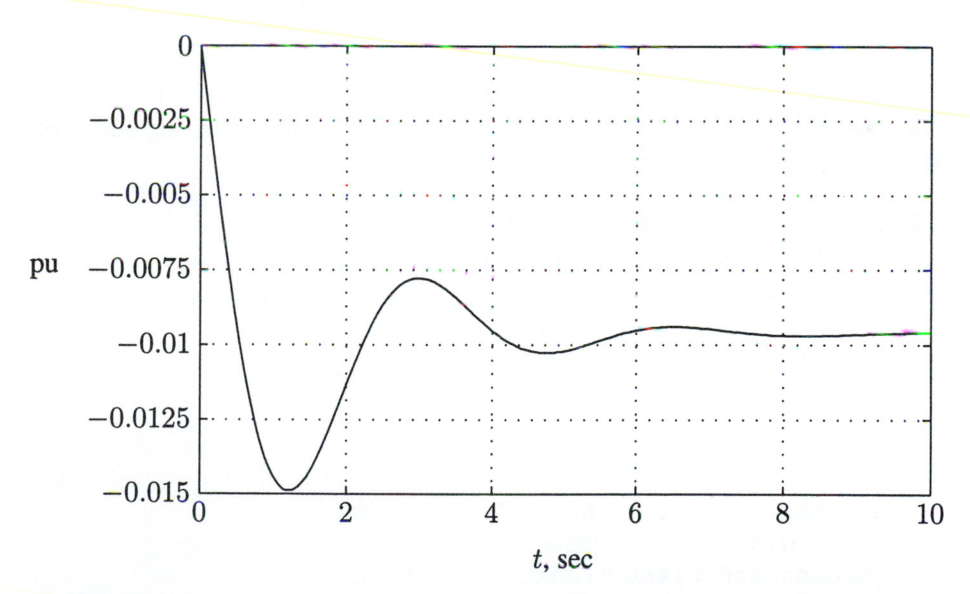

FIGURE 12.46
Uncompensated frequency deviation step response for Example 12.10.

The command **r = eig(A)** returns the roots of the characteristic equation, which are

```
r =
  -5.8863
  -0.5968 + 1.7825i
  -0.5968 - 1.7825i
```

(b) The *SIMULINK* state-space model can be used to obtain the response. A *SIMULINK* model named **sim12xxb.mdl** is constructed as shown in Figure 12.47. The state-space description dialog box is opened, and the **A**, **B**, **C**, and **D** constants are entered in the appropriate box in *MATLAB* matrix notation. The simulation

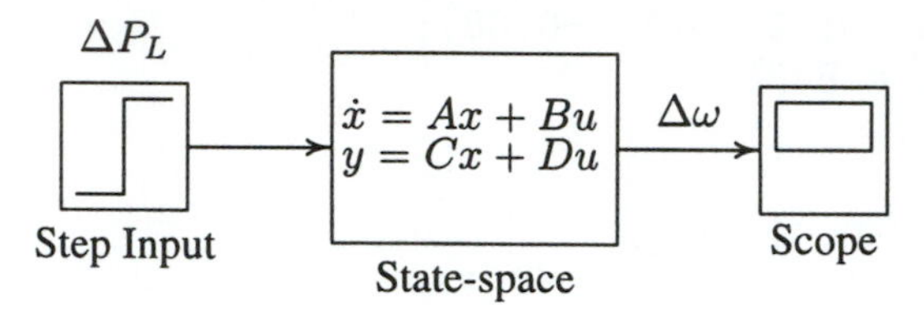

FIGURE 12.47
Simulation block diagram for Example 12.10 (b).

parameters are set to the appropriate values. The file is opened and is run in the *SIMULINK* window. The simulation results in the same response as shown in Figure 12.46.

(c) We are seeking the feedback gain vector **K** to place the roots of the system characteristic equation at $-2 \pm j6$ and -3. The following commands are added to the previous file.

```
P=[-2.0+j*6   -2.0-j*6   -3];
[K, Af] = placepol(A, B, C, P);
t=0:0.02:4;
[y, x] = step(Af, BPL, C, D, 1, t);
figure(2), plot(t, y), grid
xlabel('t,  sec'), ylabel('pu')
```

The result is

```
Feedback gain vector K
  4.2   0.8   0.8
Uncompensated Plant transfer function:
Numerator     0   -0.10   -0.70   -1.0
Denominator   1    7.08   10.56   20.8
```

```
Compensated system closed-loop transfer function:
Numerator      0    -0.1    -0.7    -1
Denominator    1     7.0    52.0   120

Compensated system matrix A - B*K
   -5.00    0.00   -100.00
    2.00   -2.00      0.00
    0.42    0.18      0.00
```

and the frequency deviation step response is shown in Figure 12.48.

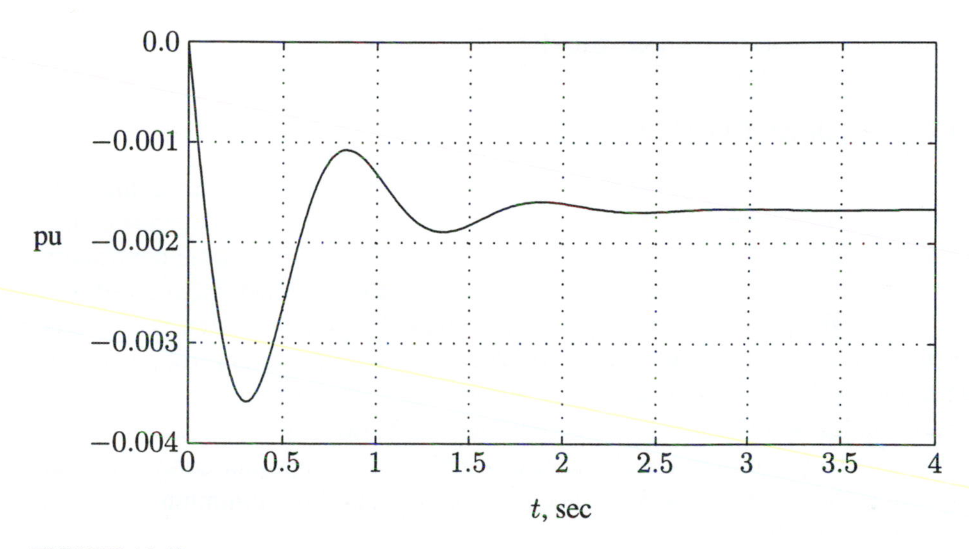

FIGURE 12.48
Compensated frequency deviation step response for Example 12.10.

Thus, the state feedback constants $K_1 = 4.2$, $K_2 = 0.8$, and $K_3 = 0.8$ result in the desired characteristic equation roots. The transient response is improved, and the response settles to a steady-state value of $\Delta_{ss} = -0.0017$ per unit in about 2.5 seconds.

(d) A *SIMULINK* model named **sim12xxd.mdl** is constructed as shown in Figure 12.49. In the state-space description dialog box, **C** is specified as an identity matrix of rank 3 to provide the three state variables as output. The file is opened and is run in the *SIMULINK* window. The simulation results in the same response as shown in Figure 12.48.

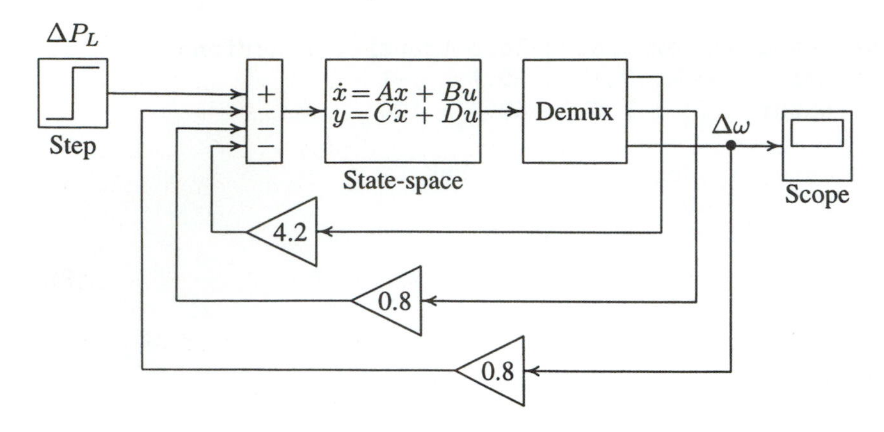

FIGURE 12.49
Simulation block diagram for Example 12.10 (d).

12.8.2 OPTIMAL CONTROL DESIGN

Optimal control is a branch of modern control theory that deals with designing controls for dynamic systems by minimizing a performance index that depends on the system variables. In this section, we will discuss the design of optimal controllers for linear systems with quadratic performance index, the so-called *linear quadratic regulator* (LQR) problem. The object of the optimal regulator design is to determine the optimal control law $\mathbf{u}^*(\mathbf{x}, t)$ which can transfer the system from its initial state to the final state such that a given performance index is minimized. The performance index is selected to give the best trade-off between performance and cost of control. The performance index that is widely used in optimal control design is known as the *quadratic performance index* and is based on minimum-error and minimum-energy criteria.

Consider the plant described by

$$\dot{\mathbf{x}}(t) = \mathbf{A}\mathbf{x}(t) + \mathbf{B}u(t) \tag{12.55}$$

The problem is to find the vector $\mathbf{K}(t)$ of the control law

$$\mathbf{u}(t) = -\mathbf{K}(t)\mathbf{x}(t) \tag{12.56}$$

which minimizes the value of a quadratic performance index $\mathbf{J}$ of the form

$$\mathbf{J} = \int_{t_0}^{t_f} (\mathbf{x}'\mathbf{Q}\mathbf{x} + \mathbf{u}'\mathbf{R}\mathbf{u})dt \tag{12.57}$$

subject to the dynamic plant equation in (12.55). In (12.57), $\mathbf{Q}$ is a positive semidefinite matrix, and $\mathbf{R}$ is a real symmetric matrix. $\mathbf{Q}$ is positive semidefinite, if all its principal minors are nonnegative. The choice of the elements of $\mathbf{Q}$ and $\mathbf{R}$ allows the relative weighting of individual state variables and individual control inputs.

To obtain a formal solution, we can use the method of Lagrange multipliers. The constraint problem is solved by augmenting (12.55) into (12.57) using an n-vector of *Lagrange multipliers*, λ. The problem reduces to the minimization of the following unconstrained function.

$$\mathcal{L}(\mathbf{x}, \lambda, u, t) = [\mathbf{x}'\mathbf{Q}\mathbf{x} + \mathbf{u}'\mathbf{R}\mathbf{u}] + \lambda'[\mathbf{A}\mathbf{x} + \mathbf{B}\mathbf{u} - \dot{\mathbf{x}}] \qquad (12.58)$$

The optimal values (denoted by the subscript $*$) are found by equating the partial derivatives to zero.

$$\frac{\partial \mathcal{L}}{\partial \lambda} = \mathbf{A}\mathbf{X}^* + \mathbf{B}\mathbf{u}^* - \dot{\mathbf{x}}^* = \mathbf{0} \quad \Rightarrow \quad \dot{\mathbf{x}}^* = \mathbf{A}\mathbf{X}^* + \mathbf{B}\mathbf{u}^* \qquad (12.59)$$

$$\frac{\partial \mathcal{L}}{\partial u} = 2\mathbf{R}\mathbf{u}^* + \lambda'\mathbf{B} = \mathbf{0} \quad \Rightarrow \quad \mathbf{u}^* = -\frac{1}{2}\mathbf{R}^{-1}\lambda'\mathbf{B} \qquad (12.60)$$

$$\frac{\partial \mathcal{L}}{\partial x} = 2\mathbf{x}'^*\mathbf{Q} + \dot{\lambda}' + \lambda'\mathbf{A} = \mathbf{0} \quad \Rightarrow \quad \dot{\lambda} = -2\mathbf{Q}\mathbf{x}^* - \mathbf{A}'\lambda \qquad (12.61)$$

Assume that there exists a symmetric, time-varying positive definite matrix $\mathbf{p}(t)$ satisfying

$$\lambda = 2\mathbf{p}(t)\mathbf{x}^* \qquad (12.62)$$

Substituting (12.42) into (12.60) gives the optimal closed-loop control law

$$\mathbf{u}^*(t) = -\mathbf{R}^{-1}\mathbf{B}'\mathbf{p}(t)\mathbf{x}^* \qquad (12.63)$$

Obtaining the derivative of (12.62), we have

$$\dot{\lambda} = 2(\dot{\mathbf{p}}\mathbf{x}^* + \mathbf{p}\dot{\mathbf{x}}^*) \qquad (12.64)$$

Finally, equating (12.61) with (12.64), we obtain

$$\dot{\mathbf{p}}(t) = -\mathbf{p}(t)\mathbf{A} - \mathbf{A}'\mathbf{p}(t) - \mathbf{Q} + \mathbf{p}(t)\mathbf{B}\mathbf{R}^{-1}\mathbf{B}'\mathbf{p}(t) \qquad (12.65)$$

The above equation is referred to as the matrix *Riccati equation*. The boundary condition for (12.65) is $\mathbf{p}(t_f) = \mathbf{0}$. Therefore, (12.65) must be integrated backward in time. Since a numerical solution is performed forward in time, a dummy time variable $\tau = t_f - t$ is replaced for time t. Once the solution to (12.65) is obtained, the solution of the state equation (12.59) in conjunction with the optimum control equation (12.63) is obtained.

The function $[\tau, \mathbf{p}, \mathbf{K}, \mathbf{t}, \mathbf{x}]$= **riccati** is developed for the time-domain solution of the Riccati equation. The function returns the solution of the matrix Riccati equation, $\mathbf{p}(\tau)$, the optimal feedback gain vector $\mathbf{k}(\tau)$, and the initial state response $\mathbf{x}(t)$. In order to use this function, the user must declare the function $[\mathbf{A}, \mathbf{B}, \mathbf{Q}, \mathbf{R}, \mathbf{t_0}, \mathbf{t_f}, \mathbf{x_0}] = \mathbf{system}(\mathbf{A}, \mathbf{B}, \mathbf{Q}, \mathbf{R}, \mathbf{t_0}, \mathbf{t_f}, \mathbf{x_0})$ containing system matrices and the performance index matrices in an M-file named **system.m**.

The optimal controller gain is a time-varying state-variable feedback. Such feedback are inconvenient to implement, because they require the storage in computer memory of time-varying gains. An alternative control scheme is to replace the time-varying optimal gain $\mathbf{K}(t)$ by its constant steady-state value. In most practical applications, the use of the steady-state feedback gain is adequate. For linear time-invariant systems, since $\dot{\mathbf{p}} = \mathbf{0}$, when the process is of infinite duration, that is $t_f = \infty$, (12.65) reduces to the algebraic Riccati equation

$$\mathbf{pA} + \mathbf{A}'\mathbf{p} + \mathbf{Q} - \mathbf{pBR}^{-1}\mathbf{B}'\mathbf{p} = \mathbf{0} \tag{12.66}$$

The *MATLAB Control System Toolbox* function **[k, p]=lqr2(A, B, Q, R)** can be used for the solution of the algebraic Riccati equation.

The LQR design procedure is in stark contrast to classical control design, where the gain matrix $\mathbf{K}$ is selected directly. To design the optimal LQR, the design engineer first selects the design parameter weight matrices $\mathbf{Q}$ and $\mathbf{R}$. Then, the feedback gain $\mathbf{K}$ is automatically given by matrix design equations and the closed-loop time responses are found by simulation. If these responses are unsuitable, new values of $\mathbf{Q}$ and $\mathbf{R}$ are selected and the design is repeated. This has the significant advantages of allowing all the control loops in a multiloop system to be closed simultaneously, while guaranteeing closed-loop stability.

Example 12.11

Design an LQR state feedback for the system described in Example 12.10.
(a) Find the optimal feedback gain vector to minimize the performance index

$$\mathbf{J} = \int_0^\infty \left(20{x_1}^2 + 15{x_2}^2 + 5{x_3}^2 + 0.15u^2\right) dt$$

The admissible states and control values are unconstrained. Obtain the frequency deviation step response for a sudden load change of $\Delta P_L = 0.2$ per unit.
(b) Construct the *SIMULINK* block diagram and obtain the frequency deviation response for the condition in part (a).

For this system we have

$$\mathbf{A} = \begin{bmatrix} -5 & 0 & -100 \\ 2 & -2 & 0 \\ 0 & 0.1 & -0.08 \end{bmatrix} \quad \mathbf{B} = \begin{bmatrix} 0 \\ 0 \\ -0.1 \end{bmatrix} \quad \mathbf{Q} = \begin{bmatrix} 20 & 0 & 0 \\ 0 & 10 & 0 \\ 0 & 0 & 5 \end{bmatrix}$$

and $R = 0.15$.

(a) We use the following commands

```
PL=0.2;
A = [-5  0  -100; 2  -2  0; 0  0.1  -0.08];
B = [0;  0;  -0.1]; BPL=PL*B;
C = [0  0  1];  D = 0;
Q = [20 0  0; 0  10 0; 0  0  5];  R = .15;
[K, P] = lqr2(A, B, Q, R)
Af = A - B*K
t=0:0.02:1;
[y, x] = step(Af, BPL, C, D, 1, t);
plot(t, y), grid, xlabel('t,  sec'), ylabel('pu')
```

The result is

```
K =
    6.4128     1.1004 -112.6003
P =
    1.5388     0.3891    -9.6192
    0.3891     2.3721    -1.6506
   -9.6192    -1.6506   168.9004
Af =
   -5.0000          0 -100.0000
    2.0000    -2.0000          0
    0.6413     0.2100   -11.3400
```

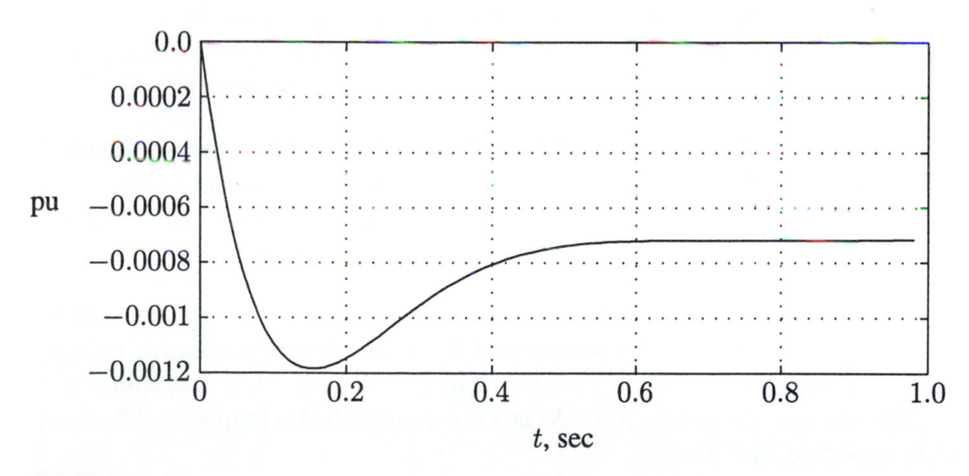

FIGURE 12.50
Frequency deviation step response for Example 12.11.

The frequency deviation step response is shown in Figure 12.50. The transient response settles to a steady-state value of $\Delta_{ss} = -0.0007$ per unit in about 0.6 second.

(b) A *SIMULINK* model named **sim12xx1.mdl** is constructed as shown in Figure 12.51. The state-space description dialog box is opened, and the **A**, **B**, **C**, and **D** constants are entered in the appropriate box in *MATLAB* matrix notation. Also, the LQR description dialog box is opened, and weighting matrix **Q** and weighting coefficient **R** are set to the given values. The simulation parameters are set to the

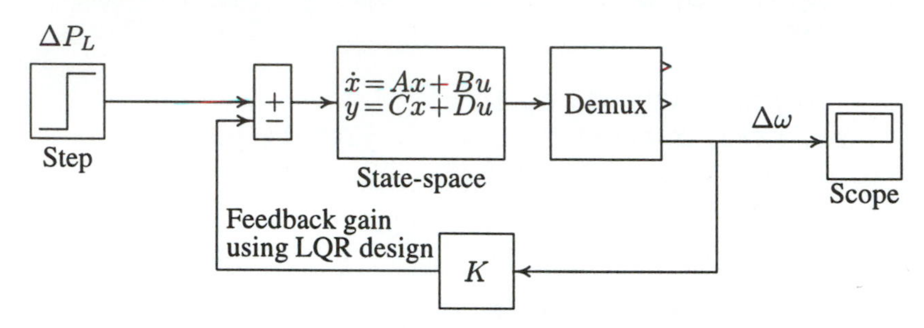

FIGURE 12.51
Simulation block diagram for Example 12.11.

appropriate values. The file is opened and is run in the *SIMULINK* window. The simulation results in the same response as shown in Figure 12.50.

PROBLEMS

12.1. A 250-MW, 60-Hz turbine generator set has a speed regulation of 5 percent based on its own rating. The generator frequency decreases from 60 Hz to a steady state value of 59.7Hz. Determine the increase in the turbine power output.

12.2. Two generating units rated for 250 MW and 400 MW have governor speed regulation of 6.0 and 6.4 percent, respectively, from no-load to full-load, respectively. They are operating in parallel and share a load of 500 MW. Assuming free governor action, determine the load shared by each unit.

12.3. A single area consists of two generating units, rated at 400 and 800 MVA, with speed regulation of 4 percent and 5 percent on their respective ratings. The units are operating in parallel, sharing 700 MW. Unit 1 supplies 200 MW and unit 2 supplies 500 MW at 1.0 per unit (60 Hz) frequency. The load is increased by 130 MW.

(a) Assume there is no frequency-dependent load, i.e., $D = 0$. Find the steady-state frequency deviation and the new generation on each unit.

(b) The load varies 0.804 percent for every 1 percent change in frequency, i.e., $D = 0.804$. Find the steady-state frequency deviation and the new generation on each unit.

12.4. An isolated power station has the LFC system as shown in Figure 12.9 with the following parameters

Turbine time constant $\tau_T = 0.5$ sec
Governor time constant $\tau_g = 0.25$ sec
Generator inertia constant $H = 8$ sec
Governor speed regulation = R per unit

The load varies by 1.6 percent for a 1 percent change in frequency, i.e., $D = 1.6$.
(a) Use the Routh-Hurwitz array (Appendix B.2.1) to find the range of R for control system stability.
(b) Use *MATLAB* **rlocus** function to obtain the root-locus plot.

12.5. The governor speed regulation of Problem 12.4 is set to $R = 0.04$ per unit. The turbine rated output is 200 MW at nominal frequency of 60 Hz. A sudden load change of 50 MW ($\Delta P_L = 0.25$ per unit) occurs.
(a) Find the steady-state frequency deviation in Hz.
(b) Obtain the closed-loop transfer function and use *MATLAB* to obtain the frequency deviation step response.
(c) Construct the *SIMULINK* block diagram and obtain the frequency deviation response.

12.6. The LFC system in Problem 12.5 is equipped with the secondary integral control loop for automatic generation control as shown in Figure 12.16.
(a) Use the *MATLAB* **step** function to obtain the frequency deviation step response for a sudden load change of $\Delta P_L = 0.25$ per unit. Set the integral controller gain to $K_I = 9$.
(b) Construct the *SIMULINK* block diagram and obtain the frequency deviation response for the condition in part (a).

12.7. The load changes of 200 MW and 150 MW occur simultaneously in areas 1 and 2 of the two-area system of Example 12.4. Modify the *SIMULINK* block diagram (sim12ex4.mdl), and obtain the frequency deviation and the power responses.

12.8. Modify the *SIMULINK* model for the two-area system of Example 12.5 with the tie-line bias control (sim12ex5.mdl) to include the load changes specified in Problem 12.7. Obtain the frequency and power response deviation for each area.

12.9. A generating unit has a simplified linearized AVR system as shown in Figure 12.52.

(a) Use the Routh-Hurwitz array (Appendix B.2.1) to find the range of K_A for control system stability.

(b) Use *MATLAB* **rlocus** function to obtain the root-locus plot.

(c) The amplifier gain is set to $K_A = 40$. Find the system closed-loop transfer function, and use *MATLAB* to obtain the step response.

(d) Construct the *SIMULINK* block diagram and obtain the step response.

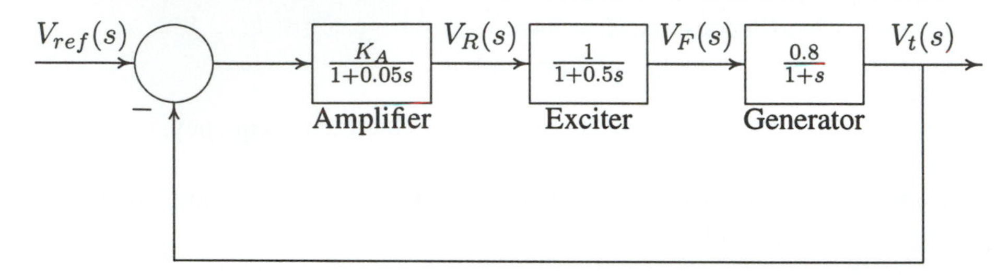

FIGURE 12.52
AVR system of Problem 12.9.

12.10. A rate feedback stabilizer is added to the AVR system of Problem 12.9 as shown in Figure 12.53. The stabilizer time constant is $\tau_F = 0.04$ second, and the derivative gain is adjusted to $K_F = 0.1$.

(a) Find the system closed-loop transfer function, and use *MATLAB* to obtain the step response.

(b) Construct the *SIMULINK* model, and obtain the step response.

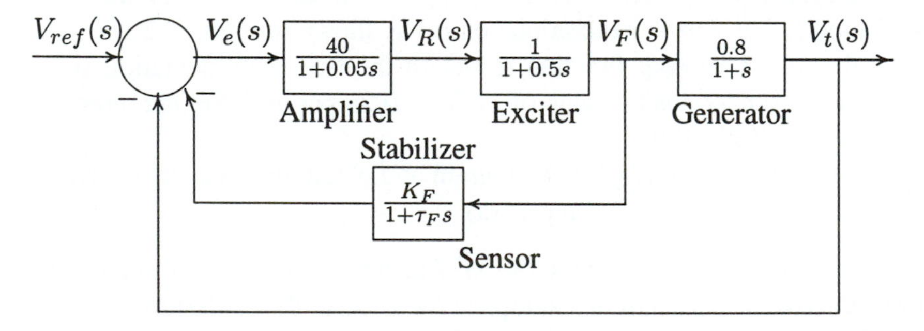

FIGURE 12.53
AVR system with rate feedback for Problem 12.10.

12.11. A PID controller is added in the forward path of the AVR system of Problem 12.9 as shown in Figure 12.54. Construct the *SIMULINK* model. Set the proportional gain K_P to 2.0, and adjust K_I and K_D until a step response with a minimum overshoot and a very small settling time is obtained (suggested values $K_I = 0.2$, and $K_P = 0.25$).

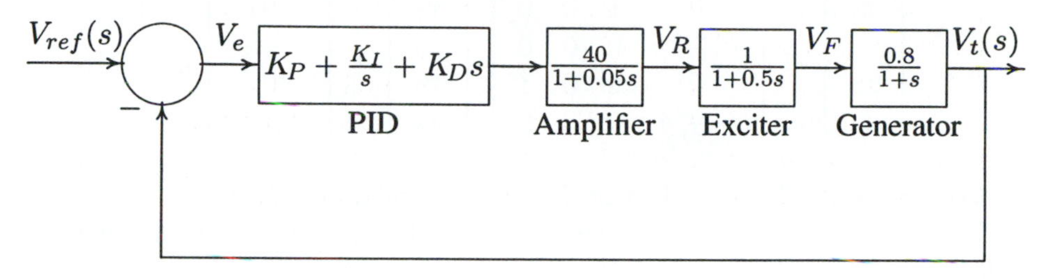

FIGURE 12.54
AVR system with PID controller for Problem 12.11.

12.12. Figure 12.55 shows an inverted pendulum of length L and mass m mounted on a cart of mass M, by means of a force u applied to the cart. This is a model of the attitude control of a space booster on takeoff. The differential equations describing the motion of the system is obtained by summing the forces on the pendulum, which result in the following nonlinear equations.

$$(M + m)\ddot{x} + mL\cos\theta\,\ddot{\theta} = mL\sin\theta\,\dot{\theta}^2 + u$$
$$mL\cos\theta\,\ddot{x} + mL^2\ddot{\theta} = mgL\sin\theta$$

(a) Linearize the above equations in the neighborhood of the zero initial

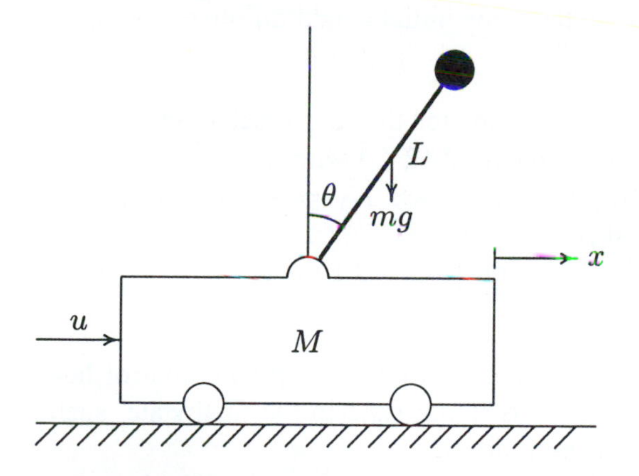

FIGURE 12.55
Inverted pendulum on a cart.

states. Hint: Substitute θ for $\sin\theta$, 1 for $\cos\theta$ and 0 for $\dot{\theta}^2$. With the state variables defined as $x_1 = \theta$, $x_2 = \dot{\theta}$, $x_3 = x$, and $x_4 = \dot{x}$, show that the linearized state equation is

$$\begin{bmatrix} \dot{x}_1 \\ \dot{x}_2 \\ \dot{x}_3 \\ \dot{x}_4 \end{bmatrix} = \begin{bmatrix} 0 & 1 & 0 & 0 \\ \frac{M+m}{ML}g & 0 & 0 & 0 \\ 0 & 0 & 0 & 1 \\ -\frac{m}{M}g & 0 & 0 & 0 \end{bmatrix} \begin{bmatrix} x_1 \\ x_2 \\ x_3 \\ x_4 \end{bmatrix} + \begin{bmatrix} 0 \\ -\frac{1}{ML} \\ 0 \\ \frac{1}{M} \end{bmatrix} u$$

Assume $M = 4$ kg, $m = 0.2$ kg, $L = 0.5$ m, and $g = 9.81$ m/s^2.

(b) Use the *MATLAB* function **eig(A)** to find the roots of the system characteristic equation.

(c) Define C as the identity matrix of rank 4, i.e., C = eye(4) and D = zeros(4, 1). Use the *MATLAB* function $[\mathbf{y}, \mathbf{x}] = \text{initial}(\mathbf{A}, \mathbf{B}, \mathbf{C}, \mathbf{D}, \mathbf{X_0}, \mathbf{t})$ to simulate the system for 20 seconds in response to an initial condition offset of $\theta(0) = 0.1$ rad, and $x(0) = 0.1$ m (i.e., $x_0 = [0.1 \quad 0 \quad 0.1 \quad 0]$). Obtain a plot of θ and x, and comment on the stability of the system.

(d) You may have found that the inverted pendulum is unstable, that is, it will fall over unless a suitable control force via state feedback is used. The purpose is to design a control system such that for a small initial disturbance the pendulum can be brought back to the vertical position ($\theta = 0$), and the cart can be brought back to the reference position ($x = 0$). A simple method is to use a state feedback gain to place the compensated closed-loop poles in the left-half of the s-plane. Use the custom made function **[K, Af] = placepol(A, B, C, P)** and design a state feedback controller to place the compensated closed-loop poles at $-2 \pm j0.5$, -4, and -5. Simulate the system for 20 seconds in response to the same initial condition offset. Obtain a plot of θ, x, and $u = -Kx'$.

12.13. Construct the *SIMULINK* block diagram for the linearized model of the inverted pendulum described in Problem 12.12 (a) with the state feedback controller. Assume the state feedback gains are $K_1 = -170.367$, $K_2 = -38.054$, $K_3 = -17.3293$, and $K_4 = -24.1081$. Obtain the response for θ, x, and u for the initial condition offset of $\theta(0) = 0.1$ rad and $x(0) = 0.1$m (i.e., $x_0 = [0.1 \quad 0 \quad 0.1 \quad 0]$).

12.14. A classical problem in control systems is to find the optimal control law which will transfer the system from its initial state to the final state, such that a given performance index is minimized.

(a) Design an LQR state feedback for the linearized model of the inverted pendulum described in Problem 12.12, and find the optimal feedback gain vector to minimize the performance index

$$\mathbf{J} = \int_0^\infty \left(10{x_1}^2 + 10{x_2}^2 + 5{x_3}^2 + +5{x_4}^2 + 0.2u^2\right) dt$$

The admissible states and control values are unconstrained.
(b) Define C as the identity matrix of rank 4, i.e., C = eye(4) and D = zeros(4, 1). Use the *MATLAB* function **[K, P] = lqr2 (A, B, Q, R)** to design a state feedback controller in response to an initial condition offset of $\theta(0) =$ 0.1 rad and $x(0) =$ 0.1m (i.e., $x_0 = [0.1 \quad 0 \quad 0.1 \quad 0]$). Use the *MATLAB* function $[\mathbf{y}, \mathbf{x}] = \text{initial}(\mathbf{A}, \mathbf{B}, \mathbf{C}, \mathbf{D}, \mathbf{X_0}, \mathbf{t})$ to simulate the system for 20 seconds. Obtain a plot of θ, x, and the control law $u = -kx'$.

12.15. Construct the *SIMULINK* block diagram for the linearized model of the inverted pendulum described in Problem 12.12(a) using the *SIMULINK* LQR model. Obtain the response for θ, x, and u for the initial condition offset described in Problem 12.14.

12.16. Obtain the state variable representation of the LFC system of Problem 12.4 with one input ΔP_L, and perform the following analysis.
(a) Use the *MATLAB* **step** function to obtain the frequency deviation step response for a sudden load change of $\Delta P_L = 0.2$ per unit.
(b) Construct the *SIMULINK* block diagram and obtain the frequency deviation response for the condition in part (a).
(c) Use **placepol(A, B, C, p)** function to place the compensated closed-loop pole at $-4 \pm j6$ and -4. Obtain the frequency deviation step response of the compensated system.
(d) Construct the *SIMULINK* block diagram and obtain the frequency deviation response for the condition in part (c).

12.17. Design a LQR state feedback for the system described in Problem 12.16.
(a) Find the optimal feedback gain vector to minimize the performance index

$$\mathbf{J} = \int_0^\infty \left(40{x_1}^2 + 20{x_2}^2 + 10{x_3}^2 + 0.2u^2\right) dt$$

The admissible states and control values are unconstrained. Obtain the frequency deviation step response for a sudden load change of $\Delta P_L = 0.2$ per unit.
(b) Construct the *SIMULINK* block diagram and obtain the frequency deviation step response for the condition in part (a).

APPENDIX A

INTRODUCTION TO *MATLAB*

MATLAB, developed by Math Works Inc., is a software package for high performance numerical computation and visualization. The combination of analysis capabilities, flexibility, reliability, and powerful graphics makes *MATLAB* the premier software package for electrical engineers.

MATLAB provides an interactive environment with hundreds of reliable and accurate built-in mathematical functions. These functions provide solutions to a broad range of mathematical problems including matrix algebra, complex arithmetic, linear systems, differential equations, signal processing, optimization, nonlinear systems, and many other types of scientific computations. The most important feature of *MATLAB* is its programming capability, which is very easy to learn and to use, and which allows user-developed functions. It also allows access to Fortran algorithms and C codes by means of external interfaces. There are several optional toolboxes written for special applications such as signal processing, control systems design, system identification, statistics, neural networks, fuzzy logic, symbolic computations, and others. *MATLAB* has been enhanced by the very powerful *SIMULINK* program. *SIMULINK* is a graphical mouse-driven program for the simulation of dynamic systems. *SIMULINK* enables students to simulate linear, as well as nonlinear, systems easily and efficiently.

The following section describes the use of *MATLAB* and is designed to give a quick familiarization with some of the commands and capabilities of *MATLAB*. For a description of all other commands, *MATLAB* functions, and many other useful features, the reader is referred to the *MATLAB User's Guide*.

A.1 INSTALLING THE TEXT TOOLBOX

The diskette included with the book contains all the developed functions and chapter examples. The M-files reside in the directories labeled Folder 1 and Folder 2. The file names for chapter examples begin with the letters **chp**. For example, the M-file for Example 11.4 is **chp11ex4**. The appendix examples begin with **exa** and **exb** and a number. The disk contains a file called **setup.exe** that automates installation of all the files. Insert the diskette in the disk drive and use Windows to view its contents. For automatic installation double click on the **setup.exe** icon to start the installation. The installation program will prompt you for the location of the *MATLAB* directory and the name of the directory where you would like the files to be installed.

Alternatively, you can copy the M-files manually. To do this, create a subdirectory, such as **power**, where the *MATLAB* toolbox resides. Copy all the files from Folder 1 and Folder 2 to the subdirectory *matlab \toolbox\power*.

In the *MATLAB* 5 **Command Window** open the Path Browser by selecting **Set Path** from the **File** menu. From the **Path** menu choose **Add to Path**. Select the directory to add, choose **Add to back** option and press OK to add to the current *MATLAB* directory area. Save before exiting the Path Browser.

If you are running *MATLAB* 4 edit the **matlabrc.m** located in the subdirectory *matlab\bin*, where the search paths are specified. Describe the subdirectory just created by adding the statement

`';C:\matlab\toolbox\power', ...` at the end of this file.

A.2 RUNNING MATLAB

MATLAB supports almost every computational platform. *MATLAB* for *WINDOWS* is started by clicking on the *MATLAB* icon. The **Command window** is launched, and after some messages such as intro, demo, help help, info, and others, the prompt " $\gg$ " is displayed. The program is in an interactive command mode. Typing **who** or **whos** displays a list of variable names currently in memory. Also, the **dir** command lists all the files on the default directory. *MATLAB* has an on-line help facility, and its use is highly recommended. The command **help** provides a list of files, built-in functions and operators for which on-line help is available. The command

`help` function name

will give information on the specified function as to its purpose and use. The command

```
help help
```

will give information as to how to use the on-line help.

MATLAB has a demonstration program that shows many of its features. The command **demo** brings up a menu of the available demonstrations. This will provide a presentation of the most important *MATLAB* facilities. Follow the instructions on the screen – it is worth trying.

MATLAB 5.2 includes a Help Desk facility that provides access to on line help topics, documentation, getting started with *MATLAB*, online reference materials, *MATLAB* functions, real-time Workshop, and several toolboxes. The online documentation is available in HTML, via either Netscape Navigator Release 3.0 or Microsoft Internet Explorer 3.0. The command **helpdesk** launches the Help Desk, or you can use the **Help** menu to bring up the Help Desk.

If an expression with correct syntax is entered at the prompt in the Command window, it is processed immediately and the result is displayed on the screen. If an expression requires more than one line, the last character of the previous line must contain three dots "...". Characters following the percent sign are ignored. The (%) may be used anywhere in a program to add clarifying comments. This is especially helpful when creating a program. The command **clear** erases all variables in the Command window.

MATLAB is also capable of executing sequences of commands that are stored in files, known as script files or *M-files*. Clicking on **File**, **New M-file**, opens the **Edit window**. A program can be written and saved in ASCII format with a filename having extension .m in the directory where *MATLAB* runs. To run the program, click on the Command window and type the filename without the *.m* extension at the *MATLAB* command "≫". You can view the text Edit window simultaneously with the Command window. That is, you can use the two windows to edit and debug a script file repeatedly and run it in the Command window without ever quitting *MATLAB*.

In addition to the Command window and Edit window are the **Graphic windows** or **Figure windows** with black (default) background. The plots created by the graphic commands appear in these windows.

Another type of M-file is a *function file*. A function provides a convenient way to encapsulate some computation, which can then be used without worrying about its implementation. In contrast to the script file, a *function file* has a name following the word "function" at the beginning of the file. The filename must be the same as the "function" name. The first line of a function file must begin with the function statement having the following syntax

function [*output arguments*] = *function name* (*input arguments*)

The output argument(s) are variables returned. A function need not return a value. The input arguments are variables passed to the function. Variables generated in function files are local to the function. The use of **global** variables make defined variables common and accessible between the main script file and other function files. For example, the statement **global R S T** declares the variables R, S, and T to be global without the need for passing the variables through the input list. This statement goes before any executable statement in the script and function files that need to access the values of the global variables.

Normally, while an M-file is executing, the commands of the file are not displayed on the screen. The command **echo** allows M-files to be viewed as they execute. **echo off** turns off the echoing of all script files. Typing **what** lists M-files and Mat-files in the default directory.

MATLAB follows conventional Windows procedure. Information from the command screen can be printed by highlighting the desired text with the mouse and then choosing the **print Selected ...** from the **File** menu. If no text is highlighted the entire Command window is printed. Similarly, selecting **print** from the Figure window sends the selected graph to the printer. For a complete list and help on general purpose commands, type **help general**.

A.3 VARIABLES

Expressions typed without a variable name are evaluated by *MATLAB*, and the result is stored and displayed by a variable called **ans**. The result of an expression can be assigned to a variable name for further use. Variable names can have as many as 19 characters (including letters and numbers). However, the first character of a variable name must be a letter. *MATLAB* is case-sensitive. Lower and uppercase letters represent two different variables. The command **casesen** makes *MATLAB* insensitive to the case. Variables in script files are global. The expressions are composed of operators and any of the available functions. For example, if the following expression is typed

```
x = exp(-0.2696*.2)*sin(2*pi*0.2)/(0.01*sqrt(3)*log(18))
```

the result is displayed on the screen as

```
x =
    18.0001
```

and is assigned to x. If a variable name is not used, the result is assigned to the variable **ans**. For example, typing the expression

```
250/sin(pi/6)
```

results in

```
ans =
     500.0000
```

If the last character of a statement is a semicolon (;), the expression is executed, but the result is not displayed. However, the result is displayed upon entering the variable name. The command **disp** may be used to display a variable without printing its name. For example, **disp(x)** displays the value of the variable without printing its name. If **x** contains a text string, the string is displayed.

A.4 OUTPUT FORMAT

While all computations in *MATLAB* are done in double precision, the default format prints results with five significant digits. The format of the displayed output can be controlled by the following commands.

MATLAB Command	Display
format	Default. Same as **format short**
format short	Scaled fixed point format with 5 digits
format long	Scaled fixed point format with 15 digits
format short e	Floating point format with 5 digits
format long e	Floating point format with 15 digits
format short g	Best of fixed or floating point with 5 digits
format long g	Best of fixed or floating point with 15 digits
format hex	Hexadecimal format
format +	The symbols +, - and blank are printed for positive, negative, and zero elements
format bank	Fixed format for dollars and cents
format rat	Approximation by ratio of small integers
format compact	Suppress extra line feeds
format loose	Puts the extra line feeds back in

For more flexibility in the output format, the command **fprintf** displays the result with a desired format on the screen or to a specified filename. The general form of this command is the following.

```
fprintf{fstr, A,...)
```

writes the real elements of the variable or matrix `A,...` according to the specifications in the string argument of `fstr`. This string can contain format characters like *ANCI C* with certain exceptions and extensions. **fprintf** is "vectorized" for the case when A is nonscalar. The format string is recycled through the elements of A (columnwise) until all the elements are used up. It is then recycled in a similar manner through any additional matrix arguments. The characters used in the format string of the commands **fprintf** are listed in the table below.

Format codes		Control characters	
%e	scientific format, lower case e	\n	new line
%E	scientific format, upper case E	\r	beginning of the line
%f	decimal format	\b	back space
%s	string	\t	tab
%u	integer	\g	new page
%i	follows the type	"	apostrophe
%x	hexadecimal, lower case	\\	back slash
%X	hexadecimal, upper case	\a	bell

A simple example of the **fprintf** is

```
fprintf('Area = %7.3f Square meters \n', pi*4.5^2)
```

The results is

```
Area =  63.617 Square meters
```

The %7.3f prints a floating point number seven characters wide, with three digits after the decimal point. The sequence \n advances the output to the left margin on the next line.

The following command displays a formatted table of the natural logarithmic for numbers 10, 20, 40, 60, and 80

```
x = [10; 20; 40; 60; 80];
y = [x,  log(x)];
fprintf('\n Number   Natural log\n')
fprintf('%4i \t %8.3f\n',y')
```

The result is

```
Number   Natural log
  10       2.303
  20       2.996
  40       3.689
  60       4.094
  80       4.382
```

An M-file can prompt for input from the keyboard. The command **input** causes the computer to request data from the keyboard. For example, the command

```
R = input('Enter radius in meter ')
```

displays the text string

```
Enter radius in meter
```

and waits for a number to be entered. If a number, say 4.5 is entered, it is assigned to variable R and displayed as

```
R =
    4.5000
```

The command **keyboard** placed in an M-file will stop the execution of the file and permit the user to examine and change variables in the file. Pressing **cntrl-z** terminates the keyboard mode and returns to the invoking file. Another useful command is **diary** A:*filename*. This command creates a file on drive A, and all output displayed on the screen is sent to that file. **diary off** turns off the diary. The contents of this file can be edited and used for merging with a word processor file. Finally, the command **save** *filename* can be used to save the expressions on the screen to a file named *filename.mat*, and the statement **load** *filename* can be used to load the file *filename.mat*.

MATLAB has a useful collection of transcendental functions, such as exponential, logarithm, trigonometric, and hyperbolic functions. For a complete list and help on operators, type **help ops**, and for elementary math functions, type **help elfun**.

A.5 CHARACTER STRING

A sequence of characters in single quotes is called a *character string* or *text variable*.

```
c ='Good'
```

results in

```
c = Good
```

A text variable can be augmented with more text variables, for example,

```
cs = [c, ' luck']
```

produces

```
cs =
     Good luck
```

A.6 VECTOR OPERATIONS

An n vector is a row or a column array of n numbers. In *MATLAB*, elements enclosed by brackets and separated by semicolons generate a column vector.

For example, the statement

```
X = [ 2; -4; 8]
```

results in

```
X =
        2
       -4
        8
```

If elements are separated by blanks or commas, a row vector is produced. Elements may be any expression. The statement

```
R = [tan(pi/4)  sqrt(9) -5]
```

results in the output

```
R =
        1.0000     3.0000    -5.0000
```

The transpose of a column vector results in a row vector, and vice versa. For example

```
Y=R'
```

will produce

```
Y =
        1.0000
        3.0000
       -5.0000
```

MATLAB has two different types of arithmetic operations. Matrix arithmetic operations are defined by the rules of linear algebra. Array arithmetic operations are carried out element-by-element. The period character (.) distinguishes the array operations from the matrix operations. However, since the matrix and array operations are the same for addition and subtraction, the character pairs .+ and .- are not used.

Vectors of the same size can be added or subtracted, where addition is performed componentwise. However, for multiplication, specific rules must be followed in order to obtain the correct resulting values. The operation of multiplying a vector X with a scalar k (scalar multiplication) is performed componentwise. For example $P = 5 * R$ produces the output

```
P =
          5.0000    15.0000    -25.0000
```

The inner product or the *dot product* of two vectors X and Y denoted by $\langle X, Y \rangle$ is a scalar quantity defined by $\sum_{i=1}^{n} x_i y_i$. If X and Y are both column vectors defined above, the inner product is given by

```
S = X'*Y
```

and results in

```
S =
        -50
```

The operator (.$*$ performs element-by-element operation. For example, for the previously defined vectors, X and Y, the statement

```
E = X.*Y
```

results in

```
E =
       2
     -12
     -40
```

The operator ./ performs element-by-element division. The two arrays must have the same size, unless one of them is a scalar. Array powers or element-by-element powers are denoted by (.^). The trigonometric functions, and other elementary mathematical functions such as **abs**, **sqrt**, **real**, and **log**, also operate element by element.

Various norms (measure of size) of a vector can be obtained. For example, the *Euclidean norm* is the square root of the inner product of the vector and itself. The command

```
N = norm(X)
```

produces the output

```
N =
      9.1652
```

The angle between two vectors X and Y is defined by $\cos\theta = \frac{\langle X,Y \rangle}{\|X\| \, \|Y\|}$. The statement

```
Theta = acos( X'*Y/(norm(X)*norm(Y)) )
```

results in the output

```
Theta =
          2.7444
```

where Theta is in radians.

The *zero vector*, also referred to as origin, is a vector with all components equal to zero. For example, to build a zero row vector of size 4, the following command

```
Z = zeros(1, 4)
```

results in

```
Z =
     0     0     0     0
```

The *one vector* is a vector with each component equal to one. To generate a one vector of size 4, use

```
I = ones(1, 4)
```

The result is

```
I =
     1     1     1     1
```

In *MATLAB*, the colon (:) can be used to generate a row vector. For example

```
x = 1:8
```

generates a row vector of integers from 1 to 8.

```
x =
      1   2   3   4   5   6   7   8
```

For increments other than unity, the following command

```
z = 0 : pi/3 : pi
```

results in

```
z =
      0000    1.0472    2.0944    3.1416
```

For negative increments

```
x = 5 : -1:1
```

results in

```
x =
         5    4    3    2    1
```

Alternatively, special vectors can be created, the command **linspace(x, y, n)** creates a vector with n elements that are spaced linearly between x and y. Similarly, the command **logspace(x, y, n)** creates a vector with n elements that are spaced in even logarithmic increments between 10^x and 10^y.

A.7 ELEMENTARY MATRIX OPERATIONS

In *MATLAB*, a matrix is created with a rectangular array of numbers surrounded by brackets. The elements in each row are separated by blanks or commas. A semicolon must be used to indicate the end of a row. Matrix elements can be any *MATLAB* expression. The statement

```
A = [ 6  1  2;  -1  8  3;  2  4  9]
```

results in the output

```
A =
          6    1    2
         -1    8    3
          2    4    9
```

If a semicolon is not used, each row must be entered in a separate line as shown below.

```
A = [ 6    1    2
     -1    8    3
      2    4    9]
```

The entire row or column of a matrix can be addressed by means of the symbol (:). For example

```
r3 =  A(3, :)
```

results in

```
r3 =
        2     4     9
```

Similarly, the statement $A(:, 2)$ addresses all elements of the second column in A.

Matrices of the same dimension can be added or subtracted. Two matrices, A and B, can be multiplied together to form the product AB if they are conformable. Two symbols are used for nonsingular matrix division. $A\backslash B$ is equivalent to $A^{-1}B$, and A/B is equivalent to AB^{-1}

Example A.1

For the matrix equation below, $AX = B$, determine the vector X.

$$\begin{bmatrix} 4 & -2 & -10 \\ 2 & 10 & -12 \\ -4 & -6 & 16 \end{bmatrix} \begin{bmatrix} x_1 \\ x_2 \\ x_3 \end{bmatrix} = \begin{bmatrix} -10 \\ 32 \\ -16 \end{bmatrix}$$

The following statements

```
A = [4  -2  -10;  2  10  -12;  -4  -6  16];
B = [-10;  32;  -16];
X = A\B
```

result in the output

```
X =
       2.0000
       4.0000
       1.0000
```

In addition to the built-in functions, numerous mathematical functions are available in the form of M-files. For the current list and their applications, see the *MATLAB User's Guide*.

Example A.2

Use the **inv** function to determine the inverse of matrix A in Example A.1 and then determine X. The following statements

```
A = [4  -2  -10;  2  10  -12;  -4  -6  16];
B = [-10;  32;  -16];
C = inv(A)
X = C*B
```

result in the output

```
C =
        2.2000    2.3000    3.1000
        0.4000    0.6000    0.7000
        0.7000    0.8000    1.1000
X =
        2.0000
        4.0000
        1.0000
```

Example A.3

Use the **lu** factorization function to express the matrix A of Example A.2 as the product of upper and lower triangular matrices, $A = LU$. Then find X from $X = U^{-1}L^{-1}B$. Typing

```
A = [ 4  -2  -10;  2  10  -12;  -4  -6  16 ]
B = [-10;  32  -16];
[L,U] = lu(A)
```

results in

```
L =
        1.0000         0         0
        0.5000    1.0000         0
       -1.0000   -0.7273    1.0000

U =
        4.0000   -2.0000  -10.0000
             0   11.0000   -7.0000
             0         0    0.9091
```

Now entering

```
X = inv(U)*inv(L)*B
```

results in

```
X =
        2.0000
        4.0000
        1.0000
```

Dimensioning is automatic in *MATLAB*. You can find the dimensions and rank of an existing matrix with the **size** and **rank** statements. For vectors, use the command **length**.

A.7.1 UTILITY MATRICES

There are many special utility matrices which are useful for matrix operations. A few examples are

eye(m, n)	Generates an m-by-n identity matrix.
zeros(m, n)	Generates an m-by-n matrix of zeros.
ones(m, n)	Generates an m-by-n matrix of ones.
diag(x)	Produces a diagonal matrix with the elements of **x** on the diagonal line.

For a complete list and help on elementary matrices and matrix manipulation, type **help elmat**. There are many other special built-in matrices. For a complete list and help on specialized matrices, type **help specmat**.

A.7.2 EIGENVALUES

If A is an n-by-n matrix, the n numbers λ that satisfy $Ax = \lambda x$ are the eigenvalues of A. They are found using **eig(A)**, which returns the eigenvalues in a column vector. Eigenvalues and eigenvectors can be obtained with a double assignment statement $[X, D] = \text{eig}(A)$. The diagonal elements of D are the eigenvalues and the columns of X are the corresponding eigenvectors such that $AX = XD$.

Example A.4

Find the eigenvalues and the eigenvectors of the matrix A given by

$$A = \begin{bmatrix} 0 & 1 & -1 \\ -6 & -11 & 6 \\ -6 & -11 & 5 \end{bmatrix}$$

```
A = [ 0  1  -1;  -6  -11  6;  -6  -11  5];
[X,D] = eig(A)
```

The eigenvalues and the eigenvectors are obtained as follows

```
X =                                       D =
        -0.7071   0.2182  -0.0921              -1     0     0
         0.0000   0.4364  -0.5523               0    -2     0
        -0.7071   0.8729  -0.8285               0     0    -3
```

A.8 COMPLEX NUMBERS

All the *MATLAB* arithmetic operators are available for complex operations. The imaginary unit $\sqrt{-1}$ is predefined by two variables i and j. In a program, if other values are assigned to i and j, they must be redefined as imaginary units, or other characters can be defined for the imaginary unit.

```
j = sqrt(-1)     or i = sqrt(-1)
```

Once the complex unit has been defined, complex numbers can be generated.

Example A.5

Evaluate the following function $V = Zc\cosh g + \sinh g/Zc$, where $Zc = 200 + j300$ and $g = 0.02 + j1.5$

```
i  = sqrt(-1); Zc = 200 +  300*i; g  = 0.02 + 1.5*i;
v  = Zc *cosh(g) + sinh(g)/Zc
```

results in the output

```
v =
        8.1672 + 25.2172i
```

It is important to note that, when complex numbers are entered as matrix elements within brackets, we avoid any blank spaces. If spaces are provided around the complex number sign, it represents two separate numbers.

Example A.6

In the circuit shown in Figure A.1, determine the node voltages V_1 and V_2 and the power delivered by each source.

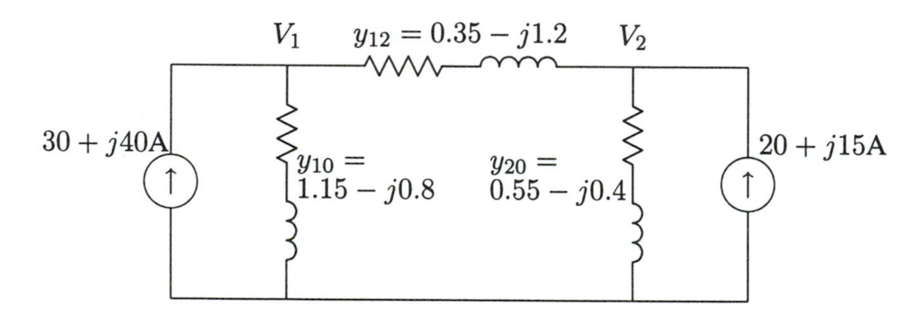

FIGURE A.1
Circuit for Example A.6.

Kirchhoff's current law results in the following matrix node equation.

$$\begin{bmatrix} 1.5 - j2.0 & -.35 + j1.2 \\ -.35 + j1.2 & 0.9 - j1.6 \end{bmatrix} \begin{bmatrix} V_1 \\ V_2 \end{bmatrix} = \begin{bmatrix} 30 + j40 \\ 20 + j15 \end{bmatrix}$$

and the complex power of each source is given by $S = VI^*$. The following program is written to yield solutions to V_1 ,V_2 and S using *MATLAB*.

```
j=sqrt(-1)                                          % Defining j
I=[30+j*40; 20+j*15]         % Column of node current phasors
Y=[1.5-j*2   -.35+j*1.2; -.35+j*1.2  .9-j*1.6]
                                   % Complex admittance matrix Y
disp('The solution is') V=inv(Y)*I  % Node voltage solution
S=V.*conj(I)                           % complex power at nodes
```

result in

```
The solution is
      V =
           3.5902 + 35.0928i
           6.0155 + 36.2212i

      S =
           1511.4 + 909.2i
            663.6 + 634.2i
```

The prime (′) transposes a real matrix; but for complex matrices, the symbol (.′) must be used to find the transpose.

A.9 POLYNOMIAL ROOTS AND CHARACTERISTIC POLYNOMIAL

If p is a row vector containing the coefficients of a polynomial, **roots(p)** returns a column vector whose elements are the roots of the polynomial. If r is a column vector containing the roots of a polynomial, **poly(r)** returns a row vector whose elements are the coefficients of the polynomial.

Example A.7

Find the roots of the following polynomial.

$$s^6 + 9s^5 + 31.25s^4 + 61.25s^3 + 67.75s^2 + 14.75s + 15$$

The polynomial coefficients are entered in a row vector in descending powers. The roots are found using **roots**.

```
p = [ 1  9  31.25  61.25  67.75  14.75  15 ]
r = roots(p)
```

The polynomial roots are obtained in column vector

```
r =
     -4.0000
     -3.0000
     -1.0000 + 2.0000i
     -1.0000 - 2.0000i
      0.0000 + 0.5000i
      0.0000 - 0.5000i
```

Example A.8

The roots of a polynomial are -1, -2, $-3 \pm j4$. Determine the polynomial equation.

Complex numbers may be entered using function i or j. The roots are then entered in a column vector. The polynomial equation is obtained using **poly** as follows

```
i = sqrt(-1)
r = [-1  -2  -3+4*i  -3-4*i ]
p = poly(r)
```

The coefficients of the polynomial equation are obtained in a row vector.

```
p =
      1   9   45   87   50
```

Therefore, the polynomial equation is

$$s^4 + 9s^3 + 45s^2 + 87s + 50 = 0$$

Example A.9

Determine the roots of the characteristic equation of the following matrix.

$$A = \begin{bmatrix} 0 & 1 & -1 \\ -6 & -11 & 6 \\ -6 & -11 & 5 \end{bmatrix}$$

The characteristic equation of the matrix is found by **poly**, and the roots of this equation are found by **roots**.

```
A = [ 0  1  -1;  -6  -11  6;  -6  -11  5];
p = poly(A)
r = roots(p)
```

The result is as follows

```
p =
        1.0000   6.0000   11.0000   6.0000
r =   -3.0000
      -2.0000
      -1.0000
```

The roots of the characteristic equation are the same as the eigenvalues of matrix A. Thus, in place of the **poly** and **roots** function, we may use

```
r = eig(A)
```

A.9.1 PRODUCT AND DIVISION OF POLYNOMIALS

The product of polynomials is the convolution of the coefficients. The division of polynomials is obtained by using the deconvolution command.

Example A.10

(a) Given $A = s^2 + 7s + 12$, and $B = s^2 + 9$, find $C = AB$.
(b) Given $Z = s^4 + 9s^3 + 37s^2 + 81s + 52$, and $Y = s^2 + 4s + 13$, find $X = \frac{Z}{Y}$.

The commands

```
A = [1   7   12]; B = [1   0   9];
C = conv(A, B)
Z = [1   9   37 81   52]; Y = [1   4   13];
[X, r] = deconv(Z, Y)
```

result in

```
C =
     1   7   21   63   108
X =
     1  5   4
r =
     0  0   0
```

A.9.2 POLYNOMIAL CURVE FITTING

In general, a polynomial fit to data in vector x and y is a function p of the form

$$p(x) = c_1 x^d + c_2 x^{d-1} + \cdots + c_n$$

The degree is d, and the number of coefficients is $n = d + 1$. Given a set of points in vectors x and y, **polyfit(x, y, d)** returns the coefficients of dth order polynomial in descending powers of x.

Example A.11

Find a polynomial of degree 3 to fit the following data

x	0	1	2	4	6	10
y	1	7	23	109	307	1231

```
x = [ 0  1  2  4  6  10];
y = [ 1  7  23  109  307  1231];
c = polyfit(x,y,3)
```

The coefficients of a third degree polynomial are found as follows

```
c =
        1.0000    2.0000    3.0000    1.0000
```

i.e., $y = x^3 + 2x^2 + 3x + 1$.

A.9.3 POLYNOMIAL EVALUATION

If c is a vector whose elements are the coefficients of a polynomial in descending powers, the **polyval(c, x)** is the value of the polynomial evaluated at **x**. For example, to evaluate the above polynomial at points 0, 1, 2, 3, and 4, use the commands

```
c = [1   2  3  1];
x = 0:1:4;
y = polyval(c, x)
```

which result in

```
y =
        7    23    55    109
```

A.9.4 PARTIAL-FRACTION EXPANSION

[r, p, k] = residue[b, a] finds the residues, poles, and direct terms of a partial fraction expansion of the ratio of two polynomials

$$\frac{P(s)}{Q(s)} = \frac{b_m s^m + b_{m-1}s^{m-1} + \cdots + b_1 s + b_0}{a_n s^n + a_{n-1}s^{n-1} + \cdots + a_1 s + a_0}$$

Vectors **b** and **a** specify the coefficients of the polynomials in descending powers of s. The residues are returned in column vector **r**, the pole locations in column vector **p**, and the direct terms in row vector **k**.

Example A.12

Determine the partial fraction expansion for

$$F(s) = \frac{2s^3 + 9s + 1}{s^3 + s^2 + 4s + 4}$$

```
      b = [ 2  0  9  1];
      a = [ 1  1  4  4];
[r,p,k] = residue(b,a)
```

The result is as follows

```
      r =
            0.0000   -0.2500i
            0.0000   +0.2500i
           -2.0000
      p =
            0.0000   +2.0000i
            0.0000   -2.0000i
           -1.0000
      K =
            2.0000
```

Therefore the partial fraction expansion is

$$2 + \frac{-2}{s+1} + \frac{j0.25}{s+j2} + \frac{-j0.25}{s-j2} = 2 + \frac{-2}{s+1} + \frac{1}{s^2+4}$$

[b, a] = residue(r, p, K) converts the partial fraction expansion back to the polynomial $P(s)/Q(s)$.

For a complete list and help on matrix analysis, linear equations, eigenvalues, and matrix functions, type **help matfun**.

A.10 GRAPHICS

MATLAB can create high-resolution, publication-quality *2-D*, *3-D*, *linear*, *semilog*, *log*, *polar*, *bar chart* and *contour plots* on plotters, dot-matrix printers, and laser printers. Some of the *2-D* graph types are **plot**, **loglog**, **semilogx**, **semi -logy**, **polar**, and **bar**. The syntax for the above plots includes the following optional symbols and colors.

COLOR SPECIFICATION		*LINE STYLE-OPTION*	
Long name	*Short name*	*Style*	*Symbol*
black	k	solid	–
blue	b	dashed	- -
cyan	c	dotted	:
green	g	dash-dot	-.
magenta	m	point	.
red	r	circle	o
white	w	x-mark	x
yellow	y	plus	+
		star	*

You have three options for plotting multiple curves on the same graph. For example,

```
plot(x1, y1,'r', x2, y2, '+b', x3, y3, '--')
```

plots `(x1, y1)` with a solid red line, `(x2, y2)` with a blue + mark, and `(x3, y3)` with a dashed line. If **X** and **Y** are matrices of the same size, **plot(X, Y)** will plot the columns of **Y** versus the column of **X**.

Alternatively, the **hold** command can be used to place new plots on the previous graph. **hold on** holds the current plot and all axes properties; subsequent plot commands are added to the existing graph. **hold off** returns to the default mode whereby a new plot command replaces the previous plot. **hold**, by itself, toggles the hold state.

Another way for plotting multiple curves on the same graph is the use of the **line** command. For example, if a graph is generated by the command `plot(x1, y1)`, then the commands

```
line(x2, y2, '+b')
line(x3, y3, '--')
```

Add curve `(x2, y2)` with a blue + mark, and `(x2, y2)` with a dashed line to the existing graph generated by the previous plot command. Multiple figure windows can be created by the **figure** command. **figure**, by itself, opens a new figure window, and returns the next available figure number, known as the figure handle. **figure(h)** makes the figure with handle h the current figure for subsequent plotting commands. Plots may be annotated with title, $x - y$ labels and grid. The command **grid** adds a grid to the graph. The commands **title**('*Graph title* ') titles the plot, and **xlabel**('*x-axis label* '), **ylabel**('*y-axis label* ') label the plot with the specified string argument. The command **text**(*x-coordinate, y-coordinate*, '*text*') can be used for placing text on the graph, where the coordinate values are taken from the current plot. For example, the statement

```
text(3.5,  1.5, 'Voltage')
```

will write Voltage at point (3.5, 1.5) in the current plot. Alternatively, you can use the **gtext**(*'text'*) command for interactive labeling. Using this command after a plot provides a crosshair in the Figure window and lets the user specify the location of the text by clicking the mouse at the desired location. Finally, the command **legend**(*string1, string2, string3, ...*) may be used to place a legend on the current plot using the specified strings as labels. This command has many optional arguments. For example, **legend**(*linetype1, string1, linetype2, string2, linetype3, string3, ...*) specifies the line types/color for each label at a suitable location. However, you can move the legend to a desired location with the mouse. **legend off** removes the legend from the current axes.

MATLAB provides automatic scaling. The command **axis([x min. x max. y min. y max.])** enforces the manual scaling. For example

```
axis([-10   40   -60  60])
```

produces an x-axis scale from -10 to 40 and a y-axis scale from -60 to 60. Typing **axis** again or **axis('auto')** resumes auto scaling. Also, the aspect ratio of the plot can be made equal to one with the command **axis('square')**. With a square aspect ratio, a line with slope 1 is at a true 45 degree angle. **axis('equal')** will make the x- and y-axis scaling factor and tic mark increments the same. For a complete list and help on general purpose graphic functions, and two- and three-dimensional graphics, see **help graphics**, **help plotxy**, and **help plotxyz**.

There are many other specialized commands for two-dimensional plotting. Among the most useful are the **semilogx** and **semilogy**, which produce a plot with an x-axis log scale and a y-axis log scale. An interesting graphic command is the **comet** plot. The command **comet(x, y)** plots the data in vectors **x** and **y** with a comet moving through the data points, and you can see the curve as it is being plotted. For a complete list and help on general purpose graphic functions and two-dimensional graphics, see **help graphics** and **help plotxy**.

A.11 GRAPHICS HARD COPY

The easiest way to obtain hard-copy printout is to make use of the Windows built-in facilities. In the Figure window, you can pull down the file menu and click on the **Print** command to send the current graph directly to the printer. You can also import a graph to your favorite word processor. To do this, select **Copy options** from the **Edit** pull-down menu, and check mark the **Invert background** option in the dialog box to invert the background. Then, use **Copy** command to copy the graph into the clipboard. Launch your word processor and use the **Paste** command to import the graph.

Some word processors may not provide the extensive support of the Windows graphics and the captured graph may be corrupted in color. To eliminate this problem use the command

```
system_dependent(14, 'on')
```

which sets the metafile rendering to the lowest common denominator. To set the metafile rendering to normal, use

```
system_dependent(14, 'off')
```

In addition *MATLAB* provides a function called **print** that can be used to produce high resolution graphic files. For example,

```
print -dhpgl [filename]
```

saves the graph under the specified *filename* with extension *hgl*. This file may be processed with an HPGL- compatible plotter. Similarly, the command

```
print -dilll [filename]
```

produces a graphic file compatible with the Adobe Illustrator'88. Another **print** option allows you to save and reload a figure. The command

```
print -dmfile [ filename ]
```

produces a MAT file and M-file to reproduce the figure again.

Example A.13

Create a linear X-Y plot for the following variables.

x	0	0.5	1.0	1.5	2.0	2.5	3.0	3.5	4.0	4.5	5.0
y	10	10	16	24	30	38	52	68	82	96	123

For a small amount of data, you can type in data explicitly using brackets.

```
x = [ 0  0.5  1.0  1.5  2.0  2.5  3.0  3.5  4.0  4.5  5.0];
y = [10   10   16   24   30   38   52   68   82   96  123];
plot(x, y), grid
xlabel('x'), ylabel('y'), title('A simple plot example')
```

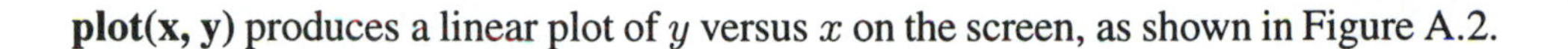

plot(x, y) produces a linear plot of y versus x on the screen, as shown in Figure A.2.

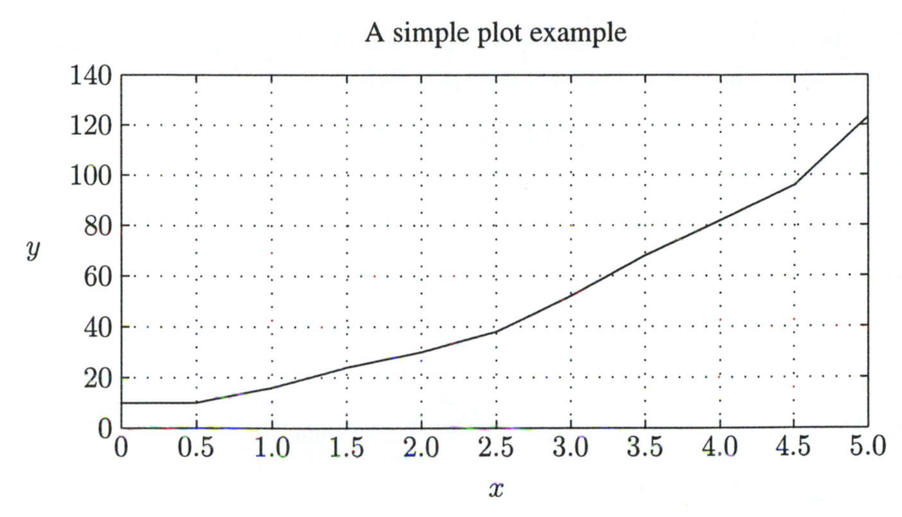

FIGURE A.2
Example of X-Y plot.

For large amounts of data, use the text editor to create a file with extension **m**. Typing the *filename* creates your data in the workspace.

Example A.14

Fit a polynomial of order 2 to the data in Example A.13. Plot the given data point with symbol x, and the fitted curve with a solid line. Place a boxed legend on the graph.

The command **p = polyfit(x, y, 2)** is used to find the coefficients of a polynomial of degree 2 that fits the data, and the command **yc = polyval(p, x)** is used to evaluate the polynomial at all values in **x**. We use the following command.

```
x = [ 0  0.5  1.0  1.5  2.0  2.5  3.0  3.5  4.0  4.5  5.0];
y = [10   10  16   24   30   38   52    68   82   96  123];
p = polyfit(x, y, 2)  % finds the coefficients of a polynomial
                            % of degree 2 that fits the data
yc = polyval(p, x);%polynomial is evaluated at all points in x
plot(x, y,'x', x, yc)%plots data with  x and fitted polynomial
xlabel('x'), ylabel('y'), grid
title('Polynomial curve fitting')
legend('Actual data', 'Fitted polynomial')
```

The result is the array of coefficients of the polynomial of degree 2, and is

```
p =
       4.0232       2.0107      9.6783
```

Thus, the parabola $4.0x^2 + 2.0x + 9.68$ is found that fits the given data in the least-square sense. The plots are shown in Figure A.3.

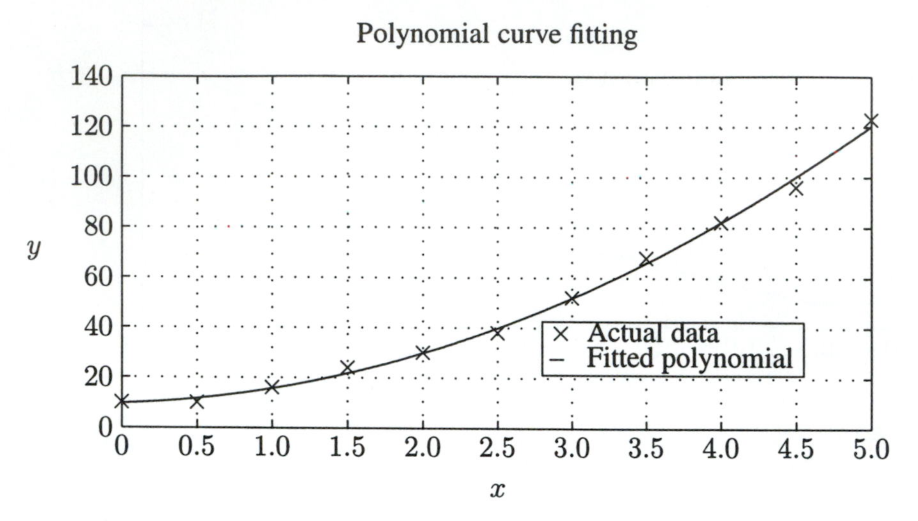

FIGURE A.3
Fitting a parabola to the data in Example A.13.

Example A.15

Plot function $y = 1 + e^{-2t}\sin(8t - \pi/2)$ from 0 to 3 seconds. Find the time corresponding to the peak value of the function and the peak value. The graph is to be labeled, titled, and have grid lines displayed.

Remember to use .* for the element-by-element multiplication of the two terms in the given equation. The command **[cp, k] = max(c)** returns the peak value and the index k corresponding to the peak time. We use the following commands.

```
t=0:.005:3; c = 1+ exp(-2*t).*sin(8*t - pi/2);
[cp, k] = max(c)  % cp is the maximum value of c at interval k
tp = t(k)                                  % tp is the peak time
plot(t, c), xlabel(' t - sec'), ylabel('c'), grid
title('Damped sine curve')
text(0.55,1.35,['cp =',num2str(cp)])%Text in quote & the value
   % of cp are printed on the graph at the specified location
text(0.55, 1.2, ['tp = ',num2str(tp)])
```

The result is

```
cp =
        1.4702
 k =
        73
tp =
        0.3600
```

and the plot is shown in Figure A.4.

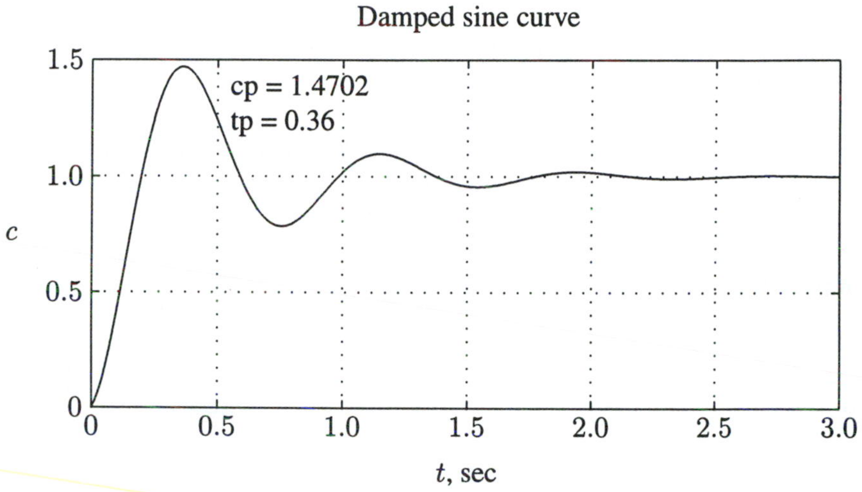

Damped sine curve

FIGURE A.4
Graph of Example A.15.

An interactive way to find the data points on the curve is by using the **ginput** command. Entering **[x, y] = ginput** will put a crosshair on the graph. Position the crosshair at the desired location on the curve, and click the mouse. You can repeat this procedure for extracting coordinates for as many points as required. When the return key is pressed, the input is terminated and the extracted data is printed on the command menu. For example, to find the peak value and the peak time for the function in Example A.15, try

```
[tp, cp] = ginput
```

A crosshair will appear. Move the crosshair to the peak position, and click the mouse. Press the return key to get

```
cp =
        1.47
tp =
        0.36
```

subplot splits the Figure window into multiple portions, in order to show several plots at the same time. The statement **subplot(m, n, p)** breaks the Figure window into an m-by-n box and uses the pth box for the subsequent plot. Thus, the command **subplot(2, 2, 3)**, **plot(x,y)** divides the Figure window into four subwindows and plots **y** versus **x** in the third subwindow, which is the first subwindow in the second row. The command **subplot(111)** returns to the default Figure window. This is demonstrated in the next example.

Example A.16

Divide the Figure window into four partitions, and plot the following functions for ωt from 0 to 3π in steps of 0.05.

1. Plot $v = 120 \sin \omega t$ and $i = 100 \sin(\omega t - \pi/4)$ versus ωt on the upper left portion.

2. Plot $p = vi$ on the upper right portion.

3. Given $F_m = 3.0$, plot $f_a = F_m \sin \omega t$, $F_b = F_m \sin(\omega t - 2\pi/3)$, and $F_c = F_m \sin(\omega t - 4\pi/3)$ versus ωt on the lower left portion.

4. For $f_R = 3F_m$, construct a circle of radius f_R on the lower right portion.

```
wt = 0: 0.05: 3*pi; v=120*sin(wt);          %Sinusoidal voltage
i = 100*sin(wt - pi/4);                      %Sinusoidal current
p = v.*i;                                   %Instantaneous power
subplot(2, 2, 1), plot(wt, v, wt, i); %Plot of v & i versus wt
title('Voltage & current'), xlabel('wt, radians');
subplot(2, 2, 2), plot(wt, p);     % Instantaneous power vs. wt
title('Power'), xlabel(' wt,  radians ')
Fm=3.0;
fa = Fm*sin(wt);                      % Three-phase mmf's fa, fb, fc
fb = Fm*sin(wt - 2*pi/3); fc = Fm*sin(wt - 4*pi/3);
subplot(2, 2, 3), plot(wt, fa, wt, fb, wt, fc)
title('3-phase mmf'), xlabel(' wt, radians ')
fR = 3/2*Fm;
subplot(2, 2, 4), plot(-fR*cos(wt), fR*sin(wt))
title('Rotating mmf'), subplot(111)
```

Example A.16 results are shown in Figure A.5.

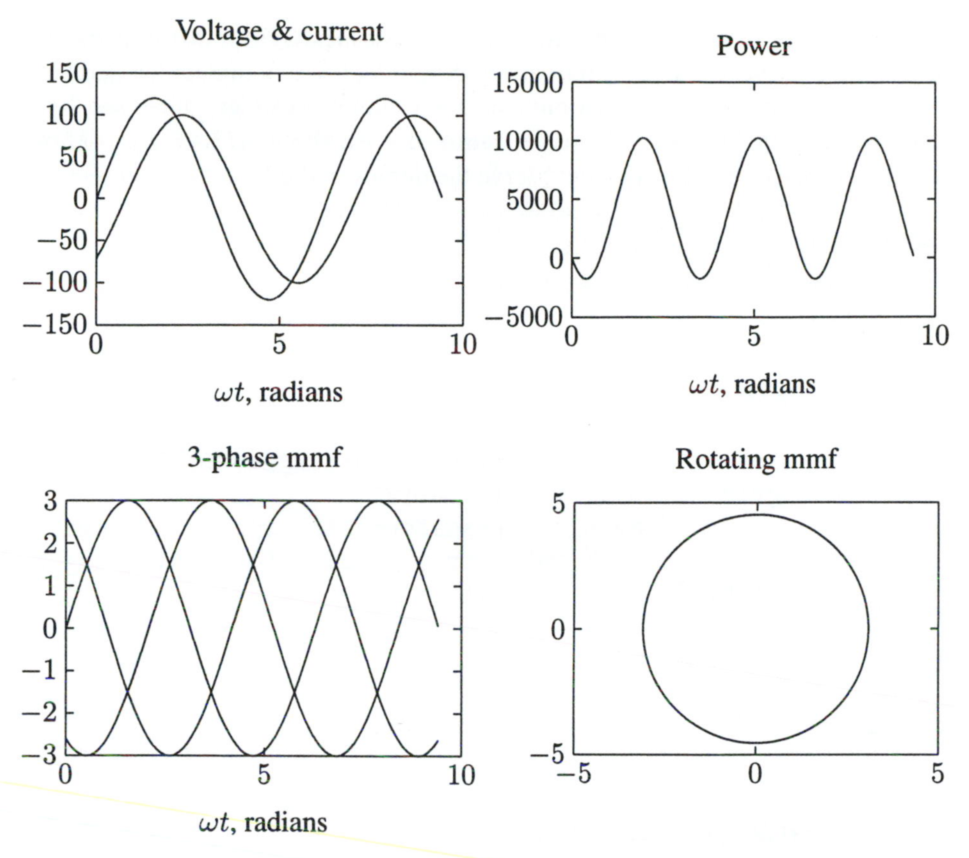

FIGURE A.5
Subplot demonstration.

A.12 THREE-DIMENSIONAL PLOTS

MATLAB provides extensive facilities for visualization of three-dimensional data. The most common are plots of curves in a three-dimensional space, mesh plots, surface plots, and contour plots. The command **plot3(x, y, z , 'style option')** produces a curve in the three-dimensional space. The viewing angle may be specified by the command **view**(*azimuth, elevation*). The arguments *azimuth*, and *elevation* specifies the horizontal and vertical rotation in degrees, respectively. The **title**, **xlabel**, **ylabel**, etc., may be used for three-dimensional plots. The **mesh** and **surf** commands have several optional arguments and are used for plotting meshes and surfaces. The **contour(z)** command creates a contour plot of matrix **z**, treating the values in **z** as heights above the plane. The statement **mesh(z)** creates a three-dimensional plot of the elements in matrix **z**. A mesh surface is defined by the **z** coordinates of points above a rectangular grid in the x-y plane. The plot is formed by joining adjacent points with straight lines. **meshgrid** transforms the domain specified by vector **x** and **y** into arrays **X** and **Y**. For a complete list and help on general purpose Graphic functions and three-dimensional graphics, see **help graphics** and **help plotxyz**. Also type **demo** to open the *MATLAB Expo Menu Map* and visit *MATLAB*. Select and observe the demos in the Visualization section.

Example A.17

Obtain the cartesian plot of the Bessel function $J_0\sqrt{x^2+y^2}$ over the range $-12 < x < 12$, $-12 < y < 12$.

Use the following commands

```
% Cartesian plot of Bessel function J0(sqrt(x^2+y^2))
[x, y] = meshgrid(-12:0.6:12, -12:.6:12);
                         % meshgrid transforms the specified domain
                              % into array x and y for evaluating z
r = sqrt(x.^2 + y.^2); z = bessel(0,r);
m = [-45 60];                                       % viewing angle
mesh(z, m)                                          % 3-D mesh plot
```

Enter **exa17** at the *MATLAB* prompt to see the result.

A.13 HANDLE GRAPHICS

It is often desirable to be able to customize the graphical output. *MATLAB* allows object-oriented programming, enabling the user to have complete control over the details of a graph. *MATLAB* provides many low-level commands known as *Handle Graphics*. These commands makes it possible to access individual objects and their properties and change any property of an object without affecting other properties or objects. Handle Graphics provides a graphical user interface (GUI) in which the user interface includes push buttons and menus. These topics are not discussed here; like *MATLAB* syntax, they are easy to follow, and we leave these topics for the interested reader to explore.

A.14 LOOPS AND LOGICAL STATEMENTS

MATLAB provides loops and logical statements for programming, like **for**, **while**, and **if** statements. The **for** statement instructs the computer to perform all subsequent expressions up to the **end** statement for a specified number of counted times. The expression may be a matrix. The following is an example of a nested loop.

```
for i = 1:n,  for j = 1:n
   expression
end, end
```

The **while** statement allows statements to be repeated an indefinite number of times under the control of a logic statement. The **if**, **else**, and **elseif** statements allow conditional execution of statements. *MATLAB* has six relational operators and four logical operators, which are defined in the following table.

Relational Operator		*Logical Operator*	
==	equal	&	logical AND
~=	not equal	|	logical OR
<	less than	~	logical complement
<=	less than or equal to	xor	exclusive OR
>	greater than		
>=	greater than or equal to		

A.15 SOLUTION OF DIFFERENTIAL EQUATIONS

Analytical solutions of linear time-invariant equations are obtained through the Laplace transform and its inversion. There are other techniques which use the state transition matrix $\phi(t)$ to provide a solution. These analytical methods are normally restricted to linear differential equations with constant coefficients. Numerical techniques solve differential equations directly in the time domain; they apply not only to linear time-invariant, but also to nonlinear and time varying differential equations. The value of the function obtained at any step is an approximation of the value which would have been obtained analytically; whereas, the analytical solution is exact. However, an analytical solution may be difficult, time consuming, or even impossible to find.

MATLAB provides two functions for numerical solutions of differential equations employing the Runge-Kutta method. These are **ode23** and **ode45**, based on the Fehlberg- second and third-order pair of formulas for medium accuracy and fourth- and fifth-order pair for high accuracy. The nth-order differential equation must be transformed into n first-order differential equations and must be placed in an M-file that returns the state derivatives of the equations. The following examples demonstrate the use of these functions.

Example A.18

Consider the simple mechanical system of Figure A.6. Three forces influence the motion of the mass, namely, the applied force, the frictional force, and the spring force.

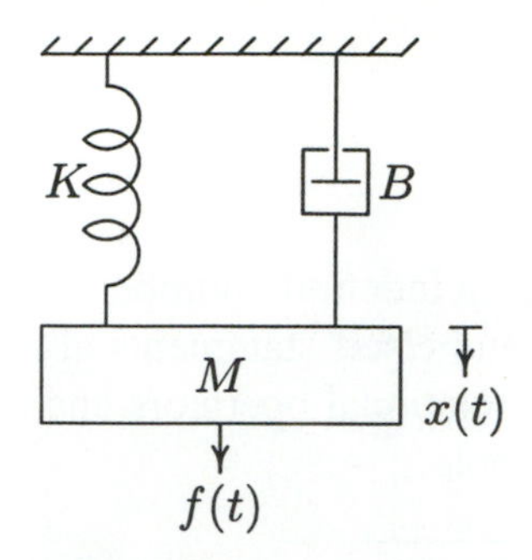

FIGURE A.6
Mechanical translational system.

Applying Newton's law of motion, the force equation of the system is

$$M\frac{d^2x}{dt^2} + B\frac{dx}{dt} + Kx = f(t)$$

Let

$$x_1 = x$$

and

$$x_2 = \frac{dx}{dt}$$

then

$$\frac{dx_1}{dt} = x_2$$
$$\frac{dx_2}{dt} = \frac{1}{M}[f(t) - Bx_2 - Kx_1]$$

With the system initially at rest, a force of 25 newtons is applied at time $t = 0$. Assume that the mass $M = 1$ kg, frictional coefficient $B = 5$ N/m/sec, and the spring constant $K = 25$ N/m. The above equations are defined in an M-file **mechsys.m** as follows.

```
function xdot = mechsys(t, x); % returns the state derivatives
F = 25;                                              % Step input
M =1; B = 5; K = 25; xdot = [x(2); 1/M*(F - B*x(2)-K*x(1) ) ];
```

The following M-file, **exa18.m**, uses **ode23** to simulate the system over an interval of 0 to 3 sec., with zero initial conditions.

```
tspan = [0, 3];                                    % time interval
x0 = [0, 0];                                  % initial conditions
```

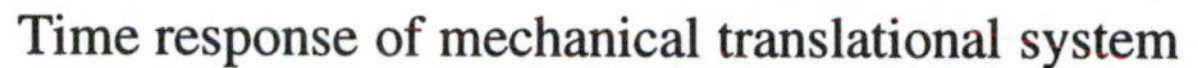

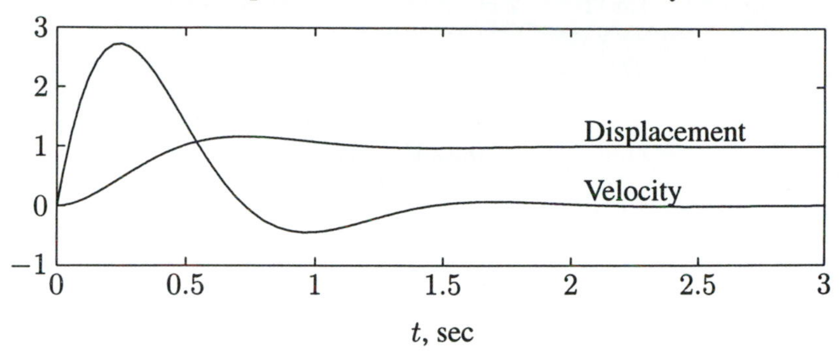

Velocity versus displacement

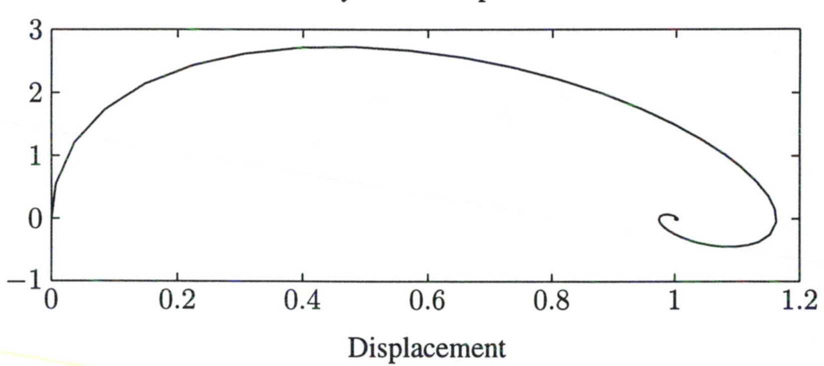

FIGURE A.7
Response of the mechanical system of Example A.18.

```
[t,x] = ode23('mechsys', tspan, x0);
subplot(2, 1, 1), plot(t, x), xlabel('t,  sec')
title('Time response of mechanical translational system')
text(2, 1.2, 'Displacement'), text(2, 0.2, 'Velocity')
d = x(:, 1);    v = x(:, 2);
subplot(2, 1, 2), plot(d, v)
title('Velocity versus displacement ')
xlabel('Displacement'), ylabel('Velocity'), subplot(111)
```

Results of the simulation are shown in Figure A.7.

Example A.19

The circuit elements in Figure A.8 are $R = 1.4\ \Omega$, $L = 2$ H, and $C = 0.32$ F. The initial inductor current is zero, and the initial capacitor voltage is 0.5 volts. A step voltage of 1 volt is applied at time $t = 0$. Determine $i(t)$ and $v(t)$ over the range $0 < t < 15$ sec. Also, obtain a plot of current versus capacitor voltage.

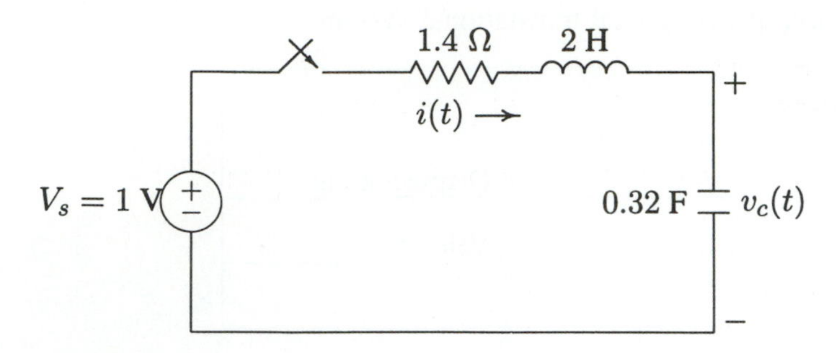

FIGURE A.8
RLC circuit for Example A.19.

Applying KVL

$$Ri + L\frac{di}{dt} + v_c = V_s$$

and

$$i = C\frac{dv_c}{dt}$$

Let

$$x_1 = v_c$$

and

$$x_2 = i$$

Then

$$\dot{x}_1 = \frac{1}{C}x_2$$

and

$$\dot{x}_2 = \frac{1}{L}(V_s - x_1 - Rx_2)$$

The above equations are defined in an M-file **electsys.m** as follows.

```
function xdot = electsys(t, x);
                                    % returns the state derivatives
V = 1;                                                  % Step input
R =1.4; L = 2; C = 0.32;
xdot = [x(2)/C; 1/L*( V - x(1) - R*x(2) )];
```

The following M-file, **exa19.m**, uses **ode23** to simulate the system over an interval of 0 to 15 sec.

```
tspan = [0, 15];                                     % time interval
x0 = [0.5, 0];                                  % initial conditions
```

```
[t,x] = ode23('electsys', tspan, x0);
subplot(2, 1, 1), plot(t, x)
title('Time response of an RLC series circuit')
xlabel('t,  sec')
text(8,1.05,'Capacitor voltage'), text(8, .05,'Current')
vc= x(:, 1);   i = x(:, 2);
subplot(2, 1, 2), plot(vc, i)
title('Current versus capacitor voltage ')
xlabel('Capacitor voltage'), ylabel('Current')
subplot(111)
```

Results of the simulation are shown in Figure A.9.

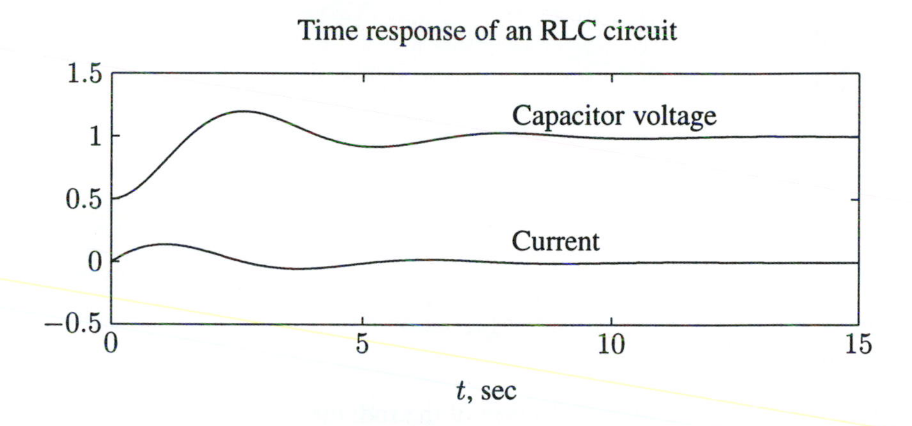

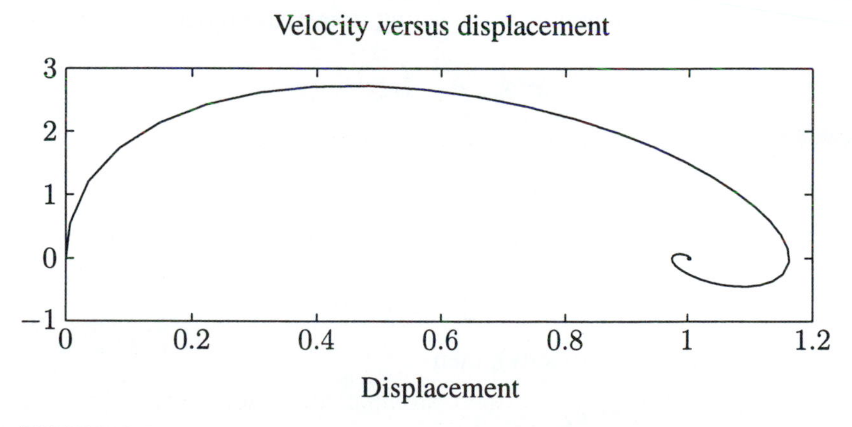

FIGURE A.9
Response of the series RLC circuit of Example A.19.

A.16 NONLINEAR SYSTEMS

A great majority of physical systems are linear within some range of the variables. However, all systems ultimately become nonlinear as the ranges are increased without limit. For the nonlinear systems, the principle of superposition does not apply. **ode23** and **ode45** simplify the task of solving a set of nonlinear differential equations, as demonstrated in Example A.20.

Example A.20

Consider the simple pendulum illustrated in Figure A.10, where a weight of $W = mg$ kg is hung from a support by a weightless rod of length L meters. While usually approximated by a linear differential equation, the system really is nonlinear and includes viscous damping with a damping coefficient of B kg/m/sec.

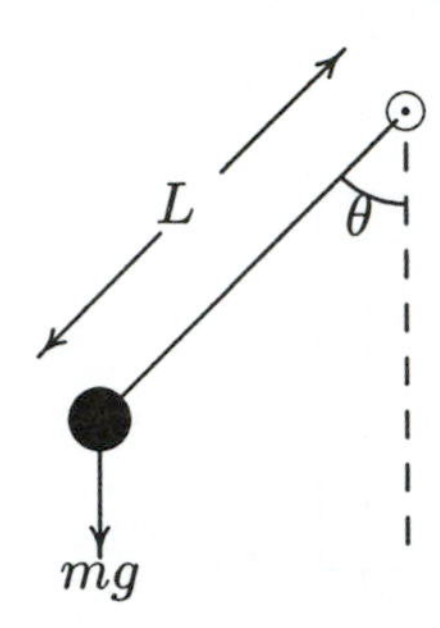

FIGURE A.10
Pendulum oscillator.

If θ in radians is the angle of deflection of the rod, the velocity of the weight at the end will be $L\dot{\theta}$ and the tangential force acting to increase the angle θ can be written as

$$F_T = -W \sin\,\theta - BL\dot{\theta}$$

From Newton's law

$$F_T = mL\ddot{\theta}$$

Combining the two equations for the force, we get

$$mL\ddot{\theta} + BL\dot{\theta} + W \sin\,\theta = 0$$

Let $x_1 = \theta$ and $x_2 = \dot{\theta}$ (angular velocity), then

$$\begin{aligned}\dot{x_1} &= x_2 \\ \dot{x_2} &= -\frac{B}{m}x_2 - \frac{W}{mL}\sin\,x_1\end{aligned}$$

The above equations are defined in an M-file **pendulum.m** as follows.

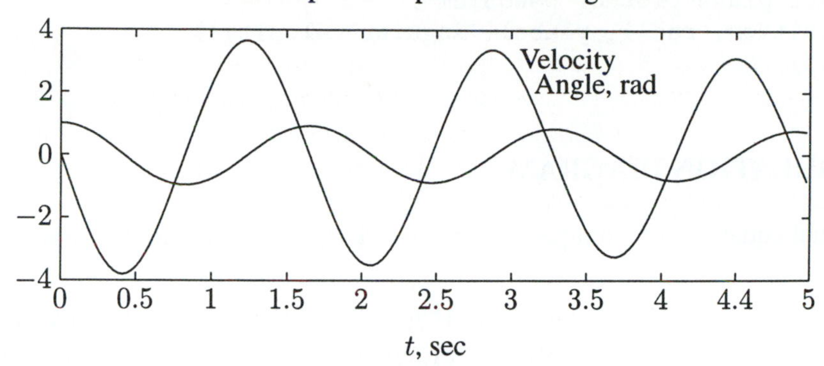

Phase plane plot of pendulum

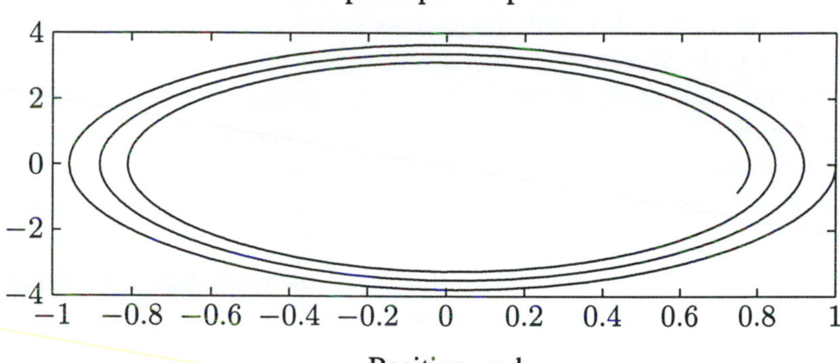

FIGURE A.11
Response of the pendulum described in Example A.20.

```
function xdot = pendulum(t,x);%returns the state derivatives
W = 2; L = .6; B = 0.02; g = 9.81;  m = W/g;
xdot = [x(2) ; -B/m*x(2)-W/(m*L)*sin(x(1)) ];
```

The following M-file, **exa20.m**, uses **ode23** to simulate the system over an interval of 0 to 5 sec. Results of the simulation are shown in Figure A.11.

```
tspan = [0, 5];                                   % time interval
x0 = [1, 0];                                 % initial conditions
[t,x] = ode23('pendulum', tspan, x0);
subplot(2, 1, 1), plot(t, x)
title('Time response of a rigid pendulum')
xlabel('t, sec')
text(3.2, 3.5, 'Velocity') , text(3.2, 1.2, 'Angle-rad.')
th= x(:, 1);    w = x(:, 2);
subplot(2, 1, 2), plot(th, w)
```

```
title('Phase plane plot of pendulum')
xlabel('Position, rad'), ylabel('Angular velocity')
subplot(111)
```

A.17 SIMULATION DIAGRAM

The differential equations of a lumped linear network can be written in the form

$$\begin{aligned}\dot{\mathbf{x}}(t) &= \mathbf{A}\mathbf{x}(t) + \mathbf{B}u(t) \\ \mathbf{y}(t) &= \mathbf{C}\mathbf{x}(t) + \mathbf{D}u(t)\end{aligned} \tag{A.1}$$

This system of first-order differential equations is known as the state equation of the system, and $\mathbf{x}$ is the state vector. One advantage of the state-space method is that the form lends itself easily to the digital and/or analog computer methods of solution. Further, the state-space method can be easily extended to analysis of nonlinear systems. State equations may be obtained from an nth-order differential equation or directly from the system model by identifying appropriate state variables.

To illustrate how we select a set of state variables, consider an nth-order linear plant model described by the differential equation

$$\frac{d^n y}{dt^n} + a_{n-1}\frac{d^{n-1}y}{dt^{n-1}} + \ldots + a_1\frac{dy}{dt} + a_0 y = u(t) \tag{A.2}$$

where $y(t)$ is the plant output and $u(t)$ is its input. A state model for this system is not unique, but depends on the choice of a set of state variables. A useful set of state variables, referred to as *phase variables*, is defined as

$$x_1 = y,\ x_2 = \dot{y},\ x_3 = \ddot{y},\ \ldots,\ x_n = y^{n-1}$$

We express $\dot{x}_k = x_{k+1}$ for $k = 1, 2, \ldots, n-1$, and then solve for $d^n y/dt^n$, and replace y and its derivatives by the corresponding state variables to give

$$\begin{aligned}\dot{x}_1 &= x_2 \\ \dot{x}_2 &= x_3 \\ &\vdots \\ \dot{x}_{n-1} &= x_n \\ \dot{x}_n &= -a_0 x_1 - a_1 x_2 - \ldots - a_{n-1}\, x_n + u(t)\end{aligned} \tag{A.3}$$

or in matrix form

$$\begin{bmatrix} \dot{x}_1 \\ \dot{x}_2 \\ \vdots \\ \dot{x}_{n-1} \\ \dot{x}_n \end{bmatrix} = \begin{bmatrix} 0 & 1 & 0 & \dots & 0 \\ 0 & 0 & 1 & \dots & 0 \\ \vdots & \vdots & \vdots & \ddots & \vdots \\ 0 & 0 & 0 & \dots & 1 \\ -a_0 & -a_1 & -a_2 & \dots & -a_{n-1} \end{bmatrix} \begin{bmatrix} x_1 \\ x_2 \\ \vdots \\ x_{n-1} \\ x_n \end{bmatrix} + \begin{bmatrix} 0 \\ 0 \\ \vdots \\ 0 \\ 1 \end{bmatrix} u(t) \tag{A.4}$$

and the output equation is

$$y = \begin{bmatrix} 1 & 0 & 0 & \dots & 0 \end{bmatrix} \mathbf{x} \tag{A.5}$$

The M-file **ode2phv.m** is developed which converts an nth-order ordinary differential equation to the state-space phase variable form. **[A, B, C] = ode2phv(ai, k)** returns the matrices **A, B, C**, where **ai** is a row vector containing coefficients of the equation in descending order, and **k** is the coefficient of the right-hand side.

Equation (A.3) indicates that state variables are determined by integrating the corresponding state equation. A diagram known as the simulation diagram can be constructed to model the given differential equations. The basic element of the simulation diagram is the integrator. The first equation in (A.3) is

$$\dot{x_1} = x_2$$

Integrating, we have

$$x_1 = \int x_2 \, dx$$

The above integral is shown by the following time-domain symbol. The integrating block is identified by symbol $\frac{1}{s}$. Adding an integrator for the remaining state variables and completing the last equation in (A.3) via a summing point and feedback paths, a simulation diagram is obtained.

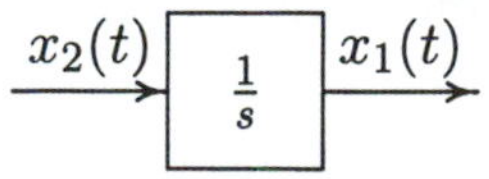

$$x_2(t) \rightarrow \boxed{\frac{1}{s}} \rightarrow x_1(t)$$

A.18 INTRODUCTION TO SIMULINK

SIMULINK is an interactive environment for modeling, analyzing, and simulating a wide variety of dynamic systems. *SIMULINK* provides a graphical user interface for constructing block diagram models using "drag-and-drop" operations. A system is configured in terms of block diagram representation from a library of standard components. *SIMULINK* is very easy to learn. A system in block diagram representation is built easily and the simulation results are displayed quickly.

Simulation algorithms and parameters can be changed in the middle of a simulation with intuitive results, thus providing the user with a ready access learning tool for simulating many of the operational problems found in the real world. *SIMULINK* is particularly useful for studying the effects of nonlinearities on the behavior of the system, and as such, it is also an ideal research tool. The key features of *SIMULINK* are

- Interactive simulations with live display.
- A comprehensive block library for creating linear, nonlinear, discrete or hybrid multi-input/output systems.
- Seven integration methods for fixed-step, variable-step, and stiff systems.
- Unlimited hierarchical model structure.
- Scalar and vector connections.
- Mask facility for creating custom blocks and block libraries.

SIMULINK provides an open architecture that allows you to extend the simulation environment:

- You can easily perform "what if" analyses by changing model parameters – either interactively or in batch mode – while your simulations are running.
- Creating custom blocks and block libraries with your own icons and user interfaces from *MATLAB*, Fortran, or C code.
- You can generate C code from *SIMULINK* models for embedded applications and for rapid prototyping of control systems.
- You can create hierarchical models by grouping blocks into subsystems. There are no limits on the number of blocks or connections.
- *SIMULINK* provides immediate access to the mathematical, graphical, and programming capabilities of *MATLAB*, you can analyze data, automate procedures, and optimize parameters directly from *SIMULINK*.
- The advanced design and analysis capabilities of the toolboxes can be executed from within a simulation using the mask facility in *SIMULINK*.

- The *SIMULINK* block library can be extended with special-purpose blocksets. The DSP Blockset can be used for DSP algorithm development, while the Fixed-Point Blockset extends *SIMULINK* for modeling and simulating digital control systems and digital filters.

A.18.1 SIMULATION PARAMETERS AND SOLVER

You set the simulation parameters and select the solver by choosing **Parameters** from the Simulation menu. *SIMULINK* displays the **Simulation Parameters** dialog box, which uses three "pages" to manage simulation parameters. **Solver**, **Workspace I/O**, and **Diagnostics**.

SOLVER PAGE

The Solver page appears when you first choose **Parameters** from the **Simulation menu** or when you select the Solver tab. The Solver page allows you to:

- Set the start and stop times – You can change the start time and stop time for the simulation by entering new values in the Start time and Stop time fields. The default start time is 0.0 seconds and the default stop time is 10.0 seconds.

- Choose the solver and specify solver parameters – The default solver provide accurate and efficient results for most problems. Some solvers may be more efficient that others at solving a particular problem; you can choose between variable-step and fixed-step solvers. Variable-step solvers can modify their step sizes during the simulation. These are **ode45**, **ode23**, **ode113**, **ode15s**, **ode23s**, and **discrete**. The default is **ode45**. For variable-step solvers, you can set the maximum and suggested initial step size parameters. By default, these parameters are automatically determined, indicated by the value auto. For fixed-step solvers, you can choose **ode5**, **ode4**, **ode3**, **ode2**, **ode1**, and **discrete**.

- Output Options – The Output options area of the dialog box enables you to control how much output the simulation generates. You can choose from three popup options. These are: Refine output, Produce additional output, and Produce specified output only.

WORKSPACE I/O PAGE

The Workspace I/O page manages the input from and the output to the *MATLAB* workspace, and allows:

- Loading input from the workspace – Input can be specified either as *MATLAB* command or as a matrix for the Import blocks.
- Saving the output to the workspace –You can specify return variables by selecting the Time, State, and/or Output check boxes in the Save to workspace area.

DIAGNOSTICS PAGE

The Diagnostics page allows you to select the level of warning messages displayed during a simulation.

A.18.2 THE SIMULATION PARAMETERS DIALOG BOX

Table below summarizes the actions performed by the dialog box buttons, which appear on the bottom of each dialog box page.

Button	Action
Apply	Applies the current parameter values and keeps the dialog box open. During a simulation, the parameter values are applied immediately.
Revert	Changes the parameter values back to the values they had when the Dialog box was most recently opened and applies the parameters.
Help	Displays help text for the dialog box page.
Close	Applies the parameter values and closes the dialog box. During a simulation, the parameter values are applied immediately.

To stop a simulation, choose Stop from the Simulation menu. The keyboard shortcut for stopping a simulation is Ctrl-T. You can suspend a running simulation by choosing Pause from the Simulation menu. When you select Pause, the menu item changes to Continue. You proceed with a suspended simulation by choosing Continue.

A.18.3 BLOCK DIAGRAM CONSTRUCTION

At the *MATLAB* prompt, type *SIMULINK*. The *SIMULINK* BLOCK LIBRARY, containing seven icons, and five pull-down menu heads, appears. Each icon contains various components in the titled category. To see the content of each category, double click on its icon. The easy-to-use pull-down menus allow you to create a *SIMULINK* block diagram, or open an existing file, perform the simulation, and make any modifications. Basically, one has to specify the model of the system (state space, discrete, transfer functions, nonlinear ode's, etc), the input (source) to the system, and where the output (sink) of the simulation of the system will go. Generally when building a model, design it first on the paper, then build it using the computer. When you start putting the blocks together into a model, add the blocks to the model window before adding the lines that connect them. This way, you can reduce how often you need to open block libraries. An introduction to *SIMULINK* is presented by constructing the *SIMULINK* diagram for the following examples.

MODELING EQUATIONS

Here are some examples that may improve your understanding of how to model equations.

Example A.21

Model the equation that converts Celsius temperature to Fahrenheit. Obtain a display of Fahrenheit-Celsius temperature graph over a range of 0 to 100°C.

$$T_F = \frac{9}{5}T_C + 32 \tag{A.6}$$

First, consider the blocks needed to build the model. These are:

- A ramp block to input the temperature signal, from the source library.
- A constant block, to define the constant of 32, also from the source library.
- A gain block, to multiply the input signal by $9/5$, from the Linear library.
- A sum block, to add the two quantities, also from the Linear library.
- A scope block to display the output, from the sink library.

To create a *SIMULINK* block diagram presentation select **new...** from the **File** menu. This provides an untitled blank window for designing and simulating a dynamic system. Copy the above blocks from the block libraries into the new window by depressing the mouse button and dragging. Assign the parameter values to the Gain and Constant blocks by opening (double clicking on) each block and entering the appropriate value. Then click on the **close** button to apply the value and close the dialog box. The next step is to connect these icons together by drawing lines connecting the icons using the left mouse button (hold the button down and drag the mouse to draw a line). You should now have the *SIMULINK* block diagram as shown in Figure A.12.

The Ramp block inputs Celsius temperature. Open this block, set the Slope to 1, Start time to 0, and the Initial output to 0. The Gain block multiplies that

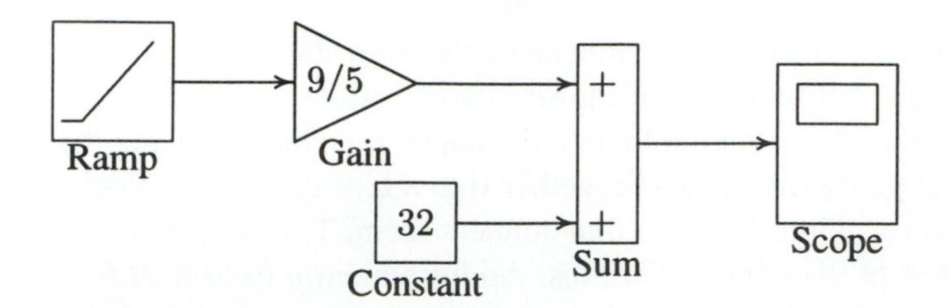

FIGURE A.12
Simulink diagram for the system of Example A.21.

temperature by the constant 9/5. The sum block adds the value 32 to the result and outputs the Fahrenheit temperature. Pull down the Simulation dialog box and select Parameters. Set the Start time to zero and the Stop Time to 100. Pull down the **File** menu and use **Save** to save the model under **simexa21** Start the simulation. Double click on the Scope, click on the **Auto Scale**, the result is displayed as shown in Figure A.13.

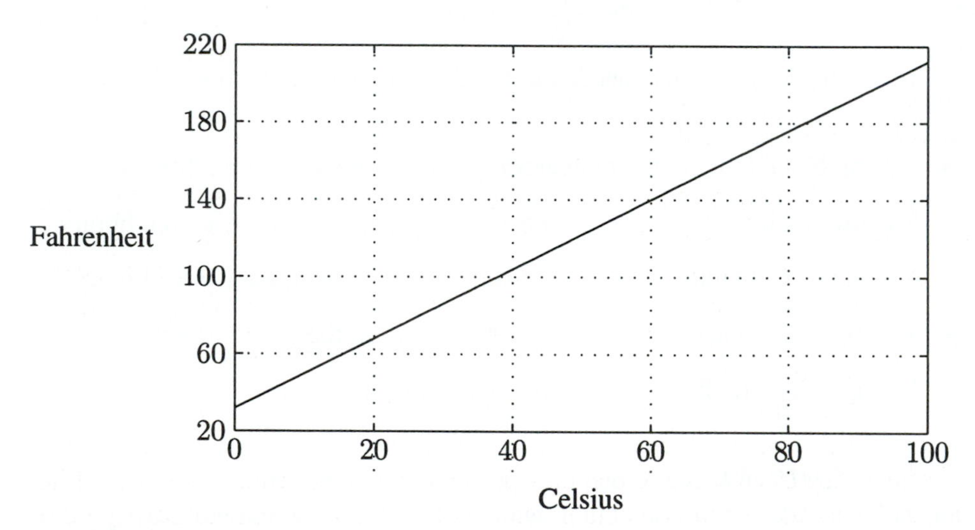

FIGURE A.13
Fahrenheit-Celsius temperature graph for Example A.21.

Example A.22

Construct a simulation diagram for the state equation described in Example A.18. Use *SIMULINK* to model and simulate the step response of this system, and display the results graphically.

State equation in Example A.18 for $M = 1$ kg, $B = 5$ N/m/sec, $K = 25$ N/m, and

$f(t) = 25u(t)$, is given by

$$\begin{aligned} \dot{x}_1 &= x_2 \\ \dot{x}_2 &= -25x_1 - 5x_2 + 25u(t) \end{aligned}$$

The simulation diagram is drawn from the above equations by inspection and is shown in Figure A.14.

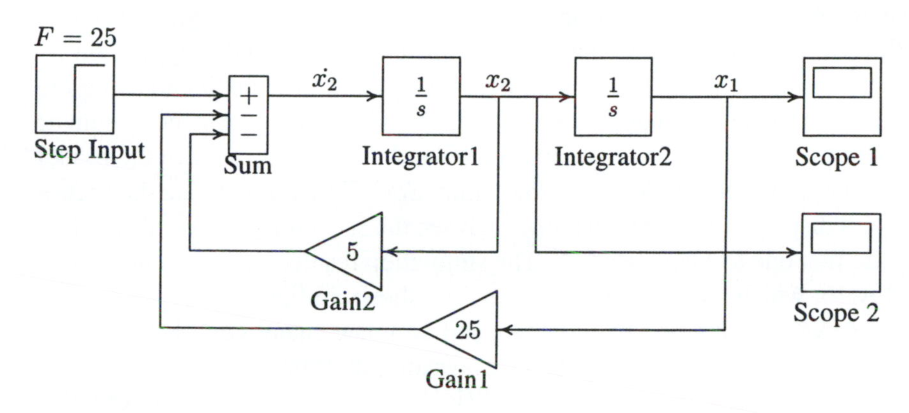

FIGURE A.14
Simulink diagram for the system of Example A.22.

To create a *SIMULINK* block diagram presentation select **new...** from the **File** menu. This provides an untitled blank window for designing and simulating a dynamic system. You can copy blocks from within any of the seven block libraries or other previously opened windows into the new window by depressing the mouse button and dragging. Open the **Source Library** and drag the Step Input block to your window. Double click on Step Input to open its dialog box. Set the step time to a large value, say 100, and set the Initial Value and the Final Value to 25 to represent the step input. Open the **Linear Library** and drag the **Sum** block to the right of the Step Input block. Open the Sum dialog box and enter + - - under List of Signs. Using the left mouse button, click and drag from the Step output port to the Summing block input port to connect them. Drag a copy of the Integrator block from the **Linear Library** and connect it to the output port of the Sum block. Click on the Integrator block once to highlight it. Use the Edit command from the menu bar to copy and paste a second Integrator. Next drag a copy of the Gain block from the **Linear Library**. Highlight the Gain block, and from the pull-down **Options** menu, click on the Flip Horizontal to rotate the Gain block by 180°. Double click on Gain block to open its dialog box and set the gain to 5. Make a copy of this block and set its gain to 25. Connect the output ports of the Gain blocks to the Sum block and their input ports to the locations shown in Figure A.14. Finally, get two Auto-Scale Graphs from the **Sink Library**, and connect them to the output of each Integrator. Before starting simulation, you must set the simulation parameters. Pull down the **Simulation** dialog box and select **Parameters**. Set the Start Time to zero, the Stop Time to 3, and for a more accurate integration, set the Maximum Step Size to 0.1. Leave the other parameters at their default values. Press OK to close the dialog box.

If you don't like some aspect of the diagram, you can change it in a variety of ways. You can move any of the icons by clicking on its center and dragging. You can move any of the lines by clicking on one of its corners and dragging. You can change the size and the shape of any of the icons by clicking and dragging on its corners. You can remove any line or icon by clicking on it to select it and using the **cut** command from the **edit** menu. You should now have exactly the same system as shown in Figure A.14. Pull down the **File** menu and use **Save as** to save the model under a file name **simexa22**. Start the simulation. *SIMULINK* will create the Figure windows and display the system responses. To see the second Figure window, click and drag the first one to a new location. The simulation results are shown in Figures A.15 and A.16, which are the same as the curves shown in Figure A.7.

SIMULINK enables you to construct and simulate many complex systems, such as control systems modeled by block diagram with transfer functions including the effect of nonlinearities. In addition, *SIMULINK* provides a number of built-in state variable models and subsystems that can be utilized easily.

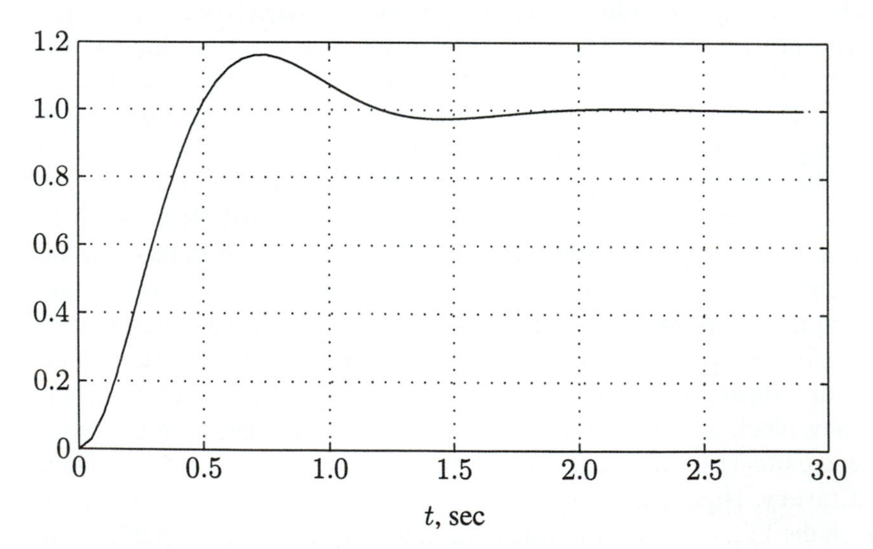

FIGURE A.15
Displacement response of the system described in Example A.22.

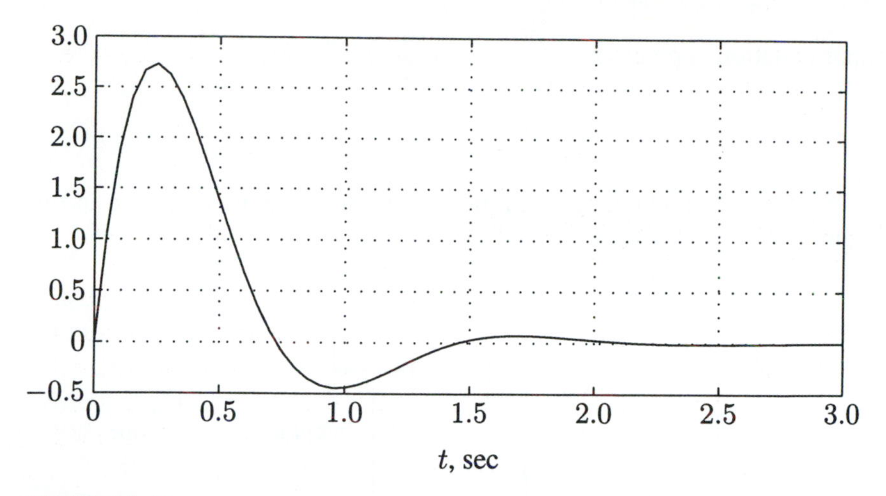

FIGURE A.16
Velocity response of the system described in Example A.22.

Example A.23

Consider the system defined by

$$2\frac{d^3y}{dt^3} + 4\frac{d^2y}{dt^2} + 8\frac{dy}{dt} + 10y = 10u(t)$$

We have a third-order system; thus there are three state variables. Let us choose the state variables as

$$\begin{aligned} x_1 &= y \\ x_2 &= \dot{y} \\ x_3 &= \ddot{y} \end{aligned}$$

Then we obtain

$$\begin{aligned} \dot{x_1} &= x_2 \\ \dot{x_2} &= x_3 \\ \dot{x_3} &= -5x_1 - 4x_2 - 2x_3 + 5u(t) \end{aligned}$$

The last of these three equations was obtained by solving the original differential equation for the highest derivative term $\dddot{y}$ and then substituting $y = x_1$, $\dot{y} = x_2$, and $\ddot{y} = x_3$ into the resulting equation. Using matrix notation, the state equation is

$$\begin{bmatrix} \dot{x_1} \\ \dot{x_2} \\ \dot{x_3} \end{bmatrix} = \begin{bmatrix} 0 & 1 & 0 \\ 0 & 0 & 1 \\ -5 & -4 & -2 \end{bmatrix} \begin{bmatrix} x_1 \\ x_2 \\ x_3 \end{bmatrix} + \begin{bmatrix} 0 \\ 0 \\ 5 \end{bmatrix} u(t)$$

and the output equation is given by

$$y = \begin{bmatrix} 1 & 0 & 0 \end{bmatrix} \mathbf{x}$$

The simulation diagram is obtained from the system differential equations and is given in Figure A.17.

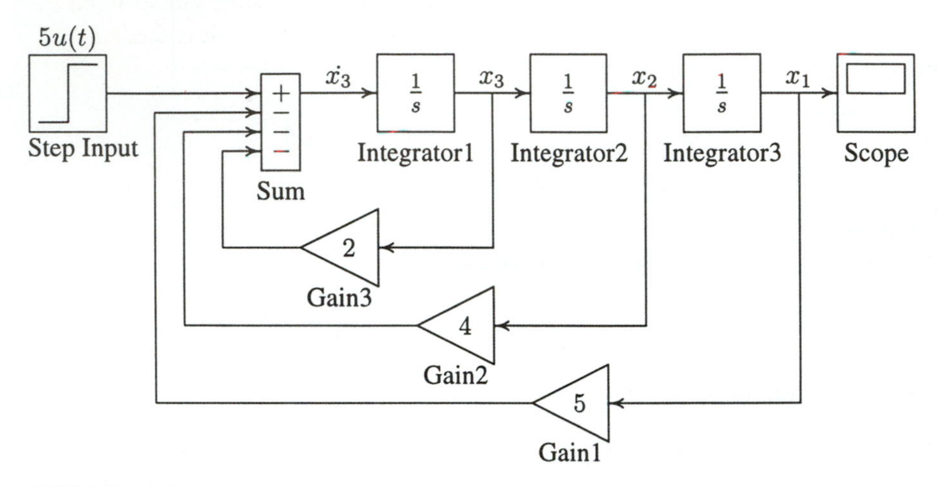

FIGURE A.17
Simulation diagram for the system of Example A.23.

A *SIMULINK* Block diagram is constructed and saved as **simexa23**. The simulation response is shown in Figure A.18.

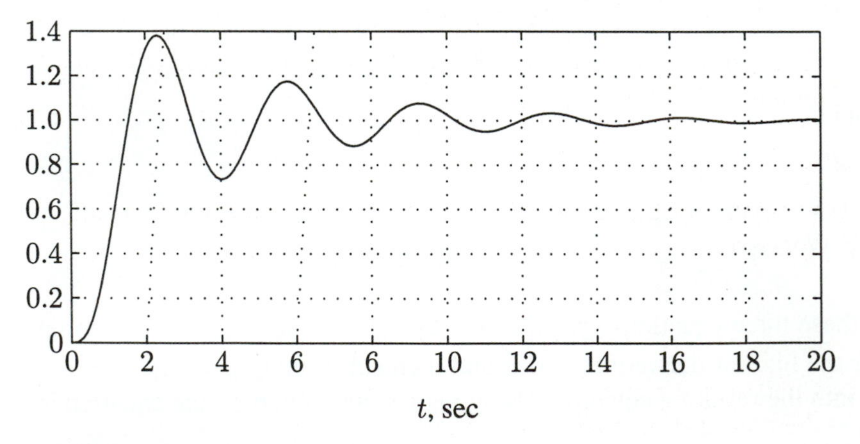

FIGURE A.18
Simulation result for the system in Example A.23.

Example A.24

Use the **state-space** model to simulate the state and output equations described in Example A.23.

The **State-Space** model provides a dialog box where the A, B, C, and D matrices can be entered in *MATLAB* matrix notation, or by variables defined in Workspace. A *SIMULINK* diagram using the **State-Space** model is constructed as shown in Figure A.19, and saved as **simexa24**. The simulation result is the same as in Figure A.18.

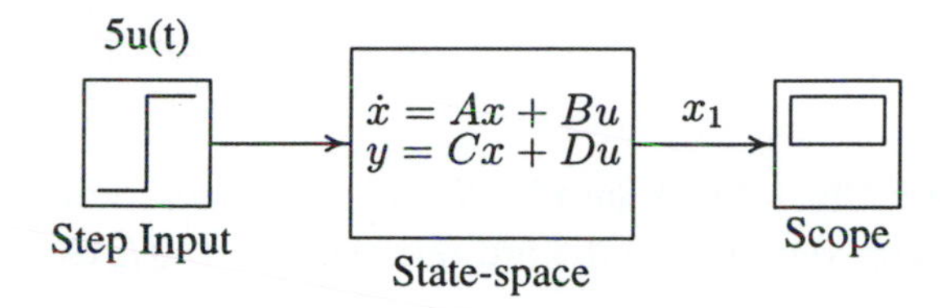

FIGURE A.19
State-space model for system in Example A.24.

Note that in this example, the output is given by $y = x_1$, and we define C as $C = [1 \quad 0 \quad 0]$. If it is desired to access all the states, then we can define C as an identity matrix, in this case a third order, i.e., $C = \text{eye}(3)$, and D as $D = \text{zeros}(3, 1)$. The output is a vector of state variables. A **DeMux** block may be added to produce individual states for graphing separately.

A.18.4 USING THE TO WORKSPACE BLOCK

The To Workspace block can be used to return output trajectories to the *MATLAB* Workspace. Example A.25 illustrates this use.

Example A.25

Obtain the step response of the following transfer function, and send the result to the *MATLAB* Workspace.

$$\frac{C(s)}{R(s)} = \frac{25}{s^2 + 2s + 25}$$

where $r(t)$ is a unit step function. The *SIMULINK* block diagram is constructed and saved in a file named **simexa25** as shown in Figure A.20.

The To Workspace block can accept a vector input, with each input element's trajectories stored as a column vector in the resulting workspace variable. To specify the variables open the To Workspace block and for the variable name enter c.

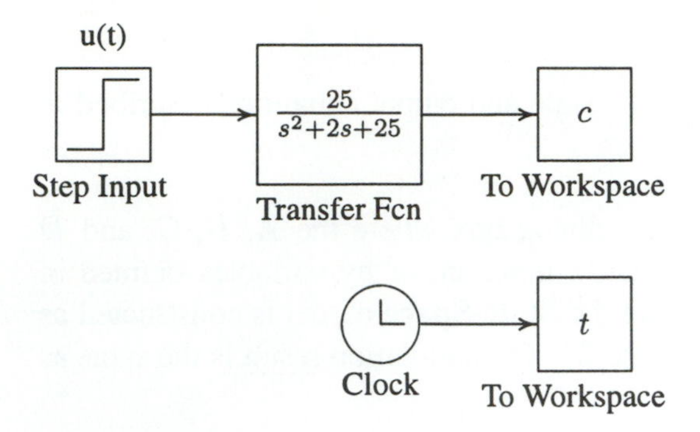

FIGURE A.20
Simulink model for system in Example A.25.

The time vector is stored by feeding a Clock block into To Workspace block. For this block variable name specify t. The vectors c and t are returned to *MATLAB* Workspace upon simulation.

A.18.5 LINEAR STATE-SPACE MODEL FROM SIMULINK DIAGRAM

SIMULINK provides the **linmod**, and **dlinmod** functions to extract linear models from the block diagram model in the form of the state-space matrices A, B, C, and D. State-space matrices describe the linear input-output relationship as

$$\dot{x}(t) = Ax(t) + Bu(t) \tag{A.7}$$

$$y(t) = Cx(t) + Du(t) \tag{A.8}$$

The following Example illustrates the use of **linmod** function. The input and outputs of the *SIMULINK* diagram must be defined using **Inport** and **Outport** blocks in place of the **Source** and **Sink** blocks.

Example A.26

Obtain the state-space model for the system represented by the block diagram shown in Figure A.21. The model is saved with a filename **simexa26**. Run the simulation and to extract the linear model of this *SIMULINK* system, in the Command Window, enter the command

```
[A,B,C,D] = linmod('simexa26')
```

The result is

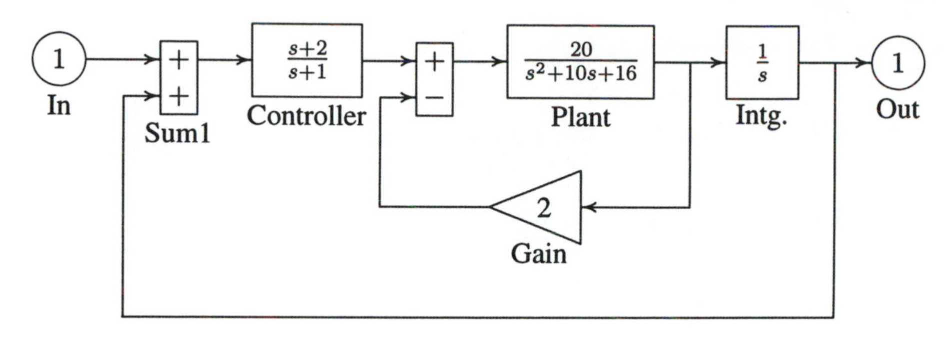

FIGURE A.21
Simulink model for system in Example A.26.

```
A =
       0     0     0    20
      -1    -1     0     0
      -1     1   -10   -56
       0     0     1     0
C =

       1     0     0     0
D =
       0
```

```
B =
       0
       1
       1
       0
```

In order to obtains the transfer function of the system from the state-space model, we use the command

```
[num, den]=ss2tf(A, B, C, D)
```

the result is

```
num =
        0.0000    0.0000    0.0000   20.0000   40.0000
den =
        1.0000   11.0000   66.0000   76.0000   40.0000
```

Thus, the transfer function model is

$$T(s) = \frac{20s + 40}{s^4 + 11s^3 + 66s^2 + 76s + 40}$$

Once the data is in the state-space form, or converted to a transfer function model, you can apply functions in Control System Toolbox for further analysis:

- Bode phase and magnitude frequency plot:

```
bode(A, B, C, D) or bode(num, den)
```

- Linearized time response:

```
step(A, B, C, D)    or  step(num, den)
lsim(A, B, C, D)    or  lsim(num, den)
impulse(A, B, C, D) or  impulse(num, den)
```

A.18.6 SUBSYSTEMS AND MASKING

SIMULINK subsystems, provide a capability within *SIMULINK* similar to subprograms in traditional programming languages.

Masking is a powerful *SIMULINK* feature that enables you to customize the dialog box and icon for a block or subsystem. With masking, you can simplify the use of your model by replacing many dialog boxes in a subsystem with a single one.

Example A.27

To encapsulate a portion of an existing *SIMULINK* model into a subsystem, consider the *SIMULINK* model of Example A.23 shown in Figure A.22, and proceed as follows:

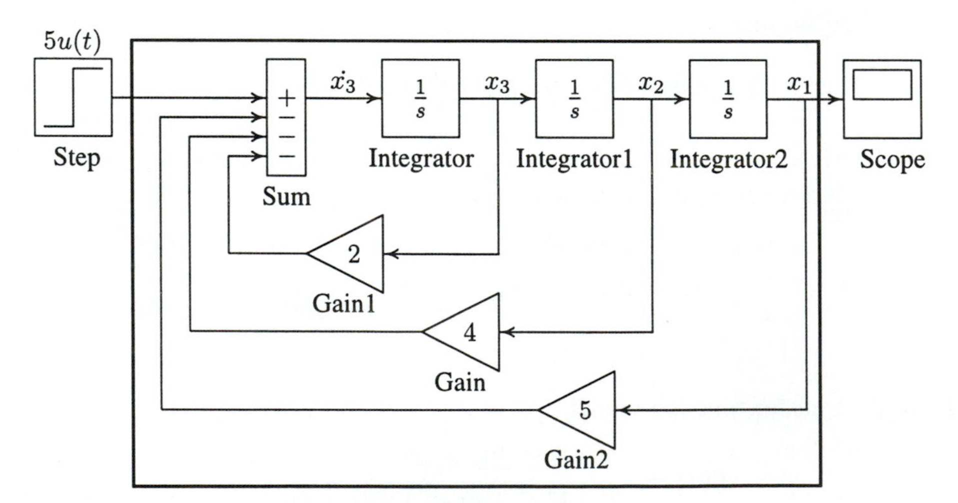

FIGURE A.22
Simulation diagram for the system of Example A.23.

1. Select all the blocks and signal lines to be included in the subsystem with the bounding box as shown.

2. Choose Edit and select Create Subsystem from the model window menu bar. *SIMULINK* will replace the select blocks with a subsystem block that has an input port for each signal entering the new subsystem and an output port for each signal leaving the new subsystem. *SIMULINK* will assign default names to the input and output ports.

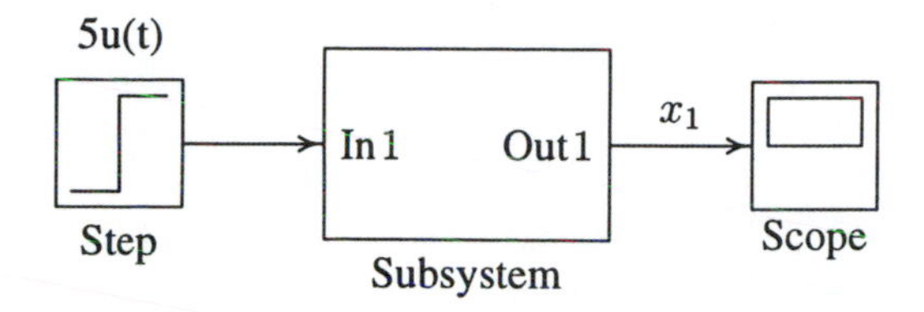

FIGURE A.23
Simulation diagram for the system of Example A.23.

To mask a block, select the block, then choose Create Mask from the Edit menu. The Mask Editor appears. The Mask Editor consists of three pages, each handling a different aspect of the mask.

- The Initialization page enables you to define and describe mask dialog box parameter prompts, name the variables associated with the parameters, and specify initialization commands.
- The Icon page enables you to define the block icon.
- The Documentation page enables you to define the mask type, and specify the block description and the block help.

In this example for icon the system transfer function is entered with command

```
dpoly([10], [2 4 8 10])
```

A short description of the system and relevant help topics can be entered in the Documentation page. The subsystem block is saved in a file named **simexa28**. Additional *SIMULINK* examples are found in Chapter 12. Also, many interesting examples are available in *SIMULINK* **demo**.

APPENDIX B

REVIEW OF FEEDBACK CONTROL SYSTEMS

B.1 THE CONTROL PROBLEM

The first step in the analysis and design of control systems is mathematical modeling of the system. The two most common methods are the transfer function approach and the state equation approach. The state equations can be applied to portray linear as well as nonlinear systems

All physical systems are nonlinear to some extent. In order to use the transfer function and linear state equations, the system must first be linearized. Thus, proper assumptions and approximations are made so that the system can be characterized by a linear mathematical model. The model may be validated by analyzing its performance for realistic input conditions and then by comparing with field test data taken from the dynamic system in its operating environment. Further analysis of the simulated model is usually necessary to obtain the model response for different feedback configurations and parameters settings. Once an acceptable controller has been designed and tested on the model, the feedback control strategy is then applied to the actual system to be controlled.

When we wish to develop a feedback control system for a specific purpose, the general procedure may be summarized as follows:

1. Choose a way to adjust the variable to be controlled; e.g., the mechanical load will be positioned with an electric motor or the temperature will be controlled by an electrical resistance heater.

2. Select suitable sensors, power supplies, amplifiers, etc., to complete the loop.

3. Determine what is required for the system to operate with the specified accuracy in steady-state and for the desired response time.

4. Analyze the resulting system to determine its stability.

5. Modify the system to provide stability and other desired operating conditions by redesigning the amplifier/controller, or by introducing additional control loops.

The objective of the control system is to control the output $c(t)$ in some prescribed manner by the input $r(t)$ through the elements of the control system. Some of the essential characteristics of feedback control systems are investigated in the following sections.

B.2 STABILITY

Consider the block diagram of a simple closed-loop control system as shown in Figure B.1 where $R(s)$ is the s-domain reference input, and $C(s)$ is the s-domain controlled output. $G(s)$ is the plant transfer function, K is a simple gain controller,

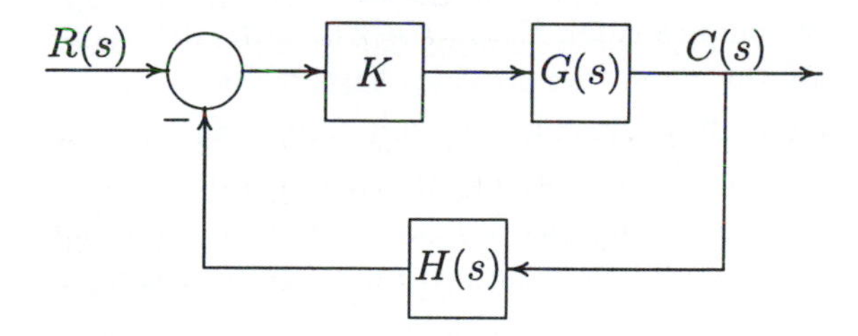

FIGURE B.1
A simple closed-loop control system.

and the feedback elements $H(s)$ represent the sensor transfer function. The closed-loop transfer function is

$$\frac{C(s)}{R(s)} = T(s) = \frac{KG(s)}{1 + KG(s)H(s)} \tag{B.1}$$

or the s-domain response is

$$C(s) = T(s)R(s) \tag{B.2}$$

The gain $KG(s)H(s)$ is commonly referred to as the *open-loop transfer function*. For a system to be usable, it must be stable. A linear time-invariant system is stable if every bounded input produces a bounded output. We call this characteristic *stability*. The denominator polynomial of the closed-loop transfer function set equal to zero is the system characteristic equation. That is, the characteristic equation is given by

$$1 + KG(s)H(s) = 0 \tag{B.3}$$

The roots of the characteristic equation are known as the poles of the closed-loop transfer function. The response is bounded if the poles of the closed-loop system are in the left-hand portion of the s-plane. Thus, a necessary and sufficient condition for a feedback system to be stable is that all the poles of the system transfer function have negative real parts.

The stability of a linear time-invariant system may be checked by using the *Control System Toolbox* function **impulse** to obtain the impulse response of the system. The system is stable if its impulse response approaches zero as time approaches infinity. One way to determine the stability of a system is by simulation. The function **lsim** can be used to observe the output for typical inputs. This is particularly useful for nonlinear systems. Alternatively, the *MATLAB* function **roots** can be utilized to obtain the roots of the characteristic equations. In the classical control theory, several techniques have been developed requiring little computation for stability analysis. One of these techniques is the *Routh-Hurwitz criterion*. Consideration of the *degree* of stability of a system often provides valuable information about its behavior. That is, if it is stable, how close is it to being unstable? This is the concept of *relative stability*. Usually, relative stability is expressed in terms of the speed of response and overshoot. Other methods frequently used for stability studies are the *Bode diagram*, *Root-locus plot*, *Nyquist criterion*, and *Lyapunov's stability criterion*.

B.2.1 THE ROUTH-HURWITZ STABILITY CRITERION

The Routh-Hurwitz criterion provides a quick method for determining absolute stability that can be applied to an nth-order characteristic equation of the form

$$a_n s^n + a_{n-1} s^{n-1} + \ldots + a_1 s + a_0 = 0 \tag{B.4}$$

The criterion is applied through the use of a *Routh table* defined as

$$\begin{array}{c|cccc}
s^n & a_n & a_{n-2} & a_{n-4} & \cdots \\
s^{n-1} & a_{n-1} & a_{n-3} & a_{n-5} & \cdots \\
s^{n-2} & b_1 & b_2 & b_3 & \cdots \\
s^{n-3} & c_1 & c_2 & c_3 & \cdots \\
\cdots & \cdots & \cdots & \cdots & \cdots
\end{array}$$

$a_n, a_{n-1}, \ldots, a_0$ are the coefficients of the characteristic equation and

$$b_1 = \frac{a_{n-1}a_{n-2} - a_n a_{n-3}}{a_{n-1}}, \quad b_2 = \frac{a_{n-1}a_{n-4} - a_n a_{n-5}}{a_{n-1}}, \quad \text{etc.}$$

$$c_1 = \frac{b_1 a_{n-3} - a_{n-1} b_2}{b_1}, \quad c_2 = \frac{b_1 a_{n-5} - a_{n-1} b_3}{b_1}, \quad \text{etc.}$$

Calculations in each row are continued until only zero elements remain. The necessary and sufficient condition that all roots of (B.4) lie in the left half of the s-plane is that the elements of the first column of the Routh-Hurwitz array have the same sign. If there are changes of signs in the elements of the first column, the number of sign changes indicates the number of roots with positive real parts.

A function called **routh(a)** is written that forms the Routh-Hurwitz array and determines if any roots have positive real parts. **a** is a row vector containing the coefficients of the characteristic equation.

If the first element in a row is zero, it is replaced by a very small positive number ϵ, and the calculation of the array is completed. If all elements in a row are zero, the system has poles on the imaginary axis, pairs of complex conjugate roots forming symmetry about the origin of the s-plane, or pairs of real roots with opposite signs. In this case, an auxiliary equation is formed from the preceding row. The all-zero row is then replaced with coefficients obtained by differentiating the auxiliary equation.

B.2.2 ROOT-LOCUS METHOD

The *root-locus method*, developed by W. R. Evans, enables us to find the closed-loop poles from the open-loop poles for all the values of the gain of the open-loop transfer function. The root locus of a system is a plot of the roots of the system characteristic equation as the gain factor K is varied. Therefore, the designer can select a suitable gain factor to achieve the desired performance criteria. If the required performance cannot be achieved, a controller can be added to the system to alter the root locus in the required manner.

Consider the feedback control system given in Figure B.1. In general, the open-loop transfer function is given by

$$KG(s)H(s) = \frac{K(s+z_1)(s+z_2)\cdots(s+z_m)}{(s+p_1)(s+p_2)\cdots(s+p_n)} \tag{B.5}$$

where m is the number of finite zeros, and n is the number of finite poles of the loop transfer function. If $n > m$, there are $(n-m)$ zeros at infinity. The characteristic equation of the closed-loop transfer function is given by (B.3); therefore

$$\frac{(s+p_1)(s+p_2)\cdots(s+p_n)}{(s+z_1)(s+z_2)\cdots(s+z_m)} = -K \tag{B.6}$$

From (B.6) it follows that for a point in the s-plane to be on the root locus, when $0 < K < \infty$, it must satisfy the following two conditions.

$$K = \frac{\text{product of vector lengths from finite poles}}{\text{product of vector lengths from finite zeros}} \tag{B.7}$$

and

$$\sum \text{angles of zeros of } GH(s) - \sum \text{angles of poles of } GH(s) = r(180)^\circ \tag{B.8}$$
$$\text{where} \quad r = \pm1, \pm3, \pm5, \cdots$$

Given a transfer function of an open-loop control system, the *Control System Toolbox* function **rlocus(num, den)** produces a root-locus plot with the gain vector automatically determined. If the open-loop system is defined in state space, we use **rlocus(A, B, C, D)**. **rlocus(num, den, K)** or **rlocus(A, B, C, D, K)** uses the user-supplied gain vector **K**. If the above commands are invoked with the left hand arguments **[r, K]**, the matrix **r** and the gain vector **K** are returned, and we need to use **plot(r, ' . ')** to obtain the plot. **rlocus** function is accurate, and we use it to obtain the root-locus. A good knowledge of the characteristics of the root loci offers insights into the effects of adding poles and zeros to the system transfer function. It is important to know how to construct the root locus by hand, so we can design a simple system and be able to understand and develop the computer-generated loci. For the basic construction rules for sketching the root locus, refer to any text on feedback control systems.

B.3 STEADY-STATE ERROR

In addition to being stable, a control system is also expected to meet a specified performance requirement when it is commanded by a set-point change or disturbed by an external force. The performance of the control system is judged not only by the transient response, but also by steady-state error. The steady-state error is the error as the transient response has decayed, leaving only the continuous response. High loop gains, in addition to sensitivity reduction, will also reduce the steady-state error. The steady-state error for a control system is classified according to its response characteristics to a polynomial input. A system may have no steady-state error to a step input, but the same system may exhibit nonzero steady-state error to a ramp input. This depends on the type of the open-loop transfer function.

Consider the system shown in Figure B.1. The closed-loop transfer function is given by (B.1). The error of the closed-loop system is

$$E(s) = R(s) - H(s)C(s) = \frac{1}{1 + KG(s)H(s)}R(s) \tag{B.9}$$

Using the final-value theorem, we have

$$e_{ss} = \lim_{s \to 0} \frac{sR(s)}{1 + KG(s)H(s)} \tag{B.10}$$

For the polynomial inputs, such as step, ramp, and parabolas, the steady-state error from the above equation will be:

Unit step input

$$e_{ss} = \frac{1}{1 + \lim_{s \to 0} KG(s)H(s)} = \frac{1}{1 + K_p} \tag{B.11}$$

Unit ramp input

$$e_{ss} = \frac{1}{\lim_{s \to 0} sKG(s)H(s)} = \frac{1}{K_v} \tag{B.12}$$

Unit parabolic input

$$e_{ss} = \frac{1}{\lim_{s \to 0} s^2 KG(s)H(s)} = \frac{1}{K_a} \tag{B.13}$$

In order to define the *system type*, the general open-loop transfer function is written in the following form.

$$KG(s)H(s) = \frac{K(1 + T_1 S)(1 + T_2 s)\dots(1 + T_m s)}{s^j (1 + T_a s)(1 + T_b s)\dots(1 + T_n s)} \tag{B.14}$$

The *type* of feedback control system refers to the *order* of the pole of $G(s)H(s)$ at $s = 0$.

Two functions, **errorzp(z,p,k)** and **errortf(num, den),** are written for computation of system steady-state error due to typical inputs, namely unit step, unit ramp, and unit parabolic. **errorzp(z,p,k)** finds the steady-state error when the system is represented by the zeros, poles, and gain. **z** is a column vector containing the transfer function zeros, **p** is a column vector containing the poles, and **k** is the gain. If the numerator power m is less than the denominator power n, then there are $(n - m)$ zeros at infinity, and vector **z** must be padded with $(n - m)$ **inf**. **errortf(num, den)** finds the steady-state error when the transfer function is expressed as the ratio of two polynomials.

B.4 STEP RESPONSE

Assessing the time-domain performance of closed-loop system models is important, because control systems are inherently time-domain systems. The performance of dynamic systems in the time domain can be defined in terms of the time response to standard test inputs. One very common input to control systems is the step function. If the response to a step input is known, it is mathematically possible to compute the response to any input. The step response for a second-order system is obtained. The standard form of the second-order transfer function is given by

$$G(s) = \frac{{\omega_n}^2}{s^2 + 2\zeta\omega_n s + {\omega_n}^2} \tag{B.15}$$

where ω_n is the natural frequency. The natural frequency is the frequency of oscillation if all of the damping is removed. Its value gives us an indication of the speed of the response. ζ is the dimensionless damping ratio. The damping ratio gives us an idea about the nature of the transient response. It gives us a feel for the amount of overshoot and oscillation that the response undergoes.

The transient response of a practical control system often exhibits damped oscillations before reaching steady-state. The underdamped response $(\zeta < 1)$ to a unit step input, subject to zero initial condition, is given by

$$c(t) = 1 - \frac{1}{\beta} e^{-\zeta\omega_n t} \sin(\beta\omega_n t + \theta) \tag{B.16}$$

where $\beta = \sqrt{1 - \zeta^2}$ and $\theta = \tan^{-1}(\beta/\zeta)$.

The performance criteria that are used to characterize the transient response to a unit step input include rise time, peak time, overshoot, and settling time. We define the rise time t_r as the time required for the response to rise from 10 percent of the final value to 90 percent of the final value. The time to reach the peak value is t_p. The swiftness of the response is measured by t_r and t_p. The similarity with which the actual response matches the step input is measured by the percent overshoot and settling time t_s. For underdamped systems, the percent overshoot $P.O.$ is defined as

$$P.O. = \frac{\text{maximum value} - \text{final value}}{\text{final value}} \tag{B.17}$$

The peak time is obtained by setting the derivative of (B.16) to zero.

$$t_p = \frac{\pi}{\omega_n\sqrt{1-\zeta^2}} \tag{B.18}$$

The peak value of the step response occurs at this time, and evaluating the response in (B.16) at $t = t_p$ yields

$$C(t_p) = M_{pt} = 1 + e^{-\zeta\pi/\sqrt{1-\zeta^2}} \tag{B.19}$$

Therefore, from (B.17), the percent overshoot is

$$P.O. = e^{-\zeta\pi/\sqrt{1-\zeta^2}} \times 100 \tag{B.20}$$

Settling time is the time required for the step response to settle within a small percent of its final value. Typically, this value may be assumed to be ± 2 percent of the final value. For the second-order system, the response remains within 2 percent after 4 time constants, that is

$$t_s = 4\tau = \frac{4}{\zeta\omega_n} \tag{B.21}$$

Given a transfer function of a closed-loop control system, the *Control System Toolbox* function **step(num, den)** produces the step response plot with the time vector automatically determined. If the closed-loop system is defined in state space, we use **step(A, B, C, D)**. **step(num, den, t)** or **step(A, B, C, D, iu, t)** uses the user-supplied time vector **t**. The scalar **iu** specifies which input is to be used for the step response. If the above commands are invoked with the left-hand arguments **[y, x, t]**, the output vector, the state response vectors, and the time vector **t** are returned, and we need to use **plot** function to obtain the plot. See also **initial** and **lsim** functions. A function called **timespec(num, den)** is written which obtains the time-domain performance specifications, $P.O.$, t_p, t_r, and t_s. **num** and **den** are the numerator and denominator of the system closed-loop transfer function.

B.5 ROOT-LOCUS DESIGN

The design specifications considered here are limited to those dealing with system accuracy and time-domain performance specifications. These performance specifications can be defined in terms of the desirable location of the dominant closed-loop poles.

The root locus can be used to determine the value of the loop gain K, which results in a satisfactory closed-loop behavior. This is called the proportional controller and provides gradual response to deviations from the set point. There are practical limits as to how large the gain can be made. In fact, very high gains lead to instabilities. If the root-locus plot is such that the desired performance cannot be achieved by the adjustment of the gain, then it is necessary to reshape the root loci by adding the additional controller $G_c(s)$ to the forward path, as shown in Figure B.2. $G_c(s)$ must be chosen so that the root locus will pass through the proper region of the s-plane.

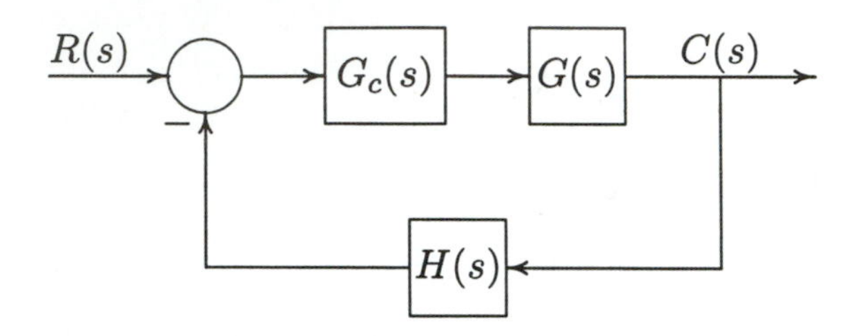

FIGURE B.2
A closed-loop control system with controller.

The proportional controller (P) has no sense of time, and its action is determined by the present value of the error. An appropriate controller must make corrections based on the past and future values. This can be accomplished by combining proportional with integral action (PI) or proportional with derivative action (PD). There is also a proportional-plus-integral-plus-derivative controller (PID).

$$G_c(s) = K_P + \frac{K_I}{s} + K_D s \tag{B.22}$$

The ideal integral and differential compensators require the use of active amplifiers.

Other compensators which can be realized with only passive network elements are lead, lag, and lead-lag compensators. A first-order compensator having a single zero and pole in its transfer function is

$$G_c(s) = \frac{K_c(s + z_0)}{s + p_0} \tag{B.23}$$

B.5.1 GAIN FACTOR COMPENSATION OR P CONTROLLER

The proportional controller is a pure gain controller. The design is accomplished by choosing a value of K_0 which results, in a satisfactory transient response.

B.5.2 PHASE-LEAD DESIGN

In (B.23) the compensator is a high-pass filter or phase-lead, if $p_0 > z_0$. The phase-lead network contributes a positive angle to the root-locus angle criterion of (B.8) and tends to shift the root locus of the plant toward the left in the s-plane. The lead network acts mainly to modify the dynamic response to raise bandwidth and to increase the speed of response. In a sense, a lead network approximates derivative control. If $p_0 < z_0$, the compensator is a low-pass filter or phase-lag. The phase-lag compensator adds a negative angle to the angle criterion and tends to shift the root locus to the right in the s-plane. The compensator angle must be small to maintain the stability of the system. The lag network is usually used to raise the low-frequency gain and thus to improve the steady-state accuracy of the system. The lag network is an approximate integral control. The DC gain of the compensator is

$$a_0 = G_c(0) = \frac{K_c z_0}{p_0} \tag{B.24}$$

For a given desired location of a closed-loop pole s_1, the design can be accomplished by trial and error. Select a proper value of z_0 and use the angle criterion of (B.8) to determine p_0. Then, the gain K_c is obtained by applying the magnitude criterion of (B.7). Alternatively, if the compensator DC gain, $a_0 = (K_c z_0)/p_0$, is specified, then for a given location of the closed-loop pole

$$s_1 = |s_1| \angle\beta \tag{B.25}$$

z_0 and p_0 are obtained such that the equation

$$1 + G_c(s_1)G_p(s_1) = 0 \tag{B.26}$$

is satisfied. It can be shown that the above parameters are found from the following equations.

$$z_0 = \frac{a_0}{a_1} \qquad p_0 = \frac{1}{b_1} \qquad \text{and} \qquad K_c = \frac{a_0 p_0}{z_0} \tag{B.27}$$

where

$$a_1 = \frac{\sin\beta + a_0 M \sin(\beta - \psi)}{|s_1|\ M \sin\psi}$$

$$b_1 = -\frac{\sin(\beta + \psi) + a_0 M \sin\beta}{|s_1|\ \sin\psi} \tag{B.28}$$

where M and ψ are the magnitude and phase angle of the open-loop plant transfer function evaluated at s_1, i.e.,

$$G_p(s_1) = M\angle\psi \tag{B.29}$$

For the case that ψ is either 0° or 180°, (B.28) is given by

$$a_1 \, |s_1| \cos\beta \pm \frac{b_1 \, |s_1|}{M} \cos\beta \pm \frac{1}{M} + a_0 = 0 \tag{B.30}$$

where the plus sign applies for $\psi = 0°$ and the minus sign applies for $\psi = 180°$. For this case, the zero of the compensator must also be assigned.

B.5.3 PHASE-LAG DESIGN

In the phase-lag control, the poles and zeros of the controller are placed very close together, and the combination is located relatively close to the origin of the s-plane. Thus, the root loci in the compensated system are shifted only slightly from their original locations. Hence, the phase-lag compensator is used when the system transient response is satisfactory but requires a reduction in the steady-state error. The function **[numo, deno, denc] = phlead(num, den,** s_1) can be used for phase-lag compensation by specifying the desired pole s_1 slightly to the right of the uncompensated pole location. Alternatively, phase-lag compensation can be obtained by assuming a DC gain of unity for the compensator based on the following approximate method.

$$a_0 = G_c(0) = \frac{K_c z_0}{p_0} = 1 \tag{B.31}$$

Therefore,

$$K_c = \frac{p_0}{z_0} \qquad \text{since } \; p_0 < z_0 \; \text{ then } \; K_c < 1 \tag{B.32}$$

If K_0 is the gain required for the desired closed-loop pole s_1, then from (B.3)

$$K_0 = -\frac{1}{G_p(s_1)} \tag{B.33}$$

If we place the pole and zero of the lag compensator very close to each other with their magnitude much smaller than s_1, then

$$G_c(s_1) = \frac{K_c(s + z_0)}{s + p_0} \simeq K_c \tag{B.34}$$

Now, the gain K required to place a closed-loop pole at approximately s_1 is given by

$$K = -\frac{1}{G_c(s_1)G_p(s_1)} \simeq -\frac{1}{K_c G_p(s_1)} \simeq \frac{K_0}{K_c} \tag{B.35}$$

Since $K_c < 1$, then $K > K_0$. Next, select the compensator zero z_0, arbitrarily small. Then from (B.31) the compensator pole is

$$p_0 = K_0 z_0 \tag{B.36}$$

The compensated system transfer function is then given by

$$K G_p G_c = K K_c \frac{s + z_0}{s + p_0} G_p \tag{B.37}$$

A lag-lead controller may be obtained by appropriately combining a lag and a lead network in series.

B.5.4 PID DESIGN

One of the most common controllers available commercially is the PID controller. Different processes are suited to different combinations of proportional, integral, and derivative control. The control engineer's task is to adjust the three gain factors to arrive at an acceptable degree of error reduction simultaneously with acceptable dynamic response. For a desired location of the closed-loop pole s_1, as given by (B.25), the following equations are obtained to satisfy (B.26).

$$K_P = \frac{-\sin(\beta + \psi)}{M \sin \beta} - \frac{2K_I \cos \beta}{|s_1|}$$

$$K_D = \frac{\sin \psi}{|s_1|\, M \sin \beta} + \frac{K_I}{|s_1|^2} \tag{B.38}$$

For PD or PI controllers, the appropriate gain is set to zero. The above equations can be used only for the complex pole s_1. For the case that s_1 is real, the zero of the PD controller ($z_0 = K_P/K_D$) and the zero of the PI controller ($z_0 = K_I/K_P$) are specified, and the corresponding gains to satisfy angle and magnitude criteria are obtained accordingly. For the PID design, the value of K_I to achieve a desired steady-state error is specified. Again, (B.38) is applied only for the complex pole s_1.

B.5.5 PD CONTROLLER

Here, both the error and its derivative are used for control, and the compensator transfer function is

$$G_c(s) = K_P + K_D s = K_D \left(s + \frac{K_P}{K_D} \right) \tag{B.39}$$

From above, it can be seen that the PD controller is equivalent to the addition of a simple zero at $s = -K_P/K_D$ to the open-loop transfer function, which improves the transient response. From a different point of view, the PD controller may also be used to improve the steady-state error, because it anticipates large errors and attempts corrective action before they occur. The function **[numo, deno, denc] = rldesign(num, den,** s_1) with option 4 is used for the PD controller design.

B.5.6 PI CONTROLLER

The integral of the error as well as the error itself is used for control, and the compensator transfer function is

$$G_c(s) = K_P + \frac{K_I}{s} = \frac{K_P(s + K_I/K_P)}{s} \tag{B.40}$$

The PI controller is common in process control or regulating systems. Integral control bases its corrective action on the cumulative error integrated over time. The controller increases the type of system by 1 and is used to reduce the steady-state errors. The function **[numo, deno, denc] = rldesign(num, den,** s_1) with option 5 is used for the PI controller design.

B.5.7 PID CONTROLLER

The PID controller is used to improve the dynamic response as well as to reduce or eliminate the steady-state error. The function **[numo, deno, denc] = rldesign(num, den,** s_1) with option 6 is used for the PID controller design.

Based on the above equations, several functions are developed for the root-locus design. These are

Function	Controller
[numo, deno, denc] = pcomp(num, den, ζ**)**	Proportional
[numo, deno, denc] = phlead(num, den, s_1**)**	Phase-Lead
[numo, deno, denc] = phlag(num, den, ζ**)**	Phase-Lag
[numo, deno, denc] = pdcomp(num, den, s_1**)**	PD
[numo, deno, denc] = picomp(num, den, s_1**)**	PI
[numo, deno, denc] = pidcomp(num, den, s_1**)**	PID

Alternatively, the function **[numo, deno, denc] = rldesign(num, den,** s_1) displays a menu with six options that allow the user to select any of the above controller designs. $s_1 = \sigma + j\omega$ is a desired pole of the closed-loop transfer function, except for the **pcomp** and **phlag** controllers, where ζ, the damping ratio of the dominant poles, is substituted for s_1. **num** and **den** are row vectors of polynomial coefficients of the uncompensated open-loop plant transfer function. The function

phlead(num, den, s1) may also be used to design phase-lag controllers. To do this, the desired pole location s_1 must be assumed slightly to the right of the uncompensated pole position. The function obtains the controller transfer function and roots of the compensated characteristic equation. Also, the function returns the open-loop and closed-loop numerators and denominators of the compensated system transfer function.

Example B.1

The block diagram of a control system is as shown in Figure B.3. $G_c(s)$ is a simple proportional controller of gain K.

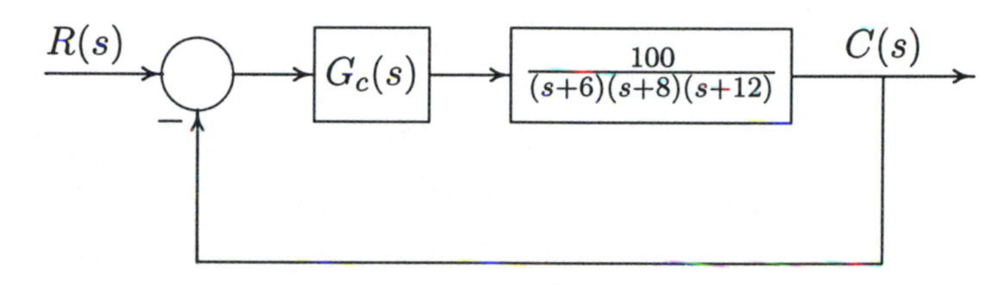

FIGURE B.3
Control system for Example B.1.

(a) Construct the Routh-Hurwitz array and determine the range of K for closed-loop stability.
(b) Find the value of K to yield a steady-state error of 0.15 for a unit step input.
(c) Use *MATLAB* **rlocus** function to obtain the root-locus plot.
(d) Use **rldesign** and option 1 to find the gain K_0 such that the dominant closed-loop poles damping ratio will be equal to 0.96. Obtain the step response curves for K_0, and the value of K found in (b).

(a) The closed-loop transfer function of the control system shown in Figure B.3 is

$$\frac{C(s)}{R(s)} = \frac{100K}{s^3 + 26s^2 + 216s + 576 + 100K}$$

The Routh-Hurwitz array for this polynomial is then (see Appendix B.2.1)

$$\begin{array}{c|ll} s^3 & 1 & 216 \\ s^2 & 26 & 576 + 100K \\ s^1 & 193.846 - 3.846K & 0 \\ s^0 & 576 + 10K & 0 \end{array}$$

From the s^1 row we see that, for control system stability, K must be less than 50.4, also from the s^0 row, K must be greater than -5.76. Thus, with positive values of

K, for control system stability, the gain must be

$$K < 50.4$$

For $K = 50.4$, the auxiliary equation from the s^2 row is

$$26s^2 + 5616 = 0$$

or $s = \pm j14.7$. That is, for $K = 50.4$, we have a pair of conjugate poles on the $j - \omega$ axis, and the control system is marginally stable.

(b) The position error constant given by (B.11) is

$$K_p = \lim_{s \to 0} G_c(s)G_p(s) = \lim_{s \to 0} \frac{100K}{(s+6)(s+8)(s+12)} = \frac{100K}{576}$$

For a unit step input

$$e_{ss} = \frac{1}{1 + K_p} = 0.15$$

Thus

$$K_p = 5.667 = \frac{100K}{576}$$

or

$$K = 32.64$$

The closed-loop transfer function for this gain is

$$\frac{C(s)}{R(s)} = \frac{(100)(32.64)}{s^3 + 26s^2 + 216s + 3840}$$

(c) The *MATLAB Control Toolbox* function **rlocus** is used to obtain the root-locus plot.

(d) To find the gain for the step response damping ratio of $\zeta = 0.96$, and the step response plots, we use the following commands.

```
num = 100;
den = [1  26  216  576];
figure(1), rlocus(num, den), grid,  axis([-20 0 -15 15]);
zeta = 0.96;                                % damping ratio
[numo, deno, denc]=rldesign(num,den,zeta);% Gain controller
```

```
t = 0:.005:4;
c1 = step(numo, denc, t); % Step response for zeta = 0.96
num2 = 100*32.64;  den2 = [1 26 216 3840];
c2 = step(num2, den2, t);   % Step response for K = 32.64
figure(2), plot(t, c1,  t, c2) , grid
xlabel('t, sec'), ylabel('c(t)')
text(3.1, 0.75, 'K = 32.64'), text(3.1, 0.1, 'K = 0.28')
timespec(num2, den2)    % Time-domain spec. for K = 32.64
```

The result is

Compensator type	Enter
Gain compensation	1
Phase-lead (or phase-lag)	2
Phase-lag (Approximate $K = K_0/K_c$)	3
PD Controller	4
PI Controller	5
PID Controller	6
To quit	0

```
Enter your choice → 1

Controller gain:          K0 = 0.28

Row vectors of polynomial coefficients of the compensated
system:

   Open-loop num.    28
   Open-loop den.     1   26   216   576
   Closed-loop den.   1   26   216   604

   Roots of the compensated characteristic equation:
   -12.8445
   -6.5778 + 1.9383i
   -6.5778 - 1.9383i

   Peak time = 0.289          Percent overshoot = 65.9
   Rise time = 0.096
   Settling time = 3.3
```

The root-locus plot is shown in Figure B.4, and the step response is shown in Figure B.5. The step response damping ratio of 0.96 resulted in a controller gain of $K_0 = 0.28$.

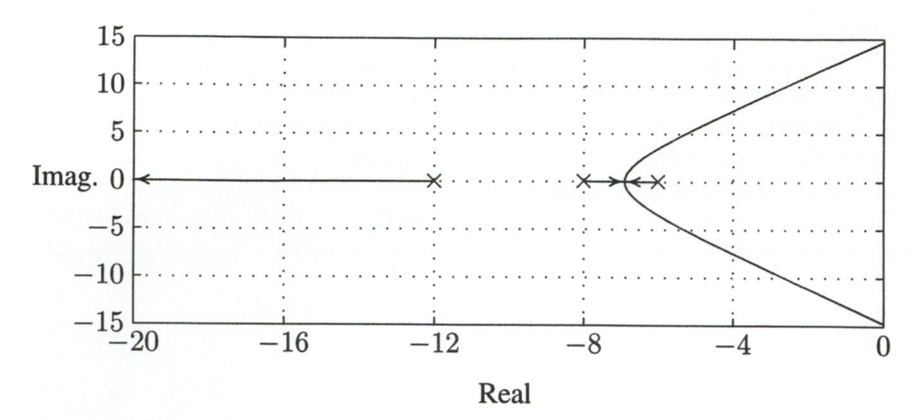

FIGURE B.4
Root-locus plot for Example B.1.

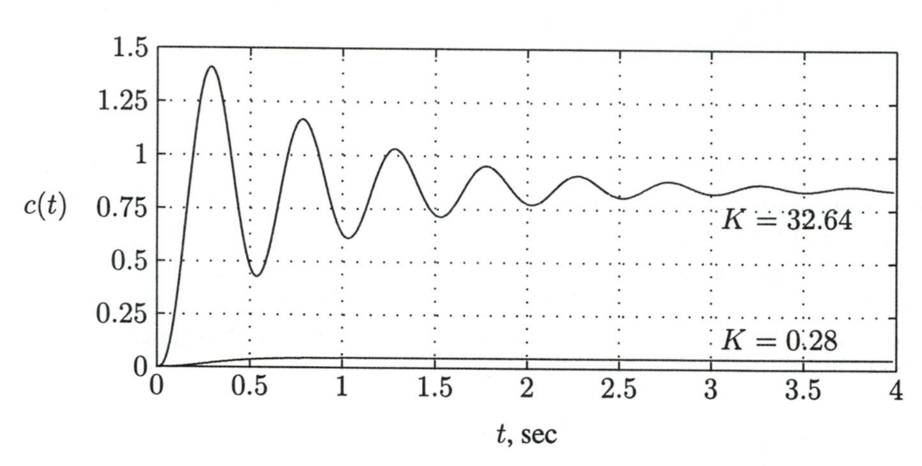

FIGURE B.5
Step response with proportional controller for Example B.1.

From Figure B.5, we see that the transient response is satisfactory, but the steady-state error given by

$$e_{ss} = \frac{1}{1 + \frac{28}{576}} = 0.9536$$

is very large, and the steady-state response is $1 - 0.9536 = 0.0464$. In order to reduce the steady-state error, the gain must be increased. The gain for a steady-state error of 0.15 was found to be 32.64, but the step response is highly oscillatory with an overshoot of 65.9 percent, which is not satisfactory.

Example B.2

For the control system in Example B.1, design a controller to meet the following specifications.

- Zero steady-state error for a step input
- Step response dominant poles damping ratio $\zeta = 0.995$
- Step response dominant pole time constant $\tau = 0.1$ second

The plant transfer function of the control system in Example B.1 is type zero. To reduce the steady-state error to zero, we must increase the system type by one. Thus, we select a PID controller, i.e.,

$$G_c(s) = K_P + \frac{K_I}{s} + K_D s$$

From the last two specifications, $\zeta\omega_n = \frac{1}{\tau} = 10$, and $\theta = \cos^{-1} 0.995 = 5.73°$. Thus, the required complex closed-loop poles are $-10 \pm j1$. The function **rldesign** with option 6 is used for a PID controller design. The user is prompted to enter a value for the integral gain K_I, and the program determines K_P and K_D. The process may be repeated for different values of K_I until a satisfactory response is obtained. For this example, use a value of 9.09 for K_I. The following commands

```
num = 100; den = [1  26  216  576];
s1= -10+j*1;         % Desired location of closed-loop poles
[numo, deno, denc]=rldesign(num, den, s1);    %PID design
t = 0:.01:4;
step(numo, denc, t), grid
xlabel('Time -sec.'), ylabel('c(t)')
```

result in

Compensator type	Enter
Gain compensation	1
Phase-lead (or phase-lag)	2
Phase-lag (Approximate $K = K_0/K_c$)	3
PD Controller	4
PI Controller	5
PID Controller	6
To quit	0

```
Enter your choice → 6
Enter the integrator gain KI → 9.09
```

```
Gc = 2.1 + 9.09/s + 0.14s

Row vectors of polynomial coefficients of the compensated
system:

  Open-loop num.    14  210  909
  Open-loop den.     1   26  230  576    0
  Closed-loop den    1   26  230  786  909

  Roots of the compensated characteristic equation:
  -10 + 1i
  -10 - 1i
  -3
  -3
```

Thus, the compensated open-loop transfer function is

$$G_c G_p = \frac{14s^2 + 210s + 909}{s(s^3 + 26s^2 + 216s + 576)}$$

and the compensated closed-loop transfer function is

$$\frac{C(s)}{R(s)} = \frac{14s^2 + 210s + 909}{s^4 + 26s^3 + 2230s^2 + 786s + 909}$$

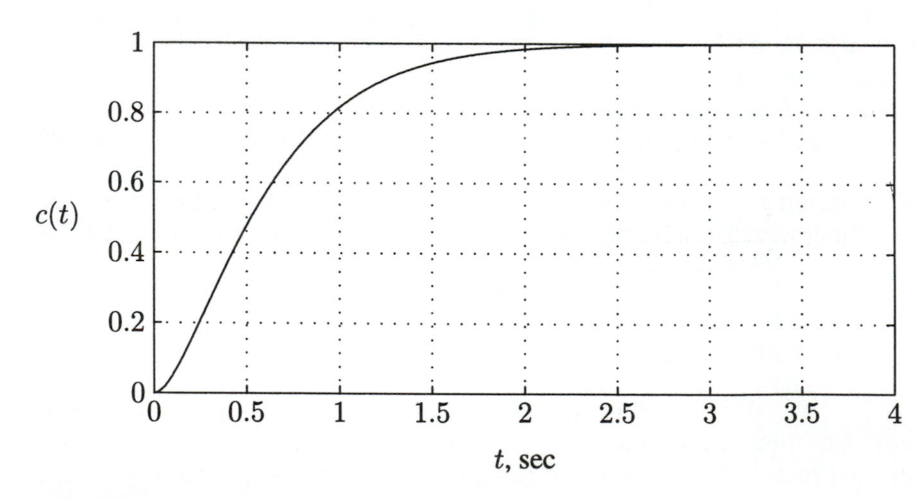

FIGURE B.6
Step response with PID controller for Example B.2.

The PID controller increases the system type by 1. That is, we have a type 1 system, and the steady-state error due to a unit step input is zero. The transient response is also improved as shown in Figure B.6.

B.6 FREQUENCY RESPONSE

The frequency response of a system is the steady-state response of the system to a sinusoidal input signal. The frequency response method and the root-locus method are simply two different ways of applying the same basic principles of analysis. These methods supplement each other, and in many practical design problems, both techniques are employed. One advantage of the frequency response method is that the transfer function of a system can be determined experimentally by frequency response tests. Furthermore, the design of a system in the frequency domain provides the designer with control over the system bandwidth and over the effect of noise and disturbance on the system response.

The response of a linear time-invariant system to sinusoidal input $r(t) = A\sin(\omega t)$ is given by

$$c(t) = A\,|G(j\omega)|\,\sin[\omega t + \theta(\omega)] \tag{B.41}$$

where the transfer function $G(j\omega)$ is obtained by substituting $j\omega$ for s in the expression for $G(s)$. The resulting transfer function may be written in *polar form* as

$$G(j\omega) = |G(j\omega)|\ \angle\theta(\omega) \tag{B.42}$$

Alternatively, the transfer function can be represented in rectangular complex form as

$$G(j\omega) = \Re\,G(j\omega) + j\Im\,G(j\omega) = R(j\omega) + jX(j\omega) \tag{B.43}$$

The most common graphical representation of a frequency response function is the *Bode plot*. Other representations of sinusoidal transfer functions are *polar plot* and *log-magnitude versus phase plot*.

B.6.1 BODE PLOT

The *Bode plot* consists of two graphs plotted on semi-log paper with linear vertical scales and logarithmic horizontal scales. The first graph is a plot of the magnitude of a frequency response function $G(j\omega)$ in decibels versus the logarithm of ω, the frequency. The second graph of a Bode plot shows the phase function $\theta(\omega)$ versus the logarithm of ω. The logarithmic representation is useful in that it shows both the low- and high-frequency characteristics of the transfer function in one diagram. Furthermore, the frequency response of a system may be approximated by a series of straight line segments.

Given a transfer function of a system, the *Control System Toolbox* function **bode(num, den)** produces the frequency response plot with the frequency vector automatically determined. If the system is defined in state space, we use **bode(A, B, C, D)**. **bode(num, den, ω)** or **bode(A, B, C, D, iu, ω)** uses the user-supplied frequency vector ω. The scalar **iu** specifies which input is to be used for the frequency response. If the above commands are invoked with the left-hand arguments **[mag, phase, ω]**, the frequency response of the system in the matrices **mag**, **phase**, and ω are returned, and we need to use **plot** or **semilogx** functions to obtain the plot.

B.6.2 POLAR PLOT

A *polar plot*, also called the *Nyquist plot*, is a graph of $\Im G(j\omega)$ versus $\Re G(j\omega)$ with ω varying from $-\infty$ to $+\infty$. The polar plot may be directly graphed from sinusoidal steady-state measurements on the components of the open-loop transfer function.

Given a transfer function of a system, the *Control System Toolbox* function **nyquist(num, den)** produces the Nyquist plot with the frequency vector automatically determined. If the system is defined in state space, we use **nyquist(A, B, C, D)**. **nyquist(num, den, ω)** or **nyquist(A, B, C, D, iu, ω)** uses the user-supplied frequency vector ω. The scalar **iu** specifies which input is to be used for the Nyquist response. If the above commands are invoked with the left-hand arguments **[re, im, ω]**, the frequency response of the system in the matrices **re**, **im**, and ω are returned, and we need to use **plot(re, im)** function to obtain the plot.

B.6.3 RELATIVE STABILITY

The closed-loop transfer function of a control system is given by

$$T(s) = \frac{C(s)}{R(s)} = \frac{KG(s)}{1 + KGH(s)} \tag{B.44}$$

For *BIBO* stability, poles of $T(s)$ must lie in the left-half s-plane. Since zeros of $1 + KGH(s)$ are poles of $T(s)$, the system is *BIBO* stable when the roots of the characteristic equation $1 + KGH(s)$ lie in the left-half s-plane. All points on the root locus satisfy the following conditions.

$$|\, KGH(s) \,| = 1 \qquad \text{and} \qquad \angle GH(s) = -180^\circ \tag{B.45}$$

The intersection of the polar plot with the negative real axis has a phase angle of -180°. The frequency ω_{pc} corresponding to this point is known as the *phase crossover frequency*. In addition, as the loop gain is increased, the polar plot crossing $(-1,0)$ point has the property described by

$$| K_c GH(j\omega_{pc}) | = 1 \qquad \text{and} \qquad \angle GH(j\omega_{pc}) = -180^\circ \tag{B.46}$$

The closed-loop response becomes marginally stable when the frequency response magnitude is unity and its phase angle is -180°. The frequency at which the polar plot intersect $(-1,0)$ point is the same frequency that the root locus crosses the $j\omega$-axis. For a still larger value of K, the polar plot will enclose the $(-1,0)$ point, and the system is unstable.

Thus, the system is stable if

$$| KGH(j\omega) | < 1 \qquad \text{at} \qquad \angle GH(j\omega_{pc}) = -180^\circ \tag{B.47}$$

The proximity of the $KGH(j\omega)$ plot in the polar coordinates to the $(-1,0)$ point gives an indication of the stability of the closed-loop system.

B.6.4 GAIN AND PHASE MARGINS

Gain margin and *phase margin* are two common design criteria related to the open-loop frequency response. The *gain margin* is the amount of gain by which the gain of a stable system must be increased for the polar plot to pass through the $(-1,0)$ point. The gain margin is defined as

$$G.M. = \frac{K_c}{K} \tag{B.48}$$

where K_c is the critical loop gain for marginal stability and K is the actual loop gain. The above ratio can be written as

$$G.M. = \frac{K_c \, |GH(j\omega_{pc})|}{K \, |GH(j\omega_{pc})|} = \frac{1}{K \, |GH(\omega_{pc})|} = \frac{1}{a} \tag{B.49}$$

In terms of decibels, the gain margin is

$$G.M._{dB} = 20 \log_{10}(G.M.) = -20 \log_{10} |KGH(j\omega_{pc})| = -20 \log_{10} a \tag{B.50}$$

The gain margin is simply the factor by which K must be changed in order to render the system unstable. The gain margin alone is inadequate to indicate relative stability when system parameters affecting the phase of $GH(j\omega)$ are subject to variation. Another measure, called *phase margin*, is required to indicate the degree of stability. Let ω_{gc}, known as the *gain crossover frequency*, be the frequency at which the open-loop frequency response magnitude is unity. The phase margin is the angle in degrees through which the polar plot must be rotated about the origin in order to intersect the $(-1, 0)$ point. The phase margin is given by

$$P.M. = \angle GH(j\omega_{gc}) - (-180^\circ) \tag{B.51}$$

For satisfactory performance, the phase margin should be between 30° and 60°, and the gain margin should be greater than 6 dB. The *MATLAB Control System Toolbox* function **[Gm, Pm,** ω_{pc} **,** ω_{gc} **] = margin(mag, phase** ω) can be used with **bode** function for evaluation of gain and phase margins, ω_{pc} and ω_{gc}.

B.6.5 NYQUIST STABILITY CRITERION

The *Nyquist stability criterion* provides a convenient method for finding the number of zeros of $1+GH(s)$ in the right-half s-plane directly from the Nyquist plot of $GH(s)$. The Nyquist stability criterion is defined in terms of the $(-1, 0)$ point on the Nyquist plot or the *zero-dB*, 180° point on the Bode plot. The Nyquist criterion is based upon a theorem of complex variable mathematics developed by Cauchy. The *Nyquist diagram* is obtained by mapping the *Nyquist path* into the complex plane via the mapping function $GH(s)$. The Nyquist path is chosen so that it encircles the entire right-half s-plane. When the s-plane locus is the Nyquist path, the Nyquist stability criterion is given by

$$Z = N + P \tag{B.52}$$

where

P = number of poles of $GH(s)$ in the right-half s-plane,

N = number of clockwise encirclements of $(-1, 0)$ point by the Nyquist diagram,

Z = number of zeros of $1 + GH(s)$ in the right-half s-plane.

For the closed-loop system to be stable, Z must be zero, that is

$$N = -P \tag{B.53}$$

B.6.6 SIMPLIFIED NYQUIST CRITERION

If the open-loop transfer function $GH(s)$ does not have poles in the right-half s-plane ($P = 0$), it is not necessary to plot the complete Nyquist diagram; the polar plot for ω increasing from 0^+ to ∞ is sufficient. Such an open-loop transfer function is called *minimum-phase* transfer function . For minimum-phase open-loop transfer functions, the closed-loop system is stable if and only if the polar plot lies to the right of $(-1, 0)$ point. For a minimum-phase open-loop transfer function, the criterion is defined in terms of the polar plot crossing with respect to $(-1, 0)$ point, as follows.

Right of	-1 stable	$\omega_{pc} > \omega_{gc}$	$\lvert GH(j\omega_{pc})\rvert < 1,\ GM_{dB} > 0,\ PM > 0°$
On	-1 marg. stable	$\omega_{pc} = \omega_{gc}$	$\lvert GH(j\omega_{pc})\rvert = 1,\ GM_{dB} = 0,\ PM = 0°$
Left of	-1 not stable	$\omega_{pc} < \omega_{gc}$	$\lvert GH(j\omega_{pc})\rvert > 1,\ GM_{dB} < 0,\ PM < 0°$

If P is not zero, the closed-loop system is stable if and only if the number of counterclockwise encirclements of the Nyquist diagram about $(-1, 0)$ point is equal to P.

The *MATLAB Control System Toolbox* function **[re, im] = nyquist(num, den, ω)** can obtain the Nyquist diagram by mapping the Nyquist path. However, the argument ω is specified as a real number. In order to map a complex number $s = a+jb$, we must specify $\omega = -js$, since the above function automatically multiplies ω by the operator j. To avoid this, the developed function **[re, im] = cnyquist(num, den, s)** can be used, where the argument s must be specified as a complex number. In defining the Nyquist path, care must be taken for the path not to pass through any poles or zeros of $GH(s)$.

B.6.7 CLOSED-LOOP FREQUENCY RESPONSE

The *closed-loop frequency response* is the frequency response of the closed-loop transfer function $T(j\omega)$. The *Control System Toolbox* function **bode**, described in Section B.6.1, is used to obtained the closed-loop frequency.

The performance specifications in terms of closed-loop frequency response are the closed-loop system *bandwidth* ω_B and the closed-loop system *resonant peak magnitude* M_p. The bandwidth ω_B is defined as the frequency at which the $|T(j\omega)|$ drops to 70.7 percent of its zero frequency value, or 3 dB down from the zero frequency value. The bandwidth indicates how well the system tracks an input sinusoid and is a measure of the speed of response. If the bandwidth is small, only signals of relatively low frequency are passed, and the response is slow; whereas, a large bandwidth corresponds to a fast rise time. Therefore, the rise time and the bandwidth are inversely proportional to each other. The frequency at which the peak occurs, the *resonant frequency* , is denoted by ω_r, and the maximum amplitude, M_p, is called the resonant peak magnitude. M_p is a measure of the relative stability of the system. A large M_p corresponds to the presence of a pair of dominant closed-loop poles with small damping ratio, which results in a large maximum overshoot of the step response in the time domain. If the gain K is set so that the open-loop frequency response $GH(j\omega)$ passes through the $(-1, 0)$ point, M_p will be infinity. In general, if M_p is kept between 1.0 and 1.7, the transient response will be acceptable. The developed function **frqspec(w, mag)** calculates M_p, ω_r, and the bandwidth ω_B from the frequency response data.

B.6.8 FREQUENCY RESPONSE DESIGN

The frequency response design provides information on the steady-state response, stability margin, and system bandwidth. The transient response performance can be estimated indirectly in terms of the phase margin, gain margin, and resonant peak magnitude. Percent overshoot is reduced with an increase in the phase margin, and the speed of response is increased with an increase in the bandwidth. Thus, the gain crossover frequency, resonant frequency, and bandwidth give a rough estimate of the speed of transient response.

A common approach to the frequency response design is to adjust the open-loop gain so that the requirement on the steady-state accuracy is achieved. This is called the *proportional controller*. If the specifications on the phase margin and gain margin are not satisfied, then it is necessary to reshape the open-loop transfer function by adding the additional controller $G_c(s)$ to the open-loop transfer function. $G_c(s)$ must be chosen so that the system has certain specified characteristics. This can be accomplished by combining proportional with integral action (PI) or proportional with derivative action (PD). There are also proportional-plus-integral-plus-derivative (PID) controllers with the following transfer function.

$$G_c(s) = K_P + \frac{K_I}{s} + K_D s \tag{B.54}$$

The ideal integral and differential compensators require the use of active amplifiers. Other compensators which can be realized with only passive network elements are lead, lag, and lead-lag compensators. A first-order compensator having a single zero and pole in its transfer function is

$$G_c(s) = \frac{K_c(s + z_0)}{s + p_0} \tag{B.55}$$

Several functions have been developed for the selection of suitable controller parameters based on the satisfaction of frequency response criteria, such as gain margin and phase margin. These functions tabulated below.

Alternatively, the function **[numo, deno, denc] = frdesign(num, den)** allows the user to select any of the above controller designs where **num** and **den** are row vectors of polynomial coefficients of the uncompensated open-loop plant transfer function. The function returns the open-loop and closed-loop numerators and denominators of the compensated system transfer function.

Function	Controller
[numo, deno, denc] = frqp(num, den)	Proportional
[numo, deno, denc] = frqlead(num, den)	Phase-lead
[numo, deno, denc] = frqlag(num, den)	Phase-lag
[numo, deno, denc] = frqpd(num, den)	PD
[numo, deno, denc] = frqpi(num, den)	PI
[numo, deno, denc] = frqpid(num,den)	PID

Example B.3

Design a PID controller for the system of Example B.1 for a compensated system phase margin of 77.8°. Choose a value of 9.09 for K_I, and select the new phase crossover frequency of 1.53 rad/s. Also, obtain the Bode plot of the compensated open-loop transfer function. The following commands:

```
num = 100;   den = [1  26  216  576];
[numo, deno, denc]=frdesign(num, den); % PID design
w = .1:.1:20;
[mag, phase] =bode(numo, deno, w); dB = 20*log10(mag);
figure(1), plot(w, dB), grid
xlabel('w, rad/sec'), ylabel('dB')
figure(2), plot(w, phase), grid
xlabel('w, rad/sec'), ylabel('Degrees')
```

result in

```
            Compensator type                              Enter
            Gain compensation                               1
            Phase-lead                                      2
            Phase-lag                                       3
            PD Controller                                   4
            PI Controller                                   5
            PID Controller                                  6
            To quit                                         0
Enter your choice → 6
Enter the integrator gain KI → 9.09
Enter desired Phase Margin → 77.8
Enter wgc → 1.53
Uncompensated control system
Gain Margin = 50.4 Gain crossover w = NaN
Phase Margin = Inf Phase crossover w = 14.7
Gc = 2.10655 + 9.09/s + 0.14074s
Row vectors of polynomial coefficients of the compensated
system:
 Open-loop num.   14.07  210.65     909
 Open-loop den.       1      26     216     576    0
 Closed-loop den      1      26  230.07  786.65  909
```

```
    Gain Margin = 30300  Gain crossover w = 1.53
    Phase Margin = 77.8  Phase crossover w = 653
    Bandwidth = 1.95
Roots of the compensated characteristic equation:
-9.9943 + 1.0097i
-9.9943 - 1.0097i
-3.1671
-2.8444
```

The PID controller increases the system type by 1. That is, we have a type 1 system, and the steady-state error due to a unit step input is zero, and the step response is similar to Figure B.6. The compensated open-loop Bode plot is shown in Figure B.7.

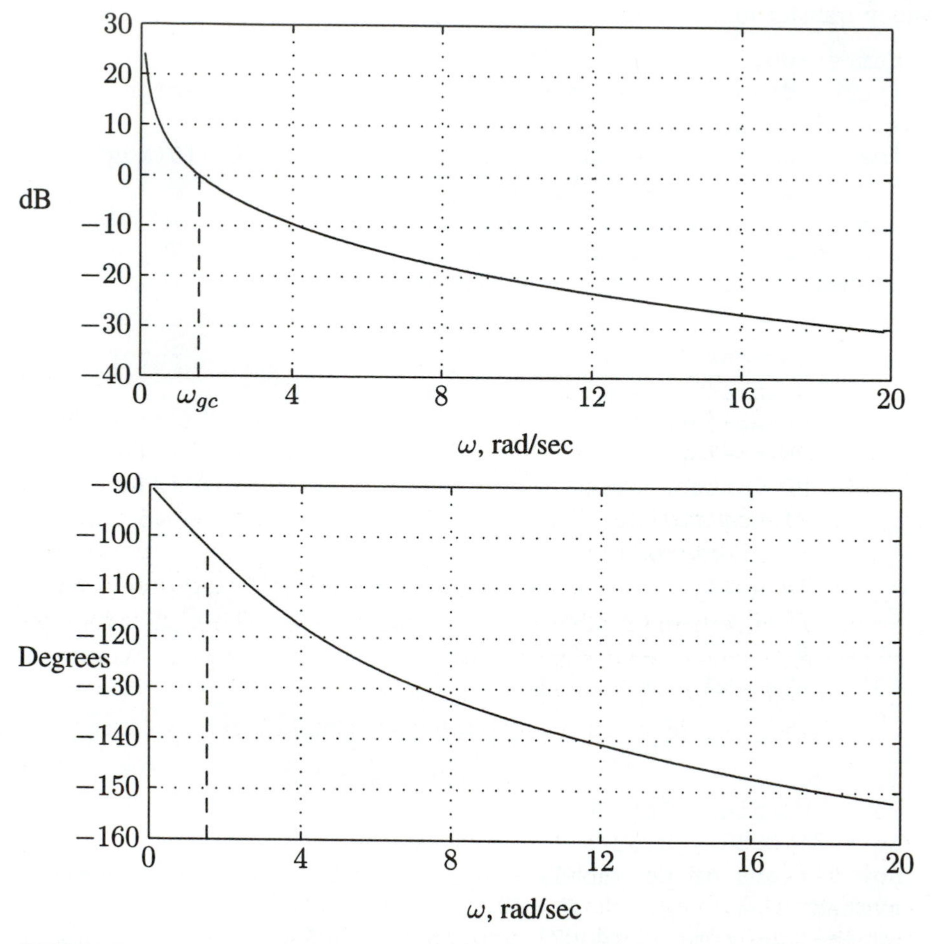

FIGURE B.7
The compensated open-loop Bode plot for Example B.3.

APPENDIX C

POWER SYSTEM TOOLBOX

The Power System Toolbox, containing a set of M-files, has been developed by the author to assist in typical power system analysis. Some of the programs, such as power flow, optimization, short-circuit and stability analysis, were originally developed by the author for a mainframe computer when working for power system consulting firms many years ago. These programs have been refined and modularized for interactive use with *MATLAB* for many problems related to the operation and analysis of power systems. The software modules are structured in such a way that the user may mix them for other power system analyses. The M-files for typical power system analyses are designed to work in synergy and communicate with each other through the use of some global variables.

The software diskette included with this book contains all the developed functions and chapter examples. Instructions for installing the Power System Toolbox can be found with the Installing the Text Toolbox described in Appendix A. We recommend that you store the files from this toolbox in a directory named *power*, where the *MATLAB* toolbox resides. Add the necessary search path to the *MATLAB* Path Browser. This appendix contains a list of all functions and script files in the Power System Toolbox developed by the author. The file names for the chapter examples are also included.

LIST OF FUNCTIONS, SCRIPT FILES, AND EXAMPLES IN THE POWER SYSTEM TOOLBOX

Load Cycle	
barcycle(data)	Plot load cycle for a given load interval

Transmission Line Parameters	
[GMD, GMRL, GMRC] = gmd	Multicircuits GMD and GMR
[L, C] = gmd2lc	Multicircuit GMD, GMRL, L, and C
acsr	Displays the ACSR characteristics

Transmission Line Performance	
lineperf	Line performance program
[r, L, C, f] = abcd2rlc(ABCD)	ABCD to rLC conversion
[Z, Y, ABCD] = abcd2pi(A, B, C)	ABCD to π model conversion
[Z, Y, ABCD] = pi2abcd(Z, Y)	π model to ABCD conversion
[Z, Y, ABCD] = rlc2abcd(r, L, C, g, f, Ln)	rLC to ABCD conversion
[Z, Y, ABCD] = zy2abcd(z, y, Ln)	zy to ABCD conversion
listmenu	Displays 8 options for analysis
givensr(ABCD)	Sending end values from receiving end power
givenss(ABCD)	Receiving end values from sending end power
givenzl(ABCD)	Sending end values from load impedance
loadabil(L, C, f)	Line loadability curves
openline((ABCD)	Open line analysis and reactor compensation
shcktlin(ABCD)	Receiving end short circuit
compmenu	Displays 3 options for capacitive compensation
sercomp(ABCD)	Series capacitor compensation
shntcomp(ABCD)	Shunt capacitor compensation
srshcomp(ABCD)	Shunt and series capacitors compensation
profmenu	Displays two options for loadabil and vprofile
pwrcirc(ABCD)	Receiving end power circle diagram
vprofile(ABCD)	Voltage curves for various loading

Optimal Dispatch of Generation	
bloss	Returns loss coefficients when followed by power flow program
dispatch	Obtains optimum dispatch of generation
gencost	Computes the total generation cost $/hr

Transformer and Induction Motor	
trans	Transformer characteristics
tperf	This script file is called by trans
[Rc, Xm] = troct(Vo, Io, Po)	Shunt branch from OC test
[Ze] = trsct(Vsc, Isc, Psc)	Obtains the series branch from SC test
[Ze1, Ze2]= trsct(E1, E2, Z1, Z2)	Winding impedances to Eq. impedance
rotfield	Revolving field demonstration
im	Equivalent circuit analysis
imchar	Torque/speed curve (called by im)
imsol	Motor performance (called by im)

Power Flow Analysis	
ybus	Obtains Y_{bus}, given R and X values
lfybus	Obtains Y_{bus}, given π model with specified *linedata* file
lfgauss	Power flow solution by the Gauss-Seidel method
lfnewton	Power flow solution by the Newton-Raphson method
decouple	Power flow solution by the Fast Decoupled method
busout	Returns the bus output result in tabular form
lineflow	Returns the line flow and losses in tabular form

Symmetrical Components	
sctm	Symmetrical Components Transformation Matrix
phasor(F)	Plots phasors expressed in rectangular or polar
F012 = abc2sc(Fabc)	Phasors to symmetrical components conversion
Fabc = sc2abc(F012)	Symmetrical components to phasors conversion
Z012 = zabc2sc(Zabc)	Impedance matrix to symmetrical components
Fr = pol2rec(Fp)	Polar phasor to rectangular phasor conversion
Fp = rec2pol(Fr)	Rectangular phasor to polar phasor conversion

Fault Analysis	
dlgfault(Z0, Zbus0, Z1, Zbus1, Z2, Zbus2, V)	Double line-to-ground fault
lgfault(Z0, Zbus0, Z1, Zbus1, Z2, Zbus2, V)	Line-to-ground fault
llfault(Z1, Zbus1, Z2, Zbus2, V)	Line-to-line fault
symfault(Z1, Zbus1, V)	Line-to-ground fault
Zbus = zbuild(zdata)	Builds the Bus Impedance Matrix
Zbus = zbuildpi(linedata, gendata, load)	Builds the Bus Impedance Matrix, compatible with load flow data

Synchronous Machine Transients	
lgshort(t,i)	Returns state derivatives of current for L-G short circuit
llshort(t, i)	Returns state derivatives of current for line-line short circuit
symshort(t, i)	Returns state derivatives of current for 3-phase short circuit

Power System Stability	
cctime	Obtains the critical clearing time for fault
eacfault(P0, E, V, X1, X2, X3)	Displays equal area criterion & finds critical clearing time of fault
eacpower(P0, E, V, X)	Displays equal area criterion & max. steady-state power
xdot = afpower(t, x)	One-machine system state derivative after fault
xdot = pfpower(t, x)	One-machine system state derivative during fault
swingmeu(Pm, E, V, X1, X2, X3, H, f, tc, tf)	One-machine swing curve, modified Euler
swingrk2(Pm, E, V, X1, X2, X3, H, f, tc, tf)	One-machine swing curve, *MATLAB* ode23
swingrk4(Pm, E, V, X1, X2, X3, H, f, tc, tf)	One-machine swing curve, *MATLAB* ode34
xot = afpek(t, x)	Multimachine system state derivative after fault
xdot = dfpek(t, x)	Multimachine system state derivative during fault
trstab	Stability analysis works in synergy with load flow
[Ybus, Ybf] = ybusbf(linedata, yload, nbus1, nbust)	Multimachine system reduced Y_{bus} before fault
Ypf = ybusbf(Ybus, nbus1, nbust, nf)	Multimachine system reduced Y_{bus} during fault
Yaf = ybusaf(linedata, yload, nbus1, nbust, nbrt)	Multimachine system reduced Y_{bus} after fault

Control System Functions	
electsys	Returns the state derivatives for Example A.19
errortf	Steady-state error, transfer function in polynomial form
errorzp	Steady-state error, transfer function in zero pole form
frcntrl	Frequency response design equations
frdesign	Frequency response design program
frqlag	Frequency response design phase-lag controller
frqlead	Frequency response design phase-lead controller
frqp	Frequency response design P controller
frqpd	Frequency response design PD controller
frqpi	Frequency response design PI controller
frqpid	Frequency response design PID controller
frqspec	Frequency response performance specifications
ghs	Returns magnitude and phase of a complex function $GH(s)$
ltstm	Laplace transform of state transition matrix
mechsys	Returns the state derivatives for Example A.18
pcomp	Root-locus design P controller
pdcomp	Root-locus design PD controller
pdlead	Root-locus design phase-lead controller
pendulum	Returns the state derivatives for Example A.20
phlag	Root-locus design phase-lag controller
picomp	Root-locus design PI controller
pidcomp	Root-locus design PID controller
placepol	Pole-placement design
pnetfdbk	Feedback compensation using passive elements
riccasim	Returns state derivative of Riccati equation
riccati	Optimal regulator design
rldesign	Root-locus design program
routh	Routh-Hurwitz array
ss2phv	Transformation to control canonical form
statesim	Returns state derivatives for use in Riccati equation
stm	Determines the state transition matrix $\phi(t)$
system	System matrices defined for use in Riccati equation
tachfdbk	Tachometer feedback control
timespec	Time-domain performance specifications

List of M-Files for Chapter Examples				
CHP1EX1	CHP5EX5	CHP7EX7	CHP10EX3	EXA1
CHP2EX1	CHP5EX6	CHP7EX8	CHP10EX4	EXA2
CHP2EX2	CHP5EX7	CHP7EX9	CHP10EX5	EXA3
CHP2EX3	CHP5EX8	CHP7EX10	CHP10EX6	EXA4
CHP2EX4	CHP5EX9	CHP7EX11	CHP10EX7	EXA5
CHP2EX5	CHP6EX1	CHP8EX1	CHP10EX8	EXA6
CHP2EX6	CHP6EX2	CHP8EX2	CHP11EX1	EXA7
CHP2EX7	CHP6EX3	CHP8EX3	CHP11EX2	EXA8
CHP2EX8	CHP6EX4	CHP8EX4	CHP11EX3	EXA9
CHP3EX1	CHP6EX5	CHP8EX5	CHP11EX4	EXA10
CHP3EX2	CHP6EX6	CHP8EX6	CHP11EX5	EXA11
CHP3EX3	CHP6EX7	CHP8EX7	CHP11EX6	EXA12
CHP3EX4	CHP6EX8	CHP8EX8	CHP11EX7	EXA13
CHP3EX5	CHP6EX9	CHP8EX9	CHP12EX1	EXA14
CHP3EX6	CHP6EX10	CHP8EX10	CHP12EX2	EXA15
CHP4EX1	CHP6EX11	CHP9EX1	CHP12EX3	EXA16
CHP4EX2	CHP6EX12	CHP9EX2	CHP12EX4	EXA17
CHP4EX3	CHP6EX13	CHP9EX3	CHP12EX5	EXA18
CHP4EX4	CHP6EX14	CHP9EX4	CHP12EX6	EXA19
CHP4EX5	CHP6EX15	CHP9EX5	CHP12EX7	EXA20
CHP4EX6	CHP7EX1	CHP9EX6	CHP12EX8	EXB1
CHP4EX7	CHP7EX2	CHP9EX7	CHP12EX9	EXB2
CHP5EX1	CHP7EX3	CHP9EX8	CHP12XX	EXB3
CHP5EX2	CHP7EX4	CHP9EX9	CHP12XX1	
CHP5EX3	CHP7EX5	CHP10EX1		
CHP5EX4	CHP7EX6	CHP10EX2		

List of SIMULINK-Files for Chapter Examples				
SIM11EX3	SIM11EX6	SIM12EX1	SIM12EX3	SIM12EX4
SIM12EX5	SIM12EX6	SIM12EX7	SIM12EX8	SIM12EX9
SIM12XXB	SIM12XXD	SIM12XX1	SIMEXA21	SIMEXA22
SIMEXA23	SIMEXA24	SIMEXA25	SIMEXA26	SIMEXA27
SIMEXA28	SIMEXB1			

If you encounter any bugs or problems please contact me at the following e-mail addresses or visit my Web sites for updates and information on this product.

Web site: http://www.msoe.edu/~ saadat
http://www.home.att.net/~saadat
e-mail: saadat@msoe.edu
saadat@worldnet.att.net

BIBLIOGRAPHY

1. Anderson, P. M., Analysis of Faulted Power Systems, IEEE Press, New York, 1973.

2. Anderson, P. M., and Fouad, A. A., Power System Control and Stability, The Iowa State University Press, Ames, Iowa, 1977.

3. Arrillaga, J., Arnold, C. P., and Harker, B. J., Computer Modeling of Electrical Power Systems, John Wiley & Sons, Inc., New York, 1986.

4. Bergen, A. R., Power Systems Analysis, Prentice-Hall, Englewood, Cliffs, New Jersey, 1970.

5. Bergseth, F. R., and Venkata, S. S., Introduction to Electric Energy Devices, Prentice-Hall, Englewood, Cliffs, New Jersey, 1987.

6. Billinton, R., Ringlee, R., and Wood, A., Power System Reliability Calculations, MIT Press, Cambridge, Massachusetts, 1973.

7. Billinton, R., Power System Reliability Evaluations, Gordon and Breach, New York, 1970.

8. Brosan, G. S., and Hayden, J. T., Advanced Electrical Power and Machines, Sir Isaac Pitman & Sons, Ltd., London, 1966.

9. Brown, H. E., Solution of Large Networks by Matrix Methods, John Wiley & Sons, Inc., New York, 1975.

10. Brown, H. E., Person, C. E., Kirchmayer, L. K., and Stagg, W. G., Digital Calculation of Three-Phase Short-Circuits by Matrix Method, AIEE Trans., Part 3, pp. 1277-1282, 1960.

11. Burchett, R. C., Happ, H. H., and Vierath, D. R., Quadratically Convergent Optimal Power Flow, IEEE Trans., PAS-103, No. 11, pp. 3267-3275, November 1984.

12. Burchett, R.C., Happ, H. H., and Wirgau, K. A., Large Scale Optimal Power Flow, IEEE Trans., PAS-101, No. 10, pp. 3722-3731, October 1982.

13. Byerly, R. T., and Kimbark, E. W., Stability of Large Electric Power Systems, IEEE Press, 1974.

14. Carson, J. R., Wave Propagation in Overhead Wires with Ground Return, Bell System Technical Journal, Vol. 5, pp. 539-554, October 1928.

15. Clarke, E., Circuit Analysis of AC Power Systems, John Wiley & Sons, Inc., New York, 1958.

16. Cohn, N., Control of Generation and Power Flow on Interconnected Systems, John Wiley & Sons, Inc., New York, 1966.

17. Concordia, C., Synchronous Machine—Theory and Performance, John Wiley & Sons, Inc., New York, 1951.

18. Crary, Power System Stability, Vol. II, John Wiley & Sons, Inc., New York, 1955.

19. Del Toro, V., Electric Power Systems, Prentice-Hall, Englewood Cliffs, New Jersey, 1992.

20. Doherty, R. E., and Nickle, C. A., Synchronous Machines, AIEE Trans., Vol. 45, pp. 912-942, 1926.

21. Dommel, H. W., and Sato, N., Fast Transient Stability Solutions, IEEE Trans., Power Apparatus and Systems, PAS-91, pp. 1643-1650, October 1972.

22. Dommel, H. W., and Tinney, W. F., Optimal Power Flow Solutions, IEEE Trans., Power Apparatus and Systems, PAS-87, pp. 1866-1876, October 1968.

23. Dopaz, J. F., Klitin, O. A., Stagg, G. W., and Watson, M., An Optimization Technique for Real and Reactive Power Allocation, Proc. of the IEEE, Vol. 55, No. 11, pp. 1877-1885, 1967.

24. Elgerd, O. I., Electric Energy Systems Theory, Second Edition, McGraw-Hill Book Company, New York, 1982.

25. El-Abiad, A. H., Digital Calculation of Line-to-Ground Short-circuits by Matrix Method, AIEE Trans., Vol. 79, pp. 323-332, 1960.

26. El-Hawary, M. E., Electrical Power Systems Design and Analysis, Reston Publishing Company, Reston, Virginia, 1983.

27. EPRI, Transmission Line Reference Book, 345 KV and Above, Electric Power Research Institute, Palo Alto, California, 1982.

28. Fink, D. G., and Beaty, H. W., Standard Handbook for Electrical Engineers, McGraw-Hill Book Company, New York, 1987.

29. Fitzgerald, A. E., Kingsley, C., and Umans, S., Electric Machinery, Fourth Edition, McGraw-Hill, New York, 1982.

30. Fortescue, C. L., Method of Symmetrical Components Applied to the Solution of Polyphase Networks, AIEE Trans., Vol. 37, pp. 1027-1140, 1918.

31. Fouad, A. A. and Stanton, S.E., Transient Stability of Multimachine Power Systems, Part I and II, IEEE Trans., Vol. PAS-100, pp. 3408-3424, July 1981.

32. Fouad, A. A., Vijay, V., Power System Transient Stability Analysis Using the Transient Energy Function Method, Prentice-Hall, Englewood Cliffs, New Jersey, 1992.

33. Gless, G. E., Direct Method of Lyapunov Applied to Transient Power System Stability, IEEE Trans., Vol. PAS-85, No. 2, pp. 159-168, February 1966.

34. Glover, J. D., Power System Analysis and Design, Second Edition, PWS Publishing Company, Boston, 1994.

35. Gönen, T., Electric Power Distribution Systems Engineering, McGraw-Hill Book Company, New York, 1986.

36. Greenwood, A., Electrical Transients in Power Systems, Wiley Interscience, New York, 1971.

37. Gross, C. A., Power System Analysis, Second Edition, John Wiley & Sons, New York, 1983.

38. Guile, A. E., and Paterson, W., Electrical Power Systems, Oliver & Boyd, Edinburgh, 1969

39. Gungor, B. R., Power Systems, Harcourt, Brace, Jovanovich, Inc., New York, 1988.

40. Happ, H., Optimal Power Dispatch—A Comprehensive Survey, IEEE Trans., Power Apparatus and Systems, PAS-96, pp. 841-854, June 1977.

41. Heydt, G. T., Computer Analysis Methods for Power Systems, Macmillan Publishing Company, New York, 1986.

42. Horton, J. S., and Grigsby, L. L., Voltage Optimization Using Combined Linear Programming and Gradient Techniques, IEEE Trans., PAS-103, No. 7, pp. 1637-1643, July 1984.

43. IEEE Standard 115, Test Procedure for Synchronous Machines.

44. Kimbark, E. W., Power System Stability, Vol. 1, Elements of Stability Calculations, John Wiley & Sons, Inc., New York, 1948.

45. Kimbark, E. W., Power System Stability, Vol. 2, Power Circuit Breakers and Protective Relays, John Wiley & Sons, Inc., New York, 1950.

46. Kimbark, E. W., Power System Stability, Vol. 3, Synchronous Machines, John Wiley & Sons, Inc., New York, 1956.

47. Kirchmayer, L. K., Economic Operation of Power Systems, John Wiley & Sons, Inc., New York, 1958.

48. Kirchmayer, L. K., Economic Control of Interconnected Systems, John Wiley & Sons, Inc., New York, 1959.

49. Knable, A., Electrical Power Systems Engineering, McGraw-Hill Book Company, New York, 1967.

50. Kron, G., Tensorial Analysis of Integrated Transmission Systems, Part I, The Six Basic Reference Frames, AIEE Trans., Vol. 70, pp. 1239-1248, 1951.

51. Kron, G., Tensorial Analysis of Integrated Transmission Systems, Part II, Off-Nominal Turn Ratios, AIEE Trans., Vol. 71, pp. 505-512, 1952.

52. Kundur, P., Power System Stability and Control, McGraw-Hill Book Company, New York, 1994.

53. Kusic, G. L., Computer Aided Power System Analysis, Prentice-Hall, Englewood Cliffs, New Jersey, 1986.

54. Lee, K. Y., Park, Y. M., and Ortiz, J. L., A United Approach to Optimal Real and Reactive Power Dispatch, IEEE Trans., PAS-104, No. 5, pp. 1147-1153, May 1985.

55. Lewis, W. A., The Principles of Synchronous Machines, IIT Press, Chicago, Illinois, 1949.

56. Lewis, W. A., A Basic Analysis of Synchronous Machines, AIEE Trans., PAS-77, pp. 436-455, 1958.

57. Lyapunov, A. M., Stability of Motion, English translation, Academic Press, Inc., New York, 1967.

58. Mamandur, K. R. C., and Chenoweth, R. D., Optimal Control of Reactive Power Flow for Improvements in Voltage Profile and Real Power Loss Minimization, IEEE Trans., PAS-100, No. 7, pp. 3185-33194, July 1981.

59. Nasar, S. A., Electric Energy Conversion and Transmission, Macmillan Publishing Company, New York, 1985.

60. Neuenswander, J. R., Modern Power Systems, International Textbook Company, Scranton, Pennsylvania, 1971.

61. Park, R. H., Two Reaction Theory of Synchronous Machines—Generalized Method of Analysis, AIEE Trans., Vol. 48, pp. 716-727, 1929.

62. Rustebakke, H. M., Electric Utility Systems and Practices, John Wiley & Sons, Inc., New York, 1983.

63. Saadat, H., Time Domain Simulation of Synchronous Machine Imbalance Faults Using PC-MATLAB, Proc. of 20th North American Power Symposium, pp. 285-291, October 1988.

64. Saadat, H., Microcomputer Applications in Electrical Power Engineering Education, Proc. of the 19th North American Power Symposium, pp. 307-314, October 1987.

65. Saadat H., Optimal Load Flow Solution by the Power Perturbation Technique, IEEE, PES Winter Meeting, A79028-2, 1979.

66. Saadat, H., Steady State Analysis of Power Systems Including the Effects of Control Devices, Vol. 2, Journal of Electric Power System Research, pp. 111-118, 1979.

67. Saadat, H., Application of Quasi-Newton Method to Load Flow Studies and Solution of Load Flow During Three-Phase Fault, Journal of Electric Power System Research, pp. 173-179, 1978.

68. Saadat, H., Computational Aids in Control Systems Using MATLAB, McGraw-Hill Book Company, New York, 1983.

69. Sauer, P. W., and Pai, M. A., Power System Dynamics and Stability, Prentice-Hall, Englewood Cliffs, New Jersey, 1998.

70. Singh, L. P., Advanced Power System Analysis and Design, Halsted Press, New York, 1983.

71. Shoults. R. R., and Sun, D. T., Optimal Power Flow Based upon P-Q Decomposition, IEEE Trans., PAS-101, No. 2, pp. 397-405, February 1982.

72. Shultz, R. D., and Smith, R. A., Introduction to Electric Power Engineering, Harper & Row Publishers, New York, 1985.

73. Stagg, G. W., and El-Abiad, A. H., Computer Methods in Power System Analysis, McGraw-Hill Book Company, New York, 1968.

74. Stevenson, W. D., and Grainger, J. J., Power System Analysis, McGraw-Hill Book Company, New York, 1994.

75. Stott, B., Decoupled Newton Load Flow, IEEE Trans. Power Apparatus and Systems, PAS-91, pp. 1955-1959, October 1972.

76. Stott, B., and Alsac, O., Fast Decoupled Load Flow, IEEE Trans. Power Apparatus and Systems, PAS-93, pp. 859-869, May-June 1974.

77. Sullivan, R., Power System Planning, McGraw-Hill Book Company, New York, 1977.

78. Sun, D. I., Ashely, B., Brewer, B., Hughes, A., and Tinney, W. F., Optimal Power Flow by Newton Approach, IEEE Trans., PAS-103, No. 10, pp. 2864-2879, October 1984.

79. Taylor, C. W., Power System Voltage Stability, McGraw-Hill Book Company, New York, 1993.

80. Tinney, W. F., Compensation Methods for Network Solutions by Optimally Ordered Triangular Factorization, IEEE Trans., PAS-91, pp. 123-127, January 1972.

81. Wadhwa, C. L., Electrical Power Systems, John Wiley & Sons, Inc., New York, 1983.

82. Wagner, C. F., and Evens, R. D., Symmetrical Components, McGraw-Hill Book Company, New York, 1933.

83. Wallach, Y., Calculations and Programs for Power System Networks, Prentice-Hall, Englewood Cliffs, New Jersey, 1986.

84. Ward, J., and Hale, H., Digital Computer Solution of Power Flow Problems, AIEE Trans., Vol. 75, pt. III, pp. 398-404, 1956.

85. Weedy, B. M., Electric Power Systems, Third Edition, John Wiley & Sons, Inc., New York, 1979.

86. Weeks, W. L., Transmission and Distribution of Electrical Energy, Harper & Row Publishers, New York, 1981.

87. Westinghouse Electric Corporation, Electric Transmission and Distribution Reference Book, East Pittsburgh, Pennsylvania, 1964.

88. Wood, A. J., and Wollenberg, B. F., Power Generation Operation and Control, John Wiley & Sons, Inc., New York, 1974.

89. Yamayee, Z. A., Electromechanical Energy Devices and Power Systems, John Wiley & Sons, Inc., New York, 1994.

90. Yu, Yao-nan, Electric Power Systems Dynamics, Academic Press, New York, 1983.

ANSWERS TO PROBLEMS

Chapter 1

1.1. 799.34 GW

1.2. 6.93%

1.3. 8 MW, 50%

Chapter 2

2.1. $S_1 = 2000$ W $+ j3464.1$ var, $S_2 = 2165.1$ W $- j1250$ var, $S_3 = 2000$ W $+ j0$ var, $S = 6165.1$ W $+ j2214.1$ var

2.2. (a) 800 W + $j600$ var, (b) $10\cos(377t - 36.87°)$ A, $7.071\angle{-36.87°}$ A, (c) $20\angle 36.87°$ Ω

2.3. 12.8 Ω, 9.6 Ω

2.4. 20Ω, 26.67Ω

2.5. 280 kW $+ j335$ kvar

2.6. (a) 60 Ω, 80 Ω, (b) $1250\angle 16.26°$ V

2.7. (a) $S_1 = 1$ kW $+ j7$ kvar, $S_2 = 1$ kW $- j2$ kvar, $S_3 = 4$ kW $+ j3$ kvar, (b) $S = 6$ kW $+ j8$ Kvar, $50\angle{-53.13°}$ A, 0.6 lagging, (c) 8 kvar, 530.5 μF, 30 A

2.8. Source 1 delivers 28 kW and receives 21 kvar, Source 2 receives 24.57 kW and delivers 32.76 kvar, 3.43 kW, 11.76 kvar

2.10. (b) 30 kW

2.11. (a) $50\angle{-36.87°}$ A, $50\angle{-156.87°}$ A, $50\angle{-276.87°}$ A, (b) 288 kW, 216 kvar

2.12. (a) $150\angle{-66.87°}$ A, $150\angle{-186.87°}$ A, $50\angle 53.13°$ A, (b) 864 kW, 648 kvar

2.13. (a) $12\angle{-53.13°}$ A, (b) 2592 W + j 3456 var, (c) 162.33 V

2.14. (a) 18 kW, 0 kvar, unity power factor, 50 A, (b) $66.9\angle{-41.63°}$ A,0.7474 lagging

2.15. (a) 360 kW + j480 kvar, 0.6 lagging, $27.78\angle{-53.13°}$ A, (b) 210 kvar, 3.58 μF, $20.835\angle{-36.87°}$ A

2.16. (a) $40\angle{-36.87°}$ A, 19.2 kW + j14.4 kvar, (b) 160 V, 277.1 V

Chapter 3

3.1. (a) 2440.80 V, 6.12 %, (b) 2200.4 V, −4.33%

3.2. (a) 36 kV, 9.59°, (b) 288 MW, (c) 547.47 A, 0.7306 lagging

3.3. (a) 12806 V, (b) 80.4 MW, (c) $3344\angle 36.73°$

3.4. (b) 16.26°, 30 kV, (c) 138.712 MW at 75°

3.5. (a) $0.4 + j0.9$ Ω, 1000 Ω, $j1500$Ω, (b) 2453.9 V, 2.247%, (c) 2387 V, −0.541%

3.6. (a) $28+j96$ Ω, 6666.67 Ω, $j5000$ Ω, (b) 21.839%, 85.97%, (c) 53.237 kVA, 86.057%, (d) 85.88%

3.7. (a) 21 kVA, (b) 96%

3.8. 13.346 kV

3.9. (a) 247.69 kV, (b) 249.72 kV

3.10. (a) 1.03205 pu, 247.69 kV, (b) 1.0405 pu, 249.72 kV

3.11. 0.926 pu, 1.0 pu

3.12. $0.122 + j0.252$ pu

3.13. $X_{G_1} = j0.1$, $X_{T_1} = j0.2$, $X_{T_2} = j0.25$, $X_{G_2} = j0.081$, $X_{Line} = j0.3$, $X_{Load} = 0.75 + j1.0$,

3.14. $X_G = j0.3$, $X_{T_1} = j0.2$, $X_{T_2} = j0.15$, $X_{T_3} = j0.16$ $X_{Line_1} = j0.25$, $X_{Line_2} = j0.35$ $X_M = j0.27$, $X_{Load} = -j10$, $Z_P = j0.06$, $Z_S = j0.18$, $Z_T = j0.12$

3.15. (a) $X_{G_1} = j0.15$, $X_{T_1} = j0.2$, $X_{T_2} = j0.2$, $X_{Line} = 0.3 + j.5$ $X_M = j0.15$, (b) 26.359 kV, 27.5 kV

3.16. 440 kV , 480 kV

3.17. 126.5 kV, 27.6 kV

Chapter 4

4.1. (a) 4.1 Ω, (b) 4.6 Ω

4.2. 0.3774 Ω/km

4.3. 1.894 cm, 556000 cmil

4.4. 0.35 cm, 46 mH

4.5. (a) $1.46r$, (b) $1.723r$

4.6. 1.486 mH/km

4.7. 10 m

4.8. (a) 1.3 mH/km, (b) 1.15 cm

4.9. 27.5% decrease, 35.25% increase

4.10. 0.88929 mH/km, 0.012658 μF/km

4.11. 0.4752 mH/km, 0.0240035 μF/km, 0.517453 mH/km, 0.0219974 μF/km

4.12. $\frac{4\pi\varepsilon}{\ln 0.866D/r}$

4.13. 5 V/km

4.14. 90 V

Chapter 5

5.1. (a) 70.508 kV, 10.17%, 58.39 MW $+j50.37$ Mvar, 95.90%,
(b) 69.0 kV, 7.83%, 127 MW $+j24.61$ Mvar, 94.465%

5.2. (a) 14.117 Mvar, 9.14μF, (b) 61.24 Mvar, 39.66μF

5.3. (a) $0.9951 + j0.000544$, $4 + j36$ Ω, $j0.0002713$ S
(b) 242.67 kV, $502.38\angle{-33.69°}$ A, 10.847%, 163.18 MW $+j134.02$ Mvar, 98.052%
(c) 230.03 kV, $799.86\angle 2.5°$ A, 5.073%, 313.74 MW $+j55.9$ Mvar, 97.53%

5.4. 141.123 Mvar, 7.734μF

5.5. $0.98182 + j0.0012447$, $4.035 + j58.947$, $j0.00061137$

5.6. 387.025 kV, $592.29\angle{-27.325°}$ A, 324.87 MW, $+j228.25$ Mvar, 14.259%, 98.5%

5.7. 345 kV, $672.54\angle{-9.633°}$ A, 401.884 MW $+j0.0$ Mvar, 2.73%, 98.743%

5.8. (a) 264.702 Ω, 29°, 4965.2 km, 2210.88 MW, 0.8746, $j128.34$ Ω, $j0.0018316$ S
(b) 896.982 kV, $1100.23\angle{-2.456°}$ A, 1600 MW $+j601.508$ Mvar, 39.536%

(c) 653.33 kV, $1748.78\angle{-43.556°}$ A, 1920 MW $+j479.33$ Mvar, 33.88%
(d) 735.13 kV, $1604.07\angle{28.98°}$ A, 2042.44 MW $+j1.32$ Mvar, 14.358%

5.9. 874.68 kV, (b) 772.13 Ω, 699.658 Mvar

5.10. $3441.47\angle{-90°}$ A, $3009.92\angle{-90°}$ A

5.11. 802.95 Mvar, 3.943 μF, $1209.46\angle{24.653°}$ A, 1600 MW $-j90.38$ Mvar, 19%

5.12. 822.677 kV, $1164.59\angle{-3.625°}$ A, 1600 MW $+j440.16$ Mvar, 21.035%

5.13. 81.464 Mvar, 51.65 μF, 563.25 Mvar, 2.765 μF, 765 kV, $1209.72\angle{16.1°}$ A, 1600 MW $-j96.32$ Mvar, 12.55%

5.14. Use **lineperf** to obtain the transmission line performance. Present a summary of the calculations along with your recommendations.

5.15. (a) 622.153 kV, $794.649\angle{-1.33°}$ A, 800 MW $+j305.408$ Mvar, 44.687%
(b) 0.96, $j39.2$, $j0.002$
(c) 530.759 kV, $891.142\angle{-5.65°}$ A, 800 MW $+j176.448$ Mvar, 10.575%

5.16. (a) 0.002 Rad/km, 500 Ω, (b) 1000 Ω, 176.4 Mvar

5.17. 400 kV

Chapter 6

6.1. $$Y_{bus} = \begin{bmatrix} 0.0 - j20.25 & 0.0 + j4.00 & 0.0 + j10.00 & 0.0 + j2.50 \\ 0.0 + j4.00 & 0.0 - j15.00 & 0.0 + j0.00 & 0.0 + j6.25 \\ 0.0 + j10.00 & 0.0 + j0.0 & 1.0 - j15.00 & 0.0 + j5.00 \\ 0.0 + j2.50 & 0.0 + j6.25 & 0.0 + j5.00 & 2.0 - j14.00 \end{bmatrix}$$

6.2. $$V_{bus} = \begin{bmatrix} 1.0293\angle{1.46°} \\ 1.0217\angle{0.99°} \\ 1.0001\angle{-0.015°} \end{bmatrix}$$

6.3. (a) $x_1 = 5.0000 \quad x_2 = 1.0000$, $x_1 = 2.0006 \quad x_2 = 3.9994$

6.4. (a) 1, (b) 1, 4, 7, 9

6.5. $$X^{(1)} = \begin{bmatrix} 4.3929 \\ 4.9286 \end{bmatrix} \quad X^{(2)} = \begin{bmatrix} 4.0222 \\ 4.9964 \end{bmatrix} \quad X^{(3)} = \begin{bmatrix} 4.0001 \\ 5.0000 \end{bmatrix}$$

6.6. (a) $V_2^{(1)} = 0.9200 - j0.1000 \quad V_2^{(2)} = 0.9024 - j0.0981$
$V_2^{(3)} = 0.9005 - j0.1000 \quad V_2^{(4)} = 0.9001 - j0.1000$
(b) $S_{12} = 300$ MW $+ j100$ Mvar
$S_{21} = -280$ MW $- j60$ Mvar
$S_L = 20$ MW $+ j40$ Mvar

6.7. (a) $V_2^{(1)} = 0.9360 - j0.0800 \quad V_3^{(1)} = 0.9602 - j0.0460$

$V_2^{(2)} = 0.9089 - j0.0974 \quad V_3^{(2)} = 0.9522 - j0.0493$

(b) $S_{12} = 300 \text{ MW} + j300 \text{ Mvar}$

$S_{21} = -300 \text{ MW} - j240 \text{ Mvar}$

$S_{L_{12}} = 0 \text{ MW} + j60 \text{ Mvar}$

$S_{13} = 400 \text{ MW} + j400 \text{ Mvar}$

$S_{31} = -400 \text{ MW} - j360 \text{ Mvar}$

$S_{L_{13}} = 0 \text{ MW} + j40 \text{ Mvar}$

$S_{23} = -100 \text{ MW} - j80 \text{ Mvar}$

$S_{32} = 100 \text{ MW} + j90 \text{ Mvar}$

$S_{L_{23}} = 0 \text{ MW} + j10 \text{ Mvar}$

$S_1 = 700 \text{ MW} + j700 \text{ Mvar}$

6.8. (a) $V_2^{(1)} = 1.0025 - j0.0500 \quad Q_3^{(1)} = 1.2360$

$V_3^{(1)} = 1.0299 + j0.0152$

$V_2^{(2)} = 1.0001 - j0.0409 \quad Q_3^{(2)} = 1.3671$

$V_3^{(2)} = 1.0298 + j0.0216$

(b) $S_{12} = 150.428 \text{ MW} + j100.159 \text{ Mvar}$

$S_{21} = -150.428 \text{ MW} - j92.387 \text{ Mvar}$

$S_{L_{12}} = 0 \text{ MW} + j7.772 \text{ Mvar}$

$S_{13} = -50.428 \text{ MW} - j9.648 \text{ Mvar}$

$S_{31} = 50.428 \text{ MW} + j10.902 \text{ Mvar}$

$S_{L_{13}} = 0 \text{ MW} + j1.255 \text{ Mvar}$

$S_{23} = -249.572 \text{ MW} - j107.613 \text{ Mvar}$

$S_{32} = 249.572 \text{ MW} + j126.034 \text{ Mvar}$

$S_{L_{23}} = 0 \text{ MW} + j18.421 \text{ Mvar}$

$S_1 = 100 \text{ MW} + j90.51 \text{ Mvar}$

6.9. $$Y_{bus} = \begin{bmatrix} -j125 & 0 & j100 & 0 \\ 0 & -j6.25 & 0 & j5 \\ j100 & 0 & -j89 & j9 \\ 0 & j5 & j9 & -j13 \end{bmatrix}$$

6.10. $|V_2^{(1)}| = 0.9100 \quad \delta_2^{(1)} = -0.1300 \text{ rad}$

$|V_2^{(2)}| = 0.8886 \quad \delta_2^{(2)} = -0.1464 \text{ rad}$

6.11. $|V_2^{(1)}| = 0.8000 \quad \delta_2^{(1)} = -0.1000 \text{ rad}$

$|V_2^{(2)}| = 0.7227 \quad \delta_2^{(2)} = -0.1350 \text{ rad}$

6.12. The bus admittance matrix in polar form is

$$Y_{bus} = \begin{bmatrix} 60\angle-\frac{\pi}{2} & 40\angle\frac{\pi}{2} & 20\angle\frac{\pi}{2} \\ 40\angle\frac{\pi}{2} & 60\angle-\frac{\pi}{2} & 20\angle\frac{\pi}{2} \\ 20\angle\frac{\pi}{2} & 20\angle\frac{\pi}{2} & 40\angle-\frac{\pi}{2} \end{bmatrix}$$

(a) Substituting for the elements of the bus admittance matrix in (6.52) and (6.53) result in the given equations.
(b)

$\delta_2^{(1)} = 0.0275$ radian $= 1.5782°$ $\delta_2^{(2)} = 0.0285$ radian $= 1.6327°$
$\delta_3^{(1)} = -0.1078$ radian $= -6.179°$ $\delta_3^{(2)} = -0.1189$ radian $= -6.816°$
$|V_3^{(1)}| = 0.9231$ pu $|V_3^{(2)}| = 0.9072$ pu
(c) The power flow program **lfnewton** is used to obtain the solution (See Example 6.9).

6.13. (a)
$\delta_2^{(1)} = 0.0262$ radian $= 1.5006°$ $\delta_2^{(2)} = 0.0277$ radian $= 1.5863°$
$\delta_3^{(1)} = -0.1119$ radian $= -6.412°$ $\delta_3^{(2)} = -0.1182$ radian $= -6.772°$
$|V_2^{(1)}| = 0.9250$ pu $|V_3^{(2)}| = 0.9088$ pu
(b) The power flow program **decouple** is used to obtain the solution (See Example 6.11).

6.14. Follow the Instruction for Data Preparation (Section 6.9) and Example 6.9.

Chapter 7

7.1. A square of side length = 1.4142, perimeter = 5.6568

For $\lambda = -2.828$, $\frac{\partial^2 L}{\partial x^2} = \frac{\partial^2 L}{\partial y^2} = -5.6568$. Second derivatives are negative. Thus, objective function is maximized.

7.2. $x = y = -\frac{4}{3}$, $\lambda = \frac{4}{3}$

$\frac{\partial^2 L}{\partial x^2} = \frac{\partial^2 L}{\partial y^2} = 2$. Second derivatives are positive. Thus, objective function is minimized.

7.3. Base = 1.732, Height = 1.5, Area = 1.299

7.4. $\zeta = 0.5$, $\omega_n = 10,000$ rad/sec, $M_{p\omega} = 1.1547$

7.5. Minimum value of the function = 12.5, at $x = 1.5$, $y = 3.2$, $\lambda = 1$

7.6. Minimum value of the function = 17, at $x = 1$, $y = 4$, $\lambda = \frac{14}{13}$

7.7. (a) $P_1 = 250$ MW, $P_2 = 300$ MW
(b) $P_1 = 500$ MW, $P_2 = 800$ MW
(c) $\beta = 6.8$, $\gamma = 0.002$

7.8. (i) $C_t = 4,849.75$ \$/h (ii) $C_t = 7,310.46$ \$/h (iii) $C_t = 12,783.04$ \$/h

7.9. (i) $P_1 = 100$ MW, $P_2 = 140$ MW, $P_3 = 210$ MW, $\lambda = 8.0$
$C_t = 4,828.70$ \$/h
(ii) $P_1 = 175$ MW, $P_2 = 260$ MW, $P_3 = 310$ MW, $\lambda = 8.6$

$C_t = 7,277.20$ \$/h
(iii) $P_1 = 325$ MW, $P_2 = 500$ MW, $P_3 = 510$ MW, $\lambda = 9.8$
$C_t = 12,705.20$ \$/h
(c) Savings: (i) 21.05 \$/h (ii) 33.26 \$/h (iii) 77.84 \$/h

7.10. (i) $P_1 = 122$ MW, $P_2 = 260$ MW, $P_3 = 68$ MW, $\lambda = 7.148$
$C_t = 4,927.13$ \$/h
(ii) $P_1 = 175$ MW, $P_2 = 260$ MW, $P_3 = 310$ MW, $\lambda = 8.6$
$C_t = 7,277.20$ \$/h
(iii) $P_1 = 350$ MW, $P_2 = 540$ MW, $P_3 = 445$ MW, $\lambda = 10$
$C_t = 12,724.38$ \$/h

7.11. $P_1 = 161.1765$ MW, $P_2 = 258.6003$ MW, $\lambda = 7.8038$ $C_t = 3,375.43$ \$/h

7.12. $P_1 = 70.360$ MW, $P_2 = 181.557$ MW, $P_7 = 97.111$ MW, $\lambda = 8.1513$
$C_t = 3,194.85$ \$/h

Chapter 8

8.1. (a) $\alpha = 75.75°$, $i(t) = 3\sin 315t$
(b) $\alpha = -14.25°$, $i(t) = 3\sin(315t - \pi/2) + 3e^{-80t}$
(c) In *MATLAB* using [Imax, k] = max(i), tmax= t(k) result in
imax = 4.37 A, tmax = 0.0096 sec.

8.2. In the file chp8ex2.m set $d = 30°$, rename the file and run the program.

8.3. In the file chp8ex3.m set $d = 30°$, rename the file and run the program.

8.5. In the file chp8ex4.m set $d = 30°$, rename the file and run the program.

8.6. (a) 0.6667 pu, 2.2222 pu, 4.0 pu
(b) $i_{ac}(t) = (2.5142e^{-25t} + 2.2e^{-0.7143t} + 0.9428)\sin\omega t$

8.7. $i_{asy}(t) = (2.5142e^{-25t} + 2.2e^{-0.7143t} + 0.9428)\sin(\omega t + \pi/2) + 5.6568e^{-3.3333t}$

8.8. $X'_d = 0.449$, $\tau'_d = 1.382$ sec, $X''_d = 0.2498$, $\tau''_d = 1.0397$ sec,

8.9. $i_{asy}(t) = (2.357e^{-25t} + 3.5355e^{-t} + 1.1785)\sin(\omega t + \pi/2) + 7.071e^{-4t}$
$I_{ac} = 5.0$ rms, $I_{dcmax} = 7.071$, $Iasy = 8.66$ rms

8.10. (a) $I''_d = 2.5$ pu, 7,216.88 A, 360.84 A
$I'_d = 2.0$ pu, 5,773.50 A, 288.68 A
$I_d = 0.6667$ pu, 1,924.5 A, 96.23 A
(b) $I_{asy} = 4.3333$ pu, 12,500 A, 625 A
(c) $i_{asy}(t) = (0.7071e^{-28.57t} + 1.8856e^{-2t} + 0.9428)\sin(\omega t + \pi/2) + 3.5355e^{-3.333t}$

8.11. $I'_g = 2.56\angle -75.53°$ pu, or $7393.69\angle -75.53°$ A

8.12. $I'_g = 3.545\angle{-78.6°}$ pu, $I'_m = 3.599\angle{-95.3°}$ pu, $I'_f = 7.068\angle{-87.03°}$ pu

Chapter 9

9.1. $2.0\angle{-90°}$ pu = $288.675\angle{-90°}$ A, 200 MVA

9.2. 1.8 Ω

9.3. (a) $j0.2$ pu, $5.0\angle{-90°}$ pu, (b) $V_1 = 0.4$ pu, $V_2 = 0.8$ pu, $V_3 = 0.7$ pu

9.4. (a) $j0.4$ pu, $2.5\angle{-90°}$ pu
(b) $V_1 = 0.925$ pu, $V_2 = 0.925$ pu, $V_3 = 0.475$ pu
$I_{12} = 0$ pu, $I_{13} = 1.5\angle{-90°}$ pu, $I_{23} = 1.0\angle{-90°}$ pu

9.5. (a) $j0.5$ pu, $2.0\angle{-90°}$ pu
(b) $V_1 = 0.60$ pu, $V_2 = 0.65$ pu, $V_3 = 0.38$ pu, $V_4 = 0$ pu
$I_{13} = 1.1\angle{-90°}$ pu, $I_{21} = 0.1\angle{-90°}$ pu, $I_{23} = 0.9\angle{-90°}$ pu
$I_{34} = 2.0\angle{-90°}$ pu
(c): (a) $j0.25$ pu, $4.0\angle{-90°}$ pu
(b) $V_1 = 0.44$ pu, $V_2 = 0.09$ pu, $V_3 = 0.3$ pu, $V_4 = 0.3$ pu
$I_{12} = 0.7\angle{-90°}$ pu, $I_{13} = 0.7\angle{-90°}$ pu, $I_{32} = 0.7\angle{-90°}$ pu

9.6. $$Z_{bus} = \begin{bmatrix} j0.2400 & j0.1400 & j0.2000 & j0.1400 \\ j0.1400 & j0.2275 & j0.1750 & j0.2275 \\ j0.2000 & j0.1750 & j0.3100 & j0.1750 \\ j0.1400 & j0.2275 & j0.1750 & j0.4175 \end{bmatrix}$$

9.7. $$Z_{bus} = \begin{bmatrix} j0.12 & j0.04 & j0.06 \\ j0.04 & j0.08 & j0.02 \\ j0.06 & j0.02 & j0.08 \end{bmatrix}$$

9.8. $$Z_{bus} = \begin{bmatrix} j0.0450 & j0.00750 & j0.0300 \\ j0.0075 & j0.06375 & j0.0300 \\ j0.0300 & j0.03000 & j0.2100 \end{bmatrix}$$

9.9. $$Z_{bus} = \begin{bmatrix} j0.32 & j0.16 & j0.28 \\ j0.16 & j0.48 & j0.24 \\ j0.28 & j0.24 & j0.42 \end{bmatrix}$$

9.10. Same as Problem 9.4

9.11. Same as Problem 9.5

9.12. $4.0\angle{-90°}$ pu
$V_1 = 0.46$ pu, $V_2 = 0.61$ pu, $V_3 = 0.16$ pu, $V_4 = 0.01$ pu
$I_{13} = 1.5\angle{-90°}$ pu, $I_{14} = 1.5\angle{-90°}$ pu, $I_{21} = 0.3\angle{-90°}$ pu,
$I_{24} = 1.0\angle{-90°}$ pu, $I_{34} = 1.5\angle{-90°}$ pu

9.13. Run chp9ex7 for a bolted fault at bus 9.

9.14. Run chp9ex8 for a bolted fault at bus 9.

9.15. Run chp9ex9 for a bolted fault at bus 9.

9.16–9.18. Make data similar to Examples 9.7–9.9.

Chapter 10

10.1. $V_a^0 = 42.265\angle{-120°}$, $V_a^1 = 193.185\angle{-135°}$, $V_a^2 = 86.947\angle{-84.896°}$

10.2. $I_a = 8.185\angle 42.216°$, $I_b = 4.0\angle{-30°}$, $I_c = 8.185\angle{-102.216°}$

10.4.

$$V_L^{012} = \begin{bmatrix} 0 \\ 763.763\angle{-10.93°} \\ 288.675\angle 30° \end{bmatrix} \quad V_{an}^{012} = \begin{bmatrix} 0 \\ 440.958\angle{-40.89°} \\ 166.667\angle 60° \end{bmatrix}$$

$$V^{abc} = \begin{bmatrix} 440.958\angle{-19.106°} \\ 600.925\angle{-166.102°} \\ 333.333\angle 60° \end{bmatrix}$$

10.5. $$I_a^{012} = \begin{bmatrix} 20\angle 90° \\ 60\angle{-90°} \\ 40\angle 90° \end{bmatrix}$$

10.6. $$I^{abc} = \begin{bmatrix} 20\angle{-90°} \\ 20\angle 150° \\ 20\angle 30° \end{bmatrix}$$

10.7. (a)

$$Z^{012} = \begin{bmatrix} 10+j50 & 0 & 0 \\ 0 & 10+j35 & 0 \end{bmatrix} \qquad (b) \quad V_a^{012} = \begin{bmatrix} 42.265\angle{-120°} \\ 193.185\angle{-135°} \\ 86.947\angle{-84.896°} \end{bmatrix}$$

$$(c) \quad I_a^{012} = \begin{bmatrix} 0.829\angle 161.31° \\ 5.307\angle 150.95° \\ 2.388\angle{-158.95°} \end{bmatrix} \qquad (d) \quad I^{abc} = \begin{bmatrix} 7.907\angle 165.46° \\ 5.819\angle 14.867° \\ 2.701\angle{-96.93°} \end{bmatrix}$$

(e) $S_{3\phi} = 1,036.8 + j3,659.6$

(f) Same as (e)

10.8.

$$V_{an}^{012} = \begin{bmatrix} 0 \\ 136.879\angle 139.933° \\ 451.105\angle 54.603° \end{bmatrix} \quad V^{abc} = \begin{bmatrix} 480.754\angle 70.560° \\ 333.338\angle 163.741° \\ 569.611\angle{-73.685°} \end{bmatrix}$$

$$I^{abc} = \begin{bmatrix} 12.993\angle 70.561° \\ 900.9\angle 163.741° \\ 15.395\angle{-73.686°} \end{bmatrix}$$

10.9. 1.8Ω

10.10. 0.825Ω

10.11. $I_a = 12\angle{-90°}pu$

10.12. $I_b = -9.116$ pu

10.13. $I_f = I_a + I_b = 12.5\angle^{9}0°$ pu

10.14. (a) $5\angle{-90°}$ pu, (b) $6\angle{-90°}$ pu, (c) $-4.33\angle{-90°}$ pu, (d) $7.5\angle 90°$ pu

10.15. (a) $4.395\angle{-90°}$ pu, (b) $4.669\angle{-90°}$ pu, (c) $-3.807\angle{-90°}$ pu, (d) $4.979\angle 90°$ pu

10.16. $I_f = 4.6693\angle{-90°}$ pu

Bus	Voltage Magnitude		
No.	Phase a	Phase b	Phase c
1	0.0000	0.9704	0.9704
2	0.5214	0.9567	0.9567
3	0.7977	0.9535	0.9535
4	0.8911	0.9739	0.9739

From	To	Line current magnitude		
Bus	Bus	Phase a	Phase b	Phase c
1	F	4.6693	0.0000	0.0000
2	1	1.4786	0.1556	0.1556
3	1	2.0234	1.0117	1.0117
4	2	1.0895	0.5447	0.5447

10.17. $I_f = -3.8067$ pu

Bus	Voltage Magnitude		
No.	Phase a	Phase b	Phase c
1	1.0000	0.5000	0.5000
2	1.0000	0.6401	0.6401
3	1.0000	0.7954	0.7954
4	1.0000	0.8871	0.8871

From	To	Line current magnitude		
Bus	Bus	Phase a	Phase b	Phase c
1	F	0.0000	3.8067	3.8067
2	1	0.0000	1.3323	1.3323
3	1	0.0000	2.4744	2.4744
4	2	0.0000	1.3323	1.3323

10.18. $I_f = 4.9793\angle 90°$ pu

Bus	Voltage Magnitude		
No.	Phase a	Phase b	Phase c
1	0.9336	0.0000	0.0000
2	0.9004	0.4965	0.4965
3	0.8921	0.7626	0.7626
4	0.9419	0.8711	0.8711

From	To	Line current magnitude		
Bus	Bus	Phase a	Phase b	Phase c
1	F	0.0000	4.5486	4.5485
2	1	0.1660	1.5076	1.5076
3	1	1.0788	2.5325	2.5325
4	2	0.5809	1.3636	1.3636

10.19.

$$Z_{bus}^{(1)} = \begin{bmatrix} 0.120 & 0.040 & 0.030 & 0.020 \\ 0.040 & 0.080 & 0.010 & 0.040 \\ 0.030 & 0.010 & 0.045 & 0.005 \\ 0.020 & 0.04070.005 & 0.045 & \end{bmatrix}$$

Total fault current = 10.8253 per unit

Bus	Voltage Magnitude		
No.	Phase a	Phase b	Phase c
1	1.0000	0.6614	0.6614
2	1.0000	0.5000	0.5000
3	1.0000	0.9079	0.9079
4	1.0000	0.6614	0.6614

From	To	Line current magnitude		
Bus	Bus	Phase a	Phase b	Phase c
1	2	0.0000	0.7217	0.7217
1	2	0.0000	1.4434	1.4434
2	F	0.0000	10.8253	10.8253
3	1	0.0000	2.1651	2.1651
4	2	0.0000	8.6603	8.6603

10.20 Run chp10ex8 for a bolted fault at bus 9.

10.21. Make data similar to Example 10.8.

Chapter 11

11.1. $M = 8.5$ MJ.rad/sec, $H = 4.0$ MJ/MVA

11.2. $600°/\text{sec}^2 = 100\text{rpm/sec}, 20°, 3620$ rpm

11.3. $376°/\text{sec}^2 = 62.667\text{rpm/sec}, 28.2°, 3609.4$ rpm

11.5. $H = 2.4$ MJ/MVA, $P_m = 0.5$ pu, $P_{max} = 2$,
$\frac{d^2\delta}{dt^2} = 4500(1 - 2\sin\delta)$, (δ is in degrees)

11.6. $E' = 1.25\angle 27.189°$, $0.03\frac{d^2\delta}{dt^2} = 0.77 - 1.65\sin\delta)$, (δ is in radians)

11.7. $0.03\frac{d^2\delta}{dt^2} = 0.77 - 0.5\sin\delta)$, (δ is in radians)

11.8. $\frac{H}{\pi f_0}\frac{d^2\Delta\delta}{dt^2} + P_s\Delta\delta = 0$, where $P_s = \left.\frac{dP}{d\delta}\right|_{\delta_0} = P_{max}\cos\delta_0 + 2P_k\cos 2\delta_0$

11.9. $\zeta = 0.6, \quad \omega_d = 4.0$ rad/sec

11.10. $\delta = 27.835° + 16.0675e^{-2.4977t}\sin(6.5059t + 69°)$

$f = 60 - 0.311e^{-2.4977t}\sin 6.5059t$ Hz

11.11.

$$A = \begin{bmatrix} 0 & 1 \\ -48.5649 & -4.9955 \end{bmatrix} \quad B = \begin{bmatrix} 0 \\ 0 \end{bmatrix} \quad c = \begin{bmatrix} 1 & 0 \\ 0 & 1 \end{bmatrix} \quad D = \begin{bmatrix} 0 \\ 0 \end{bmatrix}$$

11.12. $\delta = 27.835° + 5.8935[1 - 1.0712e^{-2.4977t}\sin(6.5059t + 69°)$

$f = 60 + 0.1222e^{-2.4977t}\sin 6.5059t$ Hz

11.13. A, B, D are the same as in Problem 11.11, $B = \begin{bmatrix} 0 \\ 4.9955 \end{bmatrix}$

11.14. (a) 0.649 pu, (b) 1.195 pu

11.15. (a) $\delta_c = 82.593°$, $t_c = 0.273$ sec (b) $\delta_c = 77.82°$

11.16. $t_c = 0.37$ sec

11.17. (a) Stable (b) Unstable (c) $t_c = 0.29$ sec

11.18. (a) Stable (b) Unstable (c) $t_c = 0.72$ sec

Chapter 12

12.1. 25 MW

12.2. $P_1 = 200$ MW, $P_2 = 300$ MW

12.3. (a) $\Delta f = -0.3$ Hz, $P_1 = 250$ MW, $P_2 = 580$ MW

(b) $\Delta f = -0.291$ Hz, $P_1 = 248.5002$ MW, $P_2 = 577.6004$ MW,
$\Delta P_L = -3.8994$

12.4. (a) $R > 0.009678$

(b) Use **rlocus** for $KG(s)H(s) = \frac{k}{2s^3+12.2s^2+17.2s+1.6}$, where $K = \frac{1}{R}$

12.5. (a) -0.5563 Hz

(b) $T(s) = \frac{0.0625s^2+0.375s+0.5}{s^3+6.1s^2+8.6s+13.3}$, use **step** function with value of -0.25 pu.

(c) Simulation response same as the response in (b)

12.6. (b) $T(s) = \frac{0.0625s^3+0.375s^2+0.5s}{s^4+6.1s^3+8.6s^2+13.3s+0.5K_I}$, use **step** function with value of -0.25

12.7. Modify sim12ex4.mdl

12.8. Modify sim12ex5.mdl

12.9. (a) $K_A < 43.3125$

(b) b) Use **rlocus** for $KG(s)H(s) = \frac{32k_A}{s^3+23s^2+62s+1320}$

(c) $T(s) = \frac{1280}{s^3+23s^2+62s+1320}$, use **step** to obtain the response

(d) Simulation response same as the response in (c)

12.10. (a) $T(s) = \frac{1280(s+25)}{s^4+48s^32+637s^2+6870s+37000}$, use **step** to obtain the response

(b) Simulation response same as the response in (a)

12.12. (b) 0, 0, ± 4.5388, (c) Unstable

(d) $K = \begin{bmatrix} -170.3666 & -38.0540 & -17.3293 & -24.1081 \end{bmatrix}$

(d) $A_f = \begin{bmatrix} 0 & 1.0000 & 0 & 0 \\ -64.5823 & -19.0270 & -8.6646 & -12.0540 \\ 0 & 0 & 0 & 1.0000 \\ 42.1012 & 9.5135 & 4.3323 & 6.0270 \end{bmatrix}$

12.14. (a) $K = \begin{bmatrix} -125.0988 & -28.6369 & -5.0000 & -10.5129 \end{bmatrix}$

(b) $A_f = \begin{bmatrix} 0 & 1.0000 & 0 & 0 \\ -41.9484 & -14.3185 & -2.5000 & -5.2565 \\ 0 & 0 & 0 & 1.0000 \\ 30.7842 & 7.1592 & 1.2500 & 2.6282 \end{bmatrix}$

12.16. (a) $A = \begin{bmatrix} -4 & 0 & -100 \\ 2 & -2 & 0 \\ 0 & 0.0625 & -0.1 \end{bmatrix}$

$B = \begin{bmatrix} 0 & 0 & -0.0625 \end{bmatrix}, BPL = B * PL, C = \begin{bmatrix} 0 & 0 & 1 \end{bmatrix}, D = 0$

(c) $K = \begin{bmatrix} 6.4 & 5.4 & -94.4 \end{bmatrix}$ $\quad A - BK = \begin{bmatrix} -4.0 & 0.0 & -100 \\ 2.0 & -2.0 & 0 \\ 0.4 & 0.4 & -6 \end{bmatrix}$

12.17. (a) $K = \begin{bmatrix} 8.3477 & 1.4637 & -162.0007 \end{bmatrix}$

$A_f = \begin{bmatrix} -4.0000 & 0 & -100.0000 \\ 2.0000 & -2.0000 & 0 \\ 0.5217 & 0.1540 & -10.2250 \end{bmatrix}$

INDEX